# METHODS IN MOLECULAR BIOLOGY™

*Series Editor*
John M. Walker
School of Life Sciences
University of Hertfordshire
Hatfield, Hertfordshire, AL10 9AB, UK



For other titles published in this series, go to
www.springer.com/series/7651

METHODS IN MOLECULAR BIOLOGY™

# Horizontal Gene Transfer: Genomes in Flux

Edited by

**Maria Boekels Gogarten, Johann Peter Gogarten**

*University of Connecticut, Storrs, CT, USA*

**and**

**Lorraine Olendzenski**

*St. Lawrence University, Canton, NY, USA*

Humana Press

*Editors*
Maria Boekels Gogarten
968 Warrenville Road
Mansfield Center CT 06250
USA
mgogarten@gmail.com

Johann Peter Gogarten
University of Connecticut
Dept. Molecular & Cell Biology
91 North Eagleville Rd.
Storrs CT 06269-3125
USA
gogarten@uconn.edu

Lorraine Olendzenski
St. Lawrence University
Biology Department
23 Romoda Drive
Canton NY 13617
USA
lolendzenski@stlawu.edu

ISSN 1064-3745 e-ISSN 1940-6029
ISBN 978-1-61737-918-5 e-ISBN 978-1-60327-853-9
DOI 10.1007/978-1-60327-853-9



*Cover illustration*: courtesy of Ford Doolittle

Printed on acid-free paper

springer.com

*This book is dedicated to our families –*
*Jacob, Jan, Jenny and Stephanie,*
*AND*
*Tim, Ryan and Brendan*

# Preface

The recombination of genetic information has long been recognized as an important process in evolution. However, genomic and metagenomic analyses, particularly of prokaryotic cells, have revealed a hitherto unexpected number of horizontal gene transfer (HGT) events. Divergent organisms exchange genes and the traits they encode; vast amounts of genetic information only sometimes reside within the chromosomes, but perhaps can be found more frequently in plasmids and phages. The importance of HGT was first recognized in pathogenic bacteria, but it plays an equal role in the evolution of non-pathogenic Bacteria and Archaea. Studies have also revealed the transfer of parasitic, selfish genes between the mitochondria of different plant and fungal species; genetic material was found to be exchanged between parasitic plants and their hosts; and many studies have identified genes that were transferred between pro- and eukaryotes, including transfers from symbionts to multicellular animals.

HGT events encompass not just a transfer of one or a few genes but the evolution of organelles by symbiosis can be viewed as an extreme instance where whole genomes were acquired by another organism. The transfer of genes from the symbiont to the host genome was crucial in eukaryogenesis, and these types of transfers continue to shape the evolution of eukaryotes, including multicellular plants and animals.

This book provides advanced undergraduates, graduate students, researchers, and professors with an overview of HGT concepts, as well as specific case histories that highlight current progress in the field. Where possible, molecular biology (wet-lab) protocols as well as overviews on *in silico* methods are provided to allow researchers to incorporate the methods of recognizing and characterizing HGT in their own research systems. This volume will be of interest for molecular biologists, geneticists, evolutionary biologists, bioinformaticians, and computational scientists.

Written by prominent contributors to the field, this volume is divided into three sections following the Introduction: Definitions, philosophy and Implications; Methods and Case Studies. Chapters provide an overview of the terminology, concepts, and implications of HGT, and current evolutionary thought and philosophy. Discussions of microbial species concepts and the possibility to reconstruct evolutionary history have frequently focused on HGT. Throughout this book, we attempt to include different and often controversial viewpoints to provide the reader with an impression of the excitement surrounding HGT. Contributions of gene transfer, often as a consequence of symbiosis, in the origin and evolution of eukaryotic cells are highlighted, as well as the role of HGT in assembling biochemical pathways, and the difficulties that HGT creates for tracing metabolic capabilities through life's history.

Methods used in exploring HGT encompass both computer and bioinformatics analyses of genomic data as well as molecular biology techniques for identifying, quantifying, and differentiating instances of HGT. These methods allow the characterization of mobile agents of HGT in modern communities as well as the identification of ancient transfers that have contributed to genome evolution.

Case Studies provide detailed accounts of how HGT has shaped evolution across the diversity of organisms and organismal lineages. In prokaryotes, transfer events in thermophiles as well as organisms that had to adapt to increased atmospheric oxygen are discussed. From a biogeochemical point of view, organisms leave a mark on the geological record in a variety of ways, including changing the isotopic composition of rocks. The metabolic pathways and their building blocks responsible for these isotopic discriminations are scattered throughout the phylogenetic web of life in distantly related lineages, suggesting their origin and distribution by HGT. The extent of plasmid ecology in marine and subsurface niches as well as mathematical modeling approaches to HGT in continuous and batch culture conditions are examined. The impacts of gene transfer on our ability to clearly define prokaryotic species and strains, our ability to accurately reconstruct phylogenetic trees, and how transferred genes or genomes undergo adaptation are also discussed. Finally, HGT has impacted evolution of eukaryotic genomes through organellar evolution. Contributions highlight the evolution of anaerobic pathogenic protozoa, plastid evolution in algae and plants, and HGT events in other eukaryotic genomes.

Taken as a whole, this book provides not just an introduction. It is a detailed compendium of how HGT has impacted genome evolution and how understanding HGT impacts our ability to accurately reconstruct and comprehend the web-like evolutionary history. It also gives a toolbox of methods to help recognize HGT events in genome analysis as well as in extant microbial communities. It is our hope that all scientists who are interested in genome evolution and microbial ecology will benefit from this new paradigm and make use of the tools and ideas presented here to further their own research.

Maria Boekels Gogarten
Johann Peter Gogarten
Lorraine Olendzenski

# Acknowledgments

The editors thank all reviewers of individual chapters:

Sonja Albers, Jan Andersson, John Archibald, Mark Bailey, Martin Bailey, Eric Bapteste, Tamar Barkay, Thomas Baum, Robert Beiko, Joe Bielawski, Robert Blankenship, Nico Boon, Maureen L. Coleman, Jonna Coombs, Joel B. Dacks, Julian Davies, Edward Feil, Michael Feldgarden, Pilar Francino, Lieve Gheysen, Beverley Green, Simonetta Gribaldo, William Hanage, Tim Harlow, Robert Hirt, John Jones, Patrick Keeling, Niels Kroer, Antonio Lazcano, Debbie Lindell, Kira Makarova, James McInerney, Cindy Nakatsu, Diana Reid Nemergut, Jan Nesvera, Maureen O'Malley, Mark Osborn, Thane Papke, Maria Poptsova, Jason Raymond, Paul Rainey, Chris Rensing, Danny Rice, Andrew Roger, Daniel Rozen, Bruce Runnegar, Janet Siefert, Barth Smets, Pat Sobecky, Ralf J. Sommer, Fredj Tekaia, Fred C. Tenover, EvaTopp. Michael Travisano, Sarah Lea Turner, Jan Dirk van Elsas, Judy Wall, Ing-Nang Wang, Linda Weigel, Jen Wernegreen, Hwan Su Yoon, and Olga Zhaxybayeva.

Special thanks to Ford Doolittle for allowing us to use his original and unique artwork (digitally manipulated scanned acrylic painting).

# Contents

## Contributors

JAN O. ANDERSSON • *Department of Cell and Molecular Biology, Uppsala University, Uppsala, Sweden*
CECILIA ALSMARK • *Institute for Cell and Molecular Biosciences, Newcastle University, Newcastle upon Tyne, UK*
MARTIN IAIN BAHL • *Department of Microbiology, University of Copenhagen, Copenhagen, Denmark*
ERIC BAPTESTE • *UPMC UMR 7138, Paris, France*
MIRIAM BARLOW • *University of California, Merced, Merced, CA, USA*
ARTURO BECERRA • *Facultad de Ciencias, Universidad Nacional Autónoma de México, México D.F., Mexico*
ROBERT G. BEIKO • *Department of Computer Science, Dalhousie University, Halifax, NS, Canada*
YAN BOUCHER • *Department of Civil and Environmental Engineering, Massachusetts Institute of Technology, Cambridge, MA, USA*
REBECCA JOSEPHINE CASE • *Harvard Center for the Environment (HUCE) and Microbiology and Molecular Genetics, Harvard Medical School, Harvard University, Boston, MA, USA*
JONNA M. COOMBS • *Department of Biology, Adelphi University, Garden City, NY, USA*
DIEGO CORTEZ • *Unité de Biologie Moléculaire du Gène chez Extremophiles Institut Pasteur, Paris, France*
LUIS DELAYE • *Facultad de Ciencias, Universidad Nacional Autónoma de México, México D.F., Mexico*
MARTIN EMBLEY • *Institute for Cell and Molecular Biosciences, Newcastle University, Newcastle upon Tyne, UK*
PETER FOSTER • *Department of Zoology, The Natural History Museum, London, UK*
GREG FOURNIER • *Department of Molecular and Cell Biology, University of Connecticut, Storrs, CT, USA*
GEORGE E. FOX • *Department of Biology and Biochemistry, University of Houston, Houston, TX, USA*
JOHANN PETER GOGARTEN • *Department of Molecular and Cell Biology, University of Connecticut, Storrs, CT, USA*
LARS HESTBJERG HANSEN • *Department of Microbiology, University of Copenhagen, Copenhagen, Denmark*
TRACY H. HAZEN • *Georgia Institute of Technology, School of Biology, Atlanta, GA, USA*
JOHANNES HELDER • *Department of Plant Sciences, Laboratory of Nematology, Wageningen University, Wageningen, Netherlands*
ROBERT HIRT • *Institute for Cell and Molecular Biosciences, Newcastle University, Newcastle upon Tyne, UK*
CHRISTOPHER H. HOUSE • *Department of Geosciences and Pennsylvania State Astrobiology Research Center, Pennsylvania State University, University Park, PA, USA*

JINLING HUANG • *Department of Biology, East Carolina University, Greenville, NC, USA*
PATRICK J. KEELING • *Botany Department, University of British Columbia, Vancouver, BC, Canada*
MAURIZIO LABBATE • *Department of Chemistry and Biomolecular Sciences, Macquarie University, Sydney, NSW, Australia*
LAURENT LARDON • *Laboratory of Environmental Biotechnology, INRA-LBE, Narbonne, France*
MAIA LARIOS-SANZ • *Department of Biology, University of St. Thomas, Houston, TX, USA*
JEFFREY G. LAWRENCE • *Department of Biological Sciences, University of Pittsburgh, Pittsburgh, PA, USA*
ANTONIO LAZCANO •*Facultad de Ciencias, Universidad Nacional Autónoma de México, México D.F., Mexico*
MICHELLE LIZOTTE-WANIEWSKI • *Department of Biology, University of Massachusetts, Amherst, MA, USA*
LYNN MARGULIS • *Department of Geosciences, University of Massachusetts, Amherst, MA, USA*
MAKEDONKA MITREVA • *Department of Genetics, Washington University School of Medicine, St. Louis, MO, USA*
KENNETH M. NOLL • *Department of Molecular and Cell Biology, University of Connecticut, Storrs, CT, USA*
LORRAINE OLENDZENSKI • *Biology Department, St. Lawrence University, Canton, NY, USA*
GUNNAR OREGAARD • *Department of Microbiology, University of Copenhagen, Copenhagen, Denmark*
R. THANE PAPKE • *Department of Molecular and Cell Biology, University of Connecticut, Storrs, CT, USA*
MARIA POPTSOVA • *Department of Molecular and Cell Biology, University of Connecticut, Storrs, CT, USA*
MARK A. RAGAN • *Institute for Molecular Bioscience and Australian Research Council Centre of Excellence in Bioinformatics, St. Lucia, Queensland, Australia*
JASON RAYMOND • *School of Natural Sciences, University of California, Merced, Merced, CA, USA*
ADAM RETCHLESS • *Department of Biological Sciences, University of Pittsburgh, Pittsburgh, PA, USA*
MARGARET A. RILEY • *Department of Biology, University of Massachusetts, Amherst, MA, USA*
THOMAS SICHERITZ-PONTEN • *Center for Biological Sequence Analysis, Department of Systems Biology, Technical University of Denmark, Lyngby, Denmark*
JANET L. SIEFERT • *Department of Statistics. Rice University, Houston, TX, USA*
GEERT SMANT • *Department of Plant Sciences, Laboratory of Nematology, Wageningen University, Wageningen, Netherlands*
BARTH F. SMETS • *Microbial Ecology Research Group, Institute of Environment & Resources DTU, Technical University of Denmark, Lyngby, Denmark*
PATRICIA A. SOBECKY • *Georgia Institute of Technology, School of Biology, Atlanta, GA, USA*

SØREN J. SØRENSEN • *Department of Biology, University of Copenhagen, Copenhagen, Denmark*
HATCH W. STOKES • *Department of Chemistry and Biomolecular Sciences, Macquarie University, Sydney, NSW, Australia*
KAMOLWAN THIRANGOON • *Department of Molecular and Cell Biology, University of Connecticut, CT, USA*
MICHAEL TRAVISANO • *Department of Ecology, Evolution and Behavior, University of Minnesota, St. Paul, MN, USA*
SHAILAJA YERRAPRAGADA • *Human Genome Sequencing Center Baylor College of Medicine, Houston, Texas, USA*
OLGA ZHAXYBAYEVA • *Department of Biochemistry and Molecular Biology, Dalhousie University, Halifax, NS, Canada*

# Color Plates

Color Plates following p. 76

| | |
|---|---|
| Color Plate 1 | Schematic representation of the GFP detection systems for HGT (*panel A*) and plasmid stability assay (*panel B*). The donor cells in the HGT detection system; (*see* complete caption and discussion on p. 79) |
| Color Plate 2 | (**A**) Segment of a bacterial colony of *Kluyvera* sp. MB101::*gfp* harboring plasmid pKJK5::$lacI^{q1}$. The *green* fluorescing patches represent plasmid-free segregant cells. (**B**) Corresponding flow cytometry data showing a small sub-population of GFP-positive, i.e., plasmid-free, cells (G1); (*see* discussion on p. 88) |

# Part I

## Introduction

# Chapter 1

# Gene Transfer: Who Benefits?

**Lorraine Olendzenski and Johann Peter Gogarten**

## Abstract

Horizontal gene and genome transfer forces us to recognize that life evolves by fusion as well as bifurcation of lineages, and necessitates the expansion of traditional views of evolution. This chapter reviews the role that horizontal gene transfer (HGT) may play in integrating selection at the gene, species, and community levels. Additionally, we provide an overview of the structure and content of this book, which reflects current thought in the dynamic field of HGT research.

**Key words:** Net of life, gene transfer agents, levels of selection, selfish genes, symbiogenesis.

## 1. Introduction: Why Study HGT?

Traditional views of evolution posit descent with modification as the origin of new species: changes passed from parent to offspring in a pattern of vertical inheritance eventually lead to divergent lineages and ultimately species *(1)*. Thus, evolutionary lineages split and evolve independently from each other, and tree-like schemes of bifurcating lineages have given us the standard imagery of a "tree of life" to describe the evolutionary history of all living organisms. Molecular analyses reinforce this view by often depicting phylogenies based on genetic sequences in the form of bifurcating trees. The same field of molecular phylogenetics has forced us to come to terms with the extent with which genes are transferred horizontally between organisms. Comparison of molecular phylogenies constructed from different molecules often yields conflicting topologies, which can be an indication of transfer between lineages *(2, 3)*.

Maria B. Gogarten et al. (eds.), *Horizontal Gene Transfer: Genomes in Flux, vol. 532*

DOI 10.1007/978-1-60327-853-9_1 Springerprotocols.com

With ever increasing data from whole genome sequences, the mark that horizontal gene transfer (HGT) makes on microbial genomic evolution becomes more apparent. The transfer of genes or genetic elements between organisms that may be closely or even distantly related causes branches on the tree of life to come together or reticulate. Symbiosis between two or more species to form a newly named "individual" likewise has the same effect on uniting formerly separate evolutionary lineages. Thus a "net of life" in which some branches fuse is a more appropriate way to envision evolutionary history *(2, 4, 5)*. Depicting life's evolutionary history only through the, perhaps tree-like history of the translation machinery does not reflect the rich history of interactions, fusions and transfers that shaped today's biosphere *(6)*.

The extent to which HGT is currently occurring and how a realization of HGT forces us to revise our understanding of speciation in Bacteria and Archaea have become hot-button issues and topics of lively debate in microbial biology *(7, 8)*. Are microbial species more than convenient but arbitrary categories to help identification of organisms and the prediction of their likely properties? To what extent do selective sweeps, random extinctions, or preferred partners for recombination each contribute to the pheno-and genotypic similarity often observed within species *(9)*?

## 2. Mechanisms of Gene Transfer

How are genes transferred among lineages? Mobile genetic elements, including plasmids, transposons, integrons, and integrative conjugative elements (ICEs) such as conjugative transposons, can provide the vehicles for HGT *(10, 11)*. These vehicles can enter the recipient cell through transformation (the uptake of DNA from the environment by competent cells), conjugation (the transfer between two cells), or by transduction (the packaging of genetic material into phage particles that deliver the DNA) *(12)*. Viruses are the most abundant biological entities and outnumber bacteria in aquatic environments by an order of magnitude *(13)*. Not only can transduction from phage result in the transfer of foreign DNA into bacteria but the evolution of virulence resistance via CRISPR elements has the ability to cause rapid genetic change in a whole community *(14)*. Additionally, gene transfer agents (GTAs) found in some alpha-proteobacteria function only in the transfer of genomic DNA between cells *(15, 16)*. GTAs are derived from prophage, but the produced phage particles no longer specifically package phage DNA and do not harm the host. If the GTAs are under purifying selection as suggested, this could only happen through their benefit to the host. Clearly, the mechanisms for transferring genes in nature are abundant; the frequency with which these elements overcome barriers to transfer to attain successful integration into new environments needs further elucidation and is still under debate *(8, 17)*.

## 3. HGT and the Units of Selection in Microbial Evolution

Darwinian evolution can be considered as changes in gene frequency in a population over time. According to Darwin's theory *(18)*, selection operates on the individual where the "fittest" organisms produce more offspring and as a consequence the traits that produce this fitness become dominant in the population. This theory was based primarily on the observations of sexually reproducing organisms that exchanged genes via sexual recombination within species, but not usually with organisms outside of their own species.

Because we now recognize that genetic material can be transferred across species via processes of HGT, the target of selection is controversial. In fact, simultaneous selection at multiple levels appears to be most likely; levels at which selection has been proposed to be working include the gene, group, or community level. Dawkins introduced the concept of selfish genes and presented a view of evolution that considers the organism as a vehicle created for the propagation of the genes, most of which cooperate in creating the organism *(19)*. In contrast, parasitic genes (e.g., inteins, homing endonucleases) do not improve the organism *(20)*. Parasitic genes have their own life cycle and are selected for efficient transfer within and between populations; they cause no or little harm to the host, thereby avoiding purifying selection at the organismal level.

Wynne-Edwards suggested that a population or a group as a whole could be the subject of selection, outcompeting groups with less optimal properties *(21)*. Since the process of evolution is itself subject to evolution *(22)*, it would not be surprising to find that lineages with better ways to evolve are favored in the long run. Conceivably, groups that adapt faster to new environments would outcompete those that adapt to new niches more slowly. Transfer of selfish genes, for example, in the quorum-sensing-mediated transfer of Ti plasmids among agrobacteria, some of which successfully transform a plant host *(23)*, is one mechanism by which a population might integrate into a selectable unit *(4)*.

Communities of organisms made up of different species can also become units of selection. Microbial communities can have access to a common shared gene pool through HGT. As such, the whole community could be considered a single individual, and could be selected as a unit *(24–27)*. The microbial community responsible for kefir production exists in the form of tightly associated granules on which selection in response to the environment can occur. Symbiosis also brings organisms together to form a unit that can be selected. This process of symbiogenesis is classically illustrated by lichens, which have been classified into over 25,000 different "species".

## 4. Book Overview: Questions and Controversies

The chapter topics of this book present areas of current research and thinking in the field by authors who are actively working to elucidate the extent and mechanisms of HGT, as well as advocating the need for incorporating HGT into our paradigms of microbiological evolution. Broadly, the chapters are grouped into sections on Definitions and Philosophy, Methods, and Case Studies. The Case Studies chapters review current research and illustrate the variety of questions being asked and methods used to answer them.

The Definitions and Philosophy chapters begin with Siefert summarizing the various mobile genetic elements that can be found in an organism's genome, which together represent the mobilome. Bahl et al. show us that selection is happening at both the level of conjugative plasmids and the organisms that contain them by investigating their mechanisms of persistence in the cell, while Labbate et al. introduce us to integrons and the importance of these gene cassettes as mobile elements in evolution. The question of how we can reconcile HGT with species definitions, particularly in Bacteria and Archaea, is carefully explored by Lawrence and Retchless, while Bapteste and Boucher argue for the adoption of the philosophical approach known as eliminative pluralism for defining bacterial species.

Phylogenetic analysis of genes, and, more recently, information contained in completely sequenced genomes, is a cornerstone of HGT research and contributes to our view of how widespread HGT must be in evolution. Interpretations of data have led to arguments that rampant HGT would erase phylogenetic history *(28)*. However, in the instances where a gene is transferred into and retained in the ancestor of a major lineage, these genes constitute a shared derived character and thus can be used to provide valuable information in phylogenetic reconstruction as shown by Huang and Gogarten in the case of red algae and green plants. House shows that despite HGT, whole genomic data create a core tree of life. Additionally, some lineages in this core tree have phenotypes recognizable in the geochemical record going back billions of years. However, Fornier's analyses of HGT in methanogens suggest that an extant lineage showing a particular phenotype may not be the one that generated a particular phenotypic signal in the biogeochemical record due to HGT of metabolic pathways. Margulis emphasizes the importance of genome transfer in contrast to single-gene transfer, provides admonitory words against poor practices that may have characterized molecular evolution analyses in the past, and reminds us that we should not ignore biology and life cycle information when reconstructing phylogenies based solely on genes.

Detection and interpretation of HGT events as well as the testing of hypotheses of how transferred genes can successfully become fixed in a population demand innovative methods; indeed development of both computer-based and experimental approaches has paralleled and fueled our burgeoning understanding in this field. Zhaxybayeva outlines the categories of methods that have been used to detect HGT using genome data, while work by Cortez et al. and Poptsova tests the efficacy of composition-based and phylogenetic methods, respectively, to detect HGT transfers. Beiko and Ragan discuss and describe automated high-throughput approaches to quantify and pinpoint transfer events from comparison between gene and organismal phylogeny. Complementing these *in silico* approaches, wet lab approaches provide the opportunity to test hypotheses of HGT in cells and to explore the conditions necessary for successful transfer. Larios-Sanz and Travisano present a model laboratory system for experimental evolution, which can track changes in the genes for sucrose catabolism after they are transferred into *Escherichia coli*. Bahl et al. detail the wet laboratory experimental approach of detecting plasmid transfer using fluorescence and flow cytometry.

Transferred genes must overcome several barriers to function successfully in new host cell environments, and additionally can undergo adaptation over time. Evidence from Noll and Thirangoon suggests that successful expression of archaeal ABC transporters transferred to *Thermotoga*, a member of the Bacteria, requires recombination downstream from functional bacterial promoters. All types of genes have undergone HGT in one lineage or another and HGT can involve genes in major metabolic pathways. Two case studies verify the importance of HGT in photosynthetic lineages. Raymond details how HGT contributed to the evolution of photosynthesis, particularly oxygenic photosynthesis, a process that radically reshaped Earth's biosphere. Furthermore, Yerrapragada and coworkers have identified a set of cyanobacterial signature genes and provide evidence for intra-lineage HGT of many of these genes.

While the analyses by Yerrapragada et al. illustrate that group-specific genes that may be used to define higher taxonomic units are not immune to HGT, Riley argues for the role of population genomics in providing evidence for the mechanisms of bacterial speciation and argues for the possibility of using a core genome to trace the evolutionary history of microbial species. Papke critiques species concepts as applied to Bacteria and Archaea and makes a case for recombination and HGT in the creation of bacterial species using haloarchaea as a model lineage in which to test hypotheses.

From its earliest detection in the role of antibiotic resistance, the role of HGT has been known to provide some selective advantage to organisms in a variety of environments. Barlow discusses the lessons learned about HGT from the study of antibiotic-resistance genes. Sobecky, Hazen, and Coombs comprehensively summarize the variety of mobile genetic elements and the evidence for HGT found in marine environments, the terrestrial subsurface, and contaminated soils, giving an environmental context for the importance of HGT to adaptations of organisms to their habitats.

Bacteria form intimate associations with eukaryotes through symbiosis. Often as a result of these associations, a variety of HGTs occur from Bacteria into Eukaryotes, including animals. **Chapter 10** provides examples of some of these associations and how they have led to symbiogenesis, the origin of new taxa through symbiotic interactions. HGT from across domains can be detected in unicellular eukaryotes, as illustrated in the genome analyses of Alsmark et al. and Andersson. The importance of gene transfer in the evolution of plastids and their hosts is discussed in **Chapter 29**. Nematodes, one of the most speciose groups of animals on Earth, seem to be acquiring genes via HGT from a variety of sources, including the bacteria with which they live. **Chapter 30** critically evaluates the impact that the acquisition of bacterial genes may have played in the evolution of plant–nematodes interactions. We hope that in reading these chapters, the reader will share our enthusiasm about HGT and the many roles this process plays in life's evolution.

## References

1. Mayr, E. (1942) *Systematics and the Origin of Species*, Columbia University Press, New York.
2. Hilario, E., Gogarten, J. P. (1993) Horizontal transfer of ATPase genes – the tree of life becomes a net of life. *Biosystems* **31**, 111–9.
3. Gogarten, J. P. (1995) The early evolution of cellular life. *Trends Ecol Evol* **10**, 147–51.
4. Gogarten, J. P., Fournier, G., Zhaxybayeva, O. (2008) Gene transfer and the reconstruction of life's early history from genomic data. *Space Sci Rev* 135, 115–31.
5. Doolittle, W. F. (1999) Phylogenetic classification and the universal tree. *Science* **284**, 2124–9.
6. Dagan, T., Martin, W. (2006) The tree of one percent. *Genome Biol* 7, 118.
7. Lawrence, J. G., Hendrickson, H. (2003) Lateral gene transfer: When will adolescence end? *Mol Microbiol* **50**, 739–49.
8. Kurland, C. G., Canback, B., Berg, O. G. (2003) Horizontal gene transfer: A critical view. *Proc Natl Acad Sci U S A* **100**, 9658–62.
9. Gevers, D., Cohan, F. M., Lawrence, J. G., Spratt, B. G., Coenye, T., Feil, E. J., Stackebrandt, E., Van De Peer, Y., Vandamme, P., Thompson, F. L., Swings, J. (2005) Re-evaluating prokaryotic species. *Nat Rev Microbiol* **3**, 733–9.
10. Frost, L. S., Leplae, R., Summers, A. O., Toussaint, A. (2005) Mobile genetic elements:

The agents of open source evolution. *Nat Rev Microbiol* **3**, 722–32.

11. Osborn, A. M., Boltner, D. (2002) When phage, plasmids, and transposons collide: Genomic islands, and conjugative- and mobilizable-transposons as a mosaic continuum. *Plasmid* **48**, 202–12.
12. Mazodier, P., Davies, J. (1991) Gene transfer between distantly related bacteria. *Annu Rev Genetics* **25**, 147–71.
13. Patel, A., Noble, R. T., Steele, J. A., Schwalbach, M. S., Hewson, I., Fuhrman, J. A. (2007) Virus and prokaryote enumeration from planktonic aquatic environments by epifluorescence microscopy with SYBR Green I. *Nat Protoc* **2**, 269–76.
14. Andersson, A. F., Banfield, J. F. (2008) Virus population dynamics and acquired virus resistance in natural microbial communities. *Science* **320**, 1047–50.
15. Lang, A. S., Beatty, J. T. (2007) Importance of widespread gene transfer agent genes in alpha-proteobacteria. *Trends Microbiol* **15**, 54–62.
16. Biers, E. J., Wang, K., Pennington, C., Belas, R., Chen, F., Moran, M. A. (2008) Occurrence and expression of gene transfer agent genes in marine bacterioplankton. *Appl Environ Microbiol* **74**, 2933–9.
17. Gogarten, J. P., Townsend, J. P. (2005) Horizontal gene transfer, genome innovation and evolution. *Nat Rev Microbiol* **3**, 679–87.
18. Darwin, C. (1859) *On the Origin of Species by Means of Natural Selection, or The Preservation of Favoured Races in the Struggle for Life*, John Murray, Albemarle Street, London.
19. Dawkins, R. (1976) *The Selfish Gene*, Oxford University Press, Oxford.
20. Gogarten, J. P., Hilario, E. (2006) Inteins, introns, and homing endonucleases: Recent revelations about the life cycle of parasitic genetic elements. *BMC Evol Biol* **6**, 94.
21. Wynne-Edwards, V. C. (1962) *Animal Dispersion in Relation to Social Behavior*, Oliver & Boyd, London.
22. Rossler, O. E. (1979) Recursive evolution. *Biosystems* **11**, 193–9.
23. White, C. E., Winans, S. C. (2007) Cell-cell communication in the plant pathogen *Agrobacterium tumefaciens. Philos Trans R Soc Lond B Biol Sci* **362**, 1135–48.
24. Sonea, S. (1988) A bacterial way of life. *Nature* **331**, 216.
25. Sonea, S., Mathieu, L. G. (2001) Evolution of the genomic systems of prokaryotes and its momentous consequences. *Int Microbiol* **4**, 67–71.
26. Caldwell, D. E., Costerton, J. W. (1996) Are bacterial biofilms constrained to Darwin's concept of evolution through natural selection? *Microbiologia* **12**, 347–58.
27. Caldwell, D. E., Wolfaardt, G. M., Korber, D. R., Lawrence, J. R. (1997) Do bacterial communities transcend Darwinism? In *Advances in Microbial Ecology* (Jones, G., ed.) Plenum Press, New York.
28. Gogarten, J. P., Doolittle, W. F., Lawrence, J. G. (2002) Prokaryotic evolution in light of gene transfer. *Mol Biol Evol* **19**, 2226–38.

# Part II

## Definitions, Philosophy, and Implications of HGT

# Chapter 2

## Defining the Mobilome


### Janet L. Siefert



### Abstract

This chapter defines the agents that provide for the movement of genetic material which fuels the adaptive potential of life on our planet. The chapter has been structured to be broadly comprehensive, arbitrarily categorizing the mobilome into four classes: (1) **transposons**, (2) **plasmids**, (3) **bacteriophage**, and (4) **self-splicing molecular parasites**.

Our increasing understanding of the mobilome is as dynamic as the mobilome itself. With continuing discovery, it is clear that nature has not confined these genomic agents of change to neat categories, but rather the classification categories overlap and intertwine. Massive sequencing efforts and their published analyses are continuing to refine our understanding of the extent of the mobilome. This chapter provides a framework to describe our current understanding of the mobilome and a foundation on which appreciation of its impact on genome evolution can be understood.



**Key words:** Mobilome, mobile genetic elements (MGEs), transposable elements (TEs), transposons, plasmids, bacteriophage, Group II introns, jumping genes, insertion sequences (ISs).


### 1. Introduction

As important as the process of horizontal gene transfer is for biology, the mobilome is the agent of change that facilitates these events. Appreciating and understanding it is critical to understanding the constraints and freedoms that dictate how, why, and when gene transfer events are likely to take place. It is not overstating the fact to say that the mobilome is a driver of evolution. Our first indication for these agents of change can be traced to Barbara McClintock. In compelling and insightful experiments, she described the mobile elements that were affecting the expressions of genes in the chromosomes of maize *(1)*. She called them "controlling elements"; her discovery would earn her the Nobel


Maria B. Gogarten et al. (eds.), *Horizontal Gene Transfer: Genomes in Flux, vol. 532*



DOI 10.1007/978-1-60327-853-9_2 Springerprotocols.com

Prize and provide a glimpse into the process of transposition. We can now appreciate, and in many cases utilize, these diverse agents of change that we recognize as mobile genetic elements.

### 1.1. What Is the Mobilome?

Simply, the mobilome consists of all mobile genetic elements (MGE) in a cell. MGEs are any type of DNA that can move around within or between genomes. MGEs can be divided into categories based on their mechanism of movement and the character of the DNA sequences. For our purposes here, I will divide MGEs into four broad categories: (1) **transposons** (synonyms: transposable elements, TE, "jumping genes"), (2) **plasmids**, (3) **bacteriophage** elements, and (4) **self-splicing molecular parasites** (**Table 2.1**). The total of all these MGEs in any cell is referred to as its mobilome.

To define MGEs it is helpful to consider the mobilome in the context of the cellular organization of prokaryotes and eukaryotes (despite the strangely fervent, well-intentioned, but largely impractical ideas to dispense with this organization in biology) *(2)*. For eukaryotes the broad category of *transposable elements* (TE) are the major mobile elements. Eukaryotes are predominantly sexually reproducing entities. In multicellular eukaryotes, the division of cell types into somatic and gametic (germ line) cells greatly limits the number of vectors that can facilitate the movement of DNA. Complicating this endeavor is the reality that for any alien DNA to be passed to the next generation, the transposition must be taking place in the germ line of the eukaryote. Within the eukaryotes, animals are the ones most likely to suffer from this barrier, as the germ line and somatic cells are separated early in development. The likelihood of inter-cell transfer by MGEs for eukaryotes is low as the alien DNA would need to overcome these hurdles. This does happen, and there are certainly dramatic examples from the past involving unicellular eukaryotes (consider mitochondria, plastids, and the diversion of genes to the host nucleus). For the most part, we can confine our estimation of the eukaryotic mobilome to intragenomic dynamics, and frequently within somatic cells.

For prokaryotes, the repertoire of MGEs is defined as a broader set of agents because movement of genetic material is not primarily confined within a genome as it is for eukaryotes, but includes elements that provide movement *between* genomes. Indeed, as essentially non-sexual entities, prokaryotic diversity is fueled by HGT. What should be appreciated is that this is a relatively new understanding for prokaryotes. In the early 1960s, Margaret Dayhoff, combining biology and extraordinary computational skill and insight, created statistical methodologies for exploring the phylogeny of DNA and protein sequences. They were based on the assumption that HGT seldom occurred, and while in practice this view was instrumental in the rich progress for the field of bioinformatics, the truth lies somewhere else. HGT does happen and in significant proportion. By 1994, Guttman and Dykhuizen demonstrated that recombination, rather than mutation, was a driving evolutionary force *(3)*. That is why you are reading this book.

The goal of this chapter is to review and describe our current understanding of the mobilome. I will include a definition and general categorization of all MGEs, and describe the current knowledge of their presence and effects on genome organization in both eukaryotes and prokaryotes.

**Table 2.1**
**Major components of the mobilome**

| Mobilome |
|---|
| **Transposons** |
| Retrotransposons |
| LTR retrotransposases |
| non-viral, non-LTR transposases |
| LINEs |
| SINEs (e.g., Alu sequences) |
| Viral-like retrotransposases |
| DNA transposons |
| Insertion sequences |
| **Plasmids** |
| Conjugative |
| Non-conjugative |
| Insertion sequences in plasmids |
| **Bacteriophages** |
| Lytic/lysogenic/prophage |
| Gene transfer agents |
| Filamentous phages |
| **Self-splicing molecular parasites** |
| Group II introns |
| Group I introns |
| Inteins |
| Homing endonucleases |

## 2. Transposons

Transposons, or "jumping genes," affect genomic dynamics within a single cell through the process of transposition and can alter gene expression by serving as promoters, enhancers, silencers, sites of epigenetic modification, or alternative splicing sites *(4)*. Sometimes transposons provide "molecular domestication" activities when their host adopts a transposon-encoded protein that can perform cellular functions *(4)*. Because copies of transposons can flood a genome, they serve as hotspots for recombination events, producing deletions, duplications, inversions, or translocations. Transposition can be disruptive, but it can have a positive role through stable integration and longlasting protein expression in cells. Through production of insertions and duplications, transposons greatly affect genome size as demonstrated by the extensive variation in nuclear genome size, or C-value, among eukaryotic species.

Based on their mechanisms of movement, transposons can be divided into three classes: **retrotransposons** (retroposons), **DNA transposons**, and **insertion sequences** (ISs).

### 2.1. Retrotransposons

Retrotransposons (also called retroposons) invade the genome by being transcribed to RNA and then back to DNA by reverse transcriptase; they are "copy and paste" reactions *(4)*. The reverse transcriptase is often, but not always, coded by the transposon itself and these MGEs can paste numerous copies into a genome, amplifying themselves in their host genome in tremendous numbers. They are ubiquitous in eukaryotic organisms, and are particularly abundant in plants. They make up between 49% and 78% of the maize genome *(5)* and 68% of the genome in wheat *(6)*. In mammals about half the genome is composed of retrotransposases and 42% of the human genome consists of remnants of retroposons *(7)*. This translates into millions of elements contained in our genomes.

There are two main categories of retrotransposons. The first group, LTR retrotransposons, encodes their own reverse transcriptase, contains long terminal repeats (LTRs), and is transcribed by RNA polymerase II. These are further classified into three subclasses: Ty1-copia-like, Ty3-gypsy-like, and Pao-BEL-like group, based on the degree of sequence similarity and the order of encoded gene products in them. Ty1-copia-like and Ty3-gypsy-like are found in high copy number in animals, fungi, protista, and plants with large genomes. Pao-BEL elements have been found only in animals so far *(8–10)*.

The second group is a non-viral, non-LTR superfamily, which is transcribed by RNA polymerase III and does not code for reverse transcriptase. There are two subtypes of non-LTR retrotransposases, long interspersed nuclear elements (LINEs) and short interspersed nuclear elements (SINEs). Originally considered "junk DNA", LINEs and SINEs play a significant role in gene evolution, structure, and determination of transcription levels. LINEs make up about 21% of the human genome and are used in forensics to generate genetic fingerprints. The non-coding SINEs are dependent upon their partner LINEs for reverse transcription *(11)*. Alu sequences are the most common SINE sequences in primates. SINEs are found in high copy numbers (up to 250,000) in the plant species and primates and make up about 13% of the human genomes *(7)*. These TEs have an impact on somatic cells. LINE 1 (L1) retrotransposases, along with Alu sequences, have been implicated in cancer through active mobilization, domestication of transposases, and by influencing genomic rearrangements *(4)*. The distribution of SINEs has been implicated in some genetic diseases and cancers. Alu sequences have been associated with breast cancer, Ewing's sarcoma, familial hypercholesterolemia, hemophilia, neurofibromatosis, and diabetes mellitus type II. SINEs and other retrotransposons are also used in forensic genomics through quantitative species-specific DNA detection, analysis of complex biomaterials, and the inference of geographic origin of human DNA samples *(12)*. Alu elements are the most abundant repetitive elements in the human genome; they emerged in the primate lineage 65 million years ago from a 5′ to 3′ fusion of the 7SL RNA gene and subsequently amplified throughout the human genome by retrotransposition to reach the present number of more than one million copies *(13)*.

A third category, viral-like retrotransposons, which are similar to retroviruses such as HIV, HIV-1 or HTLV-1, behave like retrotransposons and contain both reverse transcriptase and integrase. The integrase is the retrotransposon equivalent of the transposase of DNA-transposons. There is molecular evidence that endogenous retroviruses in the human genome may be involved in autoimmune diseases *(14)*.

### 2.2. DNA Transposons

This class of MGEs moves directly from one position to another within the genome using a transposase to "cut and paste" itself within the genome. The major difference of DNA Class II transposons from retrotransposons is that their transposition mechanism does not involve an RNA intermediate. Some DNA transposases can bind to any part of the DNA molecule and therefore the target site can be anywhere in the genome, while others bind to specific sequences. The transposase enzyme produces a staggered cut at the target site of DNA, producing sticky ends. The enzyme cuts out the transposon and ligates it into the target site, resulting in target site duplication. Insertion sites of DNA transposons may be identified by short direct repeats followed by inverted repeats (which are important for the transposon excision by transposase). Not all DNA transposons transpose through a cut and paste mechanism. In some cases, a replicative transposition has been described in which the transposon replicates itself to a new target site.

To date there have been no published accounts of naturally occurring active DNA transposases in mammalian genomes. However, there are many copies of inactive fossil DNA transposons present *(7)*, and there are genes that were domesticated from these elements in their host genomes *(15)*. An example is the recombination-activating gene RAG1 and RAG2 proteins. These are important for generating somatic diversity in the immune system because they play an indispensable role in recombination during lymphocyte development. The *Rag* gene is hypothesized to have its origin from the transposase encoded by an ancient transposon superfamily named Transib *(16, 17)*.

Analysis of the *Caenorhabditis elegans* genome indicates that approximately 12% of the *C. elegans* genome is derived from TEs. However, most of these sequences are fossil remnants that are no longer mobile but can be used by molecular archeologists to trace the interactions between parasitic "selfish DNA" *(18)* and a host genome. The investigation into transposition regulation in *C. elegans* has uncovered an unforeseen link between transposition and genome surveillance, genetic responses in the host genome involving chromatin modifications, and RNA interferences that have evolved to moderate or modify the effect of TEs *(19)*.

Tc1 and Tc3 are part of a superfamily of TEs, which is named after its two best-studied members: Tc1 and the related transposon *mariner*, which was identified in 1986 in *Drosophila mauritiana (20)*. Tc1/*mariner* elements are probably the most widespread DNA transposons in nature and can be found in fungi, plants, ciliates, and animals including ecdysozoans and vertebrates *(21)*. Molecular reconstruction of Sleeping Beauty, an ancient transposon in fish, represents a cornerstone in applying transposition-mediated gene delivery in vertebrate species, including humans *(22)*.

### 2.3. Insertion Sequences

An insertion sequence (IS, insertion sequence element, IS element) is a simple TE consisting of a short DNA sequence generally around 700–2500 bp in length. They do not carry any accessory genes and only code for proteins that are part of the transposition activity. This usually includes a transposase, which catalyses the enzymatic reaction, and a regulatory protein, which can stimulate or inhibit the activity. Inverted repeats typically flank the coding regions. ISs can be autonomous but may also be part of composite transposons. In a composite transposon (also known as a "complex transposon"), two ISs flank one or more accessory genes, such as an antibiotic-resistance gene (e.g., Tn10, Tn*5*). ISs are proficient in moving neighboring genes.

Over 2500 different ISs have been identified to date. They are integral components of the mobile, dramatically affecting bacterial genomes as they reshuffle and shape them. Massive expansion of ISs is indicated in the emergence of some pathogenic bacterial species. They also play a role in assembling genes into complex plasmid structures *(23, 24)*. Today, a minimal and incomplete repertoire deposited in the ISfinder database (http://www-is.biotoul.fr) includes 2200 different ISs in over 295 eubacterial and archaeal species *(25)*. Using shared features, ISs are classified into about 20 families. An IS family is defined as a group of ISs with related transposases, strong conservation of the catalytic site, and conservation of organization. ISs are found in most but not all eubacterial and archaeal genomes *(25)*.

Analysis of a diverse set of 18 bacterial genomes revealed that the intra-genomic sequence diversity of a given IS is very low, suggesting that most ISs in an individual genome are evolutionarily young and might have been recently acquired. This observation can be explained by a period of IS expansion followed by a series of IS "extinctions" in bacterial lineages. Wagner *(26)* explains this scenario through the transitory selective advantage of IS elements to their host through lateral gene transfers and genomic rearrangements, which might prove to be detrimental to their host in the long term. However, the loss can be explained if the element is selectively neutral or nearly so. Although ISs are usually discussed in terms of prokaryotic genomes, certain eukaryotic DNA sequences belonging to the family of Tc1/*mariner* TEs may be considered to be ISs.

## 3. Plasmids

American molecular biologist Joshua Lederberg first coined the term plasmid in 1952 *(27)*. They are extra-chromosomal DNA molecules capable of autonomous replication separate from chromosomal DNA. They are typically circular and double-stranded and occur naturally in bacteria. As naturally occurring elements, they are usually not essential elements of the cell. Sometimes they are found in eukaryotic organisms (e.g., the *2-micrometre-ring* in *Saccharomyces cerevisiae*) *(28)*. Plasmid size varies from 1 to over 400 kbp. They may exist as a single copy in a cell or hundreds or thousands of copies of the same plasmid in a single cell.

Plasmids can be categorized in a variety of ways, including by function. These include five main functional classes: fertility

(F) plasmids, resistance (R) plasmids, Col-plasmids which contain genes that code for genes that kill other bacteria, degradative plasmids which are capable of the digestion of unusual substances, and virulence plasmids which can turn bacteria into pathogens. Interestingly, plasmids can belong to more than one functional group. Plasmids are also defined as high (over 100 molecules per cell) and low copy plasmids (1–25). Plasmids can be assigned into *compatibility groups.* It is possible for plasmids of different types to coexist in a single cell, but related plasmids are often incompatible, with only one of the relative surviving in the cell line. A good example of plasmid diversity is demonstrated in *Escherichia coli*, which hosts seven different plasmids. Plasmids are active participants in horizontal gene transfer (HGT) through the transfer of genetic material during conjugation *(29)*.

### 3.1. Conjugative Plasmids

Conjugation is bacterial sex, if you define sex as cell-to-cell contact with the exchange of material. Contrary to real sex in eukaryotes, there is no fusion of gametes, creation of zygotes, or equal exchange of genetic material (the result of sexual activity in eukaryote). In bacterial sex, the donor (a bacterium with the conjugative plasmid) using a pilus initiates contact and the bacterial recipient, while in contact, receives the plasmid in the exchange. This is accomplished through the action of a suite of genes, many of which are contained on the *tra* locus of the plasmid. Sometimes the plasmids become integrated into the host genome, and when these plasmids undergo conjugation, all or part of the host genome may be transferred.

There is an interesting case of a conjugative plasmid active across domains. Plasmids in *Agrobacterium* and *Rhizobium* contain elements that transfer to plant cells *(30)*. Once the genes are transferred to the plant cell, the plant cell protein machinery is hijacked to produce opines, which are used by the bacteria for energy and cell-building sources. This inter-kingdom transfer from bacterial plasmids produces galls or root nodules in the infected plants.

### 3.2. Non-conjugative Plasmids

Non-conjugative plasmids are incapable of initiating conjugation; they are transferred when they associate with conjugative plasmids. Researchers continue to discover their mobilization potential despite this dependence. For instance, non-conjugative plasmids have been implicated in events as varied as the ripening process of goat's milk cheese *(31)* to the isolation of non-conjugative plasmids from antibiotic-resistant *Staphylococcus* strains *(32)*.

### 3.3. IS Elements in Plasmids

Siguier and co-workers *(23)* did an initial survey for IS elements in plasmids and found the incidence of IS sequences is about zero for plasmids smaller than ~20 kb. Once the IS sequence size is greater than 20 kb, there is an abrupt increase in their abundance, averaging between 5% and 15%, and reaching 20% to >40% in some cases. Siguier and co-workers hypothesize that plasmids capable of self-transfer between various species and genera are larger because they carry transfer functions and are capable of acquiring transposons, ISs, and accessory genes during passage between different host genomes. Siguier and co-workers cite evidence that certain TEs specifically target transmissible plasmids.

## 4. Bacteriophages

**Bacteriophages** (or more commonly, phages) are viruses that infect bacteria. They are typically composed of an outer protein shell surrounding the genetic material. Phages can range in size from 5 to 500 kb, can be either circular or linear in architecture, and be present as single-stranded RNA, double-stranded RNA, single-stranded DNA, or double-stranded DNA. They are estimated to be the most diverse and widely distributed entities in the biosphere, found anywhere where there are bacterial hosts. Despite their small size (between 20 and 200 nm), they are extremely effective predators and control the world's bacterial population *(33)*. Phages are major contributors to the process of HGT due to their environmental ubiquity, their large numbers, and functional affects on the hosts they infect *(34)*. In order to understand their role in HGT, we need to appreciate in basic terms, the "phage life cycle."

### 4.1. The Phage Life Cycle

**The lytic cycle** in phage depends upon chance encounters with the proper bacterial host receptors and, once the phage finds its match and attaches, with syringe-like action, injects its DNA into the host. Replication of phage genes begins almost immediately (at the expense of the host protein needs) and virion particle assembly begins. In about 15 min, separately constructed phage heads and tails will spontaneously assemble, packaged tightly with genetic material. In as little as 20 min after infection, phages can lyse their host, releasing 300 plus virus particles that can in turn infect other hosts.

With **lysogenic phage (also called temperate phage)** the infection will not initially cause lysis of the cell, but instead upon infection the phage genome integrates with the host genome (referred to as an endogenous phage or prophage). This arrangement continues until the host finds itself in deteriorating conditions. Once this happens, the dormant prophage will be excised from the host genome and the cycle resulting in lysis of the cell begins. During this process, various pieces of host DNA can be introduced into the viral genome, and as the lysis continues, host DNA is packaged into the soon-to-be-dispersed phages. Interestingly, the lysogenic cycle allows the host cell to continue to survive and reproduce while the virus remains resident in the host's genome and is reproduced in all of the cell's offspring. Furthermore, in a process known as lysogenic conversion, prophages can provide benefits to its host while they are dormant by adding new functions to the bacterial genome. The harmless strain of *Vibrio cholerae* turns into the highly virulent one that causes cholera by this process *(35)*.

### 4.2. Filamentous Phages

Filamentous phages (F-phages) are examples of how evolution produces entities that do not fit into neatly defined categories. They co-opt their host's machinery typical of phage, but never integrate nor do they ever lyse their host cell. When in their infective state, F-phages are shaped like rods, contain a single-stranded DNA genome (the wild type with ~6400 bp), and will naturally infect Gram-negative bacteria. Arguably, the most famous filamentous phage is M13. Once M13 finds an *E. coli* host that contains an F plasmid (which provided the host with a pilus that it MUST have for infection), the phage can gain entry into its host. Once there, a life cycle begins that includes a double-stranded DNA intermediate as the replicative form, which is converted to a single-stranded DNA intermediate prior to encapsidation. This is certainly a key reason why M13 is a major component of the molecular biologists' toolbox. Recombinant technologies take advantage of the fact that it never lyses its host but continually pumps phage particles into the media, yet remain discreet entities within the host (they rarely ever become lysogenic). Currently, M13 is being investigated for uses as nanostructures and nanotechnology, and in 2006 MIT researchers using a modified M13 produced a protein that would complex with cobalt ion, producing a cobalt oxide material with energy storage capacity higher than current lithium-ion batteries *(36)*.

### 4.3. Gene Transfer Agents

A virus-like agent (colloquially called "gene transfer agents") has been described in *Rhodobacter capsulatus*, which at first glance might seem to be a defective prophage *(37)*. Currently, they are found only in prokaryotes and these gene transfer agents seem to function through genomic DNA transfer between cells with no negative effects associated with the transfer to the recipient. They are widespread in alpha proteobacteria species; and Lang and Beatty *(37)* suggest that because of their regulatory mechanisms (they are expressed through histidyl-aspartyl signaling and quorum sensing genes), they should be viewed as other genuine cellular structures analogous to flagella or pili.

## 5. Self-Splicing Molecular Parasites

### 5.1. Group II Introns

Group II Introns are a class of introns found in the rRNA, tRNA, and mRNA of organelles of fungi, plants, and protists and in the mRNA of bacteria. Because of their wide distribution, their possible role in the evolution of nuclear spliceosomal introns and nuclear non-LTR-retrotransposons *(38)*, and their ability to "retrohome" and also "retrotranspose" at low frequencies into ectopic sites that resemble the normal homing site, they are a fundamental component of the mobilome. They are ribozymes that are retroelements consisting of a highly structured RNA (typically exhibiting six stem-loops known as domains) and a multifunctional intron-encoded protein (IEP). The RNA carries out the splicing and reverse splicing (integration) activities, while the IEP facilitates these reactions by stabilizing the catalytic structure of the ribozyme *(38)*. The mobility of Group II introns occurs through a target DNA-primed reverse transcriptase. The excised intron RNA reverse splices directly into a DNA target site and is then reverse transcribed by the IEP. With minimal host damage, Group II introns "retrohome" at almost 100% into specific DNA target sites, usually at unoccupied site in an intronless allele. These elements comprise more than 45% of the human genome *(38)* and sometimes they can be found internal to IS sequences.

A brief description of group II intron phylogeny and evolution can provide some insight into their current and past roles in the mobilome. There are more than 200 ORF-containing group II introns sequenced and described, and almost all encode reverse transcriptase-related proteins. Based on phylogenetic analysis, the IEPs can be divided into eight major lineages denoted mitochondrial, chloroplast-like 1 and 2, and bacterial A–E *(31, 32)* (**Fig. 2.1**).

Each lineage of IEP is associated with a distinct RNA structural subclass, implying that the IEP was likely associated with the intron RNA prior to the divergence of different group II intron lineages. Because there is no obvious exchange of IEP, it appears that these lineages have experienced little or no HGT *(39)*. The "mitochondrial" and "chloroplast" lineages include a number of bacterial group II introns (e.g., the *Lactobacillus lactis* Ll.LtrB intron belongs to the "mitochondrial" lineage), and this can be explained as a reflection that the organellar group II introns have evolved from specific bacterial lineages *(40)*. While the IEPs have experienced very little HGT, the heterogeneous phylogenetic distribution of group II intron subclasses suggests that HGT of the introns themselves is relatively common, and cross-species transfer by conjugation has been demonstrated *(38)*.

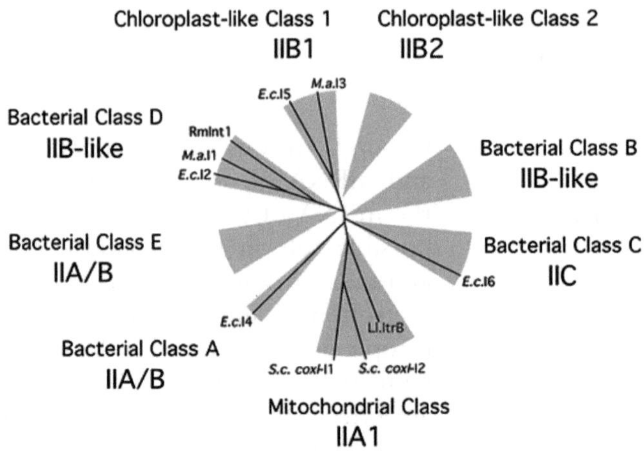


Fig. 2.1. Phylogeny of group II intron ORFs and correspondence with RNA structural classes. Phylogenetic relationships of group II intron ORFs are summarized based on neighbor-joining analyses. Group II intron ORFs are divided into eight clades, named mitochondrial, chloroplast-like 1 and 2, and bacterial A–E. Each ORF clade is associated with a distinct RNA structural class (IIA1, IIB1, IIB2, IIC, two other distinct IIB-like, and two distinct IIA/B hybrid classes) (from *(38)*, http://www.fp.ucalgary.ca/group2introns/).

Evidence for the conversion of Group II introns into spliceosomal introns is provided by degeneration of internal RNA structure, dependence on a common splicing apparatus, and the use of *trans*-acting RNAs. Because translation and transcription are separated in eukaryotes by the nuclear membrane, a Group II intron cannot bind the IEP immediately after the RNA is transcribed. This separation would favor substituting host protein in *trans*, which in turn favors the evolution of a common slicing apparatus. Mobility in early "spliceosomal" introns would be retained due to interaction with the RTs in *trans*. However, this would create a disadvantage as the number of introns grew, because the mobility would be increasingly detrimental to the host. If evolution then favored the replacement of the RT with other cellular splicing factors, then any Group II introns that had not inserted within genes would be under no selective pressure to retain splicing, enabling them to evolve into non-LTR-retrotransposons *(38)*.

### 5.2. Group I Introns, Inteins, and Homing Endonucleases

Three other classes of molecular parasites are worth mentioning. We will briefly describe Group I introns, inteins, and homing endonucleases. Group I introns are found in both eukaryotic and bacteriophage systems *(41)* and are considered part of the mobilome because they can transfer their intron to intronless alleles of the same gene. These introns catalyze their own splicing through a series of reactions initiated by a guanosine and ending in a linear segment of RNA. They can vary in size from 200 to 3000 nucleotides. Group I introns can have fairly sophisticated secondary and tertiary structure that is conserved in structure but not necessarily in sequence homology. They are workhorses in the field of ribozyme in vitro evolution studies *(42)*.

Inteins are segments of a protein that are able to excise and rejoin through a peptide bond. They usually contain an endonuclease domain that plays a role in their propagation. Inteins have been found in all three super domains of life and a nice review of the role evolution has played in the spread and maintenance of these parasites is provided by Gogarten and Hilario *(43)*. In brief, the intein can maintain a functional endonuclease over the long term despite original models which would indicate that once inteins were fixed in a population, the homing endonuclease would be lost through random processes. Certainly, inteins provide a robust platform for the consideration of how parasitic genes move and become fixed in a population.

Homing endonucleases *(44)* deserve a few sentences of mention because they are such unique entities. Basically, these are rare-cutting enzymes that are encoded by the inteins and introns. There are four families, characterized by sequence motifs LAGLI-DADG, GIY-YIG, H-N-H and His-Cys. They recognize stretches of DNA that are usually 12–40 bases long. They are considered highly invasive because they take refuge in the molecular parasites mentioned above and are therefore important elements of the mobilome community.

## 6. Conclusions/ Outlook

Our ability to sequence whole genomes in a matter of weeks provides us with opportunities to formulate and plan for strategies to quantify the mobilome in those genomes. It is an exciting prospect, especially given that microbial ecology and technology have allowed researchers to move the sequencing targets out of the laboratory and into the field. With the recognition that different bacterial lineages seemed to have different propensities for HGT *(45)*, researchers are eager to define the rules and constraints that could predict or define the transfer and the agents that accomplish them. Strides have been made. For instance, the classification and quantification of incompatibility groups of plasmids in certain lineages seems to be reaching a quantifiable threshold (e.g., 26 in the enterobacteriaceae, 14 in the pseudomonads, and 18 in the Gram-positive staphylococci) *(46)*. The burgeoning field is not without issues and they are significant ones. Current analysis schemes are hampered by the lack of large comparable data sets of MGE sequences to make gene predictions possible. The repetitive sequence, so common a feature of many MGEs, is the sequencer's nightmare. The complexity and intermixing of MGEs makes identification and categorization in large-scale and automated projects a daunting task. Physical separation of autonomous MGE elements in bacterial species so that sequencing strategies can be performed is quite problematic. Despite these hurdles, the future is bright, because to understand the mobilome means that we will get at the very heart of evolution and the adaptive process of biology. The mobilome is at our fingertips and our appreciation of the mobilome will continue to expand as more sequence data accumulate and clever schemes are devised to analyze it. There is no doubt that mobilome discovery will continue to reveal the exquisite artistry that evolution has produced in the agents of change of biology.

## References


1. McClintock, B. (1950) The origin and behavior of mutable loci in maize. *Proc Natl Acad Sci U S A* **36**, 344–355.
2. Pace, N. R. (2006) Time for a change. *Nature* **441**, 289.
3. Guttman, D. S., Dykhuizen, D. E. (1994) Clonal divergence in *Escherichia coli* as a result of recombination, not mutation. *Science* **266**, 1380–1383.
4. Collier, L. S., Largaespada, D. A. (2007) Transposable elements and the dynamic somatic genome. *Genome Biol* **8 Suppl 1**, S5.
5. SanMiguel, P., Gaut, B. S., Tikhonov, A., Nakajima,Y., Bennetzen, J. L. (1998) The paleontology of intergene retrotransposons of maize. *Nat Genet*, **20**, 43–45.
6. Li, W., Zhang, P., Fellers, J. P., Friebe, B., Gill, B. S. (2004) Sequence composition, organization, and evolution of the core Triticeae genome. *Plant J*, **40**, 500–511.
7. Lander, E. S., Linton, L. M., Birren, B., Nusbaum, C., Zody, M. C., Baldwin, J., Devon, K., Dewar, K., Doyle, M., FitzHugh, W. et al. (2001) Initial sequencing and analysis of the human genome. *Nature* **409**, 860–921.
8. King, R. D., Whelan, K. E., Jones, F. M., Reiser, P. G., Bryant, C. H., Muggleton, S. H., Kell, D. B., Oliver, S. G. (2004) Functional genomic hypothesis generation and experimentation by a robot scientist. *Nature* **427**, 247–252.
9. Copeland, C. S., Mann, V. H., Morales, M. E., Kalinna, B. H., Brindley, P. J. (2005) The Sinbad retrotransposon from the genome of the human blood fluke, Schistosoma mansoni, and the distribution of related Pao-like elements. *BMC Evol Biol* **5**, 20.
10. Wicker, T., Sabot, F., Hua-Van, A., Bennetzen, J. L., Capy, P., Chalhoub, B., Flavell, A., Leroy, P., Morgante, M., Panaud, O. et al. (2007) A unified classification system for eukaryotic transposable elements. *Nat Rev Genet* **8**, 973–982.
11. Weiner, A. M. (2002) SINEs and LINEs: the art of biting the hand that feeds you. *Curr Opin Cell Biol* **14**, 343–350.
12. Ray, D. A., Walker, J. A., Batzer, M. A. (2007) Mobile element-based forensic genomics. *Mutat Res* **616**, 24–33.
13. Hasler, J., Strub, K. (2006) Alu elements as regulators of gene expression. *Nucleic Acids Res* **34**, 5491–5497.
14. Colmegna, I., Garry, R. F. (2006) Role of endogenous retroviruses in autoimmune diseases. *Infect Dis Clin North Am* **20**, 913–929.
15. Miller, W. J., Hagemann, S., Reiter, E., Pinsker, W. (1992) P-element homologous sequences are tandemly repeated in the genome of *Drosophila guanche*. *Proc Natl Acad Sci U S A* **89**, 4018–4022.
16. van Gent, D. C., Mizuuchi, K., Gellert, M. (1996) Similarities between initiation of V(D)J recombination and retroviral integration. *Science* **271**, 1592–1594.
17. Kapitonov, V. V., Jurka, J. (2005) RAG1 core and V(D)J recombination signal sequences were derived from Transib transposons. *PLoS Biol* **3**, e181.
18. Orgel, L. E., Crick, F. H. (1980) Selfish DNA: the ultimate parasite. *Nature* **284**, 604–607.
19. Bessereau, J. L. (2006) Transposons in *C. elegans*. *WormBook* 1–13.
20. Jacobson, J. W., Medhora, M. M., Hartl, D. L. (1986) Molecular structure of a somatically

unstable transposable element in Drosophila. *Proc Natl Acad Sci U S A* **83**, 8684–8688.
21. Plasterk, R. H. (1991) The origin of footprints of the Tc1 transposon of *Caenorhabditis elegans. EMBO J* **10**, 1919–1925.
22. Ivics, Z., Izsvak, Z. (2006) Transposons for gene therapy! *Curr Gene Ther* **6**, 593–607.
23. Siguier, P., Filee, J., Chandler, M. (2006) Insertion sequences in prokaryotic genomes. *Curr Opin Microbiol* **9**, 526–531.
24. Chandler, M., Mahillon, J. (2002) Insertion sequences revisited. In *Mobile DNA vol II.* (Craig, N. L., Craigie, R., Gellernt, M., Lambowitz, A. M., eds.), ASM Press, Washington D.C., pp. 305–366.
25. Filee, J., Siguier, P., Chandler, M. (2007) Insertion sequence diversity in archaea. *Microbiol Mol Biol Rev* **71**, 121–157.
26. Wagner, A. (2006) Cooperation is fleeting in the world of transposable elements. *PLoS Comput Biol* **2**, e162.
27. Lederberg, J. (1952) Cell genetics and hereditary symbiosis. *Physiol Rev* **32**, 403–430.
28. Velmurugan, S., Mehta, S., Uzri, D., Jayaram, M. (2003) Stable propagation of 'selfish' genetic elements. *J Biosci* **28**, 623–636.
29. Thomas, C. M., Nielsen, K. M. (2005) Mechanisms of, and barriers to, horizontal gene transfer between bacteria. *Nat Rev Microbiol* **3**, 711–721.
30. Saito, K., Yamazaki, M., Murakoshi, I. (1992) Transgenic medicinal plants: *Agrobacterium*-mediated foreign gene transfer and production of secondary metabolites. *J Nat Prod* **55**, 149–162.
31. Martinez-Cuesta, M. C., Requena, T., Pelaez, C. (2001) Use of a bacteriocin-producing transconjugant as starter in acceleration of cheese ripening. *Int J Food Microbiol* **70**, 79–88.
32. Udo, E. E., Jacob, L. E. (1998) Conjugative transfer of high-level mupirocin resistance and the mobilization of non-conjugative plasmids in *Staphylococcus aureus. Microb Drug Resist* **4**, 185–193.
33. Kunin, V., He, S., Warnecke, F., Peterson, S. B., Garcia, M. H., Haynes, M., Ivanova, N., Blackall, L. L., Breitbart, M., Rohwer, F., McMahon, K.D., Hugenholtz, P. (2008) A bacterial metapopulation adapts locally to phage predation despite global dispersal. *Genome Res* **18**, 293–297.
34. Canchaya, C., Fournous, G., Brussow, H. (2004) The impact of prophages on bacterial chromosomes. *Mol Microbiol* **53**, 9–18.
35. Faruque, S. M., Nair, G. B. (2002) Molecular ecology of toxigenic Vibrio cholerae. *Microbiol Immunol* **46**, 59–66.
36. Nam, K. T., Kim, D. W., Yoo, P. J., Chiang, C. Y., Meethong, N., Hammond, P. T., Chiang, Y. M., Belcher, A. M. (2006) Virus-enabled synthesis and assembly of nanowires for lithium ion battery electrodes. *Science* **312**, 885–888.
37. Lang, A. S., Beatty, J. T. (2007) Importance of widespread gene transfer agent genes in [alpha]-proteobacteria. *Trends Microbiol* **15**, 54–62.
38. Lambowitz, A. M., Zimmerly, S. (2004) Mobile group II introns. *Annu Rev Genet* **38**, 1–35.
39. Toor, N., Hausner, G., Zimmerly, S. (2001) Coevolution of group II intron RNA structures with their intron-encoded reverse transcriptases. *RNA* 7, 1142–1152.
40. Belhocine, K., Plante, I., Cousineau, B. (2004) Conjugation mediates transfer of the Ll.LtrB group II intron between different bacterial species. *Mol Microbiol* **51**, 1459–1469.
41. Lambowitz, A. M., Belfort, M. (1993) Introns as Mobile Genetic Elements. *Annu Rev Biochem* **62**, 587–622.
42. Joyce, G. F. (2007) Forty years of in vitro evolution. *Angew Chem Int Ed Engl* **46**, 6420–6436.
43. Gogarten, J. P., Hilario, E. (2006) Inteins, introns, and homing endonucleases: recent revelations about the life cycle of parasitic genetic elements. *BMC Evol Biol* **6**, 94.
44. Gogarten, J. P., Olendzenski, L., Hilario, E., Simon, C., Holsinger, K. E. (1996) Dating the cenancester of organisms. *Science* **274**, 1750–1751.
45. Ochman, H., Lawrence, J. G., Groisman, E. A. (2000) Lateral gene transfer and the nature of bacterial innovation. *Nature* **405**, 299–304.
46. Frost, L.S., Leplae, R., Summers, A. O., Toussaint, A. (2005) Mobile genetic elements: the agents of open source evolution. *Nat Rev Micro* **3**, 722–732.

# Chapter 3

# The Interplay of Homologous Recombination and Horizontal Gene Transfer in Bacterial Speciation

**Jeffrey G. Lawrence and Adam C. Retchless**

## Abstract

Bacteria experience recombination in two ways. In the context of the Biological Species concept, allelic exchange purges genic variability within bacterial populations as gene exchange mediates selective sweeps. In contrast, horizontal gene transfer (HGT) increases the size of the population's pangenome by providing an influx of novel genetic material. Here we discuss the interplay of these two processes, with an emphasis on how they allow for the maintenance of genotypically cohesive bacterial populations, yet allow for the separation of these populations upon bacterial speciation. In populations that maintain genotypic similarity by frequent allelic exchange, horizontally transferred genes may initiate ecological barriers to genetic exchange. The resulting recombination interference allows for the accumulation of neutral mutations and, consequently, the imposition of a pre-mating barrier to gene transfer.

**Key words:** Recombination, speciation, periodic selection, recombination interference, horizontal gene transfer, cohesion, species concepts.

## 1. Introduction

The identification of organisms is a fundamental step in deciphering that organism's biology; the power of classification is the implicit understanding of what that organism is likely to do, or capable of doing, based on the past experiences with similar organisms. The species is one of the most fundamental and recognizable units of biological organization. The term is used by both biologists and non-biologists in its original, Aristotle-born meaning to encompass a group of individuals – organisms, object, or thoughts – which share commonality as well as

Maria B. Gogarten et al. (eds.), *Horizontal Gene Transfer: Genomes in Flux, vol. 532*

DOI 10.1007/978-1-60327-853-9_3 Springerprotocols.com

properties that distinguish members of one species from those of other species. In his *Scala Naturae*-based treatise on animal classification, Aristotle identified almost all of the extant species of mammals in Europe using only this general principle as a guide *(1)*, easily distinguishing between related species (e.g., three kinds of herons) while ignoring variability among individuals of the same species. While Linnaeus offered a more complex, hierarchical classification in *Systema Naturae* – placing species as the most fundamental unit of taxonomy below Genus, Family, Order, Class, Phylum, and Kingdom *(2)* – the basis for assigning organisms to species was not substantially more sophisticated than that used by Aristotle. Arguably, most schoolchildren can classify oft-encountered organisms into species with relative ease, identifying similar organisms as members of named groups and recognizing differences between groups. Indeed, children as young as four months of age are able to categorize animals such as cats and dogs in a way that distinguishes between the groups even as they recognize different individuals within each group *(3)*. So well-entrenched is the "common-sense" perception of species boundaries that the United States' Endangered Species Act of 1973 includes a definition of endangered species that fails to delineate or define a species itself *(4)*. In the court of public opinion, the term "species" lies in that nebulous group of words whose definitions may be difficult to state precisely, but which can be recognized with ease: the "I know it when I see it" phenomenon.

This has been an unsatisfying state for biologists ever since Darwin rephrased the study of biological diversity as the origination of new species from old ones *(5)*. The species played a central role in his view of life: members of the same species produced offspring that competed with one another, with only the fittest surviving. The two fundamental properties of species are evident here. First, there are genetic and ecological commonalities in members of a species in sharing the ability to reproduce with one another, as well as to compete with one another for resources. Second, there is distinction between different species, where members of different species could stably co-exist due to the lack of niche overlap. Beyond this, Darwin introduced a third concept: different species are not only distinguishable from one another but at least one must also be distinct from their "parent" species. Here, speciation is the evolutionary process by which species originate and become distinct from their ancestor. Also explicit in his model is the idea that each species originates from single ancestral species, so that descendents may be organized into hierarchical groups-within-groups, just as Linnaeus had done. Here, we examine how bacteria may be classified into species, and how HGT impacts this process.

## 2. Species

### 2.1. Why Species Matter

Beyond the purely scientific interest in defining natural groups that represent organisms with shared evolutionary trajectories, species' names impact the non-scientific world in many critical ways. Because members of a species share properties and behaviors, we rely upon their proper identification to generate an appropriate response. Nowhere, perhaps, is this more true than with microbial species. The response to finding spores of *Bacillus subtilis* versus those of *Bacillus anthracis* (the causative agent of anthrax) would be very different. Identification of bacterial species plays a central role in medical diagnosis, food safety, public health, biotechnology, and response to bioterrorism. Thus species delineation has a practical use, providing professional microbiologists with a common language to discuss the biology of important groups of microorganisms.

### 2.2. Species Definitions and Species Concepts

There are two approaches to delineating the boundaries of microbial species. First, one may define a species in any way deemed appropriate so that it encompasses strains that share phenotypes of interest. This is often the tactic taken in delineating pathogenic species: strains of a particular species are the causative agents of particular diseases. Such a description is often called a species definition. Species definitions may use broadly applied methods, such as defining bacterial species to encompass strains which share >60% similarity based on DNA hybridization, or which share >97% identity at their 16S rRNA genes *(6)*. These methods are useful for assigning names to bacteria, but the similarity of the organisms grouped this way may only passively reflect their common ancestry.

Alternatively, a species may represent a group of strains that share common features due to some active underlying biological process. These descriptions are termed species concepts, because it is the process behind the classification that is important. Here, the similarity among organisms within a species reflects a process that acts to maintain that similarity; it is not merely the reflection of common ancestry. In this way, species – as delineated by a species concept, not a species definition – are more meaningful groups than are genera, families, orders, classes, or phyla, which owe the similarity of their members to ancestry alone. As a result, the inclusion or exclusion of taxa from higher taxonomic groups is arbitrary, since no active biological process maintains the similarity among members of the group.

The biological literature is rife with the discussion of how to identify, classify, or delineate species, and what factors influence their origins. All are directed at solving the same principles outlined by Darwin: (a) how and why do members of the same species share similarity, (b) how and why do they maintain distinctiveness relative to related species and (c) how did they come into being? One biological process which promotes genotypic similarity among species members is gene exchange. Not surprisingly, homologous recombination plays a central role in many concepts of species and speciation. Mayr proposed that species' members share a common gene pool, and that high-frequency gene exchange among groups of con-specific individuals is what provided genotypic (and thus phenotypic) cohesion within species *(7, 8)*. The process of speciation, then, would amount to the imposition of barriers which prevented facile gene exchange. Paterson similarly proposed that a shared mate-recognition system would lead to genotypic cohesion *(9)*; here, speciation would entail development of different mate-recognition systems, rather than any direct barrier to gene exchange. In either case, exchange of genes among con-specific individuals works to prevent the gradual accumulation of differences that result from ongoing mutation.

Other, more broadly defined species concepts do not invoke recombination per se, but recognize the intrinsic similarly of species members. Van Valen proposed that members of the same species share an ecological niche *(10)*, necessitating a phenotypic cohesion that likely stems from genotypic similarity. Wiley proposed that members of a species share the same evolutionary trajectory *(11)*, thus also reflecting phenotypic (and thus genotypic) similarity. Most broadly, Templeton argued that species members share cohesion mechanisms, regardless of what they may be *(12)*. Below, we see that all of these ideas have bearing on the identification and delineation of groups of bacteria that share genotypic and phenotypic similarity imparted by cohesion mechanisms, and have shared ecologies and evolutionary trajectories.

## 3. Cohesion in Bacterial Species

### 3.1. Genotypic Cohesion Can Be Imparted by Periodic Selection

In bacteria, genotypic differences between strains arise every cell division as inevitable mutations arise. These differences are non-randomly distributed among strains; that is, there appears to be several mechanisms through which similarity is maintained among a larger group of strains than one would expect at random. One mechanism is periodic selection.

At the generational level, the relationships among bacteria are purely hierarchical. Reproduction in bacteria does not involve obligate reshuffling of genetic material as it does in eukaryotes that experience meiosis and syngamy to generate haploid genomes and join them to form diploid progeny. Levin *(13)*, extending ideas of Atwood et al. *(14)*, proposed periodic selection as a mechanism through which microbial population could retain similarity and avoid mutation-driven diversification. Here, when selectively beneficial mutations sweep a population, the entire chromosome "hitchhikes" with it, thus purging variability (**Fig. 3.1A**). The resulting clonal expansion of the strain bearing the beneficial allele is bounded only by its ability to outcompete similar strains lacking this allele. Cohan and colleagues have termed such lineages as ecotypes *(6, 15, 16)*, arguing that their sweeps purge genetic variability only within ecologically identical strains. Hence, periodic selection events result in genotypic cohesion in a bacterial population.

In many ways, ecotypes have properties that are associated with species. Similarity is maintained among individuals in a population by an active process, groups are clearly differentiated from one another by ecological distinctiveness, and there is a mechanism (fixation of beneficial mutations) that can lead to lineage separation. Thus ecotypes could be considered one of the most fundamental units of organization of bacterial strains. But from a conceptual standpoint, ecotype boundaries can only be established by elucidating the ecological niche of a bacterial strain. This may not be feasible, as it requires the assessment of the limits of "ecologically neutral" genetic changes. That is,

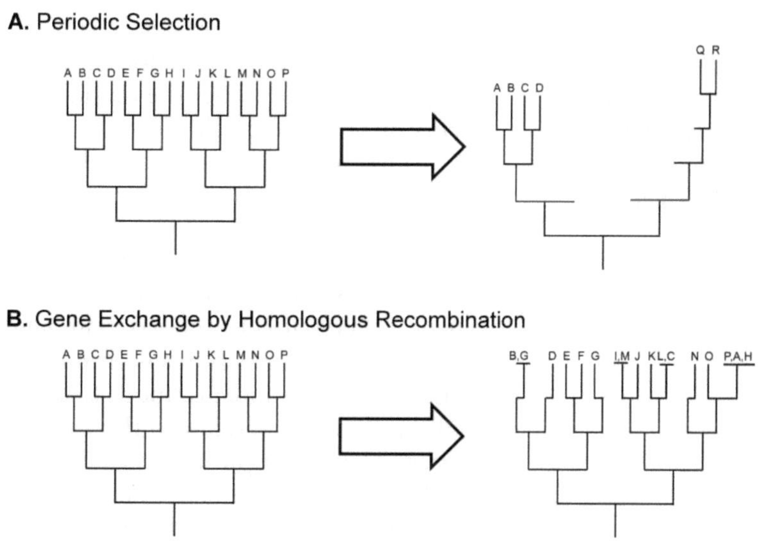


Fig. 3.1. Models for maintaining genotypic cohesion. **(A)** Periodic selection. The arrival of a beneficial mutation in lineage "P" leads to the extinction of lineages "E"–"O", thus reducing variability among strains in this population. Strains "A"–"D" represent a different ecotype and are not outcompeted in the periodic selection event. **(B)** Recombination. Here, variability is reduced at individual loci as allelic exchanges occur. Recombination events among strains B & G, I & M, L & C, and P, A & H are denoted here.

one must determine which differences among bacteria are sufficiently large to place them in different niches, and therefore be different ecotypes. However, the scope of an ecotype – that is, the boundaries of the population encompassed by a periodic selection event – is a function of the nature of the beneficial mutation driving periodic selection. Mutations of small benefit would define a narrow ecotype, whereas those with greater benefits would purge variability among a group of more diverse strains.

From a practical standpoint, it is also unlikely that ecotypes will replace current species standards. First, it has been estimated that many named bacterial species – like *Escherichia coli* – contain hundreds or thousands of ecotypes. The utility of bacterial species definitions in medical diagnosis, food safety, and bioterrorism is the rapid assessment of the biological properties of a strain based on crude estimates of its relatedness to other strains. For now, the properties of many bacteria that garner the strongest interest encompass far more strains that are found in an ecotype. Partly, this is because beneficial alleles that arise within an ecotype may also spread to a much larger, and more diverse, set of strains via homologous recombination. Second, objectively assessing "ecologically neutral" differences is impossible when considering sequence differences alone. As a result, ecotypes with particularly interesting phenotypes – e.g., *Mycobacterium tuberculosis* or *Bacillus anthracis* – may be classified as species *(6)*.

### 3.2. Genotypic Cohesion Can Be Imparted by Homologous Recombination

#### 3.2.1. Bacteria Exchange Genes

Recombination in bacteria does not involve the exchange of haploid genomes, but rather the unidirectional transfer of small fragments of DNA between donor and recipient (**Fig. 3.1B**). Here, DNA may be moved between cells by one of three mechanisms *(17)*. Transduction occurs when bacteriophages mistakenly package bacterial DNA into their capsids instead of virus DNA. When this particle finds a target cell, the DNA – limited in size to a fragment that will fit in the capsid – is injected. Transformation occurs when a bacterial cell imports fragment of naked DNA directly from the environment; this is common among bacteria that consume DNA as a source of food. Conjugation occurs when a plasmid that has integrated into the bacterial chromosome begins its process of replication and transfers into another host. Plasmid DNA is transferred directly between the cytoplasm of the donor cell into the cytoplasm of the recipient, thus requiring prolonged cell–cell contact. Conjugation can move large portions of chromosomal DNA between cells.

After the DNA has been injected into the cytoplasm of the recipient cell, it is subjected to five important processes. First, restriction endonucleases will cleave almost all incoming

DNA fragments, with the exception of DNA arriving from a cell expressing the same *hsd*-encoded restriction/modification system, whose cleavage sites have thus been protected. Given the variability in *hsd* genes within and among bacterial species *(18–20)*, this exception is rare, even within named species. Second, exonucleases will degrade the double-stranded (ds) DNA ends of the resulting fragments. These two processes act in concert to reduce the size of incoming DNA fragments and, most often, prevent the DNA from integrating into the recipient chromosome. Third, RecA-mediated homologous recombination may occur, whereby the incoming DNA fragment – reduced in size through the action of nucleases *(21)* – is integrated into the chromosome, replacing the resident allele at its cognate position. This requires nucleotide sequence identity between incoming and resident DNA, and the occurrence of mismatches reduces the probability of successful recombination *(22)*. Fourth, if no region of similarity exists between the incoming and the resident DNA, illegitimate recombination may occur, placing the arriving DNA anywhere in the chromosome or, alternatively, site-specific recombinases (e.g., phage integrases) may catalyze recombination into specific locations. Lastly, persistence of any newly acquired genes rests on the interplay of stochastic processes (which may lead to the loss of genes that are potentially advantageous) and natural selection which, as is discussed below, is the final arbiter of gene fate in bacterial genomes.

### *3.2.2. Two Kinds of Gene Exchange*

When viewed this way, HGT encompasses two distinct processes. First, genes may be added to a recipient genome after being transferred from a potentially distantly related taxa. This process is often termed horizontal (or lateral) gene transfer, and it is this process that is the topic of this book. Second, alleles may be exchanged between closely related taxa, resulting in gene conversion. This process is often called "recombination," alluding both to the role of the homologous recombination machinery in catalyzing allele replacement and to the population genetic process of reducing linkage disequilibrium *(23)*. The efficiency of the homologous recombination machinery will catalyze gene conversion events almost to the exclusion of illegitimate recombination events when the incoming DNA is highly similar to resident genes. Often the two processes are studied separately: HGT is viewed as occurring between different species, and recombination occurs primarily within species. Here, we argue that they are intimately associated, with each process affecting the scope and impact of the other; it is this association that also affects the process of lineage diversification – speciation – in bacteria.

### 3.3. Gene Transfer Among Related Organisms: A Bacterial Species Concept

Dykhuizen and Green also invoked homologous recombination in their encapsulation of Mayr's biological species concept *(23)*. Their model is a retrospective one, using the patterns of genetic diversity among individuals to delineate species boundaries. The proposal was that the relationships among individuals as inferred from different genes would not be congruent within a species, but they would be congruent between them. Thus the Dykhuizen and Green model directly invokes homologous recombination as a cohesion mechanism within bacterial species: strains of the same species exchange genes, resulting in incongruent phylogenies among different genes.

This model works well when applied to some groups of bacteria. For example, different genes among different strains of the enteric bacteria *E. coli* or *Salmonella enterica* show different relationships, reflecting homologous recombination within these groups *(23, 24)*. Yet different genes among different enteric bacteria show congruent relationships, implying that homologous recombination does not readily exchange genes across the boundaries of these named species *(25)*; *see* also **Chapter 21**. When many strains within named species are characterized by Multi-Locus Sequence Typing (MLST) – wherein alleles at a handful of shared loci are genotyped *(26)* – it is clear that many bacterial species have appreciable rates of homologous recombination among constituent strains *(27–31)*.

The contrasts between groups of organisms encompassed by homologous recombination and those delineated by periodic selection (the ecotype model) are clear. Ecotypes encompass strains with sufficiently similar ecologies that periodic selection events may purge all of the variability when beneficial mutations sweep the population, with the entire bacterial chromosome hitchhiking along. Yet this beneficial allele may also rise to high frequency among a much larger, and more diverse, set of strains via homologous recombination *(32, 33)*. Recombination among strains also works against the gradual divergence between strains that accompanies the ongoing accumulation of mutations. Here, genotypic cohesion is imparted among a much larger group of strains than a single ecotype, and their ecologies and evolutionary trajectories are closely tied to one another. The relationships among genomes at this scale are no longer dictated by lineage-specific mutations, but are instead shaped by gene exchange among strains. Thus the species concept outlined by Dykhuizen and Green resembles that defined by Mayr.

Yet while applicable to some groups of bacteria, this model fails for others such as species of the genus *Neisseria*. Here, homologous recombination allows gene exchange between named species at some loci, thereby increasing genotypic diversity within species rather than reducing it. As will be discussed further below, these cases represent points along the path establishing genetic isolation between species.

## 4. Speciation in Bacteria

### 4.1. The Barrier to Gene Exchange Between Species

Analysis of MLST data has revealed that natural populations of bacteria do generally fall into distinct sequence clusters that reflect commonly recognized microbial species, even when those species are closely related and highly recombinogenic *(34, 35)*. If recombination acts to exchange alleles within a bacterial species, what prevents transfer across species boundaries? There are three potential boundaries.

First, gene exchange may be blocked before DNA even enters the cytoplasm, thereby producing a speciation mechanism comparable to classical (whole genome) pre-mating isolation. For example, bacteriophage host ranges may be limited to closely related strains, thus mediating transduction only within species. This does not appear to be the case, as those bacteriophages assayed show variable host ranges, many of which include many named bacterial species. More importantly, bacterial species may be infected by numerous bacteriophages, each with a different host range. For example, *E. coli* is infected both by bacteriophage lambda, which has difficulty infecting other enteric bacteria due to differences in the LamB receptor protein, and bacteriophage P1, which infects many enteric bacteria. Indeed, genes encoding the P1 tail-fiber proteins have been used to create vectors for mutagenesis across numerous enteric bacterial species *(36)*. Barriers to recombination may be supplied by geographical barriers among sexual eukaryotes; if gene exchange is tied to reproduction, then the inability to find mates would curtail recombination and thus lead to allopatric speciation. But gene transfer in bacteria does not require that the donor and recipient be in the same place at the same time. Because bacteriophages can travel large distances and shelter donor DNA for long periods of time, geographic barriers – while clearly slowing down migration and/or recombination in prokaryotes *(37–39)* – is not an absolute isolating mechanism for most bacteria (excepting obligate intracellular symbionts like *Buchnera* or any taxon where vicariance in the host population represent ecological barriers to associated bacterial species).

Second, DNA moved into a recipient cell's cytoplasm may fail to undergo recombination as it has accumulated too many differences. Here, the mismatch-repair system recognizes the duplex between resident DNA and incoming DNA and prevents successful exchange. It has been shown in several bacteria that the efficiency of homologous recombination decreases exponentially with linear decreases in sequence identity *(40–44)*, and it is effectively impossible (except for very short blocks of sequence) when nucleotide sequences are greater than 5% different *(45)*. Thus, DNA is not efficiently exchanged between cells whose DNA has become too different. However, the occasional strain with a defective mismatch-repair system may have alleles converted by DNA with much lower sequence identity *(40)*; after recovery of the mismatch-repair system *(46)*, the recombinant and donor lineages may exhibit facile recombination at the converted locus, even as other loci of the recombinant lineage lack sufficient sequence identity to recombine with the donor lineage *(47)*. This type of locus-specific control over recombination may account for the "fuzzy species" boundaries inferred from MLST studies *(34)*.

Third, DNA may be exchanged, but the resulting recombinant may be counterselected if it is less fit than its two parents (although, in practical terms, it would likely only compete with its nearly isogenic maternal parent). This is comparable to post-mating isolation mechanisms in eukaryotes, while the previously mentioned mechanisms can be classified as pre-mating due to the fact that no recombinant organism is ever produced. In situations where the above pre-mating mechanisms do not reduce recombination to low rates, we believe that counterselection of recombinants will play a decisive role in establishing species boundaries by eliminating recombinants. This is discussed further below.

### 4.2. The Mutation–Recombination Balance

For recombination to act as a cohesion mechanism for bacterial species, it must purge the variability – on a locus by locus basis – faster than it accumulates by mutational processes (**Fig. 3.2**). If

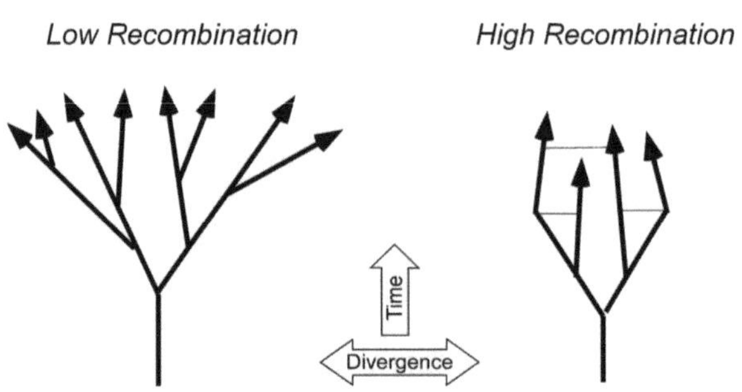


Fig. 3.2. Mutation–recombination balance. Mutations act to make individuals more different, while recombination acts to make them more similar. Under low rates of recombination, differences accumulate faster than they are eliminated by recombination. Populations will become increasingly different, and recombination will eventually cease as donor and recipient DNA become insufficiently similar to support heteroduplex formation. Under high rates of recombination, selective sweeps purge variability as quickly as it is introduced by mutation. Here, genotypic cohesion is maintained among strains over time. Branches represent ecotypes.

the rate of recombination is low, strains will diverge more quickly than recombination can make them similar; eventually, the DNA will become sufficiently different that homologous recombination cannot exchange genes between the separate lineages *(48, 49)*. Thus, "speciation" will have occurred as the inevitable result of mutation. If recombination merely deters the divergence of strains by mutation, then groups of strains that are similar via recombination are transient and, in some ways, artifactual. If speciation is the inevitable result of stochastic mutation, then there are no shared ecological properties of the individual species; species are merely groups of strains that have yet to experience sufficient numbers of mutations to prevent recombination. This is not a satisfying way to delineate species, for the selection of strains to include in a species would be arbitrary.

Alternatively, if rates of recombination are high, strains cannot accumulate sufficient numbers of mutations to develop mismatch-directed barriers to recombination *(48, 50)* (**Fig. 3.2**). That is, the high rate of recombination will prevent diversification, so that molecular barriers to recombination – those mediated by mismatch-repair systems – will never develop. As a result, strains will share genotypic similarity until recombination is prevented by another mechanism. Models of this mutation–recombination balance have indicated that the effect of sequence divergence on recombination rate is not sufficient to produce speciation on its own *(48)*. Therefore, the source of these alternative barriers to recombination is critical; without them, speciation would not occur in groups of bacteria with high rates of recombination.

### *4.3. Ecological Barriers to Recombination*

Homologous recombination moves genes between similar, but not necessarily identical, bacterial strains. The differences among strains may include those that adapt the strains to somewhat different ecological niches. For example, some strains of the gut-dwelling species *E. coli* are found more often in the intestines of herbivores, while others are found more often associated with carnivores *(51–53)*. Therefore, a barrier to gene exchange may be erected by natural selection. If recombinants are less fit than either parent, then they will be counterselected (**Fig. 3.3**). This may be the case any time gene transfer affects a locus that has an adaptive difference between the donor and the recipient strains. As a result, the effective rate of recombination at or near niche-specific genes or alleles will be lower than at genes with no adaptive differences between the donor and the recipient. This is a post-mating barrier since recombinants would be formed, but they would not persist. In contrast, recombination at other loci between these strains may not incur any selective detriment, and these recombinants would persist in the population. Therefore, this post-mating barrier to recombination will prevent selective sweeps at some loci, but not at others where recombinants suffer no detriments.

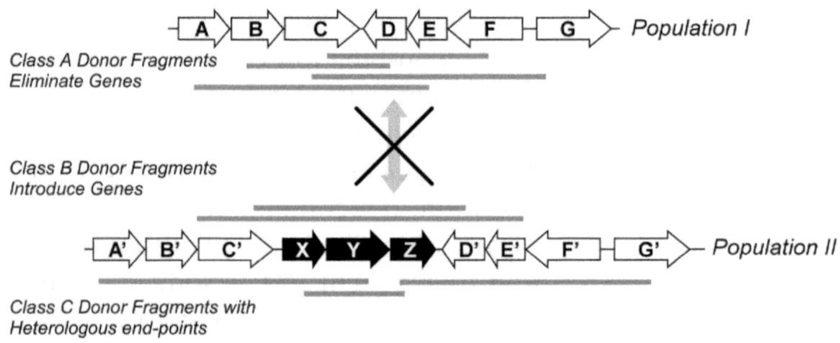


Fig. 3.3. HGT interferes with recombination between ecologically dissimilar organisms. The gain of genes X, Y, and Z in one population can interfere in several ways with recombination between populations. Movement of fragment class A would result in loss of genes X, Y, and Z from population II; movement of fragment class B would result in potentially problematic gain of genes X, Y, and Z in population I, should their expression there be counterproductive (see text). Introduction of fragment class C does not result in recombination due to a lack of homologous sequences at the dsDNA end.

This barrier – termed recombination interference – is outlined in **Fig. 3.3**; the orthologous DNA is sufficiently similar such that RecA-mediated recombination is not inhibited if strand invasion occurs. Here, we have indicated an addition of a small, adaptive gene cluster to one of the two genomes. There are several ways recombination may be inhibited in the vicinity of adaptive loci. First, recombination events that eliminate these genes would be counterselected because the recipient would lose important genes (**Fig. 3.3, class A**). Second, recombination events that move these genes to a naive host may, in fact, be detrimental (**class B**). For example, the *cadA* and *ompT* loci are present in *E. coli*, but not in the highly similar strains of *Shigella (54,55)*. Recombinants moving *Shigella* DNA lacking these genes into its cognate position in an *E. coli* strain would eliminate these genes, making a less-fit recombinant. It is also known that introduction of these genes from *E. coli* to *Shigella* interferes with virulence *(56, 57)*. Thus, recombinants in either direction may be counterselected. Third, from a purely mechanistic standpoint, fragments with dsDNA-ends within genes found only in the donor will not be effective substrates in the recipient, further reducing recombination at the shared genes flanking this adaptive locus (**class C**).

This post-mating barrier will eventually result in a pre-mating barrier *(58)*. The lack of recombination resulting from the lack of fit recombinants allows for the accumulation of neutral mutations in the region surrounding adaptive loci. Therefore, the genomes will become gradually more divergent in the genes immediately adjacent to adaptive loci because variance-purging selective sweeps cannot occur at those loci. Eventually, sufficient numbers of differences will accumulate, and DNA heteroduplexes between the two versions will be rejected by the mismatch-correction apparatus. In the end, the ecological differences that produce post-mating barriers result in robust pre-mating barriers. Given the infrequency of recombination in bacteria relative to eukaryotes, it is unlikely that pre-mating barriers would result from selection to prevent the formation of less-fit progeny.

It is clear that species of the genus *Neisseria* represent groups at the midpoint of this transition. Different species have sufficiently different ecologies that they can be readily classified as different species, yet there is also recombination across species boundaries at some loci, those that do not affect each population's niche specificity *(34)*. Given time, these species will likely develop pre-mating barriers, and the recombination across species boundaries will not necessarily lead to lineage coalescence. That is, we expect emerging species to continue to exchange genes across species boundaries so long as recombinants are not unfit. Alternatively, recombination across species boundaries may result in the coalescence of two emerging species, reuniting their gene pools *(59)*. This is occurring in two *Campylobacter* species, possibly as the result of expansion of the two populations into a shared ecological niche. Thus erasure of ecological differences between the two lineages leads naturally to the merger of the once diverging lineages *(59)*.

### 4.4. Species in Pieces

The ecological barrier to recombination discussed above works one locus at a time. This is a necessary result of the process of gene exchange in bacteria, whereby only small fragments of DNA are moved between donor and recipient. But, over time, larger fractions of the genome will experience genetic isolation as the result of two processes. First, the accumulation of additional adaptive loci will promote recombination interference in their flanking regions. Second, the accumulation of mutations in the regions experiencing recombination interference will increase the length of regions that recombine poorly. The combination of the two processes means that, over time, the entire genomes of two diverging lineages will eventually be sufficiently different that homologous recombination will not occur at any locus (**Fig. 3.4**). At this point, the genomes represent individuals of two different species *sensu* Dykhuizen and Green. Yet during this period of time, nascent species can still exchange genes at some loci; this is the case with the "fuzzy" species of *Neisseria*, wherein homologous recombination can still move genes across species boundaries at some loci.

Two questions arise from this fragmented process of speciation. First, what is the time frame over which complete genetic isolation occurs? Traditionally, the time over which lineage separation occurs is thought to be relatively short compared to the time between successive divergence events. That is, in visualizing speciation as a series of branches on a phylogenetic tree, the variance

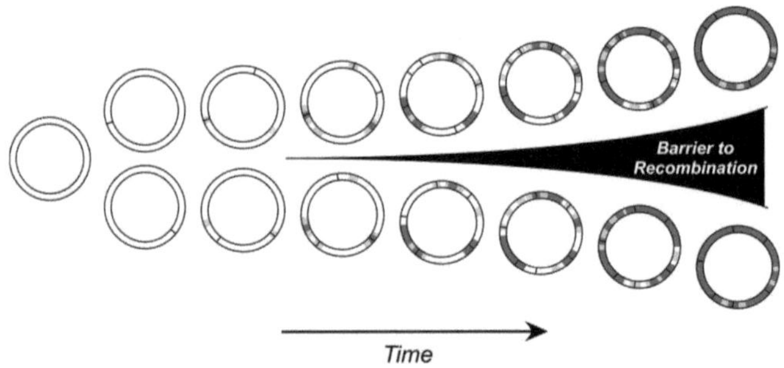


Fig. 3.4. Model for the gradual establishment of genetic isolation. An ancestral population acquires two different adaptive loci. Recombination is inhibited in the regions surrounding those loci, leading to sequence divergence in the flanking genes; this is depicted as gray regions on the chromosomal backbone. The accumulation of additional adaptive loci, as well as growth of the regions with recombination inhibition, eventually leads to the genetic isolation of all genes.

of each node is considered to be small relative to the inter-node distance. Given that a process relying on ecological barriers entails both the accumulation of adaptive differences at numerous chromosomal loci and the time to accumulate mutational differences at shared loci, this is not an insignificant time in bacteria. Second, the Dykhuizen and Green species concept groups strains with incongruent phylogenies among genes into the same species; members of different species would have congruent phylogenies because recombination between them is absent. Yet consider genomes midway through in the process outlined in **Fig. 3.4**. Here, the genomes are exchanging genes at some loci but not at others. If many such lineages were diverging simultaneously from a single parent populations, then the phylogenies of some genes may be congruent – thereby placing them in different species – where the phylogenies of others may be incongruent due to continued recombination at those loci. Thus, such lineages would be considered different species at some genes but the same species at other genes. If the process of lineage separation is rapid, then this is not an issue. But if the process takes significant amounts of time, this confounds our ability to separate strains neatly into distinct, non-overlapping species.

#### *4.4.1. Time Frame for Lineage Divergence*

To elucidate the time frame over which gene divergence takes place, one needs to calculate the time of divergence for all orthologous genes in two populations that no longer experience any recombination. One can then determine if the time frame which encompasses 95% of the divergence times (intra-node variance) is very small relative to the average time between genome divergence events (inter-node length). The average time of divergence is often inferred from the average nucleotide divergence at synonymous sites or divergence at 16S rRNA-encoding genes; rates of change at these sites have been calibrated to historical events in the fossil record *(60, 61)*. The difficult task, then, is to calculate the time of divergence for individual genes relative to the average. Nucleotide divergence is a product of the rate of divergence and the time of separation: divergence increases both with increased rate of change and with increased time of separation. Variation in nucleotide divergence among genes has traditionally been attributed to two sources: variation in rate, and error. Yet variation in time may also be a contributor if the process of speciation takes a long period of time. If divergence is corrected for variation in rate, a component of the residual variance in divergence may be attributed to variation in divergence time.

To correct for variation in rate of synonymous substitution, one may normalize divergence at synonymous sites to a time-independent metric of selection on synonymous sites, such as the degree of codon usage bias (**Fig. 3.5**). As genes experience greater selection for preferred codons, the degree of codon usage bias increases and the divergence at synonymous sites decreases *(62, 63)*. Divergence at synonymous sites is measured by comparing two sequences and noting how many substitutions affect synonymous sites *(64, 65)*. In contrast, measures of codon usage bias rely on intragenomic comparison of the degree of preferred codon usage. In this way, measures of codon usage bias reflect

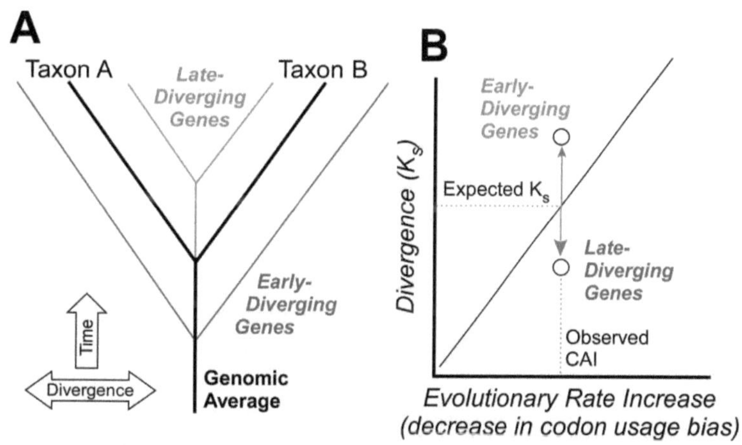


Fig. 3.5. Fragmented speciation. (**A**) Model for fragmented speciation. If the variance in divergence times among genes is large, different portions of the chromosome should contain genes with significantly different ages. As a result, sequence divergence will vary as a function of divergence times. (**B**) Testing fragmented speciation. Decreased selection on synonymous sites leads to an increase in the rate of changes and an increase in divergence. Because selection can be estimated by the degree of codon usage bias, variation from the predicted level of divergence can be attributed to differences in the time of divergence. Genes that have diverged earlier would be more divergent than expected, whereas those diverging recently would accumulate fewer differences than expected.

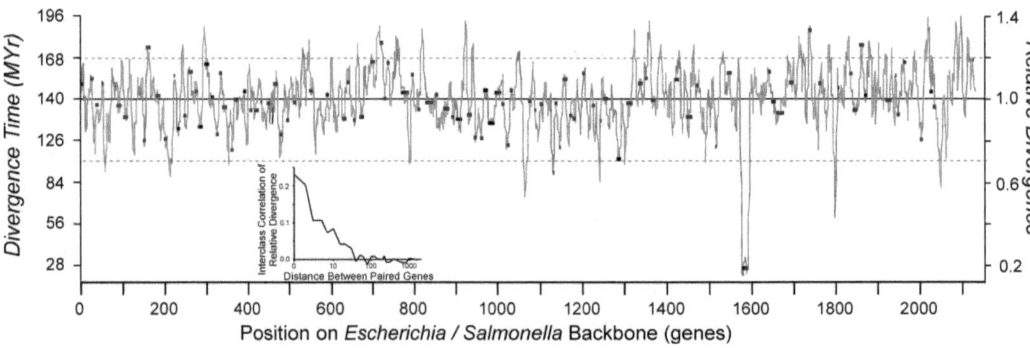


Fig. 3.6. Recombination as a function of chromosomal position. Relative times of divergence for orthologous gene pairs in *E. coli* and *Salmonella* are plotted according to position on the *E. coli* K12-MG1655 chromosome. Values are averaged across a window of seven adjacent genes, and dark horizontal lines depict divergence times for regions of seven or more genes. The region encompassing 95% of the range of divergence values is marked with dashed lines. **Inset.** The correlation of relative divergence for gene pairs as a function of their relative positions in the *E. coli* K12-MG1655 backbone *(66)*.

selection at synonymous sites that do not require comparisons with orthologous sequences. Thus, correcting the divergence at synonymous sites (e.g., Li's $K_S$) for the degree of codon usage bias (e.g., Sharp and Li's CAI) normalizes divergence for differences in the rate of substitution between genes. The remaining variation in divergence reflects a combination of error and the time of divergence. Genes diverging for longer periods of time will have greater values of normalized divergence than do genes separated for shorter periods of time.

To separate the influence of variation of divergence time from error, we follow the rules of recombination in bacteria: adjacent genes participate in the same recombination events while distantly situated genes do not. Therefore, we would expect normalized divergence of adjacent genes to be correlated (reflecting participation in the same recombination event), and this correlation should decrease with increased separation of pairs of orthologous genes *(66)*. This is what is observed in comparisons of orthologs in the *E. coli* and *Salmonella enterica* genomes (**Fig. 3.6**). Therefore, we conclude that the time to the last recombination event varies by genomic position in the *E. coli* and *Salmonella* genomes. Taking the mean time of divergence to be 140 MYr *(61, 67)*, these data suggest that individual genes diverged between 90 and 180 MYr ago, over a period of ~90 MYr (**Fig. 3.7A**).

#### *4.4.2. Models for Lineage Separation*

Regardless of the model for lineage separation, the very nature of recombination in bacteria predicts that different regions of orthologous bacterial chromosomes would show different times to the coalescent. Because gene exchange involves only a fragment of the chromosome, it cannot be true that all genes share

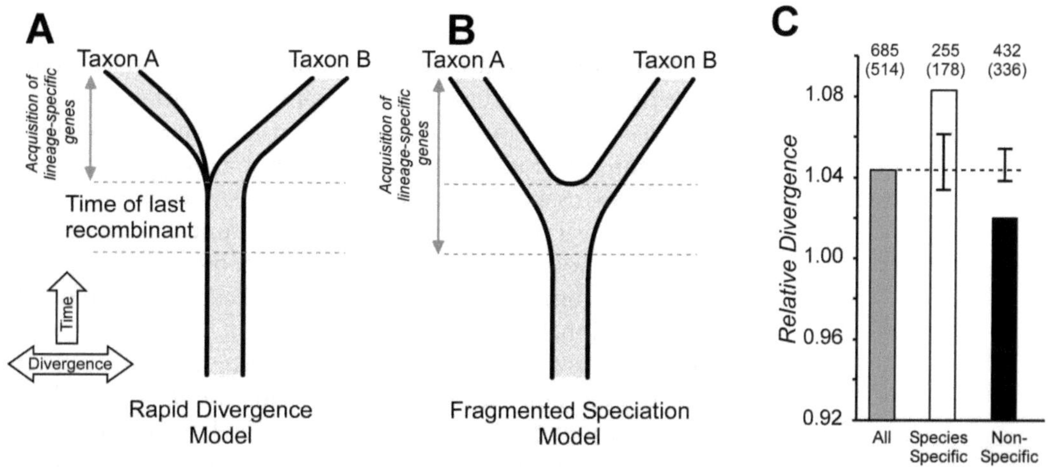


Fig. 3.7. Models for fragmented speciation. **(A)** Rapid divergence. A population divides into two, each of which adapts to a particular niche without further genetic input from the other. While recombination (cross-hatching) may produce different times of divergence between genes in taxa A and B, no cross-population recombination occurred after lineage-specific genes were acquired (*vertical arrow*). **(B)** Fragmented speciation. Ecological diversification, involving lineage-specific gene acquisition, occurs in the context of genetic exchange. **(C)** Age distribution of orthologous genes flanking lineage-specific genes in the *E. coli* and *Salmonella* genomes. The mean relative divergence of sets of orthologs (number of orthologues above bars) is shown; orthologs are classified according to adjacency to loci that distinguish genomes (number of loci in parentheses). Species-specific loci are found adjacent to genes that are present in all genomes of one species, but none of the other. Non-specific loci are found adjacent to genes that are found in some genomes of one or either species. Error bars show 1 SE for the distribution of randomized samples.

the same time of divergence. But the data described above do not discriminate between the two models discussed above. That is, even if homologous recombination were too infrequent to prevent the inevitable accumulation of mutations, different genes would still diverge at different times. Therefore, merely demonstrating that different regions of the chromosomes diverged at different times does not establish that homologous recombination acts as a cohesion mechanism for bacterial species. This is shown in **Fig. 3.7A** as the "rapid divergence" model. Here, the low rate of recombination means that few, if any, recombination events occur after lineage-specific adaptations are acquired. As a result, lineage-specific genes would be found inserted at random with respect to the age-structure of shared genes. On the other hand, if the rate of recombination is high, it would continue at loci unlinked to lineage-specific adaptations. Here, lineage-specific genes would lie primarily in the "older" regions of the genome, because regions unlinked to the insertions would continue to experience homologous recombination after acquisition (**Fig. 3.7B**). Therefore, these two models can be discriminated by examining the age distribution of different regions of the chromosome with respect to the loci that bear adaptive changes between two lineages.

#### 4.4.3. Species-Specific Genes Lie in Older Regions of the Chromosome

Adaptive changes may result from many processes, including mutation of existing genes, gene loss, and gene gain. Because the vast majority of single-base substitutions are neutral, it is very difficult to locate adaptive changes from sequence data alone. Yet gene gain and loss events are likely not neutral. Gene gain by HGT is an especially powerful mechanism whereby strains may explore novel niches using genes whose functions have been honed by selection in another organism *(68, 69)*. Therefore, we focus on the changes in gene inventory as indicators of potentially adaptive mutational events.

The results are shown in **Fig. 3.7C**. Here, the relative divergence of orthologous genes found in both the *E. coli* and the *Salmonella* genomes are summarized as several classes. In some, there is a gene interspersed or lost within either the *E. coli* or the *Salmonella* lineage so that genes which are adjacent in all strains of one species (genes AB) are interrupted by a novel locus in the other (genes AXB). We term these genes – genes AB in this example – as flanking species-specific genes. Another class of gene pairs includes those that are similarly interrupted, but in only one or a few strains in each species; these may be considered as flanking strain-specific genes, or those where the differences were acquired more recently. As shown in **Fig. 3.7C**, genes flanking species-specific insertions appear in a context of orthologous genes, which are more divergent than expected *(66)*. These data suggest that recombination has proceeded between the nascent *E. coli* and *Salmonella* lineages after the acquisition of these loci.

### 4.5. The Muddling of Bacterial Species

The protracted period of time over which lineage differentiation occurs confounds our ability to use domains of recombination to delineate species as Dykhuizen and Green proposed. It is clear that gene exchange does not occur between strains of *E. coli* and strains of *Salmonella enterica*, defining them as different species. They are also phenotypically quite distinct from each other. In addition, gene exchange does occur among strains of *E. coli* and among strains of *S. enterica*, thereby placing them into the same species by Dykhuizen and Green's criteria. But it is not clear that either group represent a genotypically or phenotypically cohesive group. Given the time taken to separate those lineages, it is clear that neither *E. coli* nor *S. enterica* represents a population of panmictic bacteria where recombining strains have an equal likelihood of exchanging genes at any locus. For example, there are commensal, uropathogenic, and enterohemorrhagic strains of *E. coli*, where each type has a distinct ecology and gene inventory *(70, 71)*. While recombination would be suppressed adjacent to niche-defining loci, it would proceed at loci unlinked to niche-specific genes. More importantly, regions of free recombination would vary depending on which sets of strains were being compared. Therefore, while phylogenies of different *E. coli* genes are not congruent, thereby placing them in the same species *sensu* Dykhuizen and Green, the strains harboring those genes represent a mosaic of partially isolated groups with a reticulate pattern of gene exchange among them.

Thus, groups of bacteria defined by recombination have muddled, indistinct boundaries; barriers to recombination can only be identified long after recombination has ceased, thereby allowing diagnostic substitutions to accumulate, differentiating the orthologous genes. This problem is shared with ecotypes, which have indistinct boundaries due to the variable magnitude of the beneficial mutations, which then drive periodic selection events to variable taxonomic breadths. This situation is inevitable given the nature of bacterial recombination, which occurs on a gene-by-gene basis. Sexual eukaryotes may be described by powerful species concepts that neatly segregate these organisms into clearly defined groups. This separation is only possible because genetic isolation may occur simultaneously for all genes in the genomes. This is simply not feasible in free-living bacteria; because genetic isolation in bacteria is imposed over a long period of time, boundaries between groups are necessarily "fuzzy." The biology of microorganisms precludes the formation of a robust species concept to delineate robustly distinct groups of bacteria.

## 5. The Interplay Between Homologous Recombination and HGT

### *5.1. Recombination Within Species Increases the Probability of Successful HGT Between Species*

Just as genes introduced by HGT affect the rate of homologous recombination between incipient species, homologous recombination within species can influence the probability of successful gene acquisition. This occurs for two reasons. First, there is the hedge against stochastic loss. After gene introgression, its frequency in the population is $1/N$; because bacterial populations are so large this initial frequency is quite small, so that the inevitable result of most gene acquisition events is rather rapid loss, even if beneficial functions are conferred. Homologous recombination acts to increase the frequency of beneficial acquired genes, thus decreasing the probability of stochastic loss. More importantly, homologous recombination places the HGT genes in different genomic contexts. Given the variability of gene inventories among different strains of the same species *(70)*, it is very unlikely that the strain into which the genes were originally inserted is the one which could maximally benefit from the new genes' functions. The success of a gene acquisition event is a function of the overall benefit conferred by the new genes, and that benefit may well be higher in another context. For example, genes allowing better scavenging of intracellular iron within eukaryotic cells would benefit pathogenic *E. coli* O157:H7 to a greater degree than commensal strains. Therefore, intraspecific gene transfer exposes a variety of strains to the new genes' functions, thus increasing the probability of long-term retention.

#### 5.2. HGT Promotes Homologous Recombination Within Species

As discussed above, laterally acquired genes may be transferred among strains of the same species by homologous recombination; such recombination events will be under selection when the newly acquired genes provide an advantageous function. The movement of these genes by homologous recombination requires gene conversion at the flanking genes that are shared between intraspecific donor and recipient (**Fig. 3.8**). Without the benefit of the novel genes at this locus, a selective sweep would be unlikely. Therefore, the effective rates of homologous recombination at some loci have been increased due to their proximity to laterally acquired genes representing an adaptive trait. As a result, selective sweeps at some loci may not have been driven by advantageous point mutations, which created beneficial alleles. Rather, the linkage to a beneficial insertion produced the accelerated rates of recombination.

#### 5.3. Does HGT Invalidate the Concept of a Bacterial Species?

Given that bacteria exchange DNA across species boundaries by HGT, what does the species boundary represent? The genomes of bacteria from many lineages show evidence for large numbers of recently acquired genes *(17, 72, 73)*. Aside from confounding bacterial phylogeny *(74)* – to the extent a phylogeny based on a small minority of constituent ("core") genes can be used to infer relationships *(75)* – the transfer of genes across species boundaries calls into question the utility of the species concept itself. If bacteria exchange genes across species boundaries, what does such a boundary define? Mayr's original concept, from which the Dykhuizen and Green formulation is derived, places high value in lack of gene exchange across species boundaries. We argue here that, from a practical standpoint, the transfer of DNA between

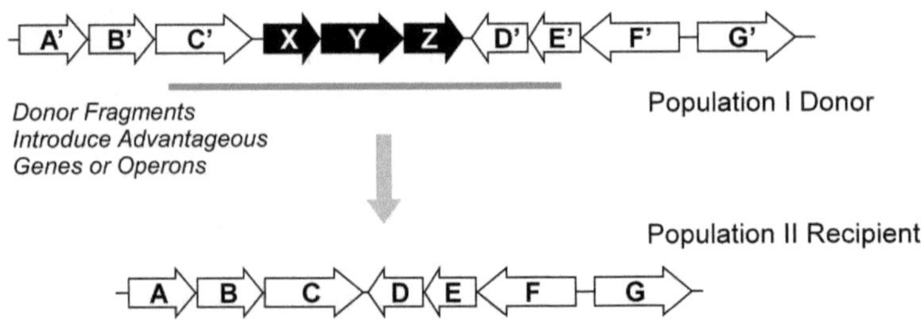


Fig. 3.8. HGT promotes homologous recombination between ecologically similar organisms. Novel genes X, Y, and Z confer an advantageous phenotype in both the donor and the recipient strain. Because recombinants are favored, the rate of recombination will be increased at the genes flanking the insertion.

species does not invalidate the utility of species definitions, nor does it interfere with a species concept.

Species definitions play an important role in microbiology in allowing the robust and consistent identification of bacteria, which play important medical, epidemiological, economic, biotechnological, or bioweapons roles. The point emphasized in this chapter is that the exchange of traits among con-specific strains provides a level of genotypic cohesion that is useful in this context. Although strains within a species may vary in phenotype, they share a core set of traits that encapsulates functionally germane behaviors. The introduction of novel traits into a species is simply a source of phenotypic variation between strains. The similarity between donor and recipient clades imparted by such transfers is minor, and does serve to confound the delineation of functionally important groups. For example, *Escherichia coli* received phosphonate utilization genes from an outside donor, likely an alpha-proteobacterium. But these genes alone do not confound the distinction between *E. coli* and *Sinorhizobium meliloti*. Lateral gene transfer does raise problems when attempting to classify organisms into strictly hierarchical schemes *(76)*, but this does not impact the delineation of the species itself when multiple traits are used for species discrimination. Moreover, lateral genes transfer is infrequent relative to the rate of homologous DNA exchange within species. Therefore, it does not confound the delineation of species as a group of strains that share genotypic cohesion as the result of allelic exchange.

### 5.4. Bacteria Without a Species Concept

From a practical standpoint, all organisms must be given names so that they can be identified and discussed. As discussed above, these names carry connotations and implications when applied to bacteria, because named organisms play important roles in disease, food safety, biowarfare, and other public health arenas. What is problematic for microbiologists is that the name of a bacterium serves a dual role as that organism's identifier and as its *species* name. That is, the simple act of assigning a name to a bacterium places it into a taxonomic group, which is useful when associated with a biological process underlying the similarity of its members. Above, we have argued that domains of homologous recombination work well in delineating some bacterial species. Yet one species concept alone is clearly insufficient to place all bacterial strains into well-ordered species. When the rate of recombination within a bacterial population is too low to prevent the inevitable diversification of strains, no cohesive group can be delineated. Any collection placing such diverse strains into a "species" is arbitrary; while this is a valid – and even useful – species definition, it is not a species concept. A group of more closely related strains – such as those formed by periodic selection events

(**Fig. 3.1**) – would represent a group whose genotypic cohesion represents a biological process.

Disturbingly, it is quite possible that many bacteria cannot be placed into any group that can be delineated with a valid species concept. That is, they may not belong to a group whose genotypic similarity is the result of a cohesion mechanism such as shared domains of gene exchange, or shared ecological niches that are subject to purifying selective sweeps. If so, then they would not belong to any species that is based on a species concept. It is unfortunate that naming conventions in bacteria demand assigning these organisms to named taxonomic groups even if there are no non-arbitrary criteria for defining their boundaries.

## 6. Conclusions and Outlook

Rates of gene exchange between a bacterium and close relatives (recombination) are intimately associated with its rate of gene acquisition from distant relatives (HGT), with each process affecting the other. This interplay sheds light on the process of lineage separation, or the creation of new bacterial species. In the model presented above, we argue that the protracted and complex process of disentangling domains of homologous gene exchange prevents the clean segregation of bacterial strains into species by virtue of their shared gene pools.

## Acknowledgments

This work was supported by grant GM078092 from the National Institutes of Health.

### References

1. Aristotle (1910) *Historia Animalium (translated by D'Arcy Wentworth Thompson)*, Clarendon Press, Oxford.
2. Linnaeus, C. (1758) *Systema naturae per regna tria naturae, secundum classes, ordines, genera, species, cum characteribus, differentiis, synonymis, locis*, Holmiae.
3. Quinn, P. C. (2002) Young infants' categorization of humans versus nonhuman animals: roles for knowledge access and perceptual process, in *Building Object Categories in Developmental Time* (Lisa Gershkoff-Stowe, D. H. R., ed.) Lawrence Erlbaum Associates, Mahwah, NJ.
4. Senate and House of Representatives of the United States of America (1973) Endangered Species Act of 1973. In. (Agency, E. P., ed.) Government of the United States of America Place.
5. Darwin, C. (1859) *On the Origin of Species by Means of Natural Selection or the Preservation of Favoured Races in the Struggle for Life*, John Murray, London.
6. Gevers, D., Cohan, F. M., Lawrence, J. G., Spratt, B. G., Coenye, T., Feil, E. J., Stackebrandt, E., Van De Peer, Y., Vandamme, P., Thompson, F. L., Swings, J. (2005) Re-evaluating prokaryotic species. *Nat Rev Microbiol* **3**, 733–9.
7. Mayr, E. (1942) *Systematics and the Origin of Species*, Columbia University Press, New York.

8. Mayr, E. (1963) *Animal Species and Evolution*, Harvard University Press, Cambridge.
9. Paterson, H. E. H. (1985) The recognition concept of species, in *Species and Speciation* (Vrba, E. S., ed.) Transvaal Museum, Pretoria, 21–9.
10. Van Valen, L. (1976) Ecological species, multispecies, and oaks. *Taxon* **25**, 223–39.
11. Wiley, E. O. (1978) The evolutionary species concept reconsidered. *Syst Zool* **27**, 17–26.
12. Templeton, A. R. (1989) The meaning of species and speciation: a genetic perspective, in *Speciation and Its Consequences* (Otte, D., Endler J. A., ed.) Sinauer Associates, Sunderland, MA, 3–27.
13. Levin, B. R. (1981) Periodic selection, infectious gene exchange and the genetic structure of E. coli populations. *Genetics* **99**, 1–23.
14. Atwood, K. C., Schneider, L. K., Ryan, F. J. (1951) Periodic selection in *Escherichia coli*. *Proc Natl Acad Sci USA* **37**, 146–55.
15. Cohan, F. M. (2001) Bacterial species and speciation. *Syst Biol* **50**, 513–24.
16. Cohan, F. M., Perry, E. B. (2007) A systematics for discovering the fundamental units of bacterial diversity. *Curr Biol* **17**, R373–86.
17. Ochman, H., Lawrence, J. G., Groisman, E. (2000) Lateral gene transfer and the nature of bacterial innovation. *Nature* **405**, 299–304.
18. O'neill, M., Chen, A., Murray, N. E. (1997) The restriction-modification genes of *Escherichia coli* K-12 may not be selfish: they do not resist loss and are readily replaced by alleles conferring different specificities. *Proc Natl Acad Sci USA* **94**, 14596–601.
19. Barcus, V. A., Titheradge, A. J., Murray, N. E. (1995) The diversity of alleles at the *hsd* locus in natural populations of *Escherichia coli*. *Genetics* **140**, 1187–97.
20. Murray, N. E. (2000) Type I restriction systems: sophisticated molecular machines (a legacy of Bertani and Weigle). *Microbiol Mol Biol Rev* **64**, 412–34.
21. Milkman, R., Raleigh, E. A., Mckane, M., Cryderman, D., Bilodeau, P., Mcweeny, K. (1999) Molecular evolution of the *Escherichia coli* chromosome. V. Recombination patterns among strains of diverse origin. *Genetics* **153**, 539–54.
22. Shen, P., Huang, H. V. (1986) Homologous recombination in *Escherichia coli*: dependence on substrate length and homology. *Genetics* **112**, 441–57
23. Dykhuizen, D. E., Green, L. (1991) Recombination in *Escherichia coli* and the definition of biological species. *J Bacteriol* **173**, 7257–68.
24. Milkman, R. (1997) Recombination and population structure in *Escherichia coli*. *Genetics* **146**, 745–50.
25. Wertz, J. E., Goldstone, C., Gordon, D. M., Riley, M. A. (2003) A molecular phylogeny of enteric bacteria and implications for a bacterial species concept. *J Evol Biol* **16**, 1236–48.
26. Maiden, M. C., Bygraves, J. A., Feil, E., Morelli, G., Russell, J. E., Urwin, R., Zhang, Q., Zhou, J., Zurth, K., Caugant, D. A., Feavers, I. M., Achtman, M., Spratt, B. G. (1998) Multilocus sequence typing: a portable approach to the identification of clones within populations of pathogenic microorganisms. *Proc Natl Acad Sci USA* **95**, 3140–5.
27. Feil, E. J., Holmes, E. C., Bessen, D. E., Chan, M. S., Day, N. P., Enright, M. C., Goldstein, R., Hood, D. W., Kalia, A., Moore, C. E., Zhou, J., Spratt, B. G. (2001) Recombination within natural populations of pathogenic bacteria: short-term empirical estimates and long-term phylogenetic consequences. *Proc Natl Acad Sci USA* **98**, 182–7.
28. Feil, E. J., Maiden, M. C., Achtman, M., Spratt, B. G. (1999) The relative contributions of recombination and mutation to the divergence of clones of *Neisseria meningitidis*. *Mol Biol Evol* **16**, 1496–502.
29. Feil, E. J., Smith, J. M., Enright, M. C., Spratt, B. G. (2000) Estimating recombinational parameters in *Streptococcus pneumoniae* from multilocus sequence typing data. *Genetics* **154**, 1439–50.
30. Feil, E. J., Spratt, B. G. (2001) Recombination and the population structures of bacterial pathogens. *Annu Rev Microbiol* **55**, 561–90.
31. Hanage, W. P., Fraser, C., Spratt, B. G. (2006) The impact of homologous recombination on the generation of diversity in bacteria. *J Theor Biol* **239**, 210–9.
32. Guttman, D. S., Dykhuizen, D. E. (1994) Detecting selective sweeps in naturally occurring *Escherichia coli*. *Genetics* **138**, 993–1003.
33. Guttman, D. S., Dykhuizen, D. E. (1994) Clonal divergence in *Escherichia coli* as a result of recombination, not mutation. *Science* **266**, 1380–3.
34. Hanage, W. P., Fraser, C., Spratt, B. G. (2005) Fuzzy species among recombinogenic bacteria. *BMC Biol* **3**, 6.
35. Hanage, W. P., Fraser, C., Spratt, B. G. (2006) Sequences, sequence clusters and bacterial species. *Philos Trans R Soc Lond B Biol Sci* **361**, 1917–27.
36. Roncero, C., Sanderson, K. E., Casadaban, M. J. (1991) Analysis of the host ranges of transposon bacteriophages Mu, MuhP1, and D108 by use of lipopolysaccharide mutants of *Salmonella typhimurium* LT2. *J Bacteriol* **173**, 5230–3.
37. Papke, R. T., Zhaxybayeva, O., Feil, E. J., Sommerfeld, K., Muise, D., Doolittle, W. F.

(2007) Searching for species in haloarchaea. *Proc Natl Acad Sci U S A* **104**, 14092–7.
38. Whitaker, R. J., Grogan, D. W., Taylor, J. W. (2003) Geographic barriers isolate endemic populations of hyperthermophilic archaea. *Science* **301**, 976–8.
39. Whitaker, R. J., Grogan, D. W., Taylor, J. W. (2005) Recombination shapes the natural population structure of the hyperthermophilic archaeon *Sulfolobus islandicus*. *Mol Biol Evol* **22**, 2354–61.
40. Vulic, M., Dionisio, F., Taddei, F., Radman, M. (1997) Molecular keys to speciation: DNA polymorphism and the control of genetic exchange in Enterobacteria. *Proc Natl Acad Sci USA* **94**, 9763–7.
41. Vulic, M., Lenski, R. E., Radman, M. (1999) Mutation, recombination, and incipient speciation of bacteria in the laboratory. *Proc Natl Acad Sci USA* **96**, 7348–51.
42. Majewski, J., Cohan, F. M. (1999) DNA sequence similarity requirements for interspecific recombination in *Bacillus*. *Genetics* **153**, 1525–33.
43. Zawadzki, P., Roberts, M. S., Cohan, F. M. (1995) The log-linear relationship between sexual isolation and sequence divergence in *Bacillus* transformation is robust. *Genetics* **140**, 917–32.
44. Springer, B., Sander, P., Sedlacek, L., Hardt, W. D., Mizrahi, V., Schar, P., Bottger, E. C. (2004) Lack of mismatch correction facilitates genome evolution in mycobacteria. *Mol Microbiol* **53**, 1601–9.
45. Matic, I., Rayssiguier, C., Radman, M. (1995) Interspecies gene exchange in bacteria: the role of SOS and mismatch repair systems in evolution of species. *Cell* **80**, 507–15.
46. Brown, E. W., Leclerc, J. E., Li, B., Payne, W. L., Cebula, T. A. (2001) Phylogenetic evidence for horizontal transfer of *mutS* alleles among naturally occurring *Escherichia coli* strains. *J Bacteriol* **183**, 1631–44.
47. Demerec, M., Ohta, N. (1964) Genetic analyses of *Salmonella typhimurium* X *Escherichia coli* hybrids. *Proc Natl Acad Sci USA* **52**, 317–23.
48. Hanage, W. P., Spratt, B. G., Turner, K. M., Fraser, C. (2006) Modelling bacterial speciation. *Philos Trans R Soc Lond B Biol Sci* **361**, 2039–44.
49. Falush, D., Torpdahl, M., Didelot, X., Conrad, D. F., Wilson, D. J., Achtman, M. (2006) Mismatch induced speciation in *Salmonella*: model and data. *Philos Trans R Soc Lond B Biol Sci* **361**, 2045–53.
50. Fraser, C., Hanage, W. P., Spratt, B. G. (2007) Recombination and the nature of bacterial speciation. *Science* **315**, 476–80.
51. Gordon, D. M., Bauer, S., Johnson, J. R. (2002) The genetic structure of *Escherichia coli* populations in primary and secondary habitats. *Microbiology* **148**, 1513–22.
52. Gordon, D. M., Cowling, A. (2003) The distribution and genetic structure of *Escherichia coli* in Australian vertebrates: host and geographic effects. *Microbiology* **149**, 3575–86.
53. Gordon, D. M., Fitzgibbon, F. (1999) The distribution of enteric bacteria from Australian mammals: host and geographical effects. *Microbiology* **145 (Pt 10)**, 2663–71.
54. Day, W. A., Jr., Fernandez, R. E., Maurelli, A. T. (2001) Pathoadaptive mutations that enhance virulence: genetic organization of the *cadA* regions of *Shigella* spp. *Infect Immun* **69**, 7471–80.
55. Nakata, N., Tobe, T., Fukuda, I., Suzuki, T., Komatsu, K., Yoshikawa, M., Sasakawa, C. (1993) The absence of a surface protease, OmpT, determines the intercellular spreading ability of *Shigella*: the relationship between the *ompT* and *kcpA* loci. *Mol Microbiol* **9**, 459–68.
56. Maurelli, A. T., Fernández, R. E., Bloch, C. A., Rode, C. K., Fasano, A. (1998) "Black holes" and bacterial pathogenicity: a large genomic deletion that enhances the virulence of *Shigella* spp. and enteroinvasive *Escherichia coli*. *Proc Natl Acad Sci, USA* **95**, 3943–8.
57. May, B. J., Zhang, Q., Li, L. L., Paustian, M. L., Whittam, T. S., Kapur, V. (2001) Complete genomic sequence of *Pasteurella multocida*, Pm70. *Proc Natl Acad Sci, USA* **98**, 3460–5.
58. Lawrence, J. G. (2002) Gene transfer in bacteria: speciation without species? *Theor Popul Biol* **61**, 449–60.
59. Sheppard, S. K., Mccarthy, N. D., Falush, D., Maiden, M. C. (2008) Convergence of *Campylobacter* species: implications for bacterial evolution. *Science* **320**, 237–9.
60. Moran, N. A., Munson, M. A., Baumann, P., Ishikawa, H. (1993) A molecular clock in endosymbiotic bacteria is calibrated using insect hosts. *Proc R Soc Lond B* **253**, 167–71.
61. Ochman, H., Wilson, A. C. (1988) Evolution in bacteria: evidence for a universal substitution rate in cellular genomes. *J Mol Evol* **26**, 74–86.
62. Sharp, P. M., Li, W.-H. (1987) The rate of synonymous substitution in enterobacterial genes is inversely related to codon usage bias. *Mol Biol Evol* **4**, 222–30.
63. Sharp, P. M., Li, W.-H. (1987) The codon adaptation index – a measure of directional synonymous codon usage bias, and its potential applications. *Nucleic Acids Res* **15**, 1281–95.

64. Li, W. H., Wu, C. I., Luo, C. C. (1985) A new method for estimating synonymous and nonsynonymous rates of nucleotide substitution considering the relative likelihood of nucleotide and codon changes. *Mol Biol Evol* **2**, 150–74.
65. Li, W. H. (1993) Unbiased estimation of the rates of synonymous and nonsynonymous substitution. *J Mol Evol* **36**, 96–9.
66. Retchless, A. C., Lawrence, J. G. (2007) Temporal fragmentation of speciation in bacteria. *Science* **317**, 1093–6.
67. Ochman, H., Wilson, A. C. (1987) Evolutionary history of enteric bacteria, in *Escherichia Coli and Salmonella Typhimurium: Cellular and Molecular Biology* (Neidhardt, F. C., Ingraham J. L., Low K. B., Magasanik B., Schaechter M., Umbarger H. E., ed.) American Society for Microbiology, Washington, D. C. 1649–54.
68. Lawrence, J. G. (1997) Selfish operons and speciation by gene transfer. *Trends Microbiol* **5**, 355–9.
69. Lawrence, J. G. (1999) Gene transfer, speciation, and the evolution of bacterial genomes. *Curr Opin Microbiol* **2**, 519–23.
70. Welch, R. A., Burland, V., Plunkett, G., 3rd, Redford, P., Roesch, P., Rasko, D., Buckles, E. L., Liou, S. R., Boutin, A., Hackett, J., Stroud, D., Mayhew, G. F., Rose, D. J., Zhou, S., Schwartz, D. C., Perna, N. T., Mobley, H. L., Donnenberg, M. S., Blattner, F. R. (2002) Extensive mosaic structure revealed by the complete genome sequence of uropathogenic *Escherichia coli*. *Proc Natl Acad Sci USA* **99**, 17020–4.
71. Kudva, I. T., Evans, P. S., Perna, N. T., Barrett, T. J., Ausubel, F. M., Blattner, F. R., Calderwood, S. B. (2002) Strains of *Escherichia coli* O157:H7 differ primarily by insertions or deletions, not single-nucleotide polymorphisms. *J Bacteriol* **184**, 1873–9.
72. Lawrence, J. G., Ochman, H. (1998) Molecular archaeology of the *Escherichia coli* genome. *Proc Natl Acad Sci USA* **95**, 9413–7.
73. Choi, I. G., Kim, S. H. (2007) Global extent of horizontal gene transfer. *Proc Natl Acad Sci USA* **104**, 4489–94.
74. Creevey, C. J., Fitzpatrick, D. A., Philip, G. K., Kinsella, R. J., O'connell, M. J., Pentony, M. M., Travers, S. A., Wilkinson, M., Mcinerney, J. O. (2004) Does a tree-like phylogeny only exist at the tips in the prokaryotes? *Proceedings* **271**, 2551–8.
75. Dagan, T., Martin, W. (2006) The tree of one percent. *Genome Biol* 7, 118.
76. Lawrence, J. G., Hatfull, G. F., Hendrix, R. W. (2002) Imbroglios of viral taxonomy: genetic exchange and failings of phenetic approaches. *J Bacteriol* **184**, 4891–905.

# Chapter 4

# Epistemological Impacts of Horizontal Gene Transfer on Classification in Microbiology

**Eric Bapteste and Yan Boucher**

## Abstract

We describe the reasons why the newly recognized process of horizontal gene transfer (HGT) forces evolutionists who study classification and microbiology to go beyond the classical Darwinian framework. We recall the importance of processes in philosophical definitions of species and for taxonomical purposes in general. More precisely, we present a brief description of a possible transition from a thinking inspired by essentialism to eliminative pluralism in the species debate and we insist on a major philosophical lesson: that processes matter and that, consequently, HGT cannot be overlooked in microbial classification. We then expand the conclusions of eliminative pluralism to microbial classification, namely (i) that species are not real and (ii) that overlapping taxonomies are equally legitimate when they are based on real natural processes. We introduce alternatives to the traditional species concept and describe what we call evolutionary units. Two types of units can be described: coherent and composite. The former are sets of co-evolving genes, pathways, or organisms, which share the same phylogenetic origin, while the latter comprise genes, pathways, or organisms with component parts from multiple phylogenetic origins. These evolutionary units are either "mostly flexible" or "mostly rigid" in their genetic composition and we discuss how this dissimilarity could profoundly affect our systematics practice. In this chapter, we illustrate how much there is to learn from the reconstruction of the complex evolutionary histories of all evolutionary units – large or small – by giving up the notion of species for recombining microbes, and suggest replacing a unique nested hierarchy of life with a comprehensive database including overlapping taxonomical groups.

**Key words:** Species concept, pluralism, systematics, horizontal gene transfer, evolutionary units, nested hierarchy.

## 1. Introduction

Panchen masterfully summarized the ideal conditions under which traditional phylogenetics could fully operate: "the natural arrangement of organisms would have the following properties:

Maria B. Gogarten et al. (eds.), *Horizontal Gene Transfer: Genomes in Flux, vol. 532*

DOI 10.1007/978-1-60327-853-9_4 Springerprotocols.com

(1) the arrangement would form an inclusive, divergent hierarchy, (2) that arrangement could therefore be represented by an "ordinally stratified hierarchical clustering", (3) all taxa at every rank (including the whole Biota) would be monophyletic in the transformed cladistic sense, (4) each taxon at every rank could be characterized by at least one apomorph and thus unique character, which, unless that taxon was terminal, would unite the two sister-taxa of immediately lower rank which that taxon included, (5) apormorph characters of every taxon could be distinguished unambiguously by (probably ontogenetic) means that did not depend on any pre-existing classification" *(1)*. However, HGT is observed at all taxonomic levels (i.e., between genera, families, domains, etc.), as well as recombination within groups of closely related organisms, which makes such a view of the world obsolete. Genetic relationships between microbial organisms (Archaea, Bacteria, and possibly some unicellular eukaryotes) are reticulated and the number of homologous characters that they can share after HGT can thwart the identification of *bona fide* synapomorphies or question their significance. In prokaryotes, because of HGT, the concept of a tree of species does not work either for systematics or for retracing the true evolutionary history. These facts lead us to reflect on some conceptual problems, which go beyond the usual concerns solely linked with improving traditional phylogenetic methods. From a philosophical point of view, acknowledgment of HGT imposes a reversal of the proposition Sober applied to phylogeny and challenges the field of systematics. This epistemologist wondered "if one wishes to reconstruct the phylogenetic pattern, how much does one need to know about the evolutionary process?" *(2)* He concluded that "ideally, a principle of 'less is more' governs this problem. The less we need to know about the evolutionary process to make an inference about pattern, the more confidence we can have in our conclusion" *(2)*. Here we intend to state the contrary. To propose a credible evolutionary scenario we need all available information concerning this process, which we can use to attempt some classification. In this context, the occurrence of significant levels of HGT can have a major impact. As Woese asks: "what does it mean, then, to speak of an organismal genealogy when nearly all the genes in the cell – genes that give it its general character – do not share a common history?" *(3)*. For him, "this question goes beyond the classical Darwinian context" *(3)*.

In this chapter, we will briefly look within philosophy for solutions to the problems raised by HGT in the fields of phylogeny and systematics. Using the concepts developed during the debate on the definition of species in biology, and the lessons inspired from the debates around the notion of emergence, we will attempt to demonstrate the importance of integrating HGT in our taxonomical concepts for the future of microbial systematics.

## 2. On the Importance of Processes in Philosophical Definitions of Species and for Taxonomical Purposes in General

It is not in our intention to retrace the entire history of the debate on species concept (interested readers are invited to refer to *(4)* and references therein). Here, we will merely recall some philosophical contributions, which have helped to identify certain difficulties traditionally associated with defining and classifying biological species. These notions could conceivably be employed as solutions to deal with microbial systematics in the presence of HGT.

### *2.1. Brief Description of a Possible Transition from Essentialism to Eliminative Pluralism in the Species Debate*

The first pre-occupation at the heart of the species problem seemed to be to move beyond an essentialist perspective. This essentialism (or property essentialism) is the view "that it is the task of pure knowledge or 'science' to discover and to describe the true nature of things; i.e. their hidden reality or essence" *(5)*. Such an approach assimilates the biological species concept to the identification of a few immutable properties, static and clearly defined, shared by a group of organisms. On the basis of these principles, "taxonomists have sought two things – a definition of 'species' which would result in real species and a unifying principle which would result in a natural classification" *(5)*. Mayr and Hull convinced most of their readers that this essentialist approach failed for 2000 years because the immutable characteristics of species it requires is in direct conflict with the changes inherent to biological evolution. Consequently, since the 1960s, major philosophers of science recommended giving up the Aristotelian definition, both for species names (i.e., *Homo sapiens* or *Mycoplasma genitalium*) and the "species" category (the presumed common essence of all species taxa). These philosophers, along with biologists, instead suggested descent with modification as a valid unifying principle for natural classification. Thus, rather than defining species using traditional property essentialism, these evolutionary units were to be defined by their history, as "chunks of the Tree of Life". Rieppel dubbed, however, such a definition of natural groups "historical essentialism", since the only essential reason why a monophyletic group is considered natural now lies in its history and the fact that all of its members share an exclusive last common ancestor *(6)*. Still, to avoid the identification of such groups from a tree being arbitrary, evolutionists had to discover how to subdivide a continuous line of descendants into distinct units in a non-artificial way. The philosopher David Hull suggested that biologists try to characterize "family resemblances", a critical combination of properties shared by a group of related organisms that could be identified as belonging to the same species. Importantly, he made clear that combinations of overlapping properties between members of the same species do not have to be identical. In fact, "several different but

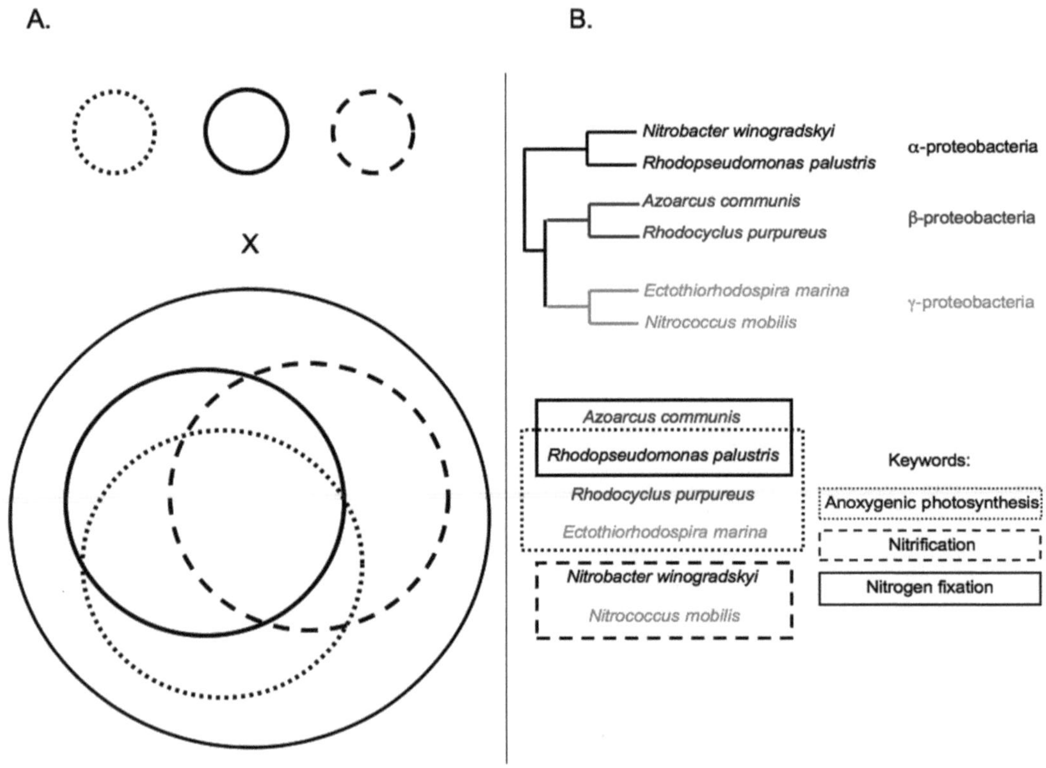


Fig. 4.1. (**A**) Family resemblance is a concept that goes back to Wittgenstein. It replaces an essentialistic standpoint with a more fluid and diversified use of concepts. Concepts do not share an essence; instead, they share overlapping features, as there are between members of a family. Here, any combination of the properties symbolized by the circles (*upper part* of the drawing) constitutes the family resemblance X (lower part of the drawing). (**B**) The tree (*upper part* of the drawing) is informative on the vertical relationships of organisms, but not necessarily on their properties. Typically, incongruent features are overlooked in such a hierarchical classification. The interactive database (*lower part* of the drawing) is structured by keywords and allows overlapping groups, where a given organism can appear in different taxonomical groups because it is naturally involved in different processes and belongs to multiple non-exclusive evolutionary units. This system preserves the information concerning vertical inheritance learned from the traditional phylogenetic approach. This information simply becomes a part and not the end of evolutionary knowledge.

overlapping sets of properties are accordingly each sufficient" *(5)* to define membership in the same natural group (**Fig. 4.1A**). This appeal to a cluster of non-essential properties could easily be integrated in the taxonomic practice because, unlike Hull's own thesis of 2000 years of stasis argued, a rigid property essentialism never had that much influence on systematics. Winsor *(7)* and Müller-Wille *(8)* demonstrated quite convincingly that the practices of pre-Darwinian taxonomists, including Linnaeus, were empirically driven rather than based on the definition of absolute types. During centuries, taxonomists had defined their higher categories using the method of exemplars or type species, by agglomerating together each new member being judged similar enough to an initial examplar in most of its characteristics. Thus, historically, taxonomical classes presented a list of properties, yet there was no need for any particular property to be always instantiated in a particular member of the group. In the philosophical debate surrounding the definition of species, one of the significant steps focused on the best means to circumscribe this "family resemblance" based on such cluster of properties (in space) and its persistence (in time).

Hull suggested that the identification of species through time could rest on the development of a metric, operating partitions in the taxonomic space, a "scale" *(9)* "to delineate evolutionary units" *(9)*. The length of this scale was not absolute but empirically derived from the observation of the phenotypic effects of some biological processes of reference (i.e., interbreeding in animals). However, the essential premise "to try to define the species and other categories in terms of the causes of evolution" *(10)*, while a variety of evolutionary processes other than interbreeding were important in nature, quickly led to regionalize the relevance of any taxonomic scale. A deeper understanding of the evolutionary biology resulted in the thesis that species are "individuals" *(10)*, and to the notion that species were "systems at various levels of integration" *(10)*, "composite wholes" *(10)*, because "constituent organisms are parts, not members" of a species *(10)*. No essential characteristic besides its particular functional integration was giving a species its reality. It was a significant breakthrough in the debate; species characterized by their status as individuals were real, regardless of divergence in their component parts. By analogy, very different people can be part of X's lab, they can even change over the years and still X's lab exists as an individual, one that publishes, organizes seminars on its favorite topic of research, competes with other labs, and so on. In other words, the species definition shifted from the identification of a set of more or less overlapping essential properties to the identification of a real causal efficacy in the evolutionary process. Ultimately, rather than leading to a unification of the general concept of species, this proposition opened the door to fairly pragmatic species concepts, known as pluralistic concepts of species taxa.

Pluralism does not mean "anything goes". It opposes monism, by endorsing the view that not only one, but several methods and theories are legitimate in an evolutionary study, because no single explanatory system can account for all the diverse phenomena of life. Of the many versions of the pluralistic species concepts proposed (from weak forms *(11)* to stronger ones *(12)* and the particularly interesting *(13)*), we will only expose one, which in our opinion describes most convincingly what is currently happening in the fields of microbiology and evolutionary biology: Ereshefsky's eliminative pluralism. In a seminal paper published in 1992, this author noted that "biologists offer various definitions of the species category" *(14)*, and recalled how the species category now played two closely connected roles in biology. First of all, "species taxa are the basal units" of traditional taxonomy: higher taxa are composed of specific taxa and constitute more inclusive units. Secondly, species taxa are "groups of organisms that evolve as units due to their exposure to common evolutionary forces" *(14)*. However, Ereshefsky demonstrated that these two roles are inconciliable, because empirical studies show that "interbreeding, ecological and monophyletic lineages do not correspond in nature" *(14)*. To the contrary, "the Tree of Life on this planet is segmented into a plurality of incompatible but equally legitimate taxonomies" *(14)*, because "different species approaches often classify the same organisms into different lineages" *(14)*, but they all rest on natural processes.

Thus Ereshefsky went on to defend that multiple approaches and taxonomies, and not only one, were relevant to classify a given organism. Unlike monists, who "insist that only one correct approach to species exists and consequently only one correct taxonomy of the organic world exists" *(14)*, he argued in favor of a "plurality of equally legitimate though incompatible taxonomies of the organic world" *(14)*. For him, "the desire for hierarchical classifications does not pose a conceptual roadblock to a realist interpretation of pluralism" *(9)*, as long as "one might be willing to ease up on the requirement that an organism belongs to only one species" *(9)*. His strong point was that the fundamental nature of an organism could be something richer than it merely being catalogued as a unique species, because "what makes an individual a single organism is the causal relations that bind its spatial and temporal parts into a single instance of life. In other words, one might give a causal account of an organism's identity" *(9)*, rather than pretending to access this identity via the arbitrary choice of a unique taxonomic label. "Given this suggestion, an organism can maintain its identity even though it belongs to different species" *(9)*, when multiple processes created it. To pursue our analogy, Mr Y, a member of X's lab, can also be the captain of an amateur soccer team and the father of a loving family; he can play a role in each one of these three groups comprising distinct members and therefore belong to all of them at the same time. Consequently, to understand evolution, there is no other choice than to "study the various types of theoretically important lineages in the world" *(14)* and the various types of causal processes, because each legitimate taxonomy offers important information. In this way, "a taxonomy of monophyletic taxa provides a framework for examining genealogy. A taxonomy of interbreeding units offers a framework for examining the effect of sex on evolution. A taxonomy of ecological units provides a structure for observing the effect of environmental selection forces" *(14)*. These different classifications should all be taken into account; "a systematic study that considers just one of these taxonomies provides an overly coarse-grained picture of evolution" *(14)* and is thus not satisfactory. For instance, Mr Y is not only a lab technician.

Finally, Ereshefsky defended the thesis that because "various taxa we call 'species' lack a common unifying feature" *(9)*, there is "reason to doubt the existence of the species category" *(9)*. In the end, he proposed to "be anti-realists when it comes to the species category" *(9)*. For instance, "merely being a genealogical entity will not suffice [to define the species category]. From an evolutionary perspective, all taxa, whether they be species, genera, or tribes are genealogical entities" *(9)*. Thus, in the absence of characteristics for the category of species, Ereshefsky proposed to "eliminate the term 'species' and replace it with a plurality of more accurate terms" *(14)* to avoid any ambiguity; for example, the terms bio-species, eco-species, and phylo-species could be used to refer to interbreeding, ecological, or phylogenetic units, respectively. This eliminative pluralist solution (the anti-realism about the species category leading to the elimination of this term and its replacement by a plurality of biologically legitimate notions) is currently widely used by biologists to deal with prokaryotic species or more exactly the lack of it *(15, 16)*.

### 2.2. A Major Philosophical Lesson: Processes Matter and HGT Cannot Be Overlooked in Microbial Classification

Attributing a primary role to biological processes in the definition of natural groups could help justify a transition from an impossible immutable property essentialism to the search of clusters of properties in historically connected organisms, and then toward eliminative pluralism. The proposition that real species are "causally efficacious, relative to some explanatory theory" *(17)* has a profound consequence. By contrast, objects "that are identified as the outcomes or artifacts of causal processes in which they do not actually function" *(17)* are characterized "as unreal or non-natural" *(17)*. Under these definitions, "higher taxa would be artifacts" *(17)*, conventional constructions. The patchiness of some characteristics of the Proteobacteria is a good example of that. *Rhodopseudomonas palustris* (an α-proteobacterium), *Rhodocyclus purpureus* (a β-proteobacterium), and *Ectothiorhodospira marina* (a γ-proteobacterium) use the same genes to realize anoxygenic photosynthesis, whereas *Nitrobacter winogradskyi* (another α-proteobacterium), *Azoarcus communis* (another β-proteobacterium), and *Nitrococcus mobilis* (another γ-proteobacterium) do not (**Fig. 4.1B**). Even worse, if we consider other functional features of the same organisms, *N. winogradskyi* and *N. mobilis* both perform the same nitrification exclusive to the other Proteobacterial species aforementioned, but only *A. communis* and *R. palustris* have the same exclusive way of fixing nitrogen. (**Fig. 4.1B**) Because there is no such thing as a real causal impact of the Proteobacteria phylum as such (i.e., there is in fact not a single physiological feature shared by all Proteobacteria, which is not a general feature of bacterial cells), the Proteobacteria cannot be considered an exclusive effective evolutionary unit when we base their definition on their role in the evolutionary process. At the very least, a microbial classification that would be satisfied with such categories as natural groups would be overly coarse-grained. Reconstructing trees that only focus on the process of vertical inheritance, from which many such taxonomic groups are derived, remains nonetheless the goal of traditional phylogenetics. However, due to HGT (and gene losses), the vast majority of the molecular characters (if not all) may have evolved along evolutionary patterns different from the vertical one. Consequently, vertical trees may have a very limited power to explain the features of extant and past microbial biodiversity. Their utility for generalizing about a species' and lineage's genomic and genetic evolution is likely close to nil at a broad evolutionary scale. Dagan and Martin notably endorsed this conclusion when they qualified an apparently magnificent Tree of Life based on 34 core genes (presumably non-transferred) as a miserable "tree of one per cent" *(18)*. A better description of the biodiversity and its history would clearly depend on a better account of how HGT binds and spreads genetic elements over time, rather than on the mere reconstruction of the vertical genealogy.

Interestingly, many philosophers acknowledge the primacy of processes over patterns, and we can thus investigate the kinds of notions they proposed when trying to develop such finer description of the evolution of life and life forms. On one hand, the fairly heterogeneous current philosophy known as "process philosophy", an alternative to entity-based philosophy, notably claims that "natural existence consists in and is best understood in terms of processes rather than things – of modes of change rather than fixed stabilities" *(19)*. In such a labile context, persistent objects (i.e., life forms in the evolutionary process) can emerge temporarily but are "no more than a statistical pattern – a stability wave in a surging sea of process" *(19)*. As a result, the biological structures we can observe and classify as taxa can be seen as ontologically secondary, as change is essential in that perspective. Following this line of thought should prevent biologists and systematicists (i) from reifying their taxonomic categories and (ii) considering that lineages present some essential aspect, persisting over time. Evolutionists should concentrate their efforts on describing this evolutionary change, stressing HGT and other diversity generating processes, rather than establishing a conservative systematics.

Emergentists' views could also bring some interesting insights to deal with the effect of HGT on the classification of evolutionary units. Emergence, a widely debated philosophical concept *(20–22)*, presupposes the existence of at least two levels of selection: a micro- and a macro-level. It states that the macro-level structure depends on the association of the micro-level components they are built upon, yet they also present emergent properties when they have their own causal power, including the possibility of affecting their own micro-level composition (by downward causation). In the evolutionary process, this downward causal effect of the higher-level structures on its components is assumed to be realized diachronically, which means over time, cycles of replication after cycles of replication. This is similar to the situation in which the presence of a selected trait (micro-level) in an individual (i.e., a recently acquired gene) can contribute to the success of a given phenotype in the population (macro-level), which in turn can affect the fitness of the carriers of the selected traits and modify the selective pressure on this micro-level character and its evolution. Importantly, such a process of emergence could give rise to identifiable evolutionary units of comparable ontological value since any new emergent higher-level structure has a causal power and plays a role in the evolutionary process. In that former example, the population (macro-level) persists while its components (individuals with original traits X) may later differ in the feature.

Interestingly, both emergentist and process-philosophy conceptions sometimes converge, as observed notably in Carl Woese's recent proposition for a new biology *(23)*. For this key player in prokaryotic phylogenetics and systematics, "the time has come to replace the purely reductionist 'eyes-down' molecular perspective with a new and genuinely holistic, 'eyes-up' view of the living world, one whose primary focus is on evolution, emergence, and biology's innate complexity" *(23)*. Importantly, one can expect two opposite phenomena (patterns) to emerge: (i) in some cases, evolution would naturally lead to some cohesion within a group of organisms, so that some "family resemblances" could help identify them as natural groups (i.e., due to interbreeding, cats look like cats generation after generation); (ii) but, on the other hand, it is also entirely possible that natural groups composed of completely disparate parts but evolving in an integrated manner could arise in nature (especially for microbes, due to HGT). In such cases, to adequately represent their natural relationships, there are no other means than to re-describe the causes of their origins and how these parts are integrated with one another in operating under a larger unit of selection.

Models where HGT plays a critical part in the evolutionary process have recently been developed and could help to account for these different emerging associations. For instance, the mix/match model integrating HGT and vertical descent proposed by Charlebois and Doolittle *(24)* seems a good way to deal with microbial evolution. In a nutshell, this model stems from the idea that cells need to fulfill various functions, but that the very genes responsible for realizing these multiple functions may differ over time. It proposes that, for a given function, the available genes can belong to different gene families (i.e., be "analogous", non-homologous markers), and that the set of genes fulfilling a given function varies during the course of evolution (due to gene and function loss). As a result, new genomic lineages would arise through mixing and matching of genes performing different functions, not only by vertical descent but also by processes of replacement. Thus, "where there are many analogous types of genes [. . .] that can perform the same general function (e.g., energy production or cell envelope formation), the living world will collectively exhibit much variability, and there will be no ubiquitous sets of genes that appear as part of any universal core. Where choices are more limited, most genes performing the needed function (some step in translation for instance) being homologous, there will appear to be little variability" *(24)*. If this model is correct, the evolutionary process would drive the emergence of multiple evolutionary units (units of selection that can interact with their environment and be replicated). Such units would share real natural genetic relationships and present a certain stability and integration at different levels of complexity.

## 3. Expanding the Conclusions of Eliminative Pluralism to Microbial Classification

The philosophical inquiries quickly evoked in the former sections, although mostly focusing on the eukaryotic world, could teach us several important lessons. First, species can be understood as evolutionary units, but then there is no such thing as species in general. Consequently, if there are no species in general, phylogenetics must redefine its (principal) objective, which can no longer be to find the universal tree of species. Phylogenetics and systematics have to revise their fundamental assumptions. It might be more relevant to assign some organisms to overlapping groups using multiple taxonomies. Second, if species taxa are evolutionary units, obviously they are not the only ones emerging from the evolutionary process. HGT may lead to the evolution of a range of evolutionary units having their own patterns (from genes, operons, plasmids, etc. to symbiotic associations). Why, then, should species taxa alone deserve the attention of microbiologists? Together, these considerations raise an important question: how could a revised enterprise of classification deal with the various evolutionary units issued from the evolutionary process? In the presence of HGT, we propose not only to embrace Ereshefsky's conclusions about species and their many legitimate taxonomies but also to extend them to all other evolutionary units and to apply them to the microbial world.

### *3.1. Multiplying "Taxa": Identification of Coherent and Composite Evolutionary Units*

The same way Ereshefsky acknowledges the equal reality of eco-species, bio-species, and phylo-species whenever their definition is based on a natural process, we propose that microbial classification should acknowledge the same ontological reality to all the evolutionary units emerging from the multilevel evolutionary process, whether this process is mostly vertical inheritance or largely involves HGT.

In its simplest form, an evolutionary unit rests on the integrated association of lower level elements held together and replicated by some biological mechanism. Phylogeneticists (and systematicists) are familiar with different evolutionary units to varying extents, depending on which biological process is responsible for the cohesion of its parts (i.e., the integration of lower level elements of the "whole" evolutionary unit). For instance, animal "species" emerge when a fairly simple reproductive process (interbreeding) causes the cohesion of a set of organisms, similar enough to interbreed, and thus produces relative persistence in the traits of their offspring across time. Based on such a process, these natural groups are comprised of phylogenetically related organisms that show some similarity (i.e., that are more similar to one another than to organisms of another interbreeding group). In that case, knowing the genealogical relationships certainly helps proposing a useful phylogenetically based taxonomy: monophyletic groups can match natural groups, thus providing a good index of biodiversity and having some explanatory power. We call such units **coherent**. Yet, due to HGT, there is no necessity for the various integrated parts to have a unique phylogenetic origin and to show some similarity with each other (like in the "species as individuals" thesis) *(10)*. In fact, and especially for microbes, whose representatives by far outnumber members of animal "species", biological processes other than interbreeding can also be responsible for the cohesion of diverse molecular parts and the emergence of more disparate – yet real – evolutionary units. We call such units **composite**.

Such disparate composite evolutionary units can be found in cases where the selective process is acting on a community of smaller evolutionary units of different origins. A clear example of this is what has been termed "the great plate count anomaly": a massive disparity between the number of microbes observed under a microscope and those that can be recovered on nutritive agar plates *(25)*. This fact suggests that, in nature, complexes of unrelated microbes, not pure isolates, are the real units of evolution. Furthermore, such evolutionary units often display emerging properties that none of their constituent parts harbor on their own. For example, syntrophic microbial consortia, composed of multiple organisms with various physiologies, are able to achieve chemical reactions that would be energetically unfavorable if carried out by a single microbe. For instance, such a relationship was uncovered between closely associated methanotrophic archaea and sulfate-reducing bacteria found in anoxic marine sediments. The archaeal partner metabolizes methane and the bacterial partner uses a resulting metabolite as an electron source *(26)*. Importantly, ecological and environmental pressures do create evolutionary units not only by associating different organisms that enable specific physiologies but also by associating different genes or pathways within organisms. Such pressures influence the rate and the nature of HGT events and consequently the genetic units composing organisms. For example, significant HGT has been detected between Sulfolobales and members of the Thermoplasmatales, two phylogenetically distant archaeal phyla that frequently share thermoacidophilic environments *(27)*. The evolution of the hyperthermophilic bacteria, *Thermotogales*, is also likely to have been shaped by their uptake of DNA from the archaea that often share their environment *(28)*.

Although coherent and composite evolutionary units cannot be studied by the same phylogenetic methods, they equally deserve to be investigated and classified by evolutionary biologists.

### 3.2. Multiplying "Taxonomies": HGT and the Era of Overlapping Rather than Nested Taxonomical Groups

The high frequency of HGT events occurring between microbes suggests that those sharing clusters of genes belong to multiple overlapping groups. We have then no reason to disregard one similarity between taxa in the name of another similarity (i.e., the sharing of a last common ancestor versus the sharing of mobile integron gene cassettes that carry drug resistance genes) and to make it the true unique diagnosis feature, which would otherwise become arbitrarily reified. Microbial biodiversity is fundamentally irreducible to a single hierarchic classification scheme, and cannot be studied with accuracy under a single model. In order to avoid overlooking any of these multiple composite evolutionary units in some overly coarse-grained perspective, it seems legitimate to propose many taxonomies describing the multiple evolutionary units and their interconnections, rather than a single taxonomy of microbial species.

Such a transition from a tree-like structure of classification to a more dynamic reticulated system is probably as shocking now as was the transition from a linear order to a series of dichotomies a thousand years ago (and in fact it is still encountering some resistance nowadays, see *(29)*). Nonetheless, such an attempt to classify microbial biodiversity, richer and more complex than traditionally proposed, will most likely prove useful in microbiology. It should be eased by the recourse to computer-based interactive databases, in which each perennial "genetic" association with a causal role in the evolutionary process could constitute an entry. To follow up on the example of the Proteobacteria described previously, the binomial nomenclature and the sole hierarchical classification were clearly poor proxies of their genetic make-up. However, thanks to keywords such as "anoxygenic photosynthesis", "nitrification", and "nitrogen fixation", describing process-based "family resemblance", the same organisms could be organized in many categories (**Fig. 4.1B**). Thus, with the recognition of widespread HGT, the statement that "because evolution seems to run in grooves, following avenues of least resistance, knowing something about one taxon gives one a very good sense of what a closely related taxon will be like" *(30)* is often inaccurate. Yet we are still able to generalize successfully about other taxonomically consistent associations, even if these "taxa" do not belong to a clade. In much the same way we can generalize about "gene replication in eukaryotes" *(31)*, we should also be able to generalize about the adaptation to high temperature in thermophiles or to the survival of halophiles at high salt concentrations (no matter if these groups comprise polyphyletic associations of archaea and bacteria), etc. This way, the extent of convergences will be better appreciated, especially in prokaryotes. For microbial classification in a context of HGT, it might well be that interactive taxonomical databases are the encyclopedia of tomorrow.

## 4. A New Critical Task for Systematics: Establishing Which Taxonomical Units Are "Mostly Flexible" or "Mostly Rigid"

Acknowledging the existence of HGT and of the reticulated genetic connections between organisms as well as the fundamental importance of change in evolution should also encourage the study of the genetic flexibility of different evolutionary units over time. At the "species" level, for instance, the study of the ontological fuzziness will likely impose itself, as "the net or web metaphor should remind us that all prokaryotic taxa are in essence imprecisely bounded and ephemeral" *(32)*. The recent transition from a microbial binomial nomenclature to a trinomial nomenclature by sheer necessity is one first instance of it. We now have to supply the strain name if we talk about the *Escherichia coli* genome. Without specifying the strain name, it is not possible to talk about what meaning should be attached to the term "*E. coli*" in the context of gene content. The recently developed notion of "bacterial pan-genome" *(33)* is another good example of that. This deliberately vague concept acknowledges two distinct circles of genes with very different diameters: the so-called "essential" core genes (i.e., the genes that are present in all the strains belonging to this group) and, in addition to it, a "cloud" comprising an undetermined number of unique adaptive genes. Such a notion sometimes advantageously replaces the one of "bacterial species", because "data clearly shows that the strategy to sequence one or two genomes per species, which has been used during the first decade of the genomic era, is not sufficient and that multiple strains need to be sequenced to understand the basics of bacterial species" *(33)*. The notions of species and pan-genome are very different, since "given that the number of unique genes is vast, the pan-genome of a bacterial species might be orders of magnitude larger than any single genome" *(33)*. In agreement with the mix/match model and the emergence of coherent and composite genetic associations in the presence of HGT, two cases are currently distinguished: "species can have an open or a closed pan-genome. An open pan-genome is typical of those species that colonize multiple environments and have multiple ways of exchanging genetic material" *(33)*. Streptococci, Meningococci, *H. pylori*, Salmonellae, and *E. coli* are notable examples of this. In these open pan-genomes, the "cloud" of genes changes according to the strains, to the environment, even according to the individual and "mathematical modeling predicts that new genes will be discovered even after sequencing hundreds of genomes per species" *(33)*. For these organisms, "the surprising conclusion from the study is that, in theory, the bacterial species will never be fully described, because new genes will be added to the genome of the species with each new genomic sequence" *(33)*. By contrast, the genes of a closed pan-genome can be exhaustively listed.

Acknowledging HGT, a new and ambitious project could thus be to investigate the evolution, composition and dynamics of each evolutionary unit, whatever its size, in order to identify which units deserve to be thought via loosely defined concepts to be correctly understood. We suggest that two notions should be developed to fulfill such a task. On one hand, if the set of genes fulfilling the selected function in a given evolutionary unit remains stable, the collection of lower level elements from which the function emerged is limited, and the genetic composition of this evolutionary unit is more or less definable, the unit is defined as **mostly rigid**. In theory, its constitutive elements can be exhaustively listed. The cat species, the closed pan-genome, or the translation machinery seem to be good instances of this type of unit, as already noted by Charlebois and Doolittle *(24)*. On the other hand, composite evolutionary units can be built from many different elements changing over time. In this case the unit is **mostly flexible**; it has a tendency to vary in the details of its make-up over long historical periods. An example of this would be the methionine biosynthesis pathway, in which the enzymes catalyzing each of the various steps can differ between organisms but still catalyze the same reaction *(34)*.

At the lineage level, we suggest naming a mostly flexible lineage where horizontal inheritance exceeds vertical inheritance over time an "open" lineage, and naming a line where vertical inheritance exceeds horizontal inheritance over time a "closed" lineage. Such a distinction would have important epistemological consequences as it will help to critically evaluate the relevance of the evolutionary scenarios about early life. It would also assist in discarding scenarios that too hastily assimilate the traditional taxonomical categories of the present time (i.e., the extant cyanobacteria) to some of their supposed common ancestors (i.e., the cyanobacteria of the past), as if the two "taxa" were composed identically. It should be clear that, in open lineages, the qualitative features associated with the extant group do not persist longer than a limited number of generations and that our power of retrodiction about them is limited. For instance, the nature of the ancestor of mitochondria might be unknowable. Nuclear genes acquired by eukaryotes from the ancestor of mitochondria should tend to reflect a single common eubacterial ancestry, but they should not necessarily belong to the known set of contemporary alpha-proteobacterial genes, regardless of how one were to define it. Rather than looking for the last common ancestor of an open lineage, we might be better inspired to look for the population of contributors that horizontally transferred these agglomerated genetic elements, a completely different viewpoint on the question of the origin.

## 5. Toward New Phylogenetic Practices

Acknowledging HGT could also provoke a profound revision of the phylogenetic practices through a fundamentally different use of the wealth of genomic data. Already, a good phylogenetic analysis of multiple markers no longer consists of the mere addition of various phylogenetic signals via concatenation of data to obtain the best unique topology. The accumulation of data under the null hypothesis that all elements used in the phylogenetic analysis share a common tree, without trying to refute this premise, suffers from a logical flaw *(35)*. In the presence of HGT, the resolution in a concatenated tree can no longer be taken as the evidence for the existence of a tree. Instead, the validity of the hypothesis concerning the existence of a single tree common to all the markers must be tested: (i) by exploring the origin of the resolution in such a super-tree, and (ii) by testing whether its support is genuine or artifactual *(36)*. To achieve this, phylogeneticists could first realize the systematic searches of sets of globally incongruent genes within a given lineage, or within a given complete genome. Good software has recently been developed to this end. For instance, the application Concaterpillar *(37)* uses a hierarchical likelihood-ratio test framework to assess both the topological congruence between gene phylogenies (i.e., whether different genes share the same evolutionary history) and the branch-length congruence (i.e., whether genes that share the same history share the same pattern of relative evolutionary rates). With such programs, it becomes possible to decide which markers belong to different incongruent gene sets, and thus to quantify whether the lineage/genome under study is closed (i.e., showing a minority of incongruent data) or open (i.e., showing a majority of incongruent data). Second, phylogeneticists could search systematically for local congruencies between a priori unrelated gene phylogenies to identify composite evolutionary units. Starting from thousands of topologies issued from metagenomic or genomic projects, analyses of split decomposition identifying common bipartitions or common embedded quartets *(38)* between trees of a same environment or between distantly related taxa should allow the discovery of co-evolving sets of genes of all sizes. If these sets of a priori unrelated genes with nonetheless common histories prove to play a role in the evolutionary process, they would likely help in discovering candidate composite evolutionary units. Third, phylogeneticists could also systematically look for the list of cosmopolitan genes, based on the detection of atypical individual gene phylogenies that do not show any correlation in their topology with the others under study. This revised practice would lead to the multiplication of phylogenetic trees and networks, each relevant for different evolutionary units, rather than to the usual proposition of a single Tree of Species.

The new phylogenetic practices described above, which emerge naturally in a theoretical framework embracing eliminative pluralism, can contribute greatly to understanding the evolution of microbial life. They are not, however, in direct opposition to more traditional analyses, but complement them. Many of these traditional analyses are still useful under a pluralist perspective, as long as they are interpreted correctly in this new context. For example, a phylogeny of Bacteria based on the 16S rRNA gene conveys some information about the histories of the organisms. The tree does not represent the true relationships of organisms, but the evolutionary history of the parts of their genomes associated with the 16S rRNA gene over a certain time span. Identifying what these parts are and how strongly they are associated is a task of new phylogenetic practices inspired by eliminative pluralism.

## 6. Conclusion

Acknowledging HGT as a major evolutionary process forces us to address questions different than those posed by traditional phylogenetics and makes it tempting to endorse the philosophical background of eliminative pluralism and the notions inspired from philosophy of emergence. In this perspective, there is certainly much to learn from the reconstruction of the complex evolutionary histories of all evolutionary units and from the discussion of the fuzziness of their contours. It could encourage us to give up the notion of species for recombining microbes and to replace it with the search for composite evolutionary units, either mostly rigid or mostly flexible. It could also facilitate giving up a unique hierarchical classification system of microbial life to the benefit of the definition of overlapping natural categories, based on the many causal roles played by microbial organisms in nature. In practice, phylogenetic methods could become pluralist and acknowledge "that different evolutionary models and representations of relationships will be appropriate, and true, for different taxa or at different scales or for different purposes" *(39)*. These fundamental changes would likely allow us to take into account a much wider part of the biodiversity and teach us that we live in a much richer and dynamic biodiversity than we previously thought, thus providing us with one of the most lively testimonies in favor of biological evolution. Until such a transition possibly occurs in phylogenetics, we all need to keep our eyes and minds open. We thus encourage our biologist readers to explore the rich literature of philosophy of biology and particularly the notion of pluralism that was only briefly introduced here.

## Acknowledgments

We thank Ford Doolittle, Pascal Tassy, Michel Morange, Armand de Ricqlès, Jean Gayon, and two anonymous referees for critical discussions. We are grateful to Jennifer Carr and Eva Boon for careful reading of the manuscript.

### References

1. Panchen, A. L. (1992) *Classification, Evolution, and the Nature of Biology*, Cambridge University Press, Cambridge.
2. Sober, E. (1991) *Reconstructing the Past: Parsimony, Evolution, and Inference*, The MIT Press, Cambridge.
3. Woese, C. R. (1987) Bacterial evolution. *Microbiol Rev* **51**, 221–71.
4. Ereshefsky, M. (2006) Species, in *The Stanford Encyclopedia of Philosophy* (Zalta, E. N., ed.) The Metaphysics Research Lab, Stanford.
5. Hull, D. L. (1965) The effect of essentialism on taxonomy – two thousands years of stasis (I). *Br J Philos Sci* **15**, 314–26.
6. Rieppel, O. (2005) The philosophy of total evidence and its relevance for phylogenetic inference. *Papéis Avulsos Zoologia* **45**, 1–31.
7. Winsor, M. P. (2003) Non-essentialist methods in pre-Darwinian taxonomy. *Biol Phil* **18**, 387–400.
8. Muller-Wille, S. (2007) Collection and collation: theory and practice of Linnaean botany. *Stud Hist Philos Biol Biomed Sci* **38**, 541–62.
9. Ereshefsky, M. (1998) Species Pluralism and Anti-Realism. *Philos Sci* **65**, 103–20.
10. Ghiselin, M. T. (1974) A radical solution to the species problem. *Systematic Zoology* **23**, 491–503.
11. Mishler, B. D., Donoghue, M. J. (1982) Species concepts: A case for pluralism. *Syst Zool* **31**, 491–503.
12. Dupre, J. (1981) Natural kinds and biological taxa. *The Philosophical Review* **90**, 66–90.

13. Kitcher, P. (1984) Species. *Philos Sci* **51**, 308–33.
14. Ereshefsky, M. (1992) Eliminative pluralism. *Philos Sci* **59**, 671–90.
15. Doolittle, W. F., Papke, R. T. (2006) Genomics and the bacterial species problem. *Genome Biol* 7, 116.
16. Gevers, D., Cohan, F. M., Lawrence, J. G., Spratt, B. G., Coenye, T., Feil, E. J., Stackebrandt, E., Van De Peer, Y., Vandamme, P., Thompson, F. L., Swings, J. (2005) Opinion: Re-evaluating prokaryotic species. *Nat Rev Microbiol* **3**, 733–9.
17. Splitter, L. J. (1988) Species and identity. *Philos Sci* **55**, 323–48.
18. Dagan, T., Martin, W. (2006) The tree of one percent. *Genome Biol* 7, 118.
19. Rescher, N. (2002) Process philosophy, in *The Stanford Encyclopedia of Philosophy* (Zalta, E. N., ed.) The Metaphysics Research Lab, Stanford.
20. Bedau, M. A. (2003) Downward causation and autonomy in weak emergence. *Principia* **6**, 5–50.
21. Brooks, D. R. (2001) Evolution in the information age: rediscovering the nature of the organism. *Semiosis, Evolution, Energy, Development* **1**, 1–29.
22. Kim, J. (1999) Making sense of emergence. *Philosophical Studies* **95**, 3–36.
23. Woese, C. R. (2004) A new biology for a new century. *Microbiol Mol Biol Rev* **68**, 173–86.
24. Charlebois, R. L., Doolittle, W. F. (2004) Computing prokaryotic gene ubiquity: rescuing the core from extinction. *Genome Res* **14**, 2469–77.
25. Staley, J. T., Konopka, A. (1985) Measurement of in situ activities of nonphotosynthetic microorganisms in aquatic and terrestrial habitats. *Annu Rev Microbiol* **39**, 321–46.
26. Nauhaus, K., Albrecht, M., Elvert, M., Boetius, A., Widdel, F. (2007) In vitro cell growth of marine archaeal-bacterial consortia during anaerobic oxidation of methane with sulfate. *Environ Microbiol* **9**, 187–96.
27. Ruepp, A., Graml, W., Santos-Martinez, M. L., Koretke, K. K., Volker, C., Mewes, H. W., Frishman, D., Stocker, S., Lupas, A. N., Baumeister, W. (2000) The genome sequence of the thermoacidophilic scavenger *Thermoplasma acidophilum*. *Nature* **407**, 508–13.
28. Nelson, K. E., Clayton, R. A., Gill, S. R., Gwinn, M. L., Dodson, R. J., Haft, D. H., Hickey, E. K., Peterson, J. D., Nelson, W. C., Ketchum, K. A., Mcdonald, L., Utterback, T. R., Malek, J. A., Linher, K. D., Garrett, M. M., Stewart, A. M., Cotton, M. D., Pratt, M. S., Phillips, C. A., Richardson, D., Heidelberg, J., Sutton, G. G., Fleischmann, R. D., Eisen, J. A., White, O., Salzberg, S. L., Smith, H. O., Venter, J. C., Fraser, C. M. (1999) Evidence for lateral gene transfer between Archaea and bacteria from genome sequence of *Thermotoga maritima*. *Nature* **399**, 323–9.
29. Gupta, R. S. (2001) The branching order and phylogenetic placement of species from completed bacterial genomes, based on conserved indels found in various proteins. *Int Microbiol* **4**, 187–202.
30. Wake, D. B. (2004) A tree grows in Manhattan, in *Assembling the Tree of Life* (Cracraft, J., ed.) Oxford University Press, New York.
31. Steel, D. (2004) Can a reductionist be a pluralist? *Biol Philos* **19**, 55–73.
32. Doolittle, W. F. (1999) Lateral genomics. *Trends Cell Biol* **9**, M5–8.
33. Medini, D., Donati, C., Tettelin, H., Masignani, V., Rappuoli, R. (2005) The microbial pan-genome. *Curr Opin Genet Dev* **15**, 589–94.
34. Gophna, U., Bapteste, E., Doolittle, W. F., Biran, D., Ron, E. Z. (2005) Evolutionary plasticity of methionine biosynthesis. *Gene* **355**, 48–57.
35. Bucknam, J., Boucher, Y., Bapteste, E. (2006) Refuting phylogenetic relationships. *Biol Direct* **1**, 26.
36. Bapteste, E., Susko, E., Leigh, J., Ruiz-Trillo, I., Bucknam, J., Doolittle, W. F. (2008) Alternative methods for concatenation of core genes indicate a lack of resolution in deep nodes of the prokaryotic phylogeny. *Mol Biol Evol* **25**, 83–91.
37. Leigh, J. W., Susko, E., Baumgartner, M., Roger, A. J. (2008) Testing congruence in phylogenomic analysis. *Syst Biol* **57**, 104–15.
38. Zhaxybayeva, O., Gogarten, J. P., Charlebois, R. L., Doolittle, W. F., Papke, R. T. (2006) Phylogenetic analyses of cyanobacterial genomes: quantification of horizontal gene transfer events. *Genome Res* **16**, 1099–108.
39. Doolittle, W. F., Bapteste, E. (2007) Pattern pluralism and the tree of life hypothesis. *Proc Natl Acad Sci U S A* **104**, 2043–9.

# Chapter 5

## Persistence Mechanisms of Conjugative Plasmids

**Martin Iain Bahl, Lars Hestbjerg Hansen, and Søren J. Sørensen**

### Abstract

Are plasmids selfish parasitic DNA molecules or an integrated part of the bacterial genome? This chapter reviews the current understanding of the persistence mechanisms of conjugative plasmids harbored by bacterial cells and populations. The diversity and intricacy of mechanisms affecting the successful propagation and long-term continued existence of these extra-chromosomal elements is extensive. Apart from the accessory genetic elements that may provide plasmid-harboring cells a selective advantage, special focus is placed on the mechanisms conjugative plasmids employ to ensure their stable maintenance in the host cell. These importantly include the ability to self-mobilize in a process termed conjugative transfer, which may occur across species barriers. Other plasmid stabilizing mechanisms include the multimer resolution system, active partitioning, and post-segregational-killing of plasmid-free cells. Finally, various molecular adaptations of plasmids to better match the genetic background of their bacterial host cell will be described.

**Key words:** Accessory elements, plasmid stability, host range, horizontal gene transfer.

### 1. Introduction – The Selective Advantage of a Plasmid Life-Style

Plasmids are ubiquitous in bacteria and constitute a driving force in exchanging genes between cells residing in the same habitat in an action termed horizontal gene transfer (HGT). This may increase the adaptability of the bacteria to changing environmental conditions, which is dramatically exemplified by the emergence and spread of multiple antibiotic-resistance plasmids in and between potentially pathogenic bacteria. This chapter aims to provide an overview of the different mechanisms and strategies that plasmids employ to ensure successful establishment and maintenance in bacterial populations and communities.

Maria B. Gogarten et al. (eds.), *Horizontal Gene Transfer: Genomes in Flux, vol. 532*

DOI 10.1007/978-1-60327-853-9_5 Springerprotocols.com

When discussing the biological reasons for the prevalence of plasmids in natural bacterial populations, it is necessary to consider the unit of Darwinian selection. The question is who is in the driving seat, the host bacterium or the plasmid. Plasmids are, due to the many useful traits associated with them, often regarded as extra-chromosomal "toolboxes" for the host bacteria, and thus may increase the fitness of the bacteria that come to carry them. This seems especially obvious in highly selective environments, such as the digestive tract of mammals during antibiotic treatment, where the carriage of appropriate antibiotic-resistance determinants ensure the survival of a subpopulation of the indigenous or invading bacteria. The ability of some plasmids to perform conjugative transfer between both closely and distantly related bacteria provides them a vehicle-like function for gene transfer. The fact that many accessory functional genes of plasmids are contained in mobile elements makes further transfer events from plasmids to the chromosome of recipient bacteria a feasible possibility that would ensure the long-term stability of the element, which would otherwise be lost together with the plasmid following unsuccessful segregation. The question is who has a selective advantage – the bacteria or the plasmid?

Some evidence suggests that conjugative plasmids basically exist as genetic parasites, ensuring self-propagation by infectious transfer. This view requires that the in situ transfer rate is higher than the effective plasmid loss rate from the bacterial population or community. The effective loss rate is a function of the rate of unsuccessful plasmid segregation and the fitness difference between plasmid-harboring and plasmid-free cells. Because of this, any mechanism that reduces this fitness difference, such as carriage of useful accessory elements, would be expected to increase the chances for the plasmid-backbone to persist over evolutionary time. The point here is that distinct mobile elements, such as plasmids, transposons, and integrons, may not be required to be associated with the genes useful to the host at all times.

## 2. Accessory Elements of Plasmids

Bacterial plasmids are typically composed of different regions or modules, which encode the functions involved in plasmid replication, stability, and transfer, which make up what is often defined as the plasmid backbone. Apart from these "household" functions plasmids very often harbor one or more accessory elements that encode useful or selectively advantageous traits to the host bacterium. The fact that most of these accessory elements are contained within transposable elements further increases their potential for dispersion in microbial communities (**Fig. 5.1**).

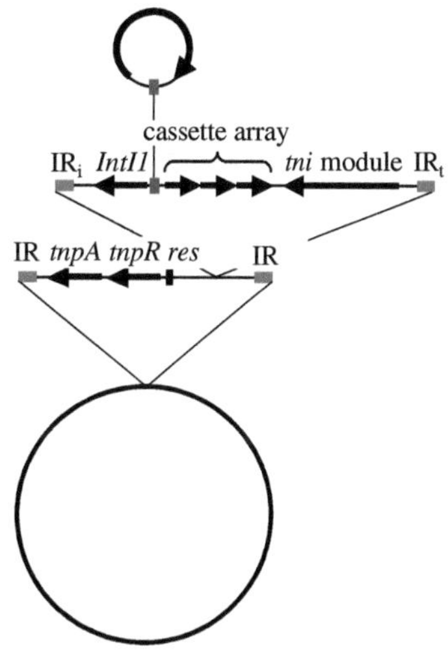


Fig. 5.1. The modular and hierarchal composition of mobile genetic elements. Gene cassettes are inserted at the *attI* site of the integron. The integron, bounded by inverted repeats, may be inserted into a composite transposon, which in turn may be inserted into a self-transmissible plasmid. This plasmid thus constitutes a large transferable package of genetic information.

One impressive example of this is Tn*21*, which is a composite transposon. Tn*21* and close relatives, known collectively as the Tn*21*-family, have been identified in numerous different large conjugative plasmids isolated from both environmental and clinical settings and have been dubbed the "flagship of the floating genome" *(1)*. This transposon belongs to the Tn*3*-familiy of transposable elements, has characteristic terminal inverted repeats, and harbors the *tnpA* and *tnpR* genes encoding a transposase and a resolvase, respectively. Tn*21* also contains a *mer* region conferring resistance toward mercury and a potentially mobile class 1 integron, which serves as a capture and expression system for gene cassettes *(2)*.

Integrons are defined as the genetic elements that encode a site-specific recombinase, known as the DNA integrase, and a corresponding recombination site, *attI*, into which gene cassettes may be specifically inserted *(3)*. Class 1 integrons are not mobile per se; however, they are typically associated with a Tn*402*-like transposon. This transposon has imperfect 25bp inverted terminal repeats and harbors either a complete or, as is often seen, a truncated form of the four-gene *tni* module, which in its complete form, as in Tn*402*, facilitates transposition. In many cases this entire potentially mobile integron/transposon element, plus possibly other inserted sequences (IS elements), has for simplicity been termed an integron *(1)*. The integron designated In2 contained in Tn*21* is an example of this. Class 1 integrons have been identified in many large conjugative plasmids including members of the IncF *(4)*, IncL/M *(5)*, IncN *(6)*, IncP (IncP-1) *(7–11)*, IncG (IncP-6) *(12)*, IncU *(13)*, IncW *(14)* incompatibility groups, and thus play an important role in their functional diversity.

Gene cassettes associated with class 1 integrons have primarily been found to carry genes conveying resistance toward different groups of antimicrobial agents including β-lactams (*bla* and *oxa*), aminoglycosides (*aad* and *aac*), trimethoprim (*dfr*), chloramphenicol (*cat* and *cml*), quaternary ammonium compounds (*qac*), rifampicin (*arr*), sulfonamides (*sul*), macrolides (*ere*), lincosamides (*lin*), streptothricin (*sat*), and quinolones (*qnr*) *(15–19)*; however, this may be due to an experimental bias toward identifying such genes and does not necessarily reflect the entire diversity or distribution of gene cassettes. Recent studies have indeed shown that class 1 integrons may also harbor gene cassettes that are not related to antibiotic resistance *(20, 21)*. Resistance genes other than those associated with gene cassettes may be found within the same integron/transposon element *(8, 22)* or as part of other accessory elements of plasmids *(9)*. These have been found to encode resistances toward a number of other antimicrobial agents such as tetracyclines *(23)*, glycopeptides *(24)*, the antibiotic growth promoter olaquindox *(25)*, as well as to toxic metal ions including $Ag^{+}$, $Cd^{2+}$, $Co^{2+}$, $CrO_4{}^{2-}$, $Cu^{2+}$, $Hg^{2+}$, $Ni^{2+}$, $Pb^{2+}$, and $Zn^{2+}$ *(26)*.

Other large groups of accessory genes found on plasmids include those encoding catabolic pathways and those encoding virulence functions. The ability of bacteria to degrade and mineralize xenobiotic compounds, here broadly defined as any compound that is not natural to the environment, is very often facilitated by the expression of specific genes residing on plasmids. A recent review lists 71 known catabolic plasmids together with their substrates and catabolic genes *(27)* and gives a picture of the impressive range of substances that may be degraded. Examples of substrates include pesticides such as 2,4-dichlorophenoxyacetic acid, aromatic hydrocarbons such as benzene, phenol, and toluene, and phthalates. Most catabolic plasmids have to date been isolated endogenously from cultured bacterial strains although a few, including the agricultural herbicide 2,4-D degrading plasmid, pEMT3, have been isolated exogenously by virtue of their conjugative transferability to a recipient strain *(28)*.

The virulence genes of pathogenic bacteria may be chromosomally located in, for example, pathogenicity islands, but are also frequently associated with plasmids. One of the more striking examples of virulence plasmids of Gram-positive bacteria are the pXO1 and pXO2 plasmids of *Bacillus anthracis*, which encode anthrax toxins and capsule proteins, respectively *(29)*. These plasmids apparently transform relatively innocent soil-dwelling bacteria into highly virulent human pathogens.

# Color Plates

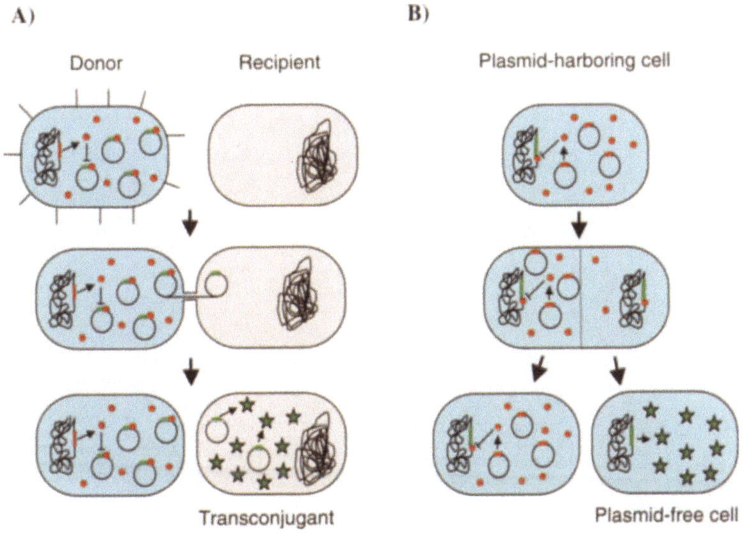


Plate 1. Schematic representation of the GFP detection systems for HGT (*panel A*) and plasmid stability assay (*panel B*). The donor cells in the HGT detection system; (*see* complete caption and discussion on p. 79)

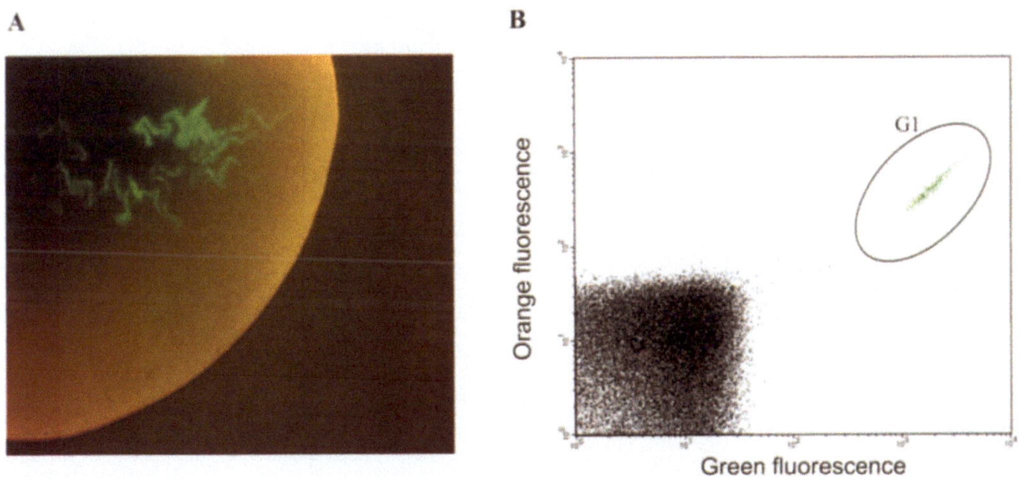


Plate 2. (**A**) Segment of a bacterial colony of *Kluyvera* sp. MB101::*gfp* harboring plasmid pKJK5::$lacI^{q1}$. The *green* fluorescing patches represent plasmid-free segregant cells. (**B**) Corresponding flow cytometry data showing a small sub-population of GFP-positive, i.e., plasmid-free, cells (G1); (*see* discussion on p. 88)

Well-characterized virulence plasmids of Gram-negative bacteria include the serovar specific plasmids of *Salmonella enterica* subsp. *enterica*. These plasmids are found in many of the serovars associated with infections in humans and farm animals, including Enteritidis, Typhimurium, and Dublin and all encode the *spvRABCD* (*Salmonella* plasmid virulence) genes. Additional virulence-associated functions of these plasmids include plasmid-encoded fimbriae, which may increase cell-adhesion during infection, and resistance to the complement killing system *(30)*.

Enterotoxigenic *Escherichia coli* (ETEC) is another serious pathogen and is the primary causative agent of food and waterborne *E. coli*-mediated diarrhea in humans worldwide *(31)*. Attachment of ETEC cells to the small intestine during colonization of the host is usually mediated by plasmid-encoded colonization factors, of which more than 20 serologically distinct types have been characterized *(31)*. One such colonization factor (CS1) causes a fimbrial morphology, and is encoded by the recently sequenced plasmid pCoo *(32)*. Interestingly, pCoo was found to have a mosaic cointegrate structure composed of two functional plasmid replication regions, namely R64-like (IncI1) and R100-like (IncFII), which increases the functional capacity and was suggested to indicate a rapid evolvement of such CS1-encoding plasmids. The enterotoxins, which are responsible for the increased fluid secretion in the intestine following ETEC infections, are frequently plasmid-encoded, and different genes may further be linked to form large mobilizable pathogenicity islands *(33)*.

One final example of plasmid-encoded virulence is the 70-kb pYV plasmid of *Yersinia* spp. This plasmid encodes the Yop virulon, which enables *Yersinia* cells to survive and multiply in the lymphoid tissues of their host. Contact to the eukaryotic cell allows injection of specialized "Yops" proteins through a secretion channel, which disturb the dynamics of the cytoskeleton and inhibit production of pro-inhibitory cytokines *(34)*. The Yop virulon represents the archetype of the type III secretion apparatus, which has also been identified in a number of other Gram-negative human pathogens including*Salmonella* spp., *Shigella* spp., *Bordetella* spp., *Pseudomonas aeruginosa*, as well as plant pathogens including *Xanthomonas* spp. and *Erwinia amylovora (35)*.

## 3. Plasmid Host Range and Tracking

The diversity of different host species in which a specific plasmid may be maintained is an important factor for its ability to spread and persist in a bacterial community and is termed its host range. Some plasmid groups have a relatively narrow host-range, e.g., the IncF and IncI plasmids of *Enterobacteriaceae*, and are consequently confined to few bacterial genera, whilst others have been shown to display an impressively broad host range (BHR). Such BHR plasmids may be either conjugative, i.e., able to facilitate their own transfer between different bacterial species, or mobilizable, i.e., transferred by employing the transfer apparatus of a co-residing conjugative plasmid or transposon. Some natural plasmids may be stably maintained in a remarkable range of both Gram-positive and Gram-negative bacterial species and are thus clearly BHR. One example of this is the plasmid pIP823, isolated from *Listeria monocytogenes* BM4293. This plasmid contains the *dfrD* gene encoding high-level trimethoprim resistance and is able to replicate in *Enterococcus faecalis, Staphylococcus aureus, Bacillus subtilis*, and *E. coli* cells *(36)*. Plasmid pIP823 is not self-transferable by conjugation, but is, in some circumstances, able to be mobilized by conjugative plasmids and by conjugative transposons from both Gram-positive and Gram-negative bacteria.

The host range of self-transferable conjugative plasmids is generally not as broad and, consequently, one widely accepted definition of BHR of such plasmids in Gram-negative bacteria is that the plasmid may replicate and be stably maintained in bacteria representing at least two different branches of the Proteobacteria, e.g., *E. coli*, γ-Proteobacteria, and *Burkholderia cepacia*, β-Proteobacteria *(37)*. Plasmids belonging to the IncP-1 incompatibility group clearly fall within the BHR definition, and have often been cited as being able to be transferred to and maintained in most Gram-negative bacterial species *(38)*, including members of the α-, β-, and γ-Proteobacteria *(39)*. It has furthermore been observed that IncP-1 plasmids may facilitate conjugation to, but not be maintained in Gram-positive bacteria, as well as to eukaryotic cells such as yeast. Conjugative BHR plasmids from Gram-positive bacteria have also been described and will here be exemplified by plasmid pIP501, originally isolated from *Streptococcus agalactiae* and conferring resistance to the macrolide, licosamide, streptogramin B (MLS) group of antibiotics as well as to chloramphenicol *(40)*. Studies have shown that pIP501 or the related plasmid pAMβ1 *(41)* may be transferred to a wide variety of Gram-positive bacteria, including streptococci, lactobacilli, lactococci, *Listeria* spp., bacilli, clostridia, and staphylococci *(42)*. Furthermore, conjugal transfer to multicellular *Streptomyces lividans* and to Gram-negative *E. coli* cells has also been demonstrated *(42)*. Such direct intergeneric plasmid transfer across the Gram-positive/Gram-negative barrier has for many years been considered virtually impossible for natural plasmids to accomplish; however, some other examples have been described in the literature *(43)*.

The traditional approach for evaluating the host range of specific bacterial plasmids is to conduct biparental matings, either in liquid broth or on the surface of a filter placed on solid agar medium, between donor cells harboring the plasmid and recipient cells. This approach, however, requires that (i) both donor and recipient are culturable, (ii) that a selective plasmid marker such as an antibiotic-resistance determinant is carried on the plasmid and able to be expressed by the recipient bacteria, and (iii) that it is possible to counter-select the donor strain. During the past few years, more advanced techniques have been employed to study the kinetics of HGT and consequently host range of conjugative plasmids *(44)*. One such approach is to incorporate a fluorescence gene-tag downstream of a repressible promoter in the plasmid of choice and the corresponding repressor-encoding gene into the chromosome of the plasmid donor strain (**Fig. 5.2A** and Color Plate 1, following p. 76) *(45)*. Such a setup using the green fluorescence protein has recently been employed to study the transfer kinetics and natural host range of the IncP-1 plasmid pKJK10 in a soil rhizosphere environment *(43)*. By using a novel flow cytometry-based cultivation-independent method, it was possible to enumerate and sort transconjugant cells and subsequently classify their phylogeny based on 16S rRNA sequencing. A high diversity of natural recipients of the IncP-1 plasmid was revealed including members of the α-, β-, and γ-Proteobacteria, which

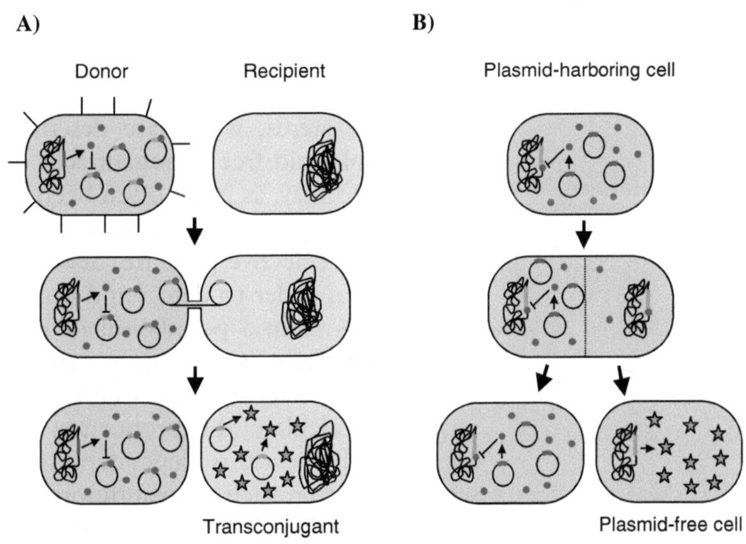


Fig. 5.2. Schematic representation of the GFP detection systems for HGT (*panel A*) and plasmid stability assay (*panel B*). The donor cells in the HGT detection system (**A**) contain a *gfp*-tagged (*green line*) conjugative plasmid. Expression of the *gfp* gene is repressed by a chromosomally encoded lac-repressor protein (*red circles*). The *gfp* gene is first expressed (*green stars*) following conjugative transfer to recipient cells. In the plasmid stability assay (**B**), the plasmid-sensor cells contain a chromosomally encoded *gfp* gene (*green line*), which is repressed by a plasmid-encoded lac-repressor protein. The *gfp* gene is expressed in plasmid-free segregant cells (*see* Color Plate 1, following p. 76).

are established hosts of IncP-1 plasmids *(46, 47)* and also, more surprisingly, an *Arthrobacter* sp. belonging to the Gram-positive *Actinobacteria*. This is significant since it demonstrates that IncP-1 plasmids may be transferred to, and their genes expressed in, phylogenetically distant Gram-positive bacteria.

Another recent study, using the *rfp*-gene-encoding red fluorescence protein as a plasmid marker, investigated the importance of the host species on the diversity of transconjugants within an activated-sludge environment *(39)*. Three strains, representing α-, β-, and γ-Proteobacteria and all harboring the IncP-1 plasmid pB10::*rfp*, were used as donor bacteria in a mating with bacteria indigenous to the activated sludge environment. It was demonstrated that the transfer range of the IncP-1 plasmid was not only determined by the plasmid itself but also affected by the host species in which it was residing. A total of 306 transconjugants belonging to 15 different genera were obtained of which 8 genera were retrieved only when the α-Proteobacterium *Sinorhizobium meliloti* was employed as donor strain. This study highlights the fact that conjugative plasmids may spread through bacterial communities in a step-stone like manner in which the plasmid may not have access to all other stones directly from the one it is currently standing on, but could well see more possibilities from the one next to it.

Prevalence of conjugative plasmids in bacterial populations is influenced by the rate of conjugative HGT and also importantly by the stability of the plasmid once hosted by the bacterial cell. An abundance of mechanisms have been reported, which enhance stable maintenance of low to medium copy-number conjugative plasmid, including active partitioning and plasmid addiction systems, which are described later on. Occasionally, however, plasmid-free segregant cells do appear in bacterial populations and may, as it has often been presumed, outcompete the slower-growing plasmid-harboring isogenic cells in the absence of a plasmid-selective pressure. Several different approaches have been used in order to detect and quantify this often low level of plasmid-loss from a population of bacteria. The most straightforward of these is to simply plate a sample of the cells on both plasmid-selective plates and non-selective plates and subsequently calculate the proportion of plasmid-free cells. This approach, however, is dependent on the plasmid harboring a selective marker, such as an antibiotic-resistance determinant, and also requires that all plasmid-harboring cells readily form colonies on the plasmid-selective plates. The latter of these requirements may well cause some problems as activity-dependent phenotypic lag or gene silencing can reduce the numbers of CFUs on a selective agar-plate, thus leading to an overestimation of plasmid loss *(48–50)*. Initial isolation on non-selective agar plates followed by replica-plating onto selective media is often a preferred method as it avoids expression lag problems and also provides a more statistical sound result.

Incorporation of fluorescence markers of biochemical tags, such as the *lacZ* gene encoding β-galactosidase, into the plasmid of interest allows a more direct discrimination between plasmid-harboring and plasmid-free cells. During the past decade the *gfp* gene, encoding green fluorescent protein, has become an extremely valuable tool for studying plasmid kinetics in bacterial populations. When employed in plasmid stability studies, it has proved useful to construct a system in which it is the plasmid-free segregants arising at a low frequency in a population of plasmid-harboring cells that express GFP and not vice versa. This has been achieved by incorporating a Lac-repressor-regulated *gfp* gene into the bacterial chromosome of *E. coli* and the corresponding *lac* repressor gene into the plasmid to be studied (**Fig. 5.2B**) *(51)*.

## 4. Plasmid Stability Mechanisms

Although bacterial plasmids may harbor an impressive diversity of "useful" genes, they per definition do not carry any absolutely essential genes for cell survival at standard conditions, as they would otherwise generally be defined as a chromosome. This fact means that plasmids are expendable and thus prone to loss, if the host-cell thereby gains a fitness advantage over isogenic plasmid-harboring cells in the same niche. Consequently, to be maintained in their host cell for an extended period of time, plasmids must either (i) contribute with functions providing a selective advantage to the host cell, (ii) have an effective system to ensure accurate segregation at cell division, or (iii) behave as a parasitic element capable of infectious HGT. The copy number of plasmids in the host cell is often considered to have a direct effect on the probability $P_0$ that a plasmid-free daughter cell will arise at cell division. In case the plasmid is distributed randomly onto the daughter cells, the probability of a plasmid-free daughter cell occurring at each cell division is given by the simple equation: $P_0 = 2^{(1-n)}$, where $n$ is the plasmid copy number *(52)*. In order to hold true, however, this requires that the plasmids are all randomly distributed and not connected in any way within the cell – a condition that is often not fulfilled, for example, due to multimer formation caused by recombination events between the identical nucleotide sequences of the plasmids. It is clear, however, that low copy number plasmids such as F and IncP-1 with copy numbers of ca. 1 and 5–7, respectively *(53, 54)*, cannot rely on random distribution alone, but must encompass active systems to circumvent plasmid loss. A number of different mechanisms that all work to

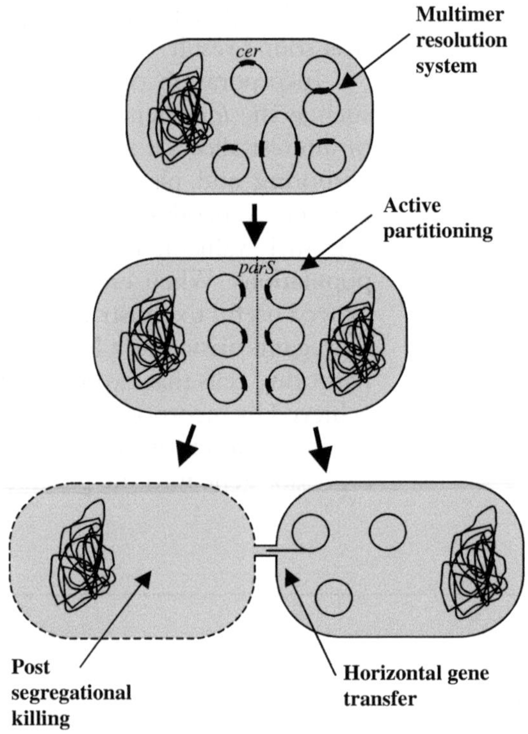


Fig. 5.3. Schematic representation of the different mechanisms during the cell cycle that may facilitate the stable maintenance of conjugative plasmids in bacterial cells. See text for further information.

stabilize plasmids in the absence of positive selection have been identified (**Fig. 5.3**) and will be presented below.

#### *4.1. Multimer Resolution Systems*

Multiple copies of plasmids within a bacterial cell are prone to host-encoded homologous recombination due to their identical DNA sequences, leading to the formation of oligomers of the plasmid. This leads to a serious stability problem, as the actual copy number of independent replicons is reduced, thus increasing the probability of the formation of plasmid-free segregant cells. The problem may be further enhanced by over-replication of already formed oligomers, an effect termed dimer catastrophe *(55)*. Many plasmids have a multimer resolution system (mrs) to circumvent multimerization. The best studied mrs is the *cer*/Xer of plasmid ColE1, in which *cer* functions as an oligomer resolution site for the host-encoded Xer site-specific recombinase. Other plasmid-encoded systems include *parCBA* of IncP-1 plasmid RK2 and the Cre-*lox* system of bacteriophage P1. Interestingly, it has been shown that the resolution site and resolvase gene of complex transposons harbored by plasmids may also provide plasmid stability by way of multimer resolution *(56)*.

#### 4.2. Plasmid Partitioning Systems

Active partitioning is an important mechanism to ensure stable maintenance of low copy number plasmids during bacterial cell division. The system is composed of an ATPase and a DNA-binding protein, which are encoded by a number of different homologous genes in various plasmids, phages, and chromosomes, and are often generically termed *parA* and *parB*, respectively. The *parA*-encoded ATPase binds to the ParB protein encoded by *parB*, which in turn binds to one or more *cis*-acting sites on the DNA strand (here termed *parS*) that function as centromere-like elements, thus forming a nucleoprotein complex. This facilitates the mitotic-like positional migration and tethering of the plasmids during cell division and ensures that both daughter cells receive at least one copy of the plasmid. Bignell and Thomas have published a comprehensive review on the bacterial ParA–ParB partitioning proteins *(57)*.

Plasmids belonging to the broad-host-range IncP-1 incompatibility group carry a partitioning system composed of the *incC* and *korB* genes, which encode a ParA family ATPase and a DNA-binding protein KorB, respectively, the latter of which also functions as a global transcriptional repressor. Palindromic target sites for the KorB protein are scattered throughout the plasmid backbone, and a total of 12 have been identified in IncP-1α plasmid RK2 ($O_B1$–$O_B12$) *(23)* of which $O_B3$ has been suggested to function as the centromere-like site *(58)*, although others may also be involved. IncP-1 plasmids have been shown to replicate and display stable maintenance in almost all Gram-negative species. The importance of the active partitioning gene *incC* for stable maintenance in such diverse bacterial hosts has recently been elucidated in a study by Siddique and Figurski *(59)*. The authors show that an in-frame deletion of the *incC* gene in an RK2-like plasmid results in a significant degree of plasmid instability in six different bacterial hosts of the α- and γ-subdivisions of the Proteobacteria. Expression of the short version of this protein (IncC2) in a *trans*-complementation analysis was sufficient to restore the stable maintenance phenotype of the plasmid *(59)*. Another study has shown that the chromosomal homologs of the ParA/ParB proteins, encoded by the *soj* and *spoOJ* genes of *Bacillus subtilis*, are able to stabilize miniF plasmids harboring the *spoOJ* gene, including one internal centromere-like site, *parS*, and the *soj* gene in the phylogenetically distant *E. coli*. This is remarkable since these two species are separated from each other at least 1,200 million years ago and indicate that the partitioning protein homologs ParA/ParB may be universally functional in a very wide range of bacterial backgrounds *(60)*.

Most bacterial chromosomes studied also contain genes, which are homologous to parA and parB, although notable exceptions to this are several species belonging to the γ-Proteobacteria, including *E. coli* and *Salmonella Typhimurium*. Whether or not such chromosomally encoded partitioning genes have any role in correct chromosome segregation at cell division has long been considered probable, due to a high similarity with their plasmid-encoded counterparts. A study by Lewis et al. has shown that different *parA* and *parB* mutations in the chromosome of *Pseudomonas putida* give rise to between 5% and 10% cells without chromosomes during transition from the exponential growth phase to the stationary growth phase when grown in M9-minimal medium *(61)*. This effect was not observed when the cells were grown in rich medium under standard conditions, which shows that the importance of the partitioning system depends on the cell's physiological state. This finding is supported by another more recent study *(62)*. The authors suggest that the Par apparatus may be redundant in exponentially growing bacteria with overlapping cell cycles, as under these conditions chromosome replication reinitiates before cell division, which may in turn facilitate symmetrical chromosome tethering on both sides of the division plane *(61)*.

A more pronounced effect of a chromosomally encoded partitioning system has recently been attributed to the smaller of the two chromosomes contained in *Vibrio cholera*, namely ChrII. Deletion of the *parAB2* genes encoded by ChrII causes the chromosome to be mislocalized within the cell and in turn frequently fails to segregate properly generating cells with only one chromosome *(63)*. Although ChrII is defined as a chromosome because it harbors essential genes for cell survival, much evidence suggests that it may well have originated as an extra-chromosomal plasmid or phage, and later acquired some essential genes through recombination or transposition events. Such evidence includes that (i) the origins of replication of the two chromosomes assume distinct subcellular positions during the cell cycle of which ChrII resembles that of low copy number plasmids such as F and P1 *(64)*, (ii) the ChrII partitioning genes are most closely related to those of plasmids, whereas the partitioning genes of the large chromosome are related to *par* genes from other chromosomes *(65)*, and lastly (iii) 13 toxin-antitoxin (TA) loci, all contained as gene cassettes in a single integron and similar to those found in plasmids, are encoded by ChrII *(66, 67)*.

### 4.3. Post-segregational Killing/Plasmid Addiction Systems

Another effective mechanism of plasmid stability is facilitated by various post-segregational killing (PSK) systems, which are also known as addiction systems or toxin–antitoxins. These systems ensure growth arrest or killing of plasmid-free cells arising following unsuccessful plasmid segregation at cell division; they have been identified in numerous different plasmids and also in many bacterial chromosomes *(66)*. PSK systems are composed of an operon of at least two essential genes encoding a stable toxin and a corresponding unstable antitoxin, respectively. The theory that PSK systems benefit plasmids by ensuring their stable maintenance is broadly accepted; however, Cooper and Heinemann have convincingly suggested that PSK may rather mediate the exclusion of competing plasmids of the same incompatibility group *(68)*. Two distinct classes of PSK systems have been defined based on the nature of the antitoxin.

#### *4.3.1. Proteic Systems*

The proteic PSK system relies on a direct protein–protein interaction between the toxin and antitoxin forming a complex, which consequently inhibits binding of the toxin to its site of action within the cell. The archetype of such a proteic PSK system is that of the F plasmid encoded by the *ccdA* and *ccdB* genes (control of cell death) *(69, 70)*. The *ccdB* gene encodes the CcdB protein, which in the absence of the unstable *ccdA* gene product binds to the *E. coli* GyrA subunit of gyrase and consequently inhibits the function of this type II topoisomerase *(71)*. A recent study has shown that resistance toward this gyrase inhibitor occurs in bacteria and may, apart from freeing the cell of the plasmid addiction system, also facilitate cross-resistance to fluoroquinolones, which also target the cellular gyrase *(72)*. Control of the *ccd* operon is negatively autoregulated at the level of transcription by a complex of both gene products *(73, 74)*. It has been proposed that the stoichiometry of the CcdA and CcdB proteins is important in determining whether or not the complex binds to the *ccd* operator–promoter region *(75)*. The CcdA antitoxin protein is prone to degradation by the Lon protease *(76)*, and consequently the intracellular concentration of CcdA decreases more rapidly than that of CcdB if segregant cells arise, which do not carry the F plasmid. This allows dissociation of the CcdA/CcdB complexes, which in turn may result in gyrase inactivation. Interestingly, the effective cell-killing function of the *ccdB* gene product has been exploited for positive selection in bacterial cloning vectors by designing these so that the *ccdB* gene is inactivated by insertion of the cloned DNA fragment *(77)*.

Other well-described plasmid-encoded proteic PSK systems includes the *parDE* system of IncP-1 plasmid RK2, in which the *parE* gene product has also been shown to inhibit the *E. coli* gyrase even though it appears to be evolutionarily unrelated to the CcdB toxin *(78)* and the identical *parD* and *pem* systems of IncFII plasmids R1 and R100 in which the Kid toxin specifically inhibits DnaB-dependent DNA replication *(79–81)*. The bacteriophage P1, which lysogenizes *E. coli* as a very stable low copy number plasmid, also harbors a proteic PSK system encoded by the *phd/doc* genes *(82)*. Two separate chromosomal homologs of the *pem* system have been identified in an *E. coli* strain and are designated *chpA* and *chpB*. This may be a result of a past recombination or insertion event with a plasmid or bacteriophage system or, conversely, the plasmid-borne system could be of chromosomal origin. The functional capability of at least some of these chromosomal genes has been confirmed by stabilization of plasmids containing these cloned sequences *(83)*.

*4.3.2. Antisense Systems*

The other class of PSK systems relies on post-transcriptional inhibition of toxin protein synthesis by anti-sense RNA. The best studied of these is the *hok/sok* (host killing/suppressor of killing) system (*parB* locus) encoded by plasmid R1 of *E. coli (84)*, which has served as a model system for general sense/antisense RNA pairing in bacteria. The *hok* gene encodes a toxic transmembrane protein, which leads to a rapid collapse of the membrane potential and arrest of respiration of the cell *(85)*. Expression of the *hok* gene is regulated post-transcriptionally by sok antisense RNA, which is complementary to the leader region of Hok mRNA, but does not overlap with the *hok* Shine-Dalgarno sequence (**Fig. 5.4**). The Sok antisense RNA does, however, overlap with the Shine-Dalgarno sequence of another gene located within the hok/sok locus, namely *mok* (mediation of killing). Translation of the part of *mok* that overlaps with the *hok* translation initiation region, although not the gene product per se, has been shown to be essential for translation of the Hok protein. Thus Sok-RNA regulates *hok* translation indirectly by regulating the translation of the *mok* gene *(86)*.

The Sok-RNA is expressed from a strong promoter, but with a half-life of less than 30 seconds it is very unstable *(87)*. Consequently, loss of the plasmid-encoded *sok* gene quickly leads to a decrease in the intracellular Sok-RNA concentration, which in turn leads to the expression of Hok protein from the much more stable *hok* mRNA (**Fig. 5.4**). A reservoir of Hok mRNAs may be present in the cells in a form that is neither translated nor bound to the antisense Sok-RNA, due to a fold-back inhibition element located at their 3′-terminal *(88)*. In a time where novel antibiotics are highly valuable due to an increase in the prevalence of antibiotic-resistant bacteria, it has been proposed that the natural host-killing proteins may be exploited for such use. One major problem is delivery of the protein into the cells. In one study

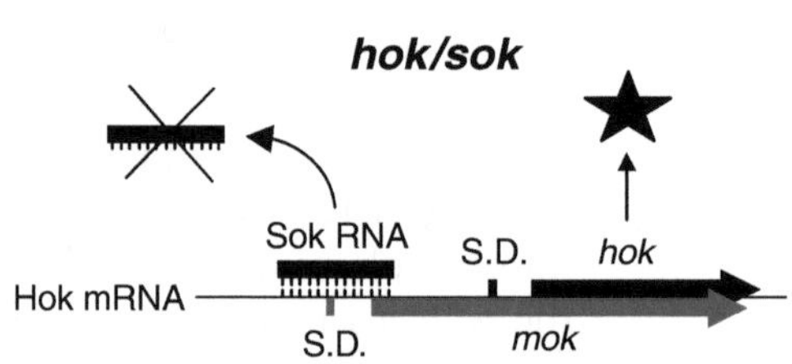


Fig. 5.4. Schematic representation of the *hok/sok* addiction system. Degradation of Sok RNA leads to translation of *mok* and in turn *hok*, which constitutes the host killing protein (*star*). S.D. represents Shine-Dalgarno boxes.

six different bacterial species were directly exposed to purified 52 amino acid Hok protein without any significant killing effect. Only if competent *E. coli* and *P. putida* cells were transformed with the Hok protein by either electroporation or heat-shock was an effect observed in *(89)*. Another approach, described in a recent study, has been to employ a competitive inhibitor of Hok mRNA::Sok-RNA interactions in the form of short anti-Sok-RNA peptide nucleic acid (PNA) oligomers. The oligomers consisted of up to 11 nucleotides homologous to the 5′-tail of Sok-RNA and an attached 10-mer peptide to enhance cell uptake. Mixing of the PNA molecules and *E. coli* cells harboring a mini-R1 *hok/sok* plasmid induced ghost cell morphology, characteristic of Hok toxin-mediated cell killing, and was shown to be more bactericidal than rifampicin *(90)*.

### 4.4. Restriction-Modification Systems

Restriction-modification (RM) systems are abundant in bacteria and almost 3800 have currently been characterized *(91)*. These systems are traditionally associated with a cellular protection mechanism against incoming foreign DNA of plasmid or bacteriophage origin. The Type II RM systems represent the largest and best-studied group and contain two components, namely a restriction enzyme, which cleaves double-stranded DNA at specific palindromic recognition sites, and a modification methylase, which methylates one or more bases within this same recognition site, thus preventing the action of the cognate restriction enzyme. If a RM system is carried by a plasmid, it may further mediate stable maintenance in much the same way as the PSK systems discussed above *(92)*. The mechanism of this is thought to be that once a plasmid-free cell arises, the concentration of both the restriction enzyme and the modification methylase decreases in the segregant cell and further descendants. Although, unlike the proteic PSK systems, both proteins appear to be equally stable, it is thought that at some critical point the concentration of the methylase becomes insufficient to methylate all recognition sites, thus allowing some DNA cleavage by the restriction enzyme *(93)*. RM have recently been regarded as very simple, selfish mobile DNA elements that function much like viruses and transposons *(94)*.

### 4.5. Horizontal Gene Transfer

Horizontal transfer of plasmids may be viewed as a specialized plasmid stability mechanism, which could be described as post-segregational re-infection by neighboring plasmid-harboring cells as opposed to the post-segregational killing mechanism described above. This view is in accordance with the notion that conjugative plasmids may behave as parasitic DNA elements or selfish DNA. Incorporation of visually detectable phenotypic markers for plasmid-carriage makes it possible to assess plasmid stability in a confined bacterial population, such as a colony growing on an agar plate, under various conditions. Sia et al. used a *lacZ* marker gene-cassette inserted into the *tetA* gene of plasmid RK2 to visualize colonies or segments of colonies with or without the plasmid by blue/white screening on plates containing X-gal (5-bromo-4-chloro-3-indolyl-$\beta$-D-galactopyranoside) *(95)*. The authors demonstrated that the loss of ability to facilitate conjugation in $Tra^-$ mutants resulted in plasmid loss within the colonies growing on solid medium, as evidenced by the formation of white sectors in otherwise blue colonies.

More recent studies have employed a flow cytometry optimized *gfp* gene as a marker for plasmid transfer and stability assessments in bacterial populations. In one such approach, a LacI-regulated *gfp* gene cassette was inserted into the chromosome of the host strain with the equivalent Lac repressor gene, $lacI^{q1}$, inserted into the plasmid of interest (**Fig. 5.2B**). Consequently, plasmid-free segregant cells arising in an otherwise plasmid-harboring bacterial population switch to a green fluorescent phenotype, detectable by both fluorescence microscopy and flow cytometry (**Fig. 5.5** and Color Plate 2, following p. 76) *(51)*. Using this approach it has been shown that conjugative transfer facilitates the stable maintenance of an IncP-1 plasmid, pKJK5, in *E. coli* cells colonizing the gastrointestinal tract of germfree rats *(48)*. Transfer-deficient pKJK5 derivatives (*traF*) were gradually lost from the *E. coli* population whereas the conjugation proficient derivative displayed a very high degree of stability and apparently was also able to re-infect plasmid-free segregant cells, which appeared shortly after inoculation of the bacteria into the animal.

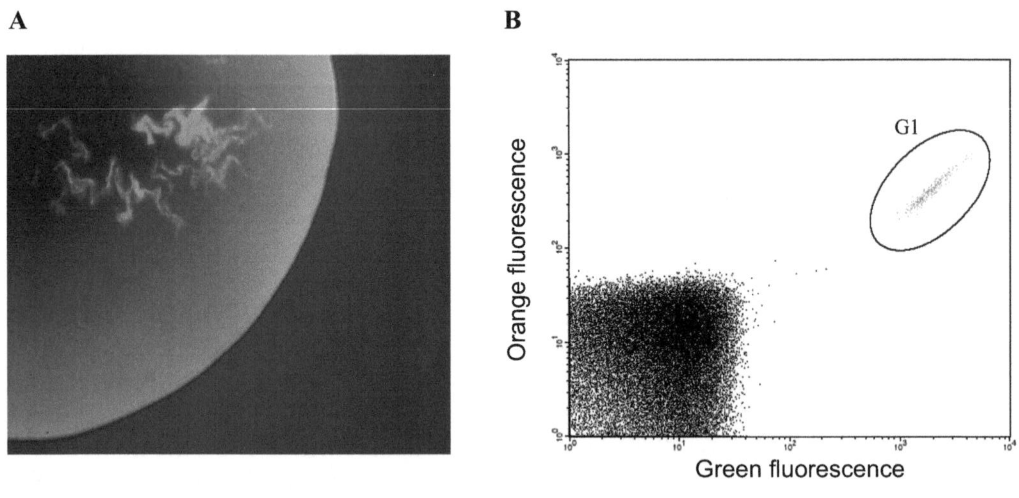


Fig. 5.5. (**A**) Segment of a bacterial colony of *Kluyvera* sp. MB101::*gfp* harboring plasmid pKJK5::$lacI^{q1}$. The *green* fluorescing patches represent plasmid-free segregant cells. (**B**) Corresponding flow cytometry data showing a small sub-population of GFP-positive, i.e., plasmid-free, cells (G1) (*see* Color Plate 2, following p. 76).

Another study using the same plasmid-sensing cell system has further demonstrated the infectious nature of plasmid pKJK5. Here mixtures of isogenic plasmid-free and plasmid-harboring cells in different ratios were applied to membrane filters placed on solid-rich agar plates. The end ratios, determined by flow cytometry after three days, revealed that plasmid pKJK5 had spread to almost 100% of both *E. coli* and *Kluyvera* sp. cells in populations initially containing as little as 10% plasmid-harboring cells *(96)*.

One important reason for the major impact of conjugation in these studies is the fact that the bacteria reside in very high cell density, nutrient-rich habitats, which are considered hot spots for conjugal transfer. It has recently been suggested that around 99% of all bacteria live in some form of close association such as a biofilm *(97)*, which suggests that conjugative transfer may well be an important mechanism to ensure the propagation and stability of plasmids within and between populations or communities of bacteria in natural settings. Furthermore, recent studies highlight the fact that some plasmids may actually facilitate enhanced aggregation and biofilm formation of the cells they reside in by expressing conjugative pili *(98)* or fimbriae *(99)*. This is a very important finding, as it indicates a two-way interaction between the mechanisms of biofilm formation and that of HGT.

## 5. Plasmid Curing

Ridding bacterial cells of residing plasmids has two main applications, namely to obtain a phenotypic characterization of the plasmid and, when resistance or virulence plasmids are concerned, revert the cell to a clinically susceptible or non-virulent phenotype. Because plasmids are defined as non-essential extra-chromosomal elements, it should theoretically be possible to cure all types of plasmids from their natural host cells. In reality, however, this has not always been possible. The various known intrinsic plasmid-stabilizing mechanisms previously described represent an obvious barrier to this and probably others remain to be discovered. A traditional approach for curing bacteria of their plasmids is the use of certain chemical agents, which cause a general stress response in the cell. Several compounds have been shown, mostly through trial and error, to be effective for this purpose. Ethidium bromide and various tricyclic compounds, including acridine orange *(100)* and phenothiazine *(101)*, are examples of such classical curing agents. These intercalating agents invade the double-stranded DNA and thereby induce conformation changes, which in turn lead to replication errors. Sodium dodecyl sulfate *(102)*, inhibitors of the cellular DNA gyrase, including quinolones and fluoroquinolones *(103)*, and growth at an elevated temperature have also been shown to induce plasmid curing. The risk of many of these approaches is a mutagenic effect on the host chromosome as well.

Transposon-mediated introduction of suicide genes into the plasmid has been suggested as an effective mechanism for plasmid curing and allows positive screening for plasmid-free cells. One example of this is the use of the *rpsL* gene, which encodes the ribosomal protein S12 and is responsible for streptomycin sensitivity. Plasmids tagged with a Tn*5*-derivative harboring this gene incur streptomycin sensitivity upon an otherwise resistant bacterial host cell. Plasmid-free cells may thus be selected for by their reversion to a resistant phenotype *(104)*. A variation of the above-mentioned approach is to exploit the fact that the transposition process itself often leads to linearization and subsequent degradation of the genetic element, in this case a plasmid, it transposes from *(105)*. The use of curing plasmids, i.e., competing plasmids from the same incompatibility group as the plasmid to be cured, has also been reported to be efficient in some circumstances *(102)*.

The last approach to be mentioned here involves the use of small molecules to interact with the regulation of replication/incompatibility systems in plasmids. Many plasmids employ small untranslated RNAs to inhibit translation of Rep protein mRNA, which in turn prevents plasmid replication. In IncB plasmids this is achieved by competitive high-affinity binding of RNAI to stem-loop I (SLI) of the Rep mRNA, thus hindering formation of a SLI/SLII pseudoknot complex, which is essential for ribosome binding *(106)*. Some aminoglycosides, including apramycine, have at sub-inhibitory concentrations been shown to mimic the binding of RNAI to SLI and consequently shut down Rep protein synthesis, resulting in dose-dependent plasmid elimination of the IncB plasmids from an *E. coli* population *(107)*.

## 6. Plasmid Adaptations

It is generally believed that plasmids impose a fitness cost on their host cell when no selectively advantageous genes are expressed *(108)*, although some studies have contradicted this *(109)*. The "excess baggage" theory predicts that under non-selective conditions, plasmid-free segregant cell will outcompete their otherwise isogenic plasmid-harboring cell, leading to loss of the plasmid. Various measures may be taken by the plasmid to reduce the intrinsic burden it imposes on its host. One important parameter is the copy number of the plasmid. A reduction in the number of plasmid copies maintained by the cell must decrease the energy consumption for DNA replication, etc. However, as discussed above, maintaining plasmids at a very low copy number necessitates very efficient mechanisms to ensure that the plasmid is not lost. The F plasmid is maintained at the lowest possible copy number of just 1–2 per cell and at the same time exhibits extremely stable maintenance. Stringent regulation of the replication machinery is achieved by a strong negative feedback loop, which ensures that replication is effectively arrested if the copy number is above the normal steady-state level *(110)*. The F plasmid encodes multiple stability functions, as described above, including active partitioning, multimer resolution, both proteic and anti-sense-regulated host-killing systems as well as the ability to facilitate conjugation and HGT. It would thus appear that the F plasmid is an extremely well-adapted plasmid.

Compensatory mutations, such as adjustment of the GC-content and codon-usage, in the plasmid to better match that of its host cell may ameliorate some if not all of the fitness burden associated with plasmid carriage. Also, genetic changes in the host chromosome may have a positive effect. Bouma and Lenski reported that 500 generations of bacteria/plasmid adaptation of a 4 kb non-conjugative plasmid and an *E. coli* host strain resulted in a significant reduction in the fitness cost of plasmid maintenance, which, after the adaptation period, surprisingly was shown to enhance the fitness of its host even in the absence of antibiotics. This adaptation was caused by alterations in the host chromosome and not in the evolved plasmid *(109)*. A similar study by Dahlberg and Chao on the fitness cost of the conjugative plasmids R1 and RP4, when harbored by an *E. coli* strain, showed that genetic changes in both the plasmids and the host chromosomes occur during 1100 generations of host/plasmid adaptation and that these changes ameliorate the cost of plasmid carriage *(108)*.

### 6.1. GC-Content

The GC-content of bacterial chromosomes is similar between closely related species, but varies from around 25 to 77% across the bacterial domain *(111)*. It has been reported that obligatory pathogens and symbionts, which rely on a host cell for survival, generally have a low GC-content *(112)*. The same was shown to hold true for plasmids, phages, and insertion sequences, which frequently display a lower GC-content than their hosts. Rocha and Danchin suggest the reasons for this are the higher energy cost and limited availability of G and C over other nucleotides *(112)*. Another reason for a discrepancy in GC-content between mobile elements and their present host could be the very fact that they are mobile and thus propagate through evolutionarily unstable relationships with many different hosts *(113)*. Identification of relative differences in GC-content between regions of DNA in the same cell is used as one approach to detect past horizontal transfer events (*see* **Chapter 11**).

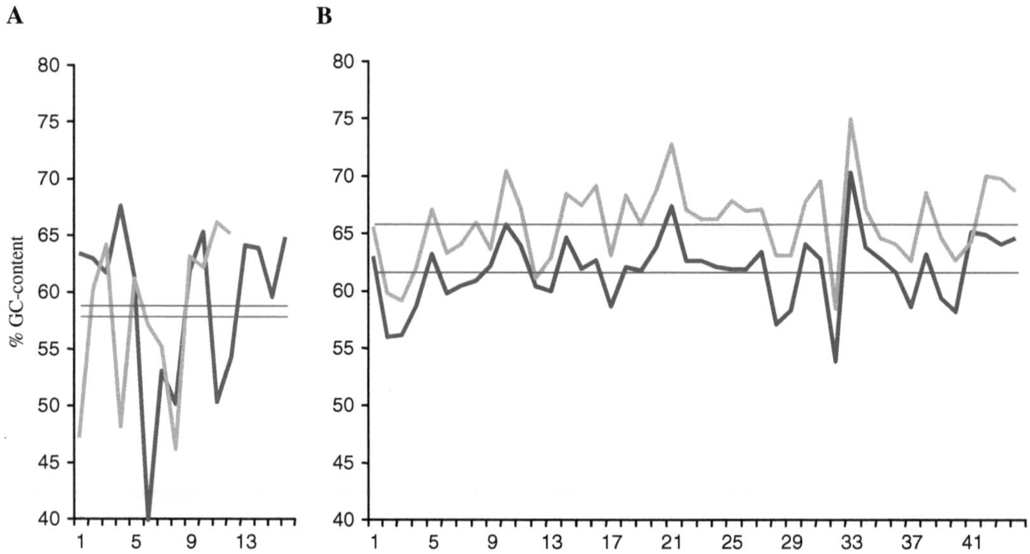


Fig. 5.6. The GC-content of genes associated with mobile accessory elements (**A**) and plasmid backbone genes (**B**) in IncP-1 plasmids R751 (*light gray*) and pKJK5 (*dark gray*). Average values are shown as *horizontal lines* in the graphs.

Plasmids belonging to the IncP-1 group have a modular backbone composition and very often contain additional mobile genetic elements, such as transposons, encoding accessory functions. The GC-content of these elements within the plasmid is typically significantly different than that of the backbone DNA, which indicates a recent acquisition of these elements *(11, 23)*. A specific example of this is the trimethoprim-resistance gene cassette harbored by a class 1 integron in IncP-1 plasmid pKJK5. The GC-content of the *dfrA1* gene is 39.7% compared to the average backbone gene content of 61.6%, which together with the placement of this gene cassette as the first in an array indicates that it was acquired recently from a low GC background (**Fig. 5.6**). Interestingly, the GC-content of all but one of 44 individual backbone genes in plasmid pKJK5, which vary between 53.7% and 70.1%, is lower than the homologous genes of the related IncP-1β plasmid R751 (**Fig. 5.6B**). This indicates that these two plasmids may have evolved for some time in different host species *(22)*.

### 6.2. Elimination of Restriction Sites

Restriction endonucleases (type II) are used extensively in molecular biology due to their ability to cleave double-stranded DNA at specific palindromic recognition sites. The biological rationale for bacteria to encode and express these enzymes is thought to be host-defense against invading DNA strands in the form of bacteriophages, plasmids, and conjugative transposons. Host-encoded specific methyltransferases ensure that cognate recognition sites are protected from the action of their own restriction enzymes. As described previously, restriction-modification systems may also function as a plasmid-stabilizing mechanism. Plasmids may be exposed to restriction enzymes following conjugative transfer from one bacterial species into another, which do not encode the same RM-systems. Depending on the presence or absence of the associated recognition sites, cleavage of the DNA may take place. One obvious way for plasmids to avoid this is to reduce the number of recognition sites in their genome. Many broad-host-range plasmids as well as phages and conjugative transposons have been shown to contain fewer than expected recognition sites, which indicates that adaptive evolution to eradicate such sites has taken place *(114–117)*.

Wilkins et al. determined the *Eco*R1 restriction index (RI), defined as the transconjugant yield to a non-restricting recipient divided by the yield to a restricting recipient, to be 31 for IncP-1α plasmid RP4 and 5400 for IncP-1β plasmid R751, which contain 1 and 5 *Eco*R1 recognition sites, respectively. Transposon-mediated incorporation of a second *Eco*R1 site into RP4 resulted in a RI-value of 209 *(115)*. Interestingly, the same study showed that the RI-values for plasmids ColIb (IncI1) and R100 (IncFII) were around 3 even though they both contain in excess of 16 *Eco*R1 sites. The reason for this difference in sensitivity is not clear; however, it emphasizes the selective pressure imposed on the IncP-1 plasmids to rid them of at least some restriction sites. Some plasmids have been found to harbor antirestriction enzymes, encoded by the *ard* genes (alleviation of restriction of DNA) and homologs, which may protect transferred DNA from different restriction enzymes particularly of type I *(118)*.

### 6.3. Gene Regulation

Tight regulation of plasmid-encoded genes reduces the intrinsic burden they confer upon their host cell and is thus an important mechanism in their pursuit in becoming energetically invisible, especially in environments not selective for plasmid-encoded traits. Some conjugative plasmids, including members of the IncP-1 incompatibility group, may express genes and functions either involved in self-replication or in conjugative transfer. This is analogous to the lysogenic phase and the lytic phase of bacteriophages. Coordinated regulation of IncP-1 genes is achieved by at least four global regulator proteins KorA, KorB, KorC, and TrbA, which have been reported to have 7 ($O_A$), 12 ($O_B$), 3 ($O_C$), and 5 ($O_T$) putative binding sites (operators), respectively, in IncP-1α plasmid RK2/RP4 *(11, 119)*. Binding of KorA to its cognate operators effectively represses the expression of genes involved in plasmid replication and stability function; and furthermore, the binding of KorA to $O_A3$ has been reported to mediate the switch in gene expression to the conjugation transfer state *(53)*. The KorC and TrbA proteins have been shown to repress transcription of the genes involved in plasmid stability function and conjugative transfer, respectively, whereas KorB represents a truly global regulator, which may repress transcription of the genes involved in nearly all the backbone functions *(53)*. Cooperative actions of the different regulators may also take place.

Functional genes contained in accessory elements, such as those encoding antibiotic resistance, harbored by plasmids may also confer a fitness burden on their host cell if constitutively expressed. This would further increase the burden of plasmid carriage in non-plasmid-selective environments. In some cases specific repressor proteins may regulate the expression. This is exemplified by the tetracycline efflux protein, TetA, which has been identified in many conjugative plasmids belonging to various incompatibility groups *(13, 23, 120, 121)*. Expression of the TetA protein is regulated by the repressor protein TetR, which inhibits transcription in the absence of bioavailable tetracycline *(122)*.

## 7. Compounds and Conditions That Affect Horizontal Gene Transfer

Some bacteria employ cell-to-cell communication to estimate the size of a population and act accordingly in a process termed quorum sensing *(123)*. A well-described example of this auto-induction mechanism is the population-size-dependent light emission (*lux* genes) of *Vibrio fischeri* colonizing the light-organ of the bobtailed squid, *Euprymna scolopes*. The biological rationale for this density-regulated expression of the *lux* genes is the fact that solitary planctonic *V. fischeri* cells would not benefit from such light-production in the darkness of the ocean, but merely pay the metabolic price.

Conjugative plasmid transfer may also be regulated by quorum sensing mechanisms as has been reported for the tumor-inducing Ti plasmids of *Agrobacterium tumefaciens (124)* or by pheromone modulated cell-to-cell communication as exemplified by the pheromone-inducible conjugative plasmids of enterococci *(125)*. In the latter example, two different cell types are involved, namely the plasmid-harboring donor cells and the plasmid-free recipient cells, in an intricate regulation of the conjugative transfer functions, which will be briefly outlined here using plasmid pCF10 as an example. The pheromones are short (7–8 amino acid) peptides, which are chromosomally encoded in both the donor and the recipients and have a high specificity for a cognate plasmid. The pheromone binds to the membrane-associated pCF10 encoded PrgZ protein and is then imported into the cytoplasm of the donor cell, where it interacts with the negative regulator PrgX, allowing expression of the conjugation functions *(126)*. This ensures that the energetically costly mechanism of conjugation is only turned on if there are potential recipients in the vicinity and may thereby stabilize the plasmid within a bacterial population. Auto-induction in the donor cell by the endogenous pheromones is impeded by several mechanisms.

Conjugative transfer may also be regulated by the presence of antimicrobial compounds at sub-inhibitory concentrations, as has been reported for the tetracycline-inducible conjugative transposons of *Bacteroides* *(127)*. Transfer of the broad-host-range Tn*916*/Tn*1545* family of conjugative transposons has also been reported to be stimulated by tetracycline *(128)*; however, some evidence suggests that other factors may be involved *(129, 130)*. Both classes of conjugative transposons appropriately harbor a corresponding tetracycline-resistance determinant. Other compounds or conditions that directly influence the frequency of transfer include temperature, pH, cations *(131)*, nutrient availability, and cell density *(132)*. The positive correlation between transfer rate and cell density acts in both directions: an increase in cell-density, such as biofilm formation, increases the rate of conjugation while expression of conjugative proteins such as pili may help to establish high-density bacterial biofilms *(98, 133)*. Recently, it has been proposed that in natural environments, antibiotics may function as inter-microbial signaling agents as opposed to antimicrobial agents *(134)*. The authors demonstrate that sub-inhibitory concentrations of tobramycin, tetracycline, and ciprofloxacin all induce increased biofilm formation of the opportunistic pathogen *Pseudomonas aeruginosa*. If this is a common feature of antibiotics, it could be speculated that this in turn may provide optimal conditions for HGT.

Most empirical and model-based studies suggest that the natural transfer rate of conjugative plasmids is not sufficiently high to ensure their persistence as genetic parasites in non-selective environments *(135)*. The recent shift in focus from viewing bacteria as isolated single-celled organisms to more socially interacting communities *(136–138)* has, however, emphasized the very important point that bacteria behave differently under laboratory conditions than in their natural habitat and consequently that predictions of bacterial behavior, such as HGT rates, should be based on in situ derived knowledge. In accordance with this, quantitative in situ determination of the conjugation rates within a defined bacterial biofilm has been reported to be in the order of 1000-fold higher than those determined by classical plating techniques *(139)*. The fact that most bacteria in complex communities elude cultivation may lead to a further underestimation of the extent of conjugation *(43)*.

## 8. Conclusion/ Outlook

The intricacy and diversity of currently known mechanisms that influence plasmid replication, stability, and transfer are truly amazing. This chapter has especially focused on the systems that enhance the intrinsic stability of conjugative plasmids within a bacterial cell or population, including multimer resolution, active partitioning, post-segregational-killing, and conjugative transfer. Although the origins of many of these functions remain elusive, it is clear that long-term plasmid–host associations have required evolution of the exceedingly well-regulated and adapted systems that are present today. Hence, the answer to the question "who has the selective advantage – Bacteria or plasmid?" is both! Plasmids should be considered as genetic symbionts rather than parasites. Historically, the association between plasmids and antibiotic-resistance genes has been a cornerstone in plasmid research (*see* **Chapter 23**) and may well have led to a bias toward discovering specific types or classes of bacterial plasmids. We predict that future studies will show that plasmids are ubiquitous as symbionts in naturally occurring bacteria. Recent meta-genomic studies of bacterial diversity have highlighted the fact that only a small fraction of all naturally occurring bacteria are presently accounted for *(140)*. This implies that many more plasmid-groups remain to be discovered. This in turn may lead to the discovery of novel mechanisms that affect the persistence of plasmids.

## Acknowledgments

The work was funded by grants from the Villum Kann Rasmussen foundation, the Danish Natural Science Council, the Natural and Accelerated Bioremediation Research (NABIR) program, Biological and Environmental Research (BER), US Department of Energy, and the Faculty of Science at the University of Copenhagen, Denmark.

### References

1. Liebert, C. A., Hall, R. M., and Summers, A. O. (1999) Transposon Tn*21*, flagship of the floating genome. *Microbiol Mol Biol Rev.* **63**, 507–522.
2. Hall, R. M., and Collis, C. M. (1995) Mobile gene cassettes and integrons: capture and spread of genes by site-specific recombination. *Mol Microbiol.* **15**, 593–600.
3. Bennett, P. M. (1999) Integrons and gene cassettes: a genetic construction kit for bacteria. *J Antimicrob Chemother.* **43**, 1–4.
4. Szczepanowski, R., Braun, S., Riedel, V., Schneiker, S., Krahn, I., Pühler, A., and Schlüter, A. (2005) The 120,592 bp IncF plasmid pRSB107 isolated from a sewage-treatment plant encodes nine different antibiotic-resistance determinants, two iron-acquisition systems and other putative virulence-associated functions. *Microbiology.* **151**, 1095–1111.
5. Villa, L., Pezzella, C., Tosini, F., Visca, P., Petrucca, A., and Carattoli, A. (2000) Multiple-antibiotic resistance mediated by structurally related IncL/M plasmids carrying an extended-spectrum beta-lactamase gene and a class 1 integron. *Antimicrob Agents Chemother.* **44**, 2911–2914.
6. Stokes, H. W., and Hall, R. M. (1992) The integron In1 in plasmid R46 includes two copies of the *oxa2* gene cassette. *Plasmid.* **28**, 225–234.

7. Tennstedt, T., Szczepanowski, R., Krahn, I., Pühler, A., and Schlüter, A. (2005) Sequence of the 68,869 bp IncP-1alpha plasmid pTB11 from a waste-water treatment plant reveals a highly conserved backbone, a Tn*402*-like integron and other transposable elements. *Plasmid.* **53**, 218–238.
8. Heuer, H., Szczepanowski, R., Schneiker, S., Pühler, A., Top, E. M., and Schlüter, A. (2004) The complete sequences of plasmids pB2 and pB3 provide evidence for a recent ancestor of the IncP-1beta group without any accessory genes. *Microbiology.* **150**, 3591–3599.
9. Schlüter, A., Heuer, H., Szczepanowski, R., Forney, L. J., Thomas, C. M., Pühler, A., and Top, E. M. (2003) The 64 508 bp IncP-1beta antibiotic multiresistance plasmid pB10 isolated from a waste-water treatment plant provides evidence for recombination between members of different branches of the IncP-1beta group. *Microbiology.* **149**, 3139–3153.
10. Schlüter, A., Heuer, H., Szczepanowski, R., Poler, S. M., Schneiker, S., Pühler, A., and Top, E. M. (2005) Plasmid pB8 is closely related to the prototype IncP-1beta plasmid R751 but transfers poorly to *Escherichia coli* and carries a new transposon encoding a small multidrug resistance efflux protein. *Plasmid.* **54**, 135–148.
11. Thorsted, P. B., Macartney, D. P., Akhtar, P., Haines, A. S., Ali, N., Davidson, P., Stafford, T., Pocklington, M. J., Pansegrau, W., Wilkins, B. M., Lanka, E., and Thomas, C. M. (1998) Complete sequence of the IncPbeta plasmid R751: implications for evolution and organisation of the IncP backbone. *J Mol Biol.* **282**, 969–990.
12. Haines, A. S., Jones, K., Cheung, M., and Thomas, C. M. (2005) The IncP-6 plasmid Rms149 consists of a small mobilizable backbone with multiple large insertions. *J Bacteriol.* **187**, 4728–4738.
13. Rhodes, G., Parkhill, J., Bird, C., Ambrose, K., Jones, M. C., Huys, G., Swings, J., and Pickup, R. W. (2004) Complete nucleotide sequence of the conjugative tetracycline resistance plasmid pFBAOT6, a member of a group of IncU plasmids with global ubiquity. *Appl Environ Microbiol.* **70**, 7497–7510.
14. Fernandez-Lopez, R., Garcillan-Barcia, M. P., Revilla, C., Lazaro, M., Vielva, L., and de la Cruz, F. (2006) Dynamics of the IncW genetic backbone imply general trends in conjugative plasmid evolution. *FEMS Microbiol Rev.* **30**, 942–966.
15. Levings, R. S., Hall, R. M., Lightfoot, D., and Djordjevic, S. P. (2006) *linG*, a new integron-associated gene cassette encoding a lincosamide nucleotidyltransferase. *Antimicrob Agents Chemother.* **50**, 3514–3515.
16. Fluit, A. C., and Schmitz, F. J. (1999) Class 1 integrons, gene cassettes, mobility, and epidemiology. *Eur J Clin Microbiol Infect Dis.* **18**, 761–770.
17. Fluit, A. C., and Schmitz, F. J. (2004) Resistance integrons and super-integrons. *Clin Microbiol Infect.* **10**, 272–288.
18. Antunes, P., Machado, J., Sousa, J. C., and Peixe, L. (2005) Dissemination of sulfonamide resistance genes (sul1, sul2, and sul3) in Portuguese Salmonella enterica strains and relation with integrons. *Antimicrob Agents Chemother.* **49**, 836–839.
19. Partridge, S. R., and Hall, R. M. (2005) Correctly identifying the streptothricin resistance gene cassette. *J Clin Microbiol.* **43**, 4298–4300.
20. Stokes, H. W., Nesbo, C. L., Holley, M., Bahl, M. I., Gillings, M. R., and Boucher, Y. (2006) Class 1 integrons potentially predating the association with tn*402*-like transposition genes are present in a sediment microbial community. *J Bacteriol.* **188**, 5722–5730.
21. Stokes, H. W., Holmes, A. J., Nield, B. S., Holley, M. P., Nevalainen, K. M., Mabbutt, B. C., and Gillings, M. R. (2001) Gene cassette PCR: sequence-independent recovery of entire genes from environmental DNA. *Appl Environ Microbiol.* **67**, 5240–5246.
22. Bahl, M. I., Hansen, L. H., Goesmann, A., and Sørensen, S. J. (2007) The multiple antibiotic resistance IncP-1 plasmid pKJK5 isolated from a soil environment is phylogenetically divergent from members of the previously established α, β and δ sub-groups. *Plasmid.* **58**(1), 31–43.
23. Pansegrau, W., Lanka, E., Barth, P. T., Figurski, D. H., Guiney, D. G., Haas, D., Helinski, D. R., Schwab, H., Stanisich, V. A., and Thomas, C. M. (1994) Complete nucleotide sequence of Birmingham IncP alpha plasmids. Compilation and comparative analysis. *J Mol Biol.* **239**, 623–663.
24. Tomita, H., Tanimoto, K., Hayakawa, S., Morinaga, K., Ezaki, K., Oshima, H., and Ike, Y. (2003) Highly conjugative pMG1-like plasmids carrying Tn*1546*-like transposons that encode vancomycin resistance in *Enterococcus faecium. J Bacteriol.* **185**, 7024–7028.
25. Sørensen, A. H., Hansen, L. H., Johannesen, E., and Sørensen, S. J. (2003) Conjugative plasmid conferring resistance to olaquindox. *Antimicrob Agents Chemother.* **47**, 798–799.
26. Silver, S. (1996) Bacterial resistances to toxic metal ions – a review. *Gene.* **179**, 9–19.

27. Nojiri, H., Shintani, M., and Omori, T. (2004) Divergence of mobile genetic elements involved in the distribution of xenobiotic-catabolic capacity. *Appl Microbiol Biotechnol.* **64**, 154–174.
28. Top, E. M., Holben, W. E., and Forney, L. J. (1995) Characterization of diverse 2,4-dichlorophenoxyacetic acid-degradative plasmids isolated from soil by complementation. *Appl Environ Microbiol.* **61**, 1691–1698.
29. Koehler, T. M. (2002) *Bacillus anthracis* genetics and virulence gene regulation. *Curr Top Microbiol Immunol.* **271**, 143–164.
30. Rychlik, I., Gregorova, D., and Hradecka, H. (2006) Distribution and function of plasmids in *Salmonella enterica*. *Vet Microbiol.* **112**, 1–10.
31. Turner, S. M., Scott-Tucker, A., Cooper, L. M., and Henderson, I. R. (2006) Weapons of mass destruction: virulence factors of the global killer enterotoxigenic *Escherichia coli*. *FEMS Microbiol Lett.* **263**, 10–20.
32. Froehlich, B., Parkhill, J., Sanders, M., Quail, M. A., and Scott, J. R. (2005) The pCoo plasmid of enterotoxigenic *Escherichia coli* is a mosaic cointegrate. *J Bacteriol.* **187**, 6509–6516.
33. Fekete, P. Z., Schneider, G., Olasz, F., Blum-Oehler, G., Hacker, J. H., and Nagy, B. (2003) Detection of a plasmid-encoded pathogenicity island in F18+enterotoxigenic and verotoxigenic *Escherichia coli* from weaned pigs. *Int J Med Microbiol.* **293**, 287–298.
34. Cornelis, G. R. (2000) Molecular and cell biology aspects of plague. *Proc Natl Acad Sci U S A.* **97**, 8778–8783.
35. Cornelis, G. R., and Van Gijsegem, F. (2000) Assembly and function of type III secretory systems. *Annu Rev Microbiol.* **54**, 735–774.
36. Charpentier, E., Gerbaud, G., and Courvalin, P. (1999) Conjugative mobilization of the rolling-circle plasmid pIP823 from *Listeria monocytogenes* BM4293 among gram-positive and gram-negative bacteria. *J Bacteriol.* **181**, 3368–3374.
37. Drønen, A., Torsvik, V., Goksøyr, J., and Top, E. (1998) Effect of mercury addition on plasmid incidence and gene mobilizing capacity in bulk soil. *FEMS Microbiol Ecol.* **27**, 381–394.
38. Droge, M., Pühler, A., and Selbitschka, W. (2000) Phenotypic and molecular characterization of conjugative antibiotic resistance plasmids isolated from bacterial communities of activated sludge. *Mol Gen Genet.* **263**, 471–482.
39. De Gelder, L., Vandecasteele, F. P., Brown, C. J., Forney, L. J., and Top, E. M. (2005) Plasmid donor affects host range of promiscuous IncP-1beta plasmid pB10 in an activated-sludge microbial community. *Appl Environ Microbiol.* **71**, 5309–5317.
40. Horodniceanu, T., Bouanchaud, D. H., Bieth, G., and Chabbert, Y. A. (1976) R plasmids in *Streptococcus agalactiae* (group B). *Antimicrob Agents Chemother.* **10**, 795–801.
41. LeBlanc, D. J., Hawley, R. J., Lee, L. N., and St Martin, E. J. (1978) "Conjugal" transfer of plasmid DNA among oral streptococci. *Proc Natl Acad Sci U S A.* **75**, 3484–3487.
42. Kurenbach, B., Bohn, C., Prabhu, J., Abudukerim, M., Szewzyk, U., and Grohmann, E. (2003) Intergeneric transfer of the *Enterococcus faecalis* plasmid pIP501 to *Escherichia coli* and *Streptomyces lividans* and sequence analysis of its *tra* region. *Plasmid.* **50**, 86–93.
43. Musovic, S., Oregaard, G., Kroer, N., and Sorensen, S. J. (2006) Cultivation-independent examination of horizontal transfer and host range of an IncP-1 plasmid among gram-positive and gram-negative bacteria indigenous to the barley rhizosphere. *Appl Environ Microbiol.* **72**, 6687–6692.
44. Sørensen, S. J., Bailey, M., Hansen, L. H., Kroer, N., and Wuertz, S. (2005) Studying plasmid horizontal transfer in situ: a critical review. *Nat Rev Microbiol.* **3**, 700–710.
45. Sørensen, S. J., Sørensen, A. H., Hansen, L. H., Oregaard, G., and Veal, D. (2003) Direct detection and quantification of horizontal gene transfer by using flow cytometry and *gfp* as a reporter gene. *Curr Microbiol.* **47**, 129–133.
46. Datta, N., and Hedges, R. W. (1972) Host ranges of R factors. *J Gen Microbiol.* **70**, 453–460.
47. Geisenberger, O., Ammendola, A., Christensen, B. B., Molin, S., Schleifer, K. H., and Eberl, L. (1999) Monitoring the conjugal transfer of plasmid RP4 in activated sludge and in situ identification of the transconjugants. *FEMS Microbiol Lett.* **174**, 9–17.
48. Bahl, M. I., Hansen, L. H., Licht, T. R., and Sørensen, S. J. (2007) Conjugative transfer facilitates stable maintenance of IncP-1 plasmid pKJK5 in *Escherichia coli* cells colonizing the gastrointestinal tract of the germfree rat. *Appl Environ Microbiol.* **73**, 341–343.
49. Enne, V. I., Delsol, A. A., Roe, J. M., and Bennett, P. M. (2006) Evidence of antibiotic resistance gene silencing in *Escherichia coli*. *Antimicrob Agents Chemother.* **50**, 3003–3010.
50. Bengtsson, G., Fossum, A., and Lindqvist, R. (2004) Persistence of plasmid RP4 in *Pseudomonas putida* and loss of its expression of

antibiotic resistance in a groundwater microcosm. *Soil Biol Biochem.* **36**, 999–1008.
51. Bahl, M. I., Sørensen, S. J., and Hansen, L. H. (2004) Quantification of plasmid loss in *Escherichia coli* cells by use of flow cytometry. *FEMS Microbiol Lett.* **232**, 45–49.
52. Turner, S. L., Bailey, M., Lilley, A. K., and Thomas, C. M. (2002) Ecological and molecular maintenance strategies of mobile genetic elements. *FEMS Microbiol Ecol.* **42**, 177–185.
53. Adamczyk, M., and Jagura-Burdzy, G. (2003) Spread and survival of promiscuous IncP-1 plasmids. *Acta Biochim Pol.* **50**, 425–453.
54. Wada, C., and Yura, T. (1984) Control of F plasmid replication by a host gene: evidence for interaction of the *mafA* gene product of *Escherichia coli* with the mini-F incC region. *J Bacteriol.* **160**, 1130–1136.
55. Summers, D. K., Beton, C. W., and Withers, H. L. (1993) Multicopy plasmid instability: the dimer catastrophe hypothesis. *Mol Microbiol.* **8**, 1031–1038.
56. Tolmasky, M. E., Colloms, S., Blakely, G., and Sherratt, D. J. (2000) Stability by multimer resolution of pJHCMW1 is due to the Tn*1331* resolvase and not to the *Escherichia coli* Xer system. *Microbiology.* **146**, 581–589.
57. Bignell, C., and Thomas, C. M. (2001) The bacterial ParA-ParB partitioning proteins. *J Biotechnol.* **91**, 1–34.
58. Williams, D. R., Macartney, D. P., and Thomas, C. M. (1998) The partitioning activity of the RK2 central control region requires only *incC, korB* and KorB-binding site O(B)3 but other KorB-binding sites form destabilizing complexes in the absence of O(B)3. *Microbiology.* **144**, 3369–3378.
59. Siddique, A., and Figurski, D. H. (2002) The active partition gene *incC* of IncP plasmids is required for stable maintenance in a broad range of hosts. *J Bacteriol.* **184**, 1788–1793.
60. Yamaichi, Y., and Niki, H. (2000) Active segregation by the *Bacillus subtilis* partitioning system in *Escherichia coli. Proc Natl Acad Sci U S A.* **97**, 14656–14661.
61. Lewis, R. A., Bignell, C. R., Zeng, W., Jones, A. C., and Thomas, C. M. (2002) Chromosome loss from par mutants of *Pseudomonas putida* depends on growth medium and phase of growth. *Microbiology.* **148**, 537–548.
62. Godfrin-Estevenon, A. M., Pasta, F., and Lane, D. (2002) The *parAB* gene products of *Pseudomonas putida* exhibit partition activity in both *P. putida* and *Escherichia coli. Mol Microbiol.* **43**, 39–49.
63. Yamaichi, Y., Fogel, M. A., and Waldor, M. K. (2006) *par* genes and the pathology of chromosome loss in *Vibrio cholerae. Proc Natl Acad Sci U S A.*
64. Fogel, M. A., and Waldor, M. K. (2005) Distinct segregation dynamics of the two *Vibrio cholerae* chromosomes. *Mol Microbiol.* **55**, 125–136.
65. Gerdes, K., Moller-Jensen, J., and Bugge Jensen, R. (2000) Plasmid and chromosome partitioning: surprises from phylogeny. *Mol Microbiol.* **37**, 455–466.
66. Pandey, D. P., and Gerdes, K. (2005) Toxin-antitoxin loci are highly abundant in free-living but lost from host-associated prokaryotes. *Nucleic Acids Res.* **33**, 966–976.
67. Heidelberg, J. F., Eisen, J. A., Nelson, W. C., Clayton, R. A., Gwinn, M. L., Dodson, R. J., Haft, D. H., Hickey, E. K., Peterson, J. D., Umayam, L., Gill, S. R., Nelson, K. E., Read, T. D., Tettelin, H., Richardson, D., Ermolaeva, M. D., Vamathevan, J., Bass, S., Qin, H., Dragoi, I., Sellers, P., McDonald, L., Utterback, T., Fleishmann, R. D., Nierman, W. C., White, O., Salzberg, S. L., Smith, H. O., Colwell, R. R., Mekalanos, J. J., Venter, J. C., and Fraser, C. M. (2000) DNA sequence of both chromosomes of the cholera pathogen *Vibrio cholerae. Nature.* **406**, 477–483.
68. Cooper, T. F., and Heinemann, J. A. (2000) Postsegregational killing does not increase plasmid stability but acts to mediate the exclusion of competing plasmids. *Proc Natl Acad Sci U S A.* **97**, 12643–12648.
69. Jaffe, A., Ogura, T., and Hiraga, S. (1985) Effects of the *ccd* function of the F plasmid on bacterial growth. *J Bacteriol.* **163**, 841–849.
70. Ogura, T., and Hiraga, S. (1983) Mini-F plasmid genes that couple host cell division to plasmid proliferation. *Proc Natl Acad Sci U S A.* **80**, 4784–4788.
71. Bernard, P., and Couturier, M. (1992) Cell killing by the F plasmid CcdB protein involves poisoning of DNA-topoisomerase II complexes. *J Mol Biol.* **226**, 735–745.
72. Ellington, M. J., and Woodford, N. (2006) Fluoroquinolone resistance and plasmid addiction systems: self-imposed selection pressure? *J Antimicrob Chemother.* **57**, 1026–1029.
73. Tam, J. E., and Kline, B. C. (1989) The F plasmid *ccd* autorepressor is a complex of CcdA and CcdB proteins. *Mol Gen Genet.* **219**, 26–32.
74. Tam, J. E., and Kline, B. C. (1989) Control of the *ccd* operon in plasmid F. *J Bacteriol.* **171**, 2353–2360.

75. Afif, H., Allali, N., Couturier, M., and Van Melderen, L. (2001) The ratio between CcdA and CcdB modulates the transcriptional repression of the *ccd* poison-antidote system. *Mol Microbiol.* **41**, 73–82.
76. Van Melderen, L., Bernard, P., and Couturier, M. (1994) Lon-dependent proteolysis of CcdA is the key control for activation of CcdB in plasmid-free segregant bacteria. *Mol Microbiol.* **11**, 1151–1157.
77. Bernard, P., Gabant, P., Bahassi, E. M., and Couturier, M. (1994) Positive-selection vectors using the F plasmid *ccdB* killer gene. *Gene.* **148**, 71–74.
78. Jiang, Y., Pogliano, J., Helinski, D. R., and Konieczny, I. (2002) ParE toxin encoded by the broad-host-range plasmid RK2 is an inhibitor of *Escherichia coli* gyrase. *Mol Microbiol.* **44**, 971–979.
79. Bravo, A., de Torrontegui, G., and Diaz, R. (1987) Identification of components of a new stability system of plasmid R1, ParD, that is close to the origin of replication of this plasmid. *Mol Gen Genet.* **210**, 101–110.
80. Tsuchimoto, S., Ohtsubo, H., and Ohtsubo, E. (1988) Two genes, *pemK* and *pemI*, responsible for stable maintenance of resistance plasmid R100. *J Bacteriol.* **170**, 1461–1466.
81. Ruiz-Echevarria, M. J., Gimenez-Gallego, G., Sabariegos-Jareno, R., and Diaz-Orejas, R. (1995) Kid, a small protein of the parD stability system of plasmid R1, is an inhibitor of DNA replication acting at the initiation of DNA synthesis. *J Mol Biol.* **247**, 568–577.
82. Lehnherr, H., Maguin, E., Jafri, S., and Yarmolinsky, M. B. (1993) Plasmid addiction genes of bacteriophage P1: *doc*, which causes cell death on curing of prophage, and *phd*, which prevents host death when prophage is retained. *J Mol Biol.* **233**, 414–428.
83. Santos-Sierra, S., Giraldo, R., and Diaz-Orejas, R. (1997) Functional interactions between homologous conditional killer systems of plasmid and chromosomal origin. *FEMS Microbiol Lett.* **152**, 51–56.
84. Gerdes, K., Rasmussen, P. B., and Molin, S. (1986) Unique type of plasmid maintenance function: postsegregational killing of plasmid-free cells. *Proc Natl Acad Sci U S A.* **83**, 3116–3120.
85. Gerdes, K., Bech, F. W., Jorgensen, S. T., Lobner-Olesen, A., Rasmussen, P. B., Atlung, T., Boe, L., Karlstrom, O., Molin, S., and von Meyenburg, K. (1986) Mechanism of postsegregational killing by the *hok* gene product of the *parB* system of plasmid R1 and its homology with the *relF* gene product of the *E. coli relB* operon. *Embo J.* **5**, 2023–2029.
86. Thisted, T., and Gerdes, K. (1992) Mechanism of post-segregational killing by the *hok/sok* system of plasmid R1. Sok antisense RNA regulates *hok* gene expression indirectly through the overlapping *mok* gene. *J Mol Biol.* **223**, 41–54.
87. Gerdes, K., Thisted, T., and Martinussen, J. (1990) Mechanism of post-segregational killing by the *hok/sok* system of plasmid R1: *sok* antisense RNA regulates formation of a *hok* mRNA species correlated with killing of plasmid-free cells. *Mol Microbiol.* **4**, 1807–1818.
88. Thisted, T., Sorensen, N. S., and Gerdes, K. (1995) Mechanism of post-segregational killing: secondary structure analysis of the entire Hok mRNA from plasmid R1 suggests a fold-back structure that prevents translation and antisense RNA binding. *J Mol Biol.* **247**, 859–873.
89. Pecota, D. C., Osapay, G., Selsted, M. E., and Wood, T. K. (2003) Antimicrobial properties of the *Escherichia coli* R1 plasmid host killing peptide. *J Biotechnol.* **100**, 1–12.
90. Faridani, O. R., Nikravesh, A., Pandey, D. P., Gerdes, K., and Good, L. (2006) Competitive inhibition of natural antisense Sok-RNA interactions activates Hok-mediated cell killing in *Escherichia coli. Nucleic Acids Res.* **34**, 5915–5922.
91. Roberts, R. J., Vincze, T., Posfai, J., and Macelis, D. (2007) REBASE – enzymes and genes for DNA restriction and modification. *Nucleic Acids Res.* **35**, D269–270.
92. Kulakauskas, S., Lubys, A., and Ehrlich, S. D. (1995) DNA restriction-modification systems mediate plasmid maintenance. *J Bacteriol.* **177**, 3451–3454.
93. Ichige, A., and Kobayashi, I. (2005) Stability of *Eco*RI restriction-modification enzymes *in vivo* differentiates the *Eco*RI restriction-modification system from other postsegregational cell killing systems. *J Bacteriol.* **187**, 6612–6621.
94. Kobayashi, I. (2001) Behavior of restriction-modification systems as selfish mobile elements and their impact on genome evolution. *Nucleic Acids Res.* **29**, 3742–3756.
95. Sia, E. A., Roberts, R. C., Easter, C., Helinski, D. R., and Figurski, D. H. (1995) Different relative importances of the *par* operons and the effect of conjugal transfer on the maintenance of intact promiscuous plasmid RK2. *J Bacteriol.* **177**, 2789–2797.
96. Bahl, M. I., Hansen, L. H., and Sørensen, S. J. (2007) Impact of conjugal transfer on the stability of IncP-1 plasmid pKJK5 in bacterial

populations. *FEMS Microbiol Lett.* **266**, 250–256.
97. Potera, C. (1996) Biofilms invade microbiology. *Science.* **273**, 1795–1797.
98. Ghigo, J. M. (2001) Natural conjugative plasmids induce bacterial biofilm development. *Nature.* **412**, 442–445.
99. Burmølle, M., Bahl, M. I., Jensen, L. B., Sørensen, S. J., and Hansen, L. H. (2008) Type 3 fimbriae, encoded by the conjugative plasmid pOLA52, enhance biofilm formation and transfer frequencies in Enterobacteriaceae strains. Microbiology **154**, 187–195.
100. Hirota, Y. (1960) The effect of acridine dyes on mating type factors in *Escherichia coli. Proc Natl Acad Sci U S A.* **46**, 57–64.
101. Molnar, A., Amaral, L., and Molnar, J. (2003) Antiplasmid effect of promethazine in mixed bacterial cultures. *Int J Antimicrob Agents.* **22**, 217–222.
102. El-Mansi, M., Anderson, K. J., Inche, C. A., Knowles, L. K., and Platt, D. J. (2000) Isolation and curing of the *Klebsiella pneumoniae* large indigenous plasmid using sodium dodecyl sulphate. *Research in Microbiology.* **151**, 201–208.
103. Brandi, L., Falconi, M., and Ripa, S. (2000) Plasmid curing effect of trovafloxacin. *FEMS Microbiol Lett.* **184**, 297–302.
104. Stojiljkovic, I., Trgovcevic, Z., and Salajsmic, E. (1991) Tn5-Rpsl – a new derivative of transposon Tn*5* useful in plasmid curing. *Gene.* **99**, 101–104.
105. Imre, A., Olasz, F., Kiss, J., and Nagy, B. (2006) A novel transposon-based method for elimination of large bacterial plasmids. *Plasmid.* **55**, 235–241.
106. Siemering, K. R., Praszkier, J., and Pittard, A. J. (1993) Interaction between the antisense and target RNAs involved in the regulation of Incb plasmid replication. *J Bacteriol.* **175**, 2895–2906.
107. DeNap, J. C. B., Thomas, J. R., Musk, D. J., and Hergenrother, P. J. (2004) Combating drug-resistant bacteria: Small molecule mimics of plasmid incompatibility as antiplasmid compounds. *J Am Chem Soc.* **126**, 15402–15404.
108. Dahlberg, C., and Chao, L. (2003) Amelioration of the cost of conjugative plasmid carriage in *Escherichia coli* K12. *Genetics.* **165**, 1641–1649.
109. Bouma, J. E., and Lenski, R. E. (1988) Evolution of a bacteria/plasmid association. *Nature.* **335**, 351–352.
110. Morrison, P. F., and Chattoraj, D. K. (2004) Replication of a unit-copy plasmid F in the bacterial cell cycle: a replication rate function analysis. *Plasmid.* **52**, 13–30.
111. Galtier, N., and Lobry, J. R. (1997) Relationships between genomic G + C content, RNA secondary structures, and optimal growth temperature in prokaryotes. *J Mol Evol.* **44**, 632–636.
112. Rocha, E. P., and Danchin, A. (2002) Base composition bias might result from competition for metabolic resources. *Trends Genet.* **18**, 291–294.
113. van Passel, M. W., Bart, A., Luyf, A. C., van Kampen, A. H., and van der Ende, A. (2006) Compositional discordance between prokaryotic plasmids and host chromosomes. *BMC Genomics.* 7, 26.
114. Rawlings, D. E., and Tietze, E. (2001) Comparative biology of IncQ and IncQ-like plasmids. *Microbiol Mol Biol Rev.* **65**, 481–496.
115. Wilkins, B. M., Chilley, P. M., Thomas, A. T., and Pocklington, M. J. (1996) Distribution of restriction enzyme recognition sequences on broad host range plasmid RP4: molecular and evolutionary implications. *J Mol Biol.* **258**, 447–456.
116. Korona, R., Korona, B., and Levin, B. R. (1993) Sensitivity of naturally occurring coliphages to type I and type II restriction and modification. *J Gen Microbiol.* **139 Pt 6**, 1283–1290.
117. Clewell, D. B., and Flannagan, S. E. (1993) in Bacterial Conjugation (Clewell, D. B., Ed.), Plenum Press, New York.
118. Wilkins, B. M. (2002) Plasmid promiscuity: meeting the challenge of DNA immigration control. *Environ Microbiol.* **4**, 495–500.
119. Vedler, E., Vahter, M., and Heinaru, A. (2004) The completely sequenced plasmid pEST4011 contains a novel IncP1 backbone and a catabolic transposon harboring tfd genes for 2,4-dichlorophenoxyacetic acid degradation. *J Bacteriol.* **186**, 7161–7174.
120. Johnson, T. J., Siek, K. E., Johnson, S. J., and Nolan, L. K. (2005) DNA sequence and comparative genomics of pAPEC-O2-R, an avian pathogenic *Escherichia coli* transmissible R plasmid. *Antimicrob Agents Chemother.* **49**, 4681–4688.
121. Gilmour, M. W., Thomson, N. R., Sanders, M., Parkhill, J., and Taylor, D. E. (2004) The complete nucleotide sequence of the resistance plasmid R478: defining the backbone components of incompatibility group H conjugative plasmids through comparative genomics. *Plasmid.* **52**, 182–202.
122. Roberts, M. C. (1996) Tetracycline resistance determinants: mechanisms of action, regulation of expression, genetic mobility, and distribution. *FEMS Microbiol Rev.* **19**, 1–24.

123. Fuqua, W. C., Winans, S. C., and Greenberg, E. P. (1994) Quorum sensing in bacteria: the LuxR-LuxI family of cell density-responsive transcriptional regulators. *J Bacteriol.* **176**, 269–275.
124. Hwang, I., Li, P. L., Zhang, L., Piper, K. R., Cook, D. M., Tate, M. E., and Farrand, S. K. (1994) TraI, a LuxI homologue, is responsible for production of conjugation factor, the Ti plasmid N-acylhomoserine lactone autoinducer. *Proc Natl Acad Sci U S A.* **91**, 4639–4643.
125. Chandler, J. R., and Dunny, G. M. (2004) Enterococcal peptide sex pheromones: synthesis and control of biological activity. *Peptides.* **25**, 1377–1388.
126. Kozlowicz, B. K., Dworkin, M., and Dunny, G. M. (2006) Pheromone-inducible conjugation in *Enterococcus faecalis*: a model for the evolution of biological complexity? *Int J Med Microbiol.* **296**, 141–147.
127. Salyers, A. A., Shoemaker, N. B., Stevens, A. M., and Li, L. Y. (1995) Conjugative transposons: an unusual and diverse set of integrated gene transfer elements. *Microbiol Rev.* **59**, 579–590.
128. Showsh, S. A., and Andrews, R. E., Jr. (1992) Tetracycline enhances Tn*916*-mediated conjugal transfer. *Plasmid.* **28**, 213–224.
129. Bahl, M. I., Sørensen, S. J., Hansen, L. H., and Licht, T. R. (2004) Effect of tetracycline on transfer and establishment of the tetracycline-inducible conjugative transposon Tn*916* in the guts of gnotobiotic rats. *Appl Environ Microbiol.* **70**, 758–764.
130. Su, Y. A., He, P., and Clewell, D. B. (1992) Characterization of the *tet*(M) determinant of Tn*916*: evidence for regulation by transcription attenuation. *Antimicrob Agents Chemother.* **36**, 769–778.
131. Khalil, T. A., and Gealt, M. A. (1987) Temperature, pH, and cations affect the ability of *Escherichia coli* to mobilize plasmids in L broth and synthetic wastewater. *Can J Microbiol.* **33**, 733–737.
132. Normander, B., Christensen, B. B., Molin, S., and Kroer, N. (1998) Effect of bacterial distribution and activity on conjugal gene transfer on the phylloplane of the bush bean (*Phaseolus vulgaris*). *Appl Environ Microbiol.* **64**, 1902–1909.
133. Molin, S., and Tolker-Nielsen, T. (2003) Gene transfer occurs with enhanced efficiency in biofilms and induces enhanced stabilisation of the biofilm structure. *Curr Opin Biotechnol.* **14**, 255–261.
134. Linares, J. F., Gustafsson, I., Baquero, F., and Martinez, J. L. (2006) Antibiotics as intermicrobial signaling agents instead of weapons. *Proc Natl Acad Sci U S A.* **103**, 19484–19489.
135. Bergstrom, C. T., Lipsitch, M., and Levin, B. R. (2000) Natural selection, infectious transfer and the existence conditions for bacterial plasmids. *Genetics.* **155**, 1505–1519.
136. Hansen, S. K., Rainey, P. B., Haagensen, J. A., and Molin, S. (2007) Evolution of species interactions in a biofilm community. *Nature.* **445**, 533–536.
137. Burmølle, M., Webb, J. S., Rao, D., Hansen, L. H., Sørensen, S. J., and Kjelleberg, S. (2006) Enhanced biofilm formation and increased resistance to antimicrobial agents and bacterial invasion are caused by synergistic interactions in multispecies biofilms. *Appl Environ Microbiol.* **72**, 3916–3923.
138. Parsek, M. R., and Greenberg, E. P. (2005) Sociomicrobiology: the connections between quorum sensing and biofilms. *Trends Microbiol.* **13**, 27–33.
139. Hausner, M., and Wuertz, S. (1999) High rates of conjugation in bacterial biofilms as determined by quantitative *in situ* analysis. *Appl Environ Microbiol.* **65**, 3710–3713.
140. Venter, J. C., Remington, K., Heidelberg, J. F., Halpern, A. L., Rusch, D., Eisen, J. A., Wu, D., Paulsen, I., Nelson, K. E., Nelson, W., Fouts, D. E., Levy, S., Knap, A. H., Lomas, M. W., Nealson, K., White, O., Peterson, J., Hoffman, J., Parsons, R., Baden-Tillson, H., Pfannkoch, C., Rogers, Y. H., and Smith, H. O. (2004) Environmental genome shotgun sequencing of the Sargasso Sea. *Science.* **304**, 66–74.

# Chapter 6

# The Integron/Gene Cassette System: An Active Player in Bacterial Adaptation

**Maurizio Labbate, Rebecca Josephine Case, and Hatch W. Stokes**

## Abstract

The integron includes a site-specific recombination system capable of integrating and expressing genes contained in structures called mobile gene cassettes. Integrons were originally identified on mobile elements from pathogenic bacteria and were found to be a major reservoir of antibiotic-resistance genes. Integrons are now known to be ancient structures that are phylogenetically diverse and, to date, have been found in approximately 9% of sequenced bacterial genomes. Overall, gene diversity in cassettes is extraordinarily high, suggesting that the integron/gene cassette system has a broad role in adaptation rather than being confined to simply conferring resistance to antibiotics. In this chapter, we provide a review of the integron/gene cassette system highlighting characteristics associated with this system, diversity of elements contained within it, and their importance in driving bacterial evolution and consequently adaptation. Ideas on the evolution of gene cassettes and gene cassette arrays are discussed.

**Key words:** Integron, gene cassette, antibiotic-resistance, *attC*.

## 1. Introduction

Horizontal gene transfer is an important process in shaping genomes over both long- and short-term time scales. In particular, in the context of the short-term evolution of prokaryotic genomes, mechanisms of non-homologous recombination are known to be crucial to adaptation in extreme environments. The most obvious example of this is the extraordinary ability of pathogens to acquire multiple antibiotic-resistance genes in historical times. The importance of transposons in facilitating the spread of resistance genes between otherwise unrelated replicons is well known. Less well known is that mechanisms of site-specific

Maria B. Gogarten et al. (eds.), *Horizontal Gene Transfer: Genomes in Flux, vol. 532*

DOI 10.1007/978-1-60327-853-9_6 Springerprotocols.com

recombination have also contributed greatly to the mobilization and spread of antibiotic resistance. Perhaps, one of the reasons for this poorer understanding is that site-specific recombination does not so obviously lend itself to the efficient radiation of DNA sequences since, in comparison to transposition, the number of target sites is more limited. Further, most site-specific recombination systems involve a very limited reaction in that any single example of a site-specific recombination reaction normally only involves recombination between two defined sequences. One site-specific system, however, the integron/gene cassette system, is atypical of site-specific recombination systems generally. Its most extraordinary feature is that at least one of the two sites participating in the recombination reaction is not a single defined site but, rather, a family of sites. This unusual departure from the norm has created a system that provides for an enormous pool of adaptive genes to be mobilized, rearranged, and disseminated amongst bacteria. Indeed, the reservoir of genes capable of being mobilized as yet has no known upper limit but must, at the very least, number in the thousands and is probably orders of magnitude higher. As a result, this system can greatly influence bacterial adaptation in ways that other systems cannot.

## 2. What Are Integrons?

The defining feature of the integron is a site-specific recombination system capable of integrating and expressing open reading frames (ORFs) contained in modular structures called mobile gene cassettes (**Fig. 6.1**). The integron platform is defined by three characteristics: an integrase gene (*intI*) whose product encodes a site-specific integrase, IntI; an attachment site (*attI*) at which point DNA sequences are inserted; and a promoter ($P_C$), although this last feature has not been functionally demonstrated for all integron classes. While not part of the site-specific recombination system, the promoter ($P_C$) expresses genes within the gene cassettes inserted at *attI (1)*. The units of insertion into integrons are mobile gene cassettes. A gene cassette is an independently mobilizable element that generally contains a promoterless ORF and an IntI-recognizable recombination site called the *attC* or 59-base element (59-be) *(2–4)* (*see* **Box 6.1**). Integrons are best known for their role in contributing to clinical antibiotic resistance, with approximately 100 different gene cassettes containing genes that confer resistance to antibiotics. Although gene cassettes are inserted independently, multiple insertion events can lead to the accumulation of many cassettes within an integron, thus contributing to multidrug resistance *(5, 6)*. In the clinical

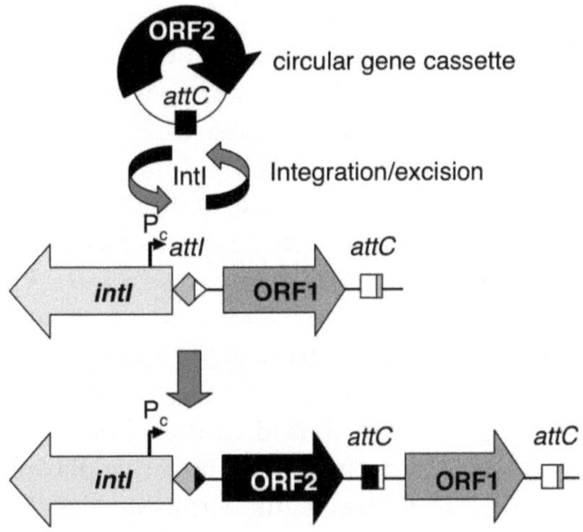


Fig. 6.1. The integron/gene cassette site-specific recombination system. The integron platform consists of the *intI* gene, *attI* site, and $P_C$ promoter. The gene cassette usually consists of a single promoterless ORF and a recombination site called *attC*. In its free form, the gene cassette is circularized and is recombined into the integron by an integrase-mediated reaction between *attI* and *attC* (or *attC* and *attC* in lower frequencies). In this figure, incorporation of the gene cassette containing ORF2 results in a two-cassette array; however, arrays can vary in size from none to over 200. Expression of incorporated gene cassettes occurs via $P_C$.

**Box 6.1. 59-base elements or *attC* sites?**

The terms 59-base element (59-be) and *attC* define the same thing – namely the gene cassette-associated recombination site recognized by an IntI integrase. The term 59-base element has historical precedent. It was coined by Ruth Hall and co-workers who were the first to identify the region adjacent to certain genes that was implicated in mobilizing such genes by site-specific recombination *(85)*. The name was meant to reflect the length of the element based on the very limited data available at that time. Although very few elements are actually 59 bases long, the term is commonly found in the literature because of its historical dominance. In recent years, the term *attC* has been introduced to describe the cassette recombination site *(86)* and is becoming favored because it is more consistent with the terminology used to describe site-specific recombination sites generally. In this case, *attC* makes reference to the fact that it is the attachment site associated with cassettes. Like many nomenclature debates, discussion from time to time has been "lively" and both terms commonly appear in the recent literature. There is however an emerging preference for the use of *attC*.

environment, there are three main groups or classes of integrons mainly associated with antibiotic resistance. These are the class 1, 2, and 3 integrons, with their classification based on their respective *intI* sequences.

The class 1 integron is the most prevalent class. They are also the most clinically important and broadly distributed integrons with 22–59% of Gram-negative clinical isolates positive for the class 1 integrase *(7–9)*. While mostly associated with Gram-negative bacteria, in recent times there have been increasing numbers of reports of Gram-positive bacteria carrying class 1 integrons *(10–14)*. The spread of integrons that convey antibiotic resistance is attributed to their linkage with mobile elements such as transposons and conjugative plasmids, allowing for their rapid transfer between and within species. Most classes of integron exist outside the clinical environment, are mostly located in chromosomes, and are not mobile, in comparison to those located in plasmids and transposons. These chromosomal integrons are relatively common, with approximately 9% of the sequenced bacterial genomes containing such integrons *(15)*. Mobilized class 1, 2, and 3 integrons contain anything from zero to six cassettes, while chromosomal integrons can be found associated with more cassettes such as *Vibrio vulnificus* CMCP6, which contains 217 gene cassettes, and *Pseudomonas alcaligenes* with 32 cassettes. However, a large array is not necessarily a signature for chromosomal integrons. The chromosomal integron of *Xanthomonas campestris* strain 85-10 contains only three cassettes while some others carry none (**Table 6.1**). The phylogenetic relationship of mobile class 1, 2, and 3 integrons with chromosomal integrons from environmental microorganisms has implications for the origins and emergence of these systems in clinical environments *(15, 16)*. This will be discussed in more detail in **Section 4**.

## 3. What Are Gene Cassettes?

Mobile gene cassettes are the smallest known mobilizable element, usually containing only a single gene and a recombination site (*attC*). Gene cassettes can exist in one of two forms. One of these is as an independent circular element in the cell (**Fig. 6.1**). Gene cassettes are not replicons and so they cannot be stably maintained during cell division when present as a free circular DNA molecule. However, circular cassettes can be inserted into an integron as a result of the integron-associated site-specific recombination reaction to create a linear form. When this happens, cleavage occurs near one end, but within, the *attC* recombination site. As a result of this, and the fact that insertion is highly orientation-specific, most of the recombination site

**Table 6.1**
**Selected list of bacterial species containing chromosomal integrons**

| Bacterial group | Bacterial strain | No. of gene cassettes | Typical length of *attC* |
|---|---|---|---|
| γ-Proteobacteria | | | |
| Vibrionaceae | *Vibrio cholerae* N16961 | 179 | 126–129 |
| | *Vibrio vulnificus* CMCP6 | 217 | 126–129 |
| | *Vibrio* sp. DAT722 | 116 | 126–128 |
| | *Vibrio parahaemolyticus* RIMD2210633 | 69 | 125–128 |
| | *Vibrio alginolyticus* 12G1 | 51 | 126–129 |
| | *Vibrio fischeri* ES114 | 38 | 125–127 |
| Pseudoalteromonas | *Pseudoalteromonas tunicata* D2 | 7 | 78–129 |
| | *Pseudoalteromonas haloplanktis* TAC125 | 5 | 98–102 |
| Xanthomonadaceae | *Xanthomonas campestris* pv. campestris ATCC 33913 | 22 | 60 |
| | *Xanthomonas campestris* pv. vesicatoria | 3 | 79–94 |
| Pseudomonadaceae | *Pseudomonas stutzeri* Q | > 7 | 76–77 |
| | *Pseudomonas alcaligenes* | 32 | 76–90 |
| Shewanellaceae | *Shewanella* sp. MR-7 | 2 | 90–92 |
| | *Shewanella denitrificans* OS217 | 0 | NA* |
| Alteromonadaceae | *Saccharophagus degradans* 2–40 | 73 | 111–140 |
| β-Proteobacteria | | | |
| | *Nitrosomonas europaea* ATCC19718 | 0 | NA |
| | *Nitrosomonas eutropha* C71 | 3 | 60–120 |
| | *Azoarcus* sp. EbN1 | 3 | 59–73 |
| | *Rubrivivax gelatinosus* PM1 | 2 | 70–126 |
| δ-Proteobacteria | | | |
| | *Geobacter metallireducens* GS-15 | 3 | 59–63 |
| Planctomycetes | *Rhodopirellula baltica* SH1 | 0 | NA |
| Spirochaetales | *Treponema denticola* ATCC35405 | 45 | 63–68 |

**NA = Not Applicable*

is located at the 3′ end of any associated gene in the integrated form of the cassette (**e.g., ORF2 in Fig. 6.1**). Also, the associated gene is oriented such as to allow expression from $P_c$ *(17)*. Since integrons are located in chromosomes or on some other replicon such as a plasmid, linear gene cassettes can be stably maintained in cell lines. Most cassettes have obvious genes embedded within them, although in many cases we do not know what the product does. However, gene cassettes with no ORF are also found, and have sufficient sequence conservation to suggest they have some biological role *(18)*. The *attC* recombination sites can range in length from 56 to 141 bp (**Table 6.1**). Recombination sites comprise an imperfect inverted repeat structure. The 20 or so bases at each end of *attC* sequences are most conserved with each end, including a site of a type generally associated with site-specific tyrosine recombinase recombination systems. Each site (called a "simple site") includes two IntI binding domains; the most right-hand site (furthest from the 3′ end of the gene in the inserted cassette; **Fig. 6.2A**) includes the point of strand cleavage *(19)*, which is located in 1R between the G and the first T residue (**Fig. 6.2B**). This core site is one of the most conserved parts of the *attC* element along with the inverse core site at the opposite end of the element. Information contained within these simple sites is important for site-specific recombination *(19, 20)*. **Figure 6.2B** also shows an alignment of *attC* sequences from environmental gene cassettes demonstrating the variability in *attC* length and conserved imperfect inverted repeats at the ends.

## 4. Evolution of Mobile Antibiotic-Resistance-Associated Integrons

Given the broad distribution of chromosomal integrons among diverse bacteria, the emergence of mobile integrons that mostly contain resistance cassettes is hypothesized to have occurred through the association of chromosomal integrons with mobile elements *(15, 16, 21)*. As stated above, such a hypothesis is supported by the close phylogenetic relationship of class 1, 2, and 3 integrases with chromosomal integrases *(15, 16)*. The phylogenetic origin of the integron classes common in pathogenic bacteria is currently unknown; however, the mobilization processes that have led to the successful radiation of class 1 integrons within clinical environments is relatively well understood. The vast majority of class 1 integrons isolated from clinical environments consist of a structure defined by conserved DNA segments flanking the gene cassette(s). The 5′-conserved segment (5′-CS) consists of the integrase gene, *intI1* and the *attI* site, while the 3′-conserved segment (3′-CS) consists of a truncated copy of *qacE*, a gene that confers resistance to quaternary ammonium

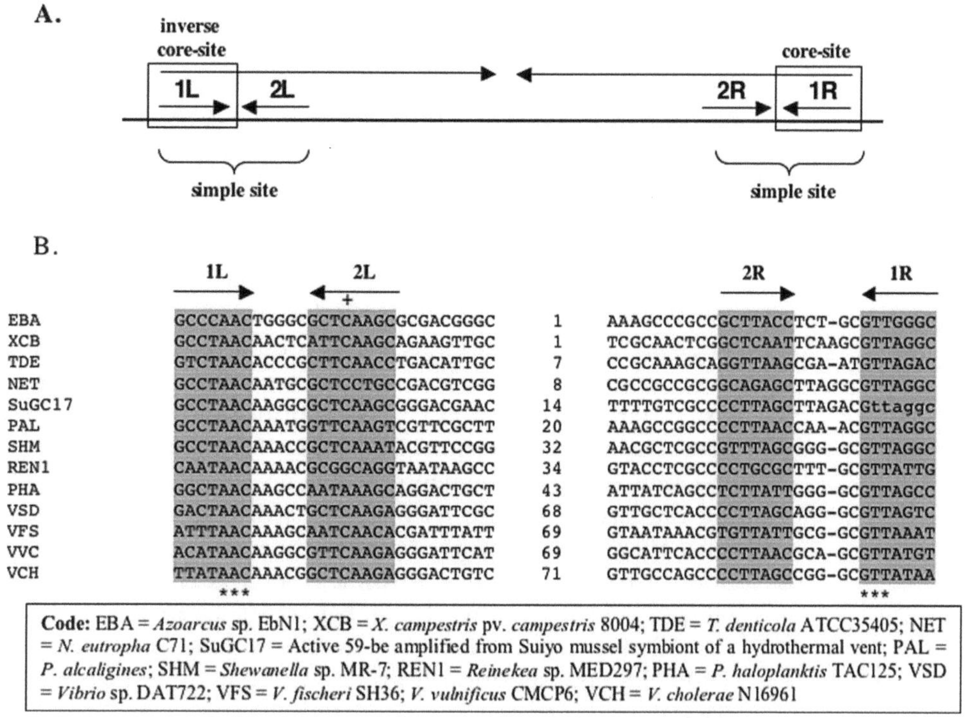


Fig. 6.2. *attC* structure and sequences. (**A**) *attC* sequences are imperfect inverted repeats (arrows demonstrate regions of symmetry) with conserved regions of approximately 20-bp at each end. These conserved regions include a simple site (*bracketed*) that facilitate IntI binding and are themselves imperfect inverted repeats when 1L is compared to 1R and 2L to 2R. The core (1R) and inverse core (1L) sites are boxed and are the most conserved regions of *attC*. The region between the two simple sites is of variable length and sequence. (**B**) Alignment of the conserved domains of *attC* sites taken from cassettes found in different bacteria. The highly variable central regions are not shown, with numbers indicating the length of the omitted sequence. Except for SuGC17, all *attC* sequences are depicted in the predicted free circular form of the cassette. SuGC17 is depicted in the linear, integrated form due to insufficient sequence data. The *lower case* sequence indicates sequence from the adjacent cassette. Putative IntI binding domains in the *left* (1L and 2L) and *right* (1R and 2R) halves of the elements are indicated by *shading* and *arrows*. 1L is an imperfect inverted repeat with respect to 1R as is 2L to 2R. In the latter case, however, 2L has an extra base ("+" symbol) that needs to be taken into account in aligning with 2R. The *asterisks* mark the highly conserved nucleotides of the core (1R) and inverse core (1L) sites. See Stokes et al., 1997 *(19)*, for a detailed discussion on important *attC* features.

compounds *(22)*, and *sulI*, a gene that confers resistance to sulfonamides *(23)*. The outer boundary of the 5′-CS is defined by a 25-bp inverted repeat sequence called IRi, of which its counterpart, IRt, is found beyond the 3′-CS when present (**Fig. 6.3**). The IRi/IRt inverted repeats are a feature of transposons that mobilize DNA using a suite of four transposition genes (*tniR*, *-Q*, *-B*, *-A*) *(24)*. However, class 1 integrons containing a 3′-CS do not contain a full complement of *tni* genes, although *tni* gene remnants are usually found adjacent to the 3′-CS (**Fig. 6.3**) *(25)*. Tn*402* typifies the structure of an integron that is also an active transposon *(26)*. It contains a full 5′-CS and a complete *tni* module downstream of the inserted gene cassettes; however, there is no 3′-CS. Such a structure supports the idea that the Tn*402*-like transposon is the ancestral mobile form of the clinical class 1 integron, and that a series of deletions, insertions, and recombination events resulted in the genesis of the 3′-CS and the partial loss of *tni* functions *(25)*. This is supported by the presence of a *qacE* gene cassette in Tn*402* that is 100% identical to the equivalent region present in the truncated *qacE* gene of the 3′-CS, suggesting a deletion event had occurred possibly as a result of the *sulI* insertion *(25, 26)*.

Class 1 integrons are associated with many different plasmids and transposons, such as those of the Tn*3* family (e.g. Tn*21* and Tn*1696*). This association is a direct result of the linkage of class 1 integrons with Tn*402*-like transposons since this transposition system targets resolution (*res*) sites of plasmids and other transposons as sites of insertion *(24, 27, 28)*. This targeting has led to a vast array of mobile elements picking up a class 1 integron, allowing the latter to make its way into a diverse range of bacteria. Since these class 1 integrons can carry many types and combinations of resistance genes, this integron class is one of the single biggest contributors to the global resistance problem. All clinically derived class 1 integrons retain evidence of at least some Tn*402*-like transposition features. A recent study identified two class 1 integrons lacking Tn*402*-like transpositional features in the chromosomes of the environmental bacteria, *Azoarcus communis* MUL2G9 and *Acidovorax* sp. MUL2G8 *(29)*. No *tni* genes are present, and the first 98-bp of the 5′-CS containing IRi are absent in both integrons. Beyond the 5′-CS divergence, both integrons contain a DNA integrase/recombinase that share significant protein similarity, implying that the integrons were recruited into their current position by a similar mechanism but one that is unrelated to Tn*402*-like transposition *(29)*. A similar general scenario is observed with class 3 integrons, which are rare in the clinical environment. Only two have been studied in detail. One is from a *Serratia marcescens* strain associated with Tn*402*-like transposition functions *(30–32)*. However, class 3 integrons recently identified in the chromosome of two *Delftia* species lack Tn*402*-like transposition functions, but are flanked by an integrase/recombinase similar to the one found flanking environmental class 1 integrons of *Azoarcus communis* MUL2G9 and *Acidovorax* sp. MUL2G8 *(33)*. Notwithstanding their widespread dispersal in clinical isolates, it appears that at least some classes of integrons are more generally mobilized in bacterial communities. The identification of non-Tn*402* associated mobilized class 1 integrons in environmentally isolated bacteria suggests that the clinically important class 1 integrons were mobilized prior to becoming associated with Tn*402*-like transposition functions. **Figure 6.3** provides a graphical representation of the likely evolution of clinically derived class 1 integrons.

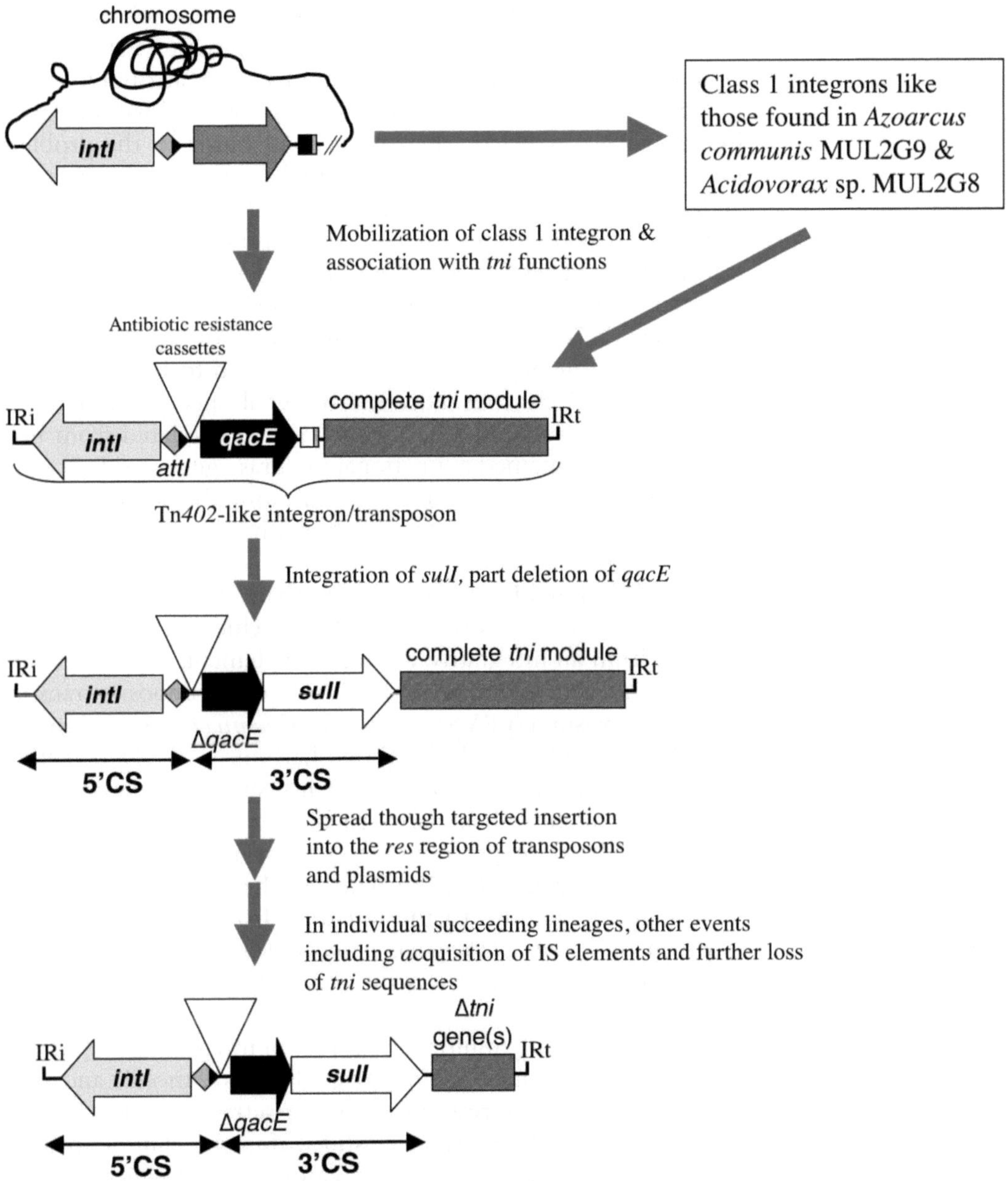


Fig. 6.3. Evolution of mobile clinically derived class 1 integrons. The origin of class 1 integrons is believed to have come from a chromosomal location. At some stage, the class 1 integron was mobilized most likely before the advent of the "antibiotic revolution". Examples of mobilized class 1 integrons have been found in the chromosome of *Azoarcus communis* MUL2G9 and *Acidovorax* sp. MUL2G8. A class 1 integron eventually became associated with *tni* functions represented by the Tn*402* structure. Partial deletion of the *qacE* cassette and insertion of *sulI* resulted in formation of the 3′-CS commonly found in clinical class 1 integrons. Rapid spread of the class 1 integron in clinical contexts occurred via the targeted insertion of the *res* region of transposons and plasmids. In individual successive lineages, events such as IS element insertion are responsible for deletions in the *tni* module creating a defective transposon.

Unlike the other mobilized integrons, most isolated class 2 integrons contain an inactive integrase gene due to a stop codon at amino acid 179 *(34)*. This mutation has been attributed to the low diversity of integrated gene cassettes observed within class 2 integrons. Nevertheless, they are a contributor to the problem of antibiotic resistance and are isolated from different bacterial and genetic backgrounds *(35–38)*. Mobility of class 2 integrons is due to an association with Tn*7*, a transposon that contains five transposition genes (*tnsA, -B, -C, -D*, and *-E*) and preferentially inserts into a unique site within bacterial chromosomes *(39, 40)*. However, Tn*7* has been shown to insert into conjugative plasmids at a lower frequency *(41)*, which is presumed to be the method by which Tn*7* transfers between bacterial species. Recently, a class 2 integron within a Tn*7* backbone was isolated from beef cattle, which contained a functional integrase gene and four non-antibiotic-resistance gene cassettes, indicating that a greater diversity of class 2 integrons may exist than what has previously been observed *(42)*.

Mobilization of integrons other than class 1, 2, or 3 has also been reported in *Vibrio* spp. These include an integron contained within an integrative conjugative element (ICE) in *Vibrio cholerae (43)* and an integron that is part of a composite transposon from a plasmid (pRVS1) in *Vibrio salmonicida* (Accession No. CAC35342; *(44)*). Both integrons harbour a *dfrA1* trimethoprim resistance cassette; however, their respective integrases have sufficient sequence dissimilarity to belong to different integron classes, and both integron-integrases are related to chromosomally located integrons *(15)*. In particular, the pRVS1 integrase has 99% sequence identity to the integrase from the chromosomal integron of *Pseudoalteromonas haloplanktis* TAC125, indicating that it or a close relative was the source for the mobilized integron in pRVS1 *(45)*.

There are clearly multiple examples of integrons being mobilized. Given that four other integron classes are mobile and associated with antibiotic-resistance gene cassettes, why has the class 1 integron been the most successful within the clinical environment? There may be a number of reasons, one of which includes the association of class 1 integrons with Tn*402*-like transposition functions. However, class 3 integrons are also associated with Tn*402*-like transposition functions but are far less prominent in a clinical environment. One other possibility is that the class 1 integrase recognizes a larger spectrum of *attC* sites and consequently is better able to access the gene cassette metagenome. In a study comparing *V. cholerae* and class 1 integrases, the class 1 integrase had a greater capability of recognizing different *attC* sites *(46)*. If it is the case that the class 1 integrase is able to recombine a wider range of *attC* sites, this may provide class 1 integrons with an important selective advantage.

## 5. Chromosomal Integrons – Beyond the Clinical Arena

It is now apparent that integrons are not only within the clinical domain but also commonly found in the chromosome of microorganisms in the general environment (**Table 6.1**). Integron sequence diversity is greater in these environmental isolates than in clinical isolates, which suggests that integrons are ancient structures in genomes and have played a general role in evolution and adaptation for a considerable period of time. As will be described later, this is further supported by the enormous diversity of integron-associated cassette genes that, in cases where a function can be ascribed, mostly encode products unrelated to antibiotic resistance. Chromosomal integrons have mainly been characterized in members of the γ-Proteobacteria and β-Proteobacteria, although chromosomal integrons have also been found among the δ-Proteobacteria, Planctomycetes, and Spirochaetales. Recent phylogenetic analyses of integrases identified three main integron clades: the soil/freshwater Proteobacteria group, the inverted integrase group, and the marine γ-Proteobacteria group. Class 1 and 3 integrases are part of the soil/freshwater Proteobacteria group while class 2 integrases are part of the marine γ-Proteobacteria group *(15, 16)*. As their name implies, the inverted integrase group have their associated *intI* gene in the inverse orientation compared to all other integrons (i.e., the *intI* gene reads toward the *attI* site). There are certain characteristics that separate chromosomal integrons from integrons found in mobile elements of pathogens. In particular, they are more likely to be "fixed" and consequently co-evolve with their host for a substantial amount of time. This can be seen with integrons from *Vibrio* spp. and *Xanthomonas* spp. whose integrases group together *(15, 16)*. Even so, phylogenetic studies suggest that at some point chromosomal integrons have been subject to lateral transfer *(15)*. For example, the chromosomal integrons found in *Pseudomonas* spp. arise from relatively recent insertions. Only some *Pseudomonas* spp. such as *P. alcaligenes (47)* and *P. stutzeri (48)* contain chromosomal integrons. No integrons are present in any of the *P. aeruginosa* genomes. Furthermore, there is also inter-strain variation as some strains of *P. stutzeri* lack chromosomal integrons *(48)*. In most cases, transposons or recombinases can be found in the gene cassette array or flanking the chromosomal integron, which provides evidence for the lateral acquisition of integrons *(15)*.

Some chromosomal integrons contain large cassette arrays. **Table 6.1** provides a selected list of chromosomal integrons and their gene cassette complement ranging from none, for example, in *Nitrosomonas europaea*, to 219 in *Vibrio vulnificus* CMCP6.

Since the integron-integrase can catalyze gene cassette excision and acquisition, the maintenance of large cassette arrays must be linked to adaptive advantage. Large arrays consist of cassettes that collectively include many diverse functions, so it is thought that multiple selection pressures driven by a rapidly changing environment maintain the large arrays. Indeed, pathogens harboring large arrays such as *Vibrio cholerae*, *V. vulnificus*, and *V. parahaemolyticus* are subject to changing environmental conditions due to their pathogenic and free-living lifestyles. Similarly, lifestyle complexity is seen in *V. fischeri* ($\sim$ 40 cassettes), which is found free-living or as a squid and fish symbiont. In the absence of selection pressure, gene cassettes encoding toxin/antitoxin (TA) systems contribute to the stability of large cassette arrays *(45)*. TA systems contain two genes encoding a stable toxin and a specific unstable antitoxin, thereby preventing growth of their host if not retained *(49)*. TA gene cassettes are common in large gene cassette arrays but absent in small arrays *(50)*. For example, the *V. cholerae* N16961 genome contains 13 TA loci, all of which are present in the integron array. Apart from single cassette excision, arrays are also subject to multi-cassette excision, with one study identifying a 38-cassette excision event in a subset of *V. cholerae* O1 El Tor strains *(51)*. Large cassette excision events are minimized by the presence of TA gene cassettes since deletions containing these genetic elements kill the host by action of the stable toxin *(45)*.

Compared to mobilized integrons, chromosomal integrons harbor cassettes with high *attC* sequence similarity. This observation has been made in the chromosomal arrays of many bacteria including *V. cholerae*, *V. fischeri*, *V. metschnikovii*, *P. alcaligenes*, *P. stutzeri*, *X. campestris*, and *Treponema denticola (21,47,52–54)*. In *V. cholerae* the *attC* sites are highly conserved, differing by less than 10% nucleotide identity. The identification of *attC* sites that are characteristic of a specific integron or species is intriguing and has led to the suggestion that gene cassettes could be synthesized by physical attachment of an *attC* site to an incoming ORF *(16)*. The mechanism of gene cassette synthesis is unknown; however, some theories will be discussed in the next section. Another explanation for the observed *attC* sequence conservation may be the *attC* site recognition range of the integrase. Biskri et al., 2005 *(46)*, determined that the *V. cholerae* integrase has a reduced ability to recognize diverse *attC* sites compared to the class 1 integrase, suggesting a tighter relationship between *V. cholerae*'s integrase and *attC* site. However, the *V. cholerae* integrase is able to recognize *attC* sites other than its own, making the issue ambiguous. It is also the case that comparing recombination efficiency between integron classes is difficult. Some progress is being made *(46, 48, 55)*, but it is an area that requires much more experimental analysis. Difficulty in comparing recombination efficiency arises since there are a number of variables to take into account. These include not just the fact that the integrases differ but that *attI* sites between classes also differ as do the *attC* sequences. In addition, some activity assays are based on insertion *(56)* while others are based on excision *(57)*, yet there is no reason to assume that the two reactions are equivalent in relation to recombination efficiency.

The fact that *attC* sites generally conform to a specific integron or species has helped determine that antibiotic-resistance gene cassettes likely arose from chromosomal integrons. For example, 12 different resistance cassettes contain an *attC* site homologous to *attC* sites in the chromosomal integron arrays of *Xanthomonas* spp. and *Vibrio* spp. *(21)*. While antibiotic-resistance cassettes circulating in clinical environments have not been identified in sequenced environmental cassette arrays, cassettes capable of providing resistance are present (e.g., *catB9*, CARB-7 and CARB-9) *(58–60)*. Furthermore, recruitment and expression of the *catB9* cassette from *V. cholerae* by a class 1 integron has been demonstrated *(60)*. In this manner, mobile class 1 integrons may sample gene cassettes from the environmental gene pool and "test" the selective advantage they convey in the clinical context. Antibiotic usage would then select for the maintenance of advantageous cassettes.

## 6. Where Do Gene Cassettes Come From?

The lack of promoters in gene cassette-containing ORFs and the fact that, overall, they generally contain very little non-coding sequence have led to the hypothesis that the formation of gene cassettes is mediated by reverse transcription of an mRNA molecule *(56, 61)*. The *attC* site is somehow added either prior to or following the reverse transcription reaction *(61)*, which is consistent with the observations that some structural genes are homologous to gene cassette ORFs *(62)*. A specific gene cassette synthesis theory has been presented based on the observation of group II introns inserted behind *attC* sites *(63, 64)*. Bacterial group II introns are novel genetic elements that are both catalytic RNA molecules and retroelements *(65)*. They are mobile via a mechanism that involves reverse splicing of an excised intron into a specific DNA target site that is then reverse transcribed by the intron-encoded protein. After DNA insertion, the introns remove themselves by protein-assisted autocatalytic RNA splicing, restarting the cycle *(65)*. Gene cassette synthesis is suggested to occur via a two-step mechanism in which an intron inserts before an *attC* site and a second intron inserts downstream of a structural gene. Recombination between the two introns would subsequently bring the structural gene and *attC* site together with the intron between them. Splicing of the intron followed by reverse transcription would result in a newly formed gene cassette *(63)*. The site-specific insertion of introns supports this hypothesis. Of the five known *attC* intron insertions encompassing two different types of group II introns, the intron is inserted identically at the left (**Fig. 6.2A**) end of the *attC (66)*. Intriguingly, some group II introns specifically insert behind or in front of rho-independent terminators *(65)*. Association of introns with integrons or in the genomes of integron-containing bacteria are rare, although it has been suggested that some environmental bacteria may act as gene cassette factories *(63)*. If this is the case, these "gene cassette factory" bacteria are quite different from the bacteria whose genomes have been sequenced since approximately 2/3 of ORFs encoded in gene cassettes have no known database homolog *(17, 18)*.

## 7. Gene Cassette Diversity, Functionality, and Role in Adaptation

Amplification of gene cassettes from metagenomic DNA has found gene cassettes to be a great source of genetic diversity and novelty *(17, 18, 67)*. Even within small physical distances, gene cassette populations differ markedly *(67)*. Approximately 80% of gene cassettes have ORFs with no known homolog or homologs of unknown function *(15, 18)*. **Figure 6.4** provides a functional distribution for 1677 gene cassette-encoded proteins from the *Vibrio* genera demonstrating the high functional diversity and large proportion of proteins with no known homolog. This distribution is typical of the gene cassettes obtained from other environments. The high level of unknown proteins encoded within gene cassettes has made it difficult to understand what the adaptive roles of the genes within these mobile units are in bacterial evolution.

Nevertheless, some cassettes have been shown to include ORFs encoding specific advantageous functions. In *V. cholerae*, three cassette-associated genes express virulence factors including a heat-stable toxin gene (*sto*), a mannose-fucose resistance hemagglutinin (*mrhA*), and a lipoprotein gene (*vlpA*) *(68–71)*. In the chromosomal integrons of *V. vulnificus*, *V. alginolyticus*, and *V. parahaemolyticus* is a cassette encoding the PAS factor. The expression of this small protein is preferentially induced during human infection by *V. vulnificus* CMCP98K *(72)* and is believed to mediate secretion of periplasmic proteins required for the in vivo survival and pathogenesis of *V. vulnificus (73)*. **Table 6.2** summarizes the ORFs within other cassettes that have been shown to encode functional proteins. The broad functional

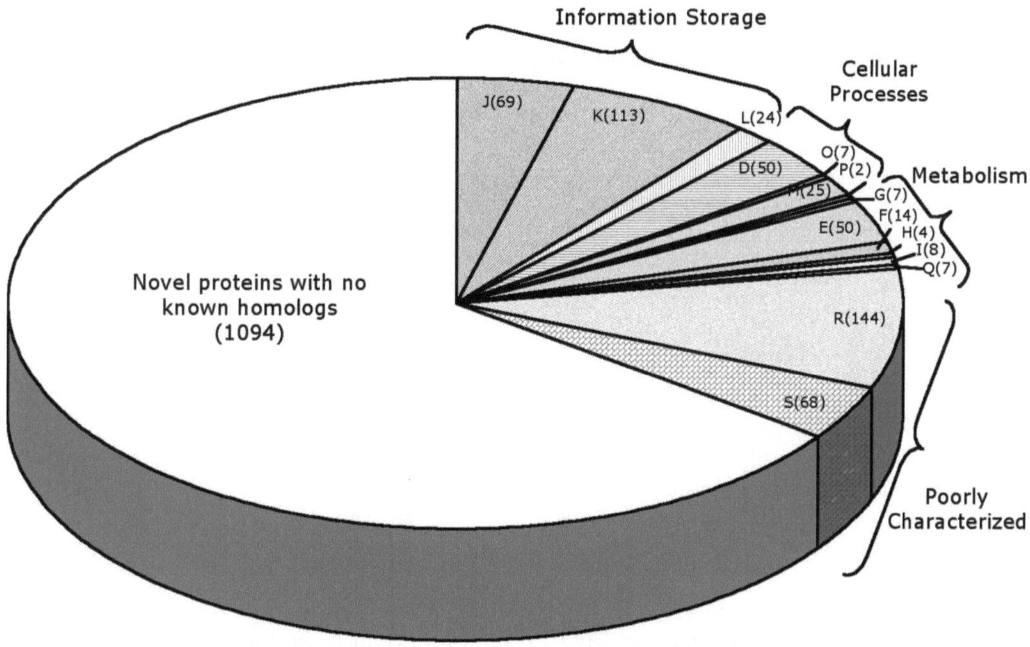


Fig. 6.4. Functional distribution for 1677 gene cassette encoded proteins from the *Vibrio* genera. These represent $> 90\%$ of all known gene cassette sequences. Functional annotation was performed using the COG database *(76)*. Proportions of proteins belonging to each broad COG functional category and subcategory are presented. The number of proteins belonging to each subcategory is indicated in *parentheses* next to its one letter abbreviation; Information storage: Translation, ribosome structure biogenesis (J), Transcription (K), and Replication recombination and repair; Cellular processes: Cell cycle control, cell division, chromosome partitioning (D), Posttranslational modification, protein turnover, chaperones (O), Cell wall/membrane/envelope biogenesis (M) and, Inorganic ion transport and metabolism (P); Metabolism: Carbohydrate transport and metabolism (G), Amino acid transport and metabolism (E), Nucleotide transport and metabolism (F), Coenzyme transport and metabolism (H), Lipid transport and metabolism (I) and, Secondary metabolites biosynthesis, transport and catabolism (Q). Finally, the hypothetical conserved broad category has within it two subcategories: general function prediction only (R) and function unknown (S). Reprinted with permission from *Trends in Microbiology (15)*.

diversity is striking and, given the novelty of many cassette genes, the potential to identify novel enzymes is great. Notable is iMazG, a dNTP pyrophosphohydrolase, which is the product of an ORF encoded within cassette 21 of the *Vibrio* sp. DAT722 integron array *(74)*. Its crystal structure and functional detail place iMazG as the first member of a new subfamily related to the MazG enzyme class. These enzymes are involved in house-cleaning activities that remove aberrant dNTPs from the metabolic pool, thereby helping to minimize DNA damage. One can speculate that the presence of additional copies of genes of this type may help the cell adapt to periods of heightened stress.

Genetic versatility provided by gene cassettes may also be environment-specific, providing bacteria with the genetic potential to adapt to specific niches. Gene cassettes amplified from deep-sea hydrothermal vents, like those seen in other environments, include many ORFs with no identifiable function

**Table 6.2**
**Examples of functional ORFs in gene cassettes**

| Source of cassette | Function | Determination of function | Reference |
|---|---|---|---|
| *Vibrio cholerae* | Sulfate-binding protein | Complementation of *E. coli* mutation | *(21)* |
| *Vibrio marinus* | Psychrophilic lipase | Active when expressed in *E. coli* at 10 °C | *(21)* |
| *Vibrio* sp. DAT722 | dNTP pyrophosphohydrolase (*i*MazG) | Crystal structure determination. Expressed in *E. coli* and enzyme activity measured | *(74)* |
| Soil metagenomic DNA | Potential transport protein | Crystal structure determination | *(83)* |
| Soil metagenomic DNA | ATPase activity | Expressed in *E. coli* and enzyme activity measured | *(84)* |
| Soil metagenomic DNA | Methyltransferase activity | Expressed in *E. coli* and enzyme activity measured | *(84)* |
| Various large integron arrays | Toxin/antitoxin genes | Demonstration that presence of toxin/antitoxin genes limits deletions in large gene cassette arrays | *(45)* |

*(75)*. However, a number of amplified cassettes from vent mussel metagenomic DNA also include ORFs consistent with symbiont-animal host metabolic processes. For example, cassette genes potentially encoding a methionine aminopeptidase, an *O*-sialoglycoprotein, and a glutamate synthase are consistent with metabolic processes the expected sulfur-oxidizing endosymbiont would exhibit. Furthermore, gene cassettes with genes encoding DNA repair proteins were also found, as expected in the strongly reducing and high-temperature hydrothermal vent environment. Potential adaptability was also seen in a cassette gene amplified from a gold mine that contained a homolog of a gene involved in the breakdown of a group of compounds used in mining *(76)*.

Members of the *Xanthomonas* genus provide an excellent example of adaptation driven by integrons. Xanthomonads are important phytopathogens causing infection in a number of different hosts including citrus, beans, cotton, and rice *(77)*. These infective Xanthomonads are surprisingly uniform, except they are believed to be capable of only causing disease on a specific plant. As a result, each plant-specific infective xanthomonad is classified as a different pathovar. Analysis of xanthomonad chromosomal integrons identified some interesting results. First, the integron is always in the same chromosomal position, indicating that it has co-evolved with its host for a substantial amount of time. Second, cassette arrays are conserved within individual pathovars. Finally, cassette arrays are fixed, at least to the extent that mutations within the integrase gene prevent cassette rearrangement by site-specific recombination *(53)*. These findings show that integrons have contributed to the diversification of the *Xanthomonas* genus into pathovars and may be responsible, at least in part, for the observed plant-specific pathogenesis, although this has yet to be shown experimentally.

## 8. Evolutionary Aspects of Gene Cassette Arrays

Gene cassette arrays are powerful "tools" in bacterial evolution as they provide the host with the ability to quickly acquire new gene functions and thereby adapt to a changing environment. This concept is supported by the faster rate at which gene cassette arrays evolve compared to the rest of the genome. *V. cholerae* pandemic strains with identical *rpoB* sequences can be separated by gene cassette composition to such a degree of sensitivity that even multi-locus sequence analysis (MLSA) cannot accomplish *(51)*. Furthermore, two *V. vulnificus* strains, with only 7-bp difference in their *rpoB* sequences, have vastly different cassette arrays *(62)*. The previous section briefly discussed the adaptive functions gene cassettes harbor; however, gene organization within the array was not discussed. Why would cassette organization in an array be of any importance? We will discuss a number of reasons as to why placement in a cassette is important and its relation to evolution of operons and gene cassette associated ORFs.

Cassette organization within an array can have an important impact on cassette gene expression. The majority of gene cassettes do not contain a promoter and therefore rely on the integron-associated promoter to drive expression of their associated genes. It is assumed, based on studies with class 1, class 3, and the *P. stutzeri* integrons, that a promoter ($P_C$) is contained in all integrons near *attI* and upstream of inserted cassette genes *(30, 78)*. However, even if $P_c$ is present in all integrons, it could at best drive expression of the first six or so cassettes. If it were assumed that only genes in cassettes close to *attI* are expressed, they would be the only genes to enhance the adaptive potential of the associated host. In this scenario, cassettes located in the distal end of large arrays would be functionally redundant and simply waiting for a shuffling event to relocate them closer to the $P_c$ promoter. This severely limits the ability of an integron to add to a host's adaptability, essentially reducing large arrays to "parking lots" for cassettes, and does not explain why large arrays are maintained.

Surprisingly, no study has been published specifically addressing gene expression in large cassette arrays. However, Yildiz et al. 2004 *(79)* did find that the majority of the cassette-associated genes were expressed under different conditions while looking at whole genome expression in *V. cholerae* N16961. This indicated that multiple promoters are present within at least this array. Several questions are then raised; are array promoters constitutive or induced under certain conditions? What is the location of these promoters? Do they make up part of other cassettes or are they within the *attC* sites? Clearly, further studies are required to identify these internal promoters, determine their strength and under what conditions, if any, they are induced. Putting aside these questions, shuffling of gene cassettes is commonly observed in the gene cassette array of *V. cholerae (51)* and could be a mechanism for "matching" a cassette or group of cassettes to a specific promoter. Such a mechanism would be a powerful tool for bacteria to induce co-expression of cassette genes under favorable conditions. Furthermore, sites in the array are free from promoters. For example, the *catB9* chloramphenicol-resistant gene cassette is silent in the *V. cholerae* N16969 array *(60)*. During the course of gene evolution, there may be intermediates that are toxic to the host cell. These "promoter free" sites could potentially be used by gene cassette ORFs to evolve through such mutations occasionally being "tested" through a shuffling event into a region of the array that allows expression.

Another interesting aspect of cassette arrays is that they often harbor multiple copies of the same cassette, with the level of cassette duplication related to cassette array size *(62)*. As an example, the large arrays of *V. cholerae* N16961 and *V. vulnificus* YJ016 contain 44 sets and 28 sets of paralogous groups, respectively *(62)*. Gene duplication is thought to play an important role in functional diversification *(80)*, as a duplicate gene copy allows the other to be freed from functional constraint, and therefore able to evolve new function. Is this occurring in gene cassette arrays?

Single gene cassettes can provide bacteria with selective advantage, such as the antibiotic-resistance genes in cassettes seen in clinical integrons or virulence genes in cassettes seen in *V. cholerae*. However, can integrons acquire cassettes with complementary functions (i.e., a gene cassette that produces an observable effect in an organism only in conjunction with another gene cassette(s)) or accumulate gene cassettes to create a complex metabolic pathway? It is difficult to answer these questions since most cassette genes encode unknown proteins, but given the functional diversity of cassettes and their acquisition, expression, and shuffling within arrays, cassettes of complementary function probably exist. It may be that complementary cassettes are physically unlinked (i.e., one cassette occupies a position in the array distant from the other cassette), but because they are co-expressed they are still able to carry out the complementary function. Intuitively, shuffling of gene cassettes is more likely in an array, allowing cassettes to be randomly linked and therefore potentially become co-regulated from one promoter. An example may exist in a pair of adjacent cassettes amplified from a hydrothermal vent *(75)*. These adjacent gene cassettes carry ORFs matching the first and second parts of the SecA-motif of nuclease-related domain proteins, suggesting a co-ordinated role. In the event that complementary cassettes become physically linked and co-expressed, their position with respect to one another would be retained by selection pressure. A permanent "lock together" of gene cassettes could be facilitated by inactivation of the *attC* site of the upstream cassette through deletion or mutation. Such a lock together has previously been observed in a class 1 integron where two resistance gene cassettes were found as a fused cassette due to deletion of the intervening *attC (81)*. Cassette fusion was also suggested for a gene cassette amplified from a hydrothermal vent containing an unprecedented four ORFs *(75)* as well as the *qacE* gene in the 3′-CS of class 1 integrons, which evolved from a cassette after deletion of the *attC* fixed its position *(25)*. Furthermore, in *V. cholerae* there are *attC* sites that are either mutated or partially deleted resulting in cassettes that are not fused, but dependent on coordinating their shuffling with adjacent cassettes *(16)*. The potential for complementary cassettes becoming locked together is an exciting prospect for gene cassette array evolution. However, understanding the extent to which complementary gene cassettes are present in gene cassette arrays will depend on the functional identification of cassette-associated ORFs. An interesting idea is that integron arrays mimic operons through orientation-specific insertion of cassettes. Have integrons played a part in operon evolution? This possibility is not hard to conceive, with pathways being constructed through successive acquisition and shuffling of gene cassettes. Operon-like structures could have been fixed through the subsequent loss of *attC* sites and then mobilized outside of the integron array. Recently, insertion sequences specifically targeting *attC* sites were identified in *P. stutzeri (82)*. Such insertion sequences would allow cassette(s) flanked by two insertion sequences to be mobilized.

## 9. Outlook

It is clear that integrons contribute to the evolution and adaption of bacteria in a changing environment. However, many more questions have arisen regarding the extent to which integrons/gene cassettes facilitate successful adaptation to new environments. Questions are posed about operon and gene evolution and the role of promoters in arrays. Furthermore, the origin of gene cassettes is still unknown. Perhaps, the biggest problem facing our understanding of the adaptive role of integrons relates to the high percentage of unknown proteins encoded within gene cassettes. This has been addressed multiple times in this chapter, but is emphasized again as a better understanding of the function of proteins encoded within gene cassettes will unveil much of their role in driving bacterial evolution.

## References


1. Collis, C., Hall, R. (1995) Expression of antibiotic resistance genes in the integrated cassettes of integrons. *Antimicrob Agents Chemother* **39**, 155–162.
2. Collis, C. M., Grammaticopoulos, G., Briton, J., Stokes, H. W., Hall, R. M. (1993) Site-specific insertion of gene cassettes into integrons. *Mol Microbiol* **9**, 41–52.
3. Recchia, G. D., Hall, R. M. (1995) Gene cassettes, a new class of mobile element. *Microbiology* **141**, 3015–3027.
4. Rowe-Magnus, D. A., Guerout, A.-M., Mazel, D. (1999) Super-integrons. *Res Microbiol* **150**, 641–651.
5. Leverstein-van Hall, M. A., Paauw, A., Box, A. T. A., Blok, H. E. M., Verhoef, J., Fluit, A. C. (2002) Presence of integron-associated resistance in the community is widespread and contributes to multidrug resistance in the hospital. *J Clin Microbiol* **40**, 3038–3040.
6. Leverstein-van Hall, M. A., Adrienne, T., A, Blok, H. E. M., Paauw, A., Fluit, A. C., Verhoef, J. (2002) Evidence of extensive interspecies transfer of integron-mediated antimicrobial resistance genes among multidrug-resistant *Enterobacteriaceae* in a clinical setting. *J Infect Dis* **186**, 49–56.
7. Maguire, A. J., Brown, D. F. J., Gray, J. J., Desselberger, U. (2001) Rapid screening technique for class 1 integrons in *Enterobacteriaceae* and nonfermenting gram-negative bacteria and its use in molecular epidemiology. *Antimicrob Agents Chemother* **45**, 1022–1029.
8. Martinez-Freijo, P., Fluit, A. C., Schmitz, F. J., Grek, V. S. C., Verhoef, J., Jones, M. E. (1998) Class 1 integrons in gram-negative isolates from different European hospitals and association with decreased susceptibility to multiple antibiotic compounds. *J Antimicrob Chemother* **42**, 689–696.
9. Sallen, B., Rajoharison, A., Desvarenne, S., Mabilat, C. (1995) Molecular epidemiology of integron-associated antibiotic resistance genes in clinical isolates of *Enterobacteriaceae. Microb Drug Resist* **1**, 195–202.
10. Tauch, A., Gotker, S., Puhler, A., Kalinowski, J., Thierbach, G. (2002) The 27.8-kb R-plasmid pTET3 from *Corynebacterium glutamicum* encodes the aminoglycoside adenyltransferase gene cassette *aadA9* and the regulated tetracycline efflux system Tet 33 flanked by active copies of the widespread insertion sequence IS*6100. Plasmid* **48**, 117–129.
11. Shi, L., Zheng, M., Xiao, Z., Asakura, M., Su, J., Li, L., Yamasaki, S. (2006) Unnoticed spread of class 1 integrons in gram-positive clinical strains isolated in Guangzhou, China. *Microbiol Immunol* **50**, 463–467.
12. Nesvera, J., Hochmannova, J., Patek, M. (1998) An integron of class 1 is present on the plasmid pCG4 from a gram-positive bacterium *Corynebacterium glutamicum. FEMS Microbiol Lett* **169**, 391–395.
13. Nandi, S., Maurer, J. J., Hofacre, C., Summers, A. O. (2004) Gram-positive bacteria are a major reservoir of class 1 antibiotic resistance integrons in poultry litter. *Proc Natl Acad Sci U S A* **101**, 7118–7122.
14. Xu, Z., Shi, L., Zhang, C., Zhang, L., Li, X., Cao, Y., Li, L., Yamasaki, S. (2007) Nosocomial infection caused by class 1 integron-carrying *Staphylococcus aureus* in a hospital in South China. *Clin Microbiol Infect* **13**, 980–984.
15. Boucher, Y., Labbate, M., Koenig, J. E., Stokes, H. (2007) Integrons: mobilizable platforms that promote genetic diversity in bacteria. *Trends Microbiol* **15**, 301–309.
16. Mazel, D. (2006) Integrons: agents of bacterial evolution. *Nat Rev Microbiol* **4**, 608–620.
17. Stokes, H. W., Holmes, A. J., Nield, B. S., Holley, M. P., Nevalainen, K. M. H., Mabbutt, B. C., Gillings, M. R. (2001) Gene cassette PCR: sequence-independent recovery of entire genes from environmental DNA. *Appl Environ Microbiol* **67**, 5240–5246.
18. Holmes, A. J., Gillings, M. R., Nield, B. S., Mabbutt, B. C., Nevalainen, K. M. H., Stokes, H. W. (2003) The gene cassette metagenome is a basic resource for bacterial genome evolution. *Environ Microbiol* **5**, 383–394.

19. Stokes, H. W., O'Gorman, D. B., Recchia, G. D., Parsekhian, M., Hall, R. M. (1997) Structure and function of 59-base element recombination sites associated with mobile gene cassettes. *Mol Microbiol* **26**, 731–745.
20. MacDonald, D., Demarre, G., Bouvier, M., Mazel, D., Gopaul, D. N. (2006) Structural basis for broad DNA-specificity in integron recombination. *Nature* **440**, 1157–1162.
21. Rowe-Magnus, D. A., Guerout, A.-M., Ploncard, P., Dychinco, B., Davies, J., Mazel, D. (2001) The evolutionary history of chromosomal super-integrons provides an ancestry for multiresistant integrons. *Proc Natl Acad Sci U S A* **98**, 652–657.
22. Paulsen, I. T., Littlejohn, T. G., Rådstrom, P., Sundström, L., Sköld, O., Swedberg, G., Skurray, R. A. (1993) The 3′ conserved segment of integrons contains a gene associated with multidrug resistance to antiseptics and disinfectants. *Antimicrob Agents Chemother* **37**, 761–768.
23. Sundström, L., Rådström, P., Swedberg, G., Sköld, O. (1988) Site-specific recombination promotes linkage between trimethoprim- and sulfonamide-resistance genes. Sequence characterization of *dhfrV* and *sulI* and a recombination active locus of Tn*21*. *Mol Gen Genet* **213**, 191–201.
24. Kholodii, G. Y., Mindlin, S. Z., Bass, I. A., Yurieva, O. V., Minakhina, S. V., Nikiforov, V. G. (1995) Four genes, two ends, and a *res* region are involved in transposition of Tn*5053*: a paradigm for a novel family of transposons carrying either a *mer* operon or an integron. *Mol Microbiol* **17**, 1189–1200.
25. Brown, H. J., Stokes, H. W., Hall, R. M. (1996) The integrons In0, In2 and In5 are defective transposon derivatives. *J Bacteriol* **178**, 4429–4437.
26. Rådstrom, P., Sköld, O., Swedberg, G., Flensburg, J., Roy, P. H., Sundström, L. (1994) Transposon Tn*5090* of plasmid R751, which carries an integron, is related to Tn*7*, Mu, and the retroelements. *J Bacteriol* **176**, 3257–3268.
27. Kamali-Moghaddam, M., Sundström, L. (2000) Transposon targeting determined by resolvase. *FEMS Microbiol. Lett.* **186**, 55–59.
28. Partridge, S. R., Recchia, G. D., Stokes, H. W., Hall, R. M. (2001) Family of class 1 integrons related to In4 from Tn*1696*. *Antimicrob Agents Chemother* **45**, 3014–3020.
29. Stokes, H. W., Nesbø, C., L, Holley, M., Bahl, I. M., Gillings, M. R., Boucher, Y. (2006) Class 1 integrons potentially predating the association with Tn*402*-like transposition genes are present in a sediment microbial community. *J Bacteriol* **188**, 5722–5730.
30. Collis, C. M., Kim, M.-J., Partridge, S. R., Stokes, H. W., Hall, R. M. (2002) Characterization of the class 3 integron and the site-specific recombination system it determines. *J Bacteriol* **184**, 3017–3026.
31. Arakawa, Y., Murakami, M., Suzuki, K., Ito, H., Wacharotayankun, E., Ohsuka, S., Kato, N., Ohta, M. (1995) A novel integron-like element carrying the metallo-B-lactamase gene $bla_{IMP}$. *Antimicrob Agents Chemother* **39**, 1612–1615.
32. Correia, M., Boavida, F., Grosso, F., Salgado, M. J., Lito, L. M., Cristino, J. M., Mendo, S., Duarte, A. (2003) Molecular characterization of a new class 3 integron in *Klebsiella pneumoniae*. *Antimicrob Agents Chemother* **47**, 2838–2843.
33. Xu, H., Davies, J., Miao, V. (2007) Molecular characterization of class 3 integrons from *Delftia spp*. *J Bacteriol* **189**, 6276–6283.
34. Hansson, K., Sundström, L., Pelletier, A., Roy, P. H. (2002) IntI2 integron integrase in Tn*7*. *J Bacteriol* **184**, 1712–1721.
35. Biskri, L., Mazel, D. (2003) Erythromycin esterase gene *ere(A)* is located in a functional gene cassette in an unusual class 2 integron. *Antimicrob Agents Chemother* **47**, 3326–3331.
36. Ramirez, M. S., Quiroga, C., Centrón, D. (2005) Novel rearrangement of a class 2 integron in two non-epidemiologically related isolates of *Acinetobacter baumanii*. *Antimicrob Agents Chemother* **49**, 5179–5181.
37. Ramirez, M. S., Vargas, L. J., Cagnoni, V., Tokumoto, M., Centrón, D. (2005) Class 2 integron with a novel cassette array in a *Burkholderia cenocepacia* isolate. *Antimicrob Agents Chemother* **49**, 4418–4420.
38. Ahmed, A. M., Nakano, H., Shimamoto, T. (2005) Molecular characterization of integrons in non-typhoid *Salmonella* serovars isolated in Japan: description of an unusual class 2 integron. *Antimicrob Agents Chemother* **55**, 371–374.
39. Heikkilä, E., Sundström, L., Skurnik, M., Houvinen, P. (1991) Analysis of genetic localization of the type 1 trimethoprim resistance gene from *Escherichia coli* isolated in Finland. *Antimicrob Agents Chemother* **35**, 1562–1569.
40. Lichtenstein, C., Brenner, S. (1982) Unique insertion site of Tn*7* in the *E. coli* chromosome. *Nature* **297**, 601–603.
41. Wolkow, C. A., DeBoy, R. T., Craig, N. L. (1996) Conjugative plasmids are preferred targets for Tn*7*. *Genes Dev* **10**, 2145–2157.
42. Barlow, R. S., Gobius, K. S. (2006) Diverse class 2 integrons in bacteria from beef

cattle sources. *J Antimicrob Chemother* **58**, 1133–1138.
43. Hochhut, B., Lotfi, Y., Mazel, D., Faruque, S. M., Woodgate, R., Waldor, M. K. (2001) Molecular analysis of antibiotic resistance gene clusters in *Vibrio cholerae* O139 and O1 SXT constins. *Antimicrob Agents Chemother* **45**, 2991–3000.
44. Sørum, H., Roberts, M. C., Crosa, J. H. (1992) Identification and cloning of a tetracycline resistance gene from the fish pathogen *Vibrio salmonicida*. *Antimicrob Agents Chemother* **36**, 611–615.
45. Szekeres, S., Dauti, M., Wilde, C., Mazel, D., Rowe-Magnus, D. A. (2007) Chromosomal toxin-antitoxin loci can diminish large-scale genome reductions in the absence of selection. *Mol Microbiol* **63**, 1588–1605.
46. Biskri, L., Bouvier, M., Guérout, A.-M., Boisnard, S., Mazel, D. (2005) Comparative study of class 1 integron and *Vibrio cholerae* superintegron integrase activities. *J Bacteriol* **187**, 1740–1750.
47. Vaisvila, R., Morgan, R. D., Posfai, J., Raleigh, E. A. (2001) Discovery and distribution of super-integrons among Pseudomonads. *Mol Microbiol* **42**, 587–601.
48. Holmes, A. J., Holley, M. P., Mahon, A., Nield, B., Gillings, M., Stokes, H. W. (2003) Recombination activity of a distinctive integron-gene cassette system associated with *Pseudomonas stutzeri* populations in soil. *J Bacteriol* **185**, 918–928.
49. Engelberg-Kulka, H., Glaser, G. (1999) Addiction modules and programmed cell death and antideath in bacteria cultures. *Ann Rev Microbiol* **53**, 43–70.
50. Prakash, D. P., Gerdes, K. (2005) Toxin-antitoxin loci are highly abundant in free-living but lost from host-associated prokaryotes. *Nucl Acids Res* **33**, 966–976.
51. Labbate, M., Boucher, Y., Joss, M. J., Michael, C. A., Gillings, M. R., Stokes, H. W. (2007) Use of chromosomal integron arrays as a phylogenetic typing system for *Vibrio cholerae* pandemic strains. *Microbiology* **153**, 1488–1498.
52. Rowe-Magnus, D. A., Guerout, A.-M., Biskri, L., Bouige, P., Mazel, D. (2003) Comparative analysis of superintegrons: Engineering extensive genetic diversity in the Vibrionaceae. *Genome Res* **13**, 428–442.
53. Gillings, M. R., Holley, M. P., Stokes, H. W., Holmes, A. J. (2005) Integrons in *Xanthomonas*: A source of species genome diversity. *Proc Nat Acad Sci U S A* **102**, 4419–4424.
54. Coleman, N., Tetu, S., Wilson, N., Holmes, A. (2004) An unusual integron in *Treponema denticola*. *Microbiology* **150**, 3524–3526.
55. Demarre, G., Frumerie, C., Gopaul, D. N., Mazel, D. (2007) Identification of key structural determinants of the IntI1 integron integrase that influence *attC* × *attI1* recombination efficiency. *Nucl Acids Res* **35**, 6475–6489.
56. Hall, R. M., Brookes, D. E., Stokes, H. W. (1991) Site-specific insertion of genes into integrons: role of the 59-base element and determination of the recombination cross-over point. *Mol Microbiol* **5**, 1941–1959.
57. Gravel, A., Messier, N., Roy, P. H. (1998) Point mutations in the integron integrase IntI1 that affect recombination and/or substrate recognition. *J Bacteriol* **180**, 5437–5442.
58. Melano, R. G., Petroni, A., Garutti, A., Saka, H., A, Mange, L., Pasterán, F., Rapoport, M., Rossi, A., Galas, M. (2002) New carbenicillin-hydrolyzing ß-lactamase (CARB-7) from *Vibrio cholerae* non-O1, non-0139 strains encoded by the VCR region of the *V. cholerae* genome. *Antimicrob Agents Chemother* **46**, 2162–2168.
59. Petroni, A., Melano, R. G., Saka, H., A, Garutti, A., Mange, L., Pasterán, F., Rapoport, M., Miranda, M., Faccone, D., Rossi, A., Hoffman, P. S., Galas, M. F. (2004) CARB-9, a carbenicillinase encoded in the VCR region of *Vibrio cholerae* non-01, non-0139 belongs to a family of cassette-encoded ß-lactamases. *Antimicrob Agents Chemother* **48**, 4042–4046.
60. Rowe-Magnus, D. A., Guerout, A.-M., Mazel, D. (2002) Bacterial resistance evolution by recruitment of super-integron gene cassettes. *Mol Microbiol* **43**, 1657–1669.
61. Recchia, G. D., Hall, R. M. (1997) Origins of the mobile gene cassettes found in integrons. *Trends Microbiol* **5**, 389–394.
62. Boucher, Y., Nesbo, C., Joss, M., Robinson, A., Mabbutt, B., Gillings, M., Doolittle, W. F., Stokes, H. (2006) Recovery and evolutionary analysis of complete integron gene cassette arrays from *Vibrio*. *BMC Evol Biol* **6**, 3.
63. Centrón, D., Roy, P. H. (2002) Presence of a group II intron in a multiresistant *Serratia marcescens* strain that harbors three integrons and a novel gene fusion. *Antimicrob Agents Chemother* **46**, 1402–1409.
64. Léon, G., Roy, P. H. (2003) Excision and integration of cassettes by an integron integrase of *Nitrosomonas europaea*. *J Bacteriol* **185**, 2036–2041.

65. Dai, L., Zimmerly, S. (2002) Compilation and analysis of group II intron insertions in bacterial genomes: evidence for retroelement behaviour. *Nucl Acids Res* **30**, 1091–1102.
66. Sunde, M. (2005) Class 1 integron with a group II intron detected in an *Escherichia coli* strain from a free-range reindeer. *Antimicrob Agents Chemother* **49**, 2512–2514.
67. Michael, C. A., Gillings, M. R., Holmes, A. J., Hughes, L., Andrew, N. R., P, H. M., Stokes, H. W. (2004) Mobile gene cassettes: a fundamental resource for bacterial evolution. *Am Nat* **164**, 1–12.
68. Ogawa, A., Takeda, T. (1993) The gene encoding the heat-stable enterotoxin of *Vibrio cholerae* is flanked by 123-bp direct repeats. *Microbiol Immunol* **37**, 607–616.
69. Franzon, V., L, Barker, A., Manning, P. (1993) Nucleotide sequence encoding the mannose-fucose-resistant hemagglutinin of *Vibrio cholerae* O1 and construction of a mutant. *Infect Immun* **61**, 3032–3037.
70. Barker, A., Clark, C. A. (1994) Identification of VCR, a repeated sequence associated with a locus encoding a hemagglutinin in *Vibrio cholerae* O1. *J Bacteriol* **176**, 5450–5458.
71. Barker, A., Manning, P. A. (1997) VlpA of *Vibrio cholerae* O1: the first bacterial member of the alpha 2-microglobulin lipocalin superfamily. *Microbiology* **143**, 1805–1813.
72. Kim, Y. R., Lee, S. E., Kim, C. M., Kim, S. Y., Shin, E. K., Shin, D. H., Chung, S. S., Choy, H. E., Progulske-Fox, A., Hillman, J. D., Handfield, M., Rhee, J. H. (2003) Characterization and pathogenic significance of *Vibrio vulnificus* antigens preferentially expressed in septicemic patients. *Infect Immun* **71**, 5461–5471.
73. Lee, J. H., Yang, S.-T., Rho, S.-H., Im, Y. J., Kim, S. Y., Kim, Y. R., Kim, M.-K., Kang, G. B., Kim, J. I., Rhee, J. H., Eom, S. H. (2006) Crystal structure and functional studies reveal that PAS factor from *Vibrio vulnificus* is a novel member of the Saposin-fold family. *J Mol Biol* **355**, 491–500.
74. Robinson, A., Guilfoyle, A. P., Harrop, S. J., Boucher, Y., Stokes, H. W., Curmi, P. M., Mabbutt, B. C. (2007) A putative house-cleaning enzyme encoded within an integron array: 1.8 Å crystal structure defines a new MazG subtype. *Mol Microbiol* **66**, 610–621.
75. Elsaied, H., Stokes, H. W., Nakamura, T., Kitamura, K., Fuse, H., Maruyama, A. (2001) Novel and diverse integron integrase genes and integron-like gene cassettes are prevalent in deep-sea hydrothermal vents. *Environ Microbiol* **9**, 2298–2312.
76. Nemergut, D. R., Martin, A. P., Schmidt, S. K. (2004) Integron diversity in heavy-metal-contaminated mine tailings and inferences about integron evolution. *Appl Environ Microbiol* **70**, 1160–1168.
77. Hayward, A. C. (1993) The Hosts of *Xanthomonas.* (J. G. Swings and E. L. Civerolo), Chapman & Hall, London, 1–17.
78. Coleman, N. V., Holmes, A. J. (2005) The native*Pseudomonas stutzeri* strain Q chromosomal integron can capture and express cassette-associated genes. *Microbiology* **151**, 1853–1864.
79. Yildiz, F. H., Liu, X. S., Heydorn, A., Schoolnik, G. K. (2004) Molecular analysis of rugosity in a *Vibrio cholerae* O1 El Tor phase variant. *Mol Microbiol* **53**, 497–515.
80. Hughes, A. L. (2005) Gene duplication and the origin of novel proteins. *Proc Natl Acad Sci U S A* **102**, 8791–8792.
81. Nass, T., Mikami, Y., Imai, T., Poirel, L., Nordmann, P. (2001) Characterization of In53, a class 1 plasmid- and composite transposon-located integron of *Escherichia coli* which carries an unusual array of gene cassettes. *J Bacteriol* **183**, 235–249.
82. Tetu, S. G., Holmes, A. J. (2008) A family of insertion sequences that impacts integrons by specific targeting of gene cassette recombination sites, the IS*1111-attc* group. *J Bacteriol* **190,** 4959–4970.
83. Robinson, A., Wu, P. S., Harrop, S. J., Schaeffer, P. M., Dosztányi, Z., Gillings, M. R., Holmes, A. J., Nevalainen, K. M., Stokes, H. W., Otting, G., Dixon, N. E., Curmi, P. M., Mabbutt, B. C. (2005) Integron-associated mobile gene cassettes code for folded proteins: the structure of Bal32a, a new member of the adaptable alpha+beta barrel family. *J Mol Biol* **346**, 1229–1241.
84. Nield, B. S., Willows, R. D., Torda, A. E., Gillings, M. R., Holmes, A. J., Nevalainen, K. M. H., Stokes, H. W., Mabbutt, B. C. (2004) New enzymes from environmental cassette arrays: Functional attributes of a phosphotransferase and an RNA-methyltransferase. *Protein Sci* **13**, 1651–1659.
85. Cameron, F. H., Groot Obbink, D. J., Ackerman, V. P., Hall, R. M. (1986) Nucleotide sequence of the AAD(2″) aminoglycoside adenyltransferase determinant *aadB*. Evolutionary relationship of this region with those surrounding *aadA* in R538-1 and *dhfrII* in R388. *Nucl Acids Res* **14**, 8625–8635.
86. Rowe-Magnus, D. A., Mazel, D. (1999) Resistance gene capture. *Curr Opin Microbiol* **2**, 483–488.

# Chapter 7

# Ancient Gene Transfer as a Tool in Phylogenetic Reconstruction


## Jinling Huang and Johann Peter Gogarten



## Abstract

Although horizontal gene transfer (HGT) is often considered as a disruptive force in reconstructing organismal phylogeny, it can also be a valuable phylogenetic tool. A gene in the net of life is often horizontally transferred to the ancestor of a major lineage. If the gene is retained in the recipient and its descendants, it will constitute a shared derived character and mark the recipient and all descendants as a monophyletic group. Additionally, phylogenetically informative HGTs also provide information about the sequence of emergence of involved taxa, because the donor organism must have emerged at least as early as the recipient. Here we review the recent applications of ancient HGT events in reconstructing organismal phylogeny as well as the promise and potential pitfalls of this approach.



**Key words:** Horizontal gene transfer, intracellular gene transfer, net of life, phylogenetic reconstruction.


## 1. Introduction

Phylogenetic classification is traditionally based on a furcating speciation process with vertical transmission from parents to offspring *(1)*. Relatedness among organisms is defined by the shared possession of evolutionary novelties (shared derived characters). Because horizontal gene transfer (HGT) transmits genes between distantly related organisms, the resulting gene phylogeny may differ significantly from the true organismal phylogeny. In fact, such incongruence between gene phylogeny and organismal phylogeny often constitutes the basis for the detection of gene transfer. Consequently, HGT is usually considered misleading in phylogenetic reconstruction *(2–4)*, creating difficulties in recovering


Maria B. Gogarten et al. (eds.), *Horizontal Gene Transfer: Genomes in Flux, vol. 532*



DOI 10.1007/978-1-60327-853-9_7 Springerprotocols.com

organismal genealogy on a single-gene basis. For example, the lack of resolution for the relationships among major eukaryotic groups is attributed partially to HGT in early eukaryotic evolution *(5,6)*. Whether HGT can ultimately overwhelm phylogenetic signals has been under vehement debate *(2, 7, 8)*.

Despite the outcry about HGT as a disruptive force, recent studies also show that HGT can be beneficial to phylogenetic reconstruction *(9–12)*. Because the requirement of resolution varies among taxonomic ranks, HGT among organisms at lower ranks will not significantly affect phylogenetic reconstruction at higher levels *(13)*. The acquired gene will also be transmitted vertically within the recipient taxon, unless a secondary gene transfer is involved. Therefore, although pervasive in prokaryotes and frequent in unicellular eukaryotes, HGT alone seldom erases the phylogenetic signal retained in a gene completely and, in many cases, traces of organismal phylogeny are still recognizable in a gene tree. On the other hand, because most acquired genes only persist in the recipient for a short period of time *(14)*, recognizable ancient HGT events actually represent rare landmarks in organismal phylogenesis. These ancient HGT events, similar to other rare genomic markers such as insertions, deletions, or rearrangements, can be valuable for the reconstruction of organismal phylogeny *(11)*.

## 2. HGT in the "Net of Life"

The potential utility of HGT in phylogenetic reconstruction was essentially realized in the 1990s and is often linked to the concept of a "net of life". This concept was largely based on frequent HGT among prokaryotes *(15, 16)* and has now gained support from genome analyses of various microbial eukaryotes *(17–20)*. In contrast to a strictly furcating speciation process, the "net of life" recognizes both vertical and lateral transmission as important factors in organismal phylogenesis. In the "net of life", vertical transmission usually dominates over the short term and phylogenesis predominantly follows a tree-like pattern *(8, 21)*, in particular for multicellular eukaryotes where HGT appears to be rare. On the other hand, the effects of gene transfer accumulate over the long term because of the gradual acquisition of foreign genes in a genome and because of the ratchet effect of homolog replacement *(22)*. Therefore, the phylogenesis pattern gradually transforms into a net such that lateral transmission intertwines with vertical transmission. The lateral transmission events also include symbioses that contributed greatly to eukaryotic evolution, including the formation of mitochondria and plastids from α-proteobacteria and cyanobacteria, respectively *(23)*. The recipient node of lateral

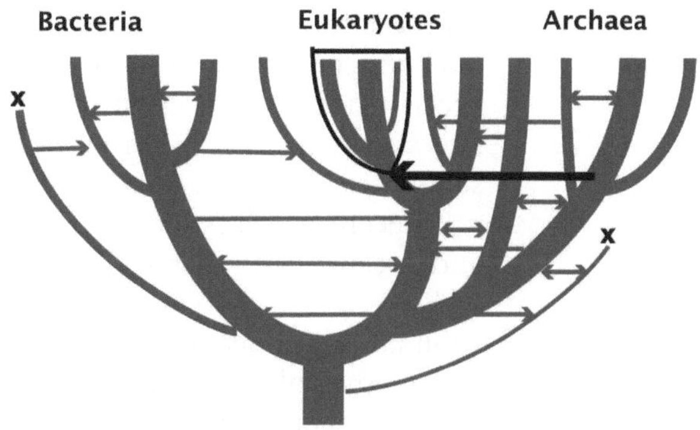


Fig. 7.1. An illustrative scheme of the net of life. Superimposed on a sketch of the phylogeny of the three domains of life, Bacteria, Eukarya, and Archaea, are cross-links that indicate gene transfers between organisms. Arrows of the links show the directions of gene transfer events. The density of the links is not scaled to the frequency of gene transfer. Crosses indicate extinct lineages. A natural group, recognizable by a transferred gene persisting in the recipient lineages, is highlighted in black. Modified from *Trends in Genetics*, 22:361–366 (with permission).

transmission in the "net of life" is often the ancestor of a major lineage, and the transmission event will implicitly mark the ancestral recipient and its descendants as monophyletic (**Fig. 7.1**) *(11)*. Additionally, such a lateral transmission event also implies that the donor must have existed prior to the divergence of the recipient *(11, 24)*. If the acquired gene is retained in all descendants, it will constitute a shared derived character in the recipient lineage.

The application of HGT as a phylogenetic marker works particularly well when a gross framework of organismal phylogeny is available and when it is combined with evidence from other aspects, such as morphology, anatomy, and other molecular characters. In the following section, we summarize a few examples that demonstrate the utility of HGT in phylogenetic reconstruction.

## 3. Case Studies

### *3.1. Relationships Between Red Algae and Green Plants*

Although both red algae and green plants contain plastids that are derived from a cyanobacterial endosymbiont, these two groups differ sharply in many anatomical and biochemical features and whether they are closely related is under debate. For example, red algae lack chlorophyll *b* and flagellated cells that are commonly found in green plants which, on the other hand, lack phycobilins that are present in red algae *(25)*. Red algae also contain Floridian starch as food reserves in the cytoplasm whereas green plants store starch in plastids. Thus far, plastid sequence data overwhelmingly support a single origin of red algae and green plants *(26)*. The results from analyses of individual nuclear genes and multigene sequence data are often ambiguous or sometimes conflicting *(27–30)*. To reconcile the incongruence between plastid and nuclear data, a hypothesis has recently been proposed based on the frequent gene transfer from plastids to the nucleus *(31)*. Such independent transfer events could have drawn red algal and green plant sequences into an apparent monophyly that does necessarily suggest a common ancestry for the carriers of the sequences (i.e., red algae and green plants) *(31)*.

Recent analyses identified multiple HGT events in red algae and green plants that support the two groups as monophyletic *(11, 30, 32)*. One of these HGT-derived genes encodes the β-subunit of topoisomerase VI (TOP6B), which has identifiable homologs only in archaea, photosynthetic eukaryotes (including red algae, green plants, cryptophytes, and haptophytes), and a few δ-proteobacteria (*Bdellovibrio* and *Anaeromyxobacter*) *(11)*. The TOP6B sequences of red algae, green plants, and other photosynthetic eukaryotes share many conserved residues with crenarchaeote homologs, and phylogenetic analyses further support a common origin of TOP6B sequences from all photosynthetic eukaryotes (**Fig. 7.2**). Because this gene is not found in cyanobacteria and α-proteobacteria, it is unlikely of plastidic or mitochondrial origin. The restricted distribution of TOP6B to photosynthetic eukaryotes also suggests that this gene is unlikely to have been present in the ancestor of all living organisms or in the ancestor of all eukaryotes. The most plausible scenario is that the top6B gene was transferred from archaea, where it is most widely distributed, to the ancestor of red algae and green plants *(11)*; the presence of this gene in other plastid-containing eukary-

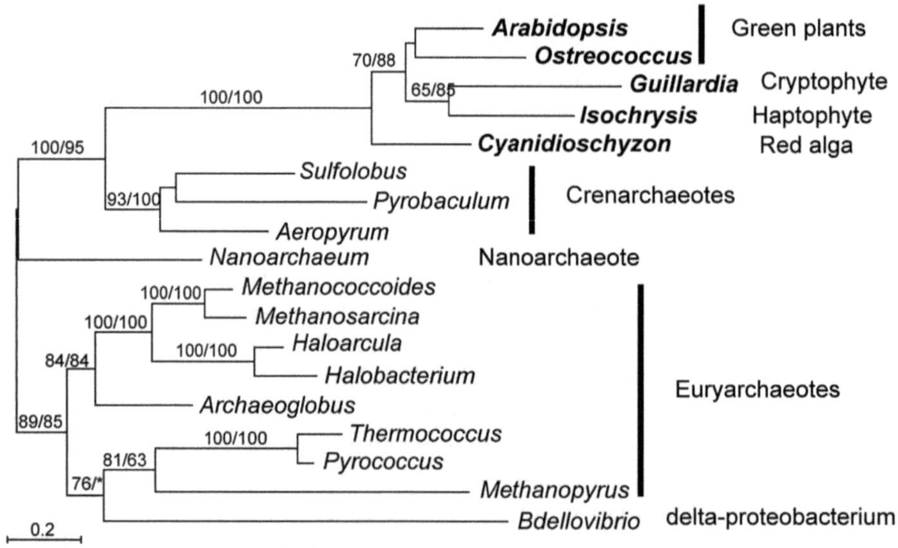


Fig. 7.2. Phylogenetic analyses of topoisomerase VI β-subunit protein sequences. Numbers above the branches show bootstrap values for maximum-likelihood analysis using PHYML *(57)* and distance analysis using *neighbor* of PHYLIP version 3.65 *(58)*, respectively. Asterisks indicate values lower than 50%. The depicted tree was calculated using PHYML based on the JTT substitution matrix and a mixed model of four gamma-distributed rate classes plus invariable sites. Bootstrap support was estimated using 100 replicates for both maximum likelihood and distance analyses.

otes might have resulted from secondary endosymbioses with an algal cell.

Another independent line of evidence for a common origin of red algae and green plants comes from chlamydiae-related genes in primary photosynthetic eukaryotes *(33, 34)*. Chlamydiae are a group of obligate intracellular bacteria commonly found in humans, insects, and *Acanthamoeba*. Although no chlamydial species has been reported in plastid-containing eukaryotes, it has been known that many chlamydial genes are similar to plant homologs that are often targeted to plastids *(35)*. Such sequence similarity between plants and chlamydiae was previously interpreted as resulting from either an ancestral relationship between the two groups or by various HGT scenarios *(33)*. In a stringent phylogenomic analysis of the red alga *Cyanidioschyzon*, 21 genes were identified as transferred between chlamydiae and primary photosynthetic eukaryotes (red algae, green plants, and glaucophytes), with the donor, in most cases, being similar to the environmental *Protochlamydia (33)*. Several of these chlamydiae-related genes lack identifiable homologs in cyanobacteria. Molecular phylogenies of most other genes indicate that chlamydial sequences form a clade distinct from cyanobacterial homologs, suggesting that the sequence similarity between chlamydiae, red algae, and green plants is not due to a close relationship between chlamydiae and cyanobacteria. Given that chlamydiae-like genes are mainly restricted to plastid-containing groups in eukaryotes and that all extant chlamydial species are obligate endosymbionts, the multiple acquired genes in red algae and green plants suggest an ancient symbiotic relationship between chlamydiae and the ancestor of primary photosynthetic eukaryotes *(33)*. This provides additional independent evidence that all primary photosynthetic eukaryotes are of a common origin.

### *3.2. Monophyly of Haptophytes and Cryptophytes*

Hapto- and cryptophytes are two of the several eukaryotic groups with chlorophyll *c*-containing plastids. Plastids of both groups are derived from an algal endosymbiont, whose remnant nucleus (the nucleomorph) still exists in cryptophytes *(36)*. Until recently, the systematic positions of cryptophytes and haptophytes were poorly understood. Analyses of plastid or plastid-derived sequence data did not suggest a specific affiliation of cryptophytes with any other group, except for a general placement within the large eukaryotic group of chromalveolates. On the other hand, haptophytes were often found to be related to heterokonts based on morphological and sequence data *(37–39)*.

A recent finding of a rare plastid gene displacement in cryptophytes and haptophytes provides evidence for a possible sister relationship between the two groups *(12)*. The *rpl36* gene of the cryptophyte *Guillardia* is only 144 bp in length, but it shares 3 indels with other cryptophytes (*Hanusia, Cryp-*

*tomonas*, and *Chroomonas*) and the haptophyte *Emiliania*. These sequences are similar to the homologs from the planctomycete *Rhodopirellula* and the $\gamma$-proteobacteria, but differ significantly from those of cyanobacteria and other plastid-containing eukaryotes. Because plastids are derived from cyanobacteria, such an apparent sequence dissimilarity between cryptophyte/haptophyte plastids and cyanobacteria would suggest a HGT in the evolution of cryptophytes and haptophytes, and this is also supported by further phylogenetic analyses *(12)*. It should be noted that, unlike nuclear and mitochondrial genomes that frequently pick up foreign genes in certain eukaryotic groups *(40)*, the acquisition of foreign genes is extremely rare in plastid genomes. Therefore, the displacement of the *rpl36* gene in cryptophytes and haptophytes constitutes a rare and shared derived character supporting the two groups as monophyletic *(12)*.

The ancient HGT of a bacterial *rpl36* gene to cryptophytes and haptophytes is the first strong evidence for a possible monophyly between the two plastid-containing groups. Such a relationship has subsequently gained independent support from more recent phylogenetic analyses using multigene sequence data *(41, 42)*.

### 3.3. Gene Transfer in Cyanobacteria

Concatenation of genes with different histories into a single dataset can result in phylogenies that reflect neither the history of individual genes nor that of the organisms. To avoid this problem, an approach based on analyses of embedded quartets was developed, which can detect conflicts and extract a plurality consensus without forcing all genes onto a single dataset *(43)* (*see* also **Chapter 11**). In an analysis of 11 complete cyanobacterial genomes, 55% of the gene families that had identifiable orthologs in at least 9 genomes contained one or more well-supported conflicts with the plurality consensus *(43)*. Despite these abundant conflicts, a clear unambiguous phylogenetic consensus also emerged: all of these plurality topologies for the different quartets are compatible with one another, and they are in perfect agreement with a single completely resolved cyanobacterial phylogeny.

Four of the analyzed genomes were from small coccoid marine cyanobacteria, including the marine *Synechococcus* WH8102 and three *Prochlorococcus marinus* strains. These marine cyanobacteria formed two clades: *P. marinus* CCMP1375 grouped with *P. marinus* MED4 and *P. marinus* MIT9313 with marine *Synechococcus* WH8102 *(43)*. Although these two clades were also recovered in an independent supertree analysis *(44)*, they are unlikely to represent organismal history. Marine *Synechococcus* and *Prochlorococcus* presumably evolved from a common phycobilisome-containing ancestor *(45)*. Many derived characters indicate that marine prochlorococci form a clade distinct from marine *Synechococcus (46)*, one of which is the presence of a threonyl-tRNA synthetase that was transferred to the ancestor of the *P. marinus* strains from a γ-proteobacterium (**Fig. 7.3**) *(43)*. One explanation for the conflict between the genome consensus and the many shared derived characters is that HGT among these genomes is so rampant that the plurality consensus no longer reflects organismal history.

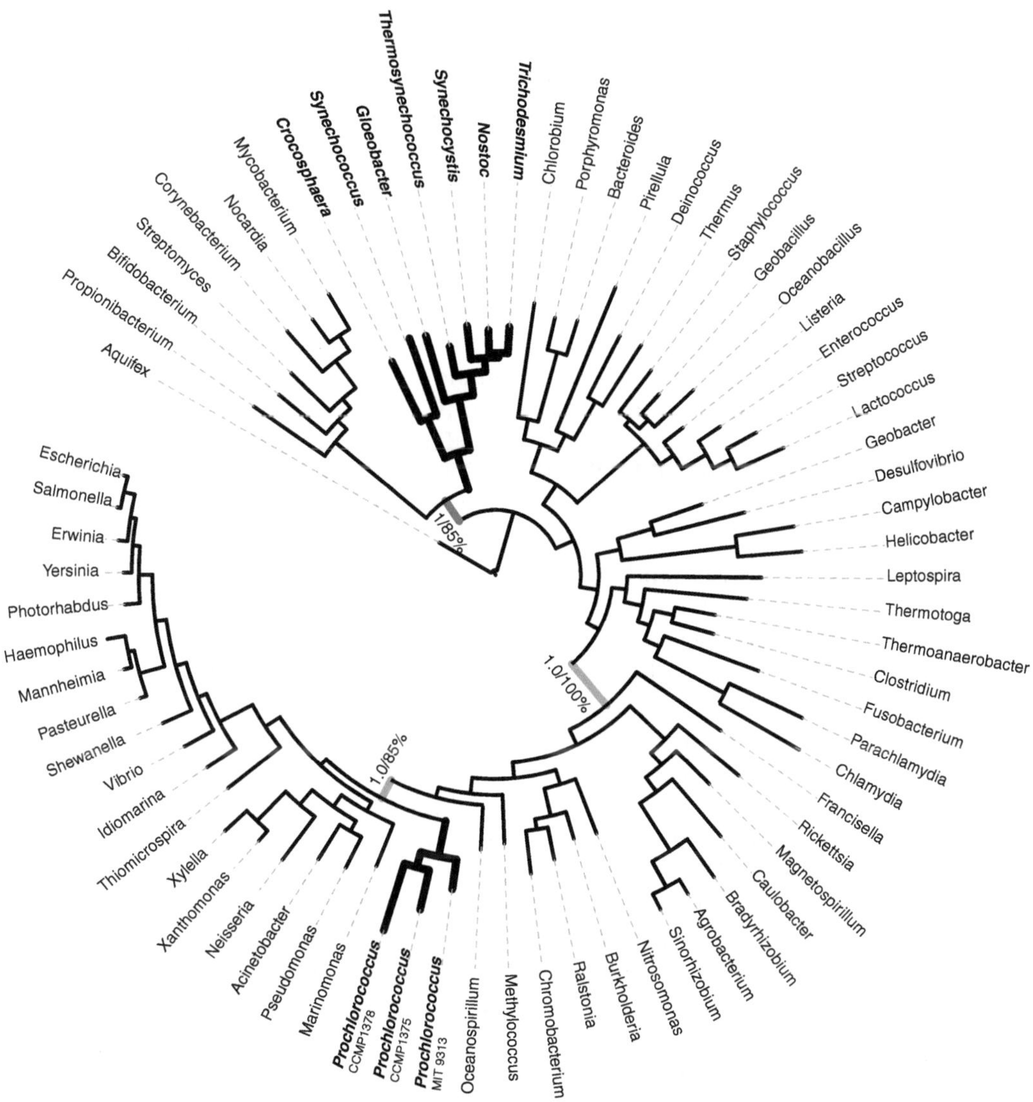


Fig. 7.3. Phylogenetic analyses of bacterial threonyl-tRNA synthetase. Sequences from cyanobacteria are indicated in *bold*. The depicted tree was calculated using MrBayes3.1 *(59)*, using model assuming the JTT substitution matrix and a gamma distribution to describe among site rate variation. MrBayes was run for 1,100,000 generations. Numbers give support values for the branches indicated in *gray*. The first gives posterior probability calculated with MrBayes, the second the percent bootstrap support calculated with PHYML using the same model. Note the sequences from the three *Prochlorococcus marinus* strains group within the gamma proteobacteria and not with the other cyanobacteria.

## 4. Promise and Pitfalls

Like all other approaches, the application of ancient HGT in phylogenetic reconstruction comes with promise and potential pitfalls. Any difficulties arising from uncertainties in identifying HGTs will potentially affect the utilization of HGT events as phylogenetic markers. In particular, taxonomic sampling, distinguishing HGT from IGT (i.e., intracellular gene transfer from organelles to the nucleus), independent gene acquisitions, and differential gene losses deserve careful consideration.

A small or biased taxonomic sample can lead to false positives for HGT identification. If the donor group is not sampled in the analysis, it is unlikely to be properly identified. For example, it was suggested that kinetoplastids once had a secondary algal endosymbiont based on the identification of "plant-like" genes in their genomes *(47)*. However, later phylogenetic analyses with larger taxonomic samples did not support this conclusion *(48–50)*. The difficulty in identifying the donor with a small or biased taxonomic sample can also translate into a problem in distinguishing HGT from IGT.

The identification of HGT is always complicated by the possibility of IGT. The existence of mitochondria and plastids in eukaryotes provides ample opportunities for continuous transfer of organellar genes into the nucleus. Because phylogenetic signals tend to fade over time, the donor of an acquired gene might be difficult to determine, and genes of organellar origin may be potentially mistaken as HGT-derived. In particular, because of the ubiquitous distribution of mitochondria in eukaryotes, it is often difficult to distinguish mitochondrion-derived genes from those that were transmitted from the ancestral eukaryotic nucleocytoplasm, or from genes that were anciently acquired from other prokaryotes. In this regard, recent analyses have used genes of archaeal origin or those encoded in the plastid genome *(9, 12, 33)*, thus effectively reducing the complications arising from IGT.

Independent gene acquisitions can lead to grouping of distantly related taxa on a gene tree. The same can be said for taxon-specific gene losses, which can always be invoked as an alternative explanation for HGT *(51)*. Independent gene losses or gains are often related to the environment and the lifestyle of an organism. If the transferred gene provides a selective advantage under certain conditions, independent gene acquisitions from the same donor or secondary gene transfer from one recipient to others can occur. For example, it has been suggested that *Entamoeba* and *Giardia* have independently acquired the same adaptations to an anaerobic lifestyle *(20)*. Additionally, HGT among eukaryotes has also been reported in multiple studies *(52–54)*.

The complications of small sample sizes and differential losses can be illustrated with the gene encoding tyrosyl-tRNA synthetase (tyrRS). In a previous analysis, two distinct tyrRS lineages were identified in eukaryotes: animal and fungal tyrRSs formed a well-supported monophyletic group with haloarchaeal homologs, whereas other eukaryotic tyrRSs formed another group *(10)*. These distinct eukaryotic tyrRS sequence lineages were interpreted as resulting from an ancient HGT from an archaeon to the ancestor of animals and fungi, providing support for a common origin of these two groups *(10)*. With the availability of additional sequences from amoebozoa and choanoflagellates, phylogenetic analyses still support the two distinct eukaryotic tyrRS groups. However, the amoebozoan tyrRS sequences are split among these groups: sequences of the archamoeba *Mastigamoeba* and the lobosean amoeba *Hartmannella* are related to animal and fungal homologs, whereas those of mycetozoa (*Dictyostelium*) and of another archamoeba, *Entamoeba*, are related to the other eukaryotic tyrRS group (**Fig. 7.4**). Such a tyrRS sequence phylogeny is in conflict with recent multigene analyses that often suggest a monophyly of amoebozoa *(55, 56)*. Additionally, the other eukaryotic tyrRS group also includes the sequence of the choanoflagellate *Monosiga* (**Fig. 7.4**), which is often considered to be related to animals.

With no eukaryotes containing copies from both tyrRS groups, it appears unlikely that the eukaryotic tyrRS gene phylogeny resulted from differential losses of duplicate gene copies in the eukaryotic ancestor. It is more likely that HGT was involved in the evolution of tyrRS. If choanoflagellates are indeed related to animals and amoebozoa are indeed monophyletic as current multigene analyses suggested, a plausible explanation is that an ancient HGT occurred at least prior to the split of amoebozoa, choanoflagellates, fungi, and animals. Under this scenario, the grouping of choanoflagellates, mycetozoa, and *Entamoeba* with the other eukaryotic tyrRS group might have resulted from losses of the acquired gene copy, whereas differential displacement of the existing tyrRS occurred in animals, fungi, and other amoebozoans.

It is unlikely that an HGT event will always create an immediate and total displacement of the existing homologs. Differential displacement of the endogenous copy occurs frequently and can lead to a scenario in conflict with the true organismal phylogeny. This has been shown in at least two chlamydiae-related genes, where some photosynthetic eukaryotes have the chlamydiae-related copy whereas others retain the plastid-derived version *(33)*. Therefore, it may be useful to treat the phylogenetic

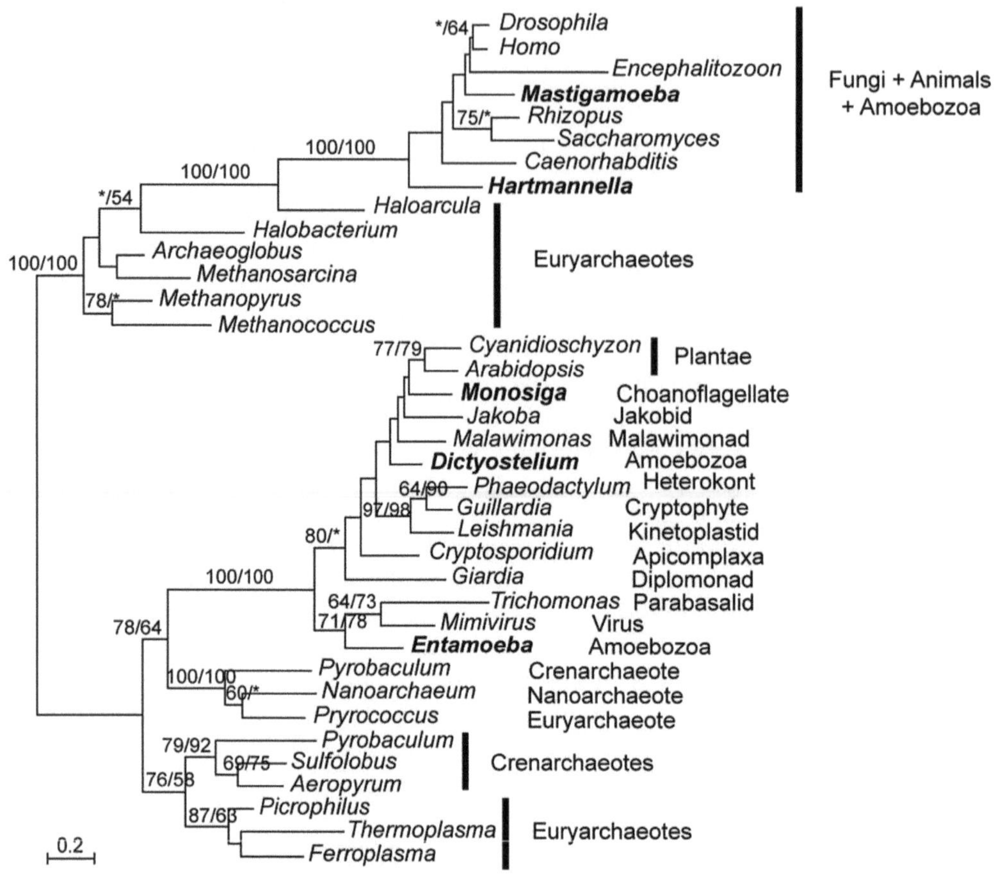


Fig. 7.4. Phylogenetic analyses of archaeal and eukaryotic tyrosyl-tRNA synthetase. Numbers above the branches show bootstrap values for maximum-likelihood analysis using PHYML and distance analysis using *neighbor* of PHYLIP version 3.65, respectively. Asterisks indicate values lower than 50%. The depicted tree was calculated using PHYML based on the JTT substitution matrix and a mixed model of 4 gamma-distributed rate classes plus invariable sites. Bootstrap support was estimated using 100 replicates for both maximum likelihood and distance analyses.

application of ancient HGT as event-based rather than sequence-based, because the acquired gene may not be retained in all descendants. Apparently, such an application of HGT events relies heavily on the availability of a gross framework of organismal phylogeny. In the case of tyrRS, if the amoebozoa indeed constitute a monophyletic group, the scenario of an ancient HGT event prior to the split of amoebozoa, animals, and fungi will provide support for a common origin of these groups. Similarly, in the case of chlamydiae-related genes in primary photosynthetic eukaryotes, it is best to treat all these genes as derived from a single endosymbiotic event rather than multiple shared derived characters. These HGT-generated characters provide independent evidence for organismal relationships, but their efficacy in reconstructing phylogenies often benefits from complementary knowledge of other aspects.

## Acknowledgments

We thank three anonymous reviewers for their comments and suggestions. This study was supported in part by a Research and Creative Activity Award from the East Carolina University to JH and through the NASA AISRP program (#NNG04GP90G) to JPG.

### References

1. Hennig, W. (1966) *Phylogenetic Systematics*, Univ. Illinois Press, Urbana, IL.
2. Doolittle, W. F. (1999) Phylogenetic classification and the universal tree. *Science* **284**, 2124–9.
3. Philippe, H., Douady, C. J. (2003) Horizontal gene transfer and phylogenetics. *Curr Opin Microbiol* **6**, 498–505.
4. Zhaxybayeva, O., Lapierre, P., Gogarten, J. P. (2004) Genome mosaicism and organismal lineages. *Trends Genet* **20**, 254–60.
5. Keeling, P. J., Burger, G., Durnford, D. G., Lang, B. F., Lee, R. W., Pearlman, R. E., Roger, A. J., Gray, M. W. (2005) The tree of eukaryotes. *Trends Ecol Evol* **20**, 670–6.
6. Parfrey, L. W., Barbero, E., Lasser, E., Dunthorn, M., Bhattacharya, D., Patterson, D. J., Katz, L. A. (2006) Evaluating support for the current classification of eukaryotic diversity. *PLoS Genet* **2**, e220.
7. Kurland, C. G., Canback, B., Berg, O. G. (2003) Horizontal gene transfer: a critical view. *Proc Natl Acad Sci U S A* **100**, 9658–62.
8. Gogarten, J. P., Doolittle, W. F., Lawrence, J. G. (2002) Prokaryotic evolution in light of gene transfer. *Mol Biol Evol* **19**, 2226–38.
9. Andersson, J. O., Sarchfield, S. W., Roger, A. J. (2005) Gene transfers from nanoarchaeota to an ancestor of diplomonads and parabasalids. *Mol Biol Evol* **22**, 85–90.
10. Huang, J., Xu, Y., Gogarten, J. P. (2005) The presence of a haloarchaeal type tyrosyl-tRNA synthetase marks the Opisthokonts as monophyletic. *Mol Biol Evol* **22**, 2142–6.
11. Huang, J., Gogarten, J. P. (2006) Ancient horizontal gene transfer can benefit phylogenetic reconstruction. *Trends Genet* **22**, 361–6.
12. Rice, D. W., Palmer, J. D. (2006) An exceptional horizontal gene transfer in plastids: gene replacement by a distant bacterial paralog and evidence that haptophyte and cryptophyte plastids are sisters. *BMC Biol* **4**, 31.
13. Woese, C. R. (2000) Interpreting the universal phylogenetic tree. *Proc Natl Acad Sci U S A* **97**, 8392–6.
14. Lawrence, J. G., Ochman, H. (1998) Molecular archaeology of the *Escherichia coli* genome. *Proc Natl Acad Sci U S A* **95**, 9413–7.
15. Hilario, E., Gogarten, J. P. (1993) Horizontal transfer of ATPase genes – the tree of life becomes a net of life. *Biosystems* **31**, 111–9.
16. Kunin, V., Goldovsky, L., Darzentas, N., Ouzounis, C. A. (2005) The net of life: reconstructing the microbial phylogenetic network. *Genome Res* **15**, 954–9.
17. Katz, L. A. (1999) The tangled web: gene genealogies and the origin of eukaryotes. *Am Nat* **154**, S137–S45.
18. Andersson, J. O., Roger, A. J. (2003) Evolution of glutamate dehydrogenase genes: evidence for lateral gene transfer within and between prokaryotes and eukaryotes. *BMC Evol Biol* **3**, 14.
19. Huang, J., Mullapudi, N., Sicheritz-Ponten, T., Kissinger, J. C. (2004) A first glimpse into the pattern and scale of gene transfer in Apicomplexa. *Int J Parasitol* **34**, 265–74.
20. Loftus, B., Anderson, I., Davies, R., Alsmark, U. C., Samuelson, J., Amedeo, P., Roncaglia, P., Berriman, M., Hirt, R. P., Mann, B. J., Nozaki, T., Suh, B., Pop, M., Duchene, M., Ackers, J., Tannich, E., Leippe, M., Hofer, M., Bruchhaus, I., Willhoeft, U., Bhattacharya, A., Chillingworth, T., Churcher, C., Hance, Z., Harris, B., Harris, D., Jagels, K., Moule, S., Mungall, K., Ormond, D., Squares, R., Whitehead, S., Quail, M. A., Rabbinowitsch, E., Norbertczak, H., Price, C., Wang, Z., Guillen, N., Gilchrist, C., Stroup, S. E., Bhattacharya, S., Lohia, A., Foster, P. G., Sicheritz-Ponten, T., Weber, C., Singh, U., Mukherjee, C., El-Sayed, N. M., Petri, W. A., Jr., Clark, C. G., Embley, T. M., Barrell, B., Fraser, C. M., Hall, N. (2005) The genome of the protist parasite *Entamoeba histolytica*. *Nature* **433**, 865–8.
21. Bapteste, E., Boucher, Y., Leigh, J., Doolittle, W. F. (2004) Phylogenetic reconstruction and lateral gene transfer. *Trends Microbiol* **12**, 406–11.

22. Doolittle, W. F. (1998) You are what you eat: a gene transfer ratchet could account for bacterial genes in eukaryotic nuclear genomes. *Trends Genet* **14**, 307–11.
23. Margulis, L. (1995) *Symbiosis in Cell Evolution: Microbial Communities in the Archean and Proterozoic Eons*, W H Freeman & Co, New York.
24. Gogarten, J. P., Murphey, R. D., Olendzenski, L. (1999) Horizontal gene transfer: pitfalls and promises. *Biol Bull* **196**, 359–61; discussion 61–2.
25. Raven, P. H., Evert, R. F., Eichhorn, S. E (1992) *Biology of Plants*, Worth Publishers, New York.
26. Palmer, J. D. (2003) The symbiotic birth and spread of plastids: how many times and whodunit? *J Phycol* **39**, 4–11.
27. Stiller, J. W., Hall, B. D. (1997) The origin of red algae: implications for plastid evolution. *Proc Natl Acad Sci U S A* **94**, 4520–5.
28. Rodriguez-Ezpeleta, N., Brinkmann, H., Burey, S. C., Roure, B., Burger, G., Loffelhardt, W., Bohnert, H. J., Philippe, H., Lang, B. F. (2005) Monophyly of primary photosynthetic eukaryotes: green plants, red algae, and glaucophytes. *Curr Biol* **15**, 1325–30.
29. Nozaki, H., Iseki, M., Hasegawa, M., Misawa, K., Nakada, T., Sasaki, N., Watanabe, M. (2007) Phylogeny of primary photosynthetic eukaryotes as deduced from slowly evolving nuclear genes. *Mol Biol Evol* **24**, 1592–5.
30. Reyes-Prieto, A., Bhattacharya, D. (2007) Phylogeny of Calvin cycle enzymes supports plantae monophyly. *Mol Phylogenet Evol* **45**, 384–91.
31. Stiller, J. W. (2007) Plastid endosymbiosis, genome evolution and the origin of green plants. *Trends Plant Sci* **12**, 391–6.
32. Richards, T. A., Dacks, J. B., Campbell, S. A., Blanchard, J. L., Foster, P. G., Mcleod, R., Roberts, C. W. (2006) Evolutionary origins of the eukaryotic shikimate pathway: gene fusions, horizontal gene transfer, and endosymbiotic replacements. *Eukaryot Cell* **5**, 1517–31.
33. Huang, J., Gogarten, P. (2007) Did an ancient chlamydial endosymbiosis facilitate the establishment of primary plastids? *Genome Biol* **8**, R99.
34. Tyra, H. M., Linka, M., Weber, A. P., Bhattacharya, D. (2007) Host origin of plastid solute transporters in the first photosynthetic eukaryotes. *Genome Biol* **8**, R212.
35. Brinkman, F. S., Blanchard, J. L., Cherkasov, A., Av-Gay, Y., Brunham, R. C., Fernandez, R. C., Finlay, B. B., Otto, S. P., Ouellette, B. F., Keeling, P. J., Rose, A. M., Hancock, R. E., Jones, S. J., Greberg, H. (2002) Evidence that plant-like genes in Chlamydia species reflect an ancestral relationship between Chlamydiaceae, cyanobacteria, and the chloroplast. *Genome Res* **12**, 1159–67.
36. Douglas, S., Zauner, S., Fraunholz, M., Beaton, M., Penny, S., Deng, L. T., Wu, X., Reith, M., Cavalier-Smith, T., Maier, U. G. (2001) The highly reduced genome of an enslaved algal nucleus. *Nature* **410**, 1091–6.
37. Cavalier-Smith, T., Allsopp, M. T., Chao, E. E. (1994) Chimeric conundra: are nucleomorphs and chromists monophyletic or polyphyletic? *Proc Natl Acad Sci U S A* **91**, 11368–72.
38. Yoon, H. S., Hackett, J. D., Pinto, G., Bhattacharya, D. (2002) The single, ancient origin of chromist plastids. *Proc Natl Acad Sci U S A* **99**, 15507–12.
39. Bachvaroff, T. R., Sanchez Puerta, M. V., Delwiche, C. F. (2005) Chlorophyll c-containing plastid relationships based on analyses of a multigene data set with all four chromalveolate lineages. *Mol Biol Evol* **22**, 1772–82.
40. Lang, B. F., Gray, M. W., Burger, G. (1999) Mitochondrial genome evolution and the origin of eukaryotes. *Annu Rev Genet* **33**, 351–97.
41. Hackett, J. D., Yoon, H. S., Li, S., Reyes-Prieto, A., Rummele, S. E., Bhattacharya, D. (2007) Phylogenomic analysis supports the monophyly of cryptophytes and haptophytes and the association of rhizaria with chromalveolates. *Mol Biol Evol* **24**, 1702–13.
42. Patron, N. J., Inagaki, Y., Keeling, P. J. (2007) Multiple gene phylogenies support the monophyly of cryptomonad and haptophyte host lineages. *Curr Biol* **17**, 887–91.
43. Zhaxybayeva, O., Gogarten, J. P., Charlebois, R. L., Doolittle, W. F., Papke, R. T. (2006) Phylogenetic analyses of cyanobacterial genomes: quantification of horizontal gene transfer events. *Genome Res* **16**, 1099–108.
44. Beiko, R. G., Harlow, T. J., Ragan, M. A. (2005) Highways of gene sharing in prokaryotes. *Proc Natl Acad Sci U S A* **102**, 14332–7.
45. Ting, C. S., Rocap, G., King, J., Chisholm, S. W. (2002) Cyanobacterial photosynthesis in the oceans: the origins and significance of divergent light-harvesting strategies. *Trends Microbiol* **10**, 134–42.
46. Partensky, F., Hess, W. R., Vaulot, D. (1999) Prochlorococcus, a marine photosynthetic prokaryote of global significance. *Microbiol Mol Biol Rev* **63**, 106–27.
47. Hannaert, V., Saavedra, E., Duffieux, F., Szikora, J. P., Rigden, D. J., Michels, P. A., Opperdoes, F. R. (2003) Plant-like traits associated with metabolism of Trypanosoma parasites. *Proc Natl Acad Sci U S A* **100**, 1067–71.

48. Huang, J., Mullapudi, N., Lancto, C. A., Scott, M., Abrahamsen, M. S., Kissinger, J. C. (2004) Phylogenomic evidence supports past endosymbiosis, intracellular and horizontal gene transfer in *Cryptosporidium parvum*. *Genome Biol* **5**, R88.
49. Rogers, M., Keeling, P. J. (2004) Lateral transfer and recompartmentalization of Calvin cycle enzymes of plants and algae. *J Mol Evol* **58**, 367–75.
50. El-Sayed, N. M., Myler, P. J., Blandin, G., Berriman, M., Crabtree, J., Aggarwal, G., Caler, E., Renauld, H., Worthey, E. A., Hertz-Fowler, C., Ghedin, E., Peacock, C., Bartholomeu, D. C., Haas, B. J., Tran, A. N., Wortman, J. R., Alsmark, U. C., Angiuoli, S., Anupama, A., Badger, J., Bringaud, F., Cadag, E., Carlton, J. M., Cerqueira, G. C., Creasy, T., Delcher, A. L., Djikeng, A., Embley, T. M., Hauser, C., Ivens, A. C., Kummerfeld, S. K., Pereira-Leal, J. B., Nilsson, D., Peterson, J., Salzberg, S. L., Shallom, J., Silva, J. C., Sundaram, J., Westenberger, S., White, O., Melville, S. E., Donelson, J. E., Andersson, B., Stuart, K. D., Hall, N. (2005) Comparative genomics of trypanosomatid parasitic protozoa. *Science* **309**, 404–9.
51. Gogarten, J. P., Townsend, J. P. (2005) Horizontal gene transfer, genome innovation and evolution. *Nat Rev Microbiol* **3**, 679–87.
52. Andersson, J. O., Sjogren, A. M., Davis, L. A., Embley, T. M., Roger, A. J. (2003) Phylogenetic analyses of diplomonad genes reveal frequent lateral gene transfers affecting eukaryotes. *Curr Biol* **13**, 94–104.
53. Richards, T. A., Dacks, J. B., Jenkinson, J. M., Thornton, C. R., Talbot, N. J. (2006) Evolution of filamentous plant pathogens: gene exchange across eukaryotic kingdoms. *Curr Biol* **16**, 1857–64.
54. Nosenko, T., Bhattacharya, D. (2007) Horizontal gene transfer in chromalveolates. *BMC Evol Biol* 7, 173.
55. Bapteste, E., Brinkmann, H., Lee, J. A., Moore, D. V., Sensen, C. W., Gordon, P., Durufle, L., Gaasterland, T., Lopez, P., Muller, M., Philippe, H. (2002) The analysis of 100 genes supports the grouping of three highly divergent amoebae: *Dictyostelium*, *Entamoeba*, and *Mastigamoeba*. *Proc Natl Acad Sci U S A* **99**, 1414–9.
56. Rodriguez-Ezpeleta, N., Brinkmann, H., Burger, G., Roger, A. J., Gray, M. W., Philippe, H., Lang, B. F. (2007) Toward resolving the eukaryotic tree: the phylogenetic positions of jakobids and cercozoans. *Curr Biol* **17**, 1420–5.
57. Guindon, S., Gascuel, O. (2003) A simple, fast, and accurate algorithm to estimate large phylogenies by maximum likelihood. *Syst Biol* **52**, 696–704.
58. Felsenstein, J. (2005) PHYLIP (Phylogeny Inference Package) Version 3.65. Distributed by the author. Department of Genome Sciences, University of Washington Place, Seattle, WA.
59. Ronquist, F., Huelsenbeck, J. P. (2003) MRBAYES 3: Bayesian phylogenetic inference under mixed models. *Bioinformatics* **19**, 1572–4.

# Chapter 8

## The Tree of Life Viewed Through the Contents of Genomes

**Christopher H. House**

### Abstract

A universal Tree of Life has been a longstanding goal of the biosciences. The most common Tree of Life, based on the small subunit rRNA gene, may or may not represent the phylogenetic history of microorganisms. The horizontal transfer of genes from one taxon to another provides a means by which each gene may tell of an independent history. When complete genomes became available, the extent to which horizontal gene transfer (HGT) has occurred became more evident. When using genomic data to study the Tree of Life, one can use any of the four broad approaches: (i) build lots of individual gene trees ("phylogenomics"), (ii) concatenate genes together for an analysis yielding one "supergene" tree, (iii) form a single tree based on the "gene content" within genomes using either orthologs or homologs, or (iv) investigate the order of genes within genomes to discern some aspects of microbial evolution. The application of whole genome tree building has suggested that there is a core tree, that such a core tree can be investigated using these varied methods, and that the results are largely similar to those of the rRNA universal Tree of Life. Some of the most interesting features of the rRNA tree, such as early diverging hyperthermophilic lineages are still uncertain, but remain a possibility. Genomic trees and geologic evidence together suggest that the vertical descent of genes and the horizontal transfer of genes between genetically similar lineages ultimately results in a core Tree of Life with at least some lineages that have phenotypic characteristics recognizable for billions of years.

**Key words:** Tree of Life, gene content, phylogenetics, phylogenomics, early Earth.

## 1. Introduction

Forming a universal Tree of Life, that is, a tree that encompasses completely or nearly all of life on Earth, has been a longstanding goal of the biological sciences. Early efforts, dating back to 1866 *(1, 2)*, include those of Ernst Haeckel (**Fig. 8.1**). Haeckel's trees are nice illustrations of the goal of representing all of evolution in a single figure, but they are dominated by the belief

Maria B. Gogarten et al. (eds.), *Horizontal Gene Transfer: Genomes in Flux, vol. 532*

DOI 10.1007/978-1-60327-853-9_8 Springerprotocols.com

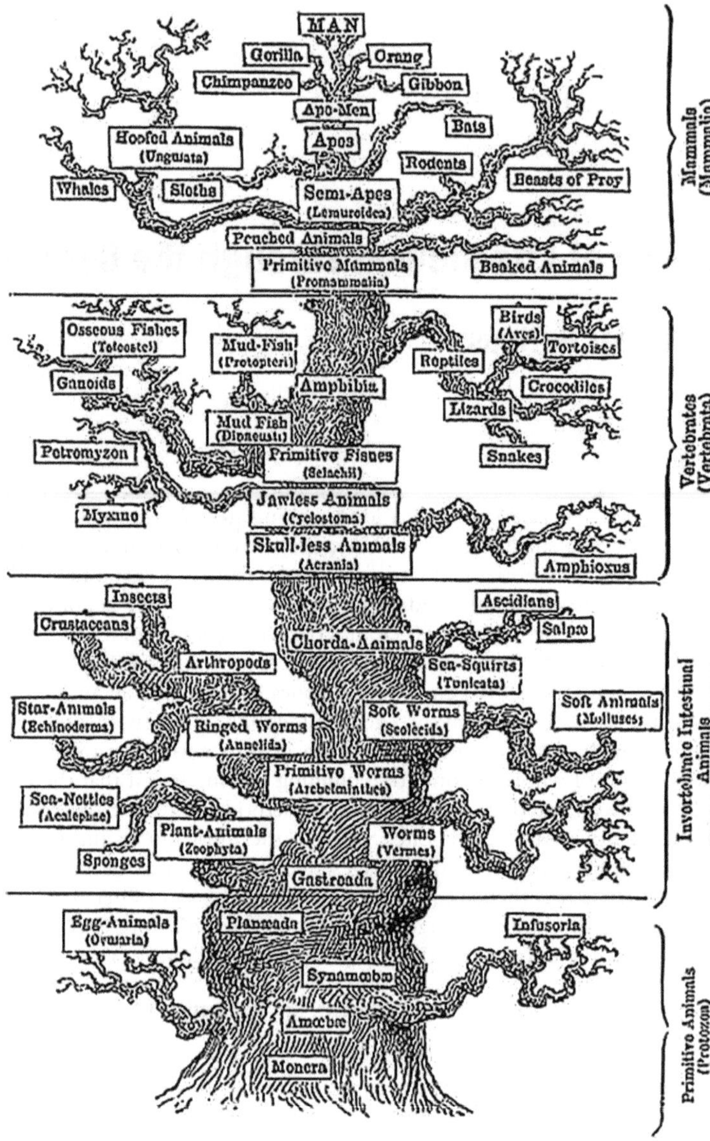


Fig. 8.1. Early Tree of Life with "Monera" at the base (*(1)*; image from later English translation *(2)*).

of directionality of evolution *(3)*, including the placement of Moneren (Prokaryotes) at the base of the tree and Humans at the top. Within Prokaryotes, there is no resolved branching. The first significant attempt to classify Prokaryotes, however, was that based on the cellular morphology proposed by Ferdinand Cohen *(4)* at about the same time Haeckel first published the tree shown in **Fig. 8.1**. By assigning known bacteria to four tribes composed of six genera, Cohen provided a useful taxonomic framework for later workers *(4, 5)*, but recognized that because his taxonomy was based on morphology rather than evolutionary relationships it probably was not a phylogeny. A phylogeny seeks to reconstruct the evolutionary relationships for a set of organisms, whereas a taxonomy is a classification of organisms, and so a taxonomy can be an attempt at phylogenetic reconstruction if, and only if, the effort is indented to recover evolutionary history. In 1909, Sigurd Orla-Jensen published a novel classification of prokaryotes *(6)*. In contrast to Cohen's earlier taxonomy, Orla-Jensen's classification (**Fig. 8.2**) was intended to represent a phylogenetic reconstruction of microorganisms *(5, 6)*. Orla-Jensen's phylogenetic tree was based on the concept that the earliest organisms on Earth would necessarily have been chemolithautotrophs, since they would have had to inhabit an environment devoid of light and organic nutrients. In his scheme, evolution proceeded from the original lithotrophic stock to give rise to bacteria having diverse nutritional requirements and increasingly complex habitats *(6)*. In contrast to the taxonomy of Cohen, Orla-Jensen's tree was based principally on physiology and used morphology only to define the highest groupings (i.e., domains).

Although in following decades acceptance of Ora-Jensen's phylogenetic reconstruction faded as the prevailing view shifted away from support for a necessarily autotrophic origin of life *(7, 8)*, phylogenetic reconstructions and/or taxonomic classifications continued to be proposed based on the combinations of morphological and physiological characters *(9, 10)*; among these were numerical (Adansonian) methods of phenetic tree construction (in which all identifiable characters are used to create a tree that therefore reflects overall similarity rather than true evolutionary relations; *(5)*). During this interval, debate among those concerned with the establishment of a true microbial phylogeny centered around which of various characters could be regarded as phylogenetically congruent and how the diverse types of relevant data could best be merged into a single, workable classification. The bacteriologist Robert G. E. Murray reasoned that because of their apparent morphological constancy and extreme complexity, characters of fine structural morphology were preferable *(11)* to the evidently more variable characters of physiology for reconstruction of phylogeny, whereas others argued that comparative biochemistry could play a useful role in classification and phylogenetic reconstruction *(12)*.

The rise of importance of protein and nucleic acid sequences for phylogenetic reconstruction includes such events as the sequencing of the protein insulin in 1955 *(13)*, the introduction of the concept of a molecular clock in 1962 *(14)*, the derivation of the Jukes-Cantor model for nucleotide substitutions in 1969 *(15)*, and Margaret O. Dayhoff's publication of the PAM substitution matrix based on the observed frequency by which different amino acids replace other amino acids during evolution *(16)*. By the late 1970s, pioneering work by Carl R. Woese and George E. Fox had shown that ribosomal RNA (rRNA) held promise for

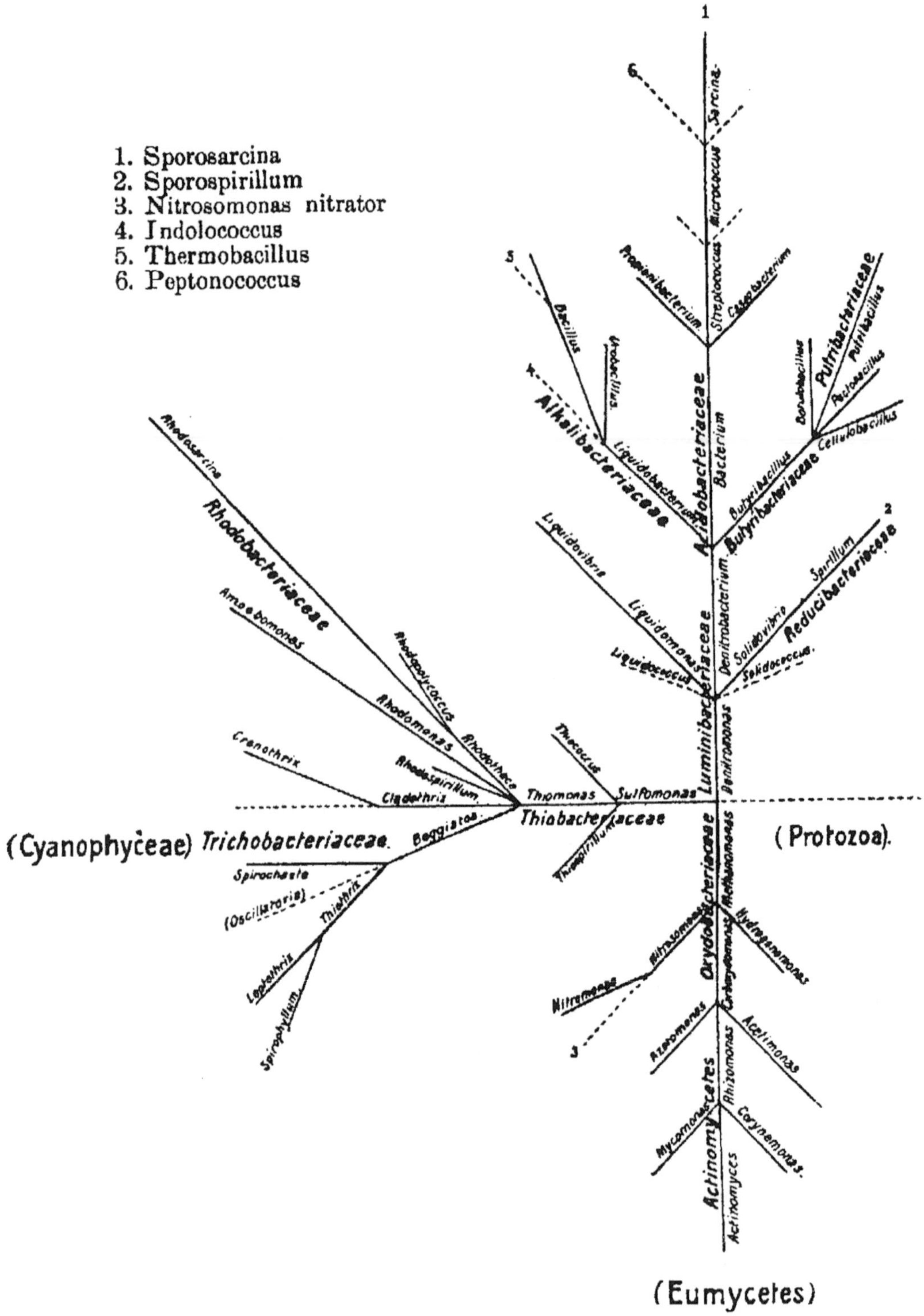


Fig. 8.2. Early 20th century microbial tree *(6)*.

phylogenetic reconstruction *(17)*, and by the end of the 1980s, analysis of universally conserved nucleic acid sequences (particularly those of the small subunit rRNA gene; *(18)*) had become a powerful tool for microbial taxonomy and phylogenetics. In particular, the analysis of small subunit rRNA genes allows for the identification of specific taxa on the basis of only a single gene sequence and provides a universal molecular phylogenetic Tree of Life (**Fig. 8.3**). In this case, the Tree of Life is built based on the evolutionary distance calculated between different gene sequences. In many ways, this universal Tree of Life has become the foundation on which rests much of our modern view of microbiology and, in particular, our view of the evolutionary history of microorganisms. The most robust conclusion that can be drawn from the rRNA universal tree is that there are three domains of life: the Bacteria, the Archaea, and the Eukaryotes *(19)*. With less certainty, the tree also shows the early diverging lineages of Bac-

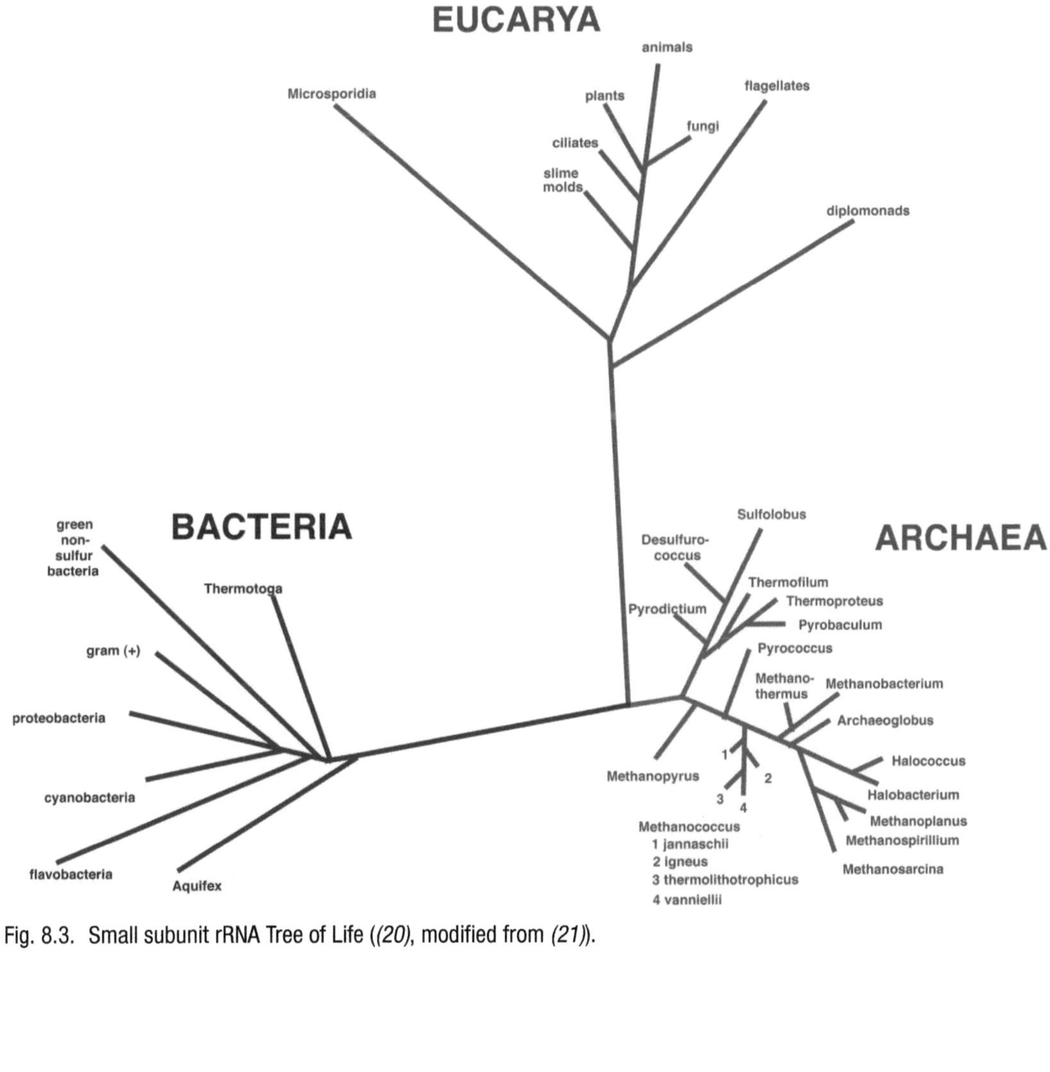


Fig. 8.3. Small subunit rRNA Tree of Life (*(20)*, modified from *(21)*).

teria and Archaea tend to be hyperthermophilic *(18, 20)*, that is, ones that have optimal growth temperatures 80°C or higher *(21)*.

In spite of its enormous impact and success, the small submit rRNA tree has inherent limitations. First, phylogenetic analyses of single gene sequences probably lack sufficient information to resolve much of the divergence pattern of the major microbial branches. Second, misalignment of sequences and differences of evolutionary rates among the various lineages can result in phylogenetic trees with the wrong topology *(22–24)*. Third, at its best, the tree really represents the evolutionary history of the small subunit rRNA gene and may or may not represent the phylogenetic history of microorganisms. The horizontal transfer of genes from one taxon to another provides a means by which each gene may tell of an independent history.

The problems created by HGT for microbial taxonomy and phylogeny were recognized early by Hartman and Goodgal, who in their review of the evidence for HGT concluded that "the ultimate solution for an enduring and stable taxonomic system would appear to reside only in thorough analysis at the genetic level; there is no reason to believe that such analysis is not possible in

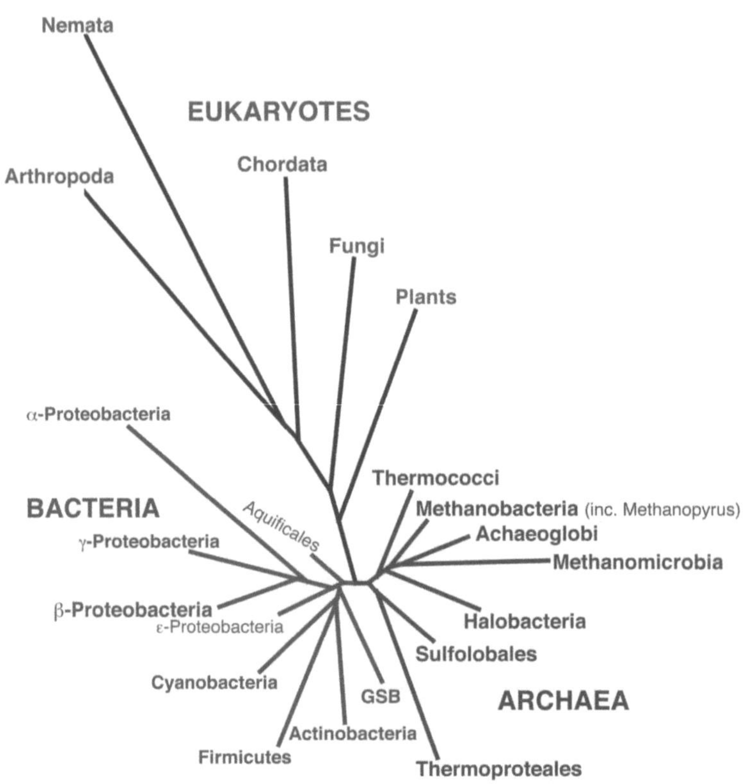


Fig. 8.4. Whole genome Tree of Life based on studies of the presence and absence of gene families (*(28)*; Data from *(29)*). The topology is based on discussions found in *(29)* and *(30)*.

any group of bacteria if looked for thoroughly and with proper techniques" *(25)*. It is likely that Hartman and Goodgal did not really anticipate the extent to which HGT might obscure taxonomic and phylogenetic signal, and it now seems that the "thorough analysis at the genetic level" that they called for is only possible with complete *(26)* or nearly complete genomic sequences *(27)*. Over the past decade, enough complete genome sequences have become available to allow for the construction of trees of life based on whole genomes. For example, **Fig. 8.4** shows a Tree of Life based on studies of the presence and absence of gene families (shown from *(28)*, based on *(29)* and *(30)*).

## 2. The Nature of Genomic Information

When complete genomes became available, the extent to which HGT has occurred became more evident. This resulted in concerns that genes are transferred with enough frequency as to seriously challenge the notion of a universal Tree of Life *(31)*. However, it was also pointed out in 2000 that the most similar archaeal genes and bacterial genes in databases are not 100% identical, but rather 74% *(32)*, suggesting that transfer of genes between these domains of life has not been rampant recently.

In any case, the movement of genes in and out of an organism over time raises the philosophical question of whether or not a lineage can represent the same organism if its parts (genes) keep changing. The question of identity in the face of changing parts has at different times been considered by philosophers such as Plutarch, Lock, and Hume *(33–35)*, and its relevance to gene flow has been noted by several molecular evolutionists *(36–38)* (Olsen, G., personal communication). In its oldest form, the philosophical debate is called Thesesus's ship *(35–37)*, named after the Greek legend where the ship of the hero Theseus is preserved, but over time each plank must one-by-one be replaced. After time, none of the original wood remains, presenting the classic paradox of identity. Is it still the Ship of Theseus, and if not, when did it cease to be? This paradox is remarkably applicable to our current debate over the Tree of Life. If gene flow is significant, then do organisms continue to be the same organism, and if not, at what point did the lineage stop representing the original organism? There is no easy solution to this paradox.

Gene transfer has happened, and presumably continues to happen. It is part of the genetic history of microbes, but so also are the organisms that have donated and received genetic material. At this point, it seems fair to establish that there can indeed be a Tree of Life, but that clearly each individual gene may have a history that is unique from other genes and is, in many cases, unique from the cellular line of descent (or the organismal history). The message for molecular biologists should be that there are many ways to treat genomic data and many ways to view the genetic history of microorganisms. The results of molecular evolutionary studies can appropriately result in either phenograms, which are branching diagrams that cluster taxa by estimates of overall similarity, or a cladogram, which technically refers to the trees constructed using cladisitc principles. Also, workers can, if needed, show gene transfers on trees *(39, 40)*, or replace the traditional tree structure with a network *(40–44)* or ring *(45)* to show the importance of HGT.

When using genomic data to study the Tree of Life, one can use any of the four broad approaches: (1) build lots of individual gene trees ("phylogenomics"), (2) concatenate genes together for an analysis yielding one "supergene" tree, (3) form a single tree based on the "gene content" within genomes, or (4) investigate the order of genes within genomes to discern some aspects of microbial evolution. Because of the complexity of molecular evolution, each of these broad methods is useful and correct, but at the same time, each of these methods has caveats and limitations.

## 3. Methods for Whole Genome Tree Building

### *3.1. Many Trees and Supertrees ("Phylogenomics")*

Because of HGT, each gene in a genome can, in principle, have a unique history that is different than the whole organism. In the phylogenomic approach, many trees are created using universal or semi-universal genes *(46–55)*. This method attempts to capture the overall picture of genetic evolution, including the individual histories of the genes studied. The method can also give the workers a sense of what the organismal history is, assuming that the plurality of genes has the same history as the overall organism. Because different genes and different sets of taxa often result in similar, but distinct trees, there are approaches for merging a number of rooted phylogenies into a single combined tree or a "supertree" *(54–57)*. The set of trees to be combined does not need to have the same taxa list and so a resultant supertree can indicate phylogenetic relationships that are not found on any single input tree. Also, supertree methods can offer a solution for cases where genes to be studied are not universally conserved or when only partial genomes are available.

The method's few drawbacks are that it is restricted to the genes common to many organisms (universal or semi-universal). The method requires reasonably good ortholog-prediction because certain paralogs, if mistaken for orthologs, will appear as horizontal transfers when they are used for tree building. Finally, some of the trees created will likely have topologies that are not correct (due to low signal in single genes or due to artifacts such as long branch attraction), and these trees may also incorrectly appear to show HGT. Interestingly, Galtier *(58)* developed a model for HGT and applied it to a bacterial and eukaryotic multigene data set. The model rejected HGT in the bacterial cases, but not in the eukaryotic cases *(58)*. This result likely is demonstrating that due to low signal in a single gene, discordant bacterial topologies are not necessarily HGT.

In application, the phylogenomic approach seems to be quite effective at recovering a core tree by assuming that the bulk of the genes follow a similar evolutionary history *(59)*. It has been noted, however, that when multiple gene markers are used for a phylogeny, inadequate sampling or the loss of lineages over time can, in principle, produce well-supported, incongruent trees that imply evolutionary events that may not have actually occurred *(42)*.

### 3.2. Concatenation of Genes ("Supergene" or "Supermatrix" Approaches)

The next approach is based on trying to maximize the amount of signal available for building a universal tree based on the sequences of universal or semi-universal genes. In this approach, one combines (concatenates) many aligned genes together to make one long sequence that can be used for phylogenetic tree building *(60–65)*. The advantage of this method is that it draws upon a high quantity of data, from which it can build a single Tree of Life. The tree formed presumably represents organismal history.

The drawbacks of this method are that it philosophically requires little or no HGT among the genes used because the signals from all of the genes are being combined. In reality, however, the signal from the bulk of the genes often overwhelms the signal of a small number of transferred genes. Like the phylogenomic approach, the supergene approach is restricted to genes common to the organisms being studied, which can be quite a small number of genes *(66)*, and the method requires good ortholog-prediction. Finally, as is true of nearly all phylogenetic methods, the resulting tree could still be incorrect due to artifacts such as long-branch attraction.

Recently, after testing several methods of genomic tree building, Dutilh et al. *(59)* concluded that the concatenation of genes (referred to superalignment in their paper) was the most successful method for recovering a known tree topology of fungal taxa. This paper also found that the multi-tree approach of phylogenomics was also quite successful, and that the gene content method tested was the least successful because it demonstrated a phenotypic bias and was also likely influenced by the large-scale duplication or loss of genes *(59)*.

### 3.3. Gene Content

The use of gene content to investigate the Tree of Life moves away from individual residues (such as nucleic bases or amino acids) as characters for tree building, opting rather for the use of genes (or even segments of protein genes) as characters for analysis *(67–80)*. The strengths of these methods are that they can draw from a high quantity of data (close to the entire genome), and that they likely target the cellular history of the microbes rather than the history of a subset of genes.

There is not a single gene content method. Rather, this is a broad category of genomic analysis that includes a wide variety of implemented methods, each with its own individual assumptions, strengths, and weaknesses. Typically, the drawbacks to these approaches are that they miss HGT events, and are usually limited to some degree by ortholog-prediction. Also, these methods usually have inherent biases created by highly derived genomes or reduced genomes *(27, 59)*. Finally, as is true of nearly all phylogenetic methods, the result tree could still be incorrect due to artifacts such as long-branch attraction.

#### 3.3.1. Ortholog Methods

The first category of gene content method is the use of the presence and absence of orthologous genes within the genomes studied *(68)*. For example, Snel et al. *(68)* used the fraction of orthologs shared between two genomes divided by the smallest genome to initially define evolutionary similarity. This can be converted to evolutionary distance either as $d = -\ln(s)$ or $d = 1 - s$ *(74)*. Also, clusters of orthologous genes have been used to construct phylogenies, first for the herpies virus *(71)*. More recently, genome trees using distance defined by mean BLAST score for orthologs shared between a pair of genomes have been constructed *(73, 80)*. By defining distance in this way, their method captures sequence divergence and is not adversely affected by genome-size differences.

In general, ortholog methods seem to be quite effective at recovering an overall average of the different phylogenetic histories for the genes in the genomes studied. Further, because in these methods "evolutionary distance" is usually based on the proportion of orthologs shared between two genomes divided by the size of the smallest genome of the two, these methods are somewhat resistant to artifacts caused by differing genome size. The drawbacks specific to ortholog methods are that reasonably good ortholog prediction is required (in most cases). Furthermore, these methods can, in principle, be greatly influenced by HGT (as recently transferred genes will appear as orthologs) and by the loss of shared genes or the duplication of unshared genes *(81)*.

The use of shared orthologs between genomes or the presence and absence of clusters of orthologs have become the most widespread gene content method used for genome tree building. It is relatively simple to implement once the necessary BLAST or FASTA searches have been performed. Furthermore, for cases where pairwise genome distances are used, the calculation for each genome pair needs only to be run once, even as more genomes are added in the future.

Tekaia and Yeramian *(78)* have recently reported the construction of trees based on gene *conservation profiles*, where a conservation profile is a binary representation of the presence and absence of an ortholog among a group of species. In their work, a conservation profile is represented by a vector of 0 and 1 corresponding respectively to the presence or absence of an ortholog in a given species. If *n* species are considered for comparisons, the conservation profile is of dimension *n*. Considering among all gene conservation profiles solely those that are distinct, in principal, corresponds to the set of different evolutionary histories that can be used to construct a tree. The main advantage of the conservation profile is that the evolutionary history is represented by a multidimensional vector that encompass jointly all considered species, whereas other methods are based on pairs of species. The initial application of this method to form a Tree of Life *(78)* looks promising with good clustering of various phyla, but also a likely artifactual clustering of greatly reduced genomes.

#### *3.3.2. Homolog Methods*

The second category of gene content method is the use of the presence and absence of gene families (homologs) within the genomes studied. This category of analysis is based on using gene families, protein domains, or protein folds as the characters on which the analysis is based *(67, 69, 70, 76, 79)*.

Homolog methods are resistant to influences of many cases of lateral gene transfer, because only the transfer of novel gene families can influence the observed distribution of characters in the tree building process. Furthermore, the expansion of certain gene families through duplication, or the reduction of the size of certain gene families through limited gene loss, has no influence on the tree building process because all gene families are treated equally regardless of their size.

Homolog methods can be adversely affected by greatly reduced genomes *(27)*. In some cases, the problem of reduced genomes adversely affecting these analyses has been addressed by adding in a weighting factor that takes into account differences in genome size *(79)*. Gene loss in greatly reduced genomes, however, is not random, but rather tends to result in genomes retaining a set of genes involved in highly conserved biological processes, and thus can produce clusters of microbial symbionts and parasites that are presumably artifacts. Interestingly, Rohwer and Edwards *(82)* used BLAST hits above a threshold shared between genomes as a measure of genome distance to look at a wide range of phage diversity *(82)*. In that case, they used a penalty for any gene found not to have a match in a genome pair to address problems that otherwise arose with the quite small size of phage genomes.

### 3.4. Gene Order

Several gene order methods depend on the presence of orthologs adjacent to each other. Watterson et al. *(83)* introduced the concept of the *breakpoint distance* between two genomes, which is the number of orthologs found paired together in one genome but separated in the other *(84)*. Recently, the presence and absence of paired genes have been used to construct trees *(62, 74)* as a gene order method similar in practice to tree building by gene content. A limitation to this approach results from the fact that small groups of laterally transferred genes will be paired after their transfer. Also, a computational method for testing phylogenetic problems using gene order has been presented by Kunisawa *(85)*. In this method, genomes are searched for cases in which the arrangement of three genes most parsimoniously suggests that a single transposition has occurred. With the use of an outgroup, the method can be used to test phylogenetic hypotheses, such as the branching order within the Proteobacteria *(85)* or Gram-positive bacteria *(86)*. The strength of this method is that it can be efficiently applied to a large dataset of genomes and that it reveals (a small number of) interesting cases of transposition. Another gene order approach often implemented is calculating the inversion distance. The *inversion distance* is the minimum possible number of inversions needed to transform one genome into the other *(87)*.

Recently, Belda et al. *(88)* have studied a subset of 244 genes universal to the genomes of 30 $\gamma$-Preotobacteria, using both the breakpoint distance and the inversion distance. They found the two distances highly correlated, suggesting that inversion was the main method of genome rearrangement for these taxa. Also, House and Fitz-Gibbon *(89)* have used a Monte Carlo approach to search for small but significant similarities in gene order. In doing so, they find some support for the uniting of the Actinobacteria and Firmicutes together into a Gram-positive clade.

## 4. Outlook

Largely, the topology of the Tree of Life and the taxonomic grouping of microorganisms, as revealed by the analysis of small subunit rRNA genes, appears quite consistent with that found by most genomic trees. When comparing genomic methods to the rRNA tree, the broad prokaryotic groups (domains, phyla, classes, etc.) largely remain intact, and the resolution of the prokaryotic phyla is usually found to be similar, but still largely uncertain with respect to branching order.

One of the most intriguing features of the current Tree of Life is the placement of the hyperthermophilic Thermotogales and Aquificales near the base of the bacterial domain *(20)*. This placement is by no means certain, but nevertheless, this placement has been one of the principal pieces of evidence cited for a hyperthermophilic last common ancestor *(90)* and potentially a hyperthermophilic origin of life (e.g., *(91–97)*). Both of these important lineages are still found at the base of the Bacteria in most genomic trees, assuming that the root of the Bacteria is correct *(98)*. Unfortunately, *Aquifex* has a relatively small genome, and so, its placement on the Tree of Life by gene content methods cannot be considered definitive. *Thermotoga*, on the other hand, does not point definitively to a hypetherthermophilic common ancestor for the Bacteria, because (i) it seems to have received considerable horizontal transfer of genes from hyperthermophilic Archaea *(99)*, and (ii) it has close relatives (*Geotoga* and *Petrotoga*) that have only thermophilic growth temperatures *(100)*. So, while genomic trees seems to support the placement of the Thermotogales and Aquificales near the base of the bacterial tree, the question of the temperature optimum for the base of the Bacteria will remain uncertain for now. If the present Tree of Life topology is correct, it can be explained through several different early Earth scenarios with largely different implications for the early biosphere and the last common ancestor (**Fig. 8.5**). The options include a hyperthermophilic origin of life (**Fig. 8.5A**), independent colonization of hyperthermophilic environments (**Fig. 8.5B**), and the cataclysmic mass extinction of non-hyperthermophiles by a major early Earth impact (**Fig. 8.5C** and **D**; *(101, 102)*). This last option can either leave a hyperthermophilic last common ancestor (**Fig. 8.5C**; *(103)*) or result in two distinct domains (Bacteria and Archaea), each descended from a different surviving lineage (**Fig. 8.5D**; *(104)*). Islas et al. *(105)* provide a fuller discussion of the possible early biosphere implications of a Tree of Life with hyperthermophilic basal lineages.

The most notable example of genomic trees suggesting a tree topology that is not found by small subunit rRNA genes is the uniting of methanogens into a derived archaeal clade *(106–108)*. This result suggests that methanogenesis was not the metabolism for the base of the Archaea. Based on these genomic trees, the early archaeal cells appear to have been sulfur *(29)* and/or Fe-reducing organisms *(109)*. The plausibility of sulfur-reduction as an early metabolism is bolstered by the fact that sulfur isotopes from Archean rocks indicate through their mass-independent signatures that elemental sulfur was being deposited from the atmosphere and being reduced to sulfide *(110–112)*. Sulfur-reduction is common in the Archaea where it is a widespread

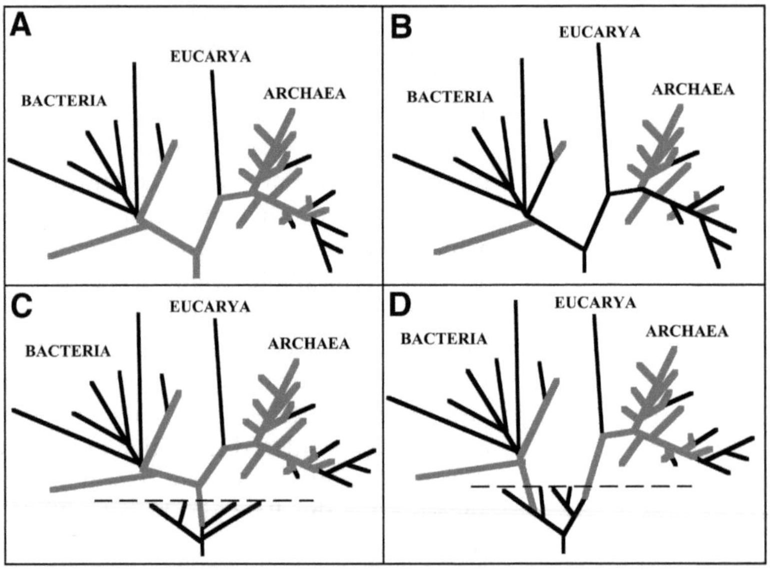


Fig. 8.5. Assuming that rRNA trees are largely correct in their topology, the various options for explaining the distribution of hyperthermophlic lineages (shown in *gray bolded lines*). (**A**) A hyperthermophilic last common ancestor followed by the evolution of non-hyperthermophilic lineages *(96)*. (**B**) Non-hyperthermophilic last common ancestor followed by colonization of hyperthermophilic environments in a number of different lineages. (**C**) Impact-induced mass extinction resulting in a hyperthermophilic last common ancestor *(103)*. (**D**) Impact-induced mass extinction leaving two distinct surviving lineages descended from a non-hyperthermophilic last common ancestor *(104)*.

chemolithotrophic metabolism in lineages of the Crenarchaeota, as well as being present in the heterotrophic euryarchaeal Pyrococcales. Furthermore, the phylogenetically uncertain *Thermoplasma* are also capable of sulfur-reduction. The widespread and basal positions of sulfur reducers support the early origin of sulfur reduction, although it is also possible that sulfur reduction genes were spread among these taxa at a later date by HGT *(113)*, and even though methanogens appear to be derived, very early cells still might have had a simple metabolism that is a direct precursor to the extant methanogenesis pathway *(114)*.

The outlook for using whole genomic information for exploring the Tree of Life (phylogenomics, supergenes, gene content, and gene order) is quite good. The work to date supports the general notion that there is a central Tree of Life and that genomic information is useful for attempting to determine the topology of such a tree. However, it is not certain, at this time, what the central Tree of Life represents. At face value, it would suggest that vertical descent of genes is considerably more common than HGT. However, it also might reflect the reality that the frequency of gene transfer between taxa is correlated with their genetic similarity. In either case, the central Tree of Life would remain a meaningful representation of phylogenetic history and diversity, and many of its branches would to a degree represent genomically stable lineages.

There are three lines of geologic evidence supporting the notion that certain lineages have been stable for a billion years or so. First, the geologic record of carbon isotopic fractionation between inorganic carbon and organic carbon (**Fig. 8.6A**) shows large-scale and largely consistent biological carbon fixation in the Earth's oceans. While this does not require the lineages responsible for the ocean carbon fixation to be constant over time, it does suggest a degree of stability in the broadest scope of microbial ecology. The most notable deviation in

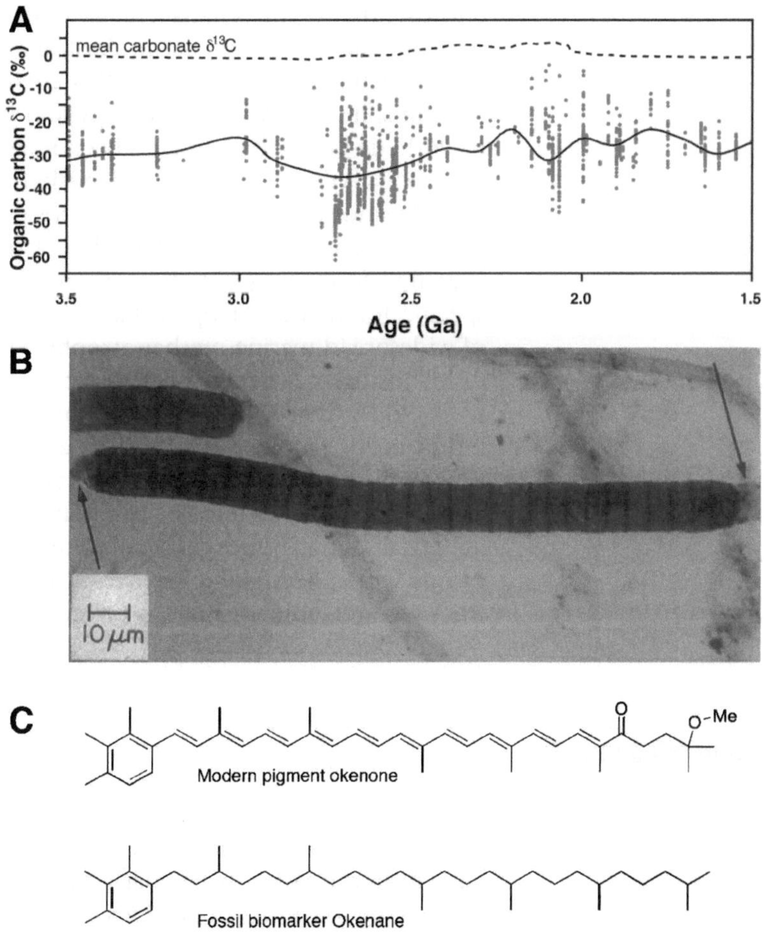


Fig. 8.6. Geologic evidence for the stability of prokaryotic lineages (**A**) Compilation of published kerogen and TOC carbon isotopic compositions ($\delta^{13}C$ values) for all sedimentary rock types. *Solid line* is a moving average for the organic carbon $\delta^{13}C$ values and the *dashed line* shows the mean carbonate $\delta^{13}C$ values through geologic time (modified from *(115)*). (**B**) *Palaeolyngbya*, a microfossil from the 950 million-year-old Lakhanda Formation that undoubtedly belongs to the Oscillatoriales (image from *(117)*). (**C**) Phototrophic purple sulfur bacteria (Chromatiaceae) pigment okenone, and its fossil biomarker okenane, which has been detected in 1.6 billion-year-old oil *(123)*.

the average observed carbon isotopic composition of sedimentary organic material is found around 2.7 billion years ago (**Fig. 8.6A**; *(115)*). This excursion represents a period during which abundant methanotrophic microorganisms were present even in shallow depositional environments *(115, 116)*. Second, the morphologies of well-preserved microfossils (particularly throughout the Neoproterozoic) are definitively related to particular living cyanobacterial groups (**Fig. 8.6B**; *(117)*), showing a remarkable consistency in those lineages for at least the past one or two billion years *(117–119)*. Third, molecular biomarkers specific to particular lineages are recognizable in ancient preserved sedimentary organic material *(120)* up to 2.7 billion years ago *(121, 122)*. In particular, okenane (**Fig. 8.6C**), a biomarker for the anaerobic photosynthetic Chromatiaceae, is found in 1.6 billion year old oil from the McArthur basin, Australia *(123)*. The detection of this particular biomarker from this euxinic basin argues strongly that, in spite of HGT, the ecology and biochemistry of the photosynthetic purple bacteria has been relatively unchanged for over 1.6 billion years. Carbonates deposited $\sim 2.7$ billion years ago provide another remarkable example of the long-term association of certain biomarkers with their associated microbial source. Eigenbrode et al. *(124)* have found methylhopane biomarkers specific for methanotrophic bacteria deposited during the time of widespread marine methanotrophy.

Considering genetic, genomic, and geologic evidence, it seems likely that a correct and complete view of microbial evolution needs to reconcile HGT as a mechanism for change over time with the observed long-term stability of lineages. The vertical descent of genes and the horizontal transfer of genes between genetically similar lineages seem to ultimately result in a core Tree of Life with at least some lineages that have phenotypic characteristics recognizable for billions of years.

## Acknowledgments

I thank all of the molecular evolutionists who I have interacted with over the years, particularly Ford Doolittle, Russ Doolittle, Peter Gogarten, Antonio Lazcano, Charles Marshall, Stanley L. Miller, and Gary Olsen. I also thank Bruce Runnegar and J. William Schopf for many useful discussions. I also thank Jennifer Eigenbrode for her compilation of carbon isotopic values, and Fredj Tekaia and Antonio Lazcano for helpful suggestions during the review of this manuscript. This work was supported by the National Aeronautics and Space Administration grant NNG05GN50G, and the NASA Astrobiology Institute.

## References


1. Haeckel, E. (1874) *Anthropogenie*. Engelmann, Leipzig.
2. Haeckel, E. (1896) *The Evolution of Man: A Popular Exposition of the Principal Points of Human Ontogeny and Phylogeny*. Appleton, New York.
3. Dayat, B. (2003) The roots of 'Phylogeny': how did Haeckel really build his trees? *System Biol* **52**, 515–27.
4. Cohen, H. J. (1872) Untersuchungen über Bacterien. I. *Beitr Biol Pfl* **1**, 127.
5. Sneath, P. H. A. (1962) Construction of taxa, in *Microbial Classification* (Cain, A. J., ed.), Cambridge University Press, London.
6. Orla-Jensen, S. (1909) Die Hauptlinien des natürlichen Bakteriensystems. *Zbl Bakt* **22**, 305.
7. Miller, S. L. (1953) A production of amino acids under possible primitive earth conditions. *Science* **117**, 528–9.
8. Oparin, A. I. (1953) *The Origin of Life*. Dover Publications Inc., New York.
9. Kluyver, A. J., van Niel, C. B. (1936) Prospects for a natural system of classification of bacteria. *Zbl Bakt* **94**, 369.
10. Stanier, R. Y., van Niel, C. B. (1941) The main outlines of bacterial classification. *J Bact* **42**, 437.
11. Murray, R. G. E. (1962) Fine structure and taxonomy of Bacteria, in *Microbial Classification* (Cain, A. J., ed.), Cambridge University Press, London.
12. De Ley, J. (1962) Comparative biochemistry and enzymology in bacterial classification, in *Microbial Classification* (Cain, A. J., ed.), Cambridge University Press, London.
13. Sanger, F., Thompson, O. P., Kitai, R. (1955) The amide groups of insulin. *Biochem J* **59**, 509–18.
14. Pauling, L., Zuckerkandl, E. (1962) Molecular paleontology. *Acta Chem Scand* **17**, S9–S16.
15. Jukes, T. H., Cantor, C. R. (1969) Evolution of protein molecules, in *Mammalian Protein Metabolism, III*. (Munro, H. N., ed.) Academic Press, New York.
16. Dayhoff, M. O. (1978) Observed frequencies of amino acid replacements between closely related proteins, in *Atlas of Protein Sequence and Structure*. National Biomedical Research Foundation, Washington, D.C.
17. Woese, C. R., Fox, G. E. (1977) Phylogenetic structure of the prokaryotic domain: the primary kingdoms. *Proc Nat Acad Sci U S A* **74**, 5088–90.
18. Pace, N. R. (1997) A molecular view of microbial diversity and the biosphere. *Science* **276**, 734–40.
19. Woese, C. R., Kandler, O., Wheelis, M. L. (1990) Towards a natural system of organisms: proposal for the domains Archaea, Bacteria, and Eucarya. *Proc Natl Acad Sci U S A* **87**, 4576–9.
20. Woese, C. R. (1987) Bacterial evolution. *Microbiol Rev* **51**, 221–71.
21. Stetter, K. O. (1996) Hyperthermophilic prokaryotes. *FEMS Microbiol Rev* **18**, 149–58.
22. Felsenstein, J. (1981) Evolutionary trees from DNA sequences: a maximum likelihood approach. *J Mol Evol* **17**, 368–76.
23. Marshall, C. R. (1997) Statistical and computational problems in reconstructing evolutionary histories from DNA data. *Comput Sci Statist* **29**, 218–26.
24. Gribaldo, S., Philippe, H. (2002) Ancient phylogenetic relationships. *Theor Popul Biol* **61**, 391–408.
25. Hartman P. E., Goodgal, S. H. (1959) Bacterial genetics (with particular reference to genetic transfer). *Annu Rev Microbiol* **13**, 465.
26. Fraser, C. M., Eisen, J. A., Salzberg, S. L. (2000) Microbial genome sequencing. *Nature*. **406**, 799–803.
27. House C. H., Fitz-Gibbon, S. T. (2002) Using homolog groups to create a whole-genomic tree of free-living organisms: an update. *J Mol Evol* **54**, 539–47.
28. House, C. H. (2007) Linking taxonomy with environmental geochemistry and why it matters to the field of geobiology. *Geobiology*. **5**, 1–3.
29. House, C. H., Runnegar, B. Fitz-Gibbon, S. T. (2003) Geobiological analysis using whole genome-based tree building applied to the bacteria, archaea, and eukarya. *Geobiology* **1**, 15–26.
30. Zerkle, A. L., House, C. H., Brantley, S. L. (2005) Biogeochemical signatures through time as inferred from whole microbial genomes. *Am J Sci* **305**, 467–502
31. Gogarten, J. P., Senejani, A. G., Zhaxybayeva, O., Olendzenski, L., Hilario, E. (2002) Inteins: structure, function, and evolution. *Annu Rev Microbiol* **56**, 263–287.
32. Doolittle, W. F. (2000) Uprooting the tree of life. *Sci Am* **282**, 90–95.
33. Perry, J. (ed.) (1975) *Personal Identity*. University of California Press, Berkeley, CA.

34. Hume, D. (1978) in *A Treatise on Human Nature*, 2nd Edition (Selby-Bigge, L. A., ed.) Clarendon Press, Oxford.
35. Clough, A (ed.), Dryden, J., and Plutarch (1992) *Plutarch's Lives, Volume 1.* Random House, Inc., New York.
36. House, C. H. (2003) Gene content based phylogenetic methods. *Amer Soc of Microb Annu Meeting.* Washington, D.C.
37. Doolittle, W. F. (2004) Q & A. *Curr Biol* **14**, R176–7.
38. Zhaxybayeva, O., Lapierre, P., Gogarten, J. P. (2004) Genome mosaicism and organismal lineages. *Trends Genet* **20**, 54–260.
39. Maddison, W. P. (1997) Gene trees in species trees. *System Biol* **46**, 523–36.
40. Philippe, H., Douady, C. J. (2003) Horizontal gene transfer and phylogenetics. *Curr Opin Microbiol* **6**, 498–505.
41. Doolittle, W. F. (1999) Phylogenetic classification and the universal tree. *Science* **284**, 2124.
42. Moret, B. M. E., Nakhleh, L., Warnow, T., Linder, C. R., Tholse, A., Padolina, A., Sun, J., Timme, R. (2004) "Phylogenetic networks: modeling, reconstructibility, and accuracy," *IEEE/ACM Trans Comput Biol Bioinf* **1**, 13–23.
43. Kunin, V., Goldovsky, L., Darzentas, N., Ouzounis, C. A. (2005) The net of life: reconstructing the microbial phylogenetic network. *Genome Res* 15, 954–59.
44. Huson, D. H., Bryant, D. (2006) Application of phylogenetic networks in evolutionary studies. *Mol Biol Evol* **23**, 254–67.
45. Rivera, M. C., Lake, J. A. (2004) The ring of life provides evidence for a genome fusion origin of eukaryotes. *Nature* **431**, 134–7.
46. Rivera, M. C., Jain, R., Moore, J. E., Lake, J. A. (1998) Genomic evidence for two functionally distinct gene classes. *Proc Natl Acad Sci U S A* **95**, 6239–44.
47. Jain, R., Rivera, M. C., Lake, J. A. (1999) Horizontal gene transfer among genomes: the complexity hypothesis. *Proc Natl Acad Sci U S A* **96**, 3801–6.
48. Daubin, V., Gouy, M., Perrière, G. (2001) Bacterial molecular phylogeny using supertree approach. *Genome Inform* **12**, 155–64.
49. Sicheritz-Ponten, T., Andersson, S. G. (2001) A phylogenomic approach to microbial evolution. *Nucleic Acids Res* **29**, 545–52.
50. Bapteste, E., Brinkmann, H., Lee, J. A., Moore, D. V., Sensen, C. W., Gordon, P., Duruflé, L. Gaasterland, T., Lopez, P., Müller, M., Philippe, H. (2002) The analysis of 100 genes supports the grouping of three highly divergent amoebae: *Dictyostelium, Entamoeba,* and *Mastigamoeba. Proc Natl Acad Sci U S A* **99**, 1414–19.
51. Daubin, V., Gouy, M., Perrière, G. (2002) A phylogenomic approach to bacterial phylogeny: evidence of a core of genes sharing a common history. *Genome Res* **12**, 1080–90.
52. Daubin, V., Moran, N. A., Ochman, H. Phylogenetics and the cohesion of bacterial genomes. *Science* 301, 829–32.
53. Canbäck, B., Tamas, I., Andersson, S. G. (2004) A phylogenomic study of endosymbiotic bacteria. *Mol Biol Evol* **21**, 1110–22.
54. Calteau, A., Daubin, V., Perrière, G. (2004) Super-tree approach for studying the phylogeny of prokaryotes: new results on completely sequenced genomes. *International Conference on Computational Science* **3039**, 700–708.
55. Sanderson, M. J., Purvis, A., Henze, C. (1998) Phylogenetic supertrees: assembling the trees of life, *Trends Ecol Evol* **13**, 105–109.
56. Baum, B. R. (1992) Combining trees as a way of combining data sets for phylogenetic inference, and the desirability of combining gene trees. *Taxon* **41**, 3–10.
57. Bininda-Emonds, O. R. P, Gittleman, J. L., Steel, M. A. (2002) The (super)tree of life: procedures, problems, and prospects. *Annu Rev Ecol Syst* **33**, 265–289.
58. Galtier, N. (2007) A model of horizontal gene transfer and the bacterial phylogeny problem. *Syst Biol* **56**, 633–42.
59. Dutilh, B. E., van Noort, V., van der Heijden, R. T., Boekhout, T., Snel, B., Huynen, M. A. (2007) Assessment of phylogenomic and orthology approaches for phylogenetic inference. *Bioinformatics* **23**, 815–24.
60. Hansmann, S., Martin, W. (2000) Phylogeny of 33 ribosomal and six other proteins encoded in an ancient gene cluster that is conserved across prokaryotic genomes: influence of excluding poorly alignable sites from analysis. *Int J Syst Evol Microbiol* **4**, 1655–63.
61. Brown, J. R., Douady, C. J., Italia, M. J., Marshall, W. E., Stanhope, M. J. (2001) Universal trees based on large combined protein sequence data sets. *Nat Genet* **28**, 281–5.
62. Wolf, Y., Rogozin, I. B., Grishin, N. V., Tatusov, R. L., Koonin, E. V. (2001) Genome trees constructed using five different approaches suggest new major bacterial clades. *BMC Evol Biol* **1**, 8.
63. Brochier, C., Bapteste, E., Moreira, D., Philippe, H. (2002) Eubacterial phylogeny

based on translational apparatus proteins. *Trends Genet* **18**, 1–5.
64. Matte-Tailliez, O., Brochier, C., Forterre, P., Philippe, H. (2002) Archaeal phylogeny based on ribosomal proteins. *Mol Biol Evol* **19**, 631–9.
65. Battistuzzi, F. U., Feijao, A, Hedges, S. B. (2004) A genomic timescale of prokaryote evolution: insights into the origin of methanogenesis, phototrophy, and the colonization of land. *BMC Evol Biol* **4**, 44.
66. Dagan, T., Martin, W. (2006) The tree of one percent. *Genome Biol* 7, 118.
67. Fitz-Gibbon, S. T., House, C. H. (1999) Whole genome-based phylogenetic analysis of free-living microorganisms. *Nucleic Acids Res* **27**, 4218–22.
68. Snel, B., Bork, P., Huynen, M. A. (1999) Genome phylogeny based on gene content. *Nat Genet* **21**, 108–10.
69. Tekaia, F., Lazcano, A., Dujon, B. (1999) The genomic tree as revealed from whole proteome comparisons. *Genome Res* **9**, 550–557.
70. Lin, J., Gerstein, M. (2000) Whole-genome trees based on the occurrence of folds and orthologs: implications for comparing genomes on different levels. *Genome Res* **10**, 808–818.
71. Montague, M. G., Hutchison, C. A. (2000) Gene content phylogeny of herpesviruses. *Proc Natl Acad Sci U S A* **97**, 5334–9.
72. Bansal, A. K., Meyer, T. E. (2002) Evolutionary analysis by whole-genome comparisons. *J Bacteriol* **184**, 2260–72.
73. Clarke, G. D., Beiko, R. G., Ragan, M. A., Charlebois, R. L. (2002) Inferring genome trees by using a filter to eliminate phylogenetically discordant sequences and a distance matrix based on mean normalized BLASTP scores. *J Bacteriol* **184**, 2072–80.
74. Korbel, J. O., Snel, B., Huynen, M. A., Bork, P. (2002) SHOT: a web server for the construction of genome phylogenies. *Trends Genet* **18**, 158–62.
75. Li, W., Fang, W., Ling, L., Wang, J., Xuan, Z., Chen, R. (2002) Phylogeny based on whole genome as inferred from complete information set analysis. *J Biol Physics* **28**, 439–47.
76. Stuart, G. W., Berry, M. W. (2003) A comprehensive whole genome bacterial phylogeny using correlated peptide motifs defined in a high dimensional vector space. *J Bioinform Comput Biol* **1**, 475–93.
77. Maruyama, O., Matsuda, A., Kuhara, S. (2005) Reconstructing phylogenetic trees of prokaryote genomes by randomly sampling oligopeptides. *Int J Bioinfo Res Applic* **1**, 429–46.
78. Tekaia, F., Yeramian, E. (2005) Genome trees from conservation profiles. *PLoS Comput Biol* **1**, e75.
79. Yang, S., Doolittle, R. F., Bourne, P. E. (2005) Phylogeny determined by protein domain content. *Proc Natl Acad Sci U S A* **102**, 373–8.
80. Gophna, U., Doolittle, W. F., Charlebois, R. L. (2005) Weighted genome trees: refinements and applications. *J Bacteriol* **187**, 1305–16.
81. Eisen, J. A. (2000) Horizontal gene transfer among microbial genomes: new insights from complete genome analysis. *Curr Opin Genet Devel* **10**, 606–11.
82. Rohwer, F., Edwards, R. (2002) The phage proteomic tree: a genome-based taxonomy for phage. *J Bacteriol* **184**, 4529–35.
83. Watterson, W. A., Ewens, W. J., Hall, T. E., Morgan, A. (1982) The chromosome inversion problem. *J Theor Biol* **99**, 1–7.
84. Blanchette, M., Kunisawa, T., Sankoff, S. (1999) Gene order breakpoint evidence in animal mitochondrial phylogeny. *J Mol Evol* **49**, 193–203.
85. Kunisawa, T. (2001) Gene arrangements and phylogeny in the class proteobacteria. *J Theor Biol* **213**, 9–19.
86. Kunisawa, T. (2003) Gene arrangements and branching orders of gram-positive bacteria. *J Theor Biol* **222**, 495–503.
87. Moret, B. M. E., Wang, L. S., Warnow, T., Wyman, S. K. (2001) New approaches for reconstructing phylogenies from gene order data. *Bioinformatics* **17** Suppl 1, S165–73.
88. Belda, E., Moya, A., Silva, F. J. (2005) Genome rearrangement distances and gene order phylogeny in gamma-proteobacteria. *Mol Biol Evol* **22**, 1456–67.
89. House, C. H., Fitz-Gibbon, S. T. (submitted) Genome-wide gene order distances support a united Gram-positive bacteria. *Geobiology.*
90. Di Giulio, M. (2003) The ancestor of the bacteria domain was a hyperthermophile. *J Theor Biol* **224**, 277–83.
91. Russell, M. J., Hall, A. J. (1988) Submarine hot springs and the origin of life. *Nature* **336**, 117.
92. Wächtershäuser, G. (1988) Pyrite formation, the first energy source for life: a hypothesis. *Syst Appl Microbiol* **10**, 207.
93. Russell, M. J., Hall, A. J. (1990) Pyrite and the origin of life. *Nature* **344**, 387.
94. Shock, E. (1990) Geochemical constraints on the origin of organic compounds in

hydrothermal systems. *Orig Life Evol Biosph* **20**, 331–67.
95. Pace, N. R. (1991) Origin of life – facing up to the physical setting. *Cell* **65**, 531–33.
96. Schwartzman, D. W., Lineweaver, C. H. (2004) The hyperthermophilic origin of life revisited. *Biochem Soc Trans* **32**, 168–71.
97. Wächtershäuser, G. (2006) From volcanic origins of chemoautotrophic life to bacteria, archaea and eukarya. *Philos Trans R Soc Lond B Biol Sci* **361**, 1787–806.
98. Forterre, P., Philippe, H. (2004) Where is the root of the universal tree of life? *Bioessays* **21**, 871–79.
99. Nelson, K. E., Clayton, R. A., Gill, S. R., Gwinn, M. L., Dodson, R. J., Haft, D. H., Hickey, E. K., Peterson, J. D., Nelson, W. C., Ketchum, K. A., McDonald, L., Utterback, T. R., Malek, J. A., Linher, K. D., Garrett, M. M., Stewart, A. M., Cotton, M. D., Pratt, M. S., Phillips, C. A., Richardson, D., Heidelberg, J., Sutton, G. G., Fleischmann, R. D., Eisen, J. A., White, O., Salzberg, S. L., Smith, H. O., Venter, J. C., Fraser, C. M. (1999) Evidence for lateral gene transfer between Archaea and Bacteria from genome sequence of *Thermotoga maritima*. *Nature* **399**, 323–9.
100. Davey, M. E., Wood, W. A., Key, R., Nakamura, K., Stahl, D. A. (1993) Isolation of three species of *Geotoga* and *Petrotoga*: Two new genera, representing a new lineage in the bacterial line of descent distantly related to the "Thermotogales". *Syst Appl Microbiol* **16**, 191–200.
101. Maher, K. A., Stevenson, D. J. (1988) Impact frustration of the origin of life. *Nature* **331**, 612–4.
102. Sleep, N. H., Zahnle, K. J., Kasting, J. F., Morowitz, H. J. (1989) Annihilation of ecosystems by large asteroid impacts on the early Earth. *Nature* **342**, 139–42.
103. Miller, S. L., Lazcano, A. (1995) The origin of life—did it occur at high temperatures? *J Mol Evol* **41**, 1432.
104. Gogarten-Boekels, M., Hilario, E., Gogarten, J. P. (1995) The effects of heavy meteorite bombardment on the early evolution – the emergence of the three domains of life. *Orig Life Evol Biosph* **25**, 251–64.
105. Islas, S., Velasco, A. M., Becerra, A., Delaye, L., Lazcano, A. (2007) Extremophiles and the origin of life, in *Physiology and Biochemistry of Extremophiles* (Gerday, C., and Glansdorff, N., ed.), ASM Press, Washington, D.C.
106. Slesarev, A. I., Mezhevaya, K. V., Makarova, K. S., Polushin, N. N., Shcherbinina, O. V., Shakhova, V. V., Belova, G. I., Aravind, L., Natale, D. A., Rogozin, I. B., Tatusov, R. L., Wolf, Y. I., Stetter, K. O., Malykh, A. G., Koonin, E. V., Kozyavkin, S. A. (2002) The complete genome of hyperthermophile *Methanopyrus kandleri* AV19 and monophyly of archaeal methanogens. *Proc Natl Acad Sci U S A* **99**, 4644–9.
107. Wolf, Y. I., Rogozin, I. B., Grishin, N. V., Koonin, E. V. (2002) *Trends Genet* **18**, 472–9.
108. Brochier, C., Forterre, P., Gribaldo, S. (2004) Archaeal phylogeny based on proteins of the transcription and translation machineries: tackling the *Methanopyrus kandleri* paradox. *Genome Biol* **5**, R17.
109. Vargas, M., Kashefi, K., Blunt-Harris, E. L., Lovley, D. R. (1998) Microbiological evidence for Fe(III) reduction on early Earth. *Nature* **395**, 65–7.
110. Farquhar, J., Bao, H., Thiemens, M. (2001) Atmospheric influence of Earth's earliest sulfur cycle. *Science* **289**, 756–8.
111. Pavlov, A. A., Kasting, J. F. (2002) Mass-independent fractionation of sulfur isotopes in Archean sediments: strong evidence for an anoxic Archean atmosphere. *Astrobiology* **2**, 27–41.
112. Ono, S., Eigenbrode, J. L., Pavlov, A. A., Kharecha, P., Rumble, D., Kasting, J. F., Freeman, K. H. (2003) New insights into Archean sulfur cycle from mass-independent sulfur isotope records from the Hamersley Basin, Australia. *Earth Planet Sc Lett* **213**, 15–30.
113. Gogarten, J. P., Doolittle, W. F., Lawrence, J. G. (2002) Prokaryotic evolution in light of gene transfer. *Mol Biol Evol* **19**, 2226–38.
114. Ferry, J. G., House, C. H. (2006) The stepwise evolution of early life driven by energy conservation. *Mol Biol Evol* **23**, 1286–92.
115. Eigenbrode, J. L., Freeman, K. H. (2006) Late Archean rise of aerobic microbial ecosystems. *Proc Natl Acad Sci U S A* **103**, 15759–64.
116. Hayes, J. M. (1994) Global methanotrophy at the Archean-Proterozoic transition, in *Early life on Earth* (S. Bengtson, ed.) Columbia University Press, New York, 220–36.
117. Schopf, J. W. (1994) Disparate rates, differing fates: tempo and mode of evolution changed from the Precambrian to the Phanerozoic. *Proc Natl Acad Sci U S A* **91**, 6735–42.
118. Golubic, S., Hofmann, H. J. (1976) Comparison of Holocene and mid-Precambrian

Entophysalidaceae (Cyanophyta) in stromatolitic algal mats; cell division and degradation. *J Paleontol* **50**, 1074–82.
119. Barghoorn, E. S., Schopf, J. W. (1965) Microorganisms from the late Precambrian of central Australia. *Science* **150**, 337–9.
120. Summons, R. E., Jahnke, L. L. (1990) Identification of the methylhopanes in sediments and petroleum. *Geochim Cosmochim Acta* **54**, 247–51.
121. Brocks, J. J., Logan, G. A., Buick, R., Summons, R. E. (1999) Archean molecular fossils and the early rise of eukaryotes. *Science* **285**, 1033–36.
122. Summons, R. E., Jahnke, L. L., Hope, J. M., Logan, G. A. (1999) 2-Methylhopanoids as biomarkers for cyanobacterial oxygenic photosynthesis. *Nature* **400**, 554–7.
123. Brocks, J. J., Love, G. D., Summons, R. E., Knoll, A. H., Logan, G. A., Bowden, S. A. (2005) Biomarker evidence for green and purple sulphur bacteria in a stratified Palaeoproterozoic sea. *Nature* **437**, 866–70.
124. Eigenbrode, J. L., Freeman, K. H., Summons, R. E. (2008) Methylhopane biomarker hydrocarbons in Hamersley Province sediments provide evidence for Neoarchean aerobiosis, *Earth Planet Sci Lett* **273**, 323–31.

# Chapter 9

# Horizontal Gene Transfer and the Evolution of Methanogenic Pathways


## Greg Fournier



## Abstract

Horizontal gene transfer (HGT) is a driving force in the evolution of metabolic pathways, allowing novel enzymatic functions that provide a selective advantage to be rapidly incorporated into an organism's physiology. Here, the role of two HGT events in the evolution of methanogenesis is described. First, the acetoclastic sub-pathway of methanogenesis is shown to have evolved via a transfer of the *ackA* and *pta* genes from a cellulolytic clostridia to a family of methanogenic archaea. Second, the system for encoding the amino acid pyrrolysine, used for the synthesis of enzymes for methanogenesis from methylamines, is shown to likely have evolved via transfer from an ancient, unknown, deeply branching organismal lineage.



**Key words:** Horizontal gene transfer, methanogenesis, acetoclastic, pyrrolysine, methanosarcina, clostridia.


## 1. Introduction

### 1.1. Evolution of Metabolic Networks

One of the primary functions of proteins in living cells is to act as enzymatic catalysts of chemical reactions. Taken together, these reactions characterize the metabolism of an organism, forming a Byzantine network of interrelated and interdependent pathways. These pathways convert substrates into biomolecules required for growth, as well as provide a source of chemical energy for driving cellular processes. Because of the diversity, ubiquity, and complexity of these pathways, understanding their history is an important part of evolutionary biology.

The most extensively studied process for the evolution of metabolic pathways is the duplication and divergence of genes encoding metabolic enzymes *(1, 2)*. In these models, gene


Maria B. Gogarten et al. (eds.), *Horizontal Gene Transfer: Genomes in Flux, vol. 532*



DOI 10.1007/978-1-60327-853-9_9 Springerprotocols.com

duplication produces a "spare" gene, relaxing purifying selection so that the extra gene has the opportunity to evolve a novel (or more specific) function. In this way, entire paralogous pathways can be generated for the biosynthesis of distinct products *(3)*. Intragenic duplications have also been shown to generate increasingly complex enzymes that catalyze additional enzymatic functions in the evolution of some biosynthesis pathways *(4)*. As the availability of genomic data and the power of computational methods have increased, the metabolic mapping of entire genomes has become a commonplace and largely automated process *(5, 6)*, making large-scale detection of these evolutionary events far easier.

Interestingly, these same computational methods have revealed that simple duplication and divergence within organismal lineages is insufficient to explain the phylogenetic signal and distribution of metabolic pathways (or parts of these pathways) in nature. Rather, observations suggest that HGT makes a significant contribution to the evolution of metabolic networks, and may even be the dominant factor *(7)*. Here the role of HGT in the evolution of a particularly interesting metabolic pathway, methanogenesis, will be discussed.

### 1.2. Methanogenesis

Several classes of the archaeal phylum *Euryarchaeota* possess the unique ability to generate energy exclusively via the reduction of single-carbon substrates to methane, a process known as methanogenesis. It is likely that the core methanogenic pathway evolved early in the evolution of the euryarchaeotes, making this an extremely ancient microbial lifestyle (**Fig. 9.1**). The paralogy of key methanogenic enzymes with those found in bacterial pathways for aerobic methane oxidation and formaldehyde detoxification suggests that the most recent common ancestor (MRCA) of bacteria and archaea may have already possessed a precursor to the methanogenic core pathway, although the ancestral function of this system is uncertain *(8, 9)*. While some derived classes of euryarchaeotes are non-methanogenic (*Thermoplasmatales, Halobacteriales*, and *Archaeoglobales*), this is most likely due to lineage-specific losses, and does not imply a polyphyletic origin or HGT of the entire pathway *(10, 11)*.

Methanogenesis can be divided into four major sub-pathways, based on the cofactors used for the transfer and/or reduction of the methyl group (**Fig. 9.2**). First is the core pathway (**Fig. 9.2**, sub-pathway I), which consists of enzymes for the final reduction of a methyl group into methane, by the formation of a disulfide bond between methyl-coenzyme M (methyl-CoM) and coenzyme B (CoB), as well as a heterodisulfide reductase complex for the re-generation of CoM and CoB. Both of these steps utilize membrane-bound protein complexes that generate an H + /Na+ gradient, which is then used for ATP production *(12)*. The

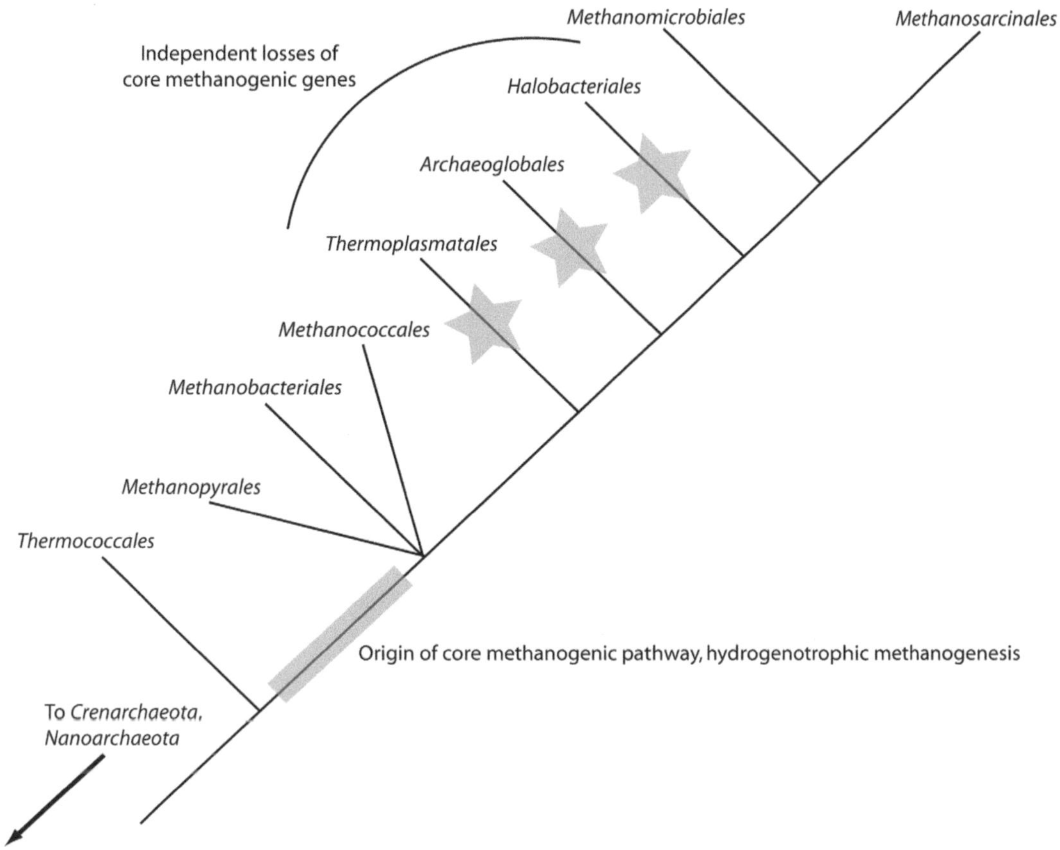


Fig. 9.1. Phylogenetic distribution of methanogens. Phylogenetic relations adopted from *(10, 11)*. This schematic illustrates the phylogenetic relationships between classes within the *Euryarchaeota*. Methanogenesis as a distinct metabolic system likely evolved after the divergence of the *Thermococcales* (*gray box*). This was followed by an independent loss of a functional methanogenic pathway in some lineages (denoted by *star symbols*).

hydrogenotrophic pathway (**Fig. 9.2**, sub-pathway II) is also likely part of the "ancestral" methanogenic route, performing the stepwise reduction of carbon dioxide, carbon monoxide, or formate using several cofactors. This joins the core pathway with the transfer of a methyl group from tetrahydromethanopterin ($H_4MPT$) to CoM.

Two alternative methanogenic sub-pathways do not require $H_2$ for energy production. Acetoclastic methanogenesis (**Fig. 9.2**, sub-pathway III) activates acetate to acetyl-CoA using either the AckA/Pta or ACS enzymes, which then transfers its methyl group to $H_4MPT$ (or $H_4SPT$ – tetrahydrosarcinapterin) via the activity of the carbon monoxide dehydrogenase enzyme complex (CODH), generating $CO_2$ in the process. The methyl group is then transferred to CoM via the same reaction as used in the hydrogenotrophic pathway. The methyl-corrinoid sub-pathway (**Fig. 9.2**, sub-pathway IV) is used for the reduction of several small molecules containing methyl groups, such as

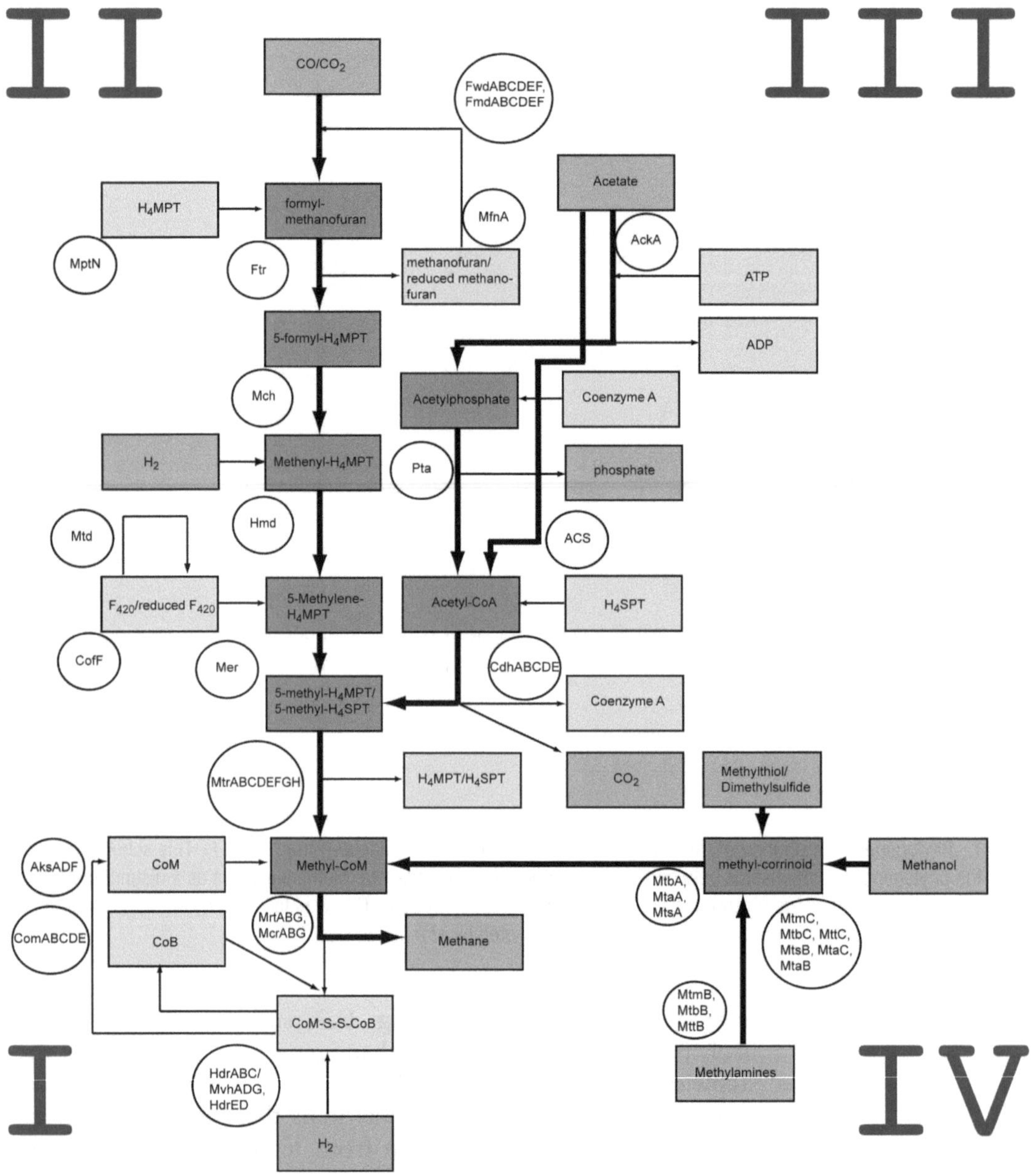


Fig. 9.2. Methanogenic pathways. Sub-pathways (data from metacyc *(5)*) are labeled with roman numerals. Enzymatic reactions are indicated by *white circles*, with their respective protein(s) listed. Cofactors are indicated by *light gray* boxes. Substrates and final products are indicated by *medium gray* boxes. *Dark gray* boxes represent the cofactors bound to single-carbon intermediates. Electron transport chain reactions are not shown. *Bold arrows* indicate the pathway of the carbon molecule being reduced to methane.

methylamines, methanol, dimethylsulfide, and methylthiols. In each case, substrate-specific methyltransferase proteins transfer the methyl group to a corrinoid protein, which then transfers the methyl group to CoM.

Methanogenic pathways utilize a large number of genes with no recognizable paralogs in nonmethanogenic pathways, as well as a large number of complex cofactors and carrier proteins, many of which are unique to methanogens. This complexity and exclusivity provides an effective barrier against HGT. In fact, no HGT of a methanogenic pathway to a nonmethanogenic lineage has ever been reported *(10)*. Therefore, since all methanogens inherited the core methanogenic pathway from a common organismal ancestor, retaining a strong phylogenetic signal to indicate their unity as a clade, it is an excellent system for detecting HGT events in the other direction, i.e., with methanogens as the recipient. Interestingly, both acetoclastic methanogenesis and methanogenesis from methylamines bear a signature of such a transfer event, albeit in quite different scenarios.

## 2. Evolution of Acetoclastic Methanogenesis

### 2.1. Physiology

Methanogenesis from acetate (acetoclastic methanogenesis, **Fig. 9.2** sub-pathway III) is the largest biogenic source of methane on Earth, and is responsible for two-thirds of all annual production *(13, 14)*. Acetoclastic methanogenesis occurs primarily in aquatic environments such as lakes, rivers, and marine sediments, wetlands, soils, and the gastrointestinal tracts of animals *(14–16)*.

However, only two genera contain pathways for acetoclastic methanogenesis, *Methanosarcina* and *Methanosaeta*, an additional member of the order *Methanosarcinales* belonging to the family *Methanosaetaceae* (**Fig. 9.1**). While *Methanosaeta* use the enzyme acetate synthase (ACS) for the generation of acetyl-CoA from acetate in a single step, *Methanosarcina* use two enzymes, acetate kinase (AckA) and phosphoacetyltransferase (Pta);

$$CH_3CO_2^- + ATP \xrightarrow{AckA} CH_3COPO_4^{2-} + ADP \quad (9.1)$$

$$CH_3COPO_4^{2-} + CoA \xrightarrow{Pta} CH_3CO - CoA + PO_4^{3-} \quad (9.2)$$

Using acetyl-CoA, the carbon monoxide dehydrogenase (CODH) complex then catalyzes the release of CoA and the transfer of $CH_3$ to tetrahydromethanopterin ($H_4MPT$) or tetrahydrosarcinapterin ($H_4SPT$). The methyl group is then transferred to Coenzyme M (CoM) by CoM methyltransferase, and reduced to $CH_4$ by methyl-CoM reductase *(17)*. CODH is also present in non-acetoclastic methanogens; however, the reaction typically proceeds in the reverse direction, synthesizing acetyl-CoA to be used in anabolic pathways. Overall, the remaining reactions of acetoclastic methanogensis can be summarized as:

$$CH_3CO - CoA + H_2O + nADP + nPO_4^{3-} \rightarrow$$
$$CH_4 + CO_2 + nATP + CoA \quad (9.3)$$

### 2.2. Phylogenetic Distribution/Analysis

Surprisingly, neither *ackA* nor *pta* has any homologs in other methanogens, or even any other archaeal species. However, these genes are widely distributed among the bacteria, and have also been reported in various eukaryotic lineages *(18)*, presumably transferred in several independent events from diverse groups of bacteria. In bacteria, AckA and Pta are used to activate acetate for use as a carbon source, or to produce acetate as an electron sink during energy production via fermentation *(19)*. In eukaryotes, AckA and/or Pta have also been reported to function in activating acetate for use as a carbon source *(18)*, assimilation of acetyl phosphate produced by the breakdown of xylulose 5-phosphate and fructose 6-phosphate *(18,20)*, and possibly protein regulation *(18, 21)*. In a situation analogous to the eukaryotic homologs of *ackA/pta*, it is likely that the archaeal homologs were also transferred from a bacterial donor lineage.

Indeed, several phylogenetic analyses support that *ackA* and *pta* were acquired by the ancestor of the *Methanosarcina* in a single HGT event, likely from within a clade of cellulolytic bacteria of the class *Clostridia (22)*.

Phylogenetic trees of AckA and Pta homologs using *Fusobacterium nucleatum* as an outgroup were generated using Bayesian inference *(23)*, neighbor joining *(24)*, and maximum likelihood *(25)*. There is strong phylogenetic support for HGT from a group of cellulolytic clostridia including *C. cellulolyticum* and *C. thermocellum*, as indicated by support values generated using the above-listed algorithms for both Pta (1.00, 82, 94) and AckA (0.90, 64, 68). The consensus tree for Pta also gives strong support for *Clostridium phytofermentans* and *Caldicellulosiruptor saccharolyticus* belonging to this same monophyletic subset within *Clostridia*. Internal nodes in the AckA consensus tree have much lower support values, although a similar grouping (including *C. saccharolyticus* but not *C. phytofermentans*) has a high posterior probability (0.63). This group also contains an AckA homolog found in *Thermoanaerobacter ethanolicus X514*. Since there is another AckA homolog present in *T. ethanolicus X514* in a highly supported grouping with other *Thermoanaerobacter* species, this gene copy is likely the result of a HGT from an organism related to *C. saccharolyticus*.

With the exception of *Moorella thermoacetica* and *Desulfitobacterium hafniense Y51*, all completed *Clostridia* genomes

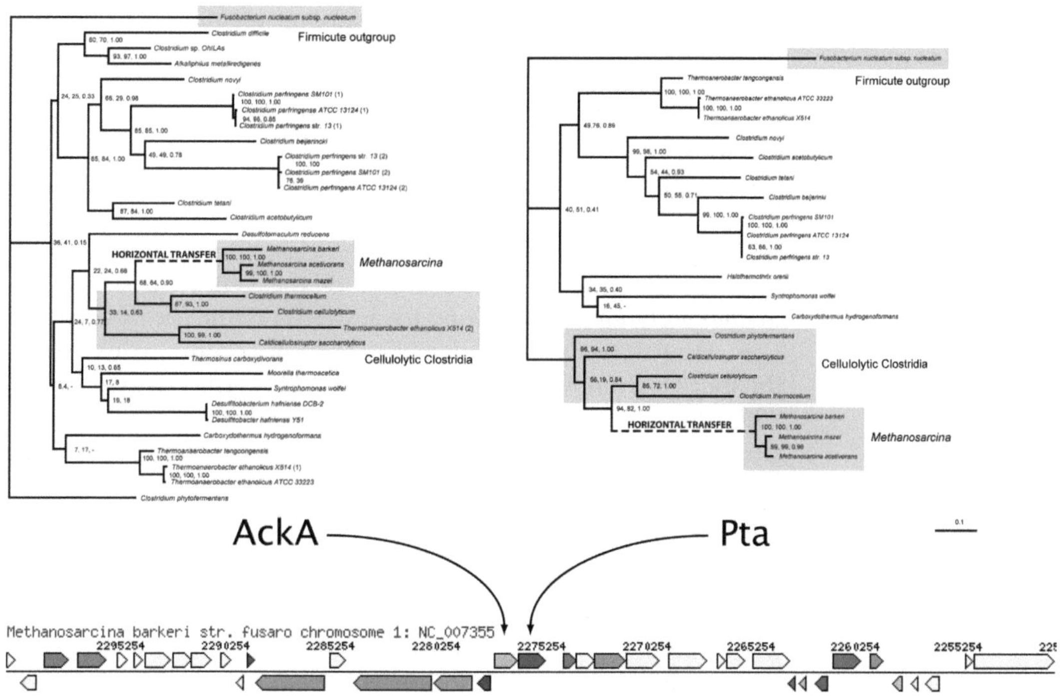


Fig. 9.3. Consensus trees of AckA and Pta homologs. Support values at nodes represent boostrap values for maximum likelihood and neighbor joining, and posterior probability for Bayesian inference, respectively. Consensus trees were generated using the PHYLIP package CONSENSE algorithm *(24)*. *Arrows* indicate the location of genes encodong AckA and Pta homologs in the *Methanosarcina barkeri* genome, showing their adjacent position.

contain both *ackA* and *pta* homologs. In all genomes with both *ackA* and *pta*, the genes are adjacent and in the same orders as the homologs in *Methanosarcina* (**Fig. 9.3**), although additional lone copies of each gene may be present elsewhere in the genome (e.g., *ackA* in *T. ethanolicus X514*). Therefore, it is likely that the absence of *ackA* or *pta* in some genomes is due to gene loss, and not incomplete sequencing. Given this assumption, we can infer that the presence of adjacent *ackA/pta* genes was very likely the ancestral state of the clade from which these genes were transferred. This agrees with the consensus phylogenies in implicating a single transfer event for the acquisition of *ackA* and *pta* in methanogenic archaea.

### 2.3. Evolutionary Considerations

HGT between cellulolytic clostridia and *Methanosarcina* is not only supported by phylogenetic analysis but also a biologically plausible event. Close physical interactions between methanogens and clostridia are commonplace *(26)*, and gene transfer between them is abundant *(27)*. Furthermore, modern representatives of *Methanosarcinaceae* and cellulolytic clostridia often co-exist in the same environment, especially in freshwater systems.

The phylogenetic relationships among members of the class *Clostridia* are largely unresolved, as this group is essentially defined as low G-C, anaerobic members of the division *Firmicutes*. Since the ancestral lineage of this group may have also given rise to organisms not classified as *Clostridia*, it is believed to be paraphyletic rather than monophyletic, and largely heterogeneous *(28)*.

According to a phylogeny of the class Clostridia published by Desvaux et al. *(29)*, the donor of the genes transferred to *Methanosarcina* would belong to cluster III of the genus *Clostridium*. However, an alternative classification scheme would place the transfer from within family #4, genus #2 of the class *Clostridia (30)*. Regardless of the specific phylogenetic relationships within clostridia, all of these schemata (including the consensus trees depicted in **Fig. 9.3**) support cellulolytic clostridia being a monophyletic group, and cellulolytic enzymatic activity as a shared derived character of the clade from which *ackA* and *pta* were transferred to *Methanosarcina*. This allows for a speculative dating of the transfer event. In fact, this gene transfer event to Methanosarcina occurred after cellulolytic clostridia had undergone significant diversification. This suggests the existence of a diverse, cellulose-rich freshwater environment, which presumably requires the availability of significant terrestrial plant biomass for fermentation, probably no earlier than the Mid-Ordovician, about 475 Mya *(31)*. Conversely, it places the diversification of the genus Methanosarcina as occurring no earlier than this time, as its members vertically inherited these transferred genes.

Interestingly, while the roles of AckA and Pta in *Methanosarcina* during the initial steps of acetoclastic methanogenesis are analogous to those in acetotrophic bacteria, cellulolytic clostridia such as *C. thermocellum* and *C. cellulolyticum* use AckA and Pta in the reverse reaction, producing acetate as a final step in the fermentation of cellulose-derived sugars *(19, 29)*. Therefore, upon transfer both genes would have immediately been required to function in a pathway in the reverse direction of that used in the donor organism.

The ancestral state of the methanogenic pathway in *Methanosarcina* before the transfer event was likely one of two possibilities: (1) the absence of an acetoclastic pathway and the use of the numerous other methanogenic substrates available to *Methanosarcina (15)*; (2) the presence of an acetoclastic pathway utilizing ACS, such as still found in *Methanosaetaceae (32)*. In the case of the second scenario, it is puzzling that the ACS pathway was not retained, since organisms with both could effectively utilize acetate at both high and low concentrations. However, the availability of alternative substrates for methanogenesis at low levels of acetate may explain such loss. Indeed, several members of *Methanosaetaceae* use only acetate for methanogenesis, apparently having lost the capacity to use any other substrate. In a similar fashion, *Methanosarcina* could have lost the ACS pathway before acquiring AckA/Pta. Further complicating the situation is the possibility that *Methanosarcina* and *Methanosaetaceae* evolved in parallel after the transfer event the ability of optimizing growth under different acetate concentrations *(32)*. Regardless of the specific evolutionary history, the evolution of the AckA/Pta acetoclastic methanogenesis pathway likely resulted in a significant increase in biogenic methane production, having global biogeochemical consequences, possibly including climate change.

## 3. Evolution of Methanogenesis from Methylamines: The Pyrrolysine System

### 3.1. Physiology

Methanogenesis from methylamines is performed through the methyl-corrinoid pathway (sub-pathway IV in **Fig. 9.2**), along with other single-carbon molecules such as methanol, dimethylsulfide, or methylthiols. Typically, each substrate is bound to a specific corrinoid protein via a specific methyltransferase enzyme. Use of methylamines for methanogenesis is limited to the family *Methanosarcinales*, using a set of specific methylamine methyltransferase proteins that allow for growth on monomethylamines, dimethylamines, or trimethylamines (MtmB, MtbB, MttB, respectively). Surprisingly, even though these enzymes perform identical functions using near-identical substrates, these proteins show very little sequence similarity, although the catalytic regions of mono- and dimethylamine methyltransferase may be homologous (Fournier, unpublished). Also, two divergent bacterial species have been identified, which possess methylamine methyltransferase genes (*Desulfitobacterium hafniense* and a delta-proteobacteria symbiont of the marine gutless worm *Olavius algarvensis*) *(33, 34)*, although these organisms are non-methanogenic.

These proteins are unique in that each requires the non-canonical amino acid pyrrolysine (Pyl) in its catalytic site, where its activity as a strong electrophile enables the transfer of a methyl group to the cobalt atom of a corrinoid carrier protein *(35)*. No other proteins have been identified that use Pyl, except for a single transposon family *(36)*.

Pyl is a non-canonical amino acid, in that (1) the vast majority of existing organisms do not have it as part of their genetic code; and (2) it requires special machinery for insertion. Unlike selenocysteine (Sec), the other known non-canonical amino acid, Pyl, has its own unique aaRS for the aminoacylation of $tRNA^{Pyl}$, and does not rely on any tRNA-dependent steps for its biosynthesis *(37)*. However, like Sec, Pyl-encoding UAG codons are associated with a downstream stem-loop structure (PYLIS) that facilitates the incorporation of Pyl *(36)*. In this way, these codons are not mistaken for termination signals. The system is not completely analogous to Sec insertion, as PYLIS does not require (and presumably does not interact with) a unique elongation factor. Furthermore, experimental evidence supports that even without the presence of a PYLIS signal, Pyl is still incorporated, albeit with less efficiency *(38)*.

### 3.2. An Evolutionary Paradox

Pyl usage has an extremely narrow phylogenetic and enzymatic distribution for a genetically encoded amino acid. How, then, did such an elaborate system evolve? In order for an organism to utilize encoded Pyl, it must have an intact Pyl biosynthesis pathway, a specific $tRNA^{Pyl}$, Pyl-aaRS, and, often, PYLIS elements. Yet, this amino acid is only used for a single type of enzymatic reaction within a handful of enzymes, none of which are essential *(15)*. Usually, biological traits with narrow phylogenetic distributions are assumed to have evolved recently, within the lineage in question. These traits are often referred to as "derived". But is usage of Pyl a highly derived trait, evidence that the genetic code is still expanding, with us having glimpsed the newest arrival in its earliest stages? One strong piece of evidence suggests otherwise. The molecular phylogeny of class II aaRS proteins places PylRS as a very deeply branching lineage within the IIb subfamily (**Fig. 9.4**). While deeply branching lineages can often be the results of misplacements due to a phylogenetic artifact known as long-branch attraction (LBA), in those cases the branch in question groups with the longest internal branch of the tree. As the internal branch leading to the class IIa aaRS is much longer than that leading to the IIb subset (where PylS roots), it is unlikely that the placement of PylS is due to an LBA artifact. This means that PylRS diverged from other aaRS sequences before the split between the bacteria and the archaea, indicating it is very ancient indeed. But if it is so ancient, why has it only been preserved in a scattering of distantly related organisms, and why is it used in such a narrow, specific manner? How could such meager utility drive the selection required to evolve and maintain the Pyl system? And, why have other genes not evolved to make use of it?

The most straightforward explanation is simply vertical inheritance with many lineage-specific independent gene losses. In this scenario, the Pyl usage machinery (as well as genes that utilize Pyl) would have been present, like other aaRS genes, in the most recent common ancestor of known life (MRCA). Then, for unknown reasons, these genes were lost in all lineages except the direct ancestors of the clades that currently possess the system. The problem with this interpretation is that it requires a bewildering number of selective losses to result in the observed distribution. For example, in order to explain the phylogenetic distribution of Pyl within the archaea alone would require at least

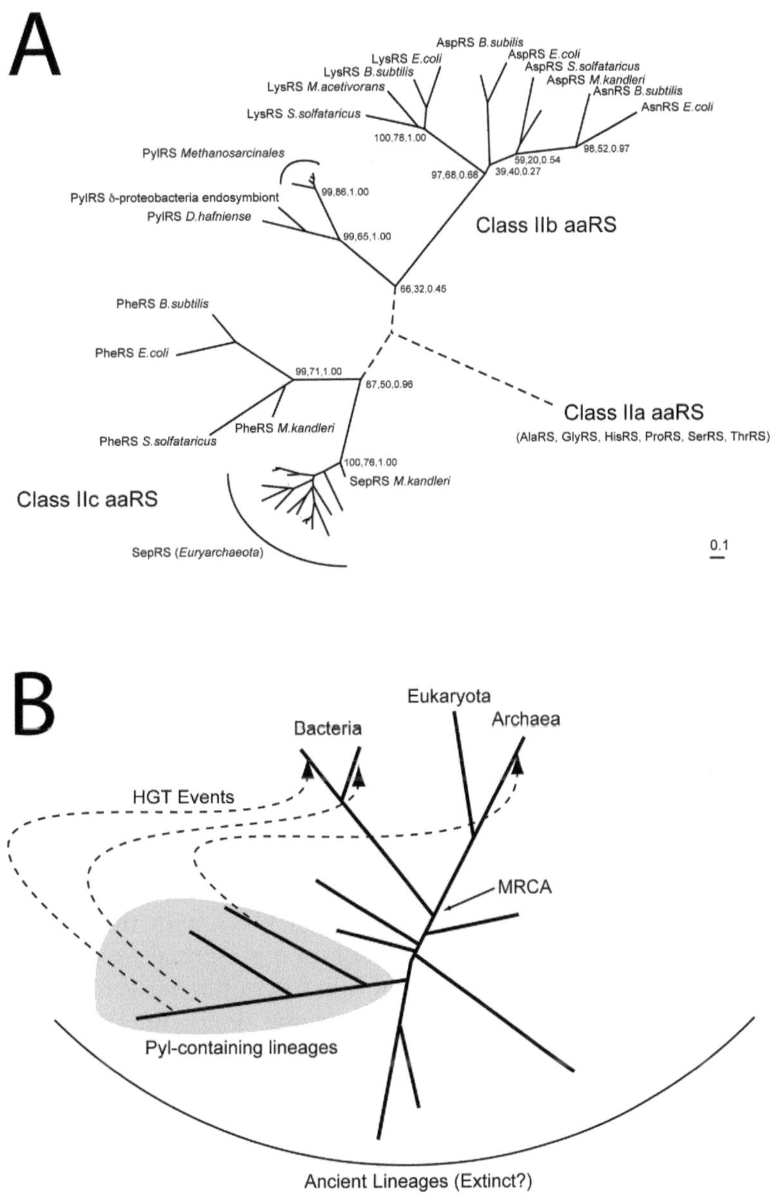


Fig. 9.4. Evolutionary history of pyrrolysine (**A**) Phylogenetic relationship of selected class II aminoacyl-tRNA synthetases. Sequences were aligned using the MUSCLE algorithm *(45)*. Consensus tree was generated using the CONSENSE algorithm in the PHYLIP package *(24)*, using trees generated via maximum likelihood *(25)*, neighbor-joining *(24)*, and Bayesian inference *(23)*. Values at nodes indicate bootstrap support/posterior probability for specific groupings of sequences using these methods, respectively. (**B**) Model for Pyl evolution. "Extinct Ancient Lineages" refers to the domains of life that diverged before the MRCA of the three known existing domains. HGT events are signified by *dotted lines*, with *arrows* pointing to their recipient lineages. The *shaded* region indicates a lineage in which Pyl originally evolved and was universally present, with extensive usage.

nine independent losses; to explain the bacterial presence in only a single delta-proteobacteria and *D. hafniense* requires many, many more. The problem is further confounded by the observation that archaeal and bacterial species both only utilize Pyl in methylamine methyltransferase enzymes. So, either the MRCA Pyl system only ever evolved to be used in this one set of enzymes (which, as previously mentioned, is highly unlikely to have provided sufficient selective pressure for its evolution) or Pyl originally evolved to be used in many enzymes, all of which were lost except for the methylamine methyltransferases. However, independent, parallel losses of entire gene families resulting in exactly the same sets of genes surviving in both bacterial and archaeal lineages seem to be an even less likely coincidence.

Clearly, vertical inheritance alone is insufficient to explain the evolution of this system. However, a solution to this paradox can be found in HGT. It has been shown experimentally that the Pyl incorporation machinery is clustered as a "cassette" that permits the synthesis and usage of Pyl in any genome to which it is transferred, as is also evidenced by its usage in *D. hafniense (39)*. Likewise, a horizontal transfer of the Pyl machinery to the ancestor of the *Methanosarcinales* would account for most of the unusual features of the system. Since PylRS is such a deeply branching protein lineage, the donor of the transfer would have to be a member of a very ancient group of organisms, one that diverged before the most recent common ancestor of the bacterial and archaeal domains (**Fig. 9.4**). Coalescence theory strongly suggests that such deeply branching lineages existed, and presumably were engaging in HGT with the ancestors of known organisms *(40)*. This solution is especially elegant in that it also explains the absence of other Pyl-utilizing genes, as, even though these probably evolved in the donor lineage, the transfer of only a single Pyl-utilizing gene would be sufficient for the retention of the Pyl system.

### 3.3. Horizontal Transfer of the Pyl System

While proposing a horizontal transfer event from a deeply branching ancient lineage resolves many of the phylogenetic problems with the evolution of the Pyl system, it also raises several new questions. Most importantly, is this donor lineage extinct, or an exotic form of life that has so far eluded discovery? Does the Pyl system represent a genetic "life raft", or a "message in a bottle"?

The PylS homologs in archaea and bacteria have a different organization. In bacteria, PylS is encoded by two adjacent genes, with one encoding the N-terminal anticodon recognition domain of the synthetase, and the other encoding the C-terminal aminoacylation domain. In archaea, PylS is present as a single gene product. This major organizational difference further diminishes support for the explanation of recent divergence followed by HGT. Additionally, phylogenetic reconstruction of the PylS gene family shows that the branch lengths between archaeal and bacterial homologs are similar to those of other vertically inheritied aaRS proteins. If these genes were recently horizontally transferred, one would expect branch lengths between the bacterial and the archaeal homologs to be much shorter than those in other synthetase families, as they are separated by less evolutionary time. Branch lengths between the bacterial homologs also show distances similar to those between other aaRS proteins found in the same species (which are presumably vertically inherited). If PylS genes were all transferred from the same donor lineage, one would expect them to be much more similar to one another than would be expected based on vertical inheritance alone *(41)*. Given this observation, one is forced to hypothesize that, assuming HGT occurred, the archaeal and bacterial genes were transferred from different donor organisms. One possible scenario satisfying this requirement is depicted in **Fig. 9.4**. A single donor lineage could possibly still account for this signal, if the "splitting" of PylS into two genes increased the rate of evolution in these homologs, increasing their branch lengths. This would explain the large pairwise distances between the archaeal and the bacterial homologs, and would create the appearance of vertical inheritance. However, this does nothing to explain the level of divergence between bacterial sequences, which should not show the same increase in branch lengths, as both have inherited the split-gene homolog. While a transfer event between bacteria followed by rapid divergence could produce the same long branch lengths, it seems implausible based on the observation that these proteins have not changed in function (implying little purifying or positive selection at work), and other HGT events involving aaRS proteins do not seem to impact their rate of evolution *(42)*. Furthermore, the branch leading to each bacterial homolog is of similar length, when asymmetrical branch lengths would result from a transfer followed by rapid divergence in the recipient lineage.

Another complication involves the timing of the transfer events. HGT indicates contemporaneous existence of both the recipient and the donor lineages. If the donor lineage is extinct, one would expect this event to be more likely earlier in the history of life. However, the donors must have persisted long enough for a HGT to the ancestor of the *Methanosarcinales* to actually take place. While the time of divergence of *Methanosarcinales* from other euryarchaeotes is not known, phylogenetic markers exist in a related group of organisms that allow for at least an approximate dating for this clade. Along with *Methanomicrobiales*, this order of methanogens forms the sister group to the *Halobacteriales* (**Fig. 9.1**). Recently, a tyrosyl-tRNA synthetase gene was identified as being transferred from within *Halobacteriales* to the ancestor of animals and fungi *(42)*. Given the estimated time of divergence of animals and fungi, this suggests *Halobacteriales* diverged from *Methanosarcinales-Methanomicrobiales* approximately 1–2 billion years ago. Therefore, the hypothesized gene transfer of the PylS system could have occurred well over a billion years ago, more than enough time necessary for massive extinctions to remove all traces of the donor lineage.

Unfortunately, this same logic does not produce a similar result for the bacterial homologs. Since *D. hafniense* is a firmicute, and the *O. algarvensis* symbiont is a delta-proteobacteria, their shared common ancestor is almost as old as the bacterial domain. However, as previously discussed, while the phylogenetic distance separating these PylS homologs suggest vertical inheritance rather than a secondary HGT event, a transfer event to their common ancestor would require massive numbers of independent gene losses, and is no more likely than simple vertical inheritance model with no HGT whatsoever. If we attempt to resolve this by hypothesizing two independent gene transfer events directly to the recipient lineages, we move too far in the other direction, with not enough geological time left for the massive extinctions required to "trim" the tree of life to its current state. One possibility is a "balanced" model in which ancient gene transfers occurred with timing similar to that discussed for the archaea. In this model, independent transfer events would have occurred in ancestral bacterial lineages, with a limited amount of gene loss along each branch to generate the sparse distribution that is currently observed. Presumably, there is a window of evolutionary time in which this could occur that would (1) plausibly correlate with the observed branch lengths; (2) avoid the invocation of an absurd number of lineage-specific gene losses; and (3) be contemporaneous with a plausible timing for the archaeal HGT events. However, this is simply an ad hoc solution, and no evidence exists to support this model.

It seems that, in the case of the bacterial homologs, the entire scenario simplifies greatly if an undiscovered, existing ancient lineage is invoked as the donor organism. Then, there is no reason to link the archaeal and bacterial HGT events in time, avoiding the above-stated convoluted scenarios. The HGT to the archaea would simply be an older transfer from the same unknown group. In order to explain the large diversity between bacterial homologs, this undiscovered lineage would need to contain a substantial amount of diversity, perhaps rivaling that currently found in the bacterial domain.

Given the large number of mass extinctions in our planet's history, HGT can allow for some gene families to survive these events, and contribute to the diversity and fitness of the surviving lineages. Since this effect would be especially pronounced early in the evolution of life, it is reasonable that some of these transferred genes are from extinct lineages that diverged before the MRCA of known living organisms *(43, 44)*. Use of pyrrolysine may be one of these systems, which happened to find a suitable niche in the methanogenic pathway. Perhaps even more compelling is the possibility that, as evidenced by the observed gene transfers of the Pyl system to highly derived, specific bacterial species, this donor lineage still survives today. Searching for other sets of phylogenetically isolated, deeply branching gene families could potentially further implicate the existence of such a cryptic lineage, and, based on their functional roles, perhaps even suggest the type of environment where these organisms might be found.

## 4. Conclusions and Outlook

HGT plays a large role in the evolution of metabolic pathways. These transfer events are most easily detected and characterized when they occur between known groups with significant sequence information. Such is the case with the acetoclastic methanogenesis pathway in *Methanosarcina*. However, HGT events can also be inferred from other information, such as anomalous phylogenetic distributions and deep rootings to paralogous sequences. These signatures are apparent in the genes for pyrrolsyine incorporation in *Methanosarcinales* and two divergent bacterial species, suggesting a transfer from an ancient, deeply rooted lineage, either extinct or hitherto undiscovered. If such transfer events are commonplace, then the physiology of modern organisms must not only be understood in the context of known biological diversity but also include those species not yet characterized, and, perhaps more importantly, the immense reservoir of extinct lineages whose only remains may be transferred genes.

## Acknowledgments

I thank J. Peter Gogarten, Pascal Lapierre, and Kristen Swithers for frequent helpful discussions and comments.

### References

1. Horowitz, N. H. (1945) On the evolution of biochemical syntheses. *Proc Natl Acad Sci U S A* **31**, 153–7.
2. Jensen, R. A. (1976) Enzyme recruitment in evolution of new function. *Annu Rev Microbiol* **30**, 409–25.
3. Fondi, M., Brilli, M., Emiliani, G., Paffetti, D., Fani, R. (2007) The primordial metabolism: an ancestral interconnection between leucine, arginine, and lysine biosynthesis. *BMC Evol Biol* 7 **Suppl 2**, S3.
4. Fani, R., Brilli, M., Fondi, M., Lio, P. (2007) The role of gene fusions in the evolution of metabolic pathways: the histidine biosynthesis case. *BMC Evol Biol* 7 **Suppl 2**, S4.
5. Caspi, R., Foerster, H., Fulcher, C. A., Kaipa, P., Krummenacker, M., Latendresse, M., Paley, S., Rhee, S. Y., Shearer, A. G.,

Tissier, C., Walk, T. C., Zhang, P., Karp, P. D. (2007) The MetaCyc Database of metabolic pathways and enzymes and the BioCyc collection of Pathway/Genome Databases. *Nucleic Acids Res* **36**, D623–31.
6. Dejongh, M., Formsma, K., Boillot, P., Gould, J., Rycenga, M., Best, A. (2007) Toward the automated generation of genome-scale metabolic networks in the SEED. *BMC Bioinformatics* **8**, 139.
7. Pal, C., Papp, B., Lercher, M. J. (2005) Adaptive evolution of bacterial metabolic networks by horizontal gene transfer. *Nat Genet* **37**, 1372–5.
8. Chistoserdova, L., Jenkins, C., Kalyuzhnaya, M. G., Marx, C. J., Lapidus, A., Vorholt, J. A., Staley, J. T., Lidstrom, M. E. (2004) The enigmatic planctomycetes may hold a key to the origins of methanogenesis and methylotrophy. *Mol Biol Evol* **21**, 1234–41.
9. Chistoserdova, L., Laukel, M., Portais, J. C., Vorholt, J. A., Lidstrom, M. E. (2004) Multiple formate dehydrogenase enzymes in the facultative methylotroph *Methylobacterium extorquens* AM1 are dispensable for growth on methanol. *J Bacteriol* **186**, 22–8.
10. Gribaldo, S., Brochier-Armanet, C. (2006) The origin and evolution of Archaea: a state of the art. *Philos Trans R Soc Lond B Biol Sci* **361**, 1007–22.
11. Bapteste, E., Brochier, C., Boucher, Y. (2005) Higher-level classification of the Archaea: evolution of methanogenesis and methanogens. *Archaea* **1**, 353–63.
12. Li, Q., Li, L., Rejtar, T., Lessner, D. J., Karger, B. L., Ferry, J. G. (2006) Electron transport in the pathway of acetate conversion to methane in the marine archaeon Methanosarcina acetivorans. *J Bacteriol* **188**, 702–10.
13. Smith, K. S., Ingram-Smith, C. (2007) *Methanosaeta*, the forgotten methanogen? *Trends Microbiol* **15**, 150–5.
14. Ferry, J. G. (1992) Methane from acetate. *J Bacteriol* **174**, 5489–95.
15. Galagan, J. E., Nusbaum, C., Roy, A., Endrizzi, M. G., Macdonald, P., Fitzhugh, W., Calvo, S., Engels, R., Smirnov, S., Atnoor, D., Brown, A., Allen, N., Naylor, J., Stange-Thomann, N., Dearellano, K., Johnson, R., Linton, L., Mcewan, P., Mckernan, K., Talamas, J., Tirrell, A., Ye, W., Zimmer, A., Barber, R. D., Cann, I., Graham, D. E., Grahame, D. A., Guss, A. M., Hedderich, R., Ingram-Smith, C., Kuettner, H. C., Krzycki, J. A., Leigh, J. A., Li, W., Liu, J., Mukhopadhyay, B., Reeve, J. N., Smith, K., Springer, T. A., Umayam, L. A., White, O., White, R. H., Conway De Macario, E., Ferry, J. G., Jarrell, K. F., Jing, H., Macario, A. J., Paulsen, I., Pritchett, M., Sowers, K. R., Swanson, R. V., Zinder, S. H., Lander, E., Metcalf, W. W., Birren, B. (2002) The genome of *M. acetivorans* reveals extensive metabolic and physiological diversity. *Genome Res* **12**, 532–42.
16. Boone, D., Whitman, W., Koga, Y. (2001) Order III. Methanosarcinales, in *Bergey's Manual of Systematic Bacteriology* (Boone, D., Castenholz G., Garrity G., ed.) Springer-Verlag, New York, 268–94.
17. Thauer, R. K. (1998) Biochemistry of methanogenesis: a tribute to Marjory Stephenson. 1998 Marjory Stephenson Prize Lecture. *Microbiology* **144 (Pt 9)**, 2377–406.
18. Ingram-Smith, C., Martin, S. R., Smith, K. S. (2006) Acetate kinase: not just a bacterial enzyme. *Trends Microbiol* **14**, 249–53.
19. White, D. (2000) *The Physiology and Biochemistry of Prokaryotes*, Oxford University Press, New York.
20. Meile, L., Rohr, L. M., Geissmann, T. A., Herensperger, M., Teuber, M. (2001) Characterization of the D-xylulose 5-phosphate/D-fructose 6-phosphate phosphoketolase gene (xfp) from *Bifidobacterium lactis*. *J Bacteriol* **183**, 2929–36.
21. Wolfe, A. J. (2005) The acetate switch. *Microbiol Mol Biol Rev* **69**, 12–50.
22. Fournier, G. P., Gogarten, J. P. (2008) Evolution of acetoclastic methanogenesis in *Methanosarcina* via horizontal gene transfer from cellulolytic Clostridia. *J Bacteriol* **190**, 1124–7.
23. Ronquist, F., Huelsenbeck, J. P. (2003) MrBayes 3: Bayesian phylogenetic inference under mixed models. *Bioinformatics* **19**, 1572–4.
24. Felsenstein, J. (2005) PHYLIP (Phylogeny Inference Package). In Distributed by the Author. Department of Genome Sciences, University of Washington, Seattle Place.
25. Guindon, S., Gascuel, O. (2003) A simple, fast, and accurate algorithm to estimate large phylogenies by maximum likelihood. *Syst Biol* **52**, 696–704.
26. Stams, A. (1994) Metabolic interactions between anaerobic bacteria in methanogenic environments. *Antonie Van Leeuwenhoek* **66**, 271–94.
27. Beiko, R. G., Harlow, T. J., Ragan, M. A. (2005) Highways of gene sharing in prokaryotes. *Proc Natl Acad Sci U S A* **102**, 14332–7.
28. Lawson, P. A., Llop-Perez, P., Hutson, R. A., Hippe, H., Collins, M. D. (1993) Towards a phylogeny of the clostridia based on 16S rRNA sequences. *FEMS Microbiol Lett* **113**, 87–92.
29. Desvaux, M. (2005) *Clostridium cellulolyticum*: model organism of mesophilic

cellulolytic clostridia. *FEMS Microbiol Rev* **29**, 741–64.
30. Collins, M. D., Lawson, P. A., Willems, A., Cordoba, J. J., Fernandez-Garayzabal, J., Garcia, P., Cai, J., Hippe, H., Farrow, J. A. (1994) The phylogeny of the genus *Clostridium*: proposal of five new genera and eleven new species combinations. *Int J Syst Bacteriol* **44**, 812–26.
31. Wellman, C. H., Osterloff, P. L., Mohiuddin, U. (2003) Fragments of the earliest land plants. *Nature* **425**, 282–5.
32. Min, H., Zinder, S. H. (1989) Kinetics of acetate utilization by two thermophilic acetotrophic methanogens: *Methanosarcina* sp. Strain CALS-1 and *Methanothrix* sp. Strain CALS-1. *Appl Environ Microbiol* **55**, 488–91.
33. Zhang, Y., Gladyshev, V. N. (2007) High content of proteins containing 21st and 22nd amino acids, selenocysteine and pyrrolysine, in a symbiotic deltaproteobacterium of gutless worm *Olavius algarvensis. Nucleic Acids Res* **35**, 4952–63.
34. Herring, S., Ambrogelly, A., Polycarpo, C. R., Soll, D. (2007) Recognition of pyrrolysine tRNA by the *Desulfitobacterium hafniense* pyrrolysyl-tRNA synthetase. *Nucleic Acids Res* **35**, 1270–8.
35. Krzycki, J. A. (2004) Function of genetically encoded pyrrolysine in corrinoid-dependent methylamine methyltransferases. *Curr Opin Chem Biol* **8**, 484–91.
36. Zhang, Y., Baranov, P. V., Atkins, J. F., Gladyshev, V. N. (2005) Pyrrolysine and selenocysteine use dissimilar decoding strategies. *J Biol Chem* **280**, 20740–51.
37. Polycarpo, C., Ambrogelly, A., Berube, A., Winbush, S. M., Mccloskey, J. A., Crain, P. F., Wood, J. L., Soll, D. (2004) An aminoacyl-tRNA synthetase that specifically activates pyrrolysine. *Proc Natl Acad Sci U S A* **101**, 12450–4.
38. Longstaff, D. G., Blight, S. K., Zhang, L., Green-Church, K. B., Krzycki, J. A. (2007) In vivo contextual requirements for UAG translation as pyrrolysine. *Mol Microbiol* **63**, 229–41.
39. Longstaff, D. G., Larue, R. C., Faust, J. E., Mahapatra, A., Zhang, L., Green-Church, K. B., Krzycki, J. A. (2007) A natural genetic code expansion cassette enables transmissible biosynthesis and genetic encoding of pyrrolysine. *Proc Natl Acad Sci U S A* **104**, 1021–6.
40. Zhaxybayeva, O., Gogarten, J. P. (2004) Cladogenesis, coalescence and the evolution of the three domains of life. *Trends Genet* **20**, 182–7.
41. Novichkov, P. S., Omelchenko, M. V., Gelfand, M. S., Mironov, A. A., Wolf, Y. I., Koonin, E. V. (2004) Genome-wide molecular clock and horizontal gene transfer in bacterial evolution. *J Bacteriol* **186**, 6575–85.
42. Huang, J., Xu, Y., Gogarten, J. P. (2005) The presence of a haloarchaeal type tyrosyl-tRNA synthetase marks the opisthokonts as monophyletic. *Mol Biol Evol* **22**, 2142–6.
43. Gogarten-Boekels, M., Hilario, E., Gogarten, J. P. (1995) The effects of heavy meteorite bombardment on the early evolution – the emergence of the three domains of life. *Orig Life Evol Biosph* **25**, 251–64.
44. Gogarten, J. P., Fournier, G., Zhaxybayeva, O. (2007) Gene transfer and the reconstruction of life's early history from genomic data. *Space Sci Rev* **135**, 115–31.
45. Edgar, R. C. (2004) MUSCLE: multiple sequence alignment with high accuracy and high throughput. *Nucleic Acids Res* **32**, 1792–7.

# Chapter 10

# Genome Acquisition in Horizontal Gene Transfer: Symbiogenesis and Macromolecular Sequence Analysis

**Lynn Margulis**


## Abstract

Phylogenetic diagrams ("trees of life") based on computer-generated analyses of nucleic acid (DNA, RNA) or protein (amino acid residues) sequences are purported to reconstruct evolutionary history of the living organisms from which the macromolecules were isolated *(1)*. "Horizontal gene transfer", an expression that refers to the ad hoc explanation of anomalous distribution of these macromolecular sequences, is an inferred past event to explain evolution that, even in principle, is not documentable. Although the diagrams ("phylogenies") help establish the details of relationships among important and widely distributed essential components of living systems (e.g., DNA of large and small replicons such as plasmids, viruses, genophores), chromatin, or protein enzymes that have conserved their function throughout the history of the evolutionary lineage (e.g., DNA that codes for polymerases or 16/18S ribosomal RNA), the HGT concept is an Alfred North Whiteheadian fallacy of misplaced concreteness *(2)*. It is deeply flawed because of sets of unstated, unwarranted assumptions accepted as fact by practitioners: genomics and proteomic experts. They tend to be zoocentric and in particular anthropocentric computer scientists. Their relative lack of familiarity with the fossil record, hard-won life histories and transmission-genetics, taxonomy, physiology, metabolism, and ecology of the communities in which the organisms invariably reside, and many other aspects of life have led to codification of systematic errors in analysis of their, often superb, molecular data. Here we point to a prodigious but little-known symbiogenesis literature that contrasts the transfer of sets of genes with HGT taken to mean one or a-very-few-genes at a time.


**Key words:** *Elysia*, karyomastigont, symbiogenesis, replicons, parasexuality, fallacy of misplaced concreteness, microbial communities as units of selection.

## 1. Introduction: Evolutionary Lineages: Divergence and Mergers

This chapter explains how, because of the limitations of the methods and their built-in assumptions, consequences of symbiogenesis may be easily confused with single-gene transfer

Maria B. Gogarten et al. (eds.), *Horizontal Gene Transfer: Genomes in Flux, vol. 532*

DOI 10.1007/978-1-60327-853-9_10 Springerprotocols.com

between partners in cyclical (parasexual) and even meiotic (sexual) associations. Some examples from an excellent, robust but underappreciated primary scientific literature that can only be mentioned here include (a) eukaryosis [symbiogenetic origin of nucleated cells from bacterial ancestors during the Proterozooic eon] *(3–5)* (**Fig. 10.1**), (b) fescue or *Lolium* grass that inherits a necrotrophic zygomycote through the seed, *(6)*, (c) *Geosiphon pyriforme* [an example of parasexuality where at approximately six-week intervals "fertilization" between the cyanobacterium *Nostoc* and a zygomycotous fungus leads to the development of unique "plant-like" offspring], *(7)*, and (d) cyclical merger where the partners [the small cephalopod *Euprymna* and its luminous bacteria *Vibrio fischeri*] daily associate *(8)*. Together they generate unique tissue morphogenesis: a ciliated, three-dimensional light organ that undergoes apoptosis. The tiny bobtail squid-luminous bacterial duo leads to counter-illumination of the shell-less cephalopod against the night sky.

Photoautrotrophy in once-heterotrophic animals provides a spectacular example of genome acquisition. Transferred from the animal's food alga (the xanthophyte *Vaucheria litorea*) is a gene (*pbo*) for a protein product (PbO) required in oxygenic

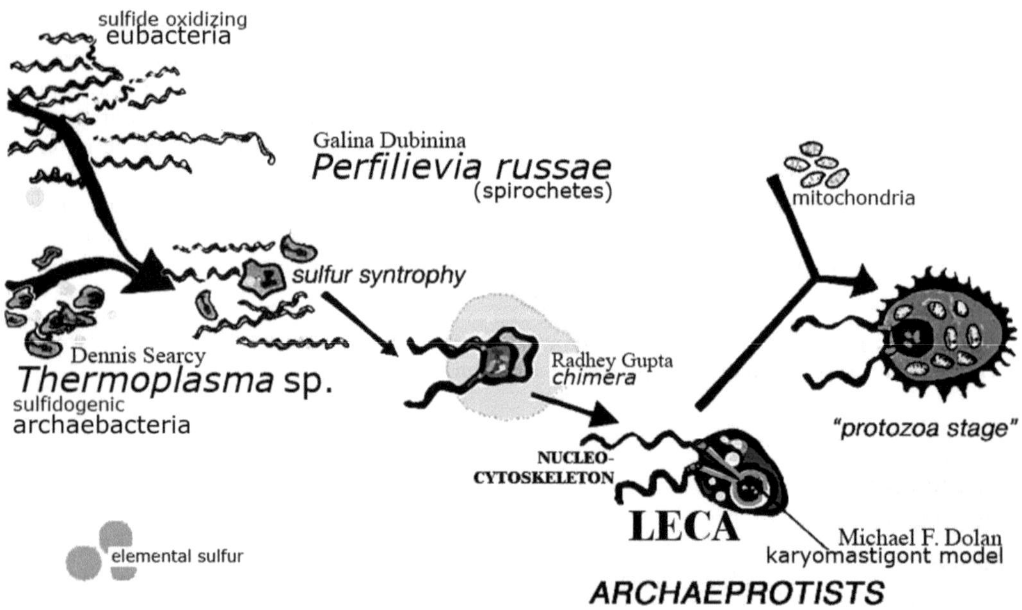


Fig. 10.1. **LECA**: Evolution of the last eukaryotic common ancestral cell by merger or joint genome acquisition. Bionts that formed the new first amitochondriate protist were sulfidogenic, fermenting thermoacidophilic *Thermoplasma*-like archaebacteria, work of Dennis Searcy and *Perfilievia* motile spirochete eubacteria (work of Galina Dubinina). The first eukaryotic cell contained a symbiogenetic organellar system: the karyomastigont with its three components undulipodium including its kinetosome, nuclear connector (labeled "rhizoplast" in old protozoological literature) described by R. Janicki in 1915. Many co-descendants of this amitochondriate protist (Phylum Archaeprotista) are extant in certain anoxic environments, e.g., parabasalids of wood-feeding roaches and termite intestines described by Michael F. Dolan and others. See Chapman and Alliegro *(5)*; Hall *(19)* for status of the cytoskeleton aspect of our symbiogenetic "karyomastigont model" *(3, 28)*.

photosynthesis. The gene, indisputably of *Vaucheria litorea* plastid origin, is detected in the nucleus of all slugs in the population *(9)*. *Elysia chlorotica*, the photosynthetic green slug, which as an adult is capable of survival on light as energy source and $CO_2$ as its only carbon source, has become an autotroph. Adventitious photosynthesis as in the North American *E. chlorotica* or the European *E. viridis* or *E. crispata* is illustrative of many examples of translucent, hungry aquatic animals or protists that secondarily became phototrophs (e.g., *Plachobranchus oscellatus, Convoluta roscoffensis, Stentor polymorphus; Paramecium busaria*). These species abandoned aspects of their original nutritional mode, which was, in these examples, near-universal heterotrophy of saccoglossan mollusks and the flatworm *Convoluta convoluta*, and the phagocytotic heterotrophy of the ciliate congenerics *Stentor coeruleus* and *Paramecium aurelia*.

Perhaps the best evidence for "horizontal gene transfer" in today's world is in the cyclical symbiosis literature. The process, especially of bacteria–insect merger, is a part of the phenomenon of partner biont merger and integration in the development of the holobiont (e.g., *Wolbachia* in insects, *Agrobacterium* in dicot plants, nitrogenase systems of legumes). Confusion and premature conclusions about evolutionary history inadvertently exemplify the "fallacies of misplaced concreteness", including the assumption-ridden phrase "horizontal gene transfer" or its equivalent "lateral gene transfer". We question the methodologically driven new science sometimes oblivious to history, especially the genetic-trait distribution pattern used to calibrate the more arcane computer methods.

Hundreds of careful observers for more than 150 years have documented and communicated with each other about life's "descent with modification", its "natural history" in at least eight countries (Germany, Japan, Russia, Scotland, Switzerland, The Netherlands, The United Kingdom and the United States *(10, 11)*). I admonish that more care be taken to recognize the ubiquity of whole-organism mergers in the study of the genes, which superficially seem to exemplify HGT. Upon closer look, the assumption that at a specific time in the past one gene at a time (by itself) entered a lineage in which that gene was not previously present should be challenged. Critical appraisal of the poorly defined hypothesis of HGT must be weighed against other modes of investigation of the merger. The goal of reconstruction of phylogeny of organisms, the reconstruction of the evolutionary past from clues taken to be representational, requires vigilant reappraisal of the validity of the clues.

The assumptions (numbered in parenthesis) and their alternatives [mentioned in brackets] that we question include (1) that "bifurcations or trifurcations of branches on a tree" represent the correct topology. [The web or net is far more justified as representative of evolutionary history]. (2) We query the inchoate typological inference that an organism's taxon name properly represents the group. [It often stands in only for a very few poorly known members of a diverse, speciose taxon]. (3) We doubt that mutational changes in macromolecular sequences are always, or even usually, a direct function of time elapsed between the divergence of the compared organisms from a common ancestor. [We suggest that rather they are differences in the molecular composition of the pair of living beings compared]. (4) We doubt that comparisons of gene sequences are necessarily revealing meaningful homologies. [Since genes are often in families, specific comparisons may be only to paralogs and therefore inappropriate. At best DNA-level gene comparisons yield only partial, not full, phylogenies of the organisms they are touted to represent]. Databases reveal sequences that in principle can code for macromolecules, i.e., hypothetical proteins, for which information about function or metabolism is limited or non-existent. As explored by H.J. Morowiz *(12)*, it is ironic that entire intermediary metabolic pathways, whose biochemical details are discerned only with great difficulty, tend to be much better conserved than the ephemeral nucleic acid sequences that code for their enzymes. For reconstruction of evolutionary history of organisms, therefore, proteomics in comparisons of multienzyme pathways (semes) are far more useful in principle than any set of single-gene sequences to discern lineages of valid taxa.

Most egregious is the systematic failure to distinguish ***horizontal gene transfer*** [**HGT** that examines one or a few genes at a time] from ***horizontal genome transfer*** [**HgenT** that, when it first occurs, usually involves from more than 500 to some 10,000 or so simultaneously transferred genes as a set that codes for the semes of whole organisms and their dormant propagules]. The way in which ***horizontal genome transfer*** is equivalent to ***symbiogenesis*** (permanent symbiosis, chronic infection) is outlined.

This chapter explores "horizontal gene transfer" from first principles of biological and paleontological science. Because the declarative statements in this chapter are supported by published evidence (or in rare cases, manuscripts in preparation), they are not developed in detail. The reader is urged to study the original sources cited in the books of our reference list. Here we outline the multi-componented process Charles Darwin called "descent with modification" that is commonly labeled "evolution", and describe its essential components. What is the minimal evolutionary system, indeed the minimal unit of life? What topologies represent evolution as understood today? To answer questions about "horizontal heredity" (equivalents of the term include "horizontal gene transfer", "lateral gene transfer", or "lateral gene flow"), the alternative assumption of "vertical heredity", both uniparental and biparental, best be contrasted.

## 2. What Evolves?

Populations of organisms embedded in their distinctive communities in specific environments at given places and times evolve. In all environments fluxes of matter (carbon compounds, mineral nutrients, etc.) and energy (light, oxidizable inorganic chemicals) that support population growth are identifiable. The places and times are distinguishable on maps of Earth and geological time scales [time-rock divisions from the lower Archean on (3600 mya) to the end of the Holocene in the last 10,000 years of the Phanerozoic eon]. "Darwinian evolution" refers to changes over time in natural populations and not changes in organisms and the cells that comprise them. Cells and organisms develop, they do not evolve.

Communities are units in time and space in which short-term changes in relative numbers of organisms that comprise them, given their specific weather regimes, can be identified as "ecological change". Evolution occurs over periods of time greater than the centuries or millenia of ecological change that correspond to climatic change. The discussion of evolution intrinsically refers to reconstruction of natural history, long-term change in populations of organisms from tens of thousands to 3000 million years. Estimates of the numbers of extant kinds of life ("species" or at least differently named less inclusive groups of organisms, i.e., "lower taxa") are: 10–20,000 types of prokaryotes; 250,000 protoctists (nucleated, eukaryotic microorganisms, and their descendants exclusive of animals, plants, and fungi. All inclusive groups of autopoietic taxa are summarized in **Table 10.1**). We scientists are advised to regard with humility the reconstruction of the history of such a prodigious and dynamic system as life on Earth. The shorthand use of "genes" to outline evolutionary history of all life must be an overgeneralization. Adequate description of total, rather than partial phylogeny, namely the history of taxa of life that equals the evolution of life on Earth, recognizes the relevance of paleontology: the geological history of life as derived directly from the sedimentary rock record.

Darwinian evolution of organisms through time is a phenomenon that intrinsically involves three processes at least: population growth, inherited variation, and natural selection. All three are required for the evolutionary process, and have been measured in myriad populations on many occasions in a huge number of laboratory and field studies. The tendency for exponential growth of populations is the fact that all populations can grow at rates that ensure they cannot be supported by their environment. The tendency toward exponential growth, biotic potential, is a species-specific measurable number: maximum number of

**Table 10.1**
**Autopoietic entities: two domain-five kingdom approach**

| Domains | Kingdoms | Basis for classification | References |
|---|---|---|---|
| PROKARYA Did not evolve symbiogenetically | PROKARYA (Eubacteria and archaebacteria) | Prokaryotic ultrastructure with genophores (large replicons) | *(20,21,24)* |
| EUKARYA Evolved by symbiogenesis | PROTOCTISTA | Eukaryotic ultrastructure, microorganisms, and their multicellular descendants exclusive of fungi, animals, and plants. Variations on mitosis and meiotic life histories | *(22,24)* |
| EUKARYA Evolved by symbiogenesis | FUNGI | Amastigote spore-formers by zygotic meiosis, haploid or dikaryotic | *(23,24)* |
| EUKARYA Evolved by symbiogenesis | ANIMALS | Anisogamonts and gametes, embryos, gametic meiosis | *(24)* |
| EUKARYA Evolved by symbiogenesis | PLANTS | Haplo-diploid life cycles, sporogenic meiosis | *(24)* |
| Did not evolve symbiogenetically | Viruses, other small replicons | NOT AUTOPOIETIC | |

offspring produced per unit time or per unit generation. For humans the biotic potential (maximum number of children per couple per generation) of 32 was reported for a Brazilian couple in 2008. *Proteus vulgaris*, a common Gram-negative bacterium can reproduce by division once every 15 min and, in principle, if not impeded, one cell could reach the weight of the Earth in less than a week. *Alternaria*, a forest fungus can produce 150,000 spores per minute for 6 months. These prodigious growth rates are invariably impeded by Darwinian natural selection. The literature on the presence of inherited variation in laboratory and field studies is vast. Evolution, therefore, is a well-documented fact.

## 3. Autopoietic Entities

The minimal living system is a cell *(12, 13)*. Cells, surrounded by water and organic molecules in solution, are autopoietic entities; they are either organisms or parts of organisms. Autopoiesis refers to the membrane-self-boundedness of cells and organisms and the maintenance of their boundaries by internally controlled metabolism. The genome is organized into a genophore. The descriptive term for the genophore, often but not always visible as a "nucleoid" in thin section electron micrographs, is a "large replicon" *(14)*. The genome of all prokaryotes consist of at least one large replicon, but many archae- and eubacteria cells harbor additional large replicons and/or one to several small replicons (e.g., plasmids, episomes, transposons, viruses, viroids, and other DNA fragments that, in principle, are too small to insure and sustain autopoiesis).

## 4. "Horizontal" vs. "Vertical" Inheritance

### *4.1. Replicons and Modes of Inheritance*

The replicons of prokaryotes, unlike those of nucleated cells, are composed of DNA in which the unit fibril is 20 Å in diameter. In eukaryotes the unit fibril, made of chromatin, is 100 Å and coated with nucleosomes composed of histone (basic, arginine- and lysine-rich) proteins. DNA molecules comprise the genomes of cells whether they are the genophores of prokaryotes or the chromosomes of eukaryotes. Both large and small replicons are made of the genes, small fractions of complete genomes, whose DNA sequences are compared in the study of both "horizontal" and "vertical" gene flow.

As this book abundantly illustrates, entire genomes are correlated with the live bodies of the individual cells and organisms in which they must reside, but component genes may flow across the boundaries of the self-maintained individuals. "Vertical" gene flow refers to "normal" reproductive and inheritance patterns from generation to generation. Vertical gene flow includes inheritance of the genome of a single parent, as in the binary fission of both archae- and eubacteria, conidia formation in fungi, soredia generation and dispersal in lichens, and many other reproductive modes of propagule formation well known to biologists. Vertical gene flow in its most widely documented form is that of genes correlated with Mendelian inheritance patterns: meiotic sexuality of animals, plants, most fungi, and many protoctists. In vertical inheritance sets of genes are "linked" on chromosomes, co-inherited in a predictable fashion.

### *4.2. Gene "Transfer" vs "Genome Transfer"*

When parents mate and conceive offspring, those offspring "vertically" acquire homologous genomes, coherent sets of genes, enzymes, and other products directly related by ancestry to their fathers and mothers. Homologous genome transfer is the apt expression to describe uniparental inheritance in genderless species such protoctists as *Amoeba proteus, Euglena gracilis*, or *Dinenympha striata*.

Symbiosis, an ecological phenomenon, refers to the protracted physical association of differently named organisms throughout most of the life history of at least one of them.

Symbiogenesis, a related evolutionary phenomenon, refers to the appearance of new behaviors, morphology, tissues, metabolic pathways, taxa including species, or other recognizable evolutionary novelties in holobionts related to long-term biont partnerships. In the most intimate of symbiogenetic associations, partners are fully integrated at all levels: topological, temporal, metabolic, gene product, and genetic. Symbiogenesis in the case of a permanent association, for example, the mitochondria and chloroplasts of plant cells, involves "genome transfer" that is the same as "heterologous genome acquisition". Professor Joel Cohen of Rockefeller University in his commentary on our book, *Acquiring Genomes (15)*, notes that in the eukaryotes, especially phagocytotic protoctists, evolution may proceed in each generation in principle "a genome at a swallow"! Branching evolution, bifurcation (dichotomy), or trifurcation (trichotomy), i.e.,

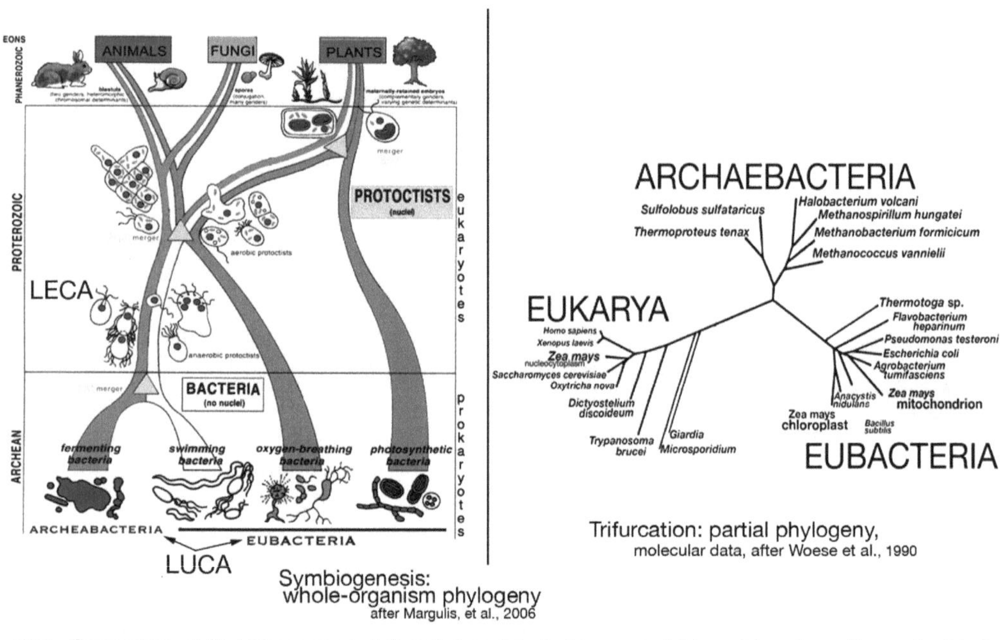


Fig. 10.2. Comparison of Carl Woese et al., trifurcated partial phylogeny model based largely on the nucleotide base sequences of small-subunit ribosomal RNA *(30)*) with our symbiogenetic whole-organism reconstruction of the evolution of early life (*left*, original drawing by K. Delisle, modified from *(29)*). **LECA**, the **l**ast **e**ukaryotic **c**ommon **a**ncestor is detailed in **Fig. 10.1**. Both depictions agree that all life emerged from **LUCA**, the **l**ast **u**niversal **c**ommon **a**ncestor, whether a theoretical construct (the "progenote") that left no trace (*right*) or the ancestor to all other prokaryotes (not just to the four bacterial types that entered eukaryosis).

The three types of ribosomal RNA in *Zea mays* (including chloroplast from free-living cyanobacteria comparable to *Anacystis nidulans*, and mitochondria from oxygen-respiring eubacteria shown near *Agrobacteium* here) support the idea that these organelles originated prior to and separate from the *Zea mays* corn plant cells. The topology at the right privileges the molecular sequence from one of the three sources that is severely limited. Because it ignores the symbiogenetic mergers (acquisition of genomes) thus permitting no anastomosis (but only divergence of branches) in the evolving system of a cell, the topology is flawed. Furthermore, it is seriously deficient as the trifurcation and various bifurcations are not plotted against time (i.e., the fossil record is ignored at the right whereas the eons named at the left provide an investigable time scale.

the "tree analogy", is an inappropriate diagramatic topology for evolution by symbiotic merger. Rather anastomosis, the fusing of lineages is more apt. Mergers of ancestral lineages form **LECA**, the Last Eukaryotic Common Ancestor (**Fig. 10.2**). This was the key event in eukaryosis as nucleated cells evolved by integration in a specific series of symbioses. Members of prokaryotic communities integrated permanently to form larger individuals at more complex levels of integration. Ivan Emmanuel Wallin wrote that microbes, primarily bacteria, as organisms "popularly associated with disease may represent the fundamental causative factor in the origin of species" *(16)* (p.8).

Autopoiesis, as self-maintenance of cells and organisms, requires many genes of different functions (from fewer than 500 to over 30,000) that code for the entire coherent system which we recognize as an organism with a name. Permanent merger, the genome acquisition of one organism by a heterologous other, is a form of symbiogenesis. Therefore, since it systematically involves transfer of far more than a single gene at a time, symbiogenesis often is systematically ignored by the methods of study optimized for gene sequence analysis. The techniques that by their very nature overlook symbiogenesis exacerbate systematic errors of evolutionary interpretation.

## 5. Speciation by "Horizontal Genome Transfer"

A few of the cases of purported origins of species and even higher taxa via symbiogenesis are listed in **Table 10.2**. All these cases proffer anastomosis in lineages that led to evolutionary innovation. Most impressive is Dr. Donald I. Williamson's claim that all holometabolism (dramatic metamorphosis) in larval

**Table 10.2**
**New species and higher taxa generated by genome acquisition (in rough order of appearance in the fossil record)**

| | |
|---|---|
| Eukaryosis, origin of nucleated cells | *(3–5)* |
| Kingdom Protoctista: *Hatena*, *Mixotricha* | *(25,26)* |
| Coleoptera (homage to Santa Rosalia: "an inordinate fondness" for insects and bacteria: *Wolbachia* and *Buchnera*) | *(27)* |
| Cyclical association of *Nostoc* sp. and a oomycota fungus | *(7)* |
| *Euprymna scolopes* and light organ eubacterium *Vibrio fischeri* | *(8)* |
| *Elysia chlorotica*, and other adventitious photosynthetic animals | *(9)* |

development of marine animals and terrestrial arthropods represent now stable ancient fusions of animal lineages *(17)*.

## 6. Conclusions: Toward Remediation by Explicit Statement of Assumptions

The unstated assumptions intrinsic to the methods of genomics and proteomics that perpetrate, especially zoocentric ignorance, may be and have been to some extent remediated by the recognition of problems mentioned here. The use of new computational approaches [e.g., *(18)* and the new algorithm "RoundTrip" developed by Hall *(19)*] have made a significant start in this direction.

The amelioration of cognitive dissonance (perpetrated by many past studies) may be accomplished in practice by research studies such as those by House, Raymond, and Fournier presented in this book. The lack of historical context so prevalent in genomics and proteomics is patently absurd as we realize that the entire goal of these fields is to reconstruct evolutionary history from clues in living autopoietic entities. The past history of life may be more appropriately and rigorously constructed if attempts are made to consider as much relevant information as possible. Such "inclusive historicity" mandates that artificial discipline boundaries be breached and that institutions recognize the relevance of many kinds of investigation (for example, in botany, biochemistry, cell biology, geochronology, infectious disease, paleontology, phycology, planetary science, ultrastructure, microbial community ecology, sedimentary geology, etc.). This admonition is subsumed under my insistence that research into life's origins and evolution on Earth, in spite of the current financial control of research funds, far transcends the borders of "evolutionary biology".

## Acknowledgments

Celeste Asikainen, Michael J. Chapman, Michael F Dolan, Victor Fet, Ricardo Guerrero, John L. Hall, Antonio Lazcano, James MacAllister, Abraham Gomel, W. E. Krumbein, Dorion Sagan, Jan Sapp, Dennis Searcy, Galina Dubinina, The Alexander von Humboldt Foundation A.I. Tauber-The Tauber Fund, Jessica H. Whiteside; the College of Natural Science and Mathematics, The Graduate School of the University of Massachusetts-Amherst and the Eastman Fellowship, Balliol College, Oxford.

## References


1. Margulis, L., Dolan, M. F., Guerrero, R. (1999) The molecular tangled bank: not seeing the phylogenies for the trees. *Bio Bull* **196**, 413–14.
2. Cobb, J. B. Jr., Ed. (2008) *Back to Darwin: A Richer Account of Evolution.* Wm. B. Eerdmans Publihing Co., Grand Rapids, MI, pp. 167–175; 176–184.
3. Margulis, L., Chapman, M., Guerrero, R., Hall. J. (2006) The last eukaryotic common ancestor (LECA): acquisition of cytoskeletal motility from aerotolerant spirochetes in the Proterozoic eon. *Proc Nat Acad Sci U S A* **103**, 13080–5.
4. Dolan, M. F. (2005) The missing piece: the microtubule cytoskeleton and the origin of eukaryotes, in *Microbial Phylogeny and Evolution: Concepts and Controversies.* (Sapp, J., ed.), Oxford University Press, New York, pp. 281–89.
5. Chapman, M. J., Alliegro, M. C. (2007). A symbiogenetic basis for the centrosome? *Symbiosis* **44**, 23–32.
6. Kozo-Polyansky, B.M. (1924) *Symbiogenesis: A New Principle of Evolution* (in Russian, translated and edited by Victor Fet., Harvard University Press, Cambridge MA, English edition, 2009, in press.)
7. Kluge, M., Mollenhauer, D., Mollenhauer, R., Kape, R. (1992) *Geosiphon pyriforme*, an endosymbiotic consortium of a fungus and a cyanobacterium *(Nostoc)*, fixes nitrogen. *Bot Acta*, **105**, 343–44.
8. McFall-Ngai, M. J. (2002) The influence of bacteria on animal development. *Dev Biol* **242**, 1–14.
9. Rumpho, M. E., Worful, J. M., Lee, J., Kannan, Tyler, M. S, Battacharya, D., Moustafa, A., Manhart, J. R. (2008) Horizontal gene transfer of the algal nuclear gene *psbO* to the photosynthetic sea slug *Elysia chlorotica. Proc Nat Acad Sci U S A* **105**, 17867–71.
10. Sapp, J. (1999) *Evolution by Association: History of Symbiosis Research.* Cambridge University Press, Cambridge, MA.
11. Sapp, J., (ed). (2005) *Molecular Phylogeny and Evolution: Concepts and Controversies.* Oxford University Press, New York.
12. Morowitz, H. J., Broyles, D., Lasus, H. (2005) The robustness of intermediary metabolism, in *Microbial Phylogeny and Evolution: Concepts and Controversies.* (Sapp, J., ed.), Oxford University Press, New York, pp. 154–9.
13. Margulis, L., Sagan, D. (2000) *What Is Life?* University of California, Berkeley CA.
14. Sonea, S., Matthieu L. G. (2001) *Prokaryotology: A Coherent View.* Les Presses De L'université de Montréal, Montréal, Canada.
15. Margulis, L., Sagan, D. (2002) *Acquiring Genomes: A Theory of the Origins of Species.* Basic Books, New York.
16. Wallin, I. E. (1926) *Symbionticism and the Origin of Species*, Williams and Wilkins, Baltimore, MD, p. 8.
17. Williamson, D. I., Vickers, S. E. (2007) The origins of larvae: mismatches between the forms of adult animals and their larvae may reflect fused genomes, expressed in sequence in complex life histories, *Am Sci* **95**, 509–17.
18. LeBlanc, M., Dyer, B. (2007) *PERL for Exploring DNA.* Oxford University Press, New York.
19. Hall, J. L. (2009) The spirochete contribution to the eukaryotic lineage: a novel genomic analysis, *Proc Natl Acad Sci U S A* (in press).
20. *Bergey's Manual of Systematic Bacteriology*, 2nd Edition, Volumes 1–5, Springer, New York.
21. Dworkin, M. M., Falkow, S., Rosenberg, E., Schleifer, K.-H. Stackebrandt, E. (eds.) *The Prokaryotes: A Handbook on the Biology of Bacteria*, 3rd Edition, Volumes 1–7, Springer-Verlag, New York.
22. Margulis, L., Corliss, J. O., Melkonian, M., Chapman, D. J. (eds.) (2010) *Handbook of Protoctists*, 2nd Edition. Jones and Bartlett Publishers, Sudbury, MA.
23. Kendrick, B. (2005) *The Fifth Kingdom*, 2nd Edition, Mycologue Press, Waterloo, Ontario, Canada.
24. Margulis, L., Chapman, M. J. (2009) *Kingdoms and Domains illustrated: Phyla of Life on Earth*, 4th Edition, Acadamic Press-Elsevier, London, UK, New York and San Diego, CA.
25. Okamoto, N., Inouye, I. (2006) *Hatena arenicola* gen. et sp. nov., a katablepharid undergoing probable plastid acquisition. *Protist* **157**, 401–19.
26. Margulis, L. (1993) *Symbiosis in Cell Evolution: Life and Its Environment on the Early Earth*, 3rd Edition, W. H. Freeman, San Francisco, CA.
27. Kostas, B., Miller, T. A. (2006) *Insect Symbiosis*, Volume 2 Contemporary Topics in Entomology, CRC Publishers, Boca Raton, FL.
28. Margulis, L., Dolan, M., Whiteside J. (2005) "Imperfections and oddities" in the origin of the nucleus. *Paleobiology* **31**, 175–91.
29. Margulis, L., Brynes. L. (1990) Hard testimony: teaching past environments with fossil foraminifera. *UNESCO Nature & Resources* **35**, 4–17. (English, French and Spanish).
30. Woese, C. R., Kandler, O., Wheelis, M. L. 1990. Towards a natural system of organisms, proposal for the domains Archaea, Bacteria, and Eucarya. *Proc Natl Acad Sci U S A* **87**, 4576–79.

# Part III

## Methods

# Chapter 11

# Detection and Quantitative Assessment of Horizontal Gene Transfer

**Olga Zhaxybayeva**


## Abstract

This chapter discusses the pros and cons of the existing computational methods for the detection of horizontal (or lateral) gene transfer and highlights the genome-wide studies utilizing these methods. The impact of horizontal gene transfer (HGT) on prokaryote genome evolution is discussed.


**Key words:** Phyletic pattern, sequence composition, phylogenetic incongruence, orthologous replacement, gene transfer rate, bipartition analysis, quartet decomposition.

## 1. Introduction

With the availability of 769 completed genomes (GOLD database *(1)*, May 2008), horizontal (or lateral) gene transfer (HGT) can be studied through the examination and comparison of genomic data from various organisms. This chapter provides an overview of the strictly computational methods currently employed for detecting instances of transferred genes, and highlights the recent genome-scale studies of HGT that use these methods. Current approaches to delineating HGT events are far from being perfect: due to varying underlying assumptions, individual methods detect transfer events at different phylogenetic distances and of different age. It is therefore not surprising that they often return disparate sets of HGT candidates in a genome *(2–5)*. In addition, all methods suffer from high error rates. Rates of false positives and false negatives can be estimated through simulations or *in silico* transfers *(6–9)*.

Maria B. Gogarten et al. (eds.), *Horizontal Gene Transfer: Genomes in Flux, vol. 532*

DOI 10.1007/978-1-60327-853-9_11 Springerprotocols.com

## 2. Overview of HGT Detection Methods

### 2.1. Compositional Methods

Since a horizontally transferred gene comes from a different genomic background, its nucleotide sequence can contain recognizable signatures of its previous "location". One group of HGT detection methods uses atypical nucleotide or oligonucleotide composition *(5, 10–12)*, atypical codon usage patterns *(13)*, or their combination *(14)* to infer which genes in a genome are instances of HGT. Because genes "ameliorate" [i.e., adopt the signatures of their new genome] *(5)*, these methods are applicable to the detection of relatively recent transfers only (e.g., in *Escherichia coli* genes ameliorating for up to about 100 million years still could be detected, although the majority of the detected transfers were introduced within the last 10 million years *(5)*). While easily applicable to completely sequenced genomes, these methods have been criticized for returning high rates of false positives and negatives *(6, 15, 16)* (*see* also the **Chapter 12**). The application of a compositional approach to 116 available genomes reveals that the number of detected, recently transferred genes ranges from 0.5% in the pea aphid endocellular symbiont *Buchnera* sp. APS to 25.2% in the anaerobic methane-producing archaeon *Methanosarcina acetivorans* C2A *(17)*. Compositional methods are also routinely utilized to aid the identification of genomic islands (e.g., *(18)*).

### 2.2. Use of Phyletic Patterns

A different way to assess whether a gene could have been transferred is to perform similarity searches of a sequence database using BLAST *(19)* or any other similarity search or clustering algorithm to find the homologs of a query gene from among a selection of available sequenced genomes or even all of GenBank. A gene family can be defined using this information and then the taxonomic distribution of members of the gene family determined. This pattern of gene presence/absence in certain organisms is known as a phyletic pattern. A significant top-scoring BLAST hit, i.e., the most similar sequence in the database, may by itself suggest a transfer event. For example, this approach was used to obtain a rough estimate (24%) for the number of horizontally transferred genes from Archaea in the *Thermotoga maritima* genome *(20)*. However, a top-scoring BLAST hit might not represent a sequence that would group with the query sequence in a phylogenetic reconstruction *(21)*. Additionally, this approach is sensitive to how broadly certain groups of organisms are represented in the database. Therefore, the phylogenetic affiliation of the top-scoring BLAST hit is often not a reliable approach to HGT detection. A modification of the top-scoring hit method *(22)*, which aims to overcome these shortcomings, re-ranks BLAST hits by taking into account taxonomic information and allowing the exclusion of hits from the organisms closely related to a query. Using this method, 11.3% of the *Thermotoga maritima* genes are predicted to be transferred from Archaea *(22)*. The lower estimate of this study is also a result of many more Bacterial genomes available for the analysis since the initial examination of the *T. maritima* genome.

As a special case, the phyletic patterns of closely related organisms can be used to judge amounts of HGT within the group. In a now classic example, three sequenced *E. coli* genomes *(23)* were each shown to harbor a substantial proportion of genes absent from the two other strains (585 genes in non-pathogenic *E. coli* K12, 1623 genes in uropathogenic *E. coli* CFT073, and 1346 genes in enterohaemorrhagic *E. coli* O157:H7); only 39.2% of their common gene pool is found in all three genomes. The genes that are present in only one of three strains are assumed to be introduced into *E. coli* through HGT. *Frankia*, a nitrogen-fixing symbiont of plants belonging to the Actinobacteria phylum, provides another example. The three *Frankia* strains studied by Normand et al. *(24)* are less than 3% divergent in their small subunit ribosomal RNA (i.e., are considered the same species by currently used taxonomy standards). But less than 20% of the genes are present in all of the studied *Frankia* genomes, and the individual genomes have 1112, 1703, and 581 genes, respectively, without any detectable homologs in the other two genomes.

Phyletic patterns can be further used to infer whether the patchy distribution of a gene is most parsimoniously explained by HGTs or by gains through duplication and divergence or losses *(25–27)*. The outcome of the inferences depends on a value of "HGT penalty" (the ratio between HGT events and gene losses), which is not known, but has to be estimated or set *a priori:* different studies disagree on what value to use. An attempt to apply this type of approach to 165 microbial genomes resulted in an inference of $\sim 40,000$ HGTs, $\sim 90,000$ gene losses, and over $\sim 600,000$ vertical transfers in all analyzed gene families *(25)*. While the numbers given above may be interpreted as showing only a limited number of HGTs among the 165 genomes, one should not forget that those estimates do not consider HGT events where an incoming gene replaces its orthologous counterpart in a genome. These instances of orthhologous replacement, sometimes also referred to as xenologous displacement, could constitute a substantial part of a genome (*cf.* **Section 2.4**). Yet another problem stems from the lack of a firm definition of "gene absence" – at what stage in its decay should a gene be declared absent? Consideration of gene remnants within a genome as absent genes can lead to systematic overestimation of within-species HGT events *(28)*.

### 2.3. Use of Substitution Rates

Recently, a class of transfer detection methods based on the comparison of gene substitution rates was proposed. In a method presented by Novichkov et al. *(29)*, pairwise distances within a gene family are compared to pairwise genomic distances (the latter are derived from a reference tree, *see* **Section 2.4**). The assumption is that if a gene evolved vertically in a lineage, then it should often accumulate substitutions at a constant rate (i.e., in a clock-like manner), and hence have distances approximately equal to corresponding genome distances. In the case of a gene that was transferred into a group of organisms from an outside source, the pairwise gene distances of a transferred gene to its untransferred homologs within the group will be approximately the same, regardless of the corresponding genome distances. A problem is that significant departures of gene distances from genome distances can be due to the violation of the clock assumption and not due to HGT. Another shortcoming that this method shares with phylogenetic approaches is that it only considers genes that have recognizable orthologs in other genomes (*see* **Section 2.4**). Because of these limitations, this approach might be best suited for sub-phylum analyses. In the case of the *Bacillus-Clostridium* group, the gamma-, and the alpha-*Proteobacteria*, the clock-like null hypothesis could not be rejected for $\sim$ 70% of the analyzed gene families *(29)*, while 13–22% of the gene families show significant deviations from the clock-like hypothesis and hence support orthologous replacements.

In a different method, Hamady et al. *(30)* propose the use of nucleotide substitution matrix changes (i.e., changes in instantaneous rates of substitution) as a proxy for determining horizontally transferred genes. These authors' reasoning is that transferred genes should have substitution rate matrices that differ from the rate matrix based on the genome as a whole. For example, this method is able to detect transferred genes for which GC content is the same, making it more powerful than compositional methods alone (*see* **Section 2.1**). The drawback of the method is that due to sampling error the method cannot accurately estimate the substitution matrix from a single gene, making it more suitable for identifying stretches of genome containing ten or more transferred genes. Another shortcoming is that the method works under the assumption that the majority of genes are not transferred (*see* also discussion of phylogenetic inference drawbacks in **Section 2.4**). Since the genome rate matrix is calculated using all genes, including the transferred genes, it is not clear how much the rate matrix is impacted by the variable rates of transferred genes.

### 2.4. Phylogenetic Incongruence

Phylogenetic incongruence methods rely on the reconstruction of phylogenetic trees for individual gene families (or, more precisely, sets of orthologous genes), assuming that trees with unexpected phylogenetic histories are results of HGT. Most of these approaches require using an expected phylogenetic history (organismal tree) as a reference tree for the comparison and identification of topological incongruence. Choices of reference trees as a proxy of organismal trees include rRNA trees, genome trees, trees derived from concatenation of selected gene families, or trees (possibly only partially resolved) supported by a plurality of gene families, such as consensus trees and supertrees. One of the earliest analyses of this kind came from the comparison of gene families from the *Aquifex aeolicus* genome to the rRNA tree, showing incongruence of individual gene trees *(31)*. Later, 205 gene families from 13 gamma-proteobacteria were compared to the phylogeny of the same genes concatenated into one dataset *(32)*. In another study, 22,432 gene families from 144 prokaryotic genomes were compared to the reference supertree constructed from strongly supported bipartitions, and families with discordant bipartitions were assumed to be instances of HGT *(33)* (*see* also **Section 3.2**). There are now tools available that consolidate information from the comparison of individual gene trees by mapping possible transfer events onto the reference topology (e.g., *(34–36)*).

Ideally, if the organismal tree is not known, all possible tree topologies should be tried as a reference tree (examples of such methodologies to analyze four and five genomes are given in *(37–39)*; a graphical evaluation of multiple selected topologies using heatmaps is found in *(40)*). However, due to the vast number of possible tree topologies, this approach is currently computationally impossible for analyses involving more than a few genomes. Alternatively, the trees to be analyzed could be broken into smaller pieces (e.g., bipartitions or quartets; see below). There is a much smaller number of possible bipartitions/quartets than trees for a given number of analyzed genomes, and therefore all possibilities can be evaluated (giving rise to bipartition *(41)* and quartet decomposition *(8)* analyses; see below).

A bipartition corresponds to a branch (or split) in a phylogenetic tree. If an edge between two nodes in a tree is removed from the tree, the tree is divided into two unconnected sub-trees. The two sets of leaves of these two sub-trees represent the *bi*partition corresponding to the edge. A given tree then can be represented as a set of bipartitions. Two bipartitions are considered compatible, if they could co-exist on a single tree, and considered incompatible otherwise. Given a tree for a gene family, support values for the different bipartitions in the tree (bootstrap support values or posterior probabilities) can be calculated *(42, 43)*. Bipartition plots (or Lento plots, after *(44)*) summarize the statistical support for bipartitions in the form of a histogram, which graphically shows how much conflicting signal exists among gene family

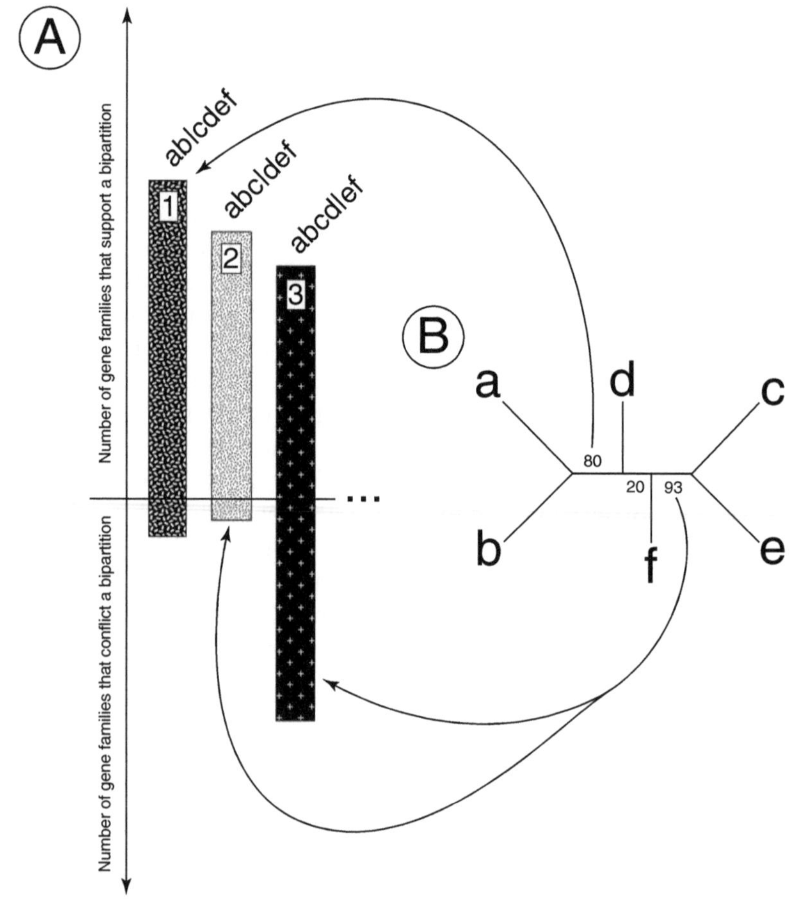


Fig. 11.1. An illustration of a bipartition analysis. **Panel A**. Histogram (Lento plot) of a hypothetical bipartition analysis of six genomes (denoted as *a, b, c, d, e,* and *f*). Each column refers to a bipartition (only three bipartitions out of possible 25 are shown). Each bipartition is labeled using a commonly used notation; for example, *ab|cdef* refers to a split dividing a tree into two sets of taxa: {*a,b*} and {*c,d,e,f*}. The height of a column corresponds to number of gene families supporting (above the X-axis) or conflicting (below the X-axis) the bipartition with a selected support cutoff (e.g., 70% of bootstrap samples). **Panel B**. An example of a gene family and its contribution to the Lento plot (shown by *arrows*). Bipartition *ab|cdef* is supported by 80% of bootstrap samples, agrees with bipartition #1 in panel A and hence counted towards the number of gene families supporting that bipartition. Bipartition *abdf|ce* is supported by 93% of bootstrap samples, contradicts both bipartitions #2 and #3 and therefore counted towards the number of gene families contradicting each of those bipartitions. Bipartition *abd|cef* is supported by only 20% of bootstrap samples (which is below the selected cutoff of 70%) and therefore not included into the Lento plot. For an example of real bipartition analysis, see *(41)*.

trees (*see* **Fig. 11.1**). Gene families giving rise to a bipartition conflicting with bipartitions supported by a majority or plurality of gene families can be considered to have experienced HGT events *(41)*.

Bipartition analyses are useful in finding gene families with conflicting phylogenetic histories without requiring a completely resolved reference tree. However, the applicability of this approach depends on at least some bipartitions being significantly supported by the plurality of gene families *(7)* (*see* also the **Chapter 13**). As more genomes are added to the analyses, the internal edges of a gene tree become shorter and their support values tend to become smaller *(45)*. It is also problematic to combine the analyses of gene families with differing numbers of taxa (*see* also **Section 3.1**). Quartet decomposition analysis attempts to resolve these problems. In this type of analysis a gene tree is "decomposed" into sets of all possible embedded quartets *(8, 43)* (*see* also **Section 6**). An embedded quartet is any subset of a tree consisting of four leaves (taxa). To calculate support for each quartet embedded in the tree, the remaining taxa on the tree are ignored and only the relationships between the four taxa constituting the quartet are evaluated (that is why the quartet is called "embedded"). As with the bipartition approach described above, the putative HGTs are determined by the examination of gene families that give rise to conflicting quartet topologies. For example, in an analysis of ten cyanobacterial genomes, 1128 gene families were present in at least eight genomes *(8)*. A quartet decomposition analysis of these families resulted in 685 gene families with phylogenetic trees incongruent with a reference tree supported by a plurality of gene families, and these were considered as candidates for HGT.

One drawback of all phylogenetic approaches (aside from artifacts of phylogenetic reconstruction, which are not discussed here, see *(46)* and the **Chapter 14**) is that horizontally transferred genes between neighboring taxa on the reference tree are invisible for these methods, because these transfers do not result in a change of tree topology. Another drawback is that weak phylogenetic signal in a set of orthologous genes often results in an unresolved (or insufficiently supported) tree. Multiple HGT events within a gene family also can produce star-like topologies *(47)*. The insufficiently supported topologies cannot be used to delineate HGT events, but they also should not be used (although unfortunately they sometimes are, e.g. *(27, 32)*) as an evidence for the absence of HGT. The third drawback is that often the choice of a reference tree may bias the results of HGT inference and quantification. This is particularly problematic when a reference tree is obtained as a plurality tree (or supertree) from the same sets of genes that are subject to HGT detection in the study. The underlying assumption is that the number of HGT events should be minimized, and that the plurality of genes therefore reflects organismal evolution but not a recurring pattern of HGT. This assumption is not always justified: in addition to the organismal history, the plurality consensus might in part reflect "highways of gene sharing" *(33)*, artifacts due to trees more frequently resulting from incomplete lineage sorting *(46, 48)*, long-branch attraction, or convergent evolution. For example, in the abovementioned analysis of cyanobacterial genomes *(8)*, the different *Prochlorococcus marinus* strains included in the study do not form a monophyletic group and the inferred relationships likely represent some of the "highways of gene sharing".

### 2.5. Scanning for Polymorphic Sites as a Method to Detect Transferred Genes in Closely Related Organisms

In the phylogenetic incongruence methods outlined in the previous section, a gene is considered to be a unit of HGT. However, due to homologous recombination, the genes themselves might be a mosaic of introduced and indigenous copies. Such conflicting, within-gene phylogenetic information often leads to unresolved or unsupported trees that are not useful for phylogenetic incongruence analyses (*see* **Section 2.4**). Mau et al. *(49)* described a method to identify the segments of genomes resulting from orthologous replacement of homologous regions from other, closely related genomes. The method employs a scan of clusters of polymorphic sites in the whole genome alignment of several genomes. For six complete *E. coli* and *Shigella flexneri* genomes, 187 segments are reported to contain polymorphic sites discordant with the predominant phylogenetic signal, spanning 311 (10%) of genes conserved across all six genomes. Several of these genes were shown to be mosaic, i.e., contain at least two significantly different phylogenetic signals. The presented method is conservative, since only highly significant segments were considered to be instances of recombination and hence the overall proportion of genomes involved in homologous recombination is underestimated.

## 3. Quantification of HGT

### 3.1. What Data Should Be Used for Quantification Analyses?

Despite the high rates of false negatives of various HGT detection methods (see also the **Chapter 13**), there are attempts to put numbers to how many transfer events happened in prokaryotic evolution. One challenge is to decide from what genes such extrapolations should be made. To compare genes across many genomes, usually ubiquitous genes (core genes, which are present in all genomes) are used. The ubiquitous core of genes shared between Archaea and Bacteria is tiny: less than 50 genes for 147 genomes *(50)*. Recently, Ciccarelli et al. *(51)* performed an analysis of 191 genomes where they aimed to find genes with "indisputable orthology" and ended up with only 31 such genes. The joint analysis of these genes results in a phylogeny that has some resolution; however, the tree's backbone is poorly resolved ( $< 80\%$), perhaps due to insufficient phylogenetic information. Dagan and Martin *(52)* refer to this tree as "the tree of 1 percent" since only 1% of the genes was determined to fit the null hypothesis of a single, strictly bifurcating tree of life. The extrapolation of results obtained from a very small subset of universal genes should be considered with caution. The relaxation of the core criteria to allow missing data can increase the amount of usable information *(50)*, and methods have been developed to deal with gene families with some missing taxa (*see* **Section 2.4**). Another problem is that many estimates to date are based on the results obtained from just one HGT detection method and hence are affected by the drawbacks of methods of HGT detection (*see* **Sections 2** and **3.4**).

### 3.2. Current Empirical Estimates

Several recent large-scale genome analyses have attempted to estimate HGT across all available sequenced genomes. Beiko et al. *(33)* analyzed the gene families from 144 prokaryotic genomes. They compared 95,194 strongly supported bipartitions from 22,432 gene family trees and found that 86.6% (82,473) of bipartitions are in agreement with bipartitions of a reference supertree (constructed from all highly supported bipartitions). This estimate should be considered biased, as it did not adjust the number of highly supported bipartitions with respect to those resulting from "trivial" relationships (such as between *E. coli* O157:H7 EDL933 and *E. coli* O157:H7). Gene sequences from such trivial relationships always group together with high statistical support values and hence inflate the total number of bipartitions congruent with the reference tree. Choi and Kim *(53)* report up to 9.7% of protein domain families with a history of transfer events, although their method, which is based on the identification of significantly atypical phylogenetic distances within a protein domain family as compared to a 16S rRNA reference tree, is extremely conservative and detects only distant transfer events. Ge et al. *(54)* analyzed 297 gene families from the gene clusters of 40 microbial genomes and compared gene family phylogenies with a genome tree, applying additional criteria to increase stringency. This resulted in 33 HGT events in 11.1% of 297 analyzed gene families. In the latter study, investigators limited themselves to the analyses of very strictly defined (ubiquitous) core genes, hence severely underestimating the number of HGT events. Analyzing larger number of gene families shows that in cyanobacteria as many as 61% of gene families may be affected by HGT *(8)*. It should also be noted that all these analyses are based on the phylogenetic incongruence approach and hence (i) ignore the pool of transitory genes and (ii) how many taxa are affected by one transfer event (*see* also **Section 3.4**), therefore even further underestimating the overall amount of HGT.

Patterns of gene presence and absence have also been used to estimate the frequency of HGT *(55)*. As discussed in **Section 2.2**, a given phyletic pattern can be explained either by vertical inheritance and gene loss, or by HGT. The former explanation alone, without consideration of gene transfer, forces one to assume that a gene present in two organisms was already present in the ancestor of the two genomes. For example, under this assumption any gene present in at least one archaeon and one bacterium would have to be assumed present in the ancestral "Genome of Eden" *(56)*. Using 190 present day genomes, Dagan and Martin *(55)* calculated the size of the genome at the base of the Bacterial domain to encode 53,658 proteins under the assumption of a complete absence of transfers. For genomes in the past to have had about the same size as today's genomes, about one gene transfer per gene family is required to have occurred. In contrast to the studies reported in the previous paragraph, this analysis ignores the transfer events that resulted in orthologous replacement.

### 3.3. Model-Based Estimates

Several studies have developed models of HGT, performed simulations under the models and attempted to estimate the rates of HGT based on fit of the real data to the model *(47, 57, 58)*. All of these models deal only with HGT events resulting in orthologous replacements and make a plethora of simplifying and often biologically unrealistic assumptions, therefore underestimating the HGT impact. For example, Galtier *(47)* reports a model where a gene tree is simulated from a known "organismal" tree by the addition of HGT events among the taxa on that tree. Multiple independent gene trees are simulated to represent different genes in an organism. Rate variation on a gene tree is introduced across branches on the tree (to simulate varying rates for different taxa on the tree) and across the whole gene tree (to simulate different overall rates for different genes), and each gene tree is further selectively pruned to mimic taxon sampling. The HGT events under this model are equiprobable and not considered to be coming from outside of the taxa present on the organismal tree; the HGT rate was assumed to be constant for all genes. The real and simulated data were compared using several tree-based statistics. One such statistic was a congruence of a gene tree to an organismal tree measured as the proportion of bipartitions not conflicting with bipartitions on the organismal tree (*see* **Section 2.4** for terminology). From the comparison between real and simulated datasets, real gene trees were less congruent to each other than the trees simulated without HGT. Such level of congruence roughly corresponded to simulations with 6–12 HGT events per gene tree. This estimate is subject to the drawbacks of the phylogenetic incongruence method for HGT detection (*see* **Section 2.4**).

### 3.4. Are These Estimates Informative?

Can we rely on the estimates outlined in **Sections 3.2** and **3.3** to answer the question of how big/small has been the impact of HGT on the evolution of prokaryotes? First, how should we quantify HGT? One way is to provide an estimate for a number of gene families affected by HGT, without looking at how many events per family had occurred. Given how imperfect HGT detection methods are, we obtain an incomplete picture at best (*see* **Fig. 11.2**). Current estimates consider vertical inheritance as a null hypothesis, counting only detectable HGT events and hence at best providing a lower bound (if false positives are ignored) for the proportion of genes subjected to HGT in the past. False positives, however, exist and potentially could affect the estimate. Currently, the impact of false positives is assessed through simulations (e.g., see *(8)*; also *see* **Chapters 12** and **13**). Obtaining upper bound estimates is probably not possible (unless

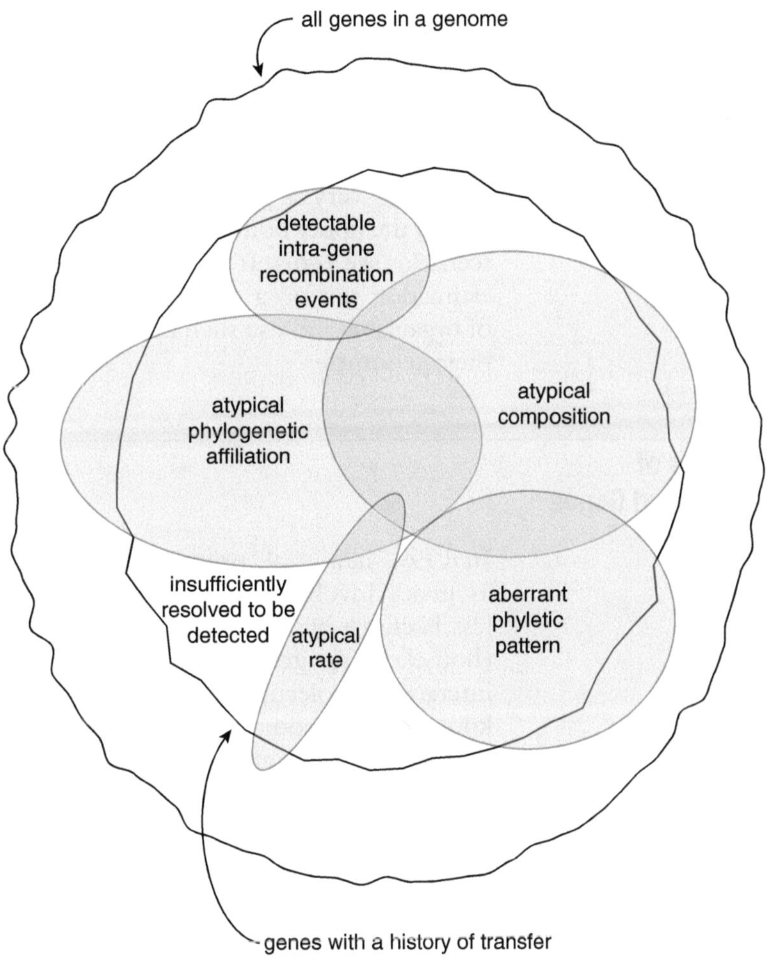


Fig. 11.2. Schematic diagram illustrating the current view of how various HGT detection methods (*grey circles*) perform (not drawn to scale; sizes of shapes meant to represent that some method classes are more general than others, and hence can identify more genes due to methodological variations employed; however, relative success of each method is not known). Most methods suffer from both false positives (the parts of gray circles that fall outside of the "genes with history of transfer" set) and false negatives to an unknown extent and all methods return different and often non-overlapping results. The figure illustrates that no single method of HGT detection covers the "true" set of genes with history of transfer, and hence the obtained estimates of number of transfer events will remain inconclusive until the performance of various methods is assessed and cross-compared, and until studies start to utilize multiple HGT detection methods.

we can prove that some gene families did not have a single transfer event in their past), so the current (but not very useful) upper bound is that all gene families could have been subject of at least one transfer event. However, just knowing the number of families affected by HGT is not sufficient to assess the impact of HGT. How many transfer events per gene family should be considered "a lot"? Even one HGT event can misplace numerous contemporary taxa on a phylogenetic tree if it occurred a long time ago in the past.

The estimation of when and where specific HGT events occurred is a very difficult task. First, an assumption of parsimony is currently employed, i.e., the inferred number of HGT events is the minimal number required to explain the observed incongruence. Aside from the tractability of inference, there is no other reason to expect that the number of HGT events should be most parsimonious. Second, even a minimum number of HGT events results in a very large number of possible scenarios (for the derivation of the upper bound of possible maximum parsimonious HGT scenarios see *(46)*). It is perhaps more useful to apply HGT event estimation not on a domain scale, but to closely-related groups of organisms and use such estimates to study processes that shape their genomes.

## 4. Types of Transferred Genes

In 1999, Jain et al. proposed that informational genes (defined as genes involved in transcription and translation) should be less likely to be transferred in comparison to operational genes (housekeeping genes) because the former are part of a highly interactive molecular assembly *(59)*. This assumption became known as "the complexity hypothesis". While genome-wide analyses indicate that all types of genes are among inferred HGTs, certain functional categories are reported to be over- or under-represented among horizontally transferred genes as compared to their genome-wide distribution. Among HGTs detected by a compositional method, Nakamura et al. detect a bias of HGTs in the functional categories of cell surface, DNA binding, and pathogenicity related genes *(17)*. In the analyses of six *E. coli* and *Shigella* genomes, Mau et al. report a bias in favor of the genes involved in motility and chemotaxis, in "DNA replication, repair and recombination", and in small molecule biosynthesis among the detected transferred genes *(49)*. Hsiao et al. report an overrepresentation of the hypothetical genes and genes involved in "DNA replication, recombination, and repair" in genomic islands *(18)*. Among the HGTs detected by a phylogenetic method, Beiko et al. *(33)* describe over-representation of "energy metabolism" and "mobile and extrachromosomal element functions" among genes with discordant bipartitions in their phylogenies (i.e., among HGT candidates). However, the transfer distance may need to be taken into account when types of transferred genes are evaluated. For example, in genome-wide analyses of HGT in cyanobacteria, no bias toward a particular functional category is found for HGTs inferred to occur within cyanobacteria, but an excess of metabolic genes and a decrease of informational genes is observed in the transfers inferred to occur between cyanobacterial and other phyla *(8)*. Also, as with estimates for amount of HGT, the analyses of types of transferred genes are sensitive to drawbacks of HGT detection methods themselves (*see* also **Section 3.4**).

## 5. Concluding Remarks

Most analyses currently utilize only one specific method of HGT detection, and the results of various analyses of even the same genomes are often incomparable directly. A framework is needed for the reconciliation of different methods of HGT detection as well as for ways to discriminate specific HGT scenarios from multiple possible ones.

In the future, the development of algorithms that use genome data to reconstruct the reticulate history of genomes, rather than providing a single phylogenetic tree only, will be an important improvement in the reconstruction of the evolutionary history of prokaryotes. In graph theory, a tree is a special case of a network structure. If supported by data, a tree-like evolutionary history can emerge from network-based inference algorithms (but not vice versa). Hence, ideally, these new methods would explore *a priori* non-tree like approaches (such as Splits-Tree and Neighbor-Net *(60)* do for single-gene phylogenies) to reconstruct the reticulate evolutionary genomic history and not begin with incompatible trees that are merged into networks only later. Such methods will open new ways to detect and assess HGT.

## 6. Methods

This section describes in detail the quartet decomposition method, which belongs to a family of methods for detecting HGT using phylogenetic incongruence (*see* **Section 2.4**). The method was originally introduced in *(8)*, hence the reader is referred there for details on assessment of the method's performance as well as its application in studying evolutionary relationships of cyanobacteria. For details on other methods described in this chapter, please refer to original studies (as cited throughout **Section 2**).

### 6.1. Selection of Gene Families

A crucial starting point for any method relying on phylogenetic incongruence is the identification of gene families that contain only orthologous genes from a collection of open-reading frames in genomes of interest. There is no unique recipe on how to select such gene families. Among commonly used approaches are families formed using mutual fully transitive reciprocal top-scoring BLAST hits (i.e., each member of such a gene family is a top-scoring BLAST hit of every other member of the gene family in respective genome BLAST searches; e.g., see *(38)*), Markov Clustering (e.g., *(61, 62)*), or in the detection of orthologous families using phylogenetic information (BRANCHCLUST; *(63)*). The difficulty of proper identification and treatment of paralogs remains a problem not satisfactorily addressed (e.g., in some cases gene families with paralogs are simply thrown out from further analyses *(61)*). Note that quartet decomposition also works with incomplete gene families, i.e., families whose members are not necessarily present in all analyzed genomes.

### 6.2. Gene Families' Alignments

Sequence alignments might be obtained by using, for example, ClustalW *(64)* or MUSCLE *(65)*. Assessment of alignment accuracy and biases is beyond the scope of this chapter; the reader is referred to some of the most recent studies *(66, 67)*. As with any large-scale phylogenetic analyses, manual inspection of alignment quality for each gene family is usually too time-consuming to be practical. Programs like GBLOCKS *(68)* automatically "clean" alignments of badly aligned regions, although such a procedure is not always recommended *(69)*.

### 6.3. Reconstruction of Phylogenetic Trees

Evolutionary history of each gene family needs to be represented by $N$ phylogenetic trees (one tree per each of $N$ bootstrap samples; posterior probabilities or any other support values can be used instead of bootstrap, if desired). Often the choice of phylogenetic method is a compromise between speed (e.g., neighbor-joining *(70)* or PhyML *(71)*) and accuracy (e.g., RAxML *(72)*) and depends on how many gene families are to be analyzed and how many taxa each family contains.

### 6.4. Evaluation of Embedded Quartets

Lists of possible embedded quartets per gene family can be generated using available combinatorial algorithms (e.g., *(73)*). The evaluation of all embedded quartets can be done either with a script provided in supplementary data accompanying ref. *(8)* or using QuartetSuite *(74)*. The result of this analysis step is a list of support values for each of three possible tree topologies per embedded quartet; such a list is generated for each gene family. These lists are used to summarize phylogenetic signal contained in all gene families (*see* **Sections 6.5** and **6.6** below).

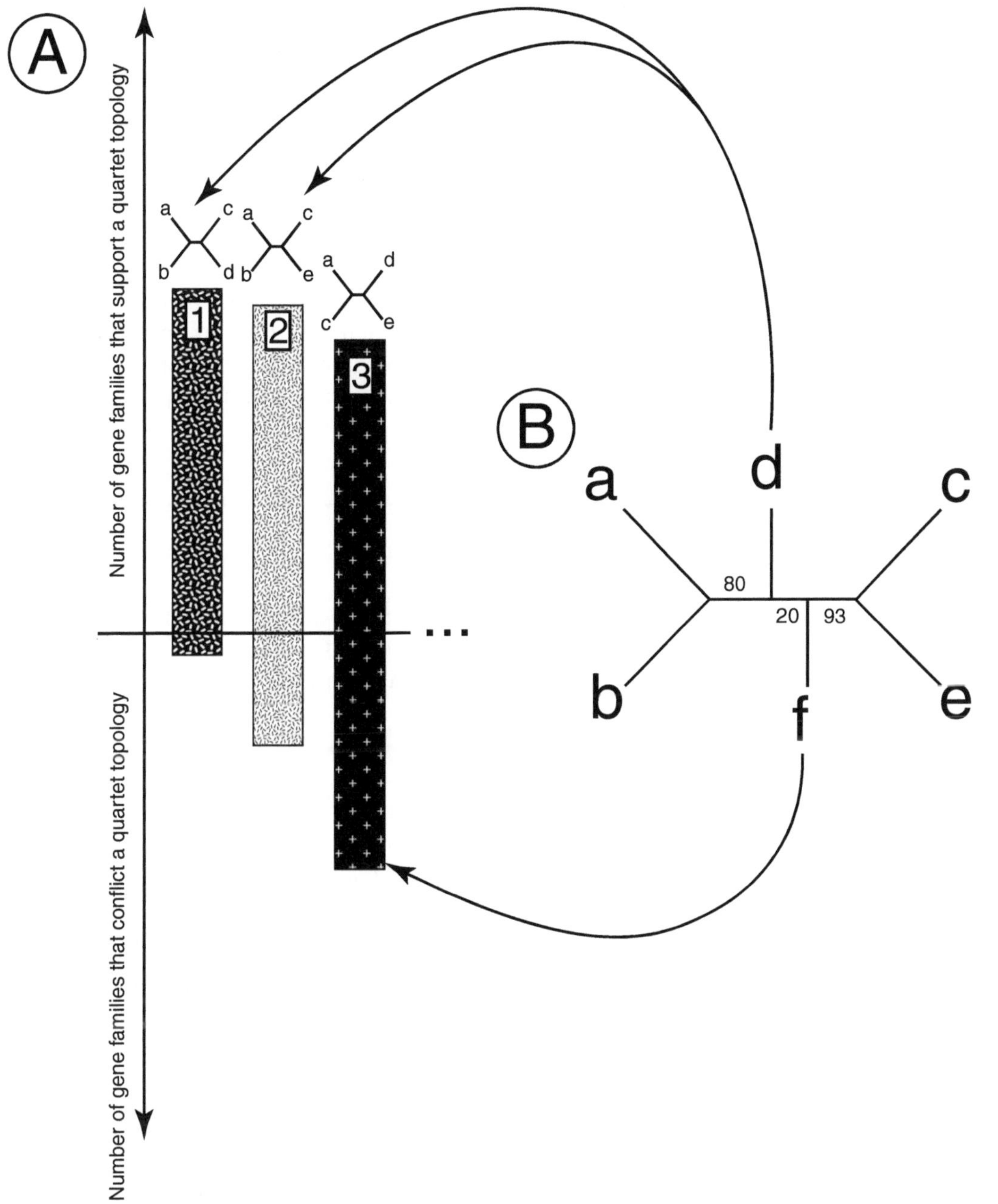


Fig. 11.3. An illustration of a quartet decomposition analysis. Analogous to a bipartition plot (*see* **Fig. 11.1**), embedded quartets can be summarized in a histogram (a spectrogram, **Panel A**). Only three quartets are shown in panel A (out of 15 possible for six genomes). **Panel B** shows an example of a gene family and its contribution to the spectrogram (*arrows*). While the gene families agree with the first two plurality quartets, it contradicts the third one. An example of the assessment of 330 quartets in 1128 gene families of 11 cyanobacterial genomes is shown in **Fig. 11.2** in *(8)*.

### *6.5. Summarizing Results in a Spectrogram*

Resulting data on how different gene families support (or conflict with) each embedded quartet can be summarized in a histogram (called a spectrogram, because it summarizes a full spectrum of possible relationships as captured in embedded quartets; *see* **Fig. 11.3**). Spectrograms are useful for gaining an overview of how much disagreement exists among different gene families and to identify the gene families that disagree (those which give rise to negative values, *see* **Fig. 11.3**) with quartets supported by a plurality of gene families.

### 6.6. Reconstruction of Plurality Topology

Information from topologies of quartets supported by a plurality of gene families can be combined into one topology, the so-called plurality topology. This is achieved either through supertree reconstruction directly from quartets (e.g., as implemented in QuartetSuite *(74)*), or using the "matrix representation using parsimony" (MRP) method *(75, 76)* (e.g., as implemented in Clann *(77)*).

## Acknowledgments

I thank W. Ford Doolittle for constructive comments and suggestions. OZ is supported through a Canadian Institute of Health Research (CIHR) postdoctoral fellowship.

### References

1. Liolios, K., Mavromatis, K., Tavernarakis, N., Kyrpides, N. C. (2007) The genomes on line database (GOLD) in 2007: Status of genomic and metagenomic projects and their associated metadata. *Nucl Acids Res* **36**, D475–D479.
2. Ragan, M. A. (2001) Detection of lateral gene transfer among microbial genomes. *Curr Opin Genet Dev* **11**, 620–6.
3. Ragan, M. A. (2002) Reconciling the many faces of lateral gene transfer. *Trends Microbiol* **10**, 4.
4. Ragan, M. A., Harlow, T. J., Beiko, R. G. (2006) Do different surrogate methods detect lateral genetic transfer events of different relative ages? *Trends Microbiol* **14**, 4–8.
5. Lawrence, J. G., Ochman, H. (1997) Amelioration of bacterial genomes: Rates of change and exchange. *J Mol Evol* **44**, 383–97.
6. Cortez, D. Q., Lazcano, A., Becerra, A. (2005) Comparative analysis of methodologies for the detection of horizontally transferred genes: A reassessment of first-order Markov models. *In Silico Biol* **5**, 0052.
7. Poptsova, M. S., Gogarten, J. P. (2007) The power of phylogenetic approaches to detect horizontally transferred genes. *BMC Evol Biol* **7**, 45.
8. Zhaxybayeva, O., Gogarten, J. P., Charlebois, R. L., Doolittle, W. F., Papke, R. T. (2006) Phylogenetic analyses of cyanobacterial genomes: Quantification of horizontal gene transfer events. *Genome Res* **16**, 1099–108.
9. Beiko, R. G., Charlebois, R. L. (2007) A simulation test bed for hypotheses of genome evolution. *Bioinformatics* **23**, 825–31.
10. Karlin, S. (2001) Detecting anomalous gene clusters and pathogenicity islands in diverse bacterial genomes. *Trends Microbiol* **9**, 335–43.
11. Putonti, C., Luo, Y., Katili, C., Chumakov, S., Fox, G. E., Graur, D., Fofanov, Y. (2006) A computational tool for the genomic identification of regions of unusual compositional properties and its utilization in the detection of horizontally transferred sequences. *Mol Biol Evol* **23**, 1863–8.
12. Van Passel, M. W. J., Bart, A., Thygesen, H. H., Luyf, A. C. M., Van Kampen, A. H. C., Van Der Ende, A. (2005) An acquisition account of genomic islands based on genome signature comparisons. *BMC Genomics* **6**, 163.

13. Lawrence, J. G., Ochman, H. (1998) Molecular archaeology of the *Escherichia coli* genome. *Proc Natl Acad Sci U S A* **95**, 9413–7.
14. Tsirigos, A., Rigoutsos, I. (2005) A new computational method for the detection of horizontal gene transfer events. *Nucl Acids Res* **33**, 922–33.
15. Koski, L. B., Morton, R. A., Golding, G. B. (2001) Codon bias and base composition are poor indicators of horizontally transferred genes. *Mol Biol Evol* **18**, 404–12.
16. Azad, R. K., Lawrence, J. G. (2005) Use of artificial genomes in assessing methods for atypical gene detection. *PLoS Comput Biol* **1**, e56.
17. Nakamura, Y., Itoh, T., Matsuda, H., Gojobori, T. (2004) Biased biological functions of horizontally transferred genes in prokaryotic genomes. *Nat Genet* **36**, 760–6.
18. Hsiao, W. W. L., Ung, K., Aeschliman, D., Bryan, J., Finlay, B. B., Brinkman, F. S. L. (2005) Evidence of a large novel gene pool associated with prokaryotic genomic islands. *PLoS Genet* **1**, e62.
19. Altschul, S. F., Madden, T. L., Schaffer, A. A., Zhang, J., Zhang, Z., Miller, W., Lipman, D. J. (1997) Gapped BLAST and PSI-BLAST: A new generation of protein database search programs. *Nucleic Acids Res* **25**, 3389–402.
20. Nelson, K. E., Clayton, R. A., Gill, S. R., Gwinn, M. L., Dodson, R. J., Haft, D. H., Hickey, E. K., Peterson, J. D., Nelson, W. C., Ketchum, K. A., Mcdonald, L., Utterback, T. R., Malek, J. A., Linher, K. D., Garrett, M. M., Stewart, A. M., Cotton, M. D., Pratt, M. S., Phillips, C. A., Richardson, D., Heidelberg, J., Sutton, G. G., Fleischmann, R. D., Eisen, J. A., White, O., Salzberg, S. L., Smith, H. O., Venter, J. C., Fraser, C. M. (1999) Evidence for lateral gene transfer between archaea and bacteria from genome sequence of *Thermotoga maritima*. *Nature* **399**, 323–9.
21. Koski, L. B., Golding, G. B. (2001) The closest BLAST hit is often not the nearest neighbor. *J Mol Evol* **52**, 540–2.
22. Podell, S., Gaasterland, T. (2007) Darkhorse: A method for genome-wide prediction of horizontal gene transfer. *Genome Biol* **8**, R16.
23. Welch, R. A., Burland, V., Plunkett, G., Iii, Redford, P., Roesch, P., Rasko, D., Buckles, E. L., Liou, S. R., Boutin, A., Hackett, J., Stroud, D., Mayhew, G. F., Rose, D. J., Zhou, S., Schwartz, D. C., Perna, N. T., Mobley, H. L. T., Donnenberg, M. S., Blattner, F. R. (2002) Extensive mosaic structure revealed by the complete genome sequence of uropathogenic *Escherichia coli*. *PNAS* **99**, 17020–4.
24. Normand, P., Lapierre, P., Tisa, L. S., Gogarten, J. P., Alloisio, N., Bagnarol, E., Bassi, C. A., Berry, A. M., Bickhart, D. M., Choisne, N., Couloux, A., Cournoyer, B., Cruveiller, S., Daubin, V., Demange, N., Francino, M. P., Goltsman, E., Huang, Y., Kopp, O. R., Labarre, L., Lapidus, A., Lavire, C., Marechal, J., Martinez, M., Mastronunzio, J. E., Mullin, B. C., Niemann, J., Pujic, P., Rawnsley, T., Rouy, Z., Schenowitz, C., Sellstedt, A., Tavares, F., Tomkins, J. P., Vallenet, D., Valverde, C., Wall, L. G., Wang, Y., Medigue, C., Benson, D. R. (2007) Genome characteristics of facultatively symbiotic *Frankia* sp. Strains reflect host range and host plant biogeography. *Genome Res* **17**, 7–15.
25. Kunin, V., Ouzounis, C. A. (2003) GeneTRACE-reconstruction of gene content of ancestral species. *Bioinformatics* **19**, 1412–6.
26. Mirkin, B. G., Fenner, T. I., Galperin, M. Y., Koonin, E. V. (2003) Algorithms for computing parsimonious evolutionary scenarios for genome evolution, the last universal common ancestor and dominance of horizontal gene transfer in the evolution of prokaryotes. *BMC Evol Biol* **3**, 2.
27. Snel, B., Bork, P., Huynen, M. A. (2002) Genomes in flux: The evolution of archaeal and proteobacterial gene content. *Genome Res* **12**, 17–25.
28. Zhaxybayeva, O., Nesbφ, C. L., Doolittle, W. F. (2007) Systematic overestimation of gene gain through false diagnosis of gene absence. *Genome Biol* **8**, 402.
29. Novichkov, P. S., Omelchenko, M. V., Gelfand, M. S., Mironov, A. A., Wolf, Y. I., Koonin, E. V. (2004) Genome-wide molecular clock and horizontal gene transfer in bacterial evolution. *J Bacteriol* **186**, 6575–85.
30. Hamady, M., Betterton, M. D., Knight, R. (2006) Using the nucleotide substitution rate matrix to detect horizontal gene transfer. *BMC Bioinformatics* **7**, 476.
31. Deckert, G., Warren, P. V., Gaasterland, T., Young, W. G., Lenox, A. L., Graham, D. E., Overbeek, R., Snead, M. A., Keller, M., Aujay, M., Huber, R., Feldman, R. A., Short, J. M., Olsen, G. J., Swanson, R. V. (1998) The complete genome of the hyperthermophilic bacterium *AQUIFEX AEOLICUS*. *Nature* **392**, 353.
32. Lerat, E., Daubin, V., Moran, N. A. (2003) From gene trees to organismal phylogeny in prokaryotes: The case of the gamma-proteobacteria. *PLoS Biol* **1**, E19.

33. Beiko, R. G., Harlow, T. J., Ragan, M. A. (2005) Highways of gene sharing in prokaryotes. *Proc Natl Acad Sci U S A* **102**, 14332–7.
34. Macleod, D., Charlebois, R., Doolittle, F., Bapteste, E. (2005) Deduction of probable events of lateral gene transfer through comparison of phylogenetic trees by recursive consolidation and rearrangement. *BMC Evolutionary Biology* **5**, 27.
35. Makarenkov, V. (2001) T-REX: Reconstructing and visualizing phylogenetic trees and reticulation networks. *Bioinformatics* **17**, 664–8.
36. Beiko, R., Hamilton, N. (2006) Phylogenetic identification of lateral genetic transfer events. *BMC Evolutionary Biology* **6**, 15.
37. Hamel, L., Zhaxybayeva, O., Gogarten, J. P. (2005) Pentaplot: A software tool for the illustration of genome mosaicism. *BMC Bioinformatics* **6**, 139.
38. Zhaxybayeva, O., Gogarten, J. (2002) Bootstrap, Bayesian probability and maximum likelihood mapping: Exploring new tools for comparative genome analyses. *BMC Genomics* **3**, 4.
39. Zhaxybayeva, O., Hamel, L., Raymond, J., Gogarten, J. (2004) Visualization of the phylogenetic content of five genomes using dekapentagonal maps. *Genome Biol* **5**, R20.
40. Bapteste, E., Susko, E., Leigh, J., Macleod, D., Charlebois, R. L., Doolittle, W. F. (2005) Do orthologous gene phylogenies really support tree-thinking? *BMC Evol Biol* **5**, 33.
41. Zhaxybayeva, O., Lapierre, P., Gogarten, J. P. (2004) Genome mosaicism and organismal lineages. *Trends Genet* **20**, 254–60.
42. Felsenstein, J. (1988) Phylogenies from molecular sequences: Inference and reliability. *Annu Rev Genet* **22**, 521–65.
43. Zhaxybayeva, O., Gogarten, J. P. (2003) An improved probability mapping approach to assess genome mosaicism. *BMC Genomics* **4**, 37.
44. Lento, G. M., Hickson, R. E., Chambers, G. K., Penny, D. (1995) Use of spectral analysis to test hypotheses on the origin of pinnipeds. *Mol Biol Evol* **12**, 28–52.
45. Wainright, P. O., Hinkle, G., Sogin, M. L., Stickel, S. K. (1993) Monophyletic origins of the metazoa: An evolutionary link with fungi. *Science* **260**, 340–2.
46. Than, C., Ruths, D., Innan, H., Nakhleh, L. (2007) Confounding factors in HGT detection: Statistical error, coalescent effects, and multiple solutions. *J Comput Biol* **14**, 517–35.
47. Galtier, N. (2007) A model of horizontal gene transfer and the bacterial phylogeny problem. *Syst Biol* **56**, 633–42.
48. Degnan, J. H., Rosenberg, N. A. (2006) Discordance of species trees with their most likely gene trees. *PLoS Genet* **2**, e68.
49. Mau, B., Glasner, J., Darling, A., Perna, N. (2006) Genome-wide detection and analysis of homologous recombination among sequenced strains of *Escherichia coli*. *Genome Biol* **7**, R44.
50. Charlebois, R. L., Doolittle, W. F. (2004) Computing prokaryotic gene ubiquity: Rescuing the core from extinction. *Genome Res* **14**, 2469–77.
51. Ciccarelli, F. D., Doerks, T., Von Mering, C., Creevey, C. J., Snel, B., Bork, P. (2006) Toward automatic reconstruction of a highly resolved tree of life. *Science* **311**, 1283–7.
52. Dagan, T., Martin, W. (2006) The tree of one percent. *Genome Biol* **7**, 118.
53. Choi, I.-G., Kim, S.-H. (2007) Global extent of horizontal gene transfer. *Proc Natl Acad Sci U S A* **104**, 4489–94.
54. Ge, F., Wang, L.-S., Kim, J. (2005) The cobweb of life revealed by genome-scale estimates of horizontal gene transfer. *PLoS Biol* **3**, e316.
55. Dagan, T., Martin, W. (2007) Ancestral genome sizes specify the minimum rate of lateral gene transfer during prokaryote evolution. *Proc Natl Acad Sci U S A* **104**, 870–5.
56. Doolittle, W. F., Boucher, Y., Nesbo, C. L., Douady, C. J., Andersson, J. O., Roger, A. J. (2003) How big is the iceberg of which organellar genes in nuclear genomes are but the tip? *Philos Trans R Soc Lond B Biol Sci* **358**, 39–57.
57. Suchard, M. A. (2005) Stochastic models for horizontal gene transfer: Taking a random walk through tree space. *Genetics* **170**, 419–31.
58. Linz, S., Radtke, A., Von Haeseler, A. (2007) A likelihood framework to measure horizontal gene transfer. *Mol Biol Evol* **24**, 1312–9.
59. Jain, R., Rivera, M. C., Lake, J. A. (1999) Horizontal gene transfer among genomes: The complexity hypothesis. *Proc Natl Acad Sci U S A* **96**, 3801–6.
60. Huson, D. H., Bryant, D. (2006) Application of phylogenetic networks in evolutionary studies. *Mol Biol Evol* **23**, 254–67.
61. Swingley, W. D., Blankenship, R. E., Raymond, J. (2008) Integrating Markov clustering and molecular phylogenetics to reconstruct the cyanobacterial species tree from conserved protein families. *Mol Biol Evol* **25**, 643–54.
62. Harlow, T. J., Gogarten, J. P., Ragan, M. A. (2004) A hybrid clustering approach to recognition of protein families in 114 microbial genomes. *BMC Bioinform* **5**, 45.

63. Poptsova, M. S., Gogarten, J. P. (2007) BranchClust: A phylogenetic algorithm for selecting gene families. *BMC Bioinform* **8**, 120.
64. Larkin, M. A., Blackshields, G., Brown, N. P., Chenna, R., Mcgettigan, P. A., Mcwilliam, H., Valentin, F., Wallace, I. M., Wilm, A., Lopez, R., Thompson, J. D., Gibson, T. J., Higgins, D. G. (2007) Clustal W and Clustal X version 2.0. *Bioinformatics* **23**, 2947–8.
65. Edgar, R. C. (2004) Muscle: Multiple sequence alignment with high accuracy and high throughput. *Nucl Acids Res* **32**, 1792–7.
66. Golubchik, T., Wise, M. J., Easteal, S., Jermiin, L. S. (2007) Mind the gaps: Evidence of bias in estimates of multiple sequence alignments. *Mol Biol Evol* **24**, 2433–42.
67. Landan, G., Graur, D. (2007) Heads or tails: A simple reliability check for multiple sequence alignments. *Mol Biol Evol* **24**, 1380–3.
68. Talavera, G., Castresana, J. (2007) Improvement of phylogenies after removing divergent and ambiguously aligned blocks from protein sequence alignments. *Syst Biol* **56**, 564–77.
69. Wong, K. M., Suchard, M. A., Huelsenbeck, J. P. (2008) Alignment uncertainty and genomic analysis. *Science* **319**, 473–6.
70. Felsenstein, J. (1993) PHYLIP (phylogeny inference package) version 3.6. Distributed by the author. *Department of Genetics, University of Washington, Seattle.*
71. Guindon, S., Gascuel, O. (2003) A simple, fast, and accurate algorithm to estimate large phylogenies by maximum likelihood. *Syst Biol* **52**, 696–704.
72. Stamatakis, A. (2006) RAxML-VI-HPC: Maximum likelihood-based phylogenetic analyses with thousands of taxa and mixed models. *Bioinformatics* **22**, 2688–90.
73. Chase, P. (1970) Algorithm 382: Combinations of m out of n objects [g6]. *Commun ACM* **13**, 368.
74. Quartet suite 1.0 (2007) (Accessed at http://genome.cs.iastate.edu/CBL/download/).
75. Baum, B. (1992) Combining trees as a way of combining data sets for phylogenetic inference, and the desirability of combining gene trees. *Taxon* **41**, 3–10.
76. Ragan, M. A. (1992) Phylogenetic inference based on matrix representation of trees. *Mol Phylogenet Evol* **1**, 53–8.
77. Creevey, C. J., Mcinerney, J. O. (2005) Clann: Investigating phylogenetic information through supertree analyses. *Bioinformatics* **21**, 390–2.

# Chapter 12

# Composition-Based Methods to Identify Horizontal Gene Transfer

**Diego Cortez, Luis Delaye, Antonio Lazcano, and Arturo Becerra**

## Abstract

The detection of horizontal gene transfer (HGT) events has become an increasingly important issue in recent years. Here we discuss a simple theoretical analysis based on the *in silico* artificial addition of known foreign genes from different prokaryotic groups into the genome of *Escherichia coli* K12 MG1655. Using this dataset as a control, we have tested the efficiency of four methodologies commonly employed to detect HGT, which are based on (a) the codon adaptation index, codon usage, and GC percentage (CAI/GC); (b) the distributional profile (DP) approach with a gene search in the closely related phylogenetic genomes; (c) the Bayesian model (BM); and (d) the first-order Markov model (MM). All methods exhibit limitations as shown here, with BM and MM giving better approximations. The MM has a better detection rate when genes from closely related organisms are evaluated. The application of the MM to detect recently transferred genes in the genomes of *E. coli* strain K12 MG1655 shows that this organism has undergone a rather significant amount of HGT, several of which have well-defined functions that appear to be involved in the direct interaction of the organisms with their environment.

**Key words:** Horizontal gene transfer, composition-based methods, methods to identify HGT.

## 1. Introduction

Horizontal gene transfer (HGT) is generally assumed to have played an important role in the innovation of genomes, especially during the early stages of biological evolution *(1–4)*. However, some authors have suggested that the role of HGT has been overestimated *(5, 6)*. The current controversy lies, in part, in the inadequate or inefficient methods to identify transfer events.

Many attempts have been made to characterize the number and types of genes involved in HGT, but as shown by ongoing

Maria B. Gogarten et al. (eds.), *Horizontal Gene Transfer: Genomes in Flux, vol. 532*

DOI 10.1007/978-1-60327-853-9_12 Springerprotocols.com

debates, as of today no infallible detection methods have been developed *(7–16)*. During the past decade, different approaches have been proposed, which can be classified in two major categories: (a) the composition-based methods and (b) anomalous phylogenetic distribution. These methods can be divided into more specific groups. Although it has been suggested that phylogenetic methods are more powerful than compositional methods to detect HGT in particular situations (i.e., when the donor is closely related to the recipient genome), they tend to be time-consuming. Therefore, the development of a compositional method with an accurate detection level of horizontally transferred genes would be an effective approach in the analysis of possible HGT in large numbers of completely sequenced genomes.

Determination of the levels of HGT depends on (i) the time of their occurrence; (ii) the phylogenetic relationship between the donor and the acceptor; and (iii) the "camouflage" of the gene; it is also affected by gene loss and the rate of divergence. Although one should not abandon all hope, it is important to keep in mind that the sequences may have undergone important signal loss.

Two examples of the controversies created by different estimates of HGT are the discrepancies in shaping the topology of the tree of life *(1, 4, 17)* and the reconstruction of the Last Common Ancestor (LCA) of all extant organisms *(18)*. The lack of congruency between different universal phylogenies may result not only from the statistical issues involved in the alignment and comparison of a large number of sequences that may have diverged more than $3.5 \times 10^9$ years ago, but also from even older additional duplications *(19)* and HGT events *(1)*, both of which may be obscuring the natural relationship between the lineages. Several authors have argued that the use of genes that are less likely to be transferred are expected to yield deep phylogenies with suitable results *(20, 21)*, and they allow not only to explore the evolution of early stages but also to backtrack the characteristics of the Last Universal Common Ancestor *(22)*. Is it generally accepted that HGT was rampant during the early evolution of genomes. Faster and better methods of HGT detection could improve our understanding of the early evolution of the life significantly.

## 2. Methodologies Used for the Detection of HGT

Current approaches for detecting HGT can be broadly divided into three major groups: (a) methods based on codon usage such as the codon adaptation index (CAI), the GC percentage analysis, Bayesian models, and higher order Markov models, all of which attempt to identify genes with anomalous compositions *(10, 12, 13, 16, 23)*; (b) comparison of the gene content of an organism with that of closely related species, based on the distributional profiles (DP) determined for every single gene in the genome *(15, 24)* – this method is based on the idea that if a gene present in the target genome is not found in any closely related genomes, given a variety of threshold values, the sequences are considered to have undergone HGT – and (c) phylogenetic reconstruction, which is based on phylogenetic conflict by alien-gene acquisition *(25)*.

Compositional approaches based on codon usage and GC content have been criticized since in some case the dissimilarities in base composition and codon usage between possible transferred genes and host sequences can be truly minor *(26, 27)*. Indeed, these results may be the outcome of compositional heterogeneity, which is now recognized as a characteristic of cellular genomes *(28, 29)*. For instance, when the compositional method developed by Lawrence and Ochman *(12)*, the high-order MM used by Hayes and Borodovsky *(10)*, the phylogenetically discordant approach followed by Clarke et al. *(30)*, and an anomalous phylogenetic distribution model by Ragan and Charlebois *(24)* were all applied to analyze the *Escherichia coli* K12 MG1655 genome, each method detected a very different set of possible horizontally transferred genes, and the intersections between these sets were less than expected by chance alone *(31)*. The development of a compositional method with an accurate detection level of horizontally transferred genes would be a powerful approach that could avoid the application of exhaustive processes and slow phylogenetic reconstructions that, moreover, might not lead to better results than the compositional methods.

In this chapter we describe a different approach based on the analysis of the significance of the most frequently used methodologies for the detection of HTG and describe a simple theoretical approach that uses the *in silico* inclusion of known foreign genes from different prokaryotes into a chosen model genome *(8)*.

### 2.1. The Accuracy of the Methodologies Based on Composition

The methodologies discussed here have been extensively employed during the past few years. However, they have usually been used without proper experimental controls that could allow an estimate of their accuracy. A simple theoretical approach was recently developed by Cortez et al. *(8)*, which consists in the *in silico* artificial inclusion of known foreign genes from different mayor prokaryotic groups into a given genome. Using this methodology we have tested the efficiency of four approaches, which are based on (a) the codon adaptation index, codon usage, and GC percentage *(12)*; (b) a distributional profile for every single gene in a genome *(15)*; (c) a Bayesian model *(16)*; and (d) a first-order Markov model we developed *(8)*. Genes were selected randomly from 30 different organisms in order to test these methodologies. The foreign genes were then randomly inserted *in silico* into the genome of *E. coli* K12 MG1655. The genes detected by each method were compared *(8)*. Our results show that the MM appears to be the most reliable approach to identify horizontally transferred genes, especially when they come from closely related species. The MM was then employed to analyze the foreign sequences acquired by *E. coli* K12 MG1655. Enterobacteria have a significant percentage of foreign genes, many of which have defined functions that might be involved in the direct interaction of the organism and its environment.

## 3. "Who Is Who" in Composition-Based Detection?

The average detection levels of artificially introduced foreign genes are shown in **Fig. 12.1**. The BM and the MM consistently detected the foreign, introduced genes. However, the MM appears to be a much better approach when genes from closely related species are studied.

The CAI/GC method appears to be a less effective approach in the detection of HGT. However, it has better detection levels when the artificially introduced foreign genes came from phylogenetically distant species, or when they belonged to

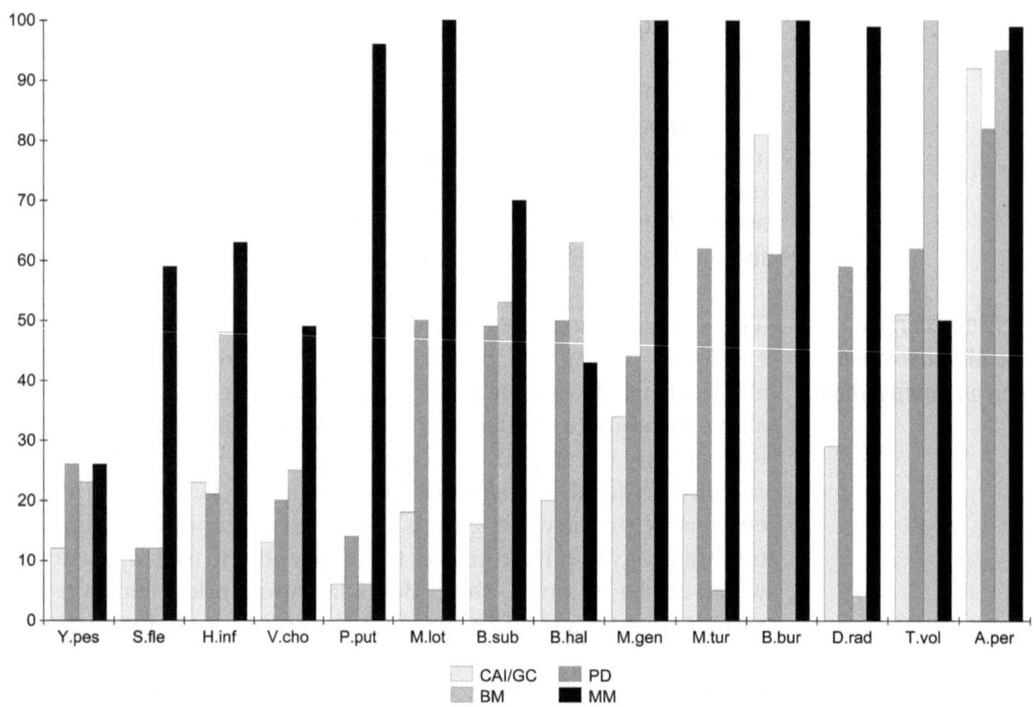


Fig. 12.1. Detection levels of 100 artificially introduced foreign genes from different organisms by the four methods discussed here: CAI/GC, PD, BM, and MM. Where: Y.pes, *Y. pestis*; S.fle, *S. flexneri*; H.inf; *H. influenzae*; V.cho, *V. cholerae*; P.put, *P. putida*; M.lot, *M. loti*; B.sub, *B. subtilis*; B.hal, *B. halodurans*; M.gen, *M. genitalium*; M.tur, *M. tuberculosis*; B.bur, *B. burgdorferi*; D.rad, *D. radiodurans*; T.vol, *T. volcanium*; A.per, *A. pernix*.

**Table 12.1**
**Functional characteristics of the four methodologies. Average detection levels of genes from close and distant prokaryotic species based on the model's performance in the experiment of *in silico* introduction of foreign genes. Probabilities of having false negatives and false positives are shown. The faculty to discriminate between gene loss and HT is shown too (for more information *see (8)*)**

| Model | Average detection levels of genes from distant species | Average detection levels of genes from Proteobacteria species | Probability of having false positives | Probability of having false negatives | Faculty to discriminate between gene loss and HT | Number of possible HTG in *E. coli* K12 |
|---|---|---|---|---|---|---|
| MM ($p < 0.01$) | High | Medium | Low | Low | YES | 422 |
| BM ($p < 0.01$) | High | Low | Low | Medium | NO | 443 |
| CAI/GC ($p < 0.33$) | Medium | Low | Medium | High | NO | 324 |
| PD ($e^{-6}$) | Medium | Low | Medium | High | NO | 545 |

genomes whose average %GC content was much lower than that of the *E. coli* K12 MG1655 (50.8% GC) genome. The DP exhibited average detection levels of approximately 50% of the introduced foreign genes. These methods (CAI/GC and DP) are efficient only when the artificially introduced foreign genes came from phylogenetically distant species. In this case, the results are phylogenetically coherent: the more related the species that the genes are taken from are to the analyzed genome, the lower the HGT detection level was. However, the DP method failed to go beyond the 80% threshold of detection, with the exception of those genes coming from *Aeropyrum pernix*, which is comparable to the CAI results observed in almost all cases.

Of all the methods analyzed, the MM appears to be the best methodology for a proper detection of HGT, and it appears to be the most reliable strategy to detect transfers from closely related species, which are believed to be the most frequent ones. The BM and the MM consistently detected the foreign, introduced genes. However, the MM appears to be more accurate when genes from closely related species are analyzed.

The functional properties of all models are summarized in **Table 12.1**. The average detection levels were obtained from the performance of the four methodologies in the experiment of the *in silico* introduction of foreign genes. The probabilities of having false negatives and positives were also analyzed.

## 4. What Kind of Genes Are Detected by Each Method (HGT and Function)?

A detailed analysis of the foreign genes found by the different methodologies shows that most of them belong to the unassigned functional category. However, a significant percentage of these putative horizontally transferred genes has well-defined functions or belongs to an assigned functional category. Few genes detected by the MM method were informational genes. In pathogenic strains, the sets of genes that have undergone HGT included a number of sequences, which are directly related to pathogenesis. Important aspects that could help to understand the HGT dynamics are: (a) most of the HGT detected belong to the unassigned functional category; these genes are mainly hypothetical insertion elements and phage-related (as shown in **Table 12.2**, most of these genes are also pseudogenes), (b) HGT belonging to all major functional categories also can be found, including a few informational genes such as a ribosomal protein, a DNA polymerase II $\varepsilon$ subunit, etc. (nevertheless, these informational genes also appear to be pseudogenes); and (c) HGT that belongs to other functional categories (for instance, amino acid metabolism, carbohydrate metabolism, membrane transport,

**Table 12.2**
**Functional categories of the transferred genes and the pseudogenes detected using the MM in the *E. coli* K12 MG1655 genome**

| Functional categories | *HGT* | *Pseudogenes* |
|---|---|---|
| Amino acid metabolism | 16 | 3 |
| Biodegradation of xenobiotics | 5 | 3 |
| Biosynthesis of secondary metabolites | 1 | 0 |
| Carbohydrate metabolism | 27 | 6 |
| Cell motility | 0 | 0 |
| Energy metabolism | 4 | 1 |
| Hypothetical | 90 | 265 |
| Insertion elements | 2 | 16 |
| Lipid metabolism | 5 | 1 |
| Membrane transport | 43 | 8 |
| Metabolism of cofactors and vitamins | 2 | 1 |
| Metabolism of complex carbohydrates | 0 | 4 |
| Metabolism of complex lipids | 0 | 0 |
| Metabolism of other amino acids | 0 | 1 |
| Nucleotide metabolism | 4 | 0 |
| Phage-related | 2 | 4 |
| Replication and repair | 0 | 1 |
| Signal transduction | 1 | 2 |
| Sorting and degradation | 0 | 4 |
| Transcription | 3 | 3 |
| Transcriptional regulator | 10 | 4 |
| Translation | 0 | 0 |

nucleotide metabolism, metabolism of cofactors and vitamins, energy metabolism, which are less likely to be pseudogenes).

## 5. Discussion

The four methodological approaches compared in this work have been widely used by different authors during the past few years *(10, 12, 13, 15, 16)*. However, they have generally been employed without proper experimental controls that could allow an accurate assessment. A control methodology based on the detection of prokaryotic genes and phage genes from two distinct genomic pools (one including all the prokaryotic genes from complete sequenced genomes and the other containing several phage genes) has been recently suggested *(23)*.

The introduction of a simple theoretical control, such as the artificial addition of prokaryotic foreign genes into a genome as discussed here, demonstrates that the accuracy level of the different methods may be low, with the exception of the BM and the MM methods that consistently detected the foreign, introduced genes and are accurate when genes from closely related species are under evaluation.

### 5.1. Different Methods Give Different Results: Is a Biological Explanation Feasible?

The proposal that massive amounts of genetic material can be promiscuously exchanged between prokaryotes raises the issue whether all the genes from a genome are equally subject to HGT *(2)*. If this is the case, foreign DNA would eventually replace all the vertically inherited genes in a few millions years *(12)*. This has led to the proposal that the history of life should not be represented as a tree but rather as a complex network *(1)*. Considerable efforts have been undertaken to discuss this hypothesis *(32–34)*. Recently, it was shown that all orthologs with non-rRNA-type phylogenies encode for unassigned proteins *(25)*. This would imply that the most essential genes from a genome, i.e., those with high selective pressure and which can be part of large complexes, such as informational genes *(35, 36)*, seem to be less likely to undergo HGT. The functional analysis of HGT confirms that most of the transferred genes encode for unassigned proteins.

However, detailed analysis shows that some of these genes have well-defined functions or belong to well-defined assigned functional categories. These genes might be encoding for the proteins involved in the direct interaction of the organism with its environment, such as membrane proteins, cytosolic enzymes, and pathogenesis-related proteins. This possibility is supported by the recognition that high levels of HGT events are recognizable in prokaryotic populations living in environments polluted with xenobiotics *(37)*. Furthermore, some badly preserved, informational genes were also detected (see results). This may imply that this sort of genes can be transferred but, for these enterobacterial species, they have not been selected in the host genome, and thus, they have experienced a sequence decay by accumulating non-sense mutations.

The results analyzed here demonstrate that it is possible to detect HGT through the compositional approaches (BM and MM) assembled with Markov chains and using Monte-Carlo simulations for statistical purposes. It is particularly interesting to observe the complementary nature of these two approaches; this suggests that the ideal detection model could be shaped using the combination of several of these approaches.

The MM exhibited its best detection levels when the transferred genes belong to distantly related organisms (**Fig. 12.1**). This implies that the detected horizontally transferred genes in *E. coli* have originated from very distant organisms, and still reflect their previous genome context. Thus it could be concluded that the barriers that limit gene exchange among prokaryotes are ecological-environmental rather than species-dependent *(36)*. As shown in **Table 12.1**, there are a high number of independent events of lateral acquisition of genes. This is consistent with the previous discussion by Ochman et al. *(2)*, Blattner *(38)*, and Perna et al. *(39)*.

During the past two years, new detection methods have been developed based on (a) the composition-based methods and (b) anomalous phylogenetic distribution (*see* **Chapters 11 and 13**). Some of these methods have a good degree of accuracy and with proper controls may yield reliable estimates. These methods include those proposed by (i) Azad and Lawrence *(40)*, which is based on a entropic clustering method and genome position; (ii) Choi and Kim *(6)*, which estimate the global extent of HGT by statistical method and analyzing protein domain families; and (iii) Linz et al. *(41)*, which propose the use of a likelihood framework to measure HGT.

In conclusion, the proper detection of HGT is affected by several, frequently occurring processes that dim the evolutionary history of the organisms and their genes. No method will be infallible, but the different approaches developed during the past few years suggest that, if we use suitable controls and consider the processes that affect the search, it is possible to generate reliable results, at least enabling us to discuss the most important hypotheses on the evolution of genomes. Furthermore, the determination of overall HGT rates and their impact on genome evolution requires that the calculation explicitly considers the rates of false positives and false negatives, and that *in silico* transfer provides an elegant approach to estimate these parameters.

## Acknowledgments

This work was supported in part by CONACYT-Mexico (Project 50520-Q) to A.L. and (Project 52226) to A.B.

## References

1. Doolittle, W. F. (1999) Phylogenetic classification and the universal tree. *Science* **284**, 2124–29.
2. Ochman, H., Lawrence, J. G., Groisman, E. A. (2000) Lateral gene transfer and the nature of bacterial innovation. *Nature* **405**, 299–304.

3. Jain, R., Rivera, M. C., Moore, J. E., Lake, J. A. (2003) Horizontal gene transfer accelerates genomes innovation and evolution. *Mol Biol Evol* **20**, 1598–1602.
4. Brown, J. R. (2003) Ancient horizontal gene transfer. *Nat Rev Genet* **4**, 121–32.
5. Galtier, N. (2007) A model of horizontal gene transfer and the bacterial phylogeny problem. *Syst Biol* **56**, 633–42.
6. Choi, I. G., Kim, S. H. (2007) Global extent of horizontal gene transfer. *Proc Natl Acad Sci U S A* **104**, 4489–94.
7. Snel, B., Bork, P., Huynen, M. A. (2002) Genomes in flux: the evolution of archaeal and proteobacterial gene content. *Genome Res* **12**, 17–25.
8. Cortez, D. Q., Lazcano, A., Becerra A. (2005) Comparative analysis of methodologies for the detection of horizontally transferred genes: A reassessment of first-order Markov models. *In silico Biology* **5**, 581–92.
9. Aravind, L., Tatusov, R. L., Wolf, Y. I., Walker, D. R., Koonin, E. V. (1998) Evidence for massive gene exchange between archaeal and bacterial hyperthermophiles. *Trends Genet* **14**, 442–4.
10. Hayes, W. S., Borodovsky, M. (1998) How to interpret an anonymous bacterial genome: machine learning approach to gene identification. *Genome Res* **8**, 1154–1171.
11. Kyrpides, N. C., Olsen, G. J. (1999) Archaeal and bacterial hyperthermophiles: horizontal gene exchange or common ancestry? *Trends Genet* **15**, 298–9.
12. Lawrence, J. G., Ochman, H. (1997) Amelioration of bacterial genomes: rates of change and exchange. *J Mol Evol* **44**, 383–97.
13. Garcia-Vallve, S., Romeu, A., Palau, J. (2000) Horizontal gene transfer in bacterial and archaeal complete genomes. *Genome Res* **10**, 1719–25.
14. Hooper, S. D., Berg, O. G. (2002) Gene import or deletion: a study of the different genes in Escherichia coli strains K12 and O157:H7. *J Mol Evol* **55**, 734–44.
15. Daubin, V., Lerat, E., Perriere, G. (2003) The source of laterally transferred genes in bacterial genomes. *Genome Biol* **4**, R57.
16. Nakamura, Y., Itoh, T., Matsuda, H., Gojobori., T. (2004) Biased biological functions of horizontally transferred genes in prokaryotic genomes. *Nature* **36**, 760–6.
17. Briones, C., Manrubia, S. C., Lázaro, E., Lazcano, A., Amils, R. (2005) Reconstructing evolutionary relationships from functional data: a consistent classification of organisms based on translation inhibition response. *Mol Phyl Evol* **34**, 371–81.
18. Delaye, L., Becerra, A., Lazcano, A (2005) The last common ancestor: what's in a name? *Orig Life Evol Biosph* **35**, 537–54.
19. Forterre, P. (1993) The great virus comeback – from an evolutionary perspective. *Res Microbiol* **154**, 223–5.
20. Philippe, H., Douady, C. J. (2003) Horizontal gene transfer and phylogetics. *Curr Opin Microbiol* **6**, 498–505.
21. Kurland, C. G., Canback, B., Berg, O. G. (2003) Horizontal gene transfer: A critical view. *Proc Natl Acad Sci U S A* **95**, 9413–7.
22. Becerra, A., Delaye, L., Islas, S., Lazcano, A. (2007) The very early stages of biological evolution related to the nature of the last common ancestor of the three major cell domains. *Annu Rev Ecol Evol Syst* **38** (in press).
23. Tsirigos, A., Rigoutsos, I. (2005). A new computational method for the detection of horizontal gene transfer events. *Nucleic Acids Res* **33**, 922–33.
24. Ragan, M. A., Charlebois, R. C. (2002) Distributional profiles of homologous open reading frames among bacterial phyla: implications for vertical and lateral transmission. *Int J Syst Evol Microbiol* **52**, 777–87.
25. Daubin, V., Moran, N. A., Ochman, H. (2003) Phylogenetics and the cohesion of bacterial genomes. *Science* **301**, 829–32.
26. Koski, L. B., Morton, R. A., Golding, G. B. (2001) Codon bias and base composition are poor indicators of horizontally transferred genes. *Mol Biol Evol* **18**, 404–12.
27. Wang, B. (2001) Limitations of compositional approach to identifying horizontally transferred genes. *J Mol Evol* **53**, 244–50.
28. Guindon, S., Perriere, G. (2001) Intragenomic base content variation is a potential source of biases when searching for horizontally transferred genes. *Mol Biol Evol* **18**, 1838–40.
29. Daubin, V., Perriere, G. (2003) G + C3 structuring along the genome: a common feature in prokaryotes. *Mol Biol Evol* **20**, 471–83.
30. Clarke, G. D., Beiko, R. G., Ragan, M. A., Charlebois, R. L. (2002) Inferring genome trees by using a filter to eliminate phylogenetically discordant sequences and a distance matrix based on mean normalized BLASTP scores. *J Bacteriol* **184**, 2072–80.
31. Ragan, M. A. (2001) On surrogate methods for detecting lateral gene transfer. *FEMS Microbiol Lett* **201**, 187–91.
32. de la Cruz, F., Davies, J. (2000) Horizontal gene transfer and the origin of species: lessons from bacteria. *Trends Microbiol* **8**, 128–33.
33. Gogarten, J. P., Doolittle, W. F., Lawrence, J. G. (2002) Prokaryotic evolution in light of gene transfer. *Mol Biol Evol* **19**, 2226–38.

34. Yang, S., Doolittle, R. F., Bourne, P. E. (2005) Phylogeny determined by protein domain content. *Proc Natl Acad Sci* **102**, 373–8.
35. Rivera, M. C., Jain, R., Moore, J. E., Lake, J. A. (1998) Genomic evidence for two functionally distinct gene classes. *Proc Natl Acad Sci* **95**, 6239–44.
36. Jain, R., Rivera, M. C., Lake, J. A. (1999) Horizontal gene transfer among genomes: the complexity hypothesis. *Proc Natl Acad Sci* **96**, 3801–6.
37. Top, E. M., Springael, D. (2003) The role of mobile genetic elements in bacterial adaptation to xenobiotic organic compounds. *Curr Opin Biotechnol* **14**, 262–9.
38. Blattner, F. R., Plunkett, G. 3rd, Bloch, C. A., Perna, N. T., Burland, V., Riley, M., Collado-Vides, J., Glasner, J. D., Rode, C. K., Mayhew, G. F., Gregor, J., et al. (1997) The complete genome sequence of Escherichia coli K-12. *Science* **277**, 1453–74.
39. Perna, N. T., Plunkett, G. 3rd, Burland, V., Mau, B., Glasner, J. D., Rose, D. J., Mayhew, G. F., Evans, P. S., Gregor, J., Kirkpatrick, H. A., Posfai, G., et al. (2001) Genome sequence of enterohaemorrhagic *Escherichia coli* O157:H7. *Nature* **409**, 529–33.
40. Azad, R. K., Lawrence, J. G. (2007) Detecting laterally transferred genes: use of entropic clustering methods and genome position. *Nucleic Acids Res* **35**, 4629–39.
41. Linz, S., Radtke, A., von Haeseler, A. (2007) A likelihood framework to measure horizontal gene transfer. *Mol Biol Evol* **24**, 1312–9.

# Chapter 13

## Testing Phylogenetic Methods to Identify Horizontal Gene Transfer

**Maria Poptsova**


### Abstract

The subject of this chapter is to describe the methodology for assessing the power of phylogenetic HGT detection methods. Detection power is defined in the framework of hypothesis testing. Rates of false positives and false negatives can be estimated by testing HGT detection methods on HGT-free orthologous sets, and on the same sets with *in silico* simulated HGT events. The whole process can be divided into three steps: obtaining HGT-free orthologous sets, *in silico* simulation of HGT events in the same set, and submitting both sets for evaluation by any of the tested methods.

Phylogenetic methods of HGT detection can be roughly divided into three types: likelihood-based tests of topologies (Kishino-Hasegawa (KH), Shimodaira-Hasegawa (SH), and Approximately Unbiased (AU) tests), tree distance methods (symmetrical difference of Robinson and Foulds (RF), and Subtree Pruning and Regrafting (SPR) distances), and genome spectral approaches (bipartition and quartet decomposition analysis). Restrictions that are inherent to phylogenetic methods of HGT detection in general and the power and precision of each method are discussed and comparative analyses of different approaches are provided, as well as some examples of assessing the power of phylogenetic HGT detection methods from a case study of orthologous sets from gamma-proteobacteria (Poptsova and Gogarten, *BMC Evol Biol* 7, 45, 2007) and cyanobacteria (Zhaxybayeva et al., *Genome Res* **16**, 1099–108, 2006).


**Key words:** Phylogenetic methods of HGT detection, power of HGT detection methods, likelihood-based tests of topologies, tree distance methods, genome spectral methods.

## 1. Introduction

Phylogenetic methods of HGT detection are, in essence, the methods of finding incongruence between a considered species (organismal, genome) tree and orthologous gene trees. It is to be understood that all phylogenetic methods for HGT detection are applicable to families of orthologs only. They cannot

Maria B. Gogarten et al. (eds.), *Horizontal Gene Transfer: Genomes in Flux, vol. 532*

DOI 10.1007/978-1-60327-853-9_13 Springerprotocols.com

consider horizontally transferred ORFans that do not have homologs in any of the considered genomes. Thus the first limitation of phylogenetic methods for HGT detection follows from the performance of the methods for the selection of orthologous gene families. Depending on the set of species under comparison, the common set of orthologs can be as small as 5–10% of the size of the genome. This number can decrease even further if one compares distant species of bacteria and archaea with the requirement that representatives from each genome must be present in orthologous families. Ideally, we would like to have orthologous sets that have one representative from each taxon. In reality, due to gene gain and gene loss, some orthologous genes may not be present in all taxa. Nevertheless, gene trees reconstructed from incomplete sets (sets that have missing taxa) can also be compared against the complete reference tree. The presence of at least one ortholog from each genome is required for the calculation of Robinson and Foulds (RF) *(1)*, Subtree Pruning and Regrafting (SPR) *(2)*, and some other tree distances such as Nearest Neighbor Interchange (NNI) and Tree Bisection and Reconnection (TBR) *(3–5)*, since they were initially defined for trees that contain equal number of taxa. However, all these tree distance metrics can be adapted to the trees with missing taxa. See *(6)* on calculation of RF and SPR distances for incomplete orthologous families. Bipartition spectral analysis *(7, 8)* was applied to trees with equal number of taxa, but principles used in *(6)* can also be extended to incomplete families. By definition, the presence of all taxa in gene trees is not required for quartet decomposition analysis *(9, 10)* since the method is based on the decomposition of either complete or incomplete families into sets of four orthologs from any of the four genomes of the analyzed set.

The second restriction on phylogenetic HGT detection methods issues from the phylogenetic reconstruction methods themselves. Unreliable phylogeny can produce wrong splits, and, consequently, lead to wrong conclusions. Statistical measures, such as bootstrap support values or Bayesian posterior probabilities, can be used to assess the reliability of each split, ultimately restricting the set that is compared against the organismal tree *(11)*. Bipartition spectral and quartet decomposition analyses explicitly take into account bootstrap support values (considering only highly supported splits) when reporting positive HGT events. In the Efficient Evaluation of Edit Paths (EEEP) algorithm *(12)* developed for the calculation of SPR distance, unsupported splits can be collapsed into multifurcating nodes and SPR distances are calculated for trees with multifurcating nodes *(6)* (*see* also **Notes 1,2**).

A separate class of phylogenetic methods relies on likelihood-based tests of topology such as the Kishino-Hasegawa (KH)

*(13, 14)*, Shimodaira-Hasegawa (SH) *(15)*, and Approximately Unbiased (AU) tests *(16)*. These tests consider relative certainty or uncertainty in maximum likelihood tree selection via the construction of a confidence set – a set of trees that are statistically no worse than the maximum likelihood tree. Trees that are not included in the confidence set are rejected. They are considered to be incompatible with the phylogeny in question, and this incompatibility serves as evidence for HGT. Limitations of such methods stem from the dependence of the results on the size of a confidence interval. If the confidence interval is too big, then conflicting trees can be included in the confidence set, thus increasing the rate of false negatives, i.e., the number of undetected HGT events. On the contrary, if the confidence interval is too small, compatible trees could be rejected, increasing the rate of false positives, i.e., the number of wrongly detected HGT events (see definition of false positives and false negatives in the framework of HGT testing at the end of this section).

The main limitation inherent to all phylogenetic methods is the reliability of the species tree. In the case of the prokaryotes, the universal species tree is still the subject of much debate. The small subunit ribosomal RNA phylogeny was used to represent organismal phylogeny *(17)*, but trees reconstructed from other conserved molecular markers were shown to contradict the small subunit ribosomal RNA tree and each other *(18–22)*. Different approaches exist for inferring an organismal phylogeny from orthologous gene trees *(23)*. Consensus methods work with trees that contain equal numbers of taxa. They include strict consensus, majority-rule consensus *(24)*, and Adams consensus *(25)* methods. Supertree approaches deal with trees from incomplete orthologous families. The varieties are matrix representation with parsimony (MRP) *(26)*, matrix representation with flipping (MRF) *(27)*, and MinCut supertrees *(28)*. Whether plurality phylogenetic signal extracted from a genome would reflect species evolution would depend on how many conflicting individual gene trees are present in the data and to which extent individual gene trees differ from the reference organismal tree. An alternative to consensus and supertree methods is the method of tree reconstruction from concatenated alignments of orthologs *(23)*. For a review of different approaches including the methods based on gene content, gene order, and evolutionary distances between orthologs, see *(23)*. The problem of a steady (bi)furcating tree as a sufficient description for species evolution is discussed in *(9, 29–31)*. Despite the uncertainty that surrounds the issue of prokaryotic species-tree reconstruction, in practice, preference is given to consensus methods and supertree approaches. The discussion of the advantages and disadvantages of each method is beyond the scope of this chapter. One must keep in mind that phylogenetic methods for HGT detection strongly depend on the reliability of a reference organismal tree.

The assessment of the power of HGT detection methods consists in the estimation of the number of false positives and false negatives, or estimation of type I (rejecting a null hypothesis that is actually true) and type II (failure to reject a null hypothesis when the alternative hypothesis is true) errors. In the framework of HGT detection, a null hypothesis is stating that the evolutionary history of orthologous genes is free from HGT, while the alternative hypothesis suggests an HGT event in the history of orthologous genes. Hence, the number of false positives is the number of times HGT events were detected in the history of orthologous genes that did not have any; and the number of false negatives is the number of times HGT events were not detected for orthologous genes with HGT events in their history. In this chapter, the technique for testing various phylogenetic methods of HGT detection using *in silico* approaches is discussed.

## 2. *In Silico* Experiments

### 2.1. Simulation of HGT by Modifying Tree Topology

Becerra and collaborators *(32)* first introduced *in silico* transfers to test the efficiency of parametric methods of HGT detection (*see* also **Chapter 12**). An extended *in silico* approach for specifically testing phylogenetic methods was implemented in *(33)*. The main idea behind *in silico* experiments is to artificially create an HGT event in a phylogenetic tree and submit the tree for evaluation by different existing methods. In terms of phylogenetic trees, the simulation of gene transfer means introducing changes to the topology of a gene tree by rearranging its branches (*see* **Fig. 13.1**). The SPR technique ideally serves this purpose. A recent transfer from species A to species E can be imitated by re-attaching, or re-grafting, the branch holding species E as a sister group of species A (**Fig. 13.1B**). An ancient transfer is more difficult to implement because we need to keep the number of species in a tree unchanged and introduce transfer events only in between the considered species. Some SPR rearrangements can imitate ancient transfers as it is the transfer from species E to the common ancestor of the group A, B, and C (**Fig. 13.1C**). Note that not all SPR rearrangements correspond to real evolutionary processes. Thus the swap between neighboring nodes, as depicted in **Fig. 13.1D**, would neither mean the transfer from species E to species D nor the converse. The constraints on SPR operations when calculating SPR distances are discussed in detail in *(12)*.

Recent *in silico* transfers were implemented in *(33)* on a test case of 13 proteobacterial genomes, which, as reported previously, were free from HGT based on SH test *(34)*. See *(35)* for a

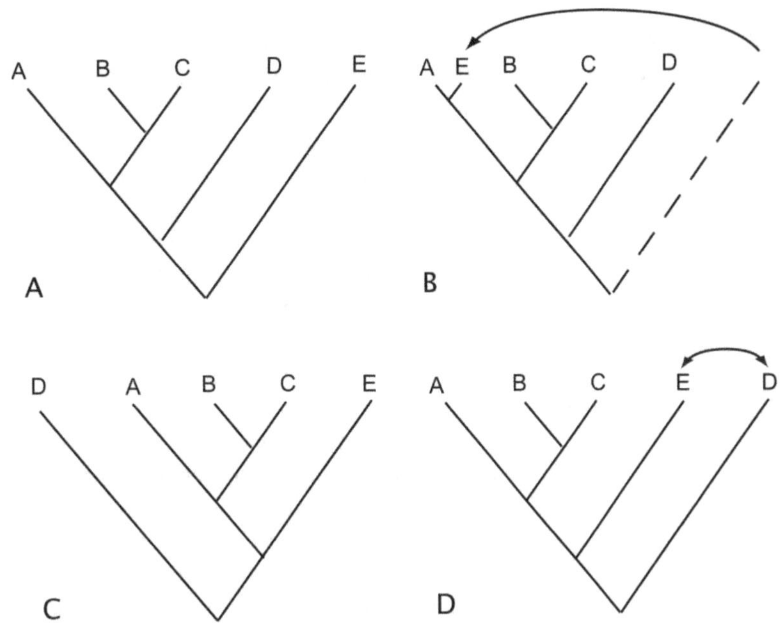


Fig. 13.1. Simulation of *in silico* transfers: (**A**) original unperturbed tree; (**B**) simulation of recent transfer (gene donation with replacement); (**C**) SPR rearrangement imitating transfer from species E to the common ancestor of group A, B, and C; (**D**) prohibited SPR rearrangement: nearest nodes interchange.

controversial discussion of this assertion. Four phylogenetic methods were tested: AU-test, RF distance, SPR distance methods, and bipartition spectral analysis (*see* **Section 3.4** for discussion). The test summary is given in **Table 13.1**.

**Table 13.1**
**Power of HGT detection methods as assessed by the case study of gamma-proteobacterial *(1)* and cyanobacterial *(2)* genomes**

| HGT detection methods | Significance level | Rate of false positives (%) | Rate of false negatives (%) |
|---|---|---|---|
| | *Gamma-proteobacteria* – in silico *transfers* | | |
| AU-test | 5% | 5 | 10 |
| RF distance | 2 SD (2.5%) | 5 | 40 |
| Bipartition spectral analysis | BP 70% | 4.2 | 6 |
| | BP 90% | 2.1 | 3 |
| | *Cyanobacteria – genomes simulation with EvolSimulator* | | |
| Quartet decomposition analysis | BP 30% (68% of fam. with HGT) | 2–5 | 45 |
| | BP 30% (40% of fam. with HGT) | | 28 |

### 2.2. Simulation of Genome Evolution

Another approach to simulate orthologous gene families whose history involves HGT is to simulate whole-genome evolution. The software package EvolSimulator *(36)* can simulate whole-genome evolution on the level of gene families with HGT included as an option. Starting with an ancestral genome containing some number of genes, one will obtain, after a number of iterations, a set of species containing genes whose evolutionary histories are exactly known. EvolSimulator uses many evolutionary models at both genome and gene levels, including lineage speciation and extinction, gene duplication, loss and horizontal transfer, combined with various sequence substitution models.

EvolSimulator was used to generate null models to assess the power of HGT detection in quartet decomposition analysis *(9)*. In the first simulation, gene evolution was carried out in the absence of gene gain, gene loss, or HGT. The resulting orthologous gene families originated strictly through vertical inheritance. Quartet decomposition analysis applied to these families allowed to estimate the number of false positives. The repeated simulation with an HGT option estimated the number of false negatives. Rate of HGT may be varied, and the estimation of false negatives can be done for different rates of HGT (*see* **Section 3.4.** for a discussion). The summary of the test is given in **Table 13.1**. The results show that the quartet decomposition method underestimated the number of HGT events.

## 3. Methods

Two types of errors deserve examination with respect to HGT detection methods: the rate of false positives and the rate of false negatives. The number of false positives can be estimated by applying HGT detection methods to HGT-free orthologous gene families. The number of false negatives will be given through testing HGT-containing gene families. The whole process can be divided into three steps: obtaining HGT-free orthologous sets, *in silico* simulation of HGT events in the same set, and submitting both sets for evaluation by the tested method.

### 3.1. Obtaining a Set of HGT-Free Phylogenetic Trees

#### 3.1.1. Real Sets of Orthologous Families

Based on the results of different HGT detection methods *(9, 34)*, some sets of orthologous gene families were reported as presumably free from HGT or as having few easily filtered HGT events. These sets can serve as null sets for introducing *in silico* changes to tree topologies.

One may consider the selection of orthologous families for particular species. Many methods are available for that purpose.

Among those, the most practical are the strict reciprocal best Blast hit method *(37–40)*, MCL Markov clustering *(41)*, and BranchClust algorithm *(42)*. To decide if the selected orthologous families are free from HGT, one must submit them to HGT detection methods, which are prone to errors. Here we risk entering a vicious circle – testing HGT detection methods on the sets selected by HGT detection methods. To refine the selection one may apply more stringent criteria and test the sets by various methods preserving the core that would agree with the results from different tests. The alternative approach described in the next section eliminates this problem: we will know the exact history of orthologous sets because we have simulated it.

#### *3.1.2. Simulated Sets of Orthologous Families*

The program EvolSimulator *(36)* (http:/bioinformatics.org.au/evolsim/) can be used to perform simulation of genome evolution. One can start with a fixed number of genes in the ancestral genome and, by balancing speciation with extinction, maintain a fixed number of lineages at any given time after a brief initial phase of "population" growth. If paralogous duplication and gene loss are disabled, then, at the end, one will get exactly the same number of genes one started with. As a result one will obtain a set of orthologous genes whose evolutionary history is exactly known. Parameters affecting sequence evolution can be adjusted to reproduce any level of sequence divergence.

For example, in *(9)* EvolSimulator software was used to assess the power of the quartet decomposition method. Seven hundreds genes were simulated for 10,000 generations at a rate of 0.015 events per generation to maintain ~50 extant lineages at any given time. Parameters affecting sequence evolution were selected to be similar to cyanobacterial species.

### 3.2. In Silico Simulation of HGT Events

#### *3.2.1. Introducing Changes to Tree Topology*

Recent transfers can be simulated by moving the branch of the recipient genome next to the donor genome (**Fig. 13.1B**). Some ancient transfers are possible to imitate by SPR rearrangements (**Fig. 13.1D**), but one has to keep in mind the constraints on all possible SPR changes so as to comply those with the real HGT events *(12)*. Such constraints include prohibition on nearest node exchange and offspring-to-parent donations.

#### *3.2.2. Simulating HGT Events with EvolSimulator*

EvolSimulator can be used to create a set of orthologous genes with HGT events in its history. One may generate the same orthologous set as described in **Section 3.1.2**, but allowing for HGT. In *(9)* the set of orthologous families with HGT was simulated with 0.5 and 1 events per generation in the entire population. The resulting sets were used to estimate the number of false negatives in quartet decomposition analysis (*see* **Table 13.1**).

### 3.3. Assessing Power of HGT Detection Methods

To obtain the number of false positives one must submit a set of trees without HGT to a tested HGT detection method and calculate the cases that were reported as containing HGT. The number of false negatives is the number of cases that were not detected by the same HGT detection method for a set of trees with simulated HGT.

#### 3.3.1. Likelihood-Based Tests of Topologies: KH, SH, and AU Tests

KH, SH, and AU tests produce *p*-value for the compared and a reference organismal tree. Trees with *p*-values lower than a certain significance level are considered as incongruent with a species tree, and this incongruence is attributed to HGT. Significance level for these tests is usually taken as 0.05 (5%) though levels of 0.01 (1%) and 0.001 (0.1%) may be considered for better precision.

The *p*-values of KH and SH tests can be calculated with TREE-PUZZLE software *(43)* or the program Consel *(44)* (*see* **Note 3**). The *p*-values of AU tests are calculated with the program Consel (*see* **Note 4** on difference between the tests).

Rate of false positives and false negatives is calculated at different significance levels, corresponding to *p*-values and equaling 0.05, 0.01, or 0.001. The results of the performance of the AU test on a particular set of 13 gamma-proteobacteria with *in silico* transfers can be found in *(33)*. *See* also **Table 13.1** for a summary.

#### 3.3.2. Distance Methods Based on the Calculation of RF or SPR Distances Between Trees

Initially, distance methods were designed to compare tree topologies only, without taking into account branch length and statistical values for splits. The RF distance gives the number of different bipartitions between trees, while the SPR distance gives the number of rearrangements necessary to convert one tree topology to another. To adapt distance methods to HGT detection, a null distribution for the original data must be obtained by calculating the distances between each gene tree and the organismal tree. For a signal generated through HGT to be considered significant, the distance is required to be at least two standard deviations larger than the mean of the distance distribution of the original data. In case of a normal distribution, this cutoff level corresponds to a significance level of 2.5%.

Symmetric difference, or Robinson and Foulds distance, can be calculated with the Treedist program from the Phylip package *(45)*. Approximation of SPR distance, the Edit Path, can be calculated with the Efficient Evaluation of Edit Paths (EEEP) software *(12)* (*see* **Note 5**).

Another approximation of SPR distances is given by the HGT detection method proposed in *(46, 47)*. An organismal tree is gradually transformed into the gene tree through a series of subtree moves, which are in essence the SPR rearrangements. To find the optimal rearrangement, a mapping procedure is applied to the distance matrix using the least-square criterion *(46)* and topological rearrangements are approximated with RF distances. The method is implemented in the program T-REX (http://www.labunix.uqam.ca/~makarenv/trex.html) *(48)*.

#### 3.3.3. Genome Spectral Approaches: Bipartition and Quartet Decomposition Analyses

Genome spectral methods, such as bipartition and quartet decomposition analysis, decompose each gene tree into small quanta of phylogenetic information such as bipartitions and quartets. The advantage of this approach is that individual bipartitions or quartets can either be compatible (they can reside in one tree) or conflicting (they cannot co-exist in one tree), and tree comparison is done by comparing small quanta of phylogenetic information. In addition, statistical support for bipartitions and quartets can be assessed through bootstrap analyses, and poorly supported bipartitions or quartets can be removed from the genome spectrum. By identifying the compatible phylogenetic entities supported by the majority of gene families, one can find a plurality consensus phylogenetic signal. Bipartitions or quartets that are in significant conflict with the plurality consensus are most likely related to a horizontal gene transfer event or to systematic artifacts of phylogenetic reconstruction.

The methods of performing bipartition spectral and quartet decomposition analysis are described in *(8, 9, 33)*. A web-based tool, Gene Phylogeny eXplorer (GPX) (http://bioinformatics.cs.uri.edu/gpx) *(49)*, was developed for automated analysis of large bipartition and quartet matrices. It uses self-organizing maps *(50)* to find structures in multidimensional phylogenetic data and reports conflicting bipartitions and quartets with their corresponding projections on the generated map *(51)*.

To test the power of phylogenetic spectral methods, one needs to have a set of trees with simulated HGT. But instead of introducing SPR rearrangements to the whole tree topology, one can introduce changes at the level of bipartitions and quartets. The latter is much easier to implement (see *(33)*) than SPR rearrangements. Calculation of the rates of false positives and false negatives proceeds in analogy with other methods by submitting sets with known evolutionary histories. The level of significance is not accessed directly but through bootstrap values of the splits.

### 3.4. Comparison of Different Approaches

Phylogenetic HGT detection methods can be roughly divided into three types: likelihood-based tests of topologies, distance-based methods, and genome spectral methods. Though in essence they are all designed to identify topological incongruence between trees, the results they generate differ and can serve different purposes.

Likelihood-based tests of topologies provide a good framework for HGT hypothesis testing by explicitly allowing to obtain probabilities (via significance levels) to accept or reject an HGT event. Likelihood-based tests assess all splits in a tree simultaneously, and changes in one part of the tree can be compensated by uncertainty in other parts of the tree. That is why likelihood-based tests fail to detect some HGT transfers *(33)*. It could also be argued that the significance level in the likelihood-based tests overestimates the false positive rate; for example, the actual rate of false positives performed with a 5% significance level might be much smaller than 5% *(33)*.

Plain tree distance (RF, SPR distances) methods without statistical support for the splits register only topological incompatibility of the trees and do not provide any information on the actual reliability of the splits. The power of detection of distance-based methods depends on the reliability of phylogenetic reconstruction for both organismal and gene trees. Some distance-based methods can incorporate statistical measures, as in the EEEP algorithm, by collapsing poorly supported splits into multifurcating nodes and adapting tree distance methodology to multifurctaing trees. The major advantage of the HGT detection methods implemented in EEEP and T-REX programs is that, in addition to HGT detection, they propose an optimal path for topological tree rearrangements, which is extremely useful in some case studies.

Spectral methods for HGT detection work with individual small phylogenetic quanta such as bipartitions or quartets. They employ bootstrap support values for individual splits that were shown to be more conservative than probability estimates *(37, 38, 52)*. The advantage of the spectral approach is that gene transfer events can be detected even in the absence of a completely resolved organismal phylogeny, because it relies on majority signal extracted from the dataset. The power of detection in spectral analyses depends on the presence of well-supported bipartitions or quartets in the majority of gene phylogenies. Individual bipartitions or quartets are compared with the majority signal, and conflicts are ascribed to HGT events. If only few majority phylogenetic signals are present in a genome spectrum, then the likelihood- and distance-based tests provide an alternative method to identify individual conflicting genes, provided a reasonable hypothesis for the organismal phylogeny can be developed.

The choice of method is also contingent on the reason why one wants to identify putatively transferred genes. If one is interested in case studies of transferred genes, then the likelihood-based tests with a high significance level, the distance-based methods, and the genome spectral methods are all effective in identifying the candidate genes. However, if one wants to remove gene families from genome-based phylogenetic analyses that might taint a combined dataset, then the AU test with a 5% significance level provides a workable solution, even though at this significance level one would remove about 5% of the gene families erroneously. At present none of the available approaches appears entirely satisfying to quantify the rates of orthologous replacement; one needs to balance unacceptable high rates of false negatives against the reliability of assessing the rates of false positives, and both types of errors need to be explicitly considered in the quantification.

## 4. Notes

1. Some of the presented methods are quite time-consuming. For example, AU test experiments for *in silico* transfers take 12 h for just one transfer test for a set of 13 species on one node of the University of Connecticut's biocluster (PowerPC G5 2.3 GHz, 2 GB RAM). To test all possible transfers, one would require 36 days of computational time on one node though the task is parallelizable, and can be possibly reduced to four days if the calculations were distributed between nine parallel nodes.

   Bootstrap analysis is another example of time-demanding calculation. For example, bootstrap analysis with Phyml can take about 20 min for one gene family of 13 species on one node of the above-mentioned biocluster. Unless running in parallel, the test will only process 72 gene families a day.

   Calculation time required for other types of analysis mentioned in this chapter is negligible compared to the time needed for likelihood-based tests of topology or bootstrap analysis.
2. Theoretically, the task of finding the minimum number of SPR operations necessary to transform one unrooted tree into another has been shown to be NP-hard *(53)*. The EEEP algorithm works with rooted trees and applies biological constraints to reduce time and complexity.
3. KH, SH, or AU tests require one multiple alignment and a tree topology, which will be compared with a maximum likelihood tree reconstructed from the multiple alignment. The reference or organismal trees, reconstructed by one of the consensus methods, does not have a multiple alignment. That is why the multiple alignment is taken for orthologous genes, and the reference (organismal) tree topology is compared against the gene tree represented by a multiple alignment.
4. The AU test is of higher-order accuracy than the KH and SH tests *(16)*. The KH test had been shown to be overconfident in the wrong trees because many trees with high likelihood fall within the confidence interval by chance. The KH test is

supposed to be biased upwards. The SH test, also known as multiple comparison test, was an improvement of the KH method, but it appeared to be biased with an increase of the number of trees to be compared *(54)*. The AU test was designed to avoid biases of KH and SH tests, and it is this test that is now frequently used for HGT detection.

5. Constrained SPR distance calculated by EEEP is similar to (or in some cases, identical to) the hybridization distance *(55,56)*. For hybrid phylogenies and how they relate to SPR operations see *(57)*.

## Acknowledgments

I thank J. Peter Gogarten for his advice and guidance, and Olga Zhaxybayeva and Robert Beiko for comments and discussion. This work was supported through NSF (MCB-0237197) and the NASA Applied Information System Research Program (NNG04GP90G) grants to J. Peter Gogarten.

### References

1. Robinson, D. R., Foulds, L. R. (1981) Comparison of phylogenetic trees. *Math Biosci* **53**, 131–47.
2. Swofford, D. L., Olsen, G. J. (1990) *Phylogeny Reconstruction in Molecular Systematics*, Sinauer Associates, Sunderland, Massachusetts.
3. Waterman, M. S., Smith, T. F. (1978) On the similarity of dendrograms. *J Theor Biol* **73**, 789–800.
4. Allen, B. L., Steel, M. (2001) Subtree transfer operations and their induced metrics on evolutionary trees. *Ann Combinatorics* **5**, 1–15.
5. Felsenstein, J. (2004) *Inferring Phylogenies*, Sinauer Associates, Sunderland, Massachusetts.
6. Beiko, R. G., Harlow, T. J., Ragan, M. A. (2005) Highways of gene sharing in prokaryotes. *Proc Natl Acad Sci U S A* **102**, 14332–7.
7. Lento, G. M., Hickson, R. E., Chambers, G. K., Penny, D. (1995) Use of spectral analysis to test hypotheses on the origin of pinnipeds. *Mol Biol Evol* **12**, 28–52.
8. Zhaxybayeva, O., Lapierre, P., Gogarten, J. P. (2004) Genome mosaicism and organismal lineages. *Trends Genet* **20**, 254–60.
9. Zhaxybayeva, O., Gogarten, J. P., Charlebois, R. L., Doolittle, W. F., Papke, R. T. (2006) Phylogenetic analyses of cyanobacterial genomes: quantification of horizontal gene transfer events. *Genome Res* **16**, 1099–108.
10. Strimmer, K., von Haeseler, A. (1996) Quartet puzzling: a quartet maximum-likelihood method for reconstructing tree topologies. *Mol. Biol. Evol.* **13**, 964–69.
11. Felsenstein, J. (1985) Confidence limits on phylogenies: an approach using the bootstrap. *Evolution* **39**, 783–91.
12. Beiko, R. G., Hamilton, N. (2006) Phylogenetic identification of lateral genetic transfer events. *BMC Evol Biol* **6**, 15.
13. Goldman, N., Anderson, J. P., Rodrigo, A. G. (2000) Likelihood-based tests of topologies in phylogenetics. *Syst Biol* **49**, 652–70.
14. Kishino, H., Hasegawa, M. (1989) Evaluation of the maximum likelihood estimate of the evolutionary tree topologies from DNA sequence data, and the branching order in Hominoidea. *J Mol Evol* **29**, 170–9.
15. Shimodaira, H., Hasegawa, M. (1999) Multiple comparisons of log-likelihoods with applications to phylogenetic inference. *Mol Biol Evol* **16**, 1114–16.
16. Shimodaira, H. (2002) An approximately unbiased test of phylogenetic tree selection. *Syst Biol* **51**, 492–508.
17. Woese, C. R., Fox, G. E. (1977) Phylogenetic structure of the prokaryotic domain: the primary kingdoms. *Proc Natl Acad Sci U S A* **74**, 5088–90.

18. Hilario, E., Gogarten, J. P. (1993) Horizontal transfer of ATPase genes – the tree of life becomes a net of life. *Biosystems* **31**, 111–9.
19. Gogarten, J. P. (1995) The early evolution of cellular life. *Trends Ecol Evol* **10**, 147–51.
20. Brown, J. R., Masuchi, Y., Robb, F. T., Doolittle, W. F. (1994) Evolutionary relationships of bacterial and archaeal glutamine synthetase genes. *J Mol Evol* **38**, 566–76.
21. Jain, R., Rivera, M. C., Lake, J. A. (1999) Horizontal gene transfer among genomes: the complexity hypothesis. *Proc Natl Acad Sci U S A* **96**, 3801–6.
22. Nesbo, C. L., Boucher, Y., Doolittle, W. F. (2001) Defining the core of nontransferable prokaryotic genes: the euryarchaeal core. *J Mol Evol* **53**, 340–50.
23. Wolf, Y. I., Rogozin, I. B., Grishin, N. V., Koonin, E. V. (2002) Genome trees and the tree of life. *Trends Genet* **18**, 472–9.
24. Margush, T., McMorris, F. R. (1981) Consensus n-trees. *Bull Math Biol* **43**, 239–44.
25. Adams, E. (1972) Consensus techniques and the comparison of taxonomic trees. *Syst Zool* **21**, 390–97.
26. Bininda-Emonds, O. R., Sanderson, M. J. (2001) Assessment of the accuracy of matrix representation with parsimony analysis supertree construction. *Syst Biol* **50**, 565–79.
27. Burleigh, J., Eulenstein, O., Fernandez-Baca, D., Sanderson, M. (2004) MRF supertrees. In *Phylogenetic Supertrees: Combining Information to Reveal the Tree of Life.* (Bininda-Emonds, O.R.P., ed.), Kluwer, Dordrecht, pp. 65–85.
28. Semple, C., Steel, M. (2000) A supertree method for rooted trees. *Discrete Appl Math* **105**, 147–58.
29. Dagan, T., Martin, W. (2006) The tree of one percent. *Genome Biol* 7, 118.
30. Ciccarelli, F. D., Doerks, T., von Mering, C., Creevey, C. J., Snel, B., Bork, P. (2006) Toward automatic reconstruction of a highly resolved tree of life. *Science* **311**, 1283–7.
31. Zhaxybayeva, O., Gogarten, J. P. (2007) Horizontal gene transfer, gene histories and the root of the tree of life. In *Astrobiology and the Origins of Life* (Pudritz, R. E., Higgs P. G., Stone J., eds.), Cambridge University Press, Cambridge.
32. Cortez, D. Q., Lazcano, A., Becerra, A. (2005) Comparative analysis of methodologies for the detection of horizontally transferred genes: a reassessment of first-order Markov models. *In Silico Biol* **5**, 581–92.
33. Poptsova, M. S., Gogarten, J. P. (2007) The power of phylogenetic approaches to detect horizontally transferred genes. *BMC Evol Biol* **7**, 45.
34. Lerat, E., Daubin, V., Moran, N. A. (2003) From gene trees to organismal phylogeny in prokaryotes: the case of the gamma-proteobacteria. *PLoS Biol* **1**, E19.
35. Bapteste, E., Boucher, Y., Leigh, J., Doolittle, W. F. (2004) Phylogenetic reconstruction and lateral gene transfer. *Trends Microbiol* **12**, 406–11.
36. Beiko, R. G., Charlebois, R. L. (2007) A simulation test bed for hypotheses of genome evolution. *Bioinformatics* **23**, 825–31.
37. Hillis, D. M., Bull, J. J. (1993) An empirical test of bootstrapping as a method for assessing confidence in phylogenetic analysis. *Syst. Biol* **42**, 182–92.
38. Zhaxybayeva, O., Gogarten, J. P. (2002) Bootstrap, Bayesian probability and maximum likelihood mapping: exploring new tools for comparative genome analyses. *BMC Genomics* **3**, 4.
39. Altschul, S. F., Gish, W., Miller, W., Myers, E. W., Lipman, D. J. (1990) Basic local alignment search tool. *J Mol Biol* **215**, 403–10.
40. Montague, M. G., Hutchison, C. A., III (2000) Gene content phylogeny of herpesviruses. *Proc Natl Acad Sci U S A* **97**, 5334–9.
41. van Dongen, S. (2000) *Graph Clustering by Flow Simulation.* University of Utrecht, Utrecht.
42. Poptsova, M. S., Gogarten, J. P. (2007) BranchClust: a phylogenetic algorithm for selecting gene families. *BMC Bioinformatics* **8**, 120.
43. Schmidt, H. A., Strimmer, K., Vingron, M., von Haeseler, A. (2002) TREE-PUZZLE: maximum likelihood phylogenetic analysis using quartets and parallel computing. *Bioinformatics* **18**, 502–4.
44. Shimodaira, H., Hasegawa, M. (2001) CONSEL: for assessing the confidence of phylogenetic tree selection. *Bioinformatics* **17**, 1246–7.
45. Felsenstein, J. (1993) PHYLIP (Phylogeny Inference Package) version 3.6 Distributed by the author. Department of Genetics, University of Washington, Seattle.
46. Boc, A., Makarenkov, V. (2003) New efficient algorithms for detection of horizontal gene transfer events. In *Algorithms in Bioinformatics.* (Benson, G., Page, R., Eds.), pp. 190–201, 3rd Workshop on Algorithms in Bioinformatics, Springer-Verlag, New York.
47. Makarenkov, V., Boc, A., Boubacar Diallo, A., Baniré Diallo, A. (2008) Algorithms for detecting complete and partial horizontal gene transfers: theory and practice. In *CRM*

*Proceedings and AMS Lecture Notes* (Pardalos, P. M., Hansen P., eds.), Vol. 45, pp. 159–79.

48. Makarenkov, V. (2001) T-REX: reconstructing and visualizing phylogenetic trees and reticulation networks. *Bioinformatics* **17**, 664–8.
49. Nahar, N., Poptsova, M. S., Hamel, L., Gogarten, J. P. (2007) GPX: a tool for the exploration and visualization of genome evolution. *Proceedings of the IEEE 7th International Symposium on Bioinformatics & Bioengineering (BIBE07) Boston*, 1338–42.
50. Kohonen, T. (2001) *Self-Organizing Maps.* Springer, New York.
51. Hamel, L., Nahar, N., Poptsova, M. S., Zhaxybayeva, O., Gogarten, J. P. (2008) Unsupervised learning in detection of gene transfer. *J Biomed Biotechnol* doi: 10.1155/2008/472719.
52. Huelsenbeck, J., Rannala, B. (2004) Frequentist properties of Bayesian posterior probabilities of phylogenetic trees under simple and complex substitution models. *Syst Biol* **53**, 904–13.
53. Hein, J., Jiang, T., Wang, L., Zhang, K. (1996) On the complexity of comparing evolutionary trees. *Discrete Appl. Math.* **71**, 153–69.
54. Strimmer, K., Rambaut, A. (2002) Inferring confidence sets of possibly misspecified gene trees. *Proc Biol Sci* **269**, 137–42.
55. Bordewich, M., Semple, C. (2007) Computing the hybridization number of two phylogenetic trees is fixed-parameter tractable. *IEEE/ACM Trans Comput Biol Bioinform* **4**, 458–66.
56. Baroni, M., Grunewald, S., Moulton, V., Semple, C. (2005) Bounding the number of hybridisation events for a consistent evolutionary history. *J Math Biol* **51**, 171–82.
57. Baroni, M., Semple, C., Steel, M. (2006) Hybrids in real time. *Syst Biol* **55**, 46–56.

# Chapter 14

# Untangling Hybrid Phylogenetic Signals: Horizontal Gene Transfer and Artifacts of Phylogenetic Reconstruction

**Robert G. Beiko and Mark A. Ragan**


## Abstract

Phylogenomic methods can be used to investigate the tangled evolutionary relationships among genomes. Building 'all the trees of all the genes' can potentially identify common pathways of horizontal gene transfer (HGT) among taxa at varying levels of phylogenetic depth. Phylogenetic affinities can be aggregated and merged with the information about genetic linkage and biochemical function to examine hypotheses of adaptive evolution via HGT. Additionally, the use of many genetic data sets increases the power of statistical tests for phylogenetic artifacts. However, large-scale phylogenetic analyses pose several challenges, including the necessary abandonment of manual validation techniques, the need to translate inferred phylogenetic discordance into inferred HGT events, and the challenges involved in aggregating results from search-based inference methods. In this chapter we describe a tree search procedure to recover the most parsimonious pathways of HGT, and examine some of the assumptions that are made by this method.


**Key words:** Phylogenetics, phylogenomics, horizontal gene transfer, subtree prune-and-regraft, bipartitions, model violation.

## 1. Introduction: Building Large Sets of Trees

Horizontal gene transfer (HGT) detection methods can be classified into two categories, depending on whether or not they rely on comparisons among genomes to identify homologous sequences. Homology-independent methods typically rely on the distribution of different compositional patterns such as G + C content *(1)* and are sometimes identified as *surrogate (2)* methods. Surrogate methods carry certain advantages: they can be applied to an entire genome, including protein-coding sequences

Maria B. Gogarten et al. (eds.), *Horizontal Gene Transfer: Genomes in Flux, vol. 532*

DOI 10.1007/978-1-60327-853-9_14 Springerprotocols.com

that have a small number of orthologs in other genomes. However, coding sequences in a genome will not all be influenced to the same degree by background signals, and the amelioration process can lead to ambiguous classifications of putatively transferred genes. Different surrogate methods have been shown to generate sets of predicted acquired genes that overlap poorly *(1, 3)*.

Homology-based approaches, particularly those based on phylogenetic analysis, can be applied only to sequences for which a sufficiently large number of reliable homologs can be identified. Such methods can be based on the observed "patchiness" of a distribution of homologous or orthologous sequences *(4, 5)*, with sparse distributions of such sequences across a reference tree of organisms constituting prima facie evidence for HGT as opposed to multiple gene loss events. Such methods are sensitive to the sparsity and bias of taxon sampling, and the unknown intrinsic rates of gene loss versus HGT. Phylogenetic or "phylogenomic" approaches are based on the consistency of phylogenetic signals across many genes: if all observed similarity relationships *(6)* or phylogenetic trees *(7)* are compatible with one another over a relatively unbiased sample of taxa, then there is taken to be little evidence of HGT. In contrast with the comparisons among surrogate methods, Beiko et al. found good agreement between a tree-based and a distribution-based approach to identifying phylogenetic discordance *(8)*. However, differences in the choice of methodology, taxon sampling, and data set (e.g., informational genes, ubiquitous genes, or whole genomes) have produced a wide range of HGT frequency estimates from different homology-based approaches *(5, 7–9)*.

Tree-based approaches potentially carry several advantages for HGT inference, including statistically consistent likelihood-based methods, the ability to specify models of sequence change, and the possibility of identifying donor–recipient relationships. Conversely, tree-based methods also have disadvantages, including sensitivities to model violation, computational demand, and challenges in interpreting and summarizing the observed patterns of discordance.

## 2. Data Set Generation/ Software

Inference of orthologs is an essential component of the tree-based approach, and there are many ways to perform this step. Protein data are typically used to compare distantly related taxa: since amino acids evolve more slowly and have more character states than nucleotides, they are less prone to the effects of substitutional saturation. Many ortholog inference methods begin with all-versus-all BLAST (or a similarly defined heuristic) to identify putative homology relationships within the set of sequences. The challenge then lies in converting the resulting graph, with sequences as vertices and similarity relationships defining edges, into a set of discrete and non-overlapping groups to be analyzed separately. Methods to do this include CD-HIT *(10)*, hybrid Markov-single linkage clustering *(11)*, and phylogenetic approaches such as BranchClust *(12)*.

There are many ways to build and trim multiple sequence alignments and infer phylogenetic trees; comprehensive discussion of these is beyond the scope of this chapter, but a key challenge in phylogenomic analysis is the high degree of reliance on automation. Techniques such as manual alignment curation are neither viable nor sufficiently consistent for very large data sets. The alignment "word-oriented" objective function WOOF *(13)* is an example of a method intended to replace manual curation by examining the same "cues" a human would look for (consistent alignment of conserved residues in the case of WOOF). In addition to being the only viable approach for large-scale analysis, such methods are more rigorous, consistent in their treatment of the data, and repeatable.

There are many approaches to quantifying the dissimilarity between phylogenetic trees. The most relevant of these to the HGT inference problem is the subtree prune-and-regraft (SPR) distance *(14)*, typically defined as the minimum number of SPR operations needed to reconcile two trees. Computing the minimum SPR distance between two unrooted trees is a NP-hard problem *(15)*, and the most frequent approach involves a search of possible intermediates to recover the final path or paths. Different constraints can be placed on the set of allowable SPR operations to eliminate the transfers that are evolutionarily impossible (such as from descendant to ancestor), and to reduce the size of the search space. Programs that implement different variants of tree comparison include LatTrans *(16)*, HorizStory *(17)*, Efficient Evaluation of Edit Paths (EEEP) *(18)*, and RIATA-HGT *(19)*. In their approaches, HorizStory and EEEP are somewhat similar, but all four methods differ in the type of trees they take as input (rooted or unrooted, strictly binary or possibly multifurcating), the evaluation of potential HGT partners, and the strategy used to restrict the size of the search space.

## 3. Assessing and Aggregating Phylogenetic Evidence for HGT

This section assumes that phylogenetic trees have already been generated through an appropriate combination of methods for orthology inference, multiple sequence alignment, alignment "trimming", and phylogenetic tree inference or calculation. The approach used in the phylogenomic project of *(8)* is described in stepwise fashion in Beiko and Ragan *(20)*, but many other approaches have been taken to generate large-scale phylogenomic data *(7,21)*. We also assume that a *reference tree* describing a plausible scenario of vertical descent of organisms is available; ideally this tree will be rooted (*see* **Section 3.1** below). This reference tree can represent the plurality signal from the inferred trees under consideration, or can represent a phylogenetic hypothesis that is extrinsic to the data (e.g., based on small-subunit ribosomal DNA, cellular ultrastructure, or other information thought to be phylogenetically informative). We refer to each member of the set of inferred trees as a *test tree*, to distinguish it from the reference tree.

### 3.1. Assessment of Phylogenetic Discordance

An efficient way to assess the degree of compatibility between a reference tree and a set of test trees is to count the incidences of reference tree features (such as bipartitions or quartets) that are topologically *congruent* or *incongruent* with individual test trees. An example of mapping concordant and discordant features is shown in **Fig. 14.1**. For the mixture of trees in **Fig. 14.1A**, the number of bipartitions that are concordant and discordant with each internal branch of the reference tree is shown above and below the corresponding branches in **Fig. 14.1B**. Although this example treats all trees as being completely resolved, the total number of trees examined should also be taken into account: a high proportion of unresolved bipartitions (i.e., with low Bayesian posterior or supporting proportion of bootstrap replicates) may suggest a difficult relationship to resolve. Bipartitions can be mapped onto a tree, but the complete spectrum of *embedded* quartets (**Fig. 14.1C**), while more difficult to summarize, confers more information about the consistency of relationships among specific taxa in the reference tree. Quartet decomposition of a phylogenetic tree *(22)* is performed by pruning away the complements of all possible $n$ choose $4 = n!/(4!(n-4)!)$ sets of taxa from a tree covering $n$ taxa.

Phylogenetic discordance can also be assessed with *consensus network* methods such as Neighbor-Net *(23)* or *super-network* methods such as Z-closure *(24)*. All of these methods can reveal cases in which the consensus signal from many trees has significant representation of specific alternative relationships, which are unlikely to be due to rare events or noisy data. Such relationships appear as reticulations in the network, with parallel edges separating groups of taxa that cluster together with significant frequency. As shown in **Fig. 14.2**, super-networks can display short-range (**Fig. 14.2B**) and long-range (**Fig. 14.2C**) transfers. Super-networks typically display a "web" of reticulations when two trees disagree due to a long-range transfer, instead of a single additional connection between donor and recipient lineages that

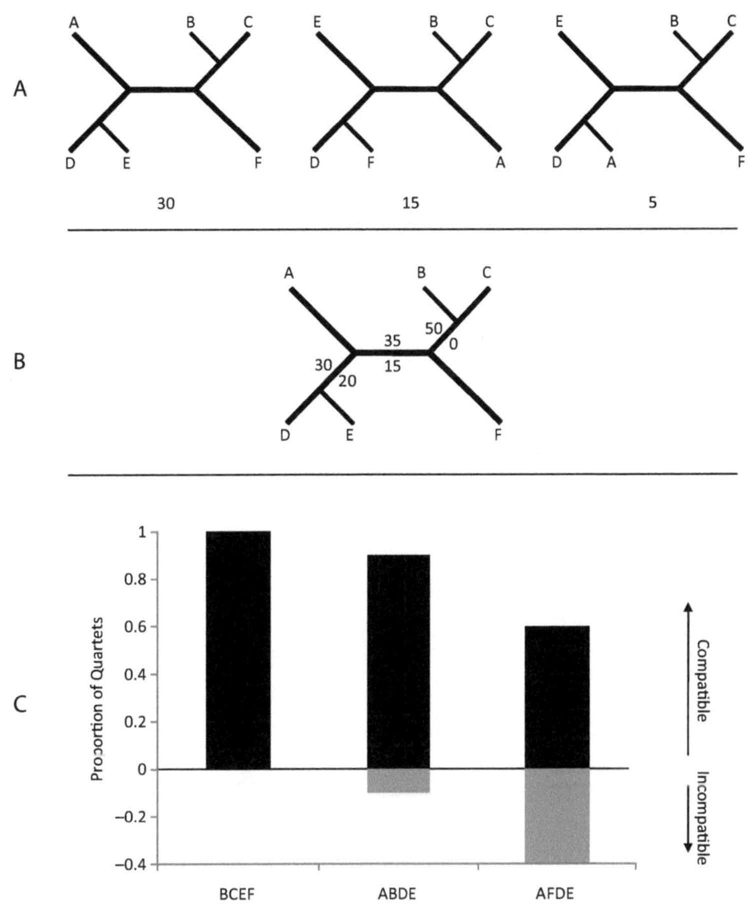


Fig. 14.1. Summarizing the relationships among a set of trees using bipartition and quartet compatibility. The trees in panel **A** represent the gene or "test" trees inferred from 50 hypothetical orthologous data sets covering taxa A–F, with numbers below indicating the number of times each tree was recovered. A hypothetical reference tree (which happens to coincide with the most frequently observed topology) is shown in panel **B** with concordant and discordant bipartitions mapped above and below the internal branches. For instance, the bipartition or split (ADE | BCF) is present in the first and third test trees, and is therefore present a total of $30 + 5 = 35$ times in the data set. The second test tree exhibits an incompatible bipartition (ABC | DEF), which accounts for the other 15 cases. Panel **C** shows a Lento plot of three distinct quartets (of the 15 possible sets of four taxa) onto the reference tree, with compatible topology frequencies shown in *black* above the 0 line and incompatible quartet frequencies aggregated below. The quartet BCEF always appears as (BC | EF) in the 50 test trees, and this topology is compatible with the reference, so a compatible proportion of 1.0 is displayed. Quartet ABDE is compatible with the reference as (AB | DE) in 90% of test trees, and the only incompatible alternative seen (AD | BE) is shown below the zero line. Finally, all three possible configurations of AFDE are observed in the set of test trees: the two incompatible versions (AD | EF) and (AE | DF) are represented together below the zero line.

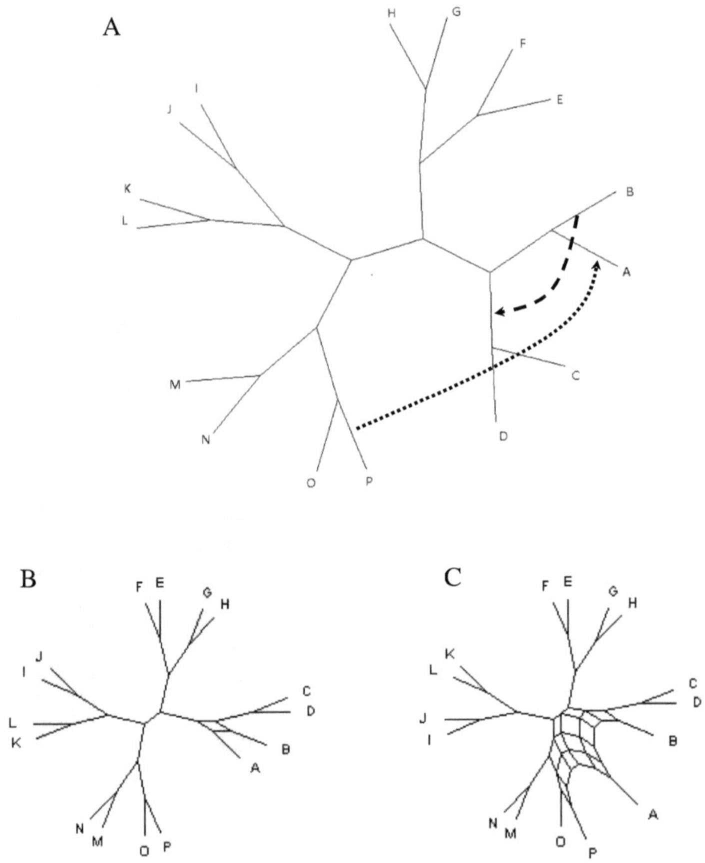


Fig. 14.2. Super-networks constructed from two trees, covering the same set of taxa A–P. Both super-networks consist of one "reference" tree as shown in panel **A**, and one additional tree. In panel **B** the second tree reflects a possible short-range transfer from B to the common ancestor of C and D (indicated with the *dashed arrow* in panel **A**), while the second tree in panel **C** reflects a transfer from an ancestor of taxon P to an ancestor of taxon A (*dotted arrow* in panel **A**). Networks were generated using the Z-closure method in SplitsTree version 4.8.

might be expected. A single such transfer can still be identified (e.g., from the reticulations shown in **Fig. 14.2C**), but two or more long-range transfers may not be distinguishable if the webs they produce intersect in the network.

### *3.2. Recovery of HGT Pathways*

If the reference tree is rooted and has accurate branch lengths that are proportional to time, then any paths of HGT could be constrained to occur only between contemporary lineages in the reference tree (but *see* **Note 1**). However, parsimony methods including the Matrix Representation with Parsimony (MRP) approach *(25)* for generating reference trees do not themselves estimate branch lengths, although an MRP tree could be fixed and branch lengths subsequently estimated from the data using a likelihood approach. If the reference tree is based on distance or likelihood analysis of, for example, the small-subunit ribosomal RNA gene or a concatenated alignment of many such genes, then branch lengths in the tree will reflect the number of substitutions per site. However, such branch lengths will be proportional to time only if the rate of sequence change has been constant through time (i.e., evolves according to a molecular clock). Since this assumption rarely holds for distantly related sets of taxa, it is customary to ignore branch lengths when inferring HGT from phylogenetic trees, and focus on differences in the branching order.

The SPR distance between two trees *(14)* considers only the branching order of taxa and is not sensitive to differences in branch lengths between two trees. Each SPR operation involves a donor and a recipient lineage, and the topological impact of SPR is to prune the recipient subtree and reattach it to the branch corresponding to the donor lineage. This is directly analogous to an HGT event in which a given gene is replaced by its ortholog from a different genome. We term a sequence of such operations an *edit path*, and the minimal-length edit path from the reference tree to a given test tree corresponds to the minimum number of HGT operations that need to be inferred in the history of the gene whose evolution is described by the test tree. Therefore, in seeking a minimum-length edit path, we are trying to recover the most parsimonious explanation for gene evolution, in terms of the number of HGT operations that need to be mapped onto the reference tree (*see* **Note 2**). Even if branch lengths are ignored, the reference tree imposes a partial ordering on lineages by specifying ancestor/descendant relationships, and we can prohibit donations of genetic material from an ancestral lineage to one of its descendants, or vice versa (but again *see* **Note 1**). This constraint on SPR operations actually produces a different distance measurement, the *hybridization distance (26, 27)*.

For a set of donor lineages $D_i$ and recipient lineages $R_i$, an edit path recovered by EEEP is of the form ($D_1 \rightarrow R_1$), ($D_2 \rightarrow R_2$), ($D_3 \rightarrow R_3$). However, unambiguous answers where only a single edit path is returned are rare: a complete set of the most parsimonious solutions may include rearrangements of the order of edit paths, inversions of the donor and recipient lineages, and alternate paths that include other lineages in place of the ones above. Questions about the extent and nature of HGT can be posed at several levels of detail, and the amount of information that can be recovered depends on several factors that are outlined below.

*What is the length (number of SPR operations) of the shortest edit path or paths?* This question, unlike the ones below, can be answered by an exact algorithm (although the computation may be limited by available time and memory). In computing the shortest distance we are making a parsimony assumption, and indeed model-based approaches to the problem have been developed in which the minimum length does not necessarily correspond to the maximum likelihood or maximum a posteriori HGT scenario (*see* **Note 2**).

*Which lineages are identified as HGT partners?* This question addresses which lineages in the tree are donor-recipient pairs, without a clear indication of the direction of transfer. Even if a reference/test tree reconciliation (achieved by choosing an appropriate donor/recipient pair) can be unambiguously attributed to lineages A and B, the direction of the implied HGT event may not be determinable from the data. The unambiguous recovery of donor–recipient pairs is not guaranteed, which is best illustrated via an HGT event that changes the topology of a four-taxon tree: if (A, B) and (C, D) are the sister-taxa pairs in the reference tree, then the phylogenetic effects of a transfer between B and C will be indistinguishable from one between A and D, unless meaningful branch lengths are present and exploited. In this case, the identity of transfer partners cannot be uniquely identified. The cases in which donor–recipient pairs *can* be recovered identifies the "partners" in HGT events; even this incomplete information can be useful in constructing and testing hypotheses about HGT.

*What is the direction of transfer?* In some cases a donor–recipient pair *does* have an unambiguous direction of transfer (i.e., A $\rightarrow$ B is present in the reconciliation path, but not B $\rightarrow$ A), in which case the identity of the donor and recipient lineages can be inferred. "Long-distance" transfers, where a gene from one organism is acquired by a very distant relative (e.g., interphylum transfers), leave the donor in its original grouping, but displace the recipient away from its canonical place in the reference tree. In addition to the information about partnerships, this information can shed light on the possible adaptive roles of the HGT event, since the ecology and metabolic capacity of the recipient genome can be compared to its closest relatives in the reference tree. A surrogate method might also be of use in identifying which genome (A or B) is the recipient, since an acquired gene that is only partially ameliorated *(1)* may still show unusual compositional patterns in the recipient genome.

*What is the order in which these transfers have taken place?* The HGT events implied by an edit path of two or more transfers can be thought of as independent if none of the donor or recipient edges is an ancestor or descendant of another. In other cases, it may be possible to assign a time ordering on successive transfers into a recipient genome if there are sufficiently many internal branches to resolve separate transfer events.

When the answers to one or both of the last two questions are ambiguous, we can merge edit paths in the solution set that differ only in the direction of transfer between donor–recipient pairs, as well as those that differ only in the ordering of successive HGT events. While the search algorithm will likely visit all of these solutions separately, considerable gains in storage efficiency and clarity of results can be achieved through this merging. For instance, a path consisting of five edits has $5! = 120$ possible orderings, but if the distinction between orderings is not of interest, then these can all be represented as a single path with an explicit indication that the ordering of elements is arbitrary. EEEP uses a compressed format to represent such paths, where a single path is shown in full, followed by numerical representations of all recovered permutations of that path.

### 3.3. Aggregating Inferred HGT Events from a Phylogenomic Analysis

If every comparison between the reference tree and a test tree yielded a single, most parsimonious edit path reconciliation, then aggregating the implied HGT events would simply require a summing up over all edits. However, the ambiguities outlined in **Section 3.2.3** were observed in a majority of the trees examined in *(8)*. Therefore, non-trivial aggregation techniques need to be built. In this section we discuss three different aggregation schemes: a *greedy* approach, *weighting* of edit paths, and *refinement* techniques.

The *greedy approach* to aggregation involves choosing the single most likely scenario from among the set of recovered paths for each test tree in turn. In the absence of models that assign different probabilities to different donor–recipient pairs, a greedy approach can be used to favor those donor–recipient pairs that are observed most frequently in the complete set of reference tree–test tree reconciliations. If lineages A and B are proposed as potential partners in the reconciliation paths of 200 out of 1000 different test trees, and if no other pair has an equal or greater frequency, then the greedy approach favors edit paths containing this pair over all possible alternatives in each of these 200 cases. In doing this, many alternative edit paths will be eliminated from consideration. With this decision made, the next most frequent pair from the set of all edit paths still under consideration is found, with the same selection and elimination process. This iterative procedure can be repeated until each test tree has a single remaining reconciliation path with the test tree (*see* **Note 3**). The greedy approach is sensitive to the order in which edit paths are chosen, and variation in rankings can lead to the recovery of different sets of edit paths. However, its robustness with respect to any data set can be tested by randomly permuting the ranking of paths many times, and comparing the sets of paths recovered from each permutation.

When aggregating the donor–recipient pairs across many test tree reconciliations, the *weighting approach* assigns fractional values to donor–recipient pairs based on their frequency in each reconciliation after path permutations have been merged. Therefore, if 50% of all reconciliation paths for a given test tree contain the donor–recipient pair (A, B), then we will add 0.5 to the total observed count of that pair (*see* **Note 4**).

*Refinement approaches* aim to decrease the ambiguity of results by reducing the precision of the phylogenetic question that is asked. For instance, a given test tree may have several mutually exclusive reconciliation paths with the reference tree, with lineages A, B, C, D, or E as possible donors and F, G, H, and I as possible recipients. Although there is no unambiguous donor/recipient pair, if A, B, C, D, and E are all part of the same grouping (e.g., the same genus or class, or clade in the reference tree), and F, G, H, and I are part of the same group (which may be the same or a different genus, class, clade, etc.), then we can say that the transfer is *obligately* between these two groups. The loss in precision is offset by the ability to address hypotheses about gene-sharing partners, even if the exact donor and recipient lineages cannot be identified. A related question is how many transfers have *possibly* been observed between two groups; in this case, all of A, B, C, D, E may not be a part of the same group, but a group that contains a subset of these lineages may be implicated in the transfer event. Summaries of obligate or possible transfers can be evaluated in light of other intersecting ecological or functional hypotheses.

### 3.4. Testing for Errors of Phylogenetic Inference

Many types of methodological error or bias can lead to incorrect inferences of HGT. Conflation of paralogs with orthologs can lead to the inferences being made on sets of sequences whose relationships are not congruent with the organismal tree, without the need to invoke HGT *(28)*. Lineage sorting is another phenomenon that can lead to mistaken inference of HGT *(19, 29)*. Errors in multiple sequence alignment can lead to incorrect phylogenetic inferences based on non-homologous residues. Sources of bias that are particularly relevant to microbial genomes include violations of the assumptions of model-based phylogenetic methods.

*Non-tree-like signals.* Inferring a phylogenetic tree from a given dataset (e.g., gene or protein sequence) makes the assumption that the underlying data have a unique, tree-like history. However, reticulate evolution events such as gene conversions can produce genes that contain multiple incongruent phylogenetic signals, and traditional statistical methods such as bootstrap resampling are not suited to the detection of clusters of mutually incompatible phylogenetic signals within an alignment. There is some evidence that HGT can affect fragments of genes in addition to whole genes and operons *(30, 31)*. Phylogenetic inference from such data may support one or the other of the correct signals, or may support a "phylogenetic compromise" that reflects neither scenario. Phylogenetic recombination detection methods, such as RecPars *(32)* and BARCE *(33)*, can be used to identify such hybrid signals, but the methods designed specifically with fragmentary HGT in mind could consider variations in sequence composition as well.

*Stochastic error.* Phylogenetic methods based on likelihood scores are statistically consistent, assuming the evolutionary model is correct. However, it has been demonstrated *(34)* that large numbers of alignment positions (relative to the length of a single gene or protein) may be necessary to yield highly accurate trees. Many "short-distance" transfers may arise as a consequence of undetected stochastic error (but *see* **Note 5**), with the influence of stochastic error increasing with decreasing alignment length, and with increasing ratios of terminal to internal branch lengths *(35)*. Supermatrix methods were developed to overcome stochastic error by concatenating many sequences from the same group of taxa. However, building a single tree from all available genes assumes that no reticulate evolution has taken place, and supermatrix methods are consequently inappropriate for the assessment of HGT. Concaterpillar *(36)* aims to group sequences based on their probable phylogenetic signals, and has the potential to approach the power of supermatrix methods while identifying the sets of phylogenetically discordant genes.

*Compositional change along branches of the tree.* Microbes have a very large range of genomic G + C composition, and amino acid usage differs both as a consequence of DNA composition *(37)* and environmental requirements *(38)*. Simulations have shown that violations of assumptions such as compositional stationarity can bias otherwise consistent phylogenetic methods in favor of incorrect trees *(39)*. Matched-site compositional tests such as Bowker's test *(40)* can be applied to diagnose the magnitude and significance of the problem, and the problem can be remedied either by using composition-insensitive methods such as LogDet distances *(41)* or by correcting for compositional differences via, for example, purine–pyrimidine recoding of nucleotides *(42)*.

*Variation in evolutionary rates.* Likelihood-based phylogenetic methods are sensitive to variations in the rate of substitution unless the substitution model used is accurate *(43)*. The best known example of this type of sensitivity is the "long-branch attraction" artifact. The relative rate test *(44)* is one widely used method to assess the degree of rate variation in the tree.

## 4. Conclusion: Estimating the Number of HGT Events

In *(8)*, we used Bayesian phylogeny and an MRP supertree to estimate the balance of vertical versus horizontal evolutionary signal in a set of 144 genomes. Based on a comparison of strongly supported (Bayesian posterior probability $\geq 0.95$) bipartitions against the reference tree, we found that 86.9% of all strongly supported bipartitions were concordant. There was considerable agreement with the reference supertree, and EEEP was used to recover frequent pathways of apparent HGT among lineages. But do these results indicate that the underlying "frequency" of HGT is 13.1%, and why does this number differ so dramatically from other estimates?

There are several reasons why our comparative approach might overestimate the true number of HGT events. Discordance may arise due to the stochastic and systematic errors outlined in **Section 3.4** above, and other phenomena such as lineage sorting may contribute when internal branches are short. A single long-distance HGT event will also disrupt more than one bipartition, since the recipient taxon will be lost from its original species, genus, etc., groupings, and disrupt bipartitions that include the donor taxon as well.

The choice of reference tree will also influence the number of recovered HGT events. Using an MRP supertree ensures that some relationship will exist between the reference and test trees. Even so, the most parsimonious supertree will not necessarily minimize the number of HGT events that need to be invoked, due to the mismatch between HGT events and disrupted bipartitions identified above. Other analyses have used different types of reference tree: for instance, Dagan and Martin *(5)* used a concatenation of three (5S, 16S, and 23S) ribosomal DNA alignments across 190 species. However, ribosomal DNA is sensitive to the mutation biases that affect the rest of the genome, and can lead to artifactual grouping of organisms based on their G + C content *(45)*. Use of an incorrect reference tree will overestimate the number of HGT events, particularly if the test data (gene trees or gene presence/absence data) are less sensitive to compositional artifacts than is the original reference tree.

Conversely, it is also probable that tree-based methods underestimate the number of HGT events that have occurred. A methodological reason for this is the need to choose a threshold of significance for inferred phylogenetic relationships. As this threshold decreases, more discordant events will appear in the dataset: for instance, while 13.1% of bipartitions were discordant at a BPP threshold of 0.95, over 23% were discordant when the threshold was reduced to 0.51. With 90% often treated as a minimum threshold of reliability for Bayesian posterior values, most of the discordant features with support in the 50–60% range are likely to be a consequence of weak phylogenetic signal or uncorrected errors.

Tree-based methods cannot detect events that do not disrupt the branching order of the recovered phylogenetic tree. The inability to detect transfers among sister taxa was mentioned above, but Ge et al. *(9)* extended this by eliminating a considerable amount of short-distance discordance that *was* detected in reference–test tree comparisons, which produced a low estimate of ~2.0%. Another way to reduce the inference of short-distance HGT events is to allow the reference tree to multifurcate, which can be done with HorizStory *(17)*. A trifurcating reference tree node imposes no ordering on the descendant lineages A, B, and C, whereas a bifurcating tree would necessarily contain a nested pair such as (A,(B,C)). The multifurcating reference tree will not be discordant with any grouping of these three lineages, so short-distance transfers among them will not be detected.

Using trees also overlooks the distributional evidence for HGT. Simple presence/absence analysis (phylogenetic profiles) reveals many orthologs with patchy distributions across groupings of taxa, including genes involved in aerobic respiration *(46)*. Indeed the analysis in *(5)* was based on the presence and absence of orthologous genes in different lineages, and led to estimates of >1 exchange per gene family, and >50% overall.

These methodological differences are all important and also highlight the importance of defining a precise question prior to inferring the networks of gene sharing. Given the evolutionary and ecological diversity of microbes and the evidence from many comparative analyses of HGT, it is clear that HGT frequency cannot be characterized with a single percentage that describes all organisms and all genes. This is reflected in the hypotheses that reflect the role of HGT in the emergence of specific functions such as photosynthesis *(47)* and variable degrees of "transferability" based on compatibility with the recipient genome and proteome, such as the Complexity Hypothesis *(48)*. These and other considerations will have a strong influence on the type of network analysis that is performed, and the consequent interpretation of phylogenetic uncertainty and discordance.

## 5. Notes

1. The requirement that donor and recipient lineages be contemporary ignores the possibility that some donor lineages are not represented in the reference tree. The sampled genomes may be descendants of recipient organisms for which the corresponding donor is not available; absence from the data set may be due to a subsequent extinction event, or to incomplete sampling of extant lineages. A "phantom donor" event may still be detectable, but the reconciling SPR operation will appear to merge the recipient lineage with one of its ancestors. Examples of this type of event are shown in **Fig. 2** in MacLeod et al. *(17)* and **Fig. 2** in Beiko and Hamilton *(18)*. HorizStory explicitly allows such events in the

reconciliation path, whereas EEEP excludes them by default, but has a command-line argument that allows the relevant ancestor-descendant SPR operations to be performed.

2. The parsimony approach seeks to minimize the number of transfers in the reconciliation path, without the use of an explicit model of HGT. HGT probabilities can be modeled for specific pairs of lineages based on their G + C compatibility, shared ecological context, phylogenetic distance, or other factors *(49,50)*. Such weightings may be useful in breaking the ties among many possible donor–recipient pairs, and may even identify cases in which longer edit paths should be favored over shorter ones based on the probability of certain lineage pairs. It is unclear, however, how phylogenetic uncertainty and the phantom donor problem should be incorporated into models of HGT.
3. The greedy approach favors the recovery of major "highways" of gene sharing among organisms, since donor–recipient pairs are concentrated wherever possible. A serious problem arises in the resolution of ties, when many possible pairs have the same frequency in the data set. The greedy approach could select one of these at random, but it would be worth evaluating the decisions that result from many possible choices, to assess the robustness of the final set of edit paths that is returned.
4. The weighted approach shares out the inferred HGT events across all possible alternative pathways. A serious problem with this approach is its sensitivity to the completeness of taxonomic sampling in the relevant gene tree. If the gene tree has few of the taxa found in the reference tree, then many possible lineage pairs will be proposed to account for any phylogenetic discordance. Consequently, adding or subtracting taxa from the gene tree can affect the number of possible alternate pairs recovered, and can have a dramatic impact on their weighting.
5. While the number of inferred short-distance transfers is likely inflated by stochastic error, such transfers are plausible given the increased likelihood of shared vectors (viruses and plasmids) that can shuttle DNA between donor and recipient cells, and a higher propensity toward homologous recombination (although see *(51)* for potential selective constraints on short-range transfers of some informational genes). Tree-based methods for detecting HGT also *underestimate* the number of short-distance transfers due to their inability to detect the transfers between lineages that are sisters in the reference tree, since an SPR operation with sister taxa as donor and recipient will not modify the branching order of a tree.

## References


1. Lawrence, J. G., Ochman, H. (1997) Amelioration of bacterial genomes: rates of change and exchange. *J Mol Evol* **44**, 383–97.
2. Ragan, M. A. (2001) On surrogate methods for detecting lateral gene transfer. *FEMS Microbiol Lett* **201**, 187–91.
3. Ragan, M. A., Harlow, T. J., Beiko, R. G. (2006) Do different surrogate methods detect lateral genetic transfer events of different relative ages? *Trends Microbiol* **14**, 4–8.
4. Ragan, M. A., Charlebois, R. L. (2002) Distributional profiles of homologous open reading frames among bacterial phyla: implications for vertical and lateral transmission. *Int J Syst Evol Microbiol* **52**, 777–87.
5. Dagan, T., Martin, W. (2007) Ancestral genome sizes specify the minimum rate of lateral gene transfer during prokaryote evolution. *Proc Natl Acad Sci USA* **104**, 870–5.
6. Clarke, G. D. P., Beiko, R. G., Ragan, M. A., Charlebois, R. L. (2002) Inferring genome trees by using a filter to eliminate phylogenetically discordant sequences and a distance matrix based on mean normalized BLASTP scores. *J Bacteriol* **184**, 2072–80.
7. Lerat, E., Daubin, V., Moran, N. A. (2003) From gene trees to organismal phylogeny in prokaryotes: the case of the gamma-Proteobacteria. *PLoS Biol* **1**, E19.
8. Beiko, R. G., Harlow, T. J., Ragan, M. A. (2005) Highways of gene sharing in prokaryotes. *Proc Natl Acad Sci USA* **102**, 14332–7.
9. Ge, F., Wang, L. S., Kim, J. (2005) The cobweb of life revealed by genome-scale estimates of horizontal gene transfer. *PLoS Biol* **3**, E16.
10. Li, W., Jaroszewski, L., Godzik, A. (2001) Clustering of highly homologous sequences to reduce the size of large protein databases. *Bioinformatics* **17**, 282–3.
11. Harlow, T. J., Gogarten, J. P., Ragan, M. A. (2004) A hybrid clustering approach to recognition of protein families in 114 microbial genomes. *BMC Bioinformatics* **5**, 45.
12. Poptsova, M. S., Gogarten, J. P. (2007) BranchClust: a phylogenetic algorithm for selecting gene families. *BMC Bioinformatics* **8**, 120.
13. Beiko, R. G., Chan, C.-X., Ragan, M. A. (2005) A word-oriented objective function for alignment validation. *Bioinformatics* **21**, 2230–9.
14. Allen, B. L., Steel, M. (2001) Subtree transfer operations and their induced metrics on evolutionary trees. *Ann. Combinatorics* **5**, 1–15.
15. Hickey, G., Dehne, F., Rau-Chaplin, A., Blouin, C. (2008) SPR distance computation for unrooted trees, *Evolutionary Bioinformatics* **4**, 17–27.
16. Hallett, M., Lagergren, J. (2001) Efficient algorithms for lateral gene transfer problems. *RECOMB 2001*, 149–56.
17. MacLeod, D., Charlebois, R. L., Doolittle, W. F., Bapteste, E. (2005) Deduction of probable events of lateral gene transfer through comparison of phylogenetic trees by recursive consolidation and rearrangement. *BMC Evol Biol* **5**, 27.
18. Beiko, R. G., Hamilton, N. (2006) Phylogenetic identification of lateral genetic transfer events. *BMC Evol Biol* **6**, 15.
19. Than, C., Ruths, D., Innan, H., Nakhleh, L. (2007) Confounding factors in HGT detection: statistical error, coalescent effects, multiple solutions. *J Comp Biol* **14**, 517–35.
20. Beiko, R. G., Ragan, M. A. (2008) Detecting lateral genetic transfer: a phylogenetic approach, in *Bioinformatics* (Keith, J. M., ed.), Humana, Totowa, NJ, 457–69.
21. Creevey, C. J., Fitzpatrick, D. A., Philip, G. K., Kinsella, R. J., O'Connell, M. J., Pentony, M. M., Travers, S. A., Wilkinson, M., McInerney, J. O. (2004) Does a tree like phylogeny only exist at the tips in the prokaryotes? *Proc Biol Sci* **271**, 2551–8.
22. Zhaxybayeva, O., Gogarten, J. P., Charlebois, R. L., Doolittle, W. F., Papke, R. T. (2006) Phylogenetic analyses of cyanobacterial genomes: quantification of horizontal gene transfer events. *Genome Res* **9**, 1099–108.
23. Bryant, D., Moulton, V. (2004) Neighbor-net: an agglomerative method for the construction of phylogenetic networks. *Mol Biol Evol* **21**, 255–65.
24. Huson, D. H., Dezulian, T., Klopper, T., Steel, M. (2004) Phylogenetic super-networks from partial trees. *IEEE Trans Comput Biol Bioinform* **1**, 151–8.
25. Ragan, M. A. (1992) Phylogenetic inference based on matrix representation of trees. *Mol Phylogenet Evol* **1**, 53–8.
26. Nakhleh, L., Warnow, T., Linder, C. R., St. John, K. (2005) Reconstructing reticulate evolution in species – theory and practice, *J Comput Biol* **12**, 796–811.
27. Bordewich, M., Semple, C. (2007) Computing the minimum number of hybridization events for a consistent evolutionary history. *Discrete Appl Math* **155**, 914–28.
28. Kurland, C. G., Canback, B., Berg, O. G. (2003) Horizontal gene transfer: a critical view. *Proc Natl Acad Sci USA* **100**, 9658–62.
29. Maddison, W. P. (1997) Gene trees in species trees. *Syst Biol* **46**, 523–46.

30. Inagaki, Y., Susko, E., Roger, A. J. (2006) Recombination between elongation factor 1α genes from distantly related archaeal lineages. *Proc Natl Acad Sci USA* **103**, 4528–33.
31. Chan, C.-X., Beiko, R. G., Ragan, M. A. (2007) A two-phase strategy for detecting recombination in nucleotide sequences. *South Africa Comp J* **38**, 20–7.
32. Hein, J. (1993) A heuristic method to reconstruct the history of sequences subject to recombination. *J Mol Evol* **36**, 396–405.
33. Husmeier, D., McGuire, G. (2002) Detecting recombination with MCMC. *Bioinformatics* **18 Suppl 1**, S345–53.
34. Swofford, D. L., Waddell, P. J., Huelsenbeck, J. P., Foster, P. G., Lewis, P. O., Rogers, J. S. (2001) Bias in phylogenetic estimation and its relevance to the choice between parsimony and likelihood methods. *Syst Biol* **50**, 525–39.
35. Philippe, H., Delsuc, F., Brinkmann, H., Lartillot, N. (2005) Phylogenomics. *Annu Rev Ecol Evol Syst* **36**, 541–62.
36. Leigh, J. W., Susko, E., Baumgartner, M., Roger, A. J. (2008) Testing congruence in phylogenomic analysis. *Syst Biol* **57**, 104–15.
37. Singer, G. A. C., Hickey, D. A. (2000) Nucleotide bias causes a genomewide bias in the amino acid composition of proteins. *Mol Biol Evol* **17**, 1581–8.
38. Fukuchi, S., Yoshimune, K., Wakayama, M., Moriguchi, M., Nishikawa, K. (2003) Unique amino acid composition of proteins in halophilic bacteria. *J Mol Biol* **327**, 347–57.
39. Ho, S. Y., Jermiin, L. S. (2004) Tracing the decay of the historical signal in biological sequence data. *Syst Biol* **53**, 623–37.
40. Jermiin, L. S., Ho, S. Y. W., Ababneh, F., Robinson, J., Larkum, A. W. D. (2004) The biasing effect of compositional heterogeneity on phylogenetic estimates may be underestimated. *Syst Biol* **53**, 638–43.
41. Lockhart, P. J., Steel, M. A., Hendy, M. D., Penny, D. (1994) Recovering evolutionary trees under a more realistic model of sequence evolution. *Mol Biol Evol* **11**, 605–12.
42. Delsuc, F., Phillips, M. J., Penny, D. (2003) Comment on "Hexapod origins: monophyletic or paraphyletic?" *Science* **301**, 1482.
43. Sullivan, J., Swofford, D. L. (1997) Are guinea pigs rodents? The importance of adequate models in molecular phylogenetics. *J Mamm Evol* **4**, 77–86.
44. Wu, C. I., Li, W. H. (1985) Evidence for higher rates of nucleotide substitution in rodents than in man. *Proc Natl Acad Sci USA* **82**, 1741–5.
45. Woese, C. R., Achenbach, L., Rouviere, P., Mandelco, L. (1991) Archaeal phylogeny: re-examination of the phylogenetic position of Archaeoglobus fulgidus in light of certain composition-induced artefacts. *Syst Appl Microbiol* **14**, 364–71.
46. Boucher, Y., Douady, C. J., Papke, R. T., Walsh, D. A., Boudreau, M. E., Nesbø, C. L., Case, R. J., Doolittle, W. F. (2003) Lateral gene transfer and the origins of prokaryotic groups. *Annu Rev Genet* **37**, 283–328.
47. Raymond, J., Zhaxybayeva, O., Gogarten, J. P., Gerdes, S. Y., Blankenship, R. E. (2002) Whole-genome analysis of photosynthetic prokaryotes. *Science* **298**, 1616–20.
48. Jain, R., Rivera, M. C., Lake, J. A. (1999) Horizontal gene transfer among genomes: the complexity hypothesis. *Proc Natl Acad Sci USA* **96**, 3801–6.
49. Galtier, N. (2007) A model of horizontal gene transfer and the bacterial phylogeny problem. *Syst Biol* **56**, 633–42.
50. Beiko, R. G., Charlebois, R. L. (2007) A simulation test bed for hypotheses of genome evolution. *Bioinformatics* **23**, 825–31.
51. Sorek, R., Zhu, Y., Creevey, C. J., Francino, M. P., Bork, P., Rubin, E. M. (2007) Genome-wide experimental determination of barriers to horizontal gene transfer. *Science* **318**, 1449–52.

# Chapter 15

## Construction and Use of Flow Cytometry Optimized Plasmid-Sensor Strains

**Martin Iain Bahl, Gunnar Oregaard, Søren J. Sørensen, and Lars Hestbjerg Hansen**

### Abstract

Determining the stability of plasmids in bacterial populations is traditionally performed by isolating a large number of clones followed by screening for the presence of plasmids by replica transfer to plasmid-selective agar plates. This is often a laborious task, especially when the intrinsic stability of the plasmid is high. The method presented here relies on a phenotypic (green fluorescence protein) marker, which is switched on if the host bacteria loses the residing plasmid. The incorporation of flow cytometry for single-cell detection and discrimination between plasmid-free and plasmid-harboring cells in a bacterial population facilitates a very high throughput of cells and thus provides excellent sensitivity and statistics toward detecting even very low levels of plasmid instability.

**Key words:** Plasmid stability, green fluorescence protein, flow cytometry, plasmid-sensor strain, lac repressor.

## 1. Introduction

Bacterial plasmids represent dynamic packages of genetic information, which frequently harbor diverse functional genes. These plasmid-encoded genes often confer "useful" accessory functional traits exemplified by antibiotic- and heavy metal resistance, and degradation of xenobiotic compounds, which thus may provide a selective advantage to the host bacterium *(1)*. Plasmid-encoded antibiotic resistance frequently complicates the treatment of bacterial infections in humans. The fact that many plasmids are self-transmissible by conjugation further increases the potential problems they may cause in health care settings *(2)*. Large

Maria B. Gogarten et al. (eds.), *Horizontal Gene Transfer: Genomes in Flux, vol. 532*

DOI 10.1007/978-1-60327-853-9_15 Springerprotocols.com

conjugative plasmids often harbor an array of different mechanisms, which collectively work to ensure their overall stability in the bacterial host and include active partitioning, post-segregational killing, and multimer resolution systems *(3)*. Although most of these systems have been comprehensively investigated and are well described at the molecular level, there is limited knowledge of the actual in situ stability of these large conjugative plasmids in natural environments. This may in part be due to the lack of efficient high-throughput methods to address this issue. In this chapter we describe in detail a high-throughput flow cytometry (FCM) based detection system, for determining plasmid stability in a bacterial population *(4)*. The plasmid-sensor system is composed of two elements, namely (i) a plasmid host strain, which is tagged with a lac-repressor-regulated and FCM-optimized green fluorescent protein gene (*gfp*) and (ii) the plasmid of interest genetically tagged with the $lacI^{q1}$ gene, which encodes the corresponding repressor protein. Plasmid-loss induces the plasmid-sensor cell to switch to a GFP-positive phenotype, which is detectable at the single-cell level by flow cytometry. Plasmid-tagging is performed using a simple in vitro Mu transposon delivery system, which, due to random insertion into the target plasmid, simultaneously generates a knock-out library of the plasmid. This may be useful in a number of applications, such as determining the genes involved in stable plasmid maintenance.

## 2. Materials

Bacterial strains and plasmids used to construct the plasmid-sensor detection system are presented in **Table 15.1**. Construction of a *gfp*-labeled host strain may be performed by tri-parental filter mating, as described below, or alternatively, *gfp*-labeled *Escherichia coli* or *Kluyvera* sp. strains may be supplied upon request.

### 2.1. Modifications in the Host Strain by Tri-parental Mating

1. Donor strain *E. coli* MT102::pir/pGS11 containing the gene cassette [$Kan^R$, $Str^R$, $P_{lac}$-*gfp*] harbored between the inverted repeats of a modified Tn*5* transposon.
2. Recipient strain of choice belonging to the $\gamma$-proteobacteria (*see* **Note 1**). This strain should be kanamycin-sensitive and not contain the *lacI* gene, as this would interfere with the plasmid detection system.
3. Helper strain *E. coli* HB101/pRK600.
4. Sterile mixed cellulose ester membrane filters (0.2 μm).
5. Luria Bertani broth.

**Table 15.1**
**Bacterial strains and plasmids**

| Strains and plasmids | Relevant phenotype | Source |
|---|---|---|
| **Strains** | | |
| *Escherichia coli* MC4100::*gfp* | [$Kan^R$, $Str^R$, $P_{lac}$-*gfp*] | *(4)* |
| *Kluyvera* Sp. MB101::*gfp* | [$Kan^R$, $Str^R$, $P_{lac}$-*gfp*] | *(5)* |
| *Escherichia coli* MT102::*pir* | Auxotroph (*leu, pro, thi*), *pir+* | *(6)* |
| *Escherichia coli* HB101 | Auxotroph (*leu, pro, thi*) | *(7)* |
| **Plasmids** | | |
| pGS11 (R6K-replicon) | [$Kan^R$, $Str^R$, $P_{lac}$-*gfp*] | *(8)* |
| pRK600 (ColE1-replicon) | $Cm^R$, *tra*+ (RP4) | *(9)* |
| pENTI$^{q1}$ (pUC19-replicon) | [$Kan^R$, *lacI*$^{q1}$] | *(4)* |

Kan=Kanamycin, Str=streptomycin, Cm=chloramphenicol.
Square brackets are used to indicate the gene cassettes flanked by transposon-inverted repeats.

6. Agar plates supplemented with Kanamycin (50 µg/ml) and counter-selective for Donor and Helper strains (e.g., minimal media).
7. Source of blue light to confirm GFP expression (*see* **Note 2**).

### 2.2. Transposon Tagging of Plasmid

1. Circular plasmid DNA extracted using e.g. Qiagen Plasmid Midi Kit (*see* **Note 3**).
2. Plasmid pENTI$^{q1}$ containing the gene cassette [$Kan^R$, *lacI*$^{q1}$] between the modified Mu transposon inverted repeats of **Fig. 15.1.**
3. Restriction enzymes *Sca*I and *Bgl*II and appropriate buffers
4. MuA transposase and specific buffer (available from Finnzymes).

### 2.3. Location of Insertion Site of the Transposons on Tagged Plasmid

If direct sequencing of the *lacI*$^{q1}$-tagged plasmid is possible, the following primers are required together with appropriate equipment and reagents for sequencing:

1. SeqB: 5′-TTATTCGGTCGAAAAGGATCC-3′
2. SeqC: 5′-ACAGTTGCGACCTGCAGGC-3′

If direct sequencing of the *lacI*$^{q1}$-tagged plasmid is not feasible (*see* **Note 4**), the following primers and reagents are needed together with appropriate equipment and reagents for sequencing.

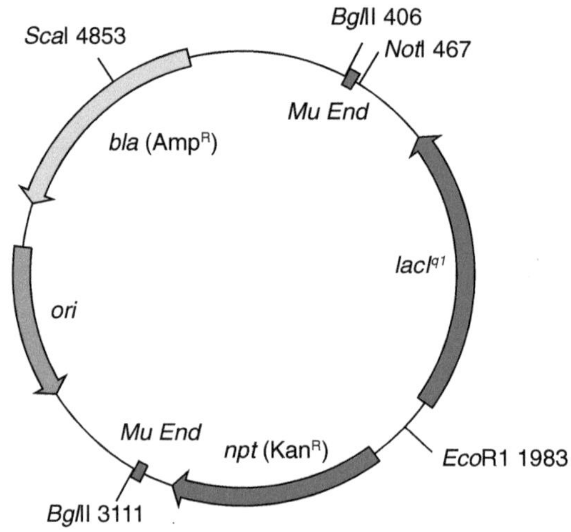


Fig. 15.1. Schematic representation of the 5.36 kb plasmid pENTI$^{q1}$ showing relevant restriction sites and functional genes. The Entranceposon::*lacI*$^{q1}$ is bounded by inverted repeats (*Mu End*) and *BglII* sites.

1. SeqC: 5′-ACAGTTGCGACCTGCAGGC-3′
2. SeqD: 5′-GGCGATTAAGTTGGGTAACGC-3′
3. *MspI* restriction enzyme and reaction buffer
4. T4 DNA ligase and reaction buffer

### 2.4. Flow Cytometry

1. Bench-top flow-cytometer (e.g., FACSCalibur from Becton Dickinson) with appropriate running fluids (sheath fluid, e.g., 0.9% NaCl, and cleaning solutions).
2. Isopropyl $\beta$-D-1-thiogalactopyranoside (IPTG) (*see* **Note 5**).
3. Nalidixic acid (e.g., 10 mg/ml stock).

## 3. Methods

### 3.1. Modifications in the Host Strain by Tri-parental Mating

1. The Donor, Recipient, and Helper strains involved in the tri-parental mating are grown to an optical density of approximately 1.0 ($\lambda = 600$ nm) in LB broth supplemented with antibiotics selecting for the plasmid.
2. One ml of each strain is centrifuged (5,000*g*, 5 min) and subsequently washed twice in preheated LB broth (25–37°C) and resuspended in 1 ml LB broth.
3. Transfer 50 μl of each cell suspension to a single microcentrifuge tube, mix gently, apply to a 0.2 μm membrane filter

placed on an LB-agar plate and incubate overnight at an appropriate temperature for the recipient (25–37°C)

4. The following day the filters are transferred to a tube containing 1–10 ml 0.9% NaCl and cells are dislodged from the filter by vortexing.
5. Tenfold dilutions of the cell suspension are spread onto transconjugant-selective plates, such as mineral salts medium amended with glucose and Kanamycin (50 µg/ml) and incubated overnight at the appropriate temperature (*see* **Note 6**).
6. As the transposon insertion in the host chromosome is not specific, it may be required to select a number of transconjugants for further use. Putative transconjugants are re-streaked twice on the same agar medium to ensure a pure culture and checked for green fluorescence by blue light excitation (e.g., in an epifluorescence microscope). The pGS11 plasmid is unstable in the recipient without the π-protein, and cells should quickly lose the plasmid without further treatment *(6)*. Absence of the narrow-host-range (*E. coli*) helper plasmid (pRK600) in the transconjugant should be confirmed by gel electrophoresis or by chloramphenicol sensitivity. If applicable, curing of the plasmid may be achieved by growth at elevated temperature (41–42°C for *E. coli*) *(6)*.
7. The *gfp*-labeled recipient is analyzed by FCM to ensure that a sufficient and easily detectable GFP signal is expressed.
8. The *gfp*-labeled strain is now ready for use and may be transformed with the plasmid of choice (see below). Preparation of electro- or chemically competent cells should be done using general guidelines for the specific species.

### 3.2. Transposon Tagging of Plasmid

1. The high-copy-number plasmid pENTI$^{q1}$, which contains the Entranceposon::*lacI*$^{q1}$, is isolated from the host *E. coli* strain by a mini plasmid preparation, such as Qiagen Plasmid Mini Kit, and the DNA concentration is determined.
2. Approximately 3 µg of the pENTI$^{q1}$ plasmid is linearized by enzymatic restriction using *Sca*I. Complete digestion is confirmed by gel electrophoresis and the 5.36 kb fragment is purified from the gel (e.g., using Qiaex® II Gel extraction kit from Qiagen) (*see* **Note 7**).
3. The 5.36 kb fragment is digested with *Bgl*II in a second endonuclease digestion to generate the 2.71 kb Entranceposon::*lacI*$^{q1}$. This fragment is excised from an agarose gel, purified using a gel extraction kit, and is ready for transposase-mediated insertion into the target plasmid.
4. The in vitro transposition reaction is setup in a microcentrifuge tube as follows, ensuring that the MuA transposase is added last:

| Reagent | Volume/amount |
|---|---|
| Deionized water | Add to 20 μl |
| Target plasmid DNA | 0.4 μg per 10 kb size |
| 5× MuA Reaction buffer | 4 μl |
| *Bgl*II-*Bgl*II entranceposon::*lacI*$^{q1}$ (2.71kb) | 50–100 ng |
| MuA Transposase | 1 μl |
| Total | 20 μl |

5. Mix the components gently and allow the transposition reaction to proceed for 1 h at 30°C.
6. Inactivate the MuA transposase by placing the microcentrifuge tube at 75°C for 10 min.
7. It is recommended to initially transform the transposon-tagged plasmid to commercially available ultra-competent *E. coli* cells (e.g., GeneHogs™ from Invitrogen) for high transformation efficiency. Because the transposon insertion is random in the target plasmid, an appropriate number of transformants should be re-streaked and saved for later use. This represents the knock-out library of the plasmid. Phenotypic screening of the transposon-tagged plasmids and/or determination of the transposon insert location (see below) may be completed in this intermediate strain before further transformation into the *gfp*-labeled strain.

### 3.3. Determination of the Insert Location in the Target Plasmid

Depending on target plasmid characteristics, direct sequencing from primer sites situated within the transposon may be performed using primers SeqB and SeqC (**Fig. 15.2**) (*see* **Note 8**). If direct sequencing is not possible (e.g., due to the large size of the plasmid), the following subcloning strategy may be used followed by sequencing to determine the insert location (**Fig. 15.3**).

1. Purify the *lacI*$^{q1}$ tagged plasmid from the host strain (e.g., using Qiagen Plasmid Mini Kit).
2. Digest the plasmid with restriction enzyme *Msp*I at 37°C for 1 h followed by heat inactivation of the enzyme at 65°C for 20 min (*see* **Note 9**).
3. Re-ligate the fragments using T4 ligase following recommendations by the supplier and heat inactivate (65°C for 20 min) (*see* **Note 10**).
4. PCR amplify the ligated DNA with primers SeqC and SeqD.
5. Purify the PCR product (*see* **Note 11**).
6. Sequence the PCR product using primer SeqC (*see* **Note 12**).

*GCGGC***TGAAGCGGCGCACGAAAAAC**GCGAAAGCGTTTCACGATAAATGCGAAAACGGATC**GCGGCCGC**AAGC
DR *Mu End* *Not*I

TTGCAT**GCCTGCAGGTCGCAACTGT**TGGGAAGGGCGATCGGTGCGGGCCTCTTCGCTATTACGCCAGCTGGC
SeqC

GAAAGGGGGATGTGCTGCAA**GGCGATTAAGTTGGGTAACGC**CAGGGTTTTCCCAGTCACGACGTTGTAAAAC
SeqD

GACGGCCAGTGAATCCGTAATCATGGTCATAGCTGTTTCCTGTGTGAAATTGTTATCCGCTCACAATTCCAC

ACAACATACGAG**CCGG**AAGCATAAAGTGTAAAGCCTGGGGTGCCTAATGAGTGAGCTAACTCACATTAATTG
*Msp*I

CGTTGCGCTCACTGCCCGCTTTCCAG------------------------GTTGTAACACTGGCAGAGCATT
~2200bp

ACGCTGACTTGACGGGACGGCGGCTTTGTTGAATAAATCGAACT**TTATTCGGTCGAAAAGGATCC**GTTTTCG
SeqB

CATTTATCGTGAAACGCTTTCGC**GTTTTTCGTGCGCCGCTTCA***GCGGC*
*Mu End* DR

Fig. 15.2. Interrupted nucleotide sequence of the 2.71 kb Entranceposon::*lacI*$^{q1}$ together with relevant restriction sites and primer-binding sites. Note the exemplified 5 bp terminal target site direct repeats (DR), which are generated during insertion into the target plasmid.

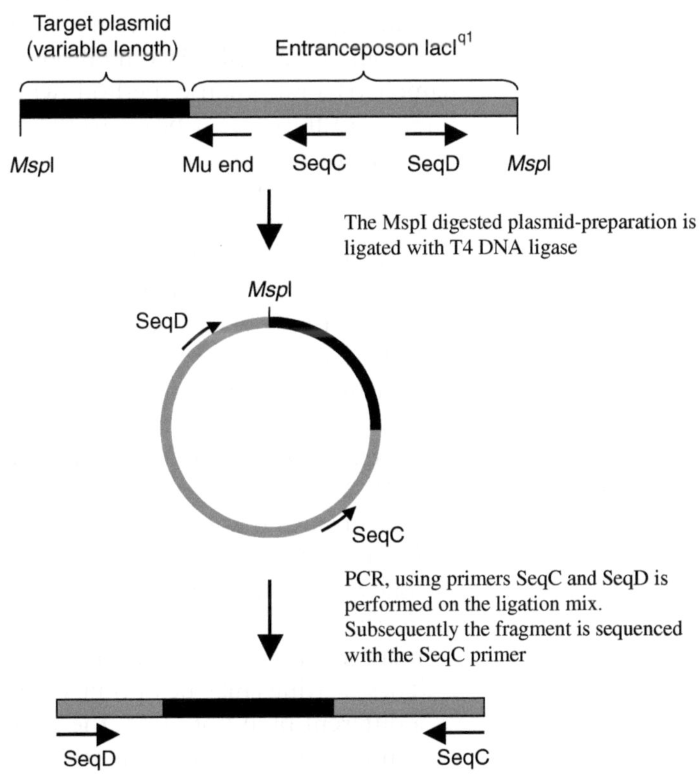


Fig. 15.3. Flow chart showing the procedure for determining the Entranceposon::*lacI*$^{q1}$ position in the target plasmid.

### 3.4. Flow Cytometry

Analysis with a bench-top flow cytometer (FCM) allows single-cell detection and discrimination between green and non-green fluorescent cells, representing plasmid-free and plasmid-harboring bacteria, respectively. In this way it is possible to determine the rate of plasmid loss from a population with high statistical accuracy. A Becton Dickinson FACSCalibur is an easy-to-use bench-top FCM, which incorporates one 488-nm air-cooled argon laser (blue-light) for excitation and three separate fluorescence detectors allowing the analysis of green (FL1, $530 \pm 15$ nm), orange (FL2, $585 \pm 21$ nm), and red fluorescence (FL3, $> 650$ nm). The recommendations in the present protocol are based on this FCM model; however, they may be easily adjusted to accommodate other FCMs. Relative cell size is correlated to the induced Forward SCatter (low-angle light scatter) and Side SCatter (90° angle light scatter), which are recorded by the FSC and SSC detectors, respectively. If possible, allow bacteria to enter an exponential growth stage prior to FCM analysis and place on ice, as this gives the best results (**Fig. 15.4**).

#### 3.4.1. Determination of FCM Flow-Rate

Unfortunately, it is not a common feature of bench-top FCMs (FACSCalibur included) to incorporate a mechanism to determine the sampled volume directly. As this flow rate is necessary to calculate the concentration of cells in a sample, it must be determined manually. Two methods may be employed to do this, namely (i) a ratiometric method, which involves adding a known number of fluorescent beads to the sample (*see* **Note 13**), or (ii) by a gravimetric volume determination, which involves determining the mass of the sample before and after a set period of time (three times 10 min is recommended) and consequently calculating the flow rate (taking the density of the sample into consideration). If the gravimetric method is used, it is recommended to determine the flow rate before and after sample analysis (*see* **Note 14**). This step may be omitted if only interested in the ratios of plasmid-free cells and not absolute numbers.

#### 3.4.2. Counting the GFP-Positive Bacteria (i.e., Plasmid-Free Cells)

1. Bacterial samples are diluted appropriately in filter-sterilized 0.9% NaCl to achieve a total event-count $< 2000$ events/s. This ensures that each cell/particle is analyzed separately.
2. If using a FACSCalibur, initially set the FCM parameters to the following values: FSC = E01 setting (tenfold signal amplification); SSC = 370 V; FL1 = 730 V; FL2 = 730 V and implement a threshold of 200 on the SSC detector (*see* **Note 15**). These settings may be optimized prior to final analysis. If using a different brand of FCM, the parameters may be set by use of control suspensions of *gfp*-positive and *gfp*-negative cells.
3. Open a bivariate acquisition dot-plot with FSC (forward-scatter) as primary parameter and SSC (side-scatter) as secondary parameter. Introduce a region around the bacterial

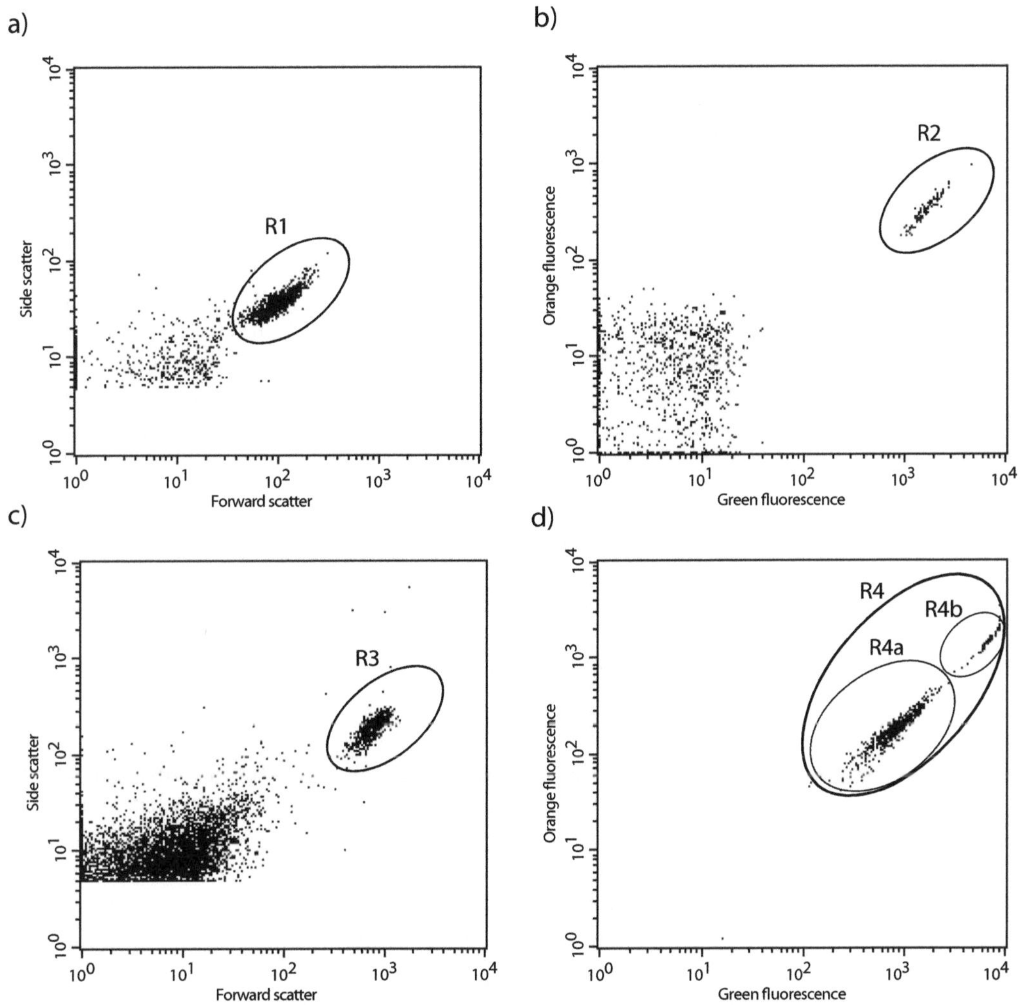


Fig. 15.4. Bivariate dot-plots showing representative FCM data and regions for enumerating GFP-positive plasmid-free bacteria and the total plasmid-sensor population. The sample contains approximately 10% plasmid-free (GFP-positive) and 90% plasmid-harboring (GFP-negative) cells. Region R1 defines the total plasmid-sensor population in a monoculture (**a**) and region R2 defines the plasmid-free, green fluorescent sub-population of cells gated in region R1 (**b**). Note that most of the cells are low fluorescent (plasmid-harboring) appearing in the lower left corner. Region R3 defines the plasmid-sensor population after nalidixic acid treatment, which increases cell size (**c**), and region R4 defines the IPTG induced GFP-positive plasmid-sensor population (**d**). Note that induction with IPTG results in two discrete sub-populations of GFP-positive cells (regions R4a and R4b), which correlate with the ratio of plasmid-free-cells in the sample. Region R2 represents 9.74% of region R1, region R4b represents 10.40% of region R4, and region R4 represents 99.41% of region R3.

population (*see* **Note 16**) and use this region to define and gate the bacterial population (**Fig. 15.4a**).

4. Open a bivariate acquisition dot-plot with FL1 (green fluorescence) as primary parameter and FL2 (orange fluorescence) as secondary parameter (*see* **Note 17**). GFP-positive cells, which should appear in a confined diagonal population in the dot-plot, are counted by introducing an elliptical user-defined region around this bacterial sub-population (**Fig. 15.4b**).

*3.4.3. Counting the Total Number of Plasmid-Sensor Cells*

If the bacterial sample to be analyzed exclusively contains plasmid-sensor cells, which are in the exponential growth phase, it is possible to discriminate these directly in bivariate dot-plot with FSC as primary parameter and SSC as secondary parameter, and consequently enumerate them by introducing a region around the total plasmid-sensor population (**Fig. 15.4a**). Under these circumstances, it is possible to obtain the ratio of plasmid-free segregant cells in a single sample-run. If on the other hand the plasmid-sensor population is part of a mixed bacterial community, determination of the total plasmid-sensor population is done by IPTG induction of the *lac* promoter, which causes all plasmid-sensor cells to express GFP. The following protocol is optimized for a nalidixic acid–susceptible *E. coli* strain:

1. Transfer 50 µl of the undiluted bacterial sample to 5 ml pre-warmed LB-broth supplemented with 1 mM IPTG (induces *gfp* expression) and 20 µg/ml nalidixic acid and incubate on a rotary shaker for 2–3 h, depending on the physiological state of the cells at 37°C. Growth in the presence of nalidixic acid results in enlarged cells and consequently an increase in both FSC and SSC signals (**Fig. 15.4c**) (*see* **Note 18**).
2. After this induction period, both plasmid-free and plasmid-harboring cells should emit green fluorescence and thus appear as a diagonal population in a bivariate dot-plot with FL1 (green fluorescence) as primary parameter and FL2 (orange fluorescence) as secondary parameter (**Fig. 15.4d**). In some circumstances, two sub-populations of GFP-positive bacteria appear, which represent plasmid-free and plasmid-harboring cells, respectively (**Fig. 15.4d**).
3. Enumerate the GFP-positive population by defining a region around these fluorescent cells (R4 in **Fig. 15.4d**), and calculate the total number of inducible plasmid-sensor cells by taking the various dilutions into consideration.
4. It is now possible to calculate the ratio of plasmid-free cells in the plasmid-sensor population.

## 4. Notes

1. Recipient bacteria other than those belonging to the $\gamma$-proteobacteria may also be used as recipients; however, we have no experience with the outcome.
2. Either bacterial colonies or single cells are inspected. Use of an epi-fluorescence microscope with an appropriate filter set (excitation wavelength approximately 488 nm) is recommended.

3. Be careful not to shear the plasmid DNA during extraction (no vortexing).
4. Direct sequencing from the insert in the plasmid may be difficult if the target plasmid is large.
5. IPTG is prepared as a 250 mM stock dissolved in water and is stored at −20°C .
6. The cell suspension may be concentrated before plating by pelleting and re-suspending in a smaller volume.
7. Initial linearization of the pENTI$^{q1}$ plasmid is necessary because *Bgl*II digestion of the circular plasmid would result in two fragments of almost identical size.
8. Due to the transposition event, 5 bp target site duplications are formed and are useful in confirming the insert location (exemplified in **Fig. 15.2**). If the target plasmid is fully sequenced, it should be easy to determine this precisely and further identify the genes that may be knocked out as a consequence of the transposition.
9. Restriction enzyme *Msp*I has a short (4 bp) recognition site (CCGG), which on average is expected to appear once every 256 bp in a random DNA strand (50% G + C-content).
10. This step is performed to circularize *Msp*I fragments, so that the region next to the insert can be amplified.
11. We typically observe many bands, which is due to incomplete *Msp*I digestion or re-ligation of more than one *Msp*I fragment. Sequencing with SeqC will thus only lead to good-quality sequence until the first *MspI* site.
12. It is recommended not to use SeqD for sequencing because more than one *Msp*I fragment may have religated.
13. Both physical size and fluorescence characteristics of the beads are useful for differentiating them from the bacterial cells.
14. In our experience the day-to-day differences in flow rate are typically less than 10%.
15. We recommend introducing a threshold on the SSC detector and not the FSC detector, as this has proved to give the best results.
16. If the plasmid-sensor cells are in an exponential growth phase, they produce a relatively high FSC and SSC signal and consequently represent a well-defined population. In this case they may be counted directly.
17. Using both these fluorescence detectors results in a superior discrimination of GFP-positive cells, compared to what is achieved by using only FL1 and a size parameter, such as SSC. GFP fluorescence is also detected in the FL2 (orange) detector.
18. Nalidixic acid is a gyrase inhibitor, which at this concentration prevents DNA replication and thus cell division, leading to elongated cells.

## Acknowledgments

The work was funded by grants from the Villum Kann Rasmussen foundation, the Danish Natural Science Council, the Natural and Accelerated Bioremediation Research (NABIR) program, Biological and Environmental Research (BER), US Department of Energy, and the Faculty of Science at the University of Copenhagen, Denmark.

### References

1. Adamczyk, M., and Jagura-Burdzy, G. (2003) Spread and survival of promiscuous IncP-1 plasmids. *Acta Biochim Pol.* **50**, 425–453.
2. Alanis, A. J. (2005) Resistance to antibiotics: are we in the post-antibiotic era? *Arch Med Res.* **36**, 697–705.
3. Summers, D. K. (1996) *The Biology of Plasmids*, Blackwell Science, Oxford.
4. Bahl, M. I., Sørensen, S. J., and Hansen, L. H. (2004) Quantification of plasmid loss in *Escherichia coli* cells by use of flow cytometry. *FEMS Microbiol Lett.* **232**, 45–49.
5. Bahl, M. I., Hansen, L. H., and Sørensen, S. J. (2007) Impact of conjugal transfer on the stability of IncP-1 plasmid pKJK5 in bacterial populations. *FEMS Microbiol Lett.* **266**, 250–256.
6. Hansen, L. H., Sørensen, S. J., and Jensen, L. B. (1997) Chromosomal insertion of the entire *Escherichia coli* lactose operon, into two strains of *Pseudomonas*, using a modified mini-Tn5 delivery system. *Gene.* **186**, 167–173.
7. Sambrook, J., Fritsch, E. F., and Maniatis, T. (1989) *Molecular Cloning: A Laboratory Manual*, Cold Spring Harbor Laboratory Press, Cold Spring Harbor, NY.
8. Sengeløv, G., Kristensen, K. J., Sørensen, A. H., Kroer, N., and Sørensen, S. J. (2001) Effect of genomic location on horizontal transfer of a recombinant gene cassette between *Pseudomonas strains* in the rhizosphere and spermosphere of barley seedlings. *Curr Microbiol.* **42**, 160–167.
9. De Vos, G. F., Walker, G. C., and Signer, E. R. (1986) Genetic manipulations in *Rhizobium meliloti* utilizing two new transposon Tn5 derivatives. *Mol Gen Genet.* **204**, 485–491.

# Chapter 16

# Experimental Evolution of an Essential *Bacillus* Gene in an *E. coli* Host

**Maia Larios-Sanz and Michael Travisano**

## Abstract

The acquisition of foreign genes by HGT potentially greatly speeds up adaptation by allowing faster evolution of beneficial traits. The evolutionary integration of novel genes into host gene expression and physiology is critical for adaptation by HGT, but remains largely unknown. We are exploring the evolutionary consequences of gene acquisition in populations of *Escherichia coli* in real time. A plasmid bearing the genes necessary for sucrose catabolism was constructed and introduced into a single *E. coli* genotype. Wild-type *E. coli* is generally incapable of utilizing sucrose, but *E. coli* transformants were able to grow on sucrose as a sole carbon and energy source, albeit poorly. Twelve replicate populations were initiated and propagated in sucrose minimal media for 300 generations. Over this time, we observed large fitness improvements in the selected environment. These results demonstrate the potential for HGT to substantially increase microbial niche breadth.

**Key words:** Experimental evolution, selection, adaptation, *Escherichia coli*.

## 1. Introduction

Comparative analyses of completed prokaryotic genomes show that bacterial genomes are mosaic, containing large amounts of genetic material acquired through the process of horizontal gene transfer (HGT) from different species and lineages *(1, 2)*. This foreign genetic material can make up 10–20% of the genome for some bacterial species *(3)*, although even these values may be a conservative estimate *(4)*. While point mutations, gene duplication *(5)*, and within-species recombination *(6)* are significant driving forces in the evolution of genome architecture, rapid microbial evolution and the acquisition of novel traits occur most

Maria B. Gogarten et al. (eds.), *Horizontal Gene Transfer: Genomes in Flux, vol. 532*

DOI 10.1007/978-1-60327-853-9_16 Springerprotocols.com

readily by the transfer of genetic information across species barriers (references *(7–15)*, and many others).

As such, HGT has been identified as a major process in the evolution of bacterial species *(16)*, symbiotic relationships *(17)*, emergence of pathogen families *(18)*, chemical defense *(19)*, interference competition *(20)*, use of novel nutrients *(21)*, and adaptation to novel environmental conditions *(22)*. The newly acquired DNA potentially allows organisms to readily expand their ecological niches into novel environments. The transfer of traits is so commonplace that long-term prokaryotic evolution is perhaps better viewed as a "mesh" of interconnecting branches *(12)* or a "net" of life *(23)* rather than a tree. There are three known modes of transfer between bacteria: conjugation, where genetic material is transferred directly from cell-to-cell; transduction, where DNA is transported from one cell to another by a virus (bacteriophage); and transformation, which involves uptake of naked DNA from the environment. While all three have been shown to have had important impacts on bacterial evolution, all modes of HGT are highly variable, rare, and are only readily observable in large bacterial populations.

Genomes are the repositories of information made up of individual genes retained due to a combination of prior selective benefit, chance, and historical contingency *(24)*. While acquisition of foreign genes by HGT potentially greatly speeds up evolution by allowing much faster attainment of beneficial genes including those encoding entirely new traits, much of the beneficial potential for the information stored in these genes requires that they be expressed and functionally integrated in the host. For successful acquisition of a novel trait, not only must the gene be acquired but also the host must be capable of appropriate expression of that gene. Although the machinery of gene expression is largely the same across bacteria, each species has its own particularities, and codon usage and rates of expression are in essence "personalized" for each.

Despite the many examples of horizontally transmitted genes, such as those for antibiotic resistance, the evolutionary integration of novel genes into host gene expression and physiology is poorly understood. It is clear that understanding the mechanisms of HGT is critical for assessing the general evolutionary processes. Many ecological and evolutionary studies have focused primarily on the consequences of selective differences among genotypes, and little emphasis has been placed on the underlying molecular causes of evolutionary and ecological effects *(25–27)*. There are several major limitations to such approaches. First, the generality of results can be difficult to evaluate since the mechanisms giving rise to the observed outcomes remain undefined. Secondly, important insights due to the interplay of cause and effect may be missed. Different genotypes, having the same selective coefficient but varying in the underlying mechanistic basis for selection, can differ in their effects on the biotic and abiotic environment and in their subsequent evolution *(24,28)*. Thirdly, predicting direct and correlated responses to selection is aided by an understanding of the possible limitations to selective response.

Functional integration of newly transferred genes can be divided into two sets of epistatic interactions: one set affecting expression of the new gene and another relating to gene–gene interactions with the host genome. Appropriate gene expression depends upon numerous species-specific factors including promoters *(29)*, ribosome binding site structure *(30)*, as well as codon bias *(31)*, and at least indirectly, GC bias *(32)*. All of these factors are likely to cause suboptimal levels of expression, as RNA polymerases and ribosomes may fail to bind or, once bound, may fall off and result in truncated mRNAs or proteins. The first two factors, promoters and ribosome binding sites, are thought to evolve rapidly in horizontally transferred genes *(33)*. In contrast, codon usage differences appear to persist for millions of generations, despite their strong impact on gene expression *(8)*.

The nature of gene–gene interactions between the newly transferred gene(s) and the host genome depends largely upon the physiological interactions between the encoded proteins. Proteins encoded by foreign genes can interfere with the existing protein function, even if the foreign genes provide an overall benefit to cell growth as might occur by the encoding of a novel catabolic pathway or resistance to an antibiotic. Amelioration of such costs can potentially occur via mutations in the plasmid, host genome, or some combination of both *(34)*. It is unclear to what extent such amelioration is likely to be host-, plasmid-, or environment-specific.

HGT has been an important laboratory technique for over 30 years, but realization of its prevalence in nature is far more recent *(35)*. Still, a substantial amount of work has been focused on the process of HGT. Most of the focus, however, has been either on the mechanisms of DNA transfer from cell to cell or on retrospective characterization of events using the genomes of extant organisms. In this regard, it is difficult to track the evolutionary dynamics of specific events and hence fully characterize both the process and the mechanisms involved. Retrospective studies are extremely valuable, but are limited in that they are indirect methods to determine the selective conditions and order of genetic changes involved with the acquisition of foreign DNA. Moreover, it can be difficult to even identify foreign DNA, especially if the transfer event is ancient or happened between closely related species. Detecting transfer events is further complicated by adaptation and alteration of the genetic language (codon usage, GC bias, etc.), which results in better integration of foreign DNA into the host's genetic machinery, but obscures the differences between initially foreign and host DNA *(8)*. Additionally, foreign DNA that does not introduce useful traits to its host can still integrate into the chromosome, as is the case with transposons, so functionality alone is not sufficient to understand the importance of HGT in nature *(36, 37)*.

## 2. Using Experimental Evolution to Examine HGT

Experimental studies to characterize the HGT process provide an ideal complement to retrospective approaches, because direct estimates of the factors affecting the transfer and functional incorporation of genetic material can be made. Experimental studies of HGT allow for quantitative examination of the mechanisms that allow these events to occur in the context of already working genomes. Integrating experimental protocols with the wealth of the information already obtained through retrospective (and largely computational) analyses allows us to probe the fundamental question: what factors are important for some genes to successfully transfer and be expressed in a host, while others are ultimately lost? Studies that can observe the progress of a new gene as it evolves within a host are essential to our understanding of the molecular processes involved in HGT. Furthermore, determination of the factors resulting in novel adaptations resulting from HGT events is essential to understand the initial stages of genome evolution, genomic constraints on adaptive evolution, and the accommodation of new genetic information into existing genomes.

Some efforts to characterize the process of HGT experimentally have been done, including, for example, a recent study focused on identifying the limitations of gene transfer using both computational and experimental approaches *(38)*. Sorek and colleagues found that there are universal transfer barriers, likely due to post-transfer gene toxicity, despite the fact that gene exchange has been previously documented for all types of genes, across all domains of life *(15)*. Phylogenomic studies supported by molecular data have also shown the importance of HGT events in the adaptation of bacterial species to novel environments. In one example, *Bacillus coahuilensis*, a strain isolated form a phosphate-limited, oligotrophic desert system, acquired genes from neighboring cyanobacterial communities as part of its genomic adaptation *(39)*. Experimental evolution with antibiotic-resistance plasmids has also shown that the costs of plasmid carriage decline, and that the decline in cost can involve epistatic interactions between host and plasmid mutations *(40, 41)*. However, these studies have not identified the specific mutations or the mechanisms of amelioration.

In this chapter, we suggest that the mechanisms of HGT and subsequent amelioration and adaptation by the host can be meticulously studied using experimental evolution with bacteria. The large amounts of complete genome data available, ease of culture, short generation times, and large population sizes make bacterial populations ideally suited for studying ecological and evolutionary processes, including HGT. An important additional advantage to such systems is that samples from evolved lines being studied can be frozen indefinitely at −80°C and re-cultured at any given time to follow the evolutionary history of each experiment. We therefore describe a series of experiments that will ultimately allow for the direct examination of newly transmitted genes over evolutionary time, using *Escherichia coli* and *E. coli* plasmids as our model system. Using selectable phenotypic markers, we have propagated populations of *E. coli* carrying foreign DNA under conditions in which the foreign DNA is essential for growth, and we show a cost-benefit analysis on the adaptation and co-adaptation of horizontally transmitted DNA and the host genome. We can further explore the evolutionary consequences of gene acquisition in populations of *E. coli* by long-term selection. This experimental system offers a tool that can be used to examine the molecular and phenotypic changes that occur over the course of selection, as well as directly observe the genetic and evolutionary constraints and the specific consequences of bacterial gene transfer.

This model system additionally offers a means to assess the factors affecting the beneficial effects of newly acquired foreign DNA on host fitness. By initially quantifying the initial fitness effects of a transfer event, we can determine the adaptive potential of foreign DNA over long-term evolution and, further, begin to analyze the molecular changes responsible. Our experimental approach involves the transfer of fully functional plasmid-borne genes into *E. coli* under selective conditions favoring carriage of foreign DNA. In this regard, our system does not address the modes or mechanisms of the transfer event itself, nor the nature of the genes being moved. Rather, it allows us to focus on the events after transfer, providing a tool for direct phenotypic characterization of the effects of the foreign DNA on the host, including the quantification of specific effects on fitness.

Our chapter is forward-looking, as analysis of the outcome of our experimental selection is still ongoing. We will ultimately obtain unique information about what is happening at the molecular level by analyzing the sequence changes in the transferred genes during the course of the evolution experiment. The experimental evolution system is easily expanded to include sophisticated analysis concerning actual integration of the transferred genes into the host genome using delivery systems such as transposon cassettes. By using a variety of standard molecular methods, one can then determine the sites of insertion of foreign genes within the host genome and the extent of amelioration or mutation on the translocated gene sequence during adaptation. The experiments are run with replicate populations (this number limited only by the experimenter), so a comprehensive analysis that incorporates all aspects of adaptation to novel traits can be easily carried out. These studies might include the determination of any constraints on functional expression due to specific site of integration (i.e., is the integration process biased toward particular spots on the host chromosome, or a result of a purely stochastic process), the characterization of the amelioration process (changes in codon usage, GC bias, etc., on the foreign sequence) throughout the process of evolution, and the global effects on gene expression of the host as adaptation occurs.

Since the bacterial lines set up for each experiment can be frozen and easily retrieved at any desired interval, it is easy to obtain snapshots of the evolutionary process as it actually happens. This will provide insight into whether the host adapts to the new gene early or if the significant changes that ensure retention of the acquired trait occur later in the process. Experimental evolution can thus be used to focus on the post-transfer events, and, in particular, in three principal aspects of the post-transfer process: the likely effects of expression of the novel gene, the molecular mechanisms likely required for functional expression of the novel trait, and the phenotypic effect it will have on the organism and the populations where the event occurs.

### 2.1. Plasmid-Borne Persistence Depends Upon Chance and Selection

The experimental system presented in this chapter involves the plasmid delivery of foreign genes that ensure survival in the experimental environment. There are numerous stochastic factors that affect plasmid persistence. These include host compatibility, which is species- and genus-specific, and compatibility with other plasmids within the cell *(42)*. Such incompatibilities greatly reduce the likelihood that the newly transmitted plasmid, and its genes, will persist in the host. Mutational loss of plasmid borne genes, like chromosomal genes, can occur, but is strongly affected by the selective effects of the genes encoded *(43)*. Selection also plays a large role in plasmid maintenance, as higher fitness costs to plasmid carriage tend to result in increased rates of plasmid loss due to chance segregation during bacterial cell division *(40)*. Gene expression is the primary determinant of the fitness costs to plasmid carriage and not the energetic costs of plasmid replication *(41)*. In previous studies, the costs of gene expression were largely explored with antibiotic resistance *(34, 43–45)*, and resistance typically involves energetically costly active antibiotic efflux or pleiotropically costly membrane alterations.

## 3. Experimental Evolution – Approach and Results

Most environmental isolates of *E. coli*, and *Enterobacteraceae* other than *Klebsiella pneumoniae*, are incapable of catabolizing sucrose, even though it is the dominant dissacharide in the environment due to its prevalence in plants *(46)*. The component sugars, glucose and fructose, are readily catabolized by *E. coli*, suggesting that the primary limitation for sucrose catabolism is uptake. *E. coli* isolates capable of growth on sucrose invariably contain at least two protein-coding regions, a sucrose inner membrane transporter, and a hydrolase, and appear to fall into general classes: phosphotransferase system (PTS) and non-PTS. PTS-mediated sucrose uptake is found in some *E. coli* isolates (e.g., enteropathogenic *E. coli* (EPEC) *(47)*) and related species (e.g., *Salmonella typhimurium*), and is typically encoded by genes on plasmids *(48)* or conjugative transposons *(49)*. Non-PTS mediated sucrose uptake in *E. coli* in frequently found in enterohemorrhagic *E. coli* (EHEC), in a chromosomal hot spot for rearrangement and site-specific phage-like integration *(50)*.

PTS sucrose transporters in *E. coli* appear to be closely related to the sucrose transport system in *Klebsiella* species, typically with 95% protein sequence similarity *(51)*. Such genes are likely to be generally well expressed and integrated into *E. coli* physiology, given the high degree of relatedness between *Klebsiella* and *E. coli*. In contrast, *Bacillus subtilis* is unrelated, but its sucrose-specific PTS transporter (*sacP*) and hydrolase (*sacA*) can confer sucrose catabolic activity to *E. coli (52)*. PTS involves a multistep phosphorylation cascade from a central metabolic intermediate (phosphoenolpyruvate), with concomitant nutrient phosphorylation and transport into the cell (**Fig. 16.1**). The tight linkage of the PTS to sucrose uptake provides an ideal system to investigate the mechanisms underlying functional integration of the genes acquired from distantly related species.

### 3.1. Cloning of sacP and sacA Genes from Bacillus subtilis into Escherichia coli Using a Plasmid System

The *sacP* and *sacA* genes from *B. subtilis* strain 168 (the first sequenced strain) were PCR-amplified from genomic DNA and cloned into the *Sac*I and *Hind*III restriction sites of plasmid pMON, a ~4.5 kb vector that contains a constitutive PSE-4 promoter (without the native lactamase structural gene) and a Tn903-derived Kanamycin selectable marker *(54)*. Two laboratory strains of *E. coli* B were used as recipients: one recipient strain (REL4536) had previously been selected in the laboratory for 10,000 generations in a glucose environment and the other recipient was its unselected ancestor *(55)*. Both were transformed with the pMON construct and with pMON without an insert (control). Transformants were initially screened for sucrose

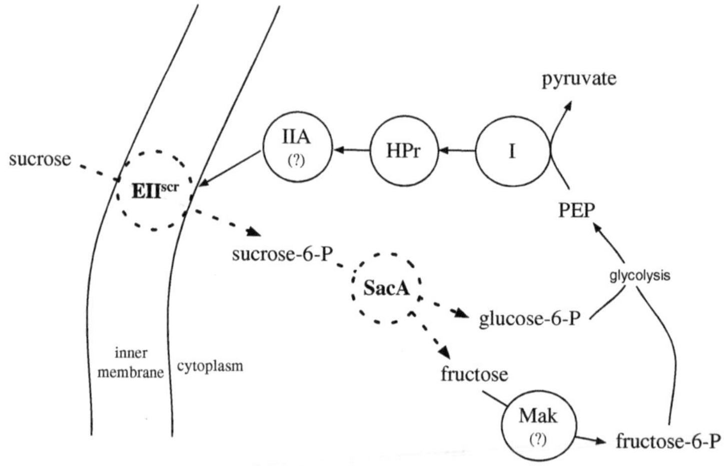


Fig. 16.1. Expected sucrose catabolic pathway for *Escherichia coli* transformants containing the two *Bacillus* genes *sacP* and *sacA*. Sucrose is transported through the inner membrane via *Bacillus* permease EII$^{scr}$ (encoded by *sacP*) concomitantly phosphorylated, and hydrolyzed by the *sacA* gene product to glucose-6-phosphate and fructose. Phosphate is obtained via the PTS phosphate transfer pathway consisting of two general PTS enzymes (I and HPr) from phosphoenolpyruvate (PEP). A IIA enzyme, currently unidentified, is almost certainly involved in the transfer of phosphate from HPr to II$^{scr}$, and is likely to be the glucose-specific IIA (IIA$^{glc}$). Glucose-6-phosphate is readily shunted into glycolysis, but fructose is typically phosphorylated during uptake due to PTS transport. In a naturally occurring sucrose utilizing *E. coli* isolate, a fructokinase is often encoded and the gene is collinear with the other genes of the sucrose operon. Our construct lacks a sucrose-associated fructokinase, and the only reported mechanism for phosphorylation is via mannokinase (Mak), encoded by a cryptic 1,032 bp ORF *(53)*.

utilization by differential growth on a Tetrazolium Sucrose (TS) plate with Kanamycin (TS plates contain 10 g tryptone, 1 g yeast extract, 5 g NaCl, 10 g sucrose, and 16 g agar in 1 L $dH_2O$, with 1 ml of filter-sterilized 5% TTC (triphenyl tetrazolium chloride) as indicator). Microorganisms that use the carbon source grow as white colonies, whereas those that do not use the carbon source grow as red colonies (**Fig. 16.2**). Expression of both genes ensured growth of transformants in Davis Minimal (DM, contains 7 g $HK_2O_4P$, 2 g $KH_2PO_4$ (anhydrous), 1 g $(NH_4)_2SO_4$, and 0.5 g $C_6H_5Na_3O_7$, carbon source to desired final concentration, 1 ml of a sterile 10% magnesium sulfate solution, and 1 ml of a filter-sterilized 0.2% thiamine (vitamin B1) solution in 1 L $dH_2O$).

The Kanamycin marker was lost in some but not all of the sucrose-selected lines during the course of evolution, despite the likely cost of kanamycin resistance. For example, after 50 days of evolution, four out of the twelve REL4536 transformant lines evolved in the sucrose environment fail to grow on Kanamycin and another four show marked decrease in growth when compared to the unevolved ancestor (**Fig. 16.3**), suggesting loss

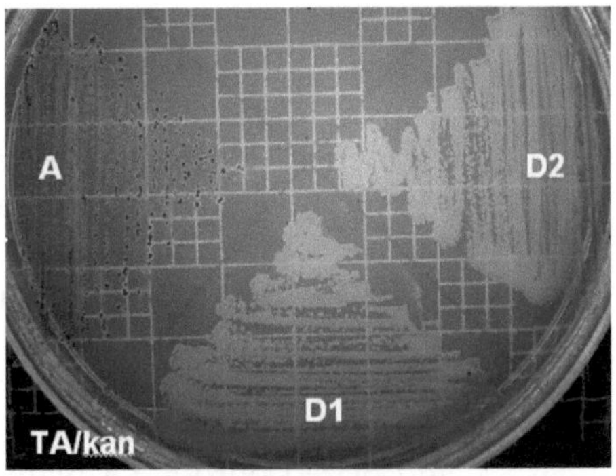


Fig. 16.2. Phenotype of ancestral colonies (A) and two derived strains (D1 and D2) on a Tetrazolium Sucrose (TS) plate. Transformants that utilize sucrose grow larger and form white colonies (D1 and D2), while those unable to utilize sucrose are smaller and grow as red (A) colonies. The color differences are due to pH differences resulting from sugar fermentation.

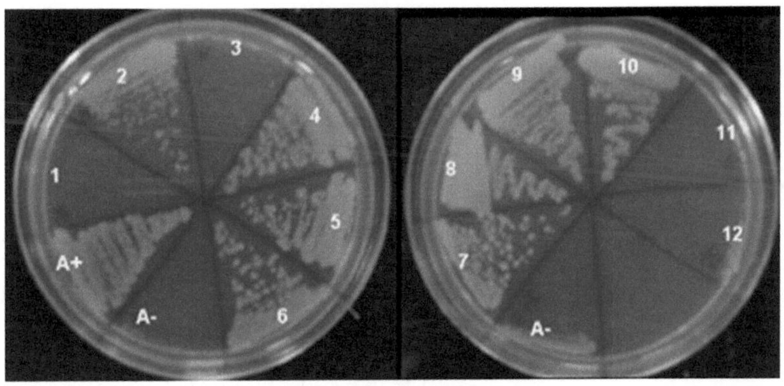


Fig. 16.3. Characterization of plasmid integrity on an LB/kan plate. "A–" is ancestral strain 4536 without sacP/sacA gene insert (kan–), "A+" is ancestral strain 4536 with sacP/sacA insert (kan+). Each replicate evolved line is denoted 1–12.

of the Kanamycin-resistance gene function. This phenomenon has been previously observed in plasmids evolved for 500 generations in environments offering different densities of available hosts *(56)*.

### 3.2. Evolution Experiments and Fitness Assays

One sucrose$^+$ clone of each recipient was used to initiate three sets of 12 replicate lines that were propagated daily in one of three environments: DM sucrose (25 μg/ml), DM glucose (25 μg/ml), and a mixture of sucrose and glucose (12. 5 μg/ml of each carbon source) for 50 days, allowing for approximately ~6.6 ($=\log_2 100$) rounds of bacterial replication every day. Glycerol stocks of all cultures were prepared and stored at −80°C every 100 generations (15 transfers).

The evolved and ancestral genotypes were compared by head-to-head competition assays under the same culture conditions as used during selection (e.g., sucrose, glucose, mixed environments, or alternating environments). Overnight cultures of sucrose-utilizing recombinants of interest and an unselected sucrose$^+$ ancestor (marked with a neutral arabinose marker, see *(57)*) are mixed in equal amounts, and cultured with shaking at 120 rpm, overnight, at 37°C. Initial population densities for each competitor are determined by plating. A 100 μl aliquot of the mixture is serially diluted 200-fold using saline solution and plated onto a TA plate (TA plates are tetrazolium indicator plates with 1% arabinose as the carbon source). The number of red (recombinant strain, sucrose$^+$) and white (ancestral strain, sucrose$^+$, not evolved) colonies are counted after 24 h (plates are incubated at 37°C) to determine the initial densities of each line ($T_0$). Final population densities are determined by plating the overnight mixed culture after a 300-fold dilution using saline solution. Plates are incubated at 37°C for 24 h and the red and white colonies are once again counted ($T_1$). **Figure 16.4** shows a schematic of this protocol.

The $T_0$ and $T_1$ densities are used determine relative fitness, $W_{ij}$; a measure of an organism's relative performance in a given environment, which is given by:

$$W_{ij} = \frac{\ln\left[{}^{N_i(1)}/_{N_i(0)}\right]}{\ln\left[{}^{N_j(1)}/_{N_j(0)}\right]},$$

where $N_i(0)$ and $N_j(0)$ are initial densities ($T_0$ densities) of the recombinant and ancestral genotypes, respectively, and $N_i(1)$ and $N_j(1)$ are the subsequent densities after 1 day ($T_1$) *(57)*. Using the replicated experimental design, we are also able to determine if there are persistent fitness effects due to the source and chromosomal location of foreign DNA *(24)*. Additionally, since we freeze samples of the lines throughout their evolution, fitness and, hence, adaptability, can be tracked and quantified for different time-points throughout the experiment, giving snap-shots of the process as it occurs.

Relative fitness values calculated using head-to-head competition assays of evolved recombinant strains (sucrose-selected lines relative to a selectively arabinose-utilization marked ancestral derivative) in both the sucrose and the glucose environments are shown in **Fig. 16.5**. Values show an improvement in both the selected and the unselected environments, suggesting a general amelioration of the cost of plasmid carriage. Clearly, there is also an interaction of the arabinose marker with fitness in sucrose. An ANOVA indicates that the fitness of the genotypes evolved in

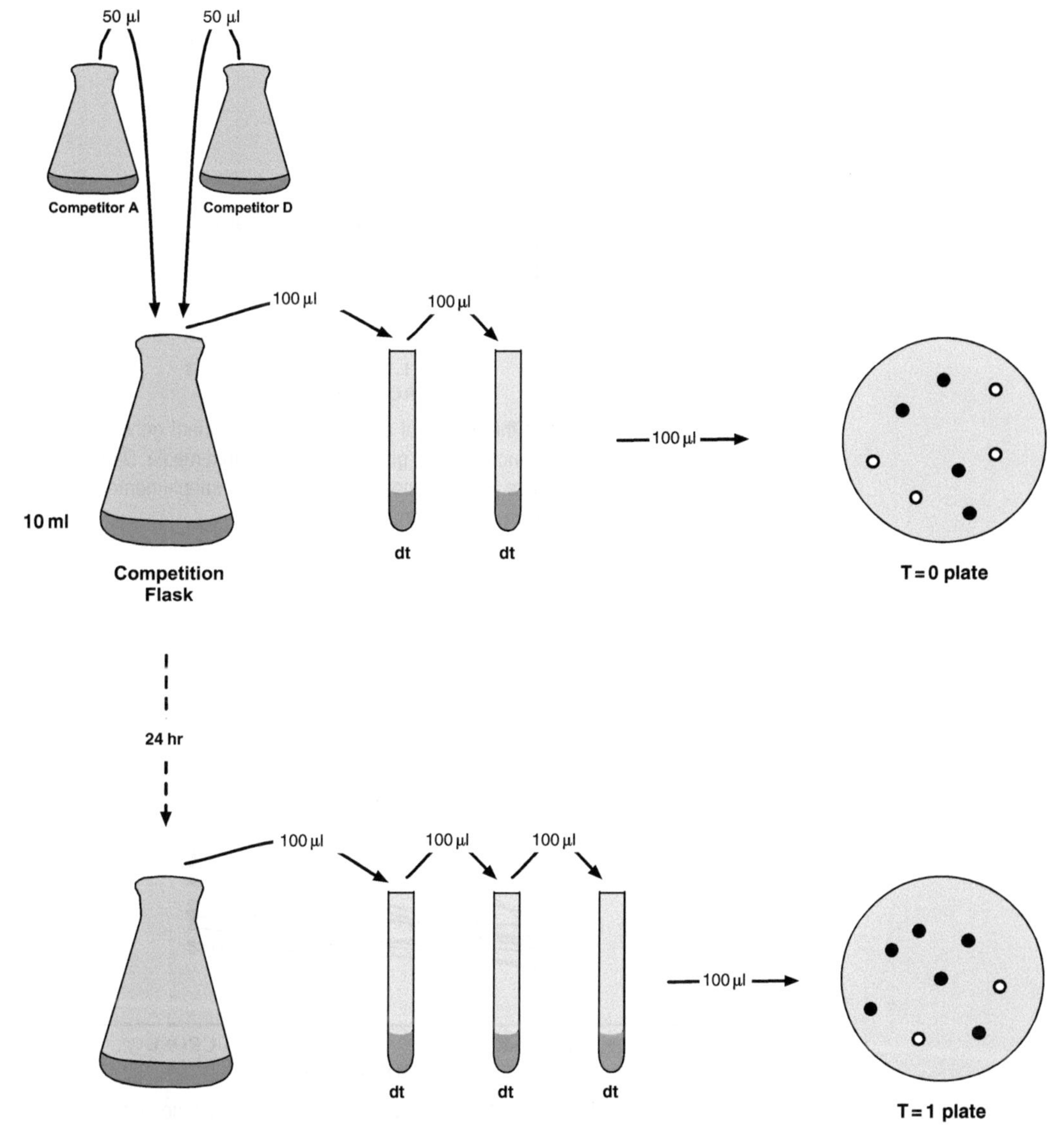


Fig. 16.4. General fitness assay protocol. For each line, cultures were plated onto fresh plates and a single colony of each replicate is used to start overnight cultures in the desired environment (e.g., sucrose or glucose). These overnight cultures are then used to seed the competition flasks. Competitions are initiated by inoculating 10 ml of the appropriate medium with 50 µl aliquots of both derived (D) and ancestor (A) cultures. Initial cell densities of each genotype are determined by plating on a TA plate, and counting red (derived recombinant strain, in this case sucrose$^+$) and white (ancestral strain, not evolved) colonies after 24 h. Competition flasks are incubated for 24 h at 37°C and plated (through two 1/100 dilutions) to obtain cell densities after competition. Each competition assay (for each of 12 lines) is run in triplicate. Cell density ratios are used to calculate the relative fitness of the evolved strains with respect to the ancestor.

sucrose is greater in sucrose than in glucose ($p < 0.0001$), that there is significant variation among lines for fitness ($p < 0.0001$), and that there are significant interactions between strains and nutrient environments ($p < 0.0001$) (**Fig. 16.6**). Fitness values for REL4536 derived strains have been more difficult to estimate,

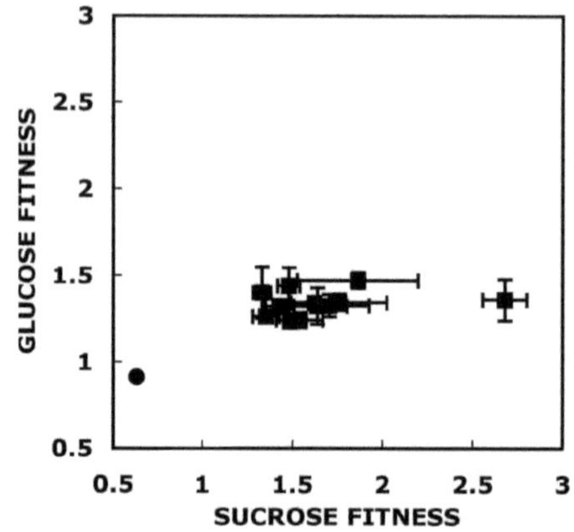


Fig. 16.5. Relative fitness of the ancestral and derived recombinant genotypes, *circle* and *squares*, respectively, in sucrose- and glucose-supplemented media. Derived genotypes were evolved for 50 days (~300 generations) in sucrose-supplemented medium. *Error bars* represent one standard deviation.

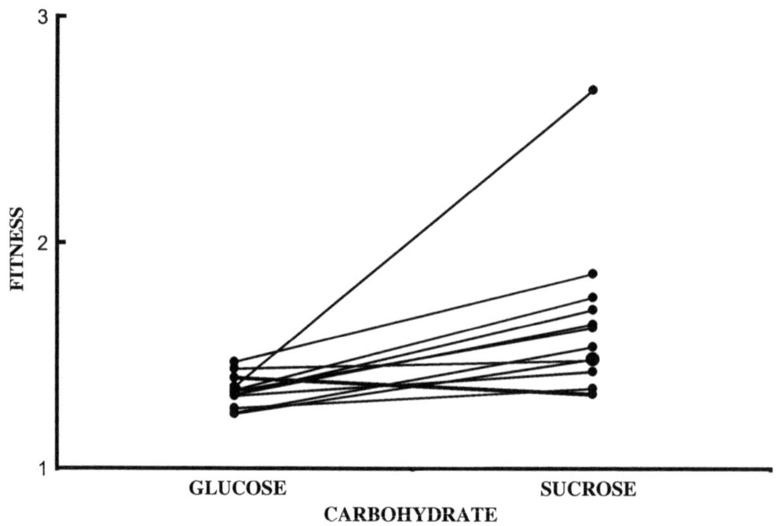


Fig. 16.6. Fitness of the selected isolates in glucose and sucrose minimal media. Fitness is generally higher in the sucrose-supplemented medium, the selected environment, and in one case, far higher. However, some isolates have higher fitness in glucose than sucrose minimal media, suggesting difficult functional integration of the newly acquired *Bacillus* sucrose transport genes. The statistically significant genotype-by-environment interaction ($p < 0.0001$) is strongly suggestive of at least transient multiple adaptive pathways.

as the original ancestral transformant grows extremely poorly relative to the sucrose-selected lines.

## 4. Characterization of the Fate of Transferred Genes in the Host – What Can Be Done with the Experimental Evolution Approach

In this chapter, we describe an experimental model system to characterize the fate of genes using bacteria once a transfer event has occurred. By using microorganisms, which are easy to manipulate, culture, and propagate, we can also get specific snap-shots of the process of evolution by freezing cultures away and re-culturing them for experiments as desired. In this way, it is possible to characterize the amelioration of the foreign sequence and what happens to the transferred genes once they are incorporated into the host's genetic environment, either by means of an autonomous plasmid maintained within the host through selection or once they incorporate into the host chromosome through a recombination event. We are presently following *E. coli* lines with a sucrose$^{+}$ phenotype. Examination of the specific changes in the foreign DNA (namely, the sucrose-utilization genes from *B. subtilis*) sequence introduced and maintained by the *E. coli* host will help characterize the pathways for adaptation of foreign genetic information to the host. Using this experimental approach, we can quantify the competitive benefits of incorporation. We can correlate these with specific sequence changes that are a result of adaptation by comparing the relative fitness of recombinants harboring the wild-type *Bacillus* sequence, and "corrected" host-equivalent (with respect to codon bias, GC content, etc.) synthetic genes. We can also assess the importance of location of insertion within the chromosome using molecular fingerprinting techniques, and determine the global effects of foreign gene expression on the rest of the genome.

Since several lines of evolution are carried out in tandem, it is possible to compare all changes and obtain a comprehensive perspective of adaptation as it occurs, in the context of both the foreign DNA and the whole genome. By analyzing different time points throughout the evolution experiment, one can directly appreciate the progression of sequence changes and determine if the evolution of the foreign sequence occurs rapidly, as well as the rate and time-frame of any specific regulatory change. Changes in foreign DNA can be correlated with adaptation using fitness data. Thus, we can determine what the specific effect of each environment is on co-adaptation of the foreign genes and the host. Initial assessments of fitness using the evolved lines versus the ancestor insert-carrying strain have already revealed a significant fitness increase with an environment effect after 300 generations (*see* **Section 3.2**).

As we stated earlier, the focus of the work discussed here is on using a plasmid to transfer the genes of interest. However, the system can easily be amended to examine chromosomal integration events, be it through a recombination event or using a delivery system such as a transposon. It is possible then to compare evolutionary trajectories of evolved lines harboring the genes in a plasmid, and those expressing the genes from the chromosome after an incorporation event. To establish the importance of location within the chromosome for functionality, an initial screening using indicator media, PCR, and protein gels to confirm the expression of each gene can be done. All recombinant lines that contain inserts which result in functional gene products, and those that carry inserts which are not functional can be retained for further study. Standard molecular techniques, such as restriction endonuclease digestion, Southern blots, and sequencing, as well as extensive statistical and bioinformatics analysis, can be used to determine the specific location of insertion in the host's chromosome. By comparing the location of inserts on the host chromosome to that of the clones that carry the insert but fail to produce functional gene products, the physical requirements for successful expression of the inserted foreign DNA can be elucidated.

### 4.1. Studying the Expression of Foreign Genes

The evolved lines generated from the experimental evolution experiment can be analyzed for gene expression at different points throughout their evolutionary history. Regulatory changes can thus be quantified in the lab, and specific questions can then be addressed. For example, we can determine if global genome expression changes occur rapidly, or later in the evolution of the bacterial lines, if there is suboptimal expression of the novel gene, if this is caused by codon bias or an inefficient promoter, if the amelioration process is a result of specific point mutations affecting codon bias or of specific regulatory changes, if there are specific epistatic interactions with other components of the host's genome, and so on. Increased expression via regulatory evolution is likely to be faster in plasmid-borne foreign genes than the identical foreign genes that have incorporated into the host chromosome. Multicopy plasmids increase the target sequence available for mutations, and increased expression is likely to be a dominant phenotype. In contrast, a mutation reducing expression on one plasmid would be expected to be largely recessive. Sequencing of evolved plasmid populations will easily reveal if such changes have incurred in the regulatory regions of the plasmid sequence.

Specific sequence changes can also be directly analyzed using the experimental evolution system. Suboptimal codon bias (which reduces gene expression) is likely to have persistent fitness consequences over evolutionary time. Since single site changes in codon use are likely to provide only small incremental fitness increases, such changes are unlikely to occur if other beneficial single mutations of greater effect are possible. However, a corollary of this is that in populations that are well adapted to their environment, changes in codon use become increasingly likely. Moreover, those codons that have the largest effect (primarily those closest to the start codon) are more likely to evolve. On the other hand, genes with more similar codon and GC bias to the host are more likely to have a greater adaptive potential over long-term evolution. Because codon and GC bias occur over the length of a gene, incremental changes would provide incremental benefits. Hence, having more similar codon and GC bias is likely to lead to more appropriate expression levels with less early termination of mRNA and protein synthesis. Additionally, in seasonal environments where expression of the acquired gene may not always be necessary, the potential for appropriate expression levels is higher, as regulatory elements will not need to compensate for the poor expression induced by codon and GC bias. The energetic costs of protein synthesis for highly expressed genes can be reduced by more than 0.025% per amino acid by encoding for less energetically costly amino acids. In *E. coli* and *B. subtilis*, highly expressed genes code for disproportionately fewer energetically costly amino acids *(58)*.

Perhaps one of the greatest strengths of the experimental evolution system is that it allows one to characterize the specific mechanisms and physical rules required for a successful incorporation event, and that one can do so at any given time-point. It can be expected that the forced incorporation of foreign genes into a host chromosome reduces the potential for their successful functional integration. This is because there are likely to be specific "hot spots" within a bacterial chromosome that determine a successful incorporation event (i.e., foreign gene is expressed and is functional), and, as has recently been suggested, there might even be lethality effects associated with particular genes *(38)*. This would support the idea that HGT into the host chromosome is often a directed mutation. Incorporation into the host genome reduces rapid loss of foreign genes in the absence of selection in comparison to plasmids. These incorporation events offer an additional step, which can help ensure that the foreign gene stays in the host – something of benefit particularly if the novel trait is advantageous. However, greater similarity between the host and foreign genes for codon and GC bias promotes plasmid stability in seasonal environments, via better regulation, than in host chromosome incorporated genes.

Acquisition of novel traits by foreign genes could, in theory, constrain subsequent evolution due to the limited potential for functional integration of the host and foreign genes *(59)*. Even though bacteria differ in modes and mechanisms of gene regulation and expression, limited evolution appears unlikely given the extant diversity of horizontally transmitted genes among diverse bacterial hosts. Instead, we expect that rapid evolution of traits encoded by foreign genes occurs. The acquisition and selection for expression of foreign genes enhances the diversity among replicates to a greater extent if populations are less well adapted to their environment. Differences in adaptive mutations early in an evolutionary sequence are more likely to lead to sustained divergence than those occurring later. Less well-adapted populations are more likely to acquire a variety of different mutations, and this variety is likely to be enhanced by the acquisition of a poorly integrated foreign gene. Note, this diversification is a function of both the adaptation of the foreign genes to the host and of the host to the environment.

The specificity of co-adaptation between host and foreign genes is more likely to be greater in populations that are initially less well adapted to their environment. As greater diversity of adaptive responses are more anticipated in initially less well-adapted populations, so too are a greater diversity of co-adaptive responses. Hence, the species specificity of horizontally transmitted genes is in part a function of the host adaptation to the environment.

## 5. Conclusion/ Outlook

Adaptation to a new sucrose environment after incorporation of foreign genes, confirmed by a threefold increase in fitness, is apparent in *E. coli* after only ~300 generations. The mechanisms of this adaptation, as well as global genetic effects of insertion of the foreign gene on the optimized expression of the rest of the genome, are being determined using further experimental techniques (e.g., microarray experiments). It has previously been established that increases in fitness in strains evolved in the laboratory are a result of very specific changes in gene expression *(60)*. One can use the evolved lines produced to determine the exact perturbations in global gene expression resulting from the successful incorporation of foreign DNA by measuring the total gene expression of ancestral and evolved strains at different time-points throughout the evolution experiment. The specific time-points are chosen based on fitness and sequencing data obtained. Microarray experimental data can then be complemented using additional methods, including quantitative real-time PCR. In both these cases, data analysis will involve the development of new statistical models to examine gene expression as a function of evolutionary time.

The experimental approach we present here is a tool that can bring together many techniques, ranging from standard microbiology and molecular techniques to sophisticated protocols involving microarray, statistical, and bioinformatics analyses. Experimental evolution affords the researcher with a basic experimental setup that generates evolved lines that can be easily studied using any of these approaches. These lines can be subsequently used to pose specific questions about gene transfer mechanisms, adaptation to novel genes, epistatic effects of the transfer event, and expression and other global effects on the host's proteome. It is clear that, in order to study the effects of a transfer event on a host, as well as the evolutionary implications such an event might have, an interdisciplinary approach is necessary. This allows for a more inclusive analysis of the complex problem of HGT.

## Acknowledgments

The authors warmly acknowledge Janet Siefert for fruitful and insightful discussions that laid the groundwork for the work presented here. We also thank two anonymous reviewers for valuable comments.

### References

1. Ochman, H., Lawrence, J.G., Groisman, E.A. (2000). Lateral gene transfer and the nature of bacterial innovation. *Nature*, **405**, 299–304.
2. Koonin, E.V., Makarova, K.S., Aravind, L. (2001). Horizontal gene transfer in prokaryotes: quantification and classification. *Annu. Rev. Microbiol.*, **55**, 709–742.
3. Lawrence, J.G., Ochman, H. (1998). Molecular archaeology of the *Escherichia coli* genome. *Proc. Natl. Acad. Sci. U. S. A*, **95**, 9413–9417.
4. Cortez, D.Q., Lazcano, A., Becerra, A. (2005). Comparative analysis of methodologies for the detection of horizontally transferred genes: a reassessment of first-order Markov models. *In Silico. Biol.*, **5**, 581–592.
5. Romero, D., Palacios, R. (1997). Gene amplification and genomic plasticity in prokaryotes. *Annu. Rev. Genet.*, **31**, 91–111.
6. Spratt, B.G., Hanage, W.P., Feil, E.J. (2001). The relative contributions of recombination and point mutation to the diversification of bacterial clones. *Curr. Opin. Microbiol.*, **4**, 602–606.
7. Koonin, E.V., Galperin, M.Y. (1997). Prokaryotic genomes: the emerging paradigm of genome-based microbiology. *Curr. Opin. Genet. Dev.*, **7**, 757–763.
8. Lawrence, J.G., Ochman, H. (1997). Amelioration of bacterial genomes: rates of change and exchange. *J. Mol. Evol.*, **44**, 383–397.
9. Lawrence, J.G. (1999). Gene transfer, speciation, and the evolution of bacterial genomes. *Curr. Opin. Microbiol.*, **2**, 519–523.
10. Aravind, L., Tatusov, R.L., Wolf, Y.I., Walker, D.R., Koonin, E.V. (1998). Evidence for massive gene exchange between archaeal and bacterial hyperthermophiles. *Trends Genet.*, **14**, 442–444.
11. Huynen, M.A., Bork, P. (1998). Measuring genome evolution. *Proc. Natl. Acad. Sci. U. S. A*, **95**, 5849–5856.
12. Doolittle, W.F. (1999). Phylogenetic classification and the universal tree. *Science*, **284**, 2124–2129.
13. Doolittle, W.F. (2000). The nature of the universal ancestor and the evolution of the proteome. *Curr. Opin. Struct. Biol.*, **10**, 355–358.
14. Woese, C.R. (2000). Interpreting the universal phylogenetic tree. *Proc. Natl. Acad. Sci. U. S. A.* **97**, 8392–8396.
15. Gogarten, J.P., Doolittle, W.F., Lawrence, J.G. (2002). Prokaryotic evolution in light of gene transfer. *Mol. Biol. Evol.*, **19**, 2226–2238.
16. de la, Cruz, F., Davies, J. (2000). Horizontal gene transfer and the origin of species: lessons from bacteria. *Trends Microbiol.*, **8**, 128–133.
17. Scholten, J.C., Culley, D.E., Brockman, F.J., Wu, G., Zhang, W. (2007). Evolution of the syntrophic interaction between *Desulfovibrio vulgaris* and *Methanosarcina barkeri*: Involvement of an ancient horizontal gene

transfer. *Biochem. Biophys. Res. Commun.*, **352**, 48–54.
18. Rosas-Magallanes, V., Deschavanne, P., Quintana-Murci, L., Brosch, R., Gicquel, B., Neyrolles, O. (2006). Horizontal transfer of a virulence operon to the ancestor of *Mycobacterium tuberculosis. Mol. Biol. Evol.*, **23**, 1129–1135.
19. Walsh, T.R. (2006). Combinatorial genetic evolution of multiresistance. *Curr. Opin. Microbiol.*, **9**, 476–482.
20. Wertz, J.E., Riley, M.A. (2004). Chimeric nature of two plasmids of *Hafnia alvei* encoding the bacteriocins alveicins A and B. *J. Bacteriol.*, **186**, 1598–1605.
21. van der Meer, J.R., Sentchilo, V. (2003). Genomic islands and the evolution of catabolic pathways in bacteria. *Curr. Opin. Biotechnol.*, **14**, 248–254.
22. O'Driscoll, J., Glynn, F., Fitzgerald, G.F., van Sinderen, D. (2006). Sequence analysis of the lactococcal plasmid pNP40: a mobile replicon for coping with environmental hazards. *J. Bacteriol.*, **188**, 6629–6639.
23. Kunin, V., Goldovsky, L., Darzentas, N., Ouzounis, C.A. (2005). The net of life: reconstructing the microbial phylogenetic network. *Genome Res.*, **15**, 954–959.
24. Travisano, M., Mongold, J.A., Bennett, A.F., Lenski, R.E. (1995). Experimental tests of the roles of adaptation, chance, and history in evolution. *Science*, **267**, 87–90.
25. Lenski, R.E., Simpson, S.C., Nguyen, T.T. (1994). Genetic analysis of a plasmid-encoded, host genotype-specific enhancement of bacterial fitness. *J. Bacteriol.*, **176**, 3140–3147.
26. Purrington, C.B., Bergelson, J. (1999). Exploring the physiological basis of costs of herbicide resistance in *Arabidopsis thaliana. The American Naturalist*, **154**, S82–S91.
27. Feder, M.E., Bennett, A.F., Huey, R.B. (2000). Evolutionary physiology. *Annu. Rev. Ecol. Syst.*, **31**, 315–341.
28. Dieckmann, U., Doebeli, M. (1999). On the origin of species by sympatric speciation. *Nature*, **400**, 354–357.
29. Goldstein, M.A., Doi, R.H. (1995). Prokaryotic promoters in biotechnology. *Biotechnol. Annu. Rev.*, **1**, 105–128.
30. Vellanoweth, R.L., Rabinowitz, J.C. (1992). The influence of ribosome-binding-site elements on translational efficiency in *Bacillus subtilis* and *Escherichia coli* in vivo. *Mol. Microbiol.*, **6**, 1105–1114.
31. Kane, J.F. (1995). Effects of rare codon clusters on high-level expression of heterologous proteins in *Escherichia coli. Curr. Opin. Biotechnol.*, **6**, 494–500.
32. Chen, S.L., Lee, W., Hottes, A.K., Shapiro, L., McAdams, H.H. (2004). Codon usage between genomes is constrained by genome-wide mutational processes. *Proc. Natl. Acad. Sci. U. S. A*, **101**, 3480–3485.
33. van Passel, M.W., van der, E.A., Bart, A. (2006). Plasmid diversity in neisseriae. *Infect. Immun.*, **74**, 4892–4899.
34. Bouma, J.E., Lenski, R.E. (1988). Evolution of a bacteria/plasmid association. *Nature*, **335**, 351–352.
35. Haubold, B., Travisano, M., Rainey, P.B., Hudson, R.R. (1998). Detecting linkage disequilibrium in bacterial populations. *Genetics*, **150**, 1341–1348.
36. Lawrence, J.G., Ochman, H., Hartl, D.L. (1992). The evolution of insertion sequences within enteric bacteria. *Genetics*, **131**, 9–20.
37. Naas, T., Blot, M., Fitch, W.M., Arber, W. (1994). Insertion sequence-related genetic variation in resting *Escherichia coli* K-12. *Genetics*, **136**, 721–730.
38. Sorek, R., Zhu, Y., Creevey, C.J., Francino, M.P., Bork, P., Rubin, E.M. (2007) Genome-wide experimental determination of barriers to horizontal gene transfer. *Science*, **318**, 1449–1452.
39. Alcaraz, L.D., Olmedo, G., Bonilla, G., Cerritos, R., Hernandez, G., Cruz, A., Ramirez, E., Putonti, C., Jimenez, B., Martinez, E., Lopez, V., Arvizu, J.L., Ayala, F., Razo, F., Caballero, J., Siefert, J., Eguiarte, L., Vielle, J.P., Martinez, O., Souza, V., Herrera-Estrella, A., Herrera-Estrella, L. (2008) The genome of *Bacillus coahuilensis* reveals adaptations essential for survival in the relic of an ancient marine environment. *Proc. Natl. Acad. Sci. U. S. A*, **105**, 5803–5808.
40. Lenski, R.E., Bouma, J.E. (1987). Effects of segregation and selection on instability of plasmid pACYC184 in *Escherichia coli* B. *J. Bacteriol.*, **169**, 5314–5316.
41. Nguyen, T.N., Phan, Q.G., Duong, L.P., Bertrand, K.P., Lenski, R.E. (1989). Effects of carriage and expression of the Tn10 tetracycline-resistance operon on the fitness of *Escherichia coli* K12. *Mol. Biol. Evol.*, **6**, 213–225.
42. Ho, T.Q., Zhong, Z., Aung, S., Pogliano, J. (2002). Compatible bacterial plasmids are targeted to independent cellular locations in *Escherichia coli. EMBO J.*, **21**, 1864–1872.
43. De Gelder, L., Ponciano, J.M., Abdo, Z., Joyce, P., Forney, L.J., Top, E.M. (2004). Combining mathematical models and statistical methods to understand and predict the dynamics of antibiotic-sensitive mutants in a population of resistant bacteria

during experimental evolution. *Genetics*, **168**, 1131–1144.

44. Lenski, R.E., Souza, V., Duong, L.P., Phan, Q.G., Nguyen, T.N., Bertrand, K.P. (1994). Epistatic effects of promoter and repressor functions of the Tn10 tetracycline-resistance operon of the fitness of *Escherichia coli. Mol. Ecol.*, **3**, 127–135.
45. Dahlberg, C., Chao, L. (2003). Amelioration of the cost of conjugative plasmid carriage in *Eschericha coli* K12. *Genetics*, **165**, 1641–1649.
46. Reid, S.J., Abratt, V.R. (2005) Sucrose utilisation in bacteria: genetic organisation and regulation. *Appl. Microbiol. Biotechnol.*, **67**, 312–321.
47. Trevino-Quintanilla, L.G., Escalante, A., Caro, A.D., Martinez, A., Gonzalez, R., Puente, J.L., Bolivar, F., Gosset, G. (2007) The phosphotransferase system-dependent sucrose utilization regulon in enteropathogenic *Escherichia coli* strains is located in a variable chromosomal region containing iap sequences. *J. Mol. Microbiol. Biotechnol.*, **13**, 117–125.
48. Cowan, P.J., Nagesha, H., Leonard, L., Howard, J.L., Pittard, A.J. (1991) Characterization of the major promoter for the plasmid-encoded sucrose genes scrY, scrA, and scrB. *J. Bacteriol.*, **173**, 7464–7470.
49. Doroshenko, V.G., Livshits, V.A. (2004) Structure and mode of transposition of Tn2555 carrying sucrose utilization genes. *FEMS Microbiol. Lett.*, **233**, 353–359.
50. Jahreis, K., Bentler, L., Bockmann, J., Hans, S., Meyer, A., Siepelmeyer, J., Lengeler, J.W. (2002) Adaptation of sucrose metabolism in the *Escherichia coli* wild-type strain EC3132. *J. Bacteriol.*, **184**, 5307–5316.
51. Titgemeyer, F., Jahreis, K., Ebner, R., Lengeler, J.W. (1996) Molecular analysis of the scrA and scrB genes from Klebsiella pneumoniae and plasmid pUR400, which encode the sucrose transport protein Enzyme II Scr of the phosphotransferase system and a sucrose-6-phosphate invertase. *Mol. Gen. Genet.*, **250**, 197–206.
52. Fouet, A., Arnaud, M., Klier, A., Rapoport, G. (1987) *Bacillus subtilis* sucrose-specific enzyme II of the phosphotransferase system: expression in *Escherichia coli* and homology to enzymes II from enteric bacteria. *Proc. Natl. Acad. Sci. U. S. A*, **84**, 8773–8777.
53. Kornberg, H.L. (2001) Routes for fructose utilization by *Escherichia coli. J. Mol. Microbiol. Biotechnol.*, **3**, 355–359.
54. Sabbagh, Y., Theriault, E., Sanschagrin, F., Voyer, N., Palzkill, T., Levesque, R.C. (1998). Characterization of a PSE-4 mutant with different properties in relation to penicillanic acid sulfones: importance of residues 216 to 218 in class A beta-lactamases. *Antimicrob. Agents Chemother.*, **42**, 2319–2325.
55. Lenski, R.E., Travisano, M. (1994). Dynamics of adaptation and diversification: a **10**, 000-generation experiment with bacterial populations. *Proc. Natl. Acad. Sci. U. S. A*, **91**, 6808–6814.
56. Turner, P.E., Cooper, V.S., Lenski, R.E. (1998). Tradeoff between horizontal and vertical modes of transmission in bacterial plasmids. *Evolution*, **52**, 315–329.
57. Travisano, M., Lenski, R.E. (1996). Long-term experimental evolution in *Escherichia coli*. IV. Targets of selection and the specificity of adaptation. *Genetics*, **143**, 15–26.
58. Akashi, H., Gojobori, T. (2002). Metabolic efficiency and amino acid composition in the proteomes of *Escherichia coli* and *Bacillus subtilis. Proc. Natl. Acad. Sci. U. S. A*, **99**, 3695–3700.
59. Mann, R.S., Carroll, S.B. (2002). Molecular mechanisms of selector gene function and evolution. *Curr. Opin. Genet. Dev.*, **12**, 592–600.
60. Cooper, T.F., Rozen, D.E., Lenski, R.E. (2003). Parallel changes in gene expression after 20,000 generations of evolution in *Escherichia coli. Proc. Natl. Acad. Sci. U. S. A*, **100**, 1072–1077.

# Chapter 17

# Mass Action Models Describing Extant Horizontal Transfer of Plasmids: Inferences and Parameter Sensitivities

**Barth F. Smets and Laurent Lardon**


## Abstract

Predicting the fate of horizontally transmissible elements in extant microbial communities might be facilitated by the availability of suitable mathematical models. Since the mid-1970s, mass action models have been introduced to describe the transfer of conjugal and mobilizable genetic elements. This chapter will summarize and explain the assumptions behind spatially homogenous models, and show the predictions by these models under typical scenarios, such as evaluating existence conditions of conjugal plasmids under chemostat or seasonal growth conditions. Special attention is given to the sensitivity of the outcomes to the various plasmid dynamic parameters. For our analysis, we developed a set of user-friendly MatLab® routines, which are deposited in the public domain. We hope that the availability of these routines will encourage the computationally untrained microbiologist to make use of these mathematical models. Finally, further permutations, as well as limitations of these mass action models in view of the structured complexity of most microbial systems are addressed.


**Key words:** Plasmid transfer, plasmid loss, plasmid kinetics, mass action, conjugation, segregation, mathematical model, MatLab® routines.

## 1. Introduction

Predicting the fate of horizontally transmissible elements in extant microbial communities might be facilitated by the availability of suitable mathematical models. Mathematical models permit the integration of all known elements of a certain process and allow one to predict outcomes under a range of scenarios without the time- and resource-consuming need to test it experimentally. This is especially true when many interdependent processes occur, which determine the response of a system. Extant (i.e., currently

Maria B. Gogarten et al. (eds.), *Horizontal Gene Transfer: Genomes in Flux, vol. 532*

DOI 10.1007/978-1-60327-853-9_17 Springerprotocols.com

ongoing) horizontal gene transfer (HGT) in microbial systems is one such example, because the response (persistence or wash out of a mobile element from a system) is the outcome of many interacting processes (relating to growth dynamics, gene transfer dynamics, as well as the interaction of the latter two with the system conditions). Since the mid-1970s, pseudo-mechanistic mass action models have been introduced, which describe the transfer of conjugal plasmids in planktonic microbial communities *(1–4)*. Their characteristics are herein reviewed, and they are solved to explore the existence conditions of plasmids under various scenarios. Plasmid transfer models applicable to biofilm-dominated systems are not reviewed.

## 2. Presentation of the Mass Action Model

### *2.1. The Governing Processes and Assumptions*

In the mass action models for plasmid transfer, typically, three cell types are recognized: donors (**D**), recipients (**R**), and transconjugants (**T**). The donor is considered the "original" host of the plasmid, while the transconjugant refers to a plasmid carrier, which is chromosomally isogenic to the recipient. The growth dynamics of all cell types are typically described by Monod-like dependencies on the limiting substrate concentrations. In addition, plasmid dynamics are explicitly considered. The model assumes the feasibility of horizontal plasmid transfer from donor to recipients (to result in transconjugants) and from transconjugants to recipients (to result in more transconjugants). No plasmid transfer between plasmid-harboring cell types is considered. Plasmid transfer follows a "mass action" model. This means that the probability of transfer is jointly proportional to the density of the participating cell types in the mating pair multiplied by a probability coefficient. This probability coefficient is a measure of the plasmid "fertility" and is typically referred to as the plasmid transfer rate coefficient (units: vol $cell^{-1}$ $time^{-1}$) *(2)*. Different values can be assumed for these rates depending on the host type (donor vs transconjugant) and sometimes even depending on the "residence time" of the plasmid in a new host (to account for the observed derepression of plasmid transfer upon infection of a new cell type) *(5)*. In addition to horizontal transfer, plasmid loss from one cell type (segregation) is considered and this process is typically considered a linear function of the cell density.

The mass action model clearly assumes that mating occurs "randomly" between the different cell types of a mating pair. This means that every cell has equal probability of making contact with another cell and that an equal fraction of these encounters results in plasmid transfer. Hence, the "structure" of the environment is not taken into account. Therefore, the use of mass action models is best limited to well-shaken or dispersed, i.e., unstructured, environments such as chemostat cultures where "zonation", "patchiness", or biofilm formation is minimal.

While the mass action model captures some of the mechanistic elements of the biology of plasmid transfer (and has been called a mechanistic model), it retains some simplifying assumptions. For example, plasmid transfer is "instantaneous": this means that a successful D to R contact results in the "immediate" creation of a T cell type, which immediately (i.e., without delay) can transfer to another R cell. The significance of the error introduced by this assumption has been examined by Lagido et al. *(6)*.

In addition, the model typically accounts for the effect of plasmid carriage on the host metabolism. It considers a kinetic and a stoichiometric effect on microbial growth, expressed as a reduction of the specific growth rate and/or reduction of the biomass/substrate growth yield. Because growth of all cell types is generally modeled via Monod-like dependencies on the limiting substrate concentrations, the metabolic burden is easily modeled as a reduction in the specific growth rate. Similarly, a gain in metabolic fitness can be considered as an increase in the specific growth rates with respect to the plasmid-free type (such situation might arise when the plasmid encoded beneficial phenotypes). Independently, the plasmid may also decrease the quantity of new cells produced per unit of substrate consumed becoming a stoichiometric burden.

### 2.2. The Governing Equations

Here, we introduce the mathematical equations used to describe the processes previously cited. We use a slightly different and more standardized nomenclature than what has been presented by Levin and collaborators in their earlier expositions *(2, 7, 8)*. In all of these processes $r$ refers to volumetric reaction rates. Because plasmid transfer occurs between "individuals", the concentration of the different cells types, $X$, is expressed in individuals-based concentrations (numbers or cells/volume) rather than in mass-based concentrations. The limiting solute concentration, $S$, remains in traditional units (mass/volume or moles/volume).

The plasmid dynamic processes can be described as follows:

$$\text{D to R plasmid transfer rate} \equiv r_{\mathrm{DR}} \equiv \gamma_{\mathrm{DR}} X_{\mathrm{D}} X_{\mathrm{R}} \quad (17.1)$$

$$\text{T to R plasmid transfer rate} \equiv r_{\mathrm{TR}} \equiv \gamma_{\mathrm{TR}} X_{\mathrm{T}} X_{\mathrm{R}} \quad (17.2)$$

$$\text{Plasmid loss rate from D} \equiv r_{\mathrm{L\_D}} \equiv -\tau_{\mathrm{D}} X_{\mathrm{D}} \quad (17.3)$$

$$\text{Plasmid loss rate from T} \equiv r_{\mathrm{L\_T}} \equiv -\tau_{\mathrm{T}} X_{\mathrm{T}} \quad (17.4)$$

With subscripts D, R, T referring to donor, recipient, and transconjugant cell types, $\gamma$ = plasmid transfer rate coefficients ($\mathrm{L\,cell^{-1}h^{-1}}$), and $\tau$ = plasmid loss rate coefficient ($\mathrm{h^{-1}}$).

The growth dynamics of the different cell types are described as follows:

$$\text{Growth rate of D} = r_{X_D} = \mu_D X_D \tag{17.5}$$

$$\text{Growth rate of R} = r_{X_R} = \mu_R X_R \tag{17.6}$$

$$\text{Growth rate of T} = r_{X_T} = \mu_T X_T \quad \mu_T = (1-\alpha)\mu_R \tag{17.7}$$

In this case, $\mu$ are specific growth rates ($h^{-1}$), which are in their turn written as functions of the limiting substrate concentration, S. In this case, $\alpha$ refers to the fractional reduction in the specific growth rate due to plasmid carriage by the transconjugant cell type.

$$\text{Specific growth rate} = \mu_D = \mu_{max,D}\frac{S}{S+K_{S,D}} \tag{17.8}$$

The stoichiometric effect is explicitly modeled by including a potential effect of the plasmid presence (expression) on the net growth yield. Therefore, a parameter $\beta$ is used and defined. This is more straightforward than in the original publication, where $\beta$ lumped both kinetic and stoichiometric effects of the plasmid *(2)*.

$$\text{Growth yield} = \Upsilon_{X_T S} = (1-\beta)\Upsilon_{X_R S}, \tag{17.9}$$

where $\Upsilon_{X_T S}$ refers to a growth yield of cell type $X_T$ on limiting substrate $S$.

Although it will not be treated in this chapter, it must be mentioned that some experimental evidence indicates that the metabolic cost incurred by plasmid carriage may be transient and ultimately disappear *(9, 10)*, which will require to model $\alpha$ or $\beta$ as time-dependent parameters.

The substrate removal and cell growth rate expressions are coupled via the growth yields:

$$\text{Substrate removal rate} = r_S = -\frac{r_{X_D}}{\Upsilon_{X_D S}} - \frac{r_{X_R}}{\Upsilon_{X_R S}} - \frac{r_{X_T}}{\Upsilon_{X_T S}} \tag{17.10}$$

Finally, these equations, describing volumetric reaction rates, are used in an appropriate mass balance to result in the governing equations for the different cell types in a given environment. For the situation of a chemostat (a well-mixed reactor, receiving a constant inflow and influent concentration of substrate, operated at a fixed dilution rate and having a constant volume), the equations become:

$$\frac{dX_D}{dt} = r_{X_D} - DX_D \tag{17.11}$$

$$\frac{dX_R}{dt} = r_{X_R} - r_{DR} - r_{TR} + r_{L\_T} - DX_R \tag{17.12}$$

$$\frac{dX_T}{dt} = r_{X_T} + r_{DR} + r_{TR} - r_{L\_T} - DX_T \tag{17.13}$$

$$\frac{dS}{dt} = D(S_0 - S) - \frac{r_{X_D}}{Y_{X_D S}} - \frac{r_{X_R}}{Y_{X_R S}} - \frac{r_{X_T}}{Y_{X_T S}} \tag{17.14}$$

with $X_{D,0}$, $X_{R,0}$, $X_{T,0}$ as initial conditions *for* $X_D$, $X_R$, $X_T$, and $S_0$ being the influent substrate concentration and $D$ the dilution rate of the chemostat (flow rate through the chemostat/volume of chemostat). In this case, the influent concentrations of all cell types are considered to be zero. However, other regimes – such as the periodic or occasional introduction of donor cell types – could also be included.

For the situation of a batch culture (assuming no inflow or outflow and constant volume), the equations are simpler:

$$\frac{dX_D}{dt} = r_{X_D} \tag{17.15}$$

$$\frac{dX_R}{dt} = r_{X_R} - r_{DR} - r_{TR} + r_{L\_T} \tag{17.16}$$

$$\frac{dX_R}{dt} = r_{X_T} + r_{DR} + r_{TR} - r_{L\_T} \tag{17.17}$$

$$\frac{dS}{dt} = -\frac{r_{X_D}}{Y_{X_D S}} - \frac{r_{X_R}}{Y_{X_R S}} - \frac{r_{X_T}}{Y_{X_T S}} \tag{17.18}$$

With $X_{D,0}$, $X_{R,0}$, $X_{T,0}$, $S_0$ as initial conditions for, respectively, $X_D$, $X_R$, $X_T$, and $S$.

The latter situation can be extended to sequencing batch growth conditions to simulate a seasonal mode. In this case, batch culture conditions are applied for a certain time $t_{Cycle}$ representing a season, allowing cell densities to increase and substrate to deplete. At the beginning of a new season, a known dilution of all cell types is considered in a new growth medium. The end-values of the cell types of the previous simulation, adjusted for the appropriate dilution factor, are used as initial conditions for the new season. This is repeated for all seasons.

### 2.3. Analytical Inferences from the Differential Equations

While the governing equations described above are not mathematically complex, their explicit analytical solution is not trivial. Nevertheless, after some clever manipulations, Stewart and Levin (S&L) were able to infer a simple "persistence" condition for a conjugal plasmid in a chemostat-grown 2-population community consisting of only a plasmid-free and a plasmid-carrying fraction *(1, 2)*.

Using the current nomenclature, their criterion reads as follows:

$$\gamma_{\mathrm{TR}}(X_{\mathrm{R}} + X_{\mathrm{T}}) \geq \alpha D + \tau_T \tag{17.19}$$

Intuitively, this S&L criterion is satisfying and can be read as: the rate of transconjugant generation must exceed the rate of transconjugant loss due to segregation and washout. The criterion is even more tangible if it is reformulated as:

$$(X_{\mathrm{R}} + X_{\mathrm{T}}) \geq \frac{\alpha D + \tau_T}{\gamma_{\mathrm{TR}}} \tag{17.20}$$

In other words, it indicates that a chemostat needs to support a minimal cell density to ensure transconjugant persistence given a certain set of plasmid dynamic parameters ($\gamma$ and $\tau$) as well as a plasmid growth effect ($\alpha$). Note that in this criterion, the stoichiometric effect of plasmid presence is not present since stoichiometric effects do not affect plasmid persistence.

In later work, based on typical plasmid dynamic parameters and cell densities relevant to *Escherichia coli*, Levin and collaborators have conjectured that this existence criterion is typically not met and that other processes must be responsible for a maintenance of a plasmid in a microbial community *(7, 8)*. They further reject the plausibility that the selective advantage caused by plasmid-encoded beneficial genes (so-called focal genes) can explain plasmid maintenance. Over *evolutionary time-scales*, it is likely that such focal genes would have moved from the plasmid to the chromosome and such chromosomal variants would no longer experience the metabolic burden of the plasmid variants, ultimately outcompeting them and the plasmids they carry. Finally, in their comprehensive and mathematical analysis, they infer that only the situation of beneficial focal genes combined with periodic invasion of fitter cell types, which can serve as plasmid hosts (selective sweeps), can explain the long-term maintenance of plasmids *(7)*.

We do not *a priori* assume that the presented persistence criterion cannot be met because this inference remains based on a rather small set of plasmid dynamic parameters, limited essentially to *E. coli (11, 12)* and a low cell density ($10^6$ ml$^{-1}$) *(8)*. This might be significantly superseded in many metabolically active microbial environments. In addition, we are not solely interested in seeking the mechanisms that explain plasmid maintenance over evolutionary time-scales, but rather under extant conditions in currently existing microbial communities. Nevertheless, we wish to point out that the routines developed and presented here could be extended to include the existence and rise of a "chromosomal" cell type as well as the periodic invasion of new "fitter" recipients to capture longer-term processes governing plasmid persistence *(7)*.

### 2.4. The Developed Software Routines

While the existence criterion, proposed by Stewart & Levin *(2)*, is elegant, it does not allow testing of existence under other conditions than the above simplified existing ones. Under the latter conditions, numerical solutions are mandatory. We have developed a set of simple MatLab® routines, which allow the computationally untrained user to solve the equations of the mass-action-based model for both chemostat and seasonal transfer conditions. The routines permit variation of all of the growth and plasmid dynamic parameters and provide graphical and tabular output for rapid or detailed results analysis. *See* **Table 17.1** for a summary of the developed MatLab® procedures.

**Table 17.1**
**Summary of the developed MatLab® procedures**

| | |
|---|---|
| Processes modelled | Three cell types: D, R, T |
| | Two growth substrates: S1, S2 |
| | Growth and dilution, plasmid-transfer, plasmid-loss |
| | D, R, and T have distinct growth rate parameters, plasmid-carrying organisms can use S2 in addition of S1 |
| | $\gamma_{DR}$ and $\gamma_{TR}$ are distinct |
| | $\gamma$ and $\tau$ are constant values, or are dependent on $\mu$ |
| | Plasmid presence can affect specific growth rate ($\alpha$) and yield ($\beta$) |
| | Influent is assumed sterile while various influent dosages for $D$ and $S_{in}$ are possible (constant influent concentration, a time-specified function, a two-dimensional ($t$, $D_{in}$) array) |
| | Chemostat (**CS**) or sequential batch (**SB**) growth conditions |
| Input options | All growth and plasmid dynamic parameters can be modified, but default conditions are set. Parameters can be read from a MS Excel sheet or from a text file |
| | Initial concentrations of D, R, T |
| | **CS**: Dilution rate (Range), influent substrate concentration (Range), total run time |
| | **SB**: Season length, dilution ratio, initial substrate concentration, total run time |
| Output options | Figures showing concentration profiles of all species and substrates during a run |
| | Figures showing concentrations, fractions, existence conditions of all species across a range of parameters |
| | Data output files of the above, immediately written in MS Excel |

## 3. Effect of Plasmid Dynamic Parameters on Existence Conditions of a Plasmid Carrying Population in a Chemostat Environment

To illustrate the use of the model, several scenarios were examined to explore the sensitivity of the population composition toward different parameters. In this case, existence conditions for the establishment of a plasmid-carrying transconjugant population are plotted. The computational experiment consisted in initiating the chemostat (or batch culture, *see above*) with a large dose of plasmid free cells (ca. $10^9\ L^{-1}$) and a small dose of plasmid containing cells ($10\ L^{-1}$). The simulation was then carried out until steady-state conditions were attained (usually for a duration of 500 h), and the cell densities recorded. Under these conditions, the establishment of a plasmid-carrying population was scored as positive if $X_{T,500\,h} > 100$ cells $L^{-1}$. In the presented simulations, for simplicity sake, no donor population was assumed (i.e., the formation of transconjugant cells from a donor to recipient transfer event occurred some time before the simulation start).

Simulations were performed with the following "default" coefficients, reflecting growth and transfer of the F plasmid under exponential growth conditions in an *E. coli* recipient/transconjugant system *(1)*:

| Parameter/symbol | Value |
|---|---|
| $[\mu_{max,D}\ \mu_{max,R}]$ | $[0\ 0.738]\ h^{-1}$ |
| $[K_{s,D}\ K_{s,R}]$ | $[0\ 4]\ mg\ L^{-1}$ |
| $[Y_{xs,D}\ Y_{xs,R}]$ | $[0\ 1.69\ 10^9]$ cell $mg^{-1}$ |
| $[\gamma_{DR}\ \gamma_{TR}]$ | $[0\ 10^{-12}]\ L\ cell^{-1}\ h^{-1}$ |
| $\tau$ | $10^{-3}\ h^{-1}$ |
| $\alpha$ | 0.1 |
| $\beta$ | 0.1 |
| D | $0–1\ h^{-1}$ |
| $S_{in}$ | $0–100\ mg\ L^{-1}$ |
| $[X_{D0}\ X_{R0}\ X_{T0}]$ | $[0\ 10^9\ 10]$ cell $L^{-1}$ |
| Simulation time | 500 h |

### 3.1. Effect of Plasmid Transfer Rate Coefficient

In a first set of results, we show existence conditions as a function of the transconjugant to recipient plasmid transfer rate coefficient, $\gamma_{TR}$, keeping all other rate parameters constant.

In this, and all following chemostat simulations, the differential equations were solved repeatedly across a range of ($D$, $S_{in}$)

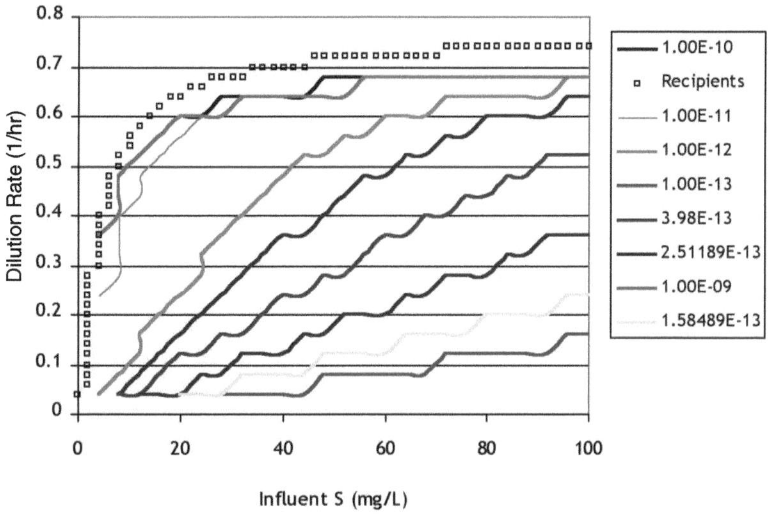


Fig. 17.1. The effect of $\gamma_{TR}$ (T-to-R plasmid transfer rate coefficient) on the existence conditions of transconjugants in chemostat growth conditions. Other parameters are $\alpha = 0.1$, $\beta = 0.1$, $\tau = 10^{-3}\,h^{-1}$.

combinations, and **Figs. 17.1, 17.2, 17.3, and 17.4** show the results at steady state. The line marked by the discrete symbols in **Figs. 17.1, 17.2, 17.3, and 17.4** indicate the *existence conditions for growth* in the chemostat: At all ($D$, $S_{in}$) combinations to the right of this line, growth in the chemostat is possible. At all ($D$, $S_{in}$) combinations to the left of this line, growth in the chemostat is not possible: dilution exceeds the maximum specific growth rate of the fastest growing cell type.

The continuous lines indicate the *existence conditions for plasmids* in the chemostat: at all ($D$, $S_{in}$) combinations to the right of this line, plasmids (transconjugants) are maintained in the chemostat. At all ($D$, $S_{in}$) combinations to the right of this line, plasmid (transconjugant) maintenance is not possible: the processes of plasmid (transconjugant) loss – dilution, segregation – exceeds those of plasmid (transconjugant) accrual – growth and *de novo* generation due to T-to-R transfer.

The results of these simulations are immediately obvious: a wider region of plasmid existence is observed as the plasmid transfer rate coefficient increases. At a $\gamma_{TR}$ value of $10^{-10}$ L cell $h^{-1}$, the plasmid nearly mirrors the existence conditions for growth (i.e., the plasmid can invade the recipient population under all conditions), and further increases if $\gamma_{TR}$ does not expand the existence region. The existence conditions steeply drop as $\gamma_{TR}$ decreases. At values below $10^{-13}$ L cell $h^{-1}$ the plasmid is no longer retained under any of the examined ($D$, $S_{in}$) conditions.

The shape of the existence regions is consistent with the S&L criterion: plasmid maintenance at a lower plasmid transfer rate

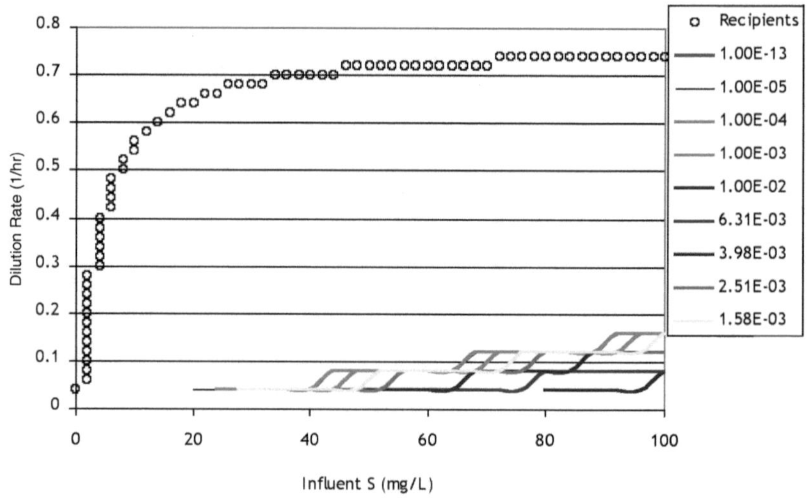


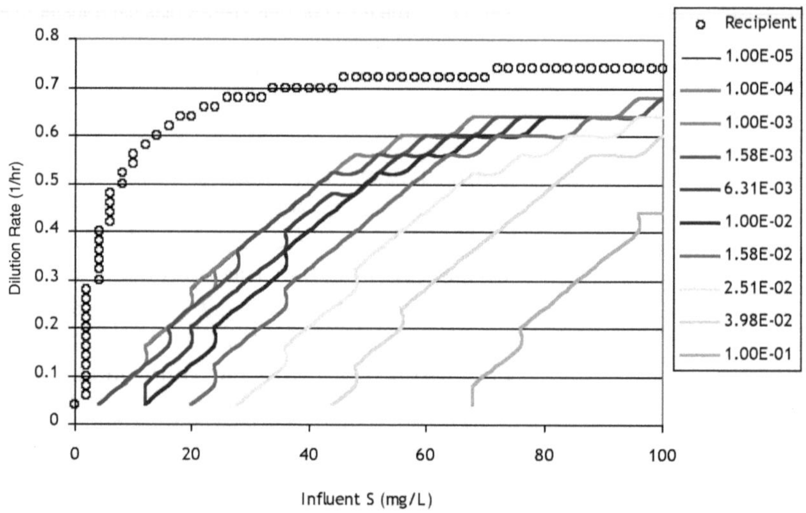


Fig. 17.2. The effect of $t$ (plasmid loss rate by $T$) on the existence conditions of transconjugants in chemostat growth conditions. Other parameters are: $\alpha = 0.1$, $\beta = 0.1$ and $\gamma_{TR}$: $10^{-13}\,\text{L cell}^{-1}\,\text{h}^{-1}$ (*top panel*) or $10^{-12}\,\text{L cell}^{-1}\,\text{h}^{-1}$ (*bottom panel*).

requires higher cell densities than at lower plasmid transfer rates, with all other parameters being equal. Indeed, cell density distributions in the $(D, S_{in})$ domain for a chemostat mirror the existence conditions shown in **Fig. 17.1** (results not shown).

### 3.2. Effect of Plasmid Loss Rate Coefficient

These simulation results indicate that a reduction in the plasmid loss rate coefficient alone cannot counteract the effect of a low plasmid transfer rate coefficient: at a $\gamma_{TR}$ value of $10^{-13}\,\text{L cell}^{-1}\,\text{h}^{-1}$, a reduction of the loss rate coefficient below $10^{-3}\,\text{h}^{-1}$ (maximal plasmid loss rate where existence is feasible,

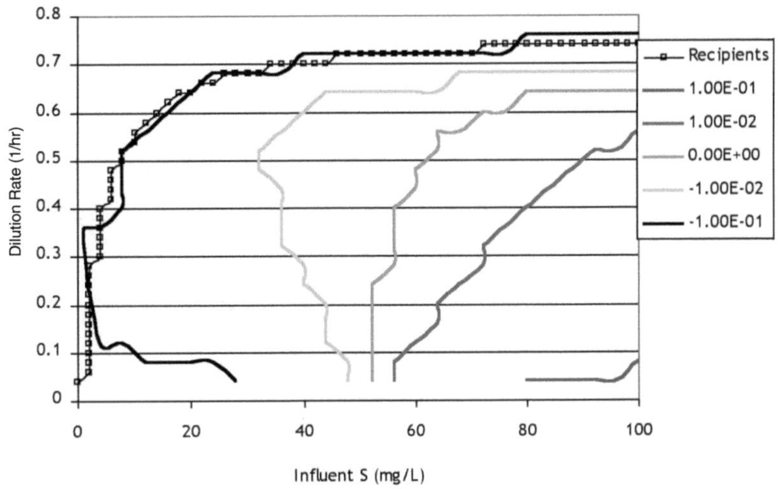


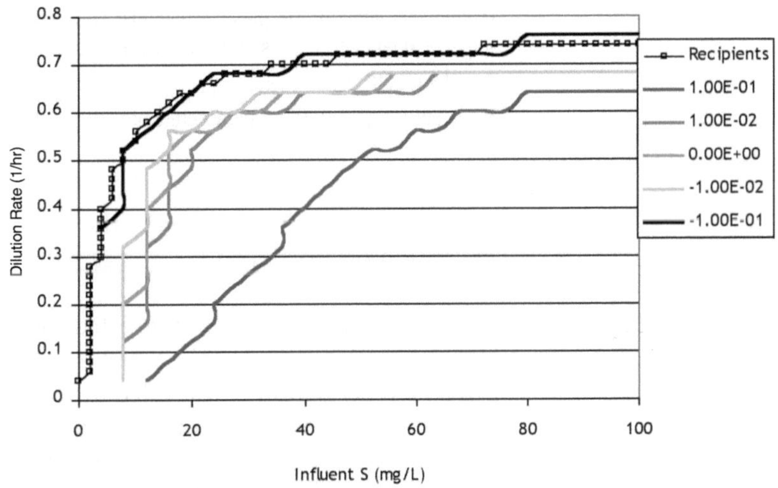


Fig. 17.3. The effect of $\alpha$ (plasmid effect on T-specific growth rate) on the existence conditions of transconjugants in chemostat growth conditions. Other parameters are: $\beta = 0.1$, $\tau = 10^{-2}\,\mathrm{h}^{-1}$, $D_{\mathrm{in}} = 0$ and $\gamma_{\mathrm{TR}}4$: $10^{-13}\,\mathrm{L\,cell^{-1}\,h^{-1}}$ (*top panel*) or $10^{-12}\,\mathrm{L\,cell^{-1}\,h^{-1}}$ (*bottom panel*).

**Fig. 17.1**) does not significantly enhance the region of plasmid existence (**Fig. 17.2, Panel A**). This is due to the fact that loss rates that are lower than the value of the dilution rate have a marginal effect on the plasmid density. Only loss rate coefficient of similar order of magnitude as dilution rate can affect the steady-state transconjugant population density.

This phenomenon can easier be observed at higher plasmid rate coefficients (**Fig. 17.2, Panel B**). At a $\gamma_{\mathrm{TR}}$ value of $10^{-13}\,\mathrm{L\,cell^{-1}\,h^{-1}}$, the plasmid existence is ensured up to $\tau$ values as high as $0.1\,\mathrm{h}^{-1}$. A significant expansion of the existence

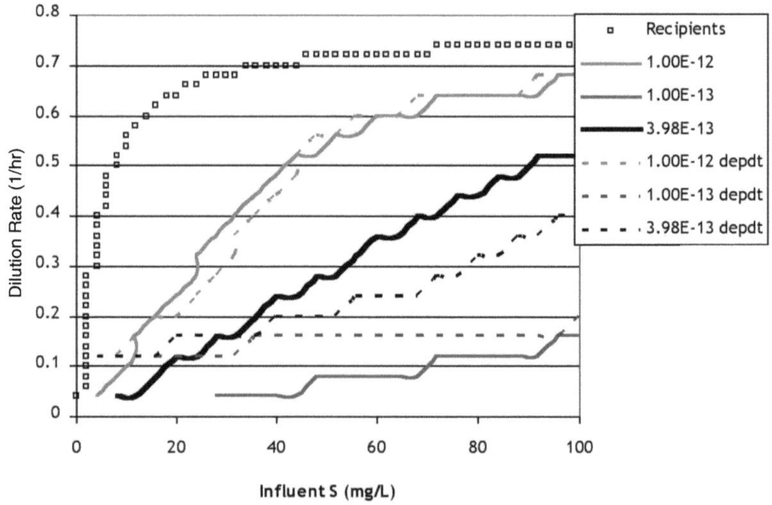


Fig. 17.4. The effect of the growth rate dependency (*dashed lines*) or independency (*continuous lines*) of the plasmid dynamic parameters on the existence conditions of transconjugants in chemostat growth conditions. $\gamma_{TR,\,max}$: varies between $10^{-13}$ and $10^{-12}$ L cell$^{-1}$ h$^{-1}$, while $\gamma_{TR,\,min}$ was set at $10^{-15}$ cell$^{-1}$ h$^{-1}$ $\tau_{max} = 10^{-3}$ h$^{-1}$ and $\tau_{min} = 0$. Other parameters are $\alpha = 0.1$, $\beta = 0.1$, $X_{Din} = 0$.

region occurs with reduction of the $t$ down to $0.01$ h$^{-1}$, but little expansion occurs beyond.

### 3.3. Effect of Fitness Loss/Gain

On the other hand, the fitness effect of plasmid carriage can have dramatic effects on plasmid existence. In case of a negative fitness effect, any reduction in this effect will result in an expansion of the existence region (e.g., **Fig. 17.3A** from 0.0 to $10^{-2}$ and from $10^{-2}$ to $10^{-1}$ to 0.0). In addition, if the fitness effect becomes positive (the plasmid encodes at least one beneficial gene), the existence region may rapidly expand to enclose the entire growth existence domain (e.g., **Fig. 17.3A** from 0.0 to $-10^{-2}$ and from $10^{-2}$ to $10^{-1}$). Further, the effect of fitness will be larger if the plasmid dynamic conditions are less favorable (**Fig. 17.3A** vs **Fig. 17.3B**). These results, again, indicate that the effect of one parameter on plasmid existence is dependent on and must consider the other parameters.

### 3.4. Effect of Parameter Dependency on Physiological State

In all the above situations, we assumed a constant value of both the plasmid transfer rate coefficient and the plasmid loss rate coefficient, which might be appropriate if cells are growing at a constant specific growth rate *(11)*. However, significant evidence accrues that these parameters may vary significantly, depending on the specific growth rate of the plasmid-harboring cells. Such dependency is intuitively obvious for $\tau$, because segregational loss is contingent on cell division and the number of cell divisions per unit time is related to the specific growth rate. Hence, an increase in plasmid transfer loss with increased specific growth rate is surmised. On the other hand, in the few studies where it has carefully been examined, it has also been observed that $\gamma_{DR}$ or $\gamma_{TR}$ vary with the donor or transconjugant population's specific growth rate with an increased value at higher specific growth rate *(1, 13–16)*. This dependency might be caused by the specific expression of conjugal transfer genes or the removal of cellular energy resources limitation to permit transfer at higher growth rates. The impact of the recipient's specific growth rate has not been rigorously examined.

Mathematically, this growth rate dependency can be incorporated readily:

$$\gamma_{DR} = \gamma_{min,DR} + \left(\gamma_{max,DR} - \gamma_{min,DR}\right) \frac{\mu_D}{\mu_{max,D}} \qquad (17.21)$$

$$\gamma_{TR} = \gamma_{min,TR} + \left(\gamma_{max,TR} - \gamma_{min,TR}\right) \frac{\mu_T}{\mu_{max,T}} \qquad (17.22)$$

$$\tau_T = \tau_{min,T} + \left(\tau_{max,T} - \tau_{min,T}\right) \frac{\mu_T}{\mu_{max,T}} \qquad (17.23)$$

In these equations, two intrinsic values for each $\gamma$ are defined. A maximum value, attained when $\mu = \mu_{max}$, and a minimal value, attained when $\mu = 0$. The actual value of $\gamma$, at any given value of $\mu$, varies linearly between these two extremes. A similar approach is used to define the $\tau/\mu$ dependency. Other dependencies can easily be formulated and incorporated in the software platform. Based on some earlier report, we set the value of $\gamma_{min}$ to 1% or 1‰ the value of $\gamma_{max}$ with $\gamma_{max}$ at $0.1 \times 10^{-12}$ *(1, 13, 14)*.

In this case, the results indicate a slightly different existence region for the plasmid (**Fig. 17.4**). While, for the highest considered $\gamma_{TR}$ value, growth dependency does not affect the existence regions, the situation varies slightly at lower $\gamma_{TR}$ values. The simulations indicate that plasmid existence is extended to lower regions in the ($D$, $S_{in}$) domain. This result suggests that the reduction in plasmid loss (due to lower values of $t$) offsets the simultaneously decreasing values of the $\gamma_{TR}$, both to lower specific growth rates at lower dilution rates. Clearly, these results confirm the need to rigorously examine the true relationship between plasmid dynamic parameters and cell physiological state.

## 4. Effect of Plasmid Dynamic Parameters on Existence Conditions of a Plasmid-Carrying Population in a Seasonal Environment

Chemostat cultivation offers very special stable conditions to microorganisms, which may be far removed from most environmental conditions, where substrate supply and the resulting growth rates vary periodically. Hence, it may need to be questioned whether dynamic environmental conditions alter the trends for existence conditions described in former paragraphs. To model seasonal conditions, constant feed and dilution are replaced by cyclic injections of substrate and biomass dilutions (**Fig. 17.5**).

To be able to compare the trends between seasonal and chemostat growth conditions, an apparent dilution rate, $\hat{D}$, is defined:

$$\hat{D} = \frac{1 - \frac{1}{f_D}}{t_{Cycle}} \tag{17.24}$$

where $t_{Cycle}$ is the length of a cycle and $f_D$ is the dilution factor.

A first observation is that the global dynamic of transconjugant existence is less permissive than under chemostat conditions (**Fig. 17.6**). Compared to existence domains obtained previously, the area is smaller. We also notice a very strong and negative effect of the growth-rate dependency on the plasmid existence region. This can be explained by the fact that during each individual cycle, the substrate concentration changes from initial values to quasi-depletion, meaning that the growth-rate range is high. As a consequence, the effect of growth-rate dependency is more visible than in the chemostat case. As expected, the growth-rate dependency affects not only the existence domain (**Fig. 17.7** in black) but the also dominance domain (**Fig. 17.7** in gray) of transconjugants.

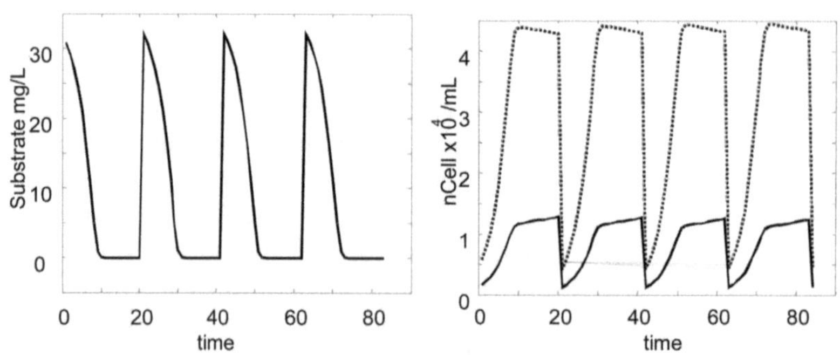


Fig. 17.5. Example of seasonal variations for substrate concentration (*left*) and cell concentrations (*dashed line*: R, *continuous line*: T).

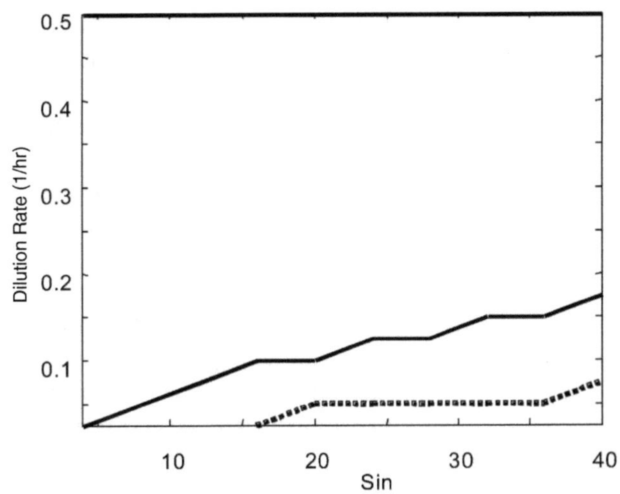


Fig. 17.6. Existence conditions (*dashed line*: constant, *continuous line*: growth-rate dependent parameters) for $\tau = 10^{-3}$ and $\gamma = 10^{-12}$. $D$ refers to the equivalent dilution rate, $\hat{D}$, obtained with a $10\times$ dilution and varying season lengths (from 1 to 10 h), $S_{\text{in}}$ is the initial concentration at each cycle.

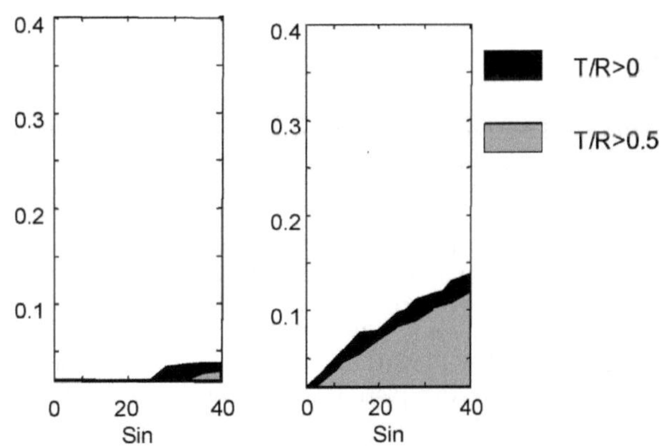


Fig. 17.7. T/R ratio with growth-rate dependent (*left*) and constant (*right*) parameters. $D$ refers to the equivalent dilution rate, $\hat{D}$, obtained with a $10\times$ dilution and a varying season length (from 1 to 10 h), $S_{\text{in}}$ is the initial concentration at each cycle.

## 5. Conclusion/ Outlook

Mass action models provide a useful pseudo-mechanistic framework to describe and predict the persistence of conjugal plasmids (as well as non-conjugal plasmids, by simple model extensions) in well-mixed model microbial communities. While solving the mass action model equation is not complex, it is outside the comfort zone of many microbiologists examining

HGT. We have developed user-friendly MatLab® routines that, we hope, can bring their use within reach of a larger community. In combination with a much larger experimental effort toward quantitative studies on extant horizontal transfer, this accessibility will allow us to validate or further refine these models. We plan to augment our routines by providing routines that allow inference of plasmid dynamic parameters (such as plasmid transfer and plasmid loss rate) from optimally designed experiments because there currently exits a true paucity of such experiments. Finally, we recognize that the mass action models presented here are unsuitable to describe and predict plasmid transfer in structured microbial communities. For such approaches, individual-based models are more appropriate, which is a subject of enquiry of both our and other research groups.

## Acknowledgments

This work was funded, in part, by EC-FP-7 as part of a Marie Curie Excellence Grant (MEXT-CT-2005-024004, RaMAda) to BFS. We wish to thank Jose Seoane for editorial assistance.

### References

1. Levin, B. R., Stewart, F. M., Rice, V. E. (1979) The kinetics of conjugative plasmid transmission: Fit of a single mass action model. *Plasmid* **2**, 247–60.
2. Stewart, F. M., Levin, B. R. (1977) The population biology of bacterial plasmids. *A priori* conditions for the existence of conjugationally transmitted factors. *Genet* **87**, 209–28.
3. Levin, B. R., Rice, V. A. (1980) The kinetics of transfer of nonconjugative plasmids by mobilizing conjugative factors. *Genet Res Camb* **35**, 241–59.
4. Levin, B. R., Stewart, F. M. (1980) The population biology of bacterial plasmids: *A priori* conditions for the existence of mobilizable nonconjugative factors. *Genet* **94**, 425–43.
5. Lundquist, P. D., Levin, B. R. (1986) Transitory depression and the maintenance of conjugative plasmids. *Genet* **113**, 483–97.
6. Lagido, C., Wilson, I. J., et al. (2003). A model for bacterial conjugal gene transfer on solid surfaces. *FEMS Microbiol Ecol* **44**(1), 67–78.
7. Bergstrom, C. T., Lipsitch, M., Levin, B. R. (2000) Natural selection, infectious transfer and the existence conditions for bacterial plasmids. *Genet* **155**, 1505–19.
8. Simonsen, L. (1991) The existence conditions for bacterial plasmids: Theory and reality. *Microb Ecol* **22**, 187–205.
9. Dahlberg, C., Chao, L. (2003) Amelioration of the cost of conjugative plasmid carriage in *eschericha coli* k12. *Genet* **165**, 1641–9.
10. Dionisio, F., Conceicao, I. C., Marques, A. C., Fernandes, L., Gordo, I. (2005) The evolution of a conjugative plasmid and its ability to increase bacterial fitness. *Biol Lett* **1**, 250–2.
11. Simonsen, L., Gordon, D. M., Stewart, F. M., Levin, B. R. (1990) Estimating the rate of plasmid transfer: An end-point method. *J Gen Microbiol* **136**, 2319–25.
12. Gordon, D. M. (1992) Rate of plasmid transfer among *escherichia coli* strains isolated from natural populations. *J Gen Microbiol* **138**, 17–21.
13. Freter, R., Freter, R. R., Brickner, H. (1983) Experimental and mathematical models of *Escherichia coli* plasmid transfer in vitro and in viro. *Infect Immun* **39**, 60–84.
14. Smets, B. F., Rittmann, B. E., Stahl, D. A. (1993) The specific growth rate of *Pseudomonas putida* paw1 influences the conjugal transfer rate of the tol plasmid. *Appl Environ Microbiol* **59**, 3430–7.

15. Smets, B. F., Rittmann, B. E., Stahl, D. A. (1995) Quantification of the effect of substrate concentration on the conjugal transfer rate of the tol plasmid in short-term batch mating experiments. *Lett Appl Microb* **21**, 167–72.

16. Macdonald, J. A., Smets, B. F., Rittmann, B. E. (1992) The effects of energy availability on the conjugative transfer kinetics of plasmid rp4. *Wat Res* **26**, 461–8.

# Part IV

## Case Studies

# Chapter 18

# Interdomain Transfers of Sugar Transporters Overcome Barriers to Gene Expression

**Kenneth M. Noll and Kamolwan Thirangoon**

## Abstract

Horizontal gene transfer (HGT) is nature's mechanism for sharing evolved physiological traits among the members of microbial communities. The extent to which such transfers can be successful is best illustrated by the fact that Archaea-derived genes are found in many bacterial genomes, particularly those in the hyperthermophile *Thermotoga maritima*. The success of these intergenomic transfers depends upon the successful transcription of the newly acquired archaeal genes using a bacterial transcription machinery that does not recognize archaeal transcriptional signals. To examine how nature solves this problem, we looked to the *T. maritima* genome for examples of interdomain transfers. Here we lay the groundwork to examine this problem by more clearly delineating the phylogenetic history of Archaea-derived transporter genes in this genome. We find that five of these polysaccharide transporters were derived from the Archaea and one came from the Archaea after that lineage inherited it from the Bacteria. These data can be used for more detailed examinations of the recombinations that allowed these transporters to be expressed in a bacterial host. This work will guide examinations of the genome sequences from other members of the Thermotogales, which will become available.

**Key words:** Thermotogales, *Thermotoga*, ABC transporters, horizontal gene transfer, promoter evolution.

## 1. Introduction

Horizontal gene transfer (HGT) is recognized as an important force in prokaryote evolution, although the magnitude of its impact is a subject of lively debate *(1, 2)*. Recognizable HGT occurs when a foreign gene integrates into a recipient's genome and is stably transmitted through subsequent generations of progeny *(3–5)*. When foreign genes are acquired by HGT, the new genes may confer selective advantages that dramatically alter

Maria B. Gogarten et al. (eds.), *Horizontal Gene Transfer: Genomes in Flux, vol. 532*

DOI 10.1007/978-1-60327-853-9_18 Springerprotocols.com

the physiology of the recipient *(6)*. To persist in the lineage, these genes must rapidly adapt to the physiology of the host, changing codons in accord with biases of the foreign translational machinery *(7)* and integrating their new functions into the pre-existing metabolism of their host.

Interdomain transfers additionally require that the new genes acquire radically different promoters and respond to new regulatory signals *(8)*. The discovery of the extensive occurrence of interdomain HGT in microbial genomes presents a challenge to discover how the newly arrived genes acquire alien transcriptional control signals. There are now numerous examples of HGT between Bacteria and Archaea *(9–12)*, yet there is no comprehensive examination of how these genes came to be expressed in their new domain. To carry out such studies, it would be useful to identify the closest living relative of the lineage that first received the gene. This is difficult for genes that are now widely distributed among lineages via vertical inheritance. In this respect, the bacterial lineage Thermotogales presents unique advantages. By several measures, it is a basal bacterial lineage and it also lives in environments largely isolated from other bacterial groups, but among archaea. Consequently, its members are likely to be the primary recipients of archaeal genes.

It was suggested that the genome sequence of the hyperthermophilic bacterium, *Thermotoga maritima*, shows evidence of extensive acquisition of genes from the Archaea *(13, 14)*. *T. maritima* is an obligately anaerobic, hyperthermophilic bacterium that can utilize simple and complex carbohydrates *(13, 15–17)* and thrives in harsh environments where archaea are relatively abundant. It is not surprising that a significant fraction of the genes that it would acquire by HGT would be derived from those archaea with which it lives.

Archaea-derived genes are clustered in 15 regions in the *T. maritima* chromosome and are not equally distributed among functional categories. The largest functional group of these genes was assigned as encoding oligopeptide ABC (ATP binding cassette) transporters *(13)*. The *T. maritima* chromosome encodes eleven putative oligopeptide ABC transporters arranged in apparent operons that, with one exception, encode three components: a substrate-binding protein (SBP), two membrane-spanning proteins (MSP), and two ATP-binding proteins (ABP) (**Fig. 18.1**). We have found that several of the Archaea-derived oligopeptide transporters are likely oligosaccharide transporters based upon the binding affinities of their encoded SBPs *(18)*, confirming the earlier demonstration of this in archaea *(19)*. We shall refer to these transporters as oligosaccharide transporters.

These transporters provide excellent examples of interdomain HGT with which to examine how such genes come to be expressed in a foreign transcriptional environment. Our working

*Archaea*-derived oligosaccharide transporter operons

| Operon | ABC loci | | | | |
|---|---|---|---|---|---|
| TM1 | *TM0031* | *TM0030* | *TM0029* | *TM0028* | *TM0027* |
| TM2 | *TM0071* | *TM0072* | *TM0073* | *TM0074* | *TM0075* |
| TM3 | *TM0300* | *TM0301* | *TM0302* | *TM0303* | *TM0304* |
| TM4 | *TM1199* | *TM1198* | *TM1197* | *TM1196* | *TM1194* |
| TM5 | *TM1223* | *TM1222* | *TM1221* | *TM1220* | *TM1219* |
| TM6 | *TM1153* | *TM1152* | *TM1151* | *TM1150* | *TM1149* |

*Bacteria*-derived oligopeptide transporter operons

| Operon | ABC loci | | | | |
|---|---|---|---|---|---|
| TM7 | *TM0060* | *TM0059* | *TM0058* | *TM0057* | *TM0056* |
| TM8 | *TM0503* | *TM0502* | *TM0501* | *TM0500* | *TM0498* |
| TM9 | | *TM0530* | *TM0531* | *TM0532* | *TM0533* |
| TM10 | *TM1067* | *TM1066* | *TM1065* | *TM1064* | *TM1063* |
| TM11 | *TM1746* | *TM1747* | *TM1748* | *TM1749* | *TM1750* |

Fig. 18.1. Gene order in *T. maritima* transporter operons. Transcription proceeds from left to right in the order shown for each operon. Genes encoding SBPs are indicated in *dark gray* boxes, MSPs are in *light gray* boxes, and ABPs in *white* boxes. The *bars* between TM1196–TM1194 and TM0500–TM0498 indicate the locations of TM1195 (a bacterial $\beta$-galactosidase) and TM0499 (a hypothetical protein of 162 amino acids), respectively. TM9 encodes only one ABP.

hypothesis is that these archaeal operons recombined downstream of promoters controlling analogous bacterial ABC transporter genes. Initially, these new operons would assume the transcriptional controls of their newly acquired promoters. In subsequent generations, those controlling elements could change, allowing positive selection for those organisms that could best utilize the new transporters. To test this working hypothesis, we need to better understand the evolutionary history of these transporters. The work described here provides this information.

BLAST analyses of these oligosaccharide transporter proteins revealed that six of these putative operons are most similar to archaeal homologs *(13, 20)*. Using this information as a starting point, we undertook a more comprehensive analysis of the evolution of these oligosaccharide transporters in *T. maritima* by examining the phylogeny of their MSPs and ABPs. SBP sequences are the least conserved of the ABC domains while those of the ABPs are the most conserved *(21, 22)*, so we used ABP and MSP sequences for this preliminary analysis of the evolutionary history of these transporters. We set out to determine if these oligosaccharide ABC transporters were inherited from the Archaea and their relationships to one another. We also sought evidence that these genes transferred as operons.

## 2. Methods

### 2.1. Sequence Acquisition and Preliminary Analyses

These analyses used amino acid sequences obtained from the Genome Database at http://www.tigr.org/. *T. maritima* sequences were searched against the non-redundant database at NCBI using BLAST *(23, 24)* to retrieve putative homologs. Four datasets, ABP1, ABP2, MSP1, and MSP2, were assembled to contain only those sequences of genes that are in apparent ABC transporter operons containing all four genes. These sequences were aligned in Clustal X version 1.83 *(25, 26)* and manually edited to remove positions with gaps and to improve the alignments. The sequences in the MSP1 dataset were concatenated with their operon partners in the MSP2 dataset in the orientation MSP1–MSP2. The final dataset is called the "concatenated MSP dataset" in this study.

### 2.2. Phylogenetic Analyses of the ABP1, ABP2, and Concatenated MSPs Datasets

Each dataset was subjected to tree construction using the following programs: PROTPARS and PROTDIST from the Phylip package *(27)*, TREE-PUZZLE, PHYML, *(28)* and MrBayes *(29)*. For the analyses using PROTPARS, PROTDIST, and TREE-PUZZLE, 100 copies of the alignments were created by SEQBOOT from the Phylip package. The most parsimonious trees were calculated in PROTPARS with 25 jumbles ($J = 25$) and 25 starting sequences (seed = 25). The ML distance matrices and distance matrices were calculated in TREE-PUZZLE and PROTDIST, respectively. Multiple ML distance matrices were calculated using PUZZLEBOOT, version 1.03 *(30)*, VT substitution model. The alpha parameter was incorporated in the analysis using TREE-PUZZLE. The coefficient of variation (CV = 1/sqrt (alpha)) was incorporated in the analysis using PRODIST. The alpha value and (CV value) of the ABP1, ABP2, and the concatenated MSPs datasets are 0.80 (1.118), 0.68 (1.213), and 1.49 (0.819), respectively. Distance matrices were used to construct trees using NEIGHBOR from the Phylip package, and consensus trees were determined by CONSENSE also from the Phylip package. The ML phylogenies of each dataset were constructed using PHYML *(28)*, JTT substitution model, and 100 replicates (bootstrapping). The same alpha values that were incorporated in this analysis were the same as the values used in TREE-PUZZLE. The ML consensus trees were determined by CONSENSE. The majority rule consensus trees of each dataset were calculated thrice in MrBayes using: 500,000 generations; four chains; gamma distribution; JTT substitution model, tree sampling every 50 trees generated; and burnin of 3124, 4073, and 1234 generations for the trees from ABP1, ABP2, and the concatenated MSPs datasets, respectively.

### 2.3. Hydropathy Plots of SBPs

Amino acid sequences of SBPs were analyzed for putative membrane-spanning domains using Kyte-Doolittle *(31)* and TMAP *(32)* analyses.

## 3. Results and Discussion

### 3.1. Archaeal-Type Oligosaccharide ABC Transporter Operons Have a Different Structure than Their Bacterial Counterparts

The genes encoding oligosaccharide and oligopeptide ABC transporters are arranged in clusters that are presumably co-transcribed as operons (**Fig. 18.1**). Six of these (TM1–6), the oligosaccharide transporters, have been said to have the most similarity to archaeal genes based upon simple BLAST analysis *(13)*. Five other apparent oligopeptide transporter operons (TM7–11) are most similar to bacterial sequences. Most of the archaeal-type operons (TM1–5) have the same order of genes, SBP-MSP1-MSP2-ABP1-ABP2, while these bacterial-type operons show little similarity of gene order. Indeed, TM8 lacks a gene encoding an SBP and TM9 has only one gene encoding an ABP, so TM8 and TM9 were not included in our analyses since they have lost genes.

### 3.2. The Archaeal-Type ABC Transporter Genes Did Not All Evolve Together in Operons

We found that the concatenated MSP tree agreed with trees derived from the individual MSP1 and MSP2 datasets, suggesting the two MSP proteins have the same evolutionary history and so evolved as a unit. To determine the evolutionary history of the transporters, the concatenated MSP, the ABP1, and the ABP2 datasets were analyzed using protein parsimony (PROTPARS), parsimony distance matrix (PROTDIST), ML distance matrix (PUZZLE), and Bayesian analysis (MrBayes) methods. All the analyses of the concatenated MSP dataset placed the TM1–TM5 sequences among those from Archaea to the exclusion of most sequences from Bacteria as seen in an unrooted, most probable concatenated tree (**Fig. 18.2**). Genes from *Thermoanaerobacter tengcongensis, Thermobifida fusca*, and *Chloroflexus aurantiacus* also clustered near archaeal-type genes, suggesting transfers to these thermophilic bacteria also occurred. This pattern of a few bacterial sequences among the branches of a largely archaeal clade supports the hypothesis that HGT of the TM1–5 genes was from the Archaea to the Thermotogales. The trees are insufficiently resolved to indicate how frequently such transfers occurred. TM2–4 consistently cluster together, suggesting they arose after an interdomain transfer by operon duplication or HGT among *Thermotoga* species. TM5 may have been acquired by a separate transfer from Archaea, but the resolution of these analyses is insufficient to strongly support this hypothesis.

By contrast, the TM6 sequences positioned with strong support within the larger bacterial clade along with sequences from

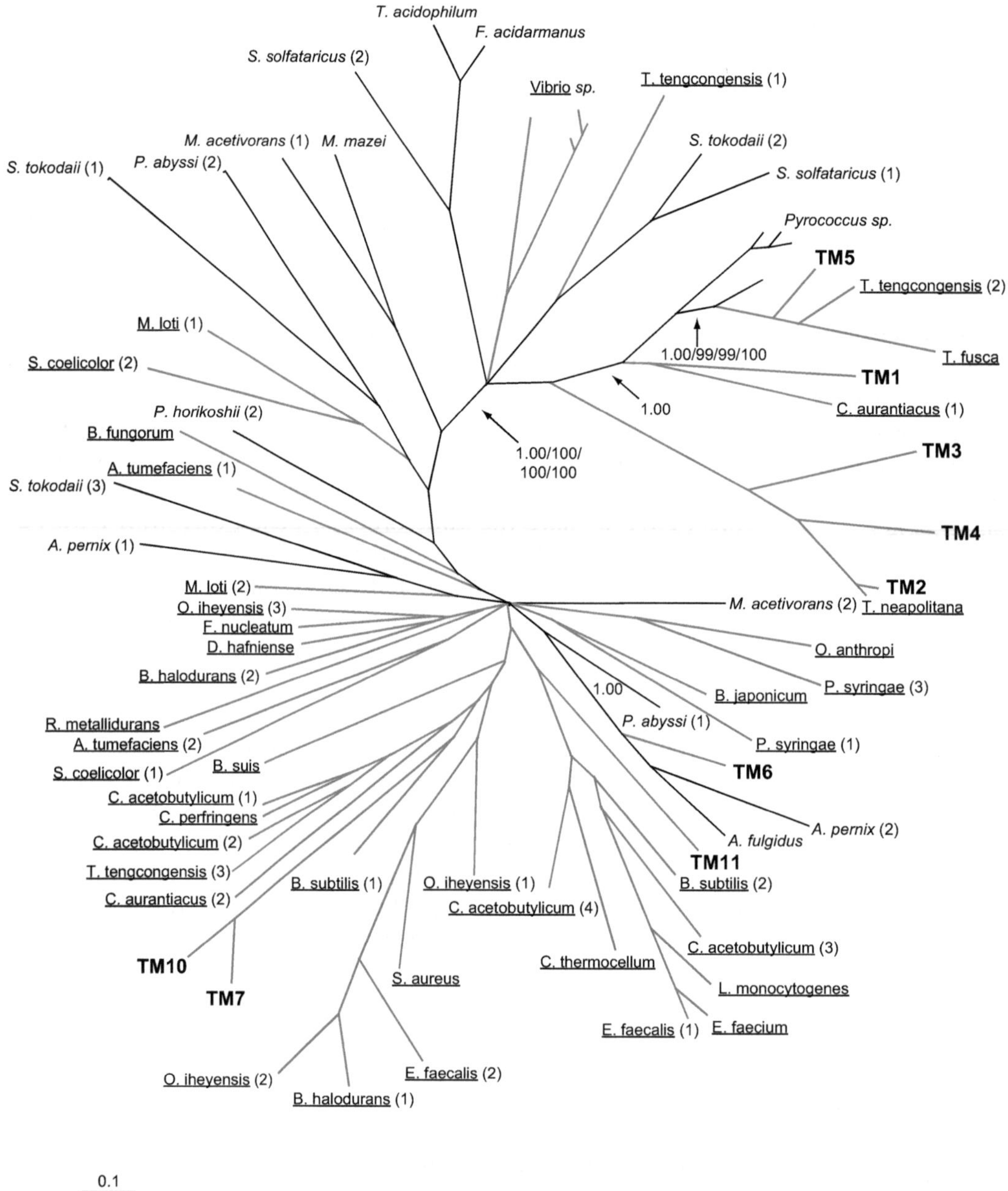


Fig. 18.2. Unrooted majority consensus concatenated MSP phylogeny. This phylogeny was calculated in MrBayes as described. Archaeal lineages are indicated by *black lines* and organisms' names are in italics. Bacterial lineages are indicated by *gray lines* and organisms' names are *underlined.* Sequences from the *T. maritima* polysaccharide ABC operons are in *bold face.* The numbers indicate posterior probabilities. Numbers in parentheses indicate the number of different sequences from the same organism used in this analysis. The same overall tree topology was obtained by parsimony analysis, parsimony, and ML distance methods in that the grouping of the sequences is the same. The support values are posterior probabilities from MrBayes, bootstrap values from TREE-PUZZLE, PROTDIST, and PROTPARS.

the archaea *Pyrococcus abyssi*, *Archaeoglobus fulgidus*, and *Aeropyrum pernix* (**Fig. 18.2**). This suggests that the Archaea received the TM6 ancestor from the Bacteria and that a *T. maritima* ancestor inherited it from an archaeon.

The grouping of the TM1–5 ABP homologs is similar to that seen with the MSPs (**Figs. 18.3 and 18.4**). This suggests that these genes may have been transferred together in operons. However, the TM6 operon shows evidence of a mosaic evolution. The TM6 ABP2 sequence grouped with the same archaeal homologs in a bacterial branch as did its MSP counterparts. In contrast, the TM6 ABP1sequence clustered with the TM1–TM5 sequences and sequences from the pyrococci. The incongruence

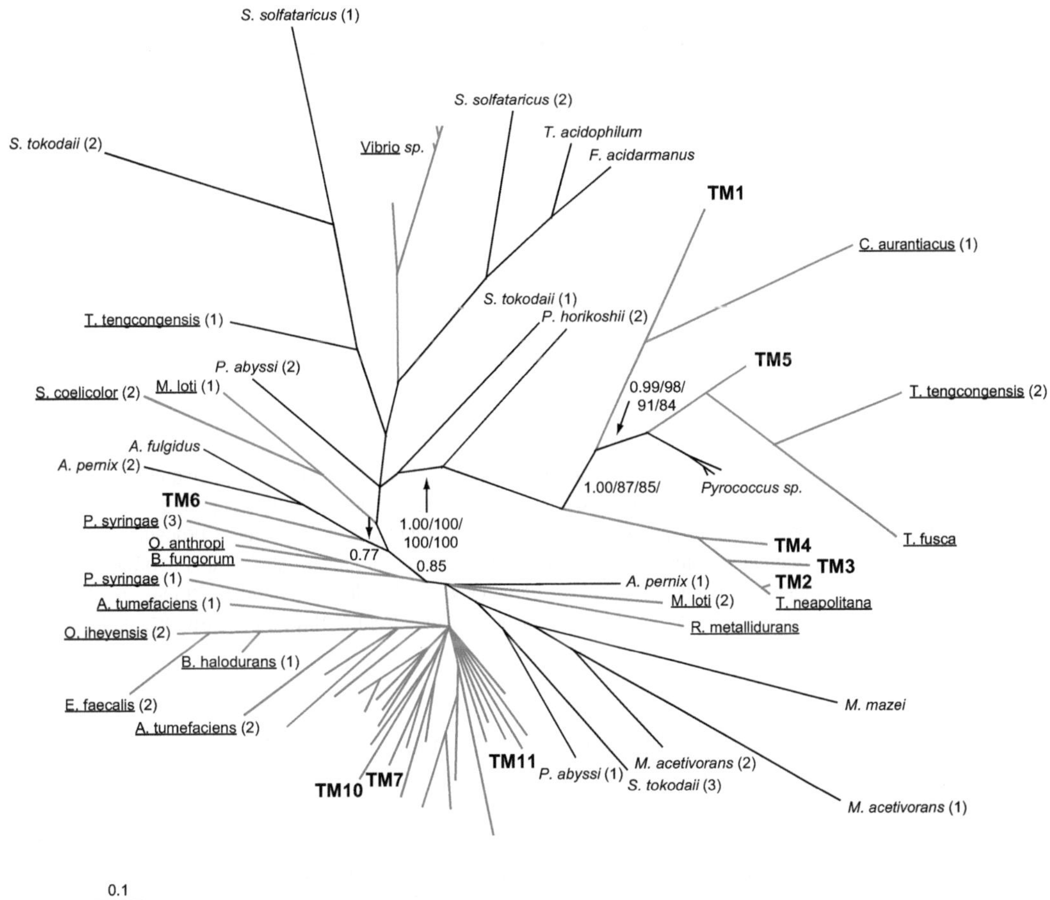


Fig. 18.3. Unrooted aligned ABP2s majority rule consensus phylogeny. This phylogeny was calculated using MrBayes as described. Archaeal lineages are indicated by *black lines* and organisms' names are in italics. Bacterial lineages are indicated by *gray lines* and organisms' names are *underlined*. The numbers indicate posterior probabilities. Numbers in parentheses indicate the number of different sequences from the same organism used in this analysis. The same overall tree topology was obtained by parsimony analysis, parsimony, and ML distance methods in that the grouping of the sequences is the same. The support values are posterior probabilities from MrBayes, bootstrap values from TREE-PUZZLE, PROTDIST, and PROTPARS.

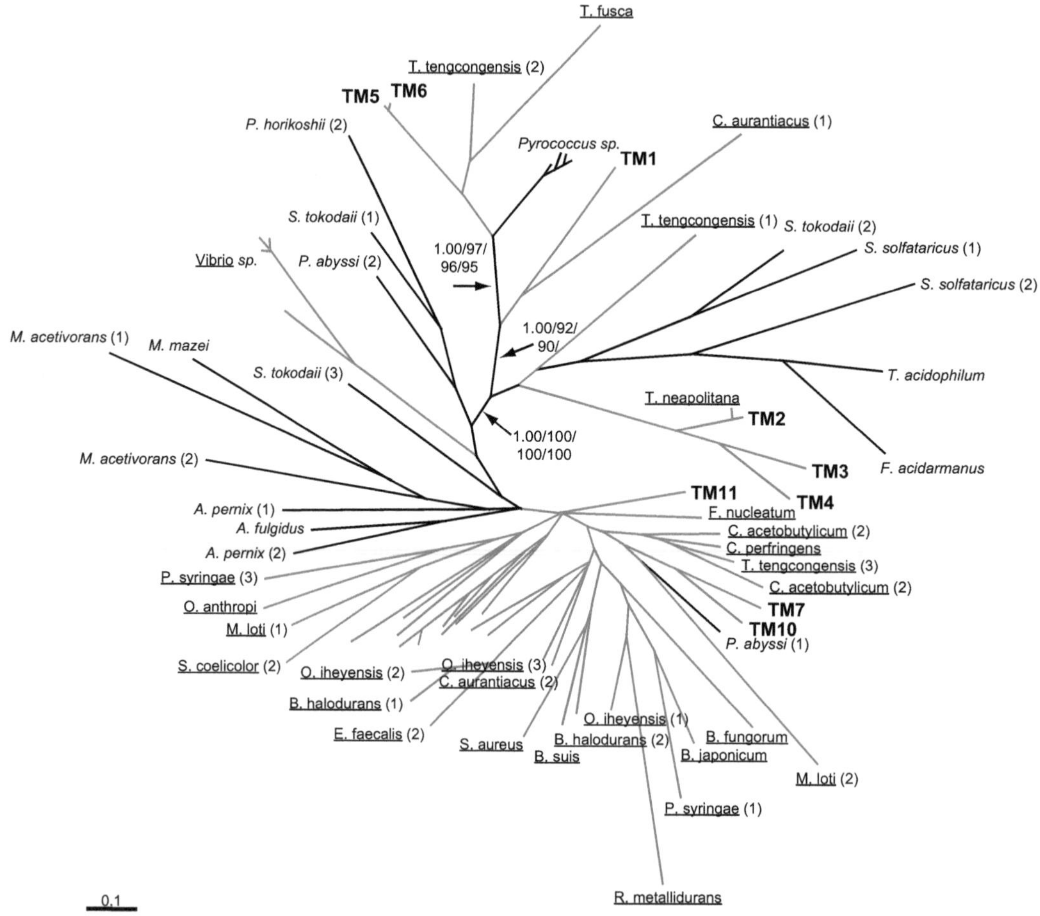


Fig. 18.4. Unrooted aligned ABP1s majority rule consensus phylogeny. This phylogeny was calculated using MrBayes as described. Archaeal lineages are indicated by *black lines* and organisms' names are in italics. Bacterial lineages are indicated by *gray lines* and organisms' names are *underlined*. The numbers indicate posterior probabilities. Numbers in parentheses indicate the number of different sequences from the same organism used in this analysis. The same overall tree topology was obtained by parsimony analysis, parsimony, and ML distance methods in that the grouping of the sequences is the same. The support values are posterior probabilities from MrBayes, bootstrap values from TREE-PUZZLE, PROTDIST, and PROTPARS.

of the TM6 sequences on the ABP1/ABP2 and concatenated MSP trees indicates evolutionary mosaicism of the TM6 operon. The gene encoding TM6 ABP1, TM1151, seems to have an archaeal origin separate from that of the genes encoding the TM6 membrane-spanning proteins, TM1149 and TM1153, and the other TM6 ATP-binding protein, ABP2, TM1152. The latter appear to have been transferred to the Thermotogales from an archaeon, but those genes were originally from a bacterium.

### 3.3. Hydropathy Plots of the SBPs Support HGT from the Archaea

Evidence in support of the acquisition of these genes from the Archaea and the more complex history of the TM6 genes is provided by the patterns of membrane anchor domains found in the cognate SBPs. Many archaeal SBPs contain a C-terminal transmembrane domain *(33)*. This domain is apparent when sequences of archaeal SBPs most closely related to the TM1–TM5 sequences are analyzed using the Kyte-Doolittle and TMAP programs (see the lower portion of **Table 18.1**). It is likely that the bacterial genes most closely related to these archaeal genes, including those encoding TM1–TM5, lost this C-terminal domain. By contrast, with one possible exception, those archaeal SBPs

**Table 18.1**
**C-terminal hydrophobic transmembrane helices predicted for archaeal SBPs.**

| Organism (operon) | Locus | Transmembrane C-terminal helix | |
|---|---|---|---|
| | | K-D | TMAP |
| *Aeropyrum pernix* (1) | APE0304 | – | – |
| *Aeropyrum pernix* (2) | APE2257 | – | – |
| *Archaeoglobus fulgidus* | AF1767 | – | – |
| *Methanococcus acetivorans* C2A (1) | MA2477 | – | – |
| *Methanococcus acetivorans* C2A | MA2482 | – | – |
| *Methanococcus acetivorans* C2A (2) | MA3461 | – | – |
| *Methanococcus mazei* Göl | MM1858 | – | + |
| *Pyrococcus abyssi* (1) | PAB0091 | – | – |
| *Pyrococcus abyssi* (3) | PAB1343 | + | + |
| *Pyrococcus furiosus* (1) | PF0357 | + | + |
| *Pyrococcus furiosus* (2) | PF1209 | + | + |
| *Pyrococcus horikoshii* (1) | PH0502 | + | + |
| *Pyrococcus horikoshii* (2) | PH0807 | + | + |
| *Sulfolobus solfataricus* (1) | SSO2669 | + | + |
| *Sulfolobus solfataricus* (2) | SSO3043 | + | + |
| *Thermoplasma acidophilum* | TA1329 | + | + |

Predictions were made using the Kyte-Doolittle *(31)* and TMAP *(32)* programs. Loci above the center line are for SBPs that cluster among the bacterial sequences on the trees derived from the ABP1, ABP2, and concatenated MSPs datasets. The "+" and "–" indicate the presence and the absence, respectively, of the predicted hydrophobic transmembrane segment at the C-terminus of the SBPs.

most closely related to TM6 lack this domain. This supports the phylogenetic data suggesting that these archaeal genes were originally inherited from a bacterium. When a member of the Thermotogales inherited that operon, it apparently already lacked that domain. It appears that the ancestor of the TM6 ABP1 (TM1151) recombined into an Archaea-derived operon, so that its evolutionary history differs from that of the other genes in TM6.

## 4. Conclusion/ Outlook

In this study we examined the hypotheses that some of the oligosaccharide ABC transporter genes in *T. maritima* originated as archaeal genes *(13)*. Our analyses suggest that at least two interdomain HGT events from the Archaea to the Thermotogales occurred, most likely from the pyrococci and from a relative of *A. pernix* or *A. fulgidus*. The positions of TM1–TM5 on both the MSP and ABP trees were very similar, so the genes encoding theses transporters appear to have been transferred as a unit containing at least the MSP and ABP domains. The TM6 proteins have archaeal homologs, but those archaea seem to have acquired that transporter from the Bacteria. TM6 appears to be a mosaic operon since at least one of its components, TM1151, has a history different than the other TM6 proteins.

Recent acquisition of genome sequence data and subsequent phylogenetic analyses provide substantial evidence of extensive HGT among microbes *(34–36)*. Unused genes are subject to negative selection and loss *(37)*, so horizontally acquired genes must assume a functional role to be maintained. Although the phenomenon of HGT is well documented using genome sequence information, the mechanisms by which transferred genes are integrated into the physiology of the host are not always apparent in these data *(38)*. The selfish operon hypothesis provides an explanation for the clustering of bacterial genes under the control of a single promoter and posits that this clustering is supported by HGT *(39)*. The authors of that hypothesis suggest that these mobilized operons are likely to come under the control of host promoters and transcription factors. However, the subsequent evolutionary "fine tuning" of this process is left unresolved. According to the selfish operon model, individual genes cannot confer selective functions *(6, 40)*. Thus, all the genes in the operon must be transferred to allow a beneficial phenotype to be maintained in the new host. Selfish acquisition of operons does not preclude subsequent recombination within those operons, recombinations that might hide the origin of the operon. Omelchenko et al. showed that intraoperon recombinations are quite common and suggested that operon structures are maintained through purifying selections *(41)*. We may have observed such a recombination in the TM6 operon, which appears to have a mosaic composition.

To function in a bacterium, HGT-acquired archaeal upstream transcriptional initiation regions would have to accumulate mutations to allow recognition by both bacterial RNA polymerase and transcription factors. Archaeal promoters look very different from bacterial promoters *(8, 42)*. Since intergenic sequences evolve relatively rapidly, evidence for an archaeal origin for these regions is not likely to be found by sequence comparisons *(43)*. Archaeal promoters do not readily function in bacterial hosts *(44)*, so fortuitous expression from archaeal promoters is unlikely, though occasionally this is observed in the laboratory *(45)* and a chromosome fragment of *Halobacterium halobium* was found to act as a promoter in all three domains *(46)*. Evolution of an archaeal to a bacterial promoter is unlikely since these upstream regions would have to acquire the necessary recognition sequences before their corresponding non-functioning genes acquired deleterious mutations *(39)*.

Archaeal ABC transporter operons probably recombined behind bacterial promoters that already controlled genes involved in metabolism of substrates similar to those transportable by the transporter gene products. This could have involved recombination between the incoming archaeal ABC transporter genes with existing ABC transporter genes or between homologous genes adjacent to the transporter operons. Adjacent genes are often involved in the catabolism of substrates of those transporters. The highly conserved ABP-encoding genes are likely sites of the necessary homeologous recombinations. Recombinational "anchors" such as this can allow introduction of large regions of DNA as would be necessary for the replacement of a resident ABC transporter operon with an archaeal operon *(47)*. Many of the archaea-acquired ABC transporters are located among genes encoding catabolism of the class of sugars accumulated by those transporters, so their co-regulation would allow their expression under conditions that are selectively advantageous.

The study here provides broad outlines of the evolution of these transporters, but provides insufficient resolution of the phylogenies to allow us to identify the transporter operons that are closest to the ancestral archaeal donor. The order of branching among the transporters differs in these trees, and there is evidence for loss of operons since the closely related species *T. neapolitana* lacks some of these operons. Horizontal transfer of these operons among the Thermotogales may have further complicated the identification of the "ancestral" operons. Fortunately, we are now acquiring more genome sequences of Thermotogales species, so we can now hope to identify those operons. These newly acquired Thermotogales genomes will provide a rich hunting ground for evidence about the phylogenetic limits to horizontal gene acquisition. As more archaeal genome sequences become available, a fuller picture of interdomain transfers will begin to emerge.

## Acknowledgments

This work was supported by funds from the NASA Exobiology program (NAG5-12367 and NNG05GN41G). The authors thank J. Peter Gogarten and Pascal Lapierre for assistance with the phylogenetic analyses.

### References

1. Lawrence, J. G., Hendrickson, H. (2003) Lateral gene transfer: When will adolescence end? *Mol Microbiol* **50**, 739–49.
2. Koonin, E. V. (2003) Horizontal gene transfer: The path to maturity. *Mol Microbiol* **50**, 725–7.
3. Eisen, J. A. (2000) Horizontal gene transfer among microbial genomes: New insights from complete genome analysis. *Curr Opin Genet Dev* **10**, 606–11.
4. Ochman, H., Lawrence, J. G., Groisman, E. A. (2000) Lateral gene transfer and the nature of bacterial innovation. *Nature Rev Genet* **405**, 299–304.
5. Philippe, H., Douady, C. J. (2003) Horizontal gene transfer and phylogenetics. *Curr Opin Microbiol* **6**, 498–505.
6. Gogarten, J. P., Doolittle, W. F., Lawrence, J. G. (2002) Prokaryotic evolution in light of gene transfer. *Mol Biol Evol* **19**, 2226–38.
7. Lawrence, J. G., Ochman, H. (1997) Amelioration of bacterial genomes: Rates of change and exchange. *Mol Evol* **44**, 383–97.
8. Bell, S. D., Magill, C. P., Jackson, S. P. (2001) Basal and regulated transcription in archaea. *Biochem Soc Trans* **29**, 392–5.
9. Brochier-Armanet, C., Forterre, P. (2007) Widespread distribution of archaeal reverse gyrase in thermophilic bacteria suggests a complex history of vertical inheritance and lateral gene transfers. *Archaea* **2**, 83–93.
10. Longstaff, D. G., Larue, R. C., Faust, J. E., Mahapatra, A., Zhang, L., Green-Church, K. B., Krzycki, J. A. (2007) A natural genetic code expansion cassette enables transmissible biosynthesis and genetic encoding of pyrrolysine. *Proc Natl Acad Sci USA* **104**, 1021–6.
11. Frigaard, N. U., Martinez, A., Mincer, T. J., Delong, E. F. (2006) Proteorhodopsin lateral gene transfer between marine planktonic bacteria and archaea. *Nature* **439**, 847–50.
12. Chistoserdova, L., Vorholt, J. A., Lidstrom, M. E. (1998) C1 transfer enzymes and coenzymes linking methylotrophic bacteria and methanogenic archaea. *Science* **281**, 99–102.
13. Nelson, K. E., Clayton, R. A., Gill, S. R., Gwinn, M. L., Dodson, R. J., Haft, D. H., Hickey, E. K., Peterson, J. D., Nelson, W. C., Ketchum, K. A., Mcdonald, L., Utterback, T. R., Malek, J. A., Linher, K. D., Garrett, M. M., Stewart, A. M., Cotton, M. D., Pratt, M. S., Phillips, C. A., Richardson, D., Heidelberg, J., Sutton, G. G., Fleischmann, R. D., Eisen, J. A., White, O., Salzberg, S. L., Smith, H. O., Venter, J. C., Fraser, C. M. (1999) Evidence for lateral gene transfer between archaea and bacteria from the genome sequence of *Thermotoga maritima*. *Nature* **399**, 323–9.
14. Logsdon, J. M., Jr (1999) Evolutionary genomics: *Thermotoga* heats up lateral gene transfer. *Curr Biol* **9**, R747–51.
15. Huber, R., Langworthy, T. A., König, H., Thomm, M., Woese, C. R., Sleytr, U. B., Stetter, K. O. (1986) *Thermotoga maritima* sp. nov. represents a new genus of unique extremely thermophilic eubacteria growing up to 90°C. *Arch Microbiol* **144**, 324–33.
16. Belkin, S., Wirsen, C. O., Jannasch, H. W. (1986) A new sulfur-reducing, extremely thermophilic eubacterium from a submarine thermal vent. *Appl Environ Microbiol* **51**, 1180–5.
17. Chhabra, S. R., Shockley, K. R., Conners, S. B., Scott, K. L., Wolfinger, R. D., Kelly, R. M. (2003) Carbohydrate-induced differential

gene expression patterns in the hyperthermophilic bacterium *Thermotoga maritima*. *J Biol Chem* **278**, 7540–52.
18. Nanavati, D. A., Thirangoon, K., Noll, K. M. (2006) Several archaeal homologs of putative oligopeptide-binding proteins encoded by *Thermotoga maritima* bind sugars. *Appl Environ Microbiol* **72**, 1336–45.
19. Elferink, M. G. L., Albers, S. V., Konings, W. N., Driessen, A. J. M. (2001) Sugar transport in *Sulfolobus solfataricus* is mediated by two families of binding protein-dependent ABC transporters. *Mol Microbiol* **39**, 1494–503.
20. Nesbo, C. L., L'haridon, S., Stetter, K. O., Doolittle, W. F. (2001) Phylogenetic analyses of two "archaeal" genes in *Thermotoga maritima* reveal multiple transfers between Archaea and Bacteria. *Mol Biol Evol* **18**, 362–75.
21. Tomii, K., Kanehisa, M. (1998) A comparative analysis of ABC transporters in complete microbial genomes. *Genome Res* **8**, 1048–59.
22. Tam, R., Saier, M. H., Jr. (1993) Structural, functional, and evolutionary relationships among extracellular solute-binding receptors of bacteria. *Microbiol Rev* **57**, 320–46.
23. Altschul, S. F., Gish, W., Miller, W., Myers, E. W., Lipman, D. J. (1990) Basic local alignment search tool. *J Mol Biol* **215**, 403–10.
24. Gish, W., States, D. J. (1993) Identification of protein coding regions by database similarity search. *Nature Genet* **3**, 266–72.
25. Higgins, D., Sharp, P. (1988) Clustal: A package for performing multiple sequence alignments on a microcomputer. *Genetics* **73**, 237–44.
26. Thompson, J., Higgins, D., Gibson, T. J. (1994) Clustal W: Improving the sensitivity of progressive multiple sequence alignment through sequence weighting, position-specific gap penalties and weight matrix choice. *Nucl Acids Res* **22**, 4673–80.
27. Felsenstein, J. (2004) Phylip (Phylogenetic Inference Package). Department of Genetics, University of Washington Seattle, Seattle.
28. Guindon, S., Gascuel, O. (2003) A simple, fast, and accurate algorithm to estimate large phylogenies by maximum likelihood. *Syst Biol* **52**, 696–704.
29. Huelsenbeck, J. P., Ronquist, F. (2001) MrBayes: Bayesian inference of phylogenetic trees. *Bioinformatics* **17**, 754–5.
30. Holder, M. E., Roger, A. J. (1999) Puzzleboot. Bioinformatic Center, The University of British Columbia Place.
31. Kyte, J., Doolittle, R. F. (1982) A simple method for displaying the hydropathic character of a protein. *J Mol Biol* **157**, 105–32.
32. Persson, B., Argos, P. (1994) Prediction of transmembrane segments in proteins utilising multiple sequence alignments. *J Mol Biol* **237**, 182–92.
33. Albers, S.-V., Van De Vossenberg, J. L. C. M., Driessen, A. J. M., Konings, W. N. (2001) Bioenergetics and solute uptake under extreme conditions. *Extremophiles* **5**, 285–94.
34. Doolittle, W. F. (1999) Phylogenetic classification and the universal tree. *Science* **284**, 2124–8.
35. Woese, C. R. (1998) The universal ancestor. *Proc Natl Acad Sci USA* **95**, 6854–9.
36. Woese, C. R. (2000) Interpreting the universal phylogenetic tree. *Proc Natl Acad Sci USA* **97**, 8392–6.
37. Frank, A. C., Amiri, H., Andersson, S. G. (2002) Genome deterioration: Loss of repeated sequences and accumulation of junk DNA. *Genetica* **115**, 1–12.
38. Snel, B., Bork, P., Huynen, M. A. (2002) Genomes in flux: The evolution of archaeal and proteobacterial gene content. *Genomic Research* **12**, 17–25.
39. Lawrence, J. G., Roth, J. R. (1996) Selfish operons: Horizontal transfer may drive the evolution of gene clusters. *Genetics* **143**, 1843–60.
40. Lawrence, J. (1999) Selfish operons: The evolutionary impact of gene clustering in prokaryotes and eukaryotes. *Curr Opin Genet Dev* **9**, 642–8.
41. Omelchenko, M. V., Makarova, K. S., Wolf, Y. I., Rogozin, I. B., Koonin, E. V. (2003) Evolution of mosaic operons by horizontal gene transfer and gene displacement *in situ*. *Genome Biol* **4**, R55. Epub.
42. Hickey, A. J., De Macario, E. C., Macario, A. J. L. (2002) Transcription in the Archaea: Basal factors, regulation, and stress-gene expression. *Crit Rev Biochem Mol Biol* **37**, 537–99.
43. Lobry, J. R., Sueoka, N. (2002) Asymmetric directional mutation pressures in bacteria. *Genome Biol* **3**, RESEARCH0058.
44. Sorek, R., Zhu, Y., Creevey, C. J., Francino, M. P., Bork, P., Rubin, E. M. (2007) Genome-wide experimental determination of barriers to horizontal gene transfer. *Science* **318**, 1449–52.
45. Wood, A. G., Redborg, A. H., Cue, D. R., Whitman, W. B., Konisky, J. (1983) Complementation of *ArgG* and *HisA* mutations of *Escherichia coli* by DNA cloned from the archaebacterium *Methanococcus voltae*. *J Bacteriol* **156**, 19–29.

46. Yang, Y., Huang, Y. P., Shen, P. (2003) The 492-bp rm07 DNA fragment from the halophilic archaea confers promoter activity in all three domains of life. *Curr Microbiol* **47**, 388–94.

47. De Vries, J., Herzfeld, T., Wackernagel, W. (2004) Transfer of plastid DNA from tobacco to the soil bacterium *Acinetobacter* sp. by natural transformation. *Mol Microbiol* **53**, 323–34.

# Chapter 19

# The Role of Horizontal Gene Transfer in Photosynthesis, Oxygen Production, and Oxygen Tolerance

Jason Raymond

## Abstract

One of the pivotal events during the early evolution of life was the advent of oxygenic photosynthesis, responsible for producing essentially all of the free oxygen in Earth's atmosphere. This molecular innovation required the development of two tandemly linked photosystems that generate a redox potential strong enough to oxidize water and then funnel those electrons ultimately to cellular processes like carbon and nitrogen fixation. The by-product of this reaction, molecular oxygen, spawned an entirely new realm of enzymatic reactions that served to mitigate its potential toxicity, as well as to take advantage of the free energy available from using $O_2$ as an electron acceptor. These ensuing events ultimately gave rise to aerobic, multicelled eukaryotes and new levels of biological complexity. Remarkably, instances of horizontal gene transfer have been identified at nearly every step in this transformation of the biosphere, from the evolution and radiation of photosynthesis to the development of biological pathways dependent on oxygen. This chapter discusses the evidence and examples of some of these occurrences that have been elucidated in recent years.

**Key words:** Photosynthesis, oxygen production, horizontal gene transfer.

## 1. Introduction

In recent years it has become clear that the evolution of life has not only been influenced by, but has itself profoundly altered the Earth's chemical and geological development. These changes are most easily observed where life's major elements are involved. Most notable are biogeochemical cycles of carbon, nitrogen, and oxygen, though effects likewise extend to minor elements including phosphorus, sulfur, and many alkali and transition metals. Understanding the natural history of these elements provides

Maria B. Gogarten et al. (eds.), *Horizontal Gene Transfer: Genomes in Flux, vol. 532*

DOI 10.1007/978-1-60327-853-9_19 Springerprotocols.com

a key opportunity to constrain the milieus in which early life originated and evolved *(1)*. However, such constraints require evidence that ties ancient organisms (or communities) to their environment. While this is quite tractable for fossils where evidence for form and/or function is often preserved in a geological context, the fossil record for Archaean life is infamously sparse *(2)*. Exciting advances from scientists studying microbial biomarkers—attempting to use geologically stable and taxonomically diagnostic biochemicals to build a molecular fossil record—are shedding new light on Precambrian microbiology *(3–5)*. Combined with traditional analyses of Earth's geological history, information from biomarkers are providing mileposts not only on what organisms were around but also on what they might have been doing (*see* also **Chapter 8**).

Importantly, molecular and macroscopic fossil evidence is limited to providing a static picture—a discrete and fixed time point—of early life and geology. While something of a prehistorical zoetrope can be reconstructed given a rich enough succession of fossils, the fossil record is notably patchy. Therefore, key insights can be gained by turning the lens of evolutionary understanding (as through phylogenetic analysis) on such studies—allowing inference of the continuum of events that underlie organismal evolution, in effect connecting the dots between geological findings. As a demonstrative example, without any knowledge of the taxonomic relationships between modern organisms (i.e., as depicted in the tree of life) one could do an extensive molecular survey of biomarkers found in extant organisms. Noticing that anytime a 2-methyl hopanoid is found in a prokaryote, that organism happens to contain chlorophyll and perform oxygenic photosynthesis, one could then focus on the occurrence of those hopanoids in the rock record as proxies for the production of oxygen from water *(5)*. However, none of the details on the origin and evolution of photosynthesis, water oxidation to molecular oxygen, or hopanoid biosynthesis are amenable to such an approach: without further knowledge of their relationships, hopanoid-containing organisms are effectively equally related to one another, and equally distant from all organisms without hopanoids. Furthermore, the usefulness of a molecular proxy for some organism or biological process is ultimately limited by our ignorance—we must assume that there are no exceptions (i.e., hopanoids in non-photosynthetic organisms) among all living organisms *yet to be studied* and, further, that there are no exceptions among all organisms *that have ever lived*. Indeed, our horizons on hopanoid biosynthesis have recently been expanded: it is now known that 2-methyl hopanoid biosynthesis is found outside of the cyanobacteria—these compounds can no longer be standalone proxies for oxygenic photosynthesis *(6)*.

While one could infer evolutionary trees based solely on biomarkers (via some scheme for quantifying how different two chemical compounds are), such an approach is far removed from the evolutionary process and could easily be misleading. This is illustrated by an example from photosynthetic organisms that will be discussed below, where classification based on the pigment bacteriochlorophyll would place the green sulfur bacteria (phylum Chlorobi) and filamentous anoxygenic phototrophs (phylum Chloroflexi) as near neighbors despite their belonging to entirely different phyla and having few commonalities outside of these photosynthetic pigments (*see* e.g. *(7)*). In fact, bacteriochlorophyll derivatives are found as visual pigments in some deep sea dragonfish; photoluminescent compounds derived from their diet *(8)*. Of course, dragonfish are most certainly not cousins of the photosynthetic bacteria, as a chemical classification scheme might suggest. The point then is that biochemical and geochemical evidence provide discrete mileposts along life's history; molecular evolutionary analyses can comprise the highways by which they are connected.

## 2. A Brief History of Oxygen

Molecular oxygen has long been at the crux of the intersection between biology and geochemistry. In the mid 19th century, Pasteur established not only that some microorganisms could grow in the absence of oxygen but also that many were inhibited by it. Facultative aerobes such as yeasts grew roughly 20 times more efficiently (biomass yield per unit of glucose) in the presence of oxygen, which also dramatically changed their metabolic end products. The details of these remarkable changes would ultimately give way to biochemists including Albert Szent-Gyorgi and Hans Krebs, whose experiments demonstrated that oxygen enabled the "total oxidation" of organic molecules—to carbon dioxide and water—netting as much as 18 times the ATP compared to anaerobic, partial oxidation that gives end products such as ethanol and acetate.

The role of oxygen in evolution has become well established only relatively recently, in large part through the insights of geologists. Though many of the early ideas have been refined on the shoulders of new data, the brilliant work of Holland *(9)*, Fischer *(10)*, Berkner and Marshall *(11)*, among others, helped establish the key role that oxygen has played in the expansion and increasing complexity of life. Early on, researchers recognized that injection of volcanic gases into Earth's early atmosphere would generate no net free oxygen. Even after the formation of primitive oceans, photolysis of water would produce insignificant quantities of free oxygen in a process that was self-limiting by the production of ozone, which attenuates ultraviolet light flux. Indeed, there is consensus in the literature that a process such as oxygenic photosynthesis is required to reach and sustain present atmospheric levels of $O_2$ *(10–13)*.

Geologists also noted the reduced state of ancient rocks, suggesting their deposition under (at least locally) anoxic conditions, and prebiotic chemists noted that oxygen was a potent inhibitor of the formation of organic compounds *(4, 14–17)*. These ideas were most convincingly borne out in the fossil record, where an utter absence of morphologically complex fossils is evident during the Archaean *(2)*; life at this early stage was entirely microscopic, and complex fossils did not appear until well after widespread geological signatures indicating an oxidizing atmosphere. The Pasteur point—the level of oxygen necessary for yeasts to switch from fermentation to aerobic respiration—as well as observed complexity of fossils, could for the first time be contrasted with the models of atmospheric $O_2$ levels *(11)*. These provocative findings linked the availability (or lack) of oxygen directly to the diversification of life.

It seems very likely that early life was energetically limited. In theory, fermentation typically yields 2–4 ATP molecules per glucose catabolized at standard biochemical state. Anaerobic respiration can optimally yield 8–12 ATP molecules per glucose, but requires strong electron acceptors such as nitrate and sulfate that may have been fleetingly present before oxygen was widely available. Importantly, these are standard biochemical state calculations; these other electron acceptors can rival and even exceed the energetic potential provided by oxygen at fleeting $O_2$ concentrations, such as those that characterized the Precambrian atmosphere *(13)*.

There are a number of other, often overlooked, factors conspiring to make oxygen well suited to its central biological role. The neutral charge means that oxygen can freely diffuse into and out of cells without requiring energy input (though many organisms have intricate mechanisms for getting oxygen in contact with cells or tissues). That oxygen is a gas is obvious but was nonetheless critical in the expansion of life onto land, where limitations of soluble, non-gaseous compounds like nitrate and phosphate can be the difference between desert and biome. Through oxygenic photosynthesis, water provides an effectively limitless source of oxygen and the fact that biology had tapped into the oxidized form of an already ubiquitous and innocuous compound ($H_2O$) was crucial. However, this is really only half the story; the key attribute of $O_2$ is its unique position in the redox cycle of water. Not only does oxygen have a universal source for its production, but its potential toxicity can be mitigated by reducing it back to $H_2O$.

This reduction turns out to be accompanied by a considerable release of free energy, something that was soon put to use by early organisms—the ancestors of modern aerobes.

### 2.1. Oxygen Production: Oxygenic Photosynthesis

As illustrated in **Fig. 19.1**, the rise of oxygen in Earth's atmosphere very likely resulted from the invention of oxygen-producing (oxygenic) photosynthesis in the ancestors of a single group of extant organisms, the cyanobacteria. These organisms, and the plants and algae that later captured their ability, can use photon energy to derive electrons from water, an otherwise incredibly difficult-to-oxidize molecule. It is generally accepted that the cyanobacteria were not among the early radiations of the bacterial tree, that is, they likely coexisted alongside many of the other major phyla of Bacteria *(18–20)*. One might imagine the ecological chaos that ensued after a primitive cyanobacterium first started pumping out molecular oxygen as a photosynthetic by-product (and managing not to destroy itself in the process). While certainly not welcomed among its neighbors, the occurrence and timing of this most significant event now provides a remarkable opportunity to study biological adaptation, especially because these adaptations occurred in parallel in the diverse organisms that existed ~2.5 billion years ago. While this parallel evolutionary processing likely occurred en masse, modern molecular evidence suggests that a handful of solutions ultimately emerged and, as might be expected, the enzymes and pathways behind these solutions have provided compelling examples of horizontal gene transfer (HGT).

The biologically instigated appearance of oxygen in Earth's atmosphere was one of the pivotal events in the evolution of life, profoundly altering the course of evolution. These events include

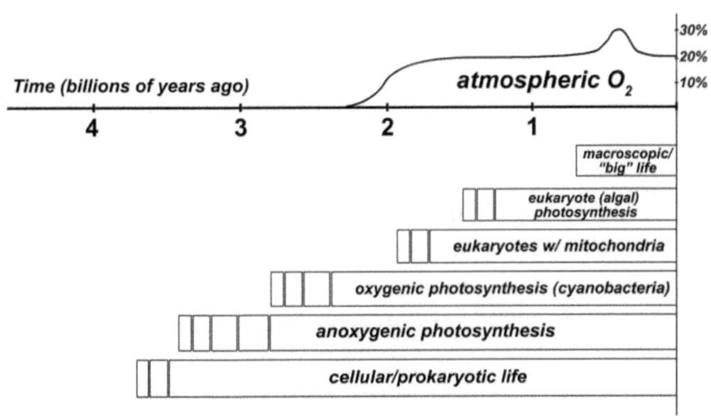


Fig. 19.1. A timeline of the history of oxygen on Earth and a few major associated biological processes. *Broken bars* around each process are meant to indicate uncertainty in dating its origin. $O_2$ percentages are given as percent of atmospheric composition leading up to the modern day value of 20.95%. Early values are not well constrained and are rough estimates at best (*see (17)*).

not only the evolution of oxygenesis by photosynthetic organisms, but also the colorful molecular repertoire with which organisms have learned to cope with oxygen. These are involved with detoxification of so-called reactive oxygen species, with using $O_2$ directly as a source of atomic oxygen (oxygenation reactions) in biosynthetic pathways, or coupling $O_2$ to energy generation as a terminal electron acceptor in prokaryotic and mitochondrial respiratory chains. Because oxygen appeared on the evolutionary scene relatively recently respective to the origin of life, the multitude of biological responses provide unique opportunities to follow the adaptive functional developments that simultaneously were occurring in many complex organisms. Indeed, one might presume that because "inventing" new capabilities is a very difficult feat, there would have been strong selection pressures toward acquiring them from other organisms through HGT, and a number of studies have borne this out. The importance of HGT in the evolution of photosynthesis, as well as across multiple molecular responses to oxygen, will be detailed here.

## 3. The Evolution of Photosynthesis

### *3.1. Archaean Origin*

The inimitable changes that have resulted from photosynthesis have fueled many ideas on how and when it came about. The geological mileposts discussed above provide a number of key insights, some of which are still actively debated. The sparse Archaean (2.5–3.8 billion years ago) fossil record yields a few clues. Kerogens—remnants of organic carbon—buried in rocks 3.8 billion years old (Ga) have carbon isotope signatures that are consistent with those produced by carbon fixation using the Calvin Cycle and have been argued to be signatures of photosynthetic carbon fixation *(21,22)*. The Calvin Cycle is found not only in photosynthetic cyanobacteria and proteobacteria but also in a number of non-photosynthetic bacteria, so its diagnostic value for photosynthesis is questionable. At about 3.5 Ga, stromatolites appear in some ancient rocks *(10, 14, 23)*. These layered structures are similar in appearance to those produced by communities of cyanobacteria in a few modern environments, though their biogenicity has been questioned because some geological processes can produce similarly laminated structures *(23)*. Also first appearing in the rock record about 3.5 Ga are putative microfossils that have been morphologically compared to cyanobacteria, though in prokaryotes morphology is generally a poor indicator either of taxonomy or of metabolism *(24)*. The hopanoid biomarkers introduced above date to about 2.7 Ga and, in modern organisms, have been found in cyanobacteria and, more recently, proteobacteria *(5, 6)*. At about the same age, geological and isotopic signatures, as well as additional biomarkers suggestive of the appearance of oxygen, are also found, providing an independent lens into the possible early appearance of oxygenic photosynthesis *(4, 17, 25–30)*. Indeed, given the speculative nature of pre-2.7 Ga features, our best evidence for when and how photosynthesis evolved comes from integrating biomarkers, isotopic, and geochemical evidence that herald the invention of oxygenic photosynthesis and arrival of $O_2$ in the atmosphere, with phylogenetic analyses of photosynthesis proteins.

### *3.2. Phylogenetics and the Role of HGT*

Importantly, phylogenetics initially proved to be more confounding than they were helpful. Even early on, 16S rRNA phylogenies showed an almost sporadic separation of photosynthetic clades around the bacterial tree, shown schematically in **Fig. 19.2**, with the Chloroflexi generally branching early on and the other four (now five) phyla with photosynthetic members appearing near the bacterial "crown", a polytomy from which most bacterial phyla emerge *(19, 31)*. Even more perplexing is that each of these groups possess a hodgepodge of photosynthetic characteristics that bear little, if any, apparent relation to their position on the tree of life *(32)*, as summarized in **Table 19.1**.

Anoxygenic phototrophs have only one of the two variants of photosynthetic reaction center (RC)—referred to as type I and type II—that couple photon energy transduction to electron transfer and use as their primary pigment bacteriochlorophyll, that absorbs at longer, lower energy wavelengths than the familiar chlorophyll *(7)*. Whereas photosynthetic Chloroflexi and proteobacteria use the type II variant, Chlorobi, Gram-positive heliobacteria, and Acidobacteria all have type I RCs *(7)*. These features are distinct from oxygenic photosynthetic organisms. In addition to using chlorophyll as their major pigment, all known oxygenic phototrophs have both types of reaction centers (referred to as photosystems I and II, based on homology to type I and II anoxygenic RCs), operating in series so that light drives

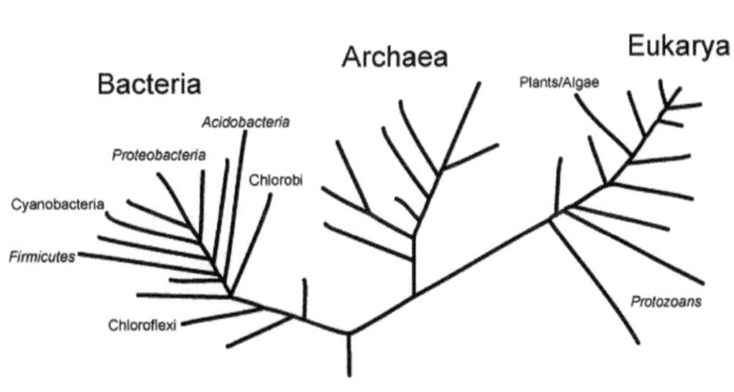


Fig. 19.2. Distribution of known photosynthetic phyla on the tree of life. Italicized phyla comprised both photosynthetic and non-photosynthetic members, invoking either HGT or loss of PS in various lineages.

**Table 19.1**
**Photosynthesis on the tree of life**

| Phylum[1] | Pigments | Reaction center | Antenna proteins[2] |
|---|---|---|---|
| *Acidobacteria* | BChl *a, c* | type I | Chlorosomes; FMO |
| Chlorobi (green sulfur bacteria) | BChl *a, c, d, e*; Chl *a* | type I | Chlorosomes; FMO |
| *Choroflexi* (filamentous anoxygenic phototrophs) | BChl *a, c* | type II | LH-like complex; Chlorosomes |
| Cyanobacteria (plants, algae, *protozoa*) | Chlorophyll (Chl) *a,b,c,d* | type I + type II | Phycobilisomes; Pcb proteins; LHC proteins |
| *Firmicutes* (heliobacteria) | BChl *g*; Chl *a* | type I | None |
| *Proteobacteria* (purple sulfur/ nonsulfur bacteria) | Bacteriochlorophyll (BChl) *a, b* | type II | LH1 + LH2 complexes |

[1]Italicized phyla comprised both photosynthetic and non-photosynthetic members; non-italicized are entirely photosynthetic.
[2]LH = Light harvesting antenna complex (1 = core, 2 = peripheral); FMO = Fenna Matthews Olson protein; Pcb = chlorophyll a/b binding protein; LHC = light harvesting proteins.

electrons from the type II RC to the type I RC—the so-called "Z scheme" *(7)*.

Though photosynthesis is seemingly sporadically distributed around the tree of life, these similar reaction centers and pigments led researchers to suspect a common evolutionary origin for oxygenic and anoxygenic photosynthesis *(32–36)*. This has been borne out by genetic analyses and protein structure data in recent studies, providing a number of important insights:

(1) Eukaryotic photosynthesis is carried out exclusively in plastids (chloroplasts in plants), derived from endosymbiosis of an ancient cyanobacterium *(37, 38)*.
(2) Photosystems I and II of oxygenic phototrophs are homologous to the type I and type II RCs, respectively, of anoxygenic phototrophs *(32, 33, 36, 37)*. Though highly diverged, type I and type II RCs appear to have evolved from a single, ancestral reaction center through gene duplication and divergence.
(3) Analysis of genes involved in bacteriochlorophyll and chlorophyll biosynthesis supports their common evolutionary

origin and, along with the much greater complexity of oxygenic photosynthesis, support that anoxygenic photosynthesis preceded oxygenic photosynthesis *(38–40)*.

HGT has played a key role in the evolution and distribution of photosynthesis *(41)*. Of course, the endosymbiotic origin of plastids is HGT en masse and stands as a remarkable example of thousands of genes, or rather a whole genome, most of which was ultimately lost, moving from the bacterial into the eukaryotic domain. A number of clear-cut examples exist for the transfer of pigment biosynthesis and RC genes both between phyla, for instance, with the Chlorobi, Chloroflexi, and Acidobacteria, and also within phyla. Despite being distant branches on the bacterial tree, having two distinct types of RCs, and having completely different growth modes (aerobic vs. obligately anaerobic, different mechanisms for assimilating carbon), the Chloroflexi, Chlorobi, and a newly discovered phototroph from the Acidobacteria share a light harvesting organelle called the chlorosome, found in no other photosynthetic organisms *(42)*. In addition, the Chlorobi and Acidobacteria both have a photosynthetic subunit known as the FMO protein, which facilitates photon energy transfer between the chlorosome and the RC *(42)*. The Chlorobi and Chloroflexi share striking similarities in their pigment biosynthesis genes, with > 70% identity in some proteins *(39)*. None of these phyla are suggested to be particularly closely related to one another, and photosynthesis is only paraphyletically distributed among the Acidobacteria (only a single photosynthetic lineage) and Chloroflexi *(42, 43)*. Taken together, these features strongly suggest that the genes for photosynthesis have been horizontally transferred between these phyla.

These present compelling examples of *inter*-phylum gene transfer have founded new ideas on how photosynthesis evolved. Likewise, strong support for *intra*-phylum HGT is found in the proteobacteria and the cyanobacteria. In proteobacteria and heliobacteria, the several dozen genes required for photosynthesis are localized in a single super-operon called the photosynthesis gene cluster, which has facilitated HGT and brought this capability to new lineages of proteobacteria as suggested by recent evidence *(44)*. Among the other photosynthetic phyla, only the Gram-positive heliobacteria have a contiguous cluster of photosynthesis genes—if this gene cluster was important in the early evolution of photosynthesis, its signature has eroded in most other lineages. More recently, comparative genomics of over 1100 cyanobacterial proteins has revealed extensive evidence of HGT, particularly within the phylum *(45)*. The recurring evidence for HGT both within and between photosynthetic phyla suggests that caution must be taken when inferring deep evolutionary events especially from a limited amount of genetic or genomic data *(46)*.

As mentioned above, another key finding borne out by phylogenetics is that anoxygenic photosynthesis indeed predates oxygenic photosynthesis. Though this is suggested by the dramatically more complex oxygenic photosynthetic apparatus, this idea received some challenge by way of the Granick hypothesis *(47, 48)*. Granick's observation was that, because chlorophyll is in fact a biosynthetic precursor to bacteriochlorophyll and is thereby "simpler" to produce, chlorophyll-utilizing organisms likely evolved first. Thus cyanobacteria would precede anoxygenic photosynthetic bacteria. However, phylogenetic analysis of biosynthetic genes has shown that the anoxygenic photosynthesis genes are in fact ancestral to those found in plants and cyanobacteria; chlorophyll came about after several end-step genes in bacteriochlorophyll biosynthesis were lost *(39)*. As mentioned above, pigment biosynthesis gene phylogenies suggested a surprisingly close relationship between Chloroflexi and Chlorobi, almost certainly a result of HGT. In addition, gene trees consistently place the heliobacteria and Cyanobacteria together. Though this might suggest HGT, a number of organismal phylogenies have the Cyanobacteria and Gram-positives as sister clades, and the basal position of the heliobacteria among the Gram + phylum suggests a possible common photosynthetic ancestry. The bacteriochlorophyll pigment used by heliobacteria in fact can spontaneously convert to chlorophyll under oxic conditions, portending a plausible route by which chlorophyll utilization may have begun. **Figure 19.3** summarizes the specific examples of HGT on the schematic tree from **Fig. 19.2** focusing only on genetic components of the photosynthetic apparatus. HGT has played a key role in the evolution and distribution of this fundamental and Earth-changing metabolic capability.

The marked differences in pigment use, serial linkage of photosystems, and the much greater subunit complexity all represent key differences between oxygenic and anoxygenic photosynthesis. However, coupling water oxidation to light absorption—required for the generation of $O_2$—represents a novel innovation of the cyanobacteria. The four manganese-containing cluster that is at the heart of the so-called oxygen evolving complex (OEC) in cyanobacterial photosystem II is absolutely required for $O_2$ production, yet has no homologous counterpart in any of the anoxygenic reaction centers. Several manganese-containing enzymes, like manganese catalase, have been invoked as plausible analogs for a transitional OEC *(49)*. Recent structural analysis indeed supports notably conserved geometry in the cofactor and ligand arrangement around the OEC and several two manganese enzymes that also act on oxygen or hydroxyl substrates *(50)*. However, it remains to be determined whether the OEC and enzymes such as manganese catalase are distant homologues, or

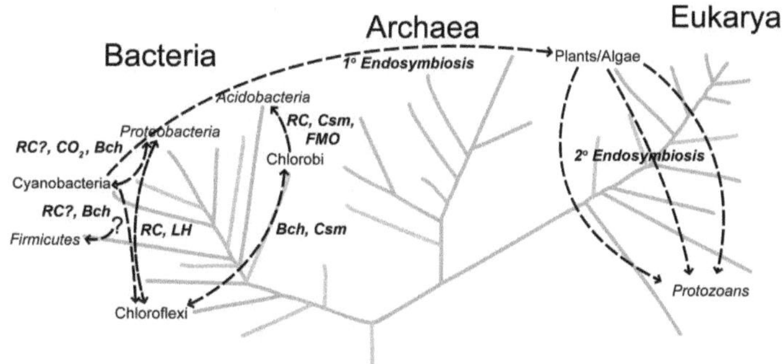


Fig. 19.3. Schematic tree from **Fig. 19.1** showing well-supported HGT events involving genetic components of photosynthesis discussed in the text and as detailed in **Table 19.1**. A number of cases of probable HGT are shown on the tree. These include the primary and secondary endosymbioses, by which eukaryotes gained (and subsequently shared) the genes for oxygenic photosynthesis; genes for reaction centers/photosystems (**RC**); antenna complexes including the chlorosome proteins (**Csm**), FMO protein (**FMO**), and light harvesting (**LH**) complexes; bacteriochlorophyll and chlorophyll pigment biosynthesis enzymes (**Bch**); and genes (e.g., rubisco large and small subunits) for the carbon-fixing Calvin Cycle ($CO_2$) that, along with several pigment biosynthesis genes, have been horizontally transferred between photosynthetic proteobacteria and marine cyanobacteria (*Prochlorococcus* and *Synechococcus* species). The origin of the tandemly linked photosystems I and II in cyanobacteria clearly originated from anoxygenic photosynthetic reaction centers, yet the origin and degree/direction of HGT are still intensely debated. These HGTs are indicated here by question marks.

if the similarities result from convergent evolution to an optimal or highly constrained valency/ligand sphere around manganese clusters.

## 4. Coping with Oxygen

While studying the origin and evolution of photosynthesis provides important insights into how the oxygen revolution came about, it is also quite feasible to approach the event from the other direction—by focusing on the changes that oxygen imparted on biology. These include the mechanisms for detoxifying reactive oxygen species (ROS) derived from oxygen, utilizing molecular oxygen in biosynthetic reactions, and taking advantage of the strong oxidizing potential of $O_2$ by using it as an electron acceptor, particularly in aerobic respiration. It seems reasonable that the capability to mitigate ROS damage was critical at the onset of oxygenic photosynthesis and, to this end, there have been proposals that enzymes including catalase, peroxidase, and superoxide dismutase played important detoxification roles. Indeed, genomic data has borne out the strong correlation between obligate anaeroby and the absence of ROS-mitigating enzymes, as originally suggested over three decades ago *(51)*.

#### 4.1. Catalase

Known catalases, often referred to as catalase-peroxidases, fall into two distinct families that appear to have arisen independently of one another. The first are the ubiquitous heme catalases, most familiar from eukaryotic peroxisomes but found in nearly all aerobic organisms. The evolution of heme catalases has been considered and recently reviewed in great detail *(52, 53)*, where three clades of heme catalases were delineated based on phylogeny. In their thorough analysis, Klotz and Loewen *(53)* note the distinct lack of congruence between monofunctional heme catalase and 16S phylogenies; bacterial lineages, for instance, are found at the base of each of the three clades that subsequently diverges into eukaryotic and bacterial branches. Fast et al. *(54)* described a heme catalase from the eukaryotic microsporidian *Nosema locustae* that groups with proteobacterial homologs. The lack of such a homolog in other microsporidia supports a bacteria-to-eukaryote HGT. Bifunctional heme catalase-peroxidases have also been extensively analyzed and their evolutionary history is most easily explained by multiple HGT events—though the timing and direction of these HGTs is debated *(53, 55)*. The non-heme catalases are much more sporadically distributed across the tree of life. The analysis by Klotz and Loewen *(53)* found only a single instance of a catalase gene outside the Bacteria, in the crenarchaeon *Pyrobaculum calidifontis (56)*. At the time of this writing, this limited archaeal distribution—almost certainly a result of HGT—has expanded slightly to include several additional *Pyrobaculum* species, a fellow crenarchaeon *Metallosphaera hakonensis*, and, uniquely, the euryarchaeon *Halorubrum lacusprofundi*. The distribution of Mn-catalases in bacteria is highly incongruent with 16S-based species trees, and its low activity relative to heme catalases led Klotz and Loewen *(53)* to suggest that the manganese-containing enzymes have been supplanted by heme-containing ones, particularly in aerobes where high catalase activity is the requisite for survival.

#### 4.2. Superoxide Dismutase

Not only has convergent evolution produced distinct and unrelated catalase enzymes, but it is quite likely that the three types of superoxide dismutase (SOD) also evolved independently. SOD isoforms include a copper/zinc-containing variety, found in a few prokaryotes, eukaryotic cytosol, and even some viruses; a manganese-containing SOD in prokaryotes and mitochondria; an iron SOD that is homologous to the MnSOD, found in prokaryotes and the plastids of some plants and algae; and a two nickel-containing SOD, recently discovered in some *Streptomyces* and cyanobacteria *(57, 58)*. With the exception of MnSOD and FeSOD, these variants show an absence of sequence and structural homology, and employ very different catalytic mechanisms *(59)*. Evolution of SOD isoforms has been studied for several decades, but has been obscured by a combination of extensive paralogy and their sporadic distribution across the tree of life. Cu/ZnSOD trees show general congruence especially with eukaryote taxonomies, though viral Cu/ZnSOD sequences cluster with their insect hosts, suggesting host-to-virus HGT *(60)*. The limited presence of Cu/ZnSODs in bacteria has been taken as evidence that HGT occurred between the bacterial and the eukaryote domains early in the evolution of the latter, but this is a difficult idea to test. Manganese and iron SOD phylogenies separate out the three domains with evidence of endosymbiotic origins of plastid and mitochondrial sequences, though a number of cases have been argued as evidence of HGT. The FeSOD of *Entamoeba histolytica* is most similar to bacterial homologs and was cited early on as an example of interdomain HGT. Wolfe-Simon et al. *(58)* elaborated the evolution of SOD in oxygenic phototrophs (algae, plants, and cyanobacteria)—in which all four SOD isoforms are found—and provide plausible routes (including renewed arguments for HGT) explaining the complex history of these enzymes.

## 5. Future Directions

The invention of oxygenic photosynthesis brought about remarkable changes in the development of the Earth and its biota. Geological markers provide important constraints on when these changes took place, and a number of studies are using these to inform molecular evolutionary analyses. This chapter has detailed evolutionary studies focused on resolving how oxygenic photosynthesis came about as well as the enzymes that would have been key in mitigating the effects of oxygen as it accumulated. These ideas have been particularly fostered by vastly increasing the availability of genomic data. The remarkable increase in prokaryotic and eukaryotic genomes, and the growing availability of environmental metagenomes, provides an unprecedented opportunity to re-evaluate evolutionary scenarios under a barrage of new data. Likewise, as understanding of the predominant conditions on early Earth become better resolved, so do insights on how factors like changing metal availability (due to redox-mediated changes in solubility) govern enzyme evolution.

## References

1. Williams, R., Frausto Da Silva, J. (2003) *The Natural Selection of the Chemical Elements*, Oxford University Press, New York.
2. Schopf, J. W. (2006) Fossil evidence of Archaean life. *Philo Trans R Soc B Biol Sci* **361**, 869–85.

3. Knoll, A. H., Javaux, E. J., Hewitt, D., Cohen, P. (2006) Eukaryotic organisms in Proterozoic oceans. *Philo Trans R Soc B Biol Sci* **361**, 1023–38.
4. Summons, R. E., Bradley, A. S., Jahnke, L. L., Waldbauer, J. R. (2006) Steroids, triterpenoids and molecular oxygen. *Philo Trans R Soc B Biol Sci* **361**, 951–68.
5. Summons, R. E., Jahnke, L. L., Hope, J. M., Logan, G. A. (1999) 2-Methylhopanoids as biomarkers for cyanobacterial oxygenic photosynthesis. *Nature* **400**, 554–7.
6. Rashby, S. E., Sessions, A. L., Summons, R. E., Newman, D. K. (2007) Biosynthesis of 2-methylbacteriohopanepolyols by an anoxygenic phototroph. *Proc Natl Acad Sci U S A* **104**, 15099–104.
7. Blankenship, R. E. (2002) *Molecular Mechanisms of Photosynthesis*, Blackwell Science, Oxford.
8. Douglas, R. H., Mullineaux, C. W., Partridge, J. C. (2000) Long-wave sensitivity in deep-sea stomiid dragonfish with far-red bioluminescence: evidence for a dietary origin of the chlorophyll-derived retinal photosensitizer of *Malacosteus niger. Philo Trans R Soc B Biol Sci* **355**, 1269–72.
9. Holland, H. D. (1962) Model for the evolution of the earth's atmosphere, in *Petrologic Studies: A Volume to Honor AF Buddington.* Geological Society of America, Washington, DC, 447–77.
10. Fischer, A. G. (1965) Fossils early life and atmospheric history. *Proc Natl Acad Sci U S A* **53**, 1205–15.
11. Berkner, L. V., Marshall, L. C. (1965) On origin and rise of oxygen concentration in Earth's Atmosphere. *J Atmos Sci* **22**, 225–61.
12. Catling, D. C., Claire, M. W. (2005) How Earth's atmosphere evolved to an oxic state: a status report. *Earth Planet Sci Lett* **237**, 1–20.
13. Catling, D. C., Glein, C. R., Zahnle, K. J., Mckay, C. P. (2005) Why $0_2$ is required by complex life on habitable planets and the concept of planetary "oxygenation time". *Astrobiology* **5**, 415–38.
14. Buick, R. (1992) The antiquity of oxygenic photosynthesis—evidence from stromatolites in sulfate-deficient Archean lakes. *Science* **255**, 74–7.
15. Kirschvink, J. L., Gaidos, E. J., Bertani, L. E., Beukes, N. J., Gutzmer, J., Maepa, L. N., Steinberger, R. E. (2000) Paleoproterozoic snowball earth: extreme climatic and geochemical global change and its biological consequences. *Proc Natl Acad Sci U S A* **97**, 1400–5.
16. Knoll, A. H. (2003) The geological consequences of evolution. *Geobiology* **1**, 3–14.
17. Kopp, R. E., Kirschvink, J. L., Hilburn, I. A., Nash, C. Z. (2005) The paleoproterozoic snowball Earth: a climate disaster triggered by the evolution of oxygenic photosynthesis. *Proc Natl Acad Sci U S A* **102**, 11131–6.
18. Des Marais, D. J. (2000) Evolution. When did photosynthesis emerge on Earth? *Science* **289**, 1703–5.
19. Woese, C. R. (1987) Bacterial evolution. *Microbiol Rev* **51**, 221–71.
20. Woese, C. R., Kandler, O., Wheelis, M. L. (1990) Towards a natural system of organisms: proposal for the domains Archaea, Bacteria, and Eucarya. *Proc Natl Acad Sci U S A* **87**, 4576–9.
21. Mojzsis, S. J., Arrhenius, G., Mckeegan, K. D., Harrison, T. M., Nutman, A. P., Friend, C. R. (1996) Evidence for life on Earth before 3,800 million years ago. *Nature* **384**, 55–9.
22. Schidlowski, M., Appel, P. W. U., Eichmann, R., Junge, C. E. (1979) Carbon isotope geochemistry of the 3.7 X 109-yr-old Isua sediments, West Greenland—Implications for the Archaean Carbon and Oxygen Cycles. *Geochimica et Cosmochimica Acta* **43**, 189–99.
23. Grotzinger, J. P., Knoll, A. H. (1999) Stromatolites in Precambrian carbonates: evolutionary mileposts or environmental dipsticks? *Ann Rev Earth Planet Sci* **27**, 313–58.
24. Schopf, J. W. (1993) Microfossils of the early Archean Apex Chert—new evidence of the antiquity of life. *Science* **260**, 640–6.
25. Brocks, J. J., Logan, G. A., Buick, R., Summons, R. E. (1999) Archean molecular fossils and the early rise of eukaryotes. *Science* **285**, 1033–6.
26. Anbar, A. D., Duan, Y., Lyons, T. W., Arnold, G. L., Kendall, B., Creaser, R. A., Kaufman, A. J., Gordon, G. W., Scott, C., Garvin, J., Buick, R. (2007) A whiff of oxygen before the great oxidation event? *Science* **317**, 1903–6.
27. Canfield, D. E. (1998) A new model for Proterozoic ocean chemistry. *Nature* **396**, 450–3.
28. Catling, D. C., Zahnle, K. J., Mckay, C. (2001) Biogenic methane, hydrogen escape, and the irreversible oxidation of early Earth. *Science* **293**, 839–43.
29. Farquhar, J., Bao, H., Thiemens, M. (2000) Atmospheric influence of Earth's earliest sulfur cycle. *Science* **289**, 756–9.
30. Tice, M. M., Lowe, D. R. (2004) Photosynthetic microbial mats in the 3,416-Myr-old ocean. *Nature* **431**, 549–52.
31. Oyaizu, H., Debrunner-Vossbrinck, B., Mandelco, L., Studier, J. A., Woese, C. R. (1987) The green non-sulfur bacteria: a deep

branching in the eubacterial line of descent. *Syst Appl Microbiol* **9**, 47–53.
32. Blankenship, R. E. (1992) Origin and early evolution of photosynthesis. *Photosynth Res* **33**, 91–111.
33. Nitschke, W., Setif, P., Liebl, U., Feiler, U., Rutherford, A. W. (1990) Reaction center photochemistry of *Heliobacterium chlorum*. *Biochemistry* **29**, 11079–88.
34. Vermaas, W. F. (1994) Evolution of heliobacteria: implications for photosynthetic reaction center complexes. *Photosynth Res* **41**, 285–94.
35. Schubert, W. D., Klukas, O., Saenger, W., Witt, H. T., Fromme, P., Krauss, N. (1998) A common ancestor for oxygenic and anoxygenic photosynthetic systems: a comparison based on the structural model of photosystem I. *J Mol Biol* **280**, 297–314.
36. Michel, H., Deisenhofer, J. (1988) Relevance of the photosynthetic reaction center from purple bacteria to the structure of photosystem II. *Biochemistry* **27**, 1–7.
37. Olson, J. M., Pierson, B. K. (1987) Evolution of reaction centers in photosynthetic prokaryotes. *Int Rev Cytol* **108**, 209–48.
38. Burke, D. H., Hearst, J. E., Sidow, A. (1993) Early evolution of photosynthesis: clues from nitrogenase and chlorophyll iron proteins. *Proc Natl Acad Sci U S A* **90**, 7134–8.
39. Xiong, J., Fischer, W. M., Inoue, K., Nakahara, M., Bauer, C. E. (2000) Molecular evidence for the early evolution of photosynthesis. *Science* **289**, 1724–30.
40. Sadekar, S., Raymond, J., Blankenship, R. E. (2006) Conservation of distantly related membrane proteins: photosynthetic reaction centers share a common structural core. *Mol Biol Evol* **23**, 2001–7.
41. Raymond, J., Zhaxybayeva, O., Gogarten, J. P., Gerdes, S. Y., Blankenship, R. E. (2002) Whole-genome analysis of photosynthetic prokaryotes. *Science* **298**, 1616–20.
42. Bryant, D. A., Costas, A. M., Maresca, J. A., Chew, A. G., Klatt, C. G., Bateson, M. M., Tallon, L. J., Hostetler, J., Nelson, W. C., Heidelberg, J. F., Ward, D. M. (2007) *Candidatus Chloracidobacterium thermophilum*: an aerobic phototrophic Acidobacterium. *Science* **317**, 523–6.
43. Klappenbach, J. A., Pierson, B. K. (2004) Phylogenetic and physiological characterization of a filamentous anoxygenic photoautotrophic bacterium '*Candidatus Chlorothrix halophila*' gen. nov., sp. nov., recovered from hypersaline microbial mats. *Arch Microbiol* **181**, 17–25.
44. Igarashi, N., Harada, J., Nagashima, S., Matsuura, K., Shimada, K., Nagashima, K. V. (2001) Horizontal transfer of the photosynthesis gene cluster and operon rearrangement in purple bacteria. *J Mol Evol* **52**, 333–41.
45. Zhaxybayeva, O., Gogarten, J. P., Charlebois, R. L., Doolittle, W. F., Papke, R. T. (2006) Phylogenetic analyses of cyanobacterial genomes: quantification of horizontal gene transfer events. *Genome Res* **16**, 1099–108.
46. Mulkidjanian, A. Y., Koonin, E. V., Makarova, K. S., Mekhedov, S. L., Sorokin, A., Wolf, Y. I., Dufresne, A., Partensky, F., Burd, H., Kaznadzey, D., Haselkorn, R., Galperin, M. Y. (2006) The cyanobacterial genome core and the origin of photosynthesis. *Proc Natl Acad Sci U S A* **103**, 13126–31.
47. Granick, S. (1965) Evolution of heme and chlorophyll, in *Evolving Genes and Proteins* (Vogel, H. J., ed.) Academic Press, New York 67–8.
48. Granick, S., Beale, S. I. (1978) Hemes, chlorophylls, and related compounds: biosynthesis and metabolic regulation. *Adv Enzymol Relat Areas Mol Biol* **46**, 33–203.
49. Blankenship, R. E., Hartman, H. (1998) The origin and evolution of oxygenic photosynthesis. *Trends Biochem Sci* **23**, 94–7.
50. Raymond, J., Blankenship, R. E. (2007) The origin of the oxygen-evolving complex. *Coord Chem Rev* **252**, 377–83.
51. Mccord, J. M., Keele, B. B., Jr., Fridovich, I. (1971) An enzyme-based theory of obligate anaerobiosis: the physiological function of superoxide dismutase. *Proc Natl Acad Sci U S A* **68**, 1024–7.
52. Klotz, M. G., Klassen, G. R., Loewen, P. C. (1997) Phylogenetic relationships among prokaryotic and eukaryotic catalases. *Mol Biol Evol* **14**, 951–8.
53. Klotz, M. G., Loewen, P. C. (2003) The molecular evolution of catalatic hydroperoxidases: evidence for multiple lateral transfer of genes between prokaryota and from bacteria into eukaryota. *Mol Biol Evol* **20**, 1098–112.
54. Fast, N. M., Law, J. S., Williams, B. A., Keeling, P. J. (2003) Bacterial catalase in the microsporidian *Nosema locustae*: implications for microsporidian metabolism and genome evolution. *Eukaryot Cell* **2**, 1069–75.
55. Faguy, D. M., Doolittle, W. F. (2000) Horizontal transfer of catalase-peroxidase genes between archaea and pathogenic bacteria. *Trends Genet* **16**, 196–7.
56. Amo, T., Atomi, H., Imanaka, T. (2002) Unique presence of a manganese catalase in a hyperthermophilic archaeon, *Pyrobaculum calidifontis* VA1. *J Bacteriol* **184**, 3305–12.

57. Choudhury, S. B., Lee, J. W., Davidson, G., Yim, Y. I., Bose, K., Sharma, M. L., Kang, S. O., Cabelli, D. E., Maroney, M. J. (1999) Examination of the nickel site structure and reaction mechanism in *Streptomyces seoulensis* superoxide dismutase. *Biochemistry* **38**, 3744–52.
58. Wolfe-Simon, F., Starovoytov, V., Reinfelder, J. R., Schofield, O., Falkowski, P. G. (2006) Localization and role of manganese superoxide dismutase in a marine diatom. *Plant Physiol* **142**, 1701–9.
59. Smith, M. W., Doolittle, R. F. (1992) A comparison of evolutionary rates of the two major kinds of superoxide dismutase. *J Mol Evol* **34**, 175–84.
60. Fink, R. C., Scandalios, J. G. (2002) Molecular evolution and structure—function relationships of the superoxide dismutase gene families in angiosperms and their relationship to other eukaryotic and prokaryotic superoxide dismutases. *Arch Biochem Biophys* **399**, 19–36.

# Chapter 20

# Horizontal Gene Transfer in Cyanobacterial Signature Genes

**Shailaja Yerrapragada, Janet L. Siefert, and George E. Fox**

## Abstract

Comparison of 15 phylogenetically diverse cyanobacterial genomes identified an updated list of 183 signature genes that are widely found in cyanobacteria but absent in non-cyanobacterial species. These signature genes comprise the unique portion of the core cyanobacterial phenotype, and their absence from other lineages implies that if they arose by horizontal gene transfer (HGT), it likely occurred before the last shared cyanobacterial ancestor. A remaining issue is whether or not these signature genes would be relatively immune to HGT within the cyanobacterial lineage. Phylogenetic trees for each signature gene were constructed and compared to cyanobacterial groupings based on 16S rRNA sequences, with clear incongruence considered indicative of HGT. Approximately 18% of the signature genes exhibited such anomalies, indicating that the incidence of inter-lineage HGT has been significant. A preliminary analysis of intra-lineage transfer was conducted using four *Synechococcus/Prochlorococcus* species. In this case, it was found that 13% of the signature genes had likely been involved in within group HGT. In order to compare this level of likely HGT to other gene types, the analysis was extended to 1380 genes shared by the four *Synechococcus/Prochlorococcus* species. Successful HGT events appear to be most frequent among genes involved in photosynthesis/respiration and genes of unknown function, many of which are signature genes. This is consistent with the hypothesis that genes that most directly effect competition and adaptation of similar species in neighboring niches would be most usefully transferred. Such genes may be more easily integrated into a new genomic environment due to close similarities in regulatory circuits. In summary, signature genes are not immune from HGT and in fact may be favored candidates for HGT among closely related cyanobacterial strains.

**Key words:** Horizontal gene transfer, signature genes, core genes, cyanobacteria, *Synechococcus/Prochlorococcus*.

## 1. Introduction

Genomic variation in species belonging to the same evolutionary lineage cannot be explained by vertical descent alone. HGT is a major evolutionary factor to be considered. HGT is the

Maria B. Gogarten et al. (eds.), *Horizontal Gene Transfer: Genomes in Flux, vol. 532*

DOI 10.1007/978-1-60327-853-9_20 Springerprotocols.com

acquisition of foreign genes by any genome either by transformation, transduction, or conjugation *(1)*. In fresh water cyanobacteria, two major modes of gene transfer have been identified: natural transformation and conjugation.

Natural transformation was first documented in *Synechococcus* sp. PCC 7942 (*Anacystis nidulans*) *(2)*, but later reported in other species of *Synechococcus (3)*, *Synechocystis (4,5) Nostoc muscorum*, or *Anabaena* sp. PCC 7120 *(6)*, and recently in *Thermosynechococcus elongatus* BP-1 *(7, 8)*. Transformation typically involves several stages including development of competence, DNA binding, DNA processing, DNA uptake, and integration of DNA into the genome of the recipient cell *(9)*. Type IV pili (Tfp) genes are involved in natural transformation among all the abovementioned cyanobacteria. No special factors required to induce competency in transformable cyanobacteria have been identified. However, light seems to play a positive influence on the ability to transform as was observed in both *Synechococcus* and *Synechocystis* species *(10)*. Exogenous DNA is taken up non-specifically during transformation and, depending on its length, may or may not be further cleaved for integration *(11)*. Homologous recombination of the incoming non-autonomously replicating plasmid DNA with the host chromosome and resident plasmids was observed in both *Synechococcus* sp. PCC 7942 and *Synechocystis* sp. PCC 6803 *(2)*. In addition, transposons may sometimes integrate foreign DNA into the host chromosome, as observed in *Synechocystis* sp. PCC 6803 *(12, 13)*.

Conjugation in cyanobacteria was first performed between several *Anabaena* strains with *(14)* and without *(15)* a cyanobacterial donor. Conjugation allows for increased genetic manipulations in cyanobacteria than natural transformation, as in natural transformation DNA may be cleaved during uptake, limiting the length of DNA that can be transferred and thereby decreasing the efficiency of homologous recombination *(16)*.

The importance of transduction in freshwater cyanobacteria remains uncertain at this point in time, though freshwater cyanophages have recently been identified *(17)*. In the case of marine cyanobacteria, many cyanophages have been found and they frequently contain important photosynthesis genes *(18–21)* as well as other genes from both archaea and bacteria. These cyanophages are thought to mediate transfer and recombination of these genes among the marine cyanobacteria *(22)*. It has also been predicted that these cyanophages serve as reservoirs for hosts and play an important role in the evolution of cyanobacterial photosynthesis.

In prokaryotic genomes, "informational" genes involved in translation, transcription, and replication were thought to be less transferable than the "operational" genes involved in metabolism

*(23)*; but any gene, even a ribosomal RNA gene, can be transferred *(24)*. Accordingly, one study found the most commonly transferred genes generally fell into three main biological functional categories: cell envelope, cell processes, and regulatory processes *(25)*. Genes belonging to the transport/binding category have also been found to be frequently subject to HGT with entire operons also being transferred *(26)*. Phylogenetic quantification of HGT events among 11 cyanobacterial genomes revealed that genes from all functional categories were subject to transfer with more transfers seen within cyanobacteria than between cyanobacteria and other phyla *(27)*. Also observed was an increase of "operational" gene transfers and a decrease in "informational" gene transfers in the inter-phylum transfers as compared to intra-phylum transfers. There is also evidence of extensive gene transfer within cyanobacteria, especially among strains belonging to the *Synechococcus/Prochlorococcus* species cluster *(28)*.

### 1.1. Methods to Detect Horizontal Gene Transfer

HGT is suggested when a gene or protein sequence from a particular organism has a stronger sequence similarity to a homolog from a very distant organism than it has with obviously closer relatives. Such gene transfers are very frequently recent events. However, one cannot always readily identify them as the result of HGT *(29, 30)*. Markers of HGT that can be used in detection include deviant GC composition, trinucleotide analysis, anomalous phylogenetic distribution, abnormal sequence similarity, and incongruent phylogenetic trees *(31, 32)*. Depending on the method, different types and relative ages of HGT events will be found *(33, 34)*. Although all these methods are subject to artifacts, phylogenetic inference is thought to be the best way to detect fixed HGT events *(35)*. The term phyletic or phylogenetic pattern was introduced with the delineation of families of orthologs to describe the pattern of species present or missing in a given cluster of orthologs *(36–38)*. Any unusual variation in the phyletic pattern of a set of orthologous genes from the phyletic pattern inferred from a 16S rRNA tree suggests probable HGT events. However, this does not always work because some genes evolve so rapidly they contain only limited conserved regions, resulting in little or no useful phylogenetic signal *(39)*. Methods that rely upon primary sequence properties such as GC content are best suited to detection of recent events, whereas phylogenetic reconstruction is more reliable in the detection of more ancient HGT events.

The main limitation on the phylogenetic approach is the reliability of the organism tree to which individual gene trees are being compared. Hence the approach can frequently detect recent events too. The most common artifact in the phylogenetic approach is long-branch attraction caused by accelerated evolution among different species *(40)*. Tree topology is a good indicator of probable HGT only when the critical nodes are strongly supported statistically, by bootstrap analysis, or by other methods *(39, 41)*.

### *1.2. The Core Hypothesis*

Cyanobacteria, as the progenitor for the endosymbiotic event leading to plastid organelle evolution, are of special interest in the study of HGT and previous studies have estimated the loss of genes to the nucleus of plants *(42)*. Various researchers have used the *Synechocystis* PCC 6803 genome to make estimations for HGT in the cyanobacteria using various methods *(43, 44)* with surprisingly disparate outcomes *(34)*. Rather than focus on all the genes, it has been argued that one should first examine the informational genes encoding transcription and translation machinery that are thought to represent an "evolutionarily stable core" of genes *(45)*. It is thought that such genes would seldom be transferred because each informational gene product is likely to interact with so many others that transfer across large evolutionary distances would be rare or impossible because the resulting hybrid informational complexes would be non-functional *(1)*. This core would then be used to define organismal phylogeny and organismal identity, and the other genes that did not conform to the same phylogeny would be inferred to have undergone HGT. However, Nesbo *et al.* *(46)* showed that even a "euryarchaeal core" from four genomes was subject to extensive HGT. Despite these potential difficulties, a phylogenetically stable core set of 323 genes, with homologs in cyanobacteria as well as other bacterial groups, was identified by comparing 13 diverse cyanobacterial genomes *(47)*.

### *1.3. Core Genomes*

Advances in DNA sequencing technology have made it possible to rapidly sequence multiple closely related genomes. This has led to the concept of a "core" or "pan" genome *(48–51)* that distinguishes genes present in essentially all members of a species cluster from those that occur sporadically. The universal set thus comprises the core genome for the species being considered and it will typically encompass the types of house-keeping genes envisioned in the original core hypothesis, but would typically include many others as well. Genes that are not universally found in the species cluster are likely to be involved in local environmental adaptation and hence thought to be more likely to be involved in HGT *(49)* than the genes in the core genome. Thus, a phylogenetic tree based on various combinations of genes in the *Synechococcus/Prochlorococcus* core genome as defined by comparison of 12 genome sequences consistently gave a single-tree topology with strong bootstrap support *(49)*. Moreover, this tree is very similar to the 16S rRNA tree.

### 1.4. Signature Genes

An alternative approach is to focus on the genes that account for the unique cyanobacterial phenotype and are therefore a characteristic signature of the group *(52)*. Such "signature genes" would be universal among cyanobacteria, but not found outside the cyanobacteria except of course in Eukaryota containing chloroplasts. These genes would be a subset of the cyanobacterial pan genome as recently defined *(48, 51)* and might also be used to study the evolutionary relationships between the cyanobacterial species. Because these genes are by definition uniquely found among the cyanobacteria, they would not have been subject to HGT from other bacterial groupings since the emergence of the cyanobacteria. Also, since they would be universal and presumably significant to aspects of the cyanobacterial life style that are shared, they may have only been laterally transferred among cyanobacteria in a limited way. Thus, one might expect that the numbers of HGT events among these signature genes would be lower than the transfer events in other gene groups.

Do such signature genes exist? An initial analysis of eight cyanobacterial genomes identified a set of 181 homologous genes common to cyanobacteria that lacked obvious homologs in other completely sequenced bacterial genomes *(53)*. Approximately 25% of the signature genes had at that point in time been associated with some function, which was in most cases related to photosynthesis. The function of the remaining genes was unknown, but it was expected that some of them would also be associated with photosynthesis.

## 2. Identification of Signature Genes

Fifteen cyanobacterial genomes were considered in the updated analysis described here. These were *Synechocystis* PCC 6803 (3.6 Mb), *Thermosynechococcus elongatus* BP-1 (2.6 Mb), *Anabaena* PCC 7120 (7.2 Mb), *Gloeobacter violaceus* PCC 7421 (4.6 Mb), *Synechococcus* sp. WH8102 (2.72 Mb), *Prochlorococcus* sp. MIT9313 (2.4 Mb), *Prochlorococcus* sp. MED4 (1.6 Mb), *Prochlorococcus marinus* SS120 (1.7 Mb), *Nostoc punctiforme* PCC 73102 (9.2 Mb), *Trichodesmium erythraeum* IMS101 (6.5 Mb), *Anabaena variabilis* ATCC 29413 (6.4 Mb), *Crocosphaera watsonii* WH8501 (6.2 Mb), *Synechococcus elongatus* PCC 7942 (2.7 Mb), *Synechococcus elongatus* PCC 6301, and *Synechococcus elongatus* PCC 7002. *Synechocystis* PCC6803 was the best annotated genome at the time of the initial study and as such was used as the reference genome for signature gene names. The following criteria were used to include and exclude genes from the signature set. Any gene that had a BLASTP search *E* value lower than $10^{-10}$ to a gene in a non-cyanobacterial species other than chloroplasts or chloroplast-containing eukaryotes was excluded from the signature set. In order to be included, genes were required to have a clear ortholog in 10 or more cyanobacterial genomes. Genes were considered to be orthologs if (a) the BLASTP results had an *E*-value cutoff below $10^{-10}$, (b) had at least 70% match to complete gene length and more than 50% amino acid identity match. The resulting signature genes were extracted and assembled into individual sequence files using the BioEdit software package (http://www.mbio.ncsu.edu/BioEdit/bioedit.html). Multiple sequence alignments were performed using ClustalW alignment tool *(54)*.

A total of 183 genes met the signature gene criteria. Of these, 112 (61%) were found in all 15 genomes, 39 (21%) were in 14 genomes, 22 (12%) were in 13 genomes, 5 (3%) were in 12 genomes, and 3 (2%) genes each were in 11 and 10 genomes. Eight additional genes, ssl3712, sll1162, slr1263, ssl2781, sll1390, slr0280, slr0208, and slr1177 (*Synechosystis* PCC 6803 gene numbers) are uniquely found in 6–8 of the 15 cyanobacteria. Approximately 49% of the signature genes (89/183) have at least a partially known function and 40% of these (36/89) are associated with photosynthesis. However, a significant portion of the signature genes, 94/183 or 51%, continue to be annotated as hypothetical proteins with no known function. Initially, most of the identified signature genes were found to be associated with photosynthesis in some manner. This likely reflected the fact that until recently, most studies on these organisms have been directed at the photosynthetic apparatus. As time has progressed, most of the genes with more recently assigned function **(Table 20.1)** are not obviously associated with photosynthesis, though in many cases the functional assignment is not well established. Twelve of the signature genes (sll0247, sll0359, sll0661, sll0822, sll0851, sll1271, sll1577, slr0042, slr0376, slr1459, slr1841, slr1908) were found to have likely paralogs in some cyanobacterial genomes. There were 29 genes removed from the initial published signature gene list because homologs are now known in other genomes. These were sll0621, sll1071, sll1109, sll1162, sll1382, sll1390, sll1654, sll1752, sll1898, slr0204, slr0208, slr0280, slr0304, slr0630, slr0732, slr0734, slr0954, slr1177, slr1263, slr1342, slr1384, slr1506, slr1535, slr1636, slr1978, slr2034, ssl0546, ssl2781, and ssl3712. Some of these might still be appropriately regarded as signature genes if it can be shown that the non-cyano homologs had in fact arisen by HGT from the cyanobacteria. Such an analysis has not been attempted.

**Table 20.1**
**Updated list of cyanobacterial signature genes**

| Gene name | Found in # of genomes | Gene function |
|---|---|---|
| all0403 | 11 | Cyanobacterial conserved hypothetical protein |
| sll0031 | 15 | **Possible Fe-S dehydrogenase** |
| sll0071 | 10 | **Possible glycosyl transferase** |
| sll0157 | 15 | **Zn-dependent hydrolase** |
| sll0169 | 15 | Cell division protein Ftn2 homolog |
| sll0199 | 13 | Plastocyanin PetE |
| sll0208 | 14 | **Ferritin** |
| sll0226 | 15 | Photosys I assembly related protein |
| sll0247 | 15 | Iron-stress chlorophyll-binding protein, homologous to PsbC (IsiA) |
| sll0258 | 13 | Cytochrome c550 PsbV |
| sll0272 | 15 | Cyanobacterial conserved hypothetical protein |
| sll0286 | 15 | **Acetyltransferase** |
| sll0288 | 15 | Septum site-determining protein MinC |
| sll0295 | 15 | **Possible phosphatase** |
| sll0350 | 15 | **Possible organic solvent tolerance protein** |
| sll0359 | 15 | Cyanobacterial conserved hypothetical protein |
| sll0372 | 14 | Cyanobacterial conserved hypothetical membrane protein |
| sll0427 | 15 | Photosys II manganese-stabilizing protein PsbO |
| sll0456 | 15 | **NAD-dependent epimerase/dehydratase** |
| sll0509 | 13 | Similar to 5′,5″-P-1,P-4-tetraphosphate phosphorylase II |
| sll0544 | 15 | **DNA-directed DNA polymerase III delta subunit HolA** |
| sll0558 | 14 | **Magnesium-protoporphyrin IX regulatory protein (GUN4)** |
| sll0584 | 14 | Cyanobacterial conserved hypothetical membrane protein |
| sll0608 | 15 | Cyanobacterial conserved hypothetical protein |
| sll0609 | 15 | Cyanobacterial conserved hypothetical membrane protein |
| **sll0610** | 15 | C-type cytochrome biogenesis protein CcdA |
| **sll0626** | 10 | Neutral invertase |
| sll0661 | 15 | Cyanobacterial conserved hypothetical outer membrane protein |
| sll0662 | 15 | 4Fe-4S type iron-sulfur protein |
| **sll0819** | 14 | Photosys I reaction center subunit III precursor PsaF |

(continued)

**Table 20.1 (continued)**

| Gene name | Found in # of genomes | Gene function |
|---|---|---|
| sll0822 | 15 | Cyanobacterial conserved hypothetical protein |
| sll0827 | 12 | Cyanobacterial conserved hypothetical membrane protein |
| sll0832 | 15 | **Holliday junction resolvase** |
| sll0851 | 15 | Photosys II CP43 protein PsbC |
| sll0854 | 14 | **Beta-lactamase-type transpeptidase** |
| sll0860 | 15 | Cyanobacterial conserved hypothetical protein |
| **sll0871** | 11 | Cyanobacterial conserved hypothetical protein |
| sll0933 | 15 | Cyanobacterial conserved hypothetical protein |
| **sll0934** | 15 | Cyanobacterial conserved hypothetical membrane protein |
| **sll0935** | 15 | Cyanobacterial conserved hypothetical protein |
| **sll1150** | 13 | Cyanobacterial conserved hypothetical membrane protein |
| sll1194 | 12 | Photosys II extrinsic protein PsbU |
| **sll1262** | 15 | **Possible NADH dehydrogenase I subunit N** |
| sll1271 | 15 | Probable porin |
| sll1317 | 15 | Apocytochrome f |
| sll1321 | 15 | ATP synthase protein 1 |
| sll1340 | 15 | Cyanobacterial conserved hypothetical protein |
| sll1372 | 14 | Cyanobacterial conserved hypothetical membrane protein |
| sll1376 | 13 | **Bifunctional sterol desaturase/short chain dehydrogenase** |
| sll1398 | 15 | Photosys II reaction center protein Psb28 |
| sll1399 | 12 | Cyanobacterial conserved hypothetical protein |
| sll1414 | 15 | **Photosystem II biogenesis protein Psb29** |
| sll1418 | 15 | Photosys II oxygen-evolving complex 23K protein PsbP2 |
| sll1507 | 15 | Salt-induced periplasmic protein |
| sll1577 | 15 | Phycocyanin beta subunit CpcB |
| sll1578 | 14 | Phycocyanin alpha subunit CpcA |
| sll1586 | 15 | Cyanobacterial conserved hypothetical protein |
| sll1632 | 15 | **Cell division protein FtsQ** |
| sll1656 | 12 | Cyanobacterial conserved hypothetical protein |
| sll1702 | 15 | Cyanobacterial conserved hypothetical protein |

(continued)

**Table 20.1 (continued)**

| Gene name | Found in # of genomes | Gene function |
|---|---|---|
| sll1737 | 15 | Cyanobacterial conserved hypothetical membrane protein |
| sll1797 | 14 | **Possible 4-hydroxybenzoate synthetase (chorismate lyase)** |
| **sll1879** | 15 | Response regulator |
| sll1902 | 14 | **Glutathione S-transferase** |
| **sll1925** | 11 | Cyanobacterial conserved hypothetical membrane protein |
| sll1926 | 14 | Cyanobacterial conserved hypothetical protein |
| sll1934 | 13 | Cyanobacterial conserved hypothetical protein |
| sll1968 | 15 | Photomixotrophic growth protein PmgA |
| sll1979 | 15 | **TPR repeat protein** |
| sll2002 | 15 | Cyanobacterial conserved hypothetical protein |
| sll2013 | 15 | Cyanobacterial conserved hypothetical protein |
| slr0013 | 13 | Cyanobacterial conserved hypothetical protein |
| slr0022 | 15 | **Possible Fe-S dehydrogenase** |
| slr0042 | 15 | Probable porin |
| slr0116 | 15 | Phycocyanobilin:ferredoxin oxidoreductase PcyA |
| slr0169 | 15 | **Possible carboxysome** |
| **slr0249** | 14 | Cyanobacterial conserved hypothetical protein |
| slr0250 | 13 | **Organic solvent tolerance protein OstA** |
| **slr0269** | 13 | Cyanobacterial conserved hypothetical membrane protein |
| slr0376 | 15 | Cyanobacterial conserved hypothetical protein |
| slr0418 | 13 | Transcription factor DevT |
| slr0438 | 13 | Cyanobacterial conserved hypothetical protein |
| slr0443 | 13 | Cyanobacterial conserved hypothetical protein |
| slr0503 | 14 | Cyanobacterial conserved hypothetical protein |
| **slr0565** | 14 | **Vitamin K epoxide reductase (VKOR)** |
| slr0575 | 15 | Cyanobacterial conserved hypothetical protein |
| slr0589 | 15 | Cyanobacterial conserved hypothetical membrane protein |
| slr0590 | 12 | Cyanobacterial conserved hypothetical protein |
| slr0598 | 14 | Cyanobacterial conserved hypothetical protein |
| **slr0613** | 15 | Cyanobacterial conserved hypothetical protein |

(continued)

**Table 20.1 (continued)**

| Gene name | Found in # of genomes | Gene function |
|---|---|---|
| slr0651 | 15 | **Possible Dihydropteroate synthase** |
| slr0728 | 14 | **O-antigen polymerase** |
| slr0730 | 15 | Cyanobacterial conserved hypothetical protein |
| slr0731 | 14 | Cyanobacterial conserved hypothetical protein |
| slr0737 | 15 | Photosys I subunit II PsaD |
| **slr0748** | 14 | Cyanobacterial conserved hypothetical protein |
| **slr0815** | 15 | Cyanobacterial conserved hypothetical protein |
| slr0816 | 13 | **Possible nucleic acid–binding protein** |
| **slr0872** | 14 | Cyanobacterial conserved hypothetical protein |
| slr0906 | 15 | Photosys II core light harvesting protein PsbB |
| **slr0921** | 14 | Cyanobacterial conserved hypothetical protein |
| **slr0923** | 14 | **Plastid-specific ribosomal protein** |
| slr0941 | 15 | **Polyketide cyclase/dehydrase** |
| **slr0948** | 15 | Cyanobacterial conserved hypothetical protein |
| slr1034 | 15 | **Single-stranded DNA-binding protein** |
| **slr1052** | 14 | Cyanobacterial conserved hypothetical protein |
| slr1122 | 14 | Cyanobacterial conserved hypothetical protein |
| slr1160 | 15 | Cyanobacterial conserved hypothetical protein |
| slr1195 | 15 | Cyanobacterial conserved hypothetical protein |
| slr1206 | 13 | Cyanobacterial conserved hypothetical protein |
| slr1220 | 14 | **Possible photosynthetic reaction centre subunit H** |
| slr1273 | 14 | Cyanobacterial conserved hypothetical protein |
| slr1287 | 15 | Cyanobacterial conserved hypothetical protein |
| slr1306 | 14 | Cyanobacterial conserved hypothetical protein |
| **slr1311** | 15 | Photosystem II D1 protein PsbA2 |
| slr1459 | 15 | Phycobilisome core component ApcF |
| slr1470 | 15 | **Possible Aldo/keto reductase** |
| slr1495 | 15 | **Rad3-related DNA helicase DinG** |
| slr1530 | 15 | **Segregation and condensation protein ScpA** |
| slr1557 | 15 | Cyanobacterial conserved hypothetical protein |

(continued)

**Table 20.1 (continued)**

| Gene name | Found in # of genomes | Gene function |
|---|---|---|
| slr1579 | 13 | Cyanobacterial conserved hypothetical protein |
| slr1596 | 11 | Light-induced proton extrusion protein PxcA |
| slr1599 | 14 | Cyanobacterial conserved hypothetical membrane protein |
| slr1600 | 15 | Cyanobacterial conserved hypothetical protein |
| slr1623 | 15 | **NADH dehydrogenase I subunit M** |
| slr1638 | 15 | Cyanobacterial conserved hypothetical protein |
| slr1645 | 14 | Photosys II protein PsbZ |
| slr1649 | 14 | Cyanobacterial conserved hypothetical protein |
| slr1655 | 14 | Photosys I subunit XI PsaL |
| slr1660 | 14 | Cyanobacterial conserved hypothetical protein |
| slr1677 | 15 | **Possible Lipid A disaccharide synthetase** |
| slr1699 | 14 | Cyanobacterial conserved hypothetical protein |
| slr1702 | 15 | Cyanobacterial conserved hypothetical protein |
| slr1780 | 13 | Cyanobacterial conserved hypothetical protein |
| slr1796 | 15 | **Thioredoxin** |
| slr1800 | 13 | Cyanobacterial conserved hypothetical protein |
| slr1834 | 15 | P700 apoprotein subunit Ia PsaA |
| slr1841 | 15 | Probable porin |
| slr1896 | 15 | Cyanobacterial conserved hypothetical protein |
| slr1900 | 15 | Cyanobacterial conserved hypothetical protein |
| slr1908 | 15 | Probable porin |
| slr1915 | 15 | Cyanobacterial conserved hypothetical protein |
| slr1918 | 15 | Cyanobacterial conserved hypothetical protein |
| slr1926 | 15 | Cyanobacterial conserved hypothetical protein |
| slr1946 | 13 | **Zinc-dependent metalloprotease** |
| slr1949 | 14 | Cyanobacterial conserved hypothetical protein |
| slr1990 | 15 | Cyanobacterial conserved hypothetical membrane protein |
| slr2032 | 15 | Cyanobacterial conserved hypothetical protein |
| slr2049 | 15 | **Phycoerythrin linker protein CpeS** |
| slr2144 | 15 | Cyanobacterial conserved hypothetical protein |

(continued)

**Table 20.1 (continued)**

| Gene name | Found in # of genomes | Gene function |
|---|---|---|
| **sml0011** | 15 | Cyanobacterial conserved hypothetical protein |
| smr0006 | 15 | Cytochrome b559 beta subunit PsbF |
| smr0009 | 15 | Photosys II protein PsbN |
| **ssl0105** | 14 | Cyanobacterial conserved hypothetical protein |
| ssl0352 | 15 | Cyanobacterial conserved hypothetical protein |
| ssl0353 | 15 | Cyanobacterial conserved hypothetical protein |
| ssl0461 | 15 | Cyanobacterial conserved hypothetical protein |
| ssl0563 | 15 | Photosys I subunit VII PsaC |
| ssl1263 | 15 | Cyanobacterial conserved hypothetical protein |
| ssl1417 | 15 | Cyanobacterial conserved hypothetical protein |
| **ssl1690** | 15 | Cyanobacterial conserved hypothetical protein |
| ssl1972 | 14 | Cyanobacterial conserved hypothetical protein |
| ssl2598 | 15 | Photosys II protein PsbH |
| ssl3364 | 14 | CP12 protein |
| ssl3379 | 15 | **ATP-dependent protease adaptor protein ClpS** |
| ssl3451 | 15 | **Possible regulator of histidine kinase NblS** |
| **ssl3829** | 15 | Cyanobacterial conserved hypothetical protein |
| ssr0109 | 13 | Cyanobacterial conserved hypothetical protein |
| **ssr0332** | 15 | Cyanobacterial conserved hypothetical protein |
| ssr0390 | 14 | Photosys I reaction center subunit X PsaK1 |
| ssr0657 | 13 | Cyanobacterial conserved hypothetical protein |
| ssr1041 | 15 | **Ferredoxin** |
| ssr1425 | 15 | Cyanobacterial conserved hypothetical protein |
| **ssr1528** | 10 | Cyanobacterial conserved hypothetical protein |
| ssr1698 | 13 | **Heme iron utilization protein** |
| ssr1789 | 15 | High light-inducible protein HliD |
| **ssr2551** | 15 | Cyanobacterial conserved hypothetical protein |
| ssr2595 | 15 | High light-inducible protein HliD |
| ssr2831 | 15 | Photosys I subunit IV PsaE |
| ssr2843 | 14 | **Tryptophan-rich protein** |

(continued)

**Table 20.1 (continued)**

| Gene name | Found in # of genomes | Gene function |
|---|---|---|
| ssr2998 | 14 | Cyanobacterial conserved hypothetical protein |
| **ssr3188** | 13 | Cyanobacterial conserved hypothetical protein |
| ssr3451 | 15 | Cytochrome b559 alpha subunit PsbE |

The table includes the genetic nomenclature for each gene based on the *Synechocystis* sp. PCC 6803 genome. The genes whose names appear in bold were not included on the earlier signature gene list, which was based on far fewer organisms *(53)*. Gene function appears in bold for those genes that were previously considered hypothetical but which have now been assigned a function.

**Table 20.2** lists the phyletic distribution of the signature gene absent in each genome. *Gloeobacter violaceus*, by far, has the highest number of missing genes. This likely reflects the fact that photosynthesis takes place in the cytoplasmic membrane of *Gloeobacter* and not in the thylakoid membranes as in all other cyanobacteria *(55)*. There are 29/183 (16%) signature genes that are not found in *Gloeobacter violaceus* and these are listed along with their function and size in **Table 20.3**. Most of the missing signature genes are hypothetical proteins that do not appear to code for membrane proteins. Also, only a few of the missing genes are very short as would typically be the case if they were regulators. The clustering of the signature genes in the genomes was also examined in order to identify possible regulatory groupings, e.g., operons. Generally, few such clusters were found. One exception was a cluster containing two genes, sll0608 and sll0609, found to be conserved in all 15 organisms: one of the genes in this cluster codes for a 4Fe-4S-type iron–sulfur protein while the function of the second gene is still unknown. The putative operon for the cytochrome b559 alpha and beta subunits is found conserved in 14 cyanobacterial genomes, but absent in *Synechocystis* sp. PCC 6803.

## 3. Inter-lineage HGT and Signature Genes

With the exception of the most deeply branching species, traditional phylogenetic reconstruction did not readily provide definitive resolution of the branching order of the individual lineages within the cyanobacterial radiation *(56)*. However, the monophyly of individual lineages has been robustly supported. An example of this is monophyly of the "Nostoc" group, containing *Nostoc* and two species of *Anabaena*, that is supported by phylogenetic inference as well as the phenotypic characteristics

**Table 20.2**
**Phyletic distribution of gene losses in each cyanobacterial genome**

| Organism | Gene(s) absent | # of genes absent | # of Uniquely missing genes |
|---|---|---|---|
| *Anabaena* PCC 7120 | 0 | 0 | 0 |
| *Anabaena variabilis* | slr0249, **sll1150** | 2 | 1 |
| *Crocosphaera watsonii* WH8501 | slr1122, slr1220, sll0286, slr0921 | 4 | 4 |
| *Gleobacter violaceus* PCC 7421 | **sll1399**, slr0731, ssr0390, ssr2998, **slr1800**, **sll1656**, slr1306, slr1596, slr1599, **sll1934**, slr1699, sll1797, sll0854, sll1376, **slr0438**, **slr0418**, sll1372, slr0204, sll0509, **slr1579**, slr1660, slr1052, **sll1150**, slr0923, slr0748 | 24 | 17 |
| *Nostoc punctiforme* | ssl1972, slr1384, ssr0109, sll0819 | 4 | 4 |
| *Prochlorococcus marinus* sp MED4 | **slr0250**, **slr1800**, **sll1194**, sll1578, sll1926, slr1206, ssl3364, **slr0816**, slr1649, **slr0443**, **sll0258**, **slr1946**, **slr0013**, **sll0071**, slr0630, **sll0827**, **slr1579**,**slr0590**. Slr0872 | 19 | 7 |
| *Prochlorococcus marinus* sp. MIT 9313 | **slr1780**, **sll1934**, **slr0418**, sll0584, slr0598, **slr0269** | 6 | 2 |
| *Prochlorococcus marinus* sp. SS120 | **slr0250**, **slr1780**, sll1654, **sll1194** **slr0816**, slr1273, **slr0443**, **sll0258**, **slr1946**, **slr0013**, **slr0590**, **slr0269** | 12 | 2 |
| *Synechococcus* sp. WH8102 | 0 | 0 | 0 |
| *Synechococcus elongatus* PCC 7942 | **sll1656**,**slr0438**,**sll0827**, ssr3188 | 4 | 1 |
| 2 *Synechococcus elongatus* PCC 9311 | **slr0613**, ssl3451, **slr0443**, **ssr0657**, **slr0590** | 5 | 1 |
| *Synechococcus elongatus* CC 7002 | | 139 | 85 |
| 2 *Thermosynechococcus elongatus* BP-1 | **sll1399**, **ssr1698**, ssr2843, sll0199, | 5 | 3 |
| *Trichodesmium erythraeum* | **ssr1698**, **sll0071**, ssl0105 | 3 | 1 |
| | | Total = 83 | Total = 42 |

Missing cyanobacterial signature genes in each organism are indicated by their *Synechocystis* PCC 6803 gene number in Column 2. Genes in bold are genes missing in more than one cyanobacterial genome.

**Table 20.3**
**List of 29 signature genes absent in *Gloeobacter violaceus***

| PCC 6803 gene # | Found in | Function | Protein length (aa) |
|---|---|---|---|
| sll0558 | 14 | Magnesium-protoporphyrin IX regulatory protein (GUN4) | 233 |
| sll1399 | 12 | Cyanobacterial conserved hypothetical protein | 379 |
| slr0731 | 14 | Cyanobacterial conserved hypothetical protein | 310 |
| ssr0390 | 14 | Photosys I reaction center subunit X PsaKl | 86 |
| ssr2998 | 14 | Cyanobacterial conserved hypothetical protein | 62 |
| slr1800 | 13 | Cyanobacterial conserved hypothetical protein | 329 |
| sll1656 | 12 | Cyanobacterial conserved hypothetical protein | 176 |
| slr1306 | 14 | Cyanobacterial conserved hypothetical protein | 464 |
| slr1599 | 14 | Cyanobacterial conserved hypothetical membrane protein | 229 |
| slr1052 | 14 | Cyanobacterial conserved hypothetical protein | 170 |
| sll1934 | 13 | Cyanobacterial conserved hypothetical protein | 138 |
| slr1699 | 14 | Cyanobacterial conserved hypothetical protein | 272 |
| sll1797 | 14 | Possible 4-hydroxybenzoate synthetase (chorismate lyase) | 169 |
| sll0854 | 14 | Beta-lactamase-type transpeptidase | 306 |
| sll1150 | 13 | Cyanobacterial conserved hypothetical membrane protein | 252 |
| sll1376 | 13 | Bifunctional sterol desaturase/short-chain dehydrogenase | 272 |
| sll1372 | 14 | Cyanobacterial conserved hypothetical membrane protein | 204 |
| slr1645 | 14 | Photosys II protein PsbZ | 131 |
| ssr0657 | 13 | Cyanobacterial conserved hypothetical protein | 73 |
| slr1949 | 14 | Cyanobacterial conserved hypothetical protein | 195 |
| slr0438 | 13 | Cyanobacterial conserved hypothetical protein | 101 |
| slr0418 | 13 | Transcription factor DevT | 294 |
| sll0372 | 14 | Cyanobacterial conserved hypothetical membrane protein | 195 |
| slr0923 | 14 | Plastid-specific ribosomal protein | 112 |
| slr0503 | 14 | Cyanobacterial conserved hypothetical protein | 331 |
| sll0509 | 13 | Similar to 5′,5″-P-1, P-4-tetraphosphate phosphorylase II | 275 |
| slr1579 | 13 | Cyanobacterial conserved hypothetical protein | 522 |
| slr0748 | 14 | Cyanobacterial conserved hypothetical protein | 230 |
| slr1660 | 14 | Cyanobacterial conserved hypothetical protein | 192 |

The first column is the *Synechocystis* PCC6803 signature gene #; second column shows how many genomes the gene is present in, third column is gene function; the fourth column lists the protein length in amino acids as in *Synechocystis* PCC6803.

such as heterocysts, which are diagnostic of the group *(57, 58)*. Another robustly monophyletic group includes members of the marine *Synechococcus/Prochlorococcus* group and related strains *(59)*. *Gloeobacter* represents one of the most deeply branching species of cyanobacteria, which likely diverged before the plastids *(60)*. Its expected deep placement is not readily explained by long-branch attraction, as the cyanobacterial lineage does not appear to have excessive sequence divergence *(61)*. A 16S rRNA tree (**Fig. 20.1**) preserves these core groupings and also agrees well with the topology obtained recently using large numbers of protein families *(62)*.

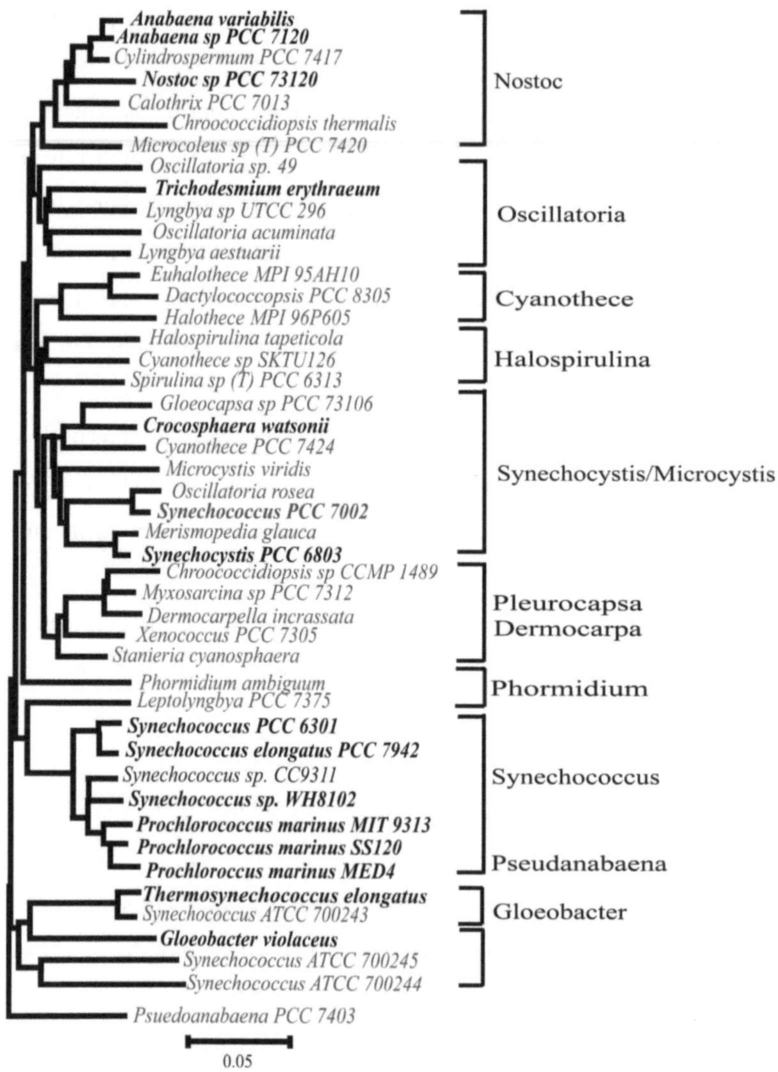


Fig. 20.1. Small subunit ribosomal RNA (16S rRNA) tree of cyanobacterial groups. Neighbor joining tree illustrating the major cyanobacterial groups *(56)*. Species in bold were used in constructing the signature gene set.

Signature genes likely define much of what was unique to the cyanobacterial genotype at the point of the group's last common ancestor and may carry meaningful information about the groups' subsequent evolutionary history, if they were not subject to significant lateral transfer. In order to assess lateral transfer, amino acid alignments were made for each signature gene using the Bioedit software *(54)*. Phylogenetic trees were then constructed using maximum parsimony *(63)*. The resulting trees were manually inter-compared for congruence in topology among themselves and with the 16S rRNA tree.

Many of the individual signature genes are short and, as single genes, it is unreasonable to expect each to exhibit precisely the same tree topology. Instead, each was examined for consistency in the presence of well-established subgroups that occur in the 16S rRNA tree and with one minor exception in the recent protein tree *(62)* as well. Possible lateral transfer events were inferred by the presence of aberrant subgroup composition without reference to how the subgroups were interconnected at the lower levels, as these lower level joins are variable from tree to tree due to the limited phylogenetic signal in the individual genes. The key groupings were the *Synechococcus/Prochlorococcus* cluster containing *Prochlorococcus* sp. MIT 9313, *Prochlorococcus* sp. MED4, *Prochlorococcus marinus* SS120, and *Synechococcus* sp. WH8102. Second, a group of ten species clustered into varying individual lineages commonly showing: (a) *Anabaena* sp. PCC 7120 with *Anabaena variabilis* ATCC29413 and *Nostoc punctiforme*; (b) *Synechocystis* PCC 6803, with *Crocosphaera watsonii* WH8501 and sometimes *Synechococcus elongatus* PCC 7002; (c) *Synechococcus elongatus* PCC 7942 with *Synechococcus elongatus* PCC 6301 – consistent with the 16S rRNA tree, **Fig. 20.1**, strains 7942 and 6301 are sometimes associated with the *Synechococcus/Prochlorococcus* group and this placement was not considered an anomaly; (d) *Trichodesmium erythraeum* and *Thermosynechococcus elongates* together or each alone; and third, *Gloeobacter violaceus* always clustered alone as the most divergent cyanobacterial species when it had a homolog of the gene being considered. The signature gene trees were always unrooted as by definition there is no outgroup. When only the presence/absence of these subgroupings is considered, most of the individual 15 species gene trees were topologically consistent with the 16S rRNA tree and likely have not been subjected to extensive lateral transfer. **Figure 20.2** is a representative example.

Within the major clusters, some variation in local topology of the sub clusters did occur. In some instances, as illustrated in **Fig. 20.3**, trees are obtained, which clearly differ from the usual tree, suggesting that HGT has occurred. In this example, *Crocosphaera* is associated with members of the *Nostoc* group instead of being a close relative with *Synechocystis* 6803 as is

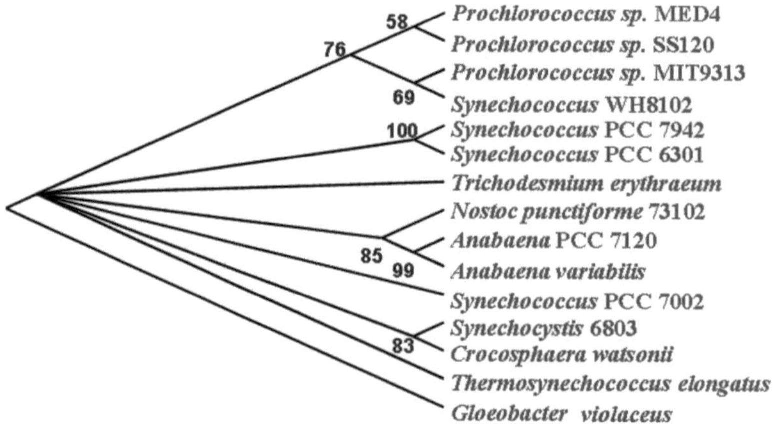


Fig. 20.2. Example of a common signature gene tree. An unrooted bootstrapped maximum parsimony phylogenetic tree of cyanobacterial signature gene sll1398 (a photosystem II reaction center 13 kDa protein) from *Synechocystis* PCC 6803 and its 14 homologs is shown. The figure illustrates the typical pattern seen with *Gloeobacter* as the most divergent member. Numbers at the nodes indicate bootstrap values.

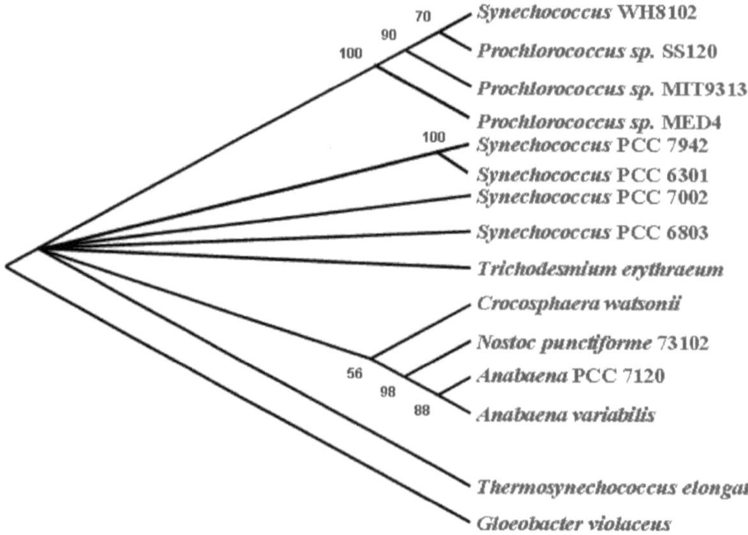


Fig. 20.3. Example of an atypical signature gene tree. An unrooted bootstrapped maximum parsimony phylogenetic tree of cyanobacterial signature gene sll1968 (a photomixotrophic growth related protein) *Synechocystis* PCC 6803 and its 14 homologs is shown. *Synechococcus/Prochlorococcus marinus* SS120 and *Crocosphaera* are highlighted in bold. The position of these two strains is highly aberrant as indicated in the tree and verified by inspection of the sequence alignments and actual changes seen. The placement strongly suggests HGT has occurred in the past.

usual. This topology suggests HGT occurred from the *Nostoc* group. The arrangement of organisms within the *Synechococcus/Prochlorococcus* cluster is also anomalous in this gene, suggesting intra-group HGT may have also occurred. The particular anomaly is of course potentially different for each specific gene. In those cases where a clearly aberrant tree was obtained, a detailed examination of the sequence alignment was undertaken to verify that the phylogenetic placements were reasonable and not an artifact of tree construction or bad alignment. Since only well-evidenced HGT events were considered, it is likely that the actual number of HGT events is being underestimated.

Overall, evidence of HGT between major clusters was found in 34 of the 183 (18.5%) signature genes. Ten of these examples involve *Synechococcus/Prochlorococcus* species, and in nine of the cases a larger cluster is created that typically includes *Synechococcus elongatus* PCC7942 and *Synechococcus* PCC 6301 and another species. Inclusion of the 7942/6301 grouping, although rare, was not regarded as anomalous but the other organism, which was frequently *T. elongatus*, was. Although the direction of transfer cannot be reliably inferred, of the 34 examples, the source appeared to be the *Nostoc* cluster in 16 cases.

## 4. Intra-lineage HGT and Signature Genes

In order to examine intra-lineage transfer, one must have larger organism groupings and hence in the present case this was only attempted with the four *Synechococcus/Prochlorococcus* species. Three of these species (MED4, MIT9313, and WH8102) had previously been shown to be closely related using rRNA, rpoC (a subunit of DNA-dependent RNA polymerase), and ribosomal DNA internal transcribed spacer sequences *(64–66)*. The fourth species, *Prochlorococcus marinus* SS120, has a very small genome like MED4 and was recently shown to also be closely related to the other three as well *(67)*. Both MIT9313 and SS120 are low light-adapted strains. Examination of the WH8102 strain genome *(68)* revealed that this marine cyanobacterium has been greatly influenced by HGT. This transfer appears to have been mediated in part by phages and especially involves genes that modify the cell surface and the ability for a unique kind of swimming motility. The abundance of phages in aquatic ecosystems *(69–71)* has implicated them as major factors in their own genomic diversity and as likely contributors to the genome dynamics of the ecosystem's microbial inhabitants as well. Within the *Synechococcus/Prochlorococcus* lineage, 151/183 or 82% of the signature genes produce a topology that clusters MIT9313 with WH8102 and MED4 with SS120. However, 25 genes (14%) show MIT9313 pairing with SS120 or MED4 and WH8102 with MED4 or SS120.

#### *4.1. What Types of Genes Are Most Susceptible to Intra-lineage HGT?*

Given that HGT is clearly occurring among signature genes within the *Synechococcus/Prochlorococcus* cluster, it is reasonable to ask how the rate of HGT among signature genes compares to that for other genes. A preliminary assessment of this was undertaken. The goal was to obtain an assessment of the relative frequency of HGT events in various categories of genes, not to identify all of the events or even to unequivocally prove any particular event.

All 1713 genes from MED4, which has the smallest genome of the four marine *Synechococcus/Prochlorococcus* species considered here, were inter-compared in order to find orthologs in MIT9313, SS120, and WH8102. Genes involved in photosynthesis/respiration and those of unknown function (hypothetical proteins) were additionally compared to orthologs from the freshwater *Synechococcus* species 6301 and 7942. Translated gene sequences for MED4, MIT9313, SS120, and WH8102 were retrieved from the Cyanobase website http://www.kazusa.or.jp/cyano/cyano.html. BLASTP scores with an $E$ value of $10^{-10}$ or lower were used as the cut-off point. All sequences were aligned using the ClustalW multiple sequence alignment tool *(54, 63)*.

333 of the 1713 MED4 genes (19%) had orthologs in less than 4 of the *Synechococcus/Prochlorococcus* species and were therefore not used in the phylogenetic analysis. For the 1380 genes with orthologs in all 4 species, phylogenetic trees were constructed by unrooted bootstrapped maximum parsimony. 90% of the genes (1237/1380) had the typical arrangement of MED4 with SS120 and MIT9313 pairing with WH8102. There were 143/1380 genes (10%) that showed two variant arrangements such as MED4/WH8102+MIT9313+SS120 or MIT9313/SS120+MED4+WH8102 (79%) and MIT9313/MED4+SS120+WH8102 or SS120/WH8102+MIT9313+MED4 (21%). Thus, the majority of the likely intra-lineage transfers appear to be between the low light-adapted species *Synechococcus/Prochlorococcus* sp. MIT9313 and *Synechococcus/Prochlorococcus* marinus SS120.

The genes were also grouped together by functional category as shown in **Table 20.4**.

It is apparent that likely HGT events among the *Synechococcus/Prochlorococcus* species are not uniformly distributed in various functional categories. Instead, likely HGT events were dramatically more common in genes that are known to be associated with photosynthesis/respiration as well as among the genes of unknown function. These are precisely the categories that encompass most signature genes, and of the 124 signature genes that occur in all four organisms, 13% appear to have been subjected

**Table 20.4**
**Possible HGT events among four marine *Synechococcus/Prochlorococcus* species by functional category**

| Gene category | Total genes | Genes <4 | Frequent pattern | Variant pattern *a* | Variant pattern *b* | TOTAL variants |
|---|---|---|---|---|---|---|
| Aminoacid biosynthesis genes | 77 | 3 | 72 (94%) | 2 | 0 | 2 |
| Biosynthesis of cofactor genes | 101 | 1 | 96 (95%) | 3 | 1 | 4 |
| Cell envelope genes | 36 | 4 | 29 (81%) | 3 | 0 | 3 |
| Cellular processes genes | 48 | 1 | 46 (96%) | 1 | 0 | 1 |
| Central intermediary metabolism genes | 13 | 7 | 5 (38%) | 1 | 0 | 1 |
| DNA replication, recombination genes | 56 | 8 | 47 (84%) | 1 | 0 | 1 |
| Energy metabolism genes | 21 | 2 | 18 (86%) | 1 | 0 | 1 |
| Fatty acid metabolism genes | 35 | 2 | 31 (86%) | 1 | 1 | 2 |
| Hypothetical genes | 534 | 145 | 335 (63%) | 45 (8%) | 9 (2%) | 54 (10%) |
| Other category genes | 274 | 92 | 167 (61%) | 8 | 7 | 15 |
| Photosynthesis & Respiratory genes | 86 | 3 | 61 (71%) | 20 (24%) | 2 (2%) | 22 (26%) |
| Purines, pyrimidines, nucleosides and nucleotides | 35 | 1 | 33 (94%) | 0 | 1 | 1 |
| Regulatory functions genes | 139 | 39 | 92 (66%) | 8 | 0 | 8 |
| Transcription genes | 30 | 3 | 26 (87%) | 1 | 0 | 1 |
| Translation genes | 131 | 3 | 115 (88%) | 11 | 2 | 13 |
| Transport genes | 97 | 18 | 65 (67%) | 11 | 3 | 14 |
| Total | 1713 | 332 (19%) | 1238 (72%) | 117 (7%) | 26 (2%) | 143 (9%) |
| Signature genes | 192 | 43 (22%) | 124 (65%) | 19 (10%) | 6 (3%) | 25 (13%) |

Usual pattern = MED4+SS120 or MIT9313+WH8102.
Variant pattern a = MIT9313+SS120 or MED4+WH8102.
Variant pattern b = MIT9313+MED4 or SS120+WH8102.

**Table 20.5**
**Relationship between gene sequence length and possible HGT events in *Synechococcus/Prochlorococcus* sp. MED4**

| Sequence length_aa | All MED4 genes | Typical pattern | Variant pattern | Photosynthesis/respiratory genes | | | Hypothetical genes | | |
|---|---|---|---|---|---|---|---|---|---|
| | | | | # of genes | Typical pattern | Atypical pattern | # of genes | Typical pattern | Atypical pattern |
| 0–99 | 150 | 119 (79%) | 31 (21%) | 24 | 17(71%) | 7(29%) | 81 | 65(80) | 16(20%) |
| 100–199 | 311 | 258 (83%) | 53 (17%) | 25 | 17(68%) | 8(32%) | 149 | 119(80%) | 30(20%) |
| 200–299 | 302 | 277 (92%) | 25 (8%) | 8 | 6(75%) | 2(25%) | 76 | 71(93%) | 5(7%) |
| 300–399 | 273 | 258 (95%) | 15 (5%) | 11 | 10(91%) | 1(9%) | 47 | 47(100%) | |
| 400–499 | 171 | 166 (97%) | 5 (3%) | 5 | 4(80%) | 1(20%) | 21 | 20(95%) | 1(5%) |
| 500–599 | 79 | 76 (96%) | 3 (4%) | 6 | 6(100%) | | 10 | 10(100%) | |
| 600–699 | 40 | 39 (98%) | 1 (2%) | 3 | 3(100%) | | 3 | 3(100%) | |
| 700–799 | 15 | 15 (100%) | 0 | 0 | | | 1 | 1(100%) | |
| 800–899 | 15 | 14 (93%) | 1 (7%) | 0 | | | 0 | | |
| 900–999 | 10 | 10 (100%) | 0 | 0 | | | 0 | | |
| > 1000 | 14 | 14 (100%) | 0 | 0 | | | 1 | 1(100%) | |

to HGT. Thus, instead of being immune from HGT, the signature genes may in fact be hot spots for HGT. All 1380 genes were further examined to determine the effect of the sequence length on the probability of a HGT event. The results are summarized in **Table 20.5**. It is clear that hypothetical HGT events were more likely to be found in genes with less than 200 amino acids. This may be a reflection of limited phylogenetic signal in the smaller genes. It would be useful to confirm these likely HGT events with a second approach such as examinations of sequence composition.

### 4.2. Future Opportunities for Studies of Intra-lineage HGT

In order to examine intra-lineage transfer, one must have larger groupings and hence in the present case this could only be attempted with the four *Synechococcus/Prochlorococcus* species whose genomes were available when this work was undertaken. Given the rapid increase in the number of available closely related genomes, it would already be possible to do this type of analysis for multiple strains from several widely disparate taxonomic groups including the lactic acid bacteria *(49)* and *Streptococcus (48)* strains as well as the rapidly expanding *Synechococcus/Prochlorococcus* group with others on the way as well. It thus should soon become clear which types of genes are most likely to be effected by HGT and how this mix is or is not changed by the nature of the organism.

## 5. Conclusion

At the outset, it seemed reasonable to suppose that signature genes would be more immune to HGT transfer than other genes. By their very definition, the signature genes are uniquely shared by most cyanobacteria and likely to have been present in the group for 2 billion years or longer. Under such conditions, they would have many opportunities to be integrated into the regulatory machinery in complex ways. In contrast, a new arrival by HGT, with identical function would be an orphan gene competing with the well-integrated original version and, hence, quite likely to be the version that gets eliminated. The probability of a successful HGT event involving a signature gene might, however, increase significantly in instances where the original version had been lost or damaged.

This proved not to be the case. Initial examination of the signature genes revealed that 18% of the signature genes may have experienced inter-lineage HGT within the modest set of genomes being considered. Although no analysis was undertaken to establish that the rate is actually more or less than that associated with other gene groups, it seems unlikely that this is an unusually low rate. When intra-lineage HGT was examined using four *Synechococcus/Prochlorococcus* genomes approximately 15% of the signature genes appear to have been involved in lateral transfer. A preliminary analysis was then undertaken to determine how this rate compares to that of other gene categories. It is clear from this analysis that genes involved in photosynthesis directly or indirectly as well as the genes of unknown function were far more likely to be involved in HGT than any other gene category. Most of the signature genes are in one of these two categories, and although the analysis was simplistic, the rate of intra-lineage HGT of signature genes is far greater than that seen for most other gene groups. So the initial speculation that signature genes might be relatively immune to HGT is clearly unreasonable. Because signature genes will typically be included in the "pan-genome" as currently defined, it might be appropriate to exclude them when using the pan genome to construct trees of relationship between closely related species. However, the inclusion of signature genes in the pan-genome in the case of *Prochlorococcus (51)* seems not to have been detrimental, likely because the number of signature genes is quite small compared to all the genes considered.

Most of the likely HGT events seen in the present analysis were between the two low light-adapted strains, MED4 and MIT. Clearly, ecological proximity and mechanisms for transfer, e.g., the large numbers of integrases in WH8102 *(68)*, are the necessary conditions for HGT. However, they need not be sufficient to produce a high frequency of successful HGT. Successful HGT between closely related strains will likely be favored if the transferred variant gene is advantageous to the donor strain. Closely related organisms would favor this because the substantial overlap in ecological niches and similar regulatory circuits is likely to make the variant gene advantageous in the acceptor strain too. Hence, a direct swap, by recombination perhaps, would have a good chance of being successful if the transfer allowed extension of the recipient organism's habitable zone. But why would signature genes be more likely to be successfully transferred than other genes? The answer likely resides in the fact that by their very definition, the signature genes are involved in the biochemical processes that make cyanobacteria unique. It is these processes that must continuously BE adapted to ecological conditions. In view of the preliminary results described here, and the current availability of at least 12 genomes from the *Synechococcus/Prochlorococcus* group now available *(51)* and more in process, future detailed studies of intra-lineage HGT in this group are likely to be fruitful.

## Acknowledgments

This work was supported in part by grants from the NSF LExEn program (#0085562) to JLS and the Institute of Space Systems Operation and the NASA Exobiology program (NNG05GN75G) to GEF.

### References


1. Jain, R., Rivera, M. C., Moore, J. E., Lake, J. A. (2002) Horizontal gene transfer in microbial genome evolution. *Theor Popul Biol* **61**, 489–95.
2. Nakasugi, K., Svenson, C. J., Neilan, B. A. (2006) The competence gene, comF, from *Synechocystis* sp. strain PCC 6803 is involved in natural transformation, phototactic motility and piliation. *Microbiology* **152**, 3623–31.
3. Golden, S. S., Sherman, L. A. (1984) Optimal conditions for genetic transformation of the cyanobacterium *Anacystis nidulans* R2. *J Bacteriol* **158**, 36–42.
4. Grigorieva, G., Shestakov, S. (1982) Transformation in cyanobacterium *Synechocystis* sp. 6803. *FEMS Microbiol Lett* **13**, 367–70.
5. Stevens-Jr, S. E., Porter, R. D. (1986) Heterospecific transformation among cyanobacteria. *J Bacteriology* **167**, 1074–6.
6. Trehan, K., Sinha, U. (1981) Genetic transfer in a nitrogen-fixing filamentous cyanobacterium. *J Gen Microbiology* **124**, 349–52.
7. Iwai, M., Katoh, H., Katayama, M., Ikeuchi, M. (2004) Improved genetic transformation of the thermophilic cyanobacterium, *Thermosynechococcus elongatus* BP-1. *Plant Cell Physiol* **45**, 171–5.
8. Onai, K., Morishita, M., Kaneko, T., Tabata, S., Ishiura, M. (2004) Natural transformation of the thermophilic cyanobacterium *Thermosynechococcus elongatus* BP-1: A simple and efficient method for gene transfer. *Mol Genet Genomics* **271**, 50–9.
9. Smith, H. O., Danner, D. B., Deich, R. A. (1981) Genetic transformation. *Annu Rev Biochem* **50**, 41–68.
10. Morrison, S. S., Mullineaux, C. W., Ashby, M. K. (2005) The influence of acetyl phosphate on DspA signalling in the cyanobacterium *Synechocystis* sp. PCC6803. *BMC Microbiol* **5**, 47.
11. Notani, N. K., Joshi, V.P., Kanade, R.P. (1984) Genetic transformation in bacteria. *J Biosci* **6**, 525–33.
12. Okamoto, S., Ikeuchi, M., Ohmori, M. (1999) Experimental analysis of recently transposed insertion sequences in the cyanobacterium *Synechocystis* sp. PCC 6803. *DNA Res* **6**, 265–73.
13. Ochman, H., Lawrence, J. G., Groisman, E. A. (2000) Lateral gene transfer and the nature of bacterial innovation. *Nature* **405**, 299–304.
14. Muro-Pastor, A. M., Kuritz, T., Flores, E., Herrero, A., Wolk, C. P. (1994) Transfer of a genetic marker from a megaplasmid of *Anabaena* sp. strain PCC 7120 to a megaplasmid of a different *Anabaena* strain. *J Bacteriol* **176**, 1093–8.
15. Wolk, C. P., Vonshak, A., Kehoe, P., Elhai, J. (1984) Construction of shuttle vectors capable of conjugative transfer from *Escherichia coli* to nitrogen-fixing filamentous cyanobacteria. *Proc Natl Acad Sci U S A* **81**, 1561–5.
16. Elhai, J., Wolk, C. P. (1988) Conjugal transfer of DNA to cyanobacteria. *Methods Enzymol* **167**, 747–54.
17. Matteson, W. S. A. A. (2008) Freshwater and marine virioplankton: A brief overview of commonalities and differences. *Freshw Biol* **53**, 1076–89.
18. Mann, N. H., Clokie, M. R., Millard, A., Cook, A., Wilson, W. H., Wheatley, P. J., Letarov, A., Krisch, H. M. (2005) The genome of S-PM2, a "photosynthetic" T4-type bacteriophage that infects marine *Synechococcus* strains. *J Bacteriol* **187**, 3188–200.
19. Lindell, D., Sullivan, M. B., Johnson, Z. I., Tolonen, A. C., Rohwer, F., Chisholm, S. W. (2004) Transfer of photosynthesis genes to and from *Prochlorococcus* viruses. *Proc Natl Acad Sci U S A* **101**, 11013–8.
20. Sullivan, M. B., Coleman, M. L., Weigele, P., Rohwer, F., Chisholm, S. W. (2005) Three *Prochlorococcus* cyanophage genomes: Signature features and ecological interpretations. *PLoS Biol* **3**, e144.
21. Weigele, P. R., Pope, W. H., Pedulla, M. L., Houtz, J. M., Smith, A. L., Conway, J. F., King, J., Hatfull, G. F., Lawrence, J. G., Hendrix, R. W. (2007)

Genomic and structural analysis of Syn9, a cyanophage infecting marine *Prochlorococcus* and *Synechococcus*. *Environ Microbiol* **9**, 1675–95.

22. Zeidner, G., Bielawski, J. P., Shmoish, M., Scanlan, D. J., Sabehi, G., Beja, O. (2005) Potential photosynthesis gene recombination between *Prochlorococcus* and *Synechococcus* via viral intermediates. *Environ Microbiol* **7**, 1505–13.
23. Rivera, M. C., Jain, R., Moore, J. E., Lake, J. A. (1998) Genomic evidence for two functionally distinct gene classes. *Proc Natl Acad Sci U S A* **95**, 6239–44.
24. Yap, W. H., Zhang, Z., Wang, Y. (1999) Distinct types of rRNA operons exist in the genome of the actinomycete *Thermomonospora chromogena* and evidence for horizontal transfer of an entire rRNA operon. *J Bacteriol* **181**, 5201–9.
25. Nakamura, Y., Itoh, T., Matsuda, H., Gojobori, T. (2004) Biased biological functions of horizontally transferred genes in prokaryotic genomes. *Nat Genet* **36**, 760–6.
26. Shi, S. Y., Cai, X. H., Ding, D. F. (2005) Identification and categorization of horizontally transferred genes in prokaryotic genomes. *Acta Biochim Biophys Sin (Shanghai)* **37**, 561–6.
27. Zhaxybayeva, O., Gogarten, J. P., Charlebois, R. L., Doolittle, W. F., Papke, R. T. (2006) Phylogenetic analyses of cyanobacterial genomes: Quantification of horizontal gene transfer events. *Genome Res* **16**, 1099–108.
28. Beiko, R. G., Harlow, T. J., Ragan, M. A. (2005) Highways of gene sharing in prokaryotes. *Proc Natl Acad Sci U S A* **102**, 14332–7.
29. Hayashi, T., Makino, K., Ohnishi, M., Kurokawa, K., Ishii, K., Yokoyama, K., Han, C. G., Ohtsubo, E., Nakayama, K., Murata, T., Tanaka, M., Tobe, T., Iida, T., Takami, H., Honda, T., Sasakawa, C., Ogasawara, N., Yasunaga, T., Kuhara, S., Shiba, T., Hattori, M., Shinagawa, H. (2001) Complete genome sequence of enterohemorrhagic *Escherichia coli* O157:H7 and genomic comparison with a laboratory strain K-12. *DNA Res* **8**, 11–22.
30. Welch, R. A., Burland, V., Plunkett, G., 3rd, Redford, P., Roesch, P., Rasko, D., Buckles, E. L., Liou, S. R., Boutin, A., Hackett, J., Stroud, D., Mayhew, G. F., Rose, D. J., Zhou, S., Schwartz, D. C., Perna, N. T., Mobley, H. L., Donnenberg, M. S., Blattner, F. R. (2002) Extensive mosaic structure revealed by the complete genome sequence of uropathogenic *Escherichia coli*. *Proc Natl Acad Sci U S A* **99**, 17020–4.
31. Eisen, J. A. (2000) Horizontal gene transfer among microbial genomes: New insights from complete genome analysis. *Curr Opin Genet Dev* **10**, 606–11.
32. Koonin, E. V., Makarova, K. S., Aravind, L. (2001) Horizontal gene transfer in prokaryotes: Quantification and classification. *Annu Rev Microbiol* **55**, 709–42.
33. Ragan, M. A. (2001) Detection of lateral gene transfer among microbial genomes. *Curr Opin Genet Dev* **11**, 620–6.
34. Lawrence, J. G., Ochman, H. (2002) Reconciling the many faces of lateral gene transfer. *Trends Microbiol* **10**, 1–4.
35. Syvanen, M. (1994) Horizontal gene transfer: Evidence and possible consequences. *Annu Rev Genet* **28**, 237–61.
36. Gaasterland, T., Ragan, M. A. (1998) Microbial genescapes: Phyletic and functional patterns of ORF distribution among prokaryotes. *Microb Comp Genomics* **3**, 199–217.
37. Tatusov, R. L., Galperin, M. Y., Natale, D. A., Koonin, E. V. (2000) The COG database: A tool for genome-scale analysis of protein functions and evolution. *Nucleic Acids Res* **28**, 33–6.
38. Tatusov, R. L., Koonin, E. V., Lipman, D. J. (1997) A genomic perspective on protein families. *Science* **278**, 631–7.
39. Brown, J. R., Doolittle, W. F. (1999) Gene descent, duplication, and horizontal transfer in the evolution of glutamyl- and glutaminyl-tRNA synthetases. *J Mol Evol* **49**, 485–95.
40. Moreira, D., Philippe, H. (2000) Molecular phylogeny: pitfalls and progress. *Int Microbiol* **3**, 9–16.
41. Efron, B., Halloran, E., Holmes, S. (1996) Bootstrap confidence levels for phylogenetic trees. *Proc Natl Acad Sci U S A* **93**, 13429–34.
42. Martin, W., Rujan, T., Richly, E., Hansen, A., Cornelsen, S., Lins, T., Leister, D., Stoebe, B., Hasegawa, M., Penny, D. (2002) Evolutionary analysis of *Arabidopsis*, cyanobacterial, and chloroplast genomes reveals plastid phylogeny and thousands of cyanobacterial genes in the nucleus. *Proc Natl Acad Sci U S A* **99**, 12246–51.
43. Mrazek, J., Bhaya, D., Grossman, A. R., Karlin, S. (2001) Highly expressed and alien genes of the *Synechocystis* genome. *Nucleic Acids Res* **29**, 1590–601.
44. Garcia-Vallve, S., Romeu, A., Palau, J. (2000) Horizontal gene transfer in bacterial and archaeal complete genomes. *Genome Res* **10**, 1719–25.
45. Makarova, K. S., Aravind, L., Galperin, M. Y., Grishin, N. V., Tatusov, R. L., Wolf, Y. I., Koonin, E. V. (1999) Comparative genomics of the Archaea (Euryarchaeota): Evolution of

conserved protein families, the stable core, and the variable shell. *Genome Res* **9**, 608–28.
46. Nesbo, C. L., Boucher, Y., Doolittle, W. F. (2001) Defining the core of nontransferable prokaryotic genes: The euryarchaeal core. *J Mol Evol* **53**, 340–50.
47. Shi, T., Falkowski, P. G. (2008) Genome evolution in cyanobacteria: The stable core and the variable shell. *Proc Natl Acad Sci U S A* **105**, 2510–5.
48. Tettelin, H., Masignani, V., Cieslewicz, M. J., Donati, C., Medini, D., Ward, N. L., Angiuoli, S. V., Crabtree, J., Jones, A. L., Durkin, A. S., Deboy, R. T., Davidsen, T. M., Mora, M., Scarselli, M., Margarit Y Ros, I., Peterson, J. D., Hauser, C. R., Sundaram, J. P., Nelson, W. C., Madupu, R., Brinkac, L. M., Dodson, R. J., Rosovitz, M. J., Sullivan, S. A., Daugherty, S. C., Haft, D. H., Selengut, J., Gwinn, M. L., Zhou, L., Zafar, N., Khouri, H., Radune, D., Dimitrov, G., Watkins, K., O'connor, K. J., Smith, S., Utterback, T. R., White, O., Rubens, C. E., Grandi, G., Madoff, L. C., Kasper, D. L., Telford, J. L., Wessels, M. R., Rappuoli, R., Fraser, C. M. (2005) Genome analysis of multiple pathogenic isolates of Streptococcus agalactiae: Implications for the microbial "pan-genome". *Proc Natl Acad Sci U S A* **102**, 13950–5.
49. Makarova, K., Slesarev, A., Wolf, Y., Sorokin, A., Mirkin, B., Koonin, E., Pavlov, A., Pavlova, N., Karamychev, V., Polouchine, N., Shakhova, V., Grigoriev, I., Lou, Y., Rohksar, D., Lucas, S., Huang, K., Goodstein, D. M., Hawkins, T., Plengvidhya, V., Welker, D., Hughes, J., Goh, Y., Benson, A., Baldwin, K., Lee, J. H., Diaz-Muniz, I., Dosti, B., Smeianov, V., Wechter, W., Barabote, R., Lorca, G., Altermann, E., Barrangou, R., Ganesan, B., Xie, Y., Rawsthorne, H., Tamir, D., Parker, C., Breidt, F., Broadbent, J., Hutkins, R., O'sullivan, D., Steele, J., Unlu, G., Saier, M., Klaenhammer, T., Richardson, P., Kozyavkin, S., Weimer, B., Mills, D. (2006) Comparative genomics of the lactic acid bacteria. *Proc Natl Acad Sci U S A* **103**, 15611–6.
50. Mulkidjanian, A. Y., Koonin, E. V., Makarova, K. S., Mekhedov, S. L., Sorokin, A., Wolf, Y. I., Dufresne, A., Partensky, F., Burd, H., Kaznadzey, D., Haselkorn, R., Galperin, M. Y. (2006) The cyanobacterial genome core and the origin of photosynthesis. *Proc Natl Acad Sci U S A* **103**, 13126–31.
51. Kettler, G. C., Martiny, A. C., Huang, K., Zucker, J., Coleman, M. L., Rodrigue, S., Chen, F., Lapidus, A., Ferriera, S., Johnson, J., Steglich, C., Church, G. M., Richardson, P., Chisholm, S. W. (2007) Patterns and implications of gene gain and loss in the evolution of *Prochlorococcus. PLoS Genet* **3**, e231.
52. Graham, D. E., Overbeek, R., Olsen, G. J., Woese, C. R. (2000) An archaeal genomic signature. *Proc Natl Acad Sci U S A* **97**, 3304–8.
53. Martin, K. A., Siefert, J. L., Yerrapragada, S., Lu, Y., Mcneill, T. Z., Moreno, P. A., Weinstock, G. M., Widger, W. R., Fox, G. E. (2003) Cyanobacterial signature genes. *Photosynth Res* **75**, 211–21.
54. Thompson, J. D., Higgins, D. G., Gibson, T. J. (1994) CLUSTAL W: Improving the sensitivity of progressive multiple sequence alignment through sequence weighting, position-specific gap penalties and weight matrix choice. *Nucleic Acids Res* **22**, 4673–80.
55. Nakamura, Y., Kaneko, T., Sato, S., Mimuro, M., Miyashita, H., Tsuchiya, T., Sasamoto, S., Watanabe, A., Kawashima, K., Kishida, Y., Kiyokawa, C., Kohara, M., Matsumoto, M., Matsuno, A., Nakazaki, N., Shimpo, S., Takeuchi, C., Yamada, M., Tabata, S. (2003) Complete genome structure of *Gloeobacter violaceus* PCC 7421, a cyanobacterium that lacks thylakoids (supplement). *DNA Res* **10**, 181–201.
56. Turner, S., Pryer, K. M., Miao, V. P., Palmer, J. D. (1999) Investigating deep phylogenetic relationships among cyanobacteria and plastids by small subunit rRNA sequence analysis. *J Eukaryot Microbiol* **46**, 327–38.
57. Castenholz, R. W. (1989) Order Oscillatoriales. In *Bergey's Manual of Systematic Bacteriology*, Vol. **3**, Williams & Wilkins, London, pp. 1771–80.
58. Rippka, R., Deruelles, J., Waterbury, J.B., Herdman, M., Stanier, R.Y. (1979) Generic assignments, strain histories and properties of pure cultures of cyanobacteria. *J Gen Microbiology* **111**, 1–61.
59. Waterbury, J. B., Watson, S.W., Valois, F.W., Franks, D.G. (1986) Biological and ecological characterization of the maine unicellular cyanobacterium *Synechococcus. Can Bull Fish Aquat Sci* **214**, 71–120.
60. Nelissen, B., Van De Peer, Y., Wilmotte, A., De Wachter, R. (1995) An early origin of plastids within the cyanobacterial divergence is suggested by evolutionary trees based on complete 16S rRNA sequences. *Mol Biol Evol* **12**, 1166–73.
61. Garrity, G., Boone, D. R., Castenholz, R.W. (2001) Phylum BX. Cyanobacteria, oxygenic photosynthetic bacteria. *Bergey's Manual of Systematic Bacteriology*, Vol., **1**, Springer, New York.
62. Swingley, W. D., Blankenship, R. E., Raymond, J. (2008) Integrating Markov clustering and molecular phylogenetics to recon-

struct the cyanobacterial species tree from conserved protein families. *Mol Biol Evol* **25**, 643–54.
63. Kumar, S., Tamura, K., Nei, M. (1994) MEGA: Molecular Evolutionary Genetics Analysis software for microcomputers. *Comput Appl Biosci* **10**, 189–91.
64. Palenik, B., Haselkorn, R. (1992) Multiple evolutionary origins of prochlorophytes, the chlorophyll b-containing prokaryotes. *Nature* **355**, 265–7.
65. Urbach, E., Robertson, D. L., Chisholm, S. W. (1992) Multiple evolutionary origins of prochlorophytes within the cyanobacterial radiation. *Nature* **355**, 267–70.
66. Rocap, G., Distel, D. L., Waterbury, J. B., Chisholm, S. W. (2002) Resolution of *Prochlorococcus* and *Synechococcus* ecotypes by using 16S-23S ribosomal DNA internal transcribed spacer sequences. *Appl Environ Microbiol* **68**, 1180–91.
67. Fuhrman, J. (2003) Genome sequences from the sea. *Nature* **424**, 1001–2.
68. Palenik, B., Brahamsha, B., Larimer, F. W., Land, M., Hauser, L., Chain, P., Lamerdin, J., Regala, W., Allen, E. E., Mccarren, J., Paulsen, I., Dufresne, A., Partensky, F., Webb, E. A., Waterbury, J. (2003) The genome of a motile marine *Synechococcus*. *Nature* **424**, 1037–42.
69. Wommack, K. E., Colwell, R. R. (2000) Virioplankton: Viruses in aquatic ecosystems. *Microbiol Mol Biol Rev* **64**, 69–114.
70. Pedulla, M. L., Ford, M. E., Houtz, J. M., Karthikeyan, T., Wadsworth, C., Lewis, J. A., Jacobs-Sera, D., Falbo, J., Gross, J., Pannunzio, N. R., Brucker, W., Kumar, V., Kandasamy, J., Keenan, L., Bardarov, S., Kriakov, J., Lawrence, J. G., Jacobs, W. R., Jr., Hendrix, R. W., Hatfull, G. F. (2003) Origins of highly mosaic mycobacteriophage genomes. *Cell* **113**, 171–82.
71. Suttle, C. A. (2007) Marine viruses – major players in the global ecosystem. *Nat Rev Microbiol* **5**, 801–12.

# Chapter 21

## Population Genomics and the Bacterial Species Concept

**Margaret A. Riley and Michelle Lizotte-Waniewski**

### Abstract

In recent years, the importance of horizontal gene transfer (HGT) in bacterial evolution has been elevated to such a degree that many bacteriologists now question the very existence of bacterial species. If gene transfer is as rampant as comparative genomic studies have suggested, how could bacterial species survive such genomic fluidity? And yet, most bacteriologists recognize, and name, as species, clusters of bacterial isolates that share complex phenotypic properties. The Core Genome Hypothesis (CGH) has been proposed to explain this apparent paradox of fluid bacterial genomes associated with stable phenotypic clusters. It posits that there is a core of genes responsible for maintaining the species-specific phenotypic clusters observed throughout bacterial diversity and argues that, even in the face of substantial genomic fluidity, bacterial species can be rationally identified and named.

**Key words:** Bacterial species concept, core genome hypothesis, population genomics.

## 1. Introduction

The impact of molecular systematics on bacterial classification has been profound. Indeed, phylogenies based on highly conserved molecules, such as ribosomal RNA, have fundamentally changed our view of biological diversity *(1, 2)*. These molecular phylogenies have confirmed the existence of three primary divisions of life (Archaea, Bacteria, and Eukarya), rather than the five that had emerged from phenotype studies (Animalia, Plantae, Fungi, Protista, and Monera), and reveal that microbes comprise, by far, the greatest amount of biological diversity *(1, 3–5)*. Further, as additional highly conserved genes are examined, such as elongation factor-1α, actin, α-tubulin, and β-tubulin, we gain confidence that

Maria B. Gogarten et al. (eds.), *Horizontal Gene Transfer: Genomes in Flux, vol. 532*

DOI 10.1007/978-1-60327-853-9_21 Springerprotocols.com

molecular-based phylogenies can provide a fairly robust description of the major evolutionary lineages *(6, 7)*.

Just as molecules appear to have solved some of the outstanding phylogenetic questions, their application has generated an entirely new and unexpected controversy. They have revealed that horizontal gene transfer (HGT) may play an important and unexpectedly large role in evolution (**Fig. 21.1**) *(8–11)*. Recent observations of putative gene transfer events between some of the deepest branches in the 16S ribosomal RNA-based tree of life have raised the question of whether we should employ networks, rather than dichotomously branching trees, to represent the relationships of evolutionary lineages over time *(12)*. In fact, the importance of HGT in bacterial evolution has been elevated to such a degree that numerous bacteriologists now question the very existence of bacterial species *(13–15)*. If gene transfer is as rampant as some propose, how could bacterial species survive such genomic fluidity?

Traditional bacterial species designations were based upon extensive phenotypic characterization of a large number of isolates. Although current methods now require the use of 16S rRNA sequence comparisons to identify the closest relatives of a proposed species, phenotype still remains the primary criterion by which species are identified *(16)*. The emerging phylo-phenetic bacterial species concept posits that a bacterial species is "a monophyletic and genomically coherent cluster of individual organisms that show a high degree of overall similarity with respect to many independent characteristics, and is diagnosable by a discriminative phenotypic property." *(16)*.

Numerous studies have revealed clusters of bacterial isolates that share complex phenotypes and these clusters are often designated as species *(17–21)*. In fact, Cohan uses the mere existence of

Fig. 21.1. **The impact of gene transfer on molecular phylogenies.** The tree on the *left* represents a dichotomously branching set of lineages. Over time the lineages diverge and result in extensive biological diversity. The tree on the *right* shows the impact of extensive HGT, which serves to limit divergence and the resulting diversity. Such lineages are homogenized by HGT and appear to have a shortened evolutionary history.

these clusters as prima facie evidence of the existence of bacterial species. He notes “Bacterial species exist ... bacterial diversity is organized into discrete phenotypic and genetic clusters, which are separated by large phenotypic and genetic gaps, and these clusters are recognized as species.” *(22)*.

## 2. Molecular-Based Species Distinctions

More recent investigations into microbial species distinctions have sought to incorporate estimates of molecular diversity into the process of species identification. The assumption is that this molecular diversity will fall into discrete clusters that correspond with observed phenotype-based species clusters. Can sequence variability be employed to inform the division of a genus into species, to distinguish among similar species, or to address whether bacterial species even exist *(23–28)*?

Early attempts to use molecular data to delineate bacterial species began with the introduction of DNA–DNA hybridization, in which bacterial species were defined as those isolates sharing at least 70% hybridization under standardized conditions *(29)*. Given the enormous range of genetic variation detected in different clearly recognized species, it became clear that a variability cutoff, such as is imposed with hybridization methods, was not appropriate. Levels of variability will vary over the lifetime of a species and will reflect aspects of its life history, particularly the process by which it adapts to its range of habitats.

The use of DNA–DNA hybridization has largely been replaced by the use of 16S rRNA sequences to determine the closest relatives of an isolate, combined with extensive phenotypic data. A disturbingly large number of publications now report species diversity based solely upon the quantity of novel 16S rRNA sequences detected *(30–32)*. This approach has no valid systematic basis and should be avoided at all cost *(16, 33)*. Variability in 16S rRNA sequences can provide a valid estimate of molecular diversity, but that estimate cannot necessarily be equated with species diversity.

One of the first gene-based investigations into the microbial species concept was conducted in 2003 by Wertz et al., who sequenced six housekeeping genes from a sample of environmental bacteria representing seven species of Enterobacteriaceae *(34, 35)*. Molecular phylogenies for each of the genes were inferred (**Fig. 21.2**) and the branching patterns of the resulting trees compared. In almost every case, isolates from a species formed a well-supported monophyletic group, which corresponded precisely with the clusters identified by phenotypic data, and upon which species distinctions were initially delineated

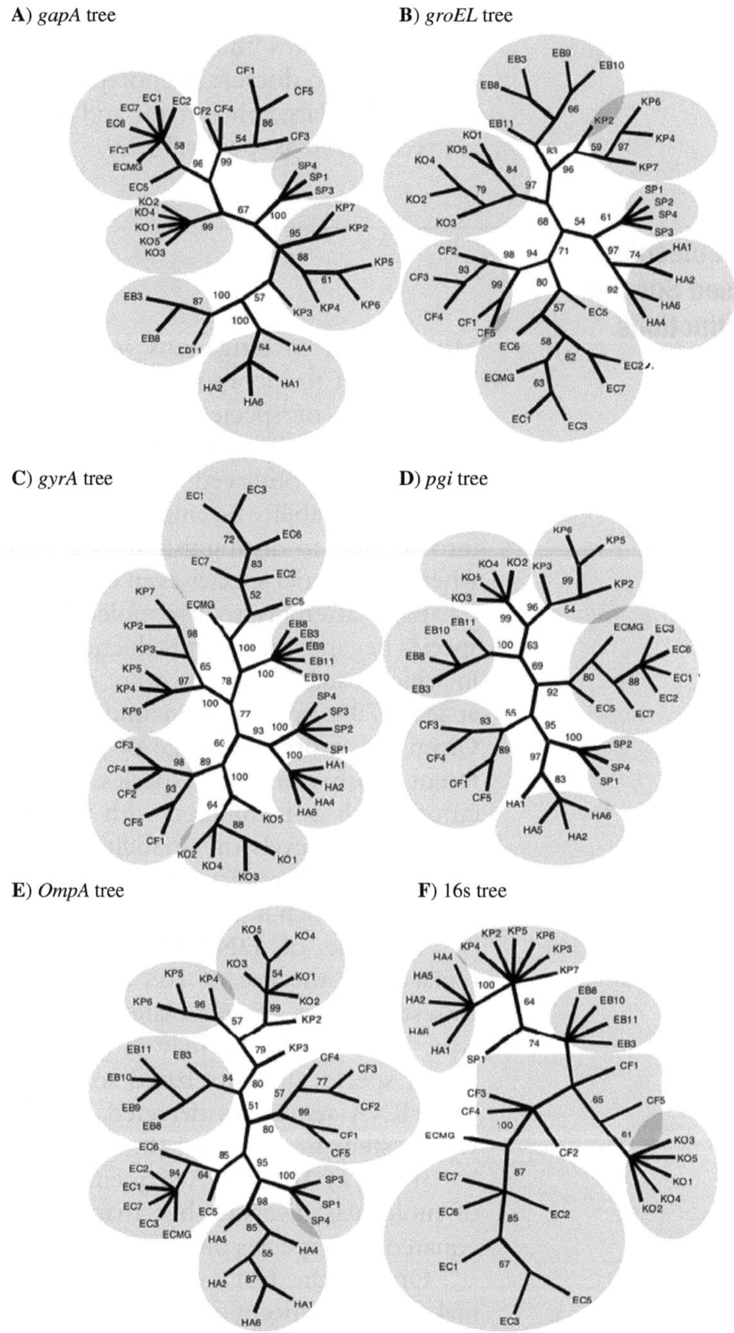


Fig. 21.2. **Molecular phylogenies of six housekeeping genes for six species of enteric bacteria**. Each molecular phylogeny was inferred from the DNA sequences of one housekeeping gene using maximum likelihood. Cladograms were based on maximum likelihood for six putative core genes. (**a**) gapA, (**b**) groEL, (**c**) gyrA, (**d**) pgi, (**e**) ompA, and (**f**) 16S rRNA. Bootstrap values less than 50% (of 500 replicates) were omitted. Confidence in the branching patterns is indicated along the branches. *Shading* denotes a cluster of lineages that correspond to a named species. Taxa abbreviations are CF: *Citrobacter freundii*; EB: *Enterobacter cloacae*; EC: *Escherichia coli*; HA: *Hafnia alvei*; KO: *Klebsiella oxytoca*; KP: *Klebsiella pneumoniae*, and SP: *Serratia plymuthica*. ECMG refers to *E. coli* strain MG1655; see *(34)*.

*(16,36)*. A molecular-based enteric species phylogeny was inferred from the composite data by concatenating the sequences of the six genes *(34)*. The concatenated sequence contained enough phylogenetic signal to resolve all of the interspecies nodes and thus provided a robust estimate of the enteric phylogeny, which corresponded with the existing phenotype-based phylogeny. The authors concluded that, at least for these enteric bacteria, bacterial species clearly do exist and identical species designations emerge from both phenotypic and genotypic data.

The use of multi-locus sequence typing (MLST), in which portions of, generally, seven housekeeping genes are sequenced, has become the norm for characterizing genetic diversity within a bacterial species *(37, 38)*. This method permits the analysis of large numbers of related bacterial isolates, which is essential to the determination of species designations *(24, 38)*. Such studies have confirmed that species designations based upon phenotypic criteria have a corresponding underlying MLST-based genotypic clustering *(23, 26, 39, 40)*.

## 3. Population Genomics Informs Species Designations

The availability of whole genome sequences for multiple isolates of *E. coli* provided our first glimpse into the dynamic nature of a species genome. Glasner and Perna *(41)* and Mau et al. *(42)* compared six complete genomes of *E. coli* (including *Shigella flexneri*) and revealed a highly conserved genomic backbone of more than 3,000 genes, each with an average of greater than 98% sequence similarity among the isolates. Further, Mau et al. *(42)* detected a high level of homologous recombination among these shared genes, confirming earlier studies by Roger Milkman *(43, 44)*. More surprising was the observation that this conserved backbone was punctuated with hundreds of "sequence islands" specific to one strain or another.

Edwards et al. *(45)* provided a similar comparison of two complete and three draft genome sequences of *Salmonella enterica*. As in *E. coli*, a backbone of highly conserved genes was identified, each with an average of greater than 99% sequence similarity and a similar pattern of unique sequence islands specific to one strain or another. This pattern of shared and unique sequences appears to be common among many bacterial species *(46–48)*.

Studies with subtractive hybridization and comparative genome hybridization revealed that for *Helicobacter pylori*, *E. coli*, and *Staphylococcus aureus*, strains within a species share roughly 75–85% of their genome. A comparison of eight genomes of group B Streptococci revealed a core of 1,806 genes present in every genome and 907 genes absent in one or more genomes.

A similar comparison between five genomes of *Streptococcus pyogenes* revealed a comparable level of genomic diversity and predicted that each new genome added approximately 27 strain-specific genes to the species total pan-genome. In contrast, eight genomes of *Bacillus anthracis* revealed very few strain-specific genes. In fact, after the addition of four genomes to the comparison, no new unique genes were identified. The general pattern that emerges is that members of a bacterial species share some large fraction of their genomes, but often carry unique, strain-specific sequences. The fraction of the genome that is shared versus unique varies greatly from one bacterial species to the next.

## 4. The Core Genome Hypothesis and the Bacterial Species Concept

Lan and Reeves were the first to recognize the potential link between the *observation* of shared versus unique sequences in bacterial genomes and its *implication* for discriminating bacterial species *(49)*. They proposed the Core Genome Hypothesis (CGH), which distinguishes between that fraction of the genome (the core) shared by all members of a species and the fraction found in only a subset of the population (the auxiliary). Core genes encode essential metabolic housekeeping and informational processing functions *(50)*. They are ubiquitous in a species and define the species-specific characteristics. Auxiliary genes may or may not be present in a strain and are generally genes that encode supplementary biochemical pathways, are associated with phage or other mobile elements, or encode products that serve to interact with the external environment. Thus, auxiliary genes serve in the adaptation of strains to local competitive or environmental pressures *(14)*. They are likely to encode antibiotic resistance, novel metabolic functions, toxin production, and the like *(51–53)*.

The CGH has dramatically influenced how bacteriologists think about the nature of bacterial species. Prior to the CGH, the strongest argument against the recognition of bacterial "species" was the simple observation of HGT between bacterial lineages. The fact that bacterial species gene pools may not be tightly closed was enough reason for many microbiologists to conclude bacterial species could not survive such exchange. This contradicts the clearly demonstrated fact that bacteria exist in phenotypic clusters, which many microbiologists recognize as species. Even more compelling, it is becoming clear that these well-defined phenotypic clusters correspond to underlying genotype clusters *(26, 39, 48, 54)*.

Some have argued that it is futile to expect a bacterial species to ever be characterized fully at the genome level, particularly since as more genome sequences are obtained, the pan-genomes (i.e., the sum of all genes identified within a species) of numerous species continues to grow *(55, 56)*. In fact, some predict that hundreds of thousands of genome sequences are required to fully define certain bacterial species *(55)*. Others suggest that the wide range of intra-species variation observed for bacterial species reflects the lack of a universal and meaningful species definition *(50)*.

Many ecological and evolutionary factors will impact how many unique genes a species pan-genome may encode and how much genetic variation it harbors. There is no "one size fits all" concept that can, or should, be applied. In fact, no existing species definition requires that either the pan-genome or the level of genetic variation be known in order to delineate members of a species.

One of the more commonly accepted species concepts is the Biological Species Concept (BSC) proposed by Ernst Mayr *(57)*. Mayr proposed that a biological species is comprised of groups of actually or potentially interbreeding natural populations, which are reproductively isolated from other such groups *(57)*. Although Mayr developed this definition specifically for eukaryotes, it can be modified to apply to bacteria. However, it is important to note that at this juncture, the BSC should not be taken to imply any particular process of speciation, merely that the observation of more gene "sharing" (via recombination and/or LGT) is observed within versus between putative bacterial species. The Core Genome Hypothesis provides a perfect backdrop from which to articulate this bacterial-based modification of the BSC. According to the CGH, a bacterial species is comprised of groups of strains that frequently exchange, or could exchange, core genes, but which are relatively restricted from such exchange with other groups.

The CGH predicts that a subset of genes, the core, is present in all, or nearly all, individuals within a species. These are the genes that provide the defining characteristics of a species and are assumed to experience primarily purifying selection, to remove deleterious mutations, and to maintain existing functions. As a species evolves, its core genome will evolve as a complex of co-evolved functions. When transferred between species, such genes will most likely experience a selective disadvantage, as this will disrupt co-evolved functions. Such transfer will rarely survive. Thus, core genes will diverge as the species diverge (**Fig. 21.1**).

In contrast, auxiliary genes will be found in only a subset of individuals within a species. The CGH predicts that these genes experience intermittent positive selection, when their function enhances survival in a varied and ever-changing environment. When such genes are exchanged between species, their functions will often provide a selective advantage to the recipient. Frequent successful transfers between species will serve to limit the divergence of auxiliary genes, relative to the core (**Fig. 21.1**).

The most specific prediction that emerges from the CGH concerns the rate at which core and auxiliary genes accumulate variability. Core genes will, on average, display a neutral rate of evolution, while auxiliary genes will experience a variety of selective pressures, including diversifying selection (acting to increase levels of variation), directional selection (acting to decrease levels of variation), balancing selection (acting to maintain particular alleles in the species), and purifying selection (acting to weed out deleterious mutations). Thus, the average rate of evolution for auxiliary genes could be just about anything, and the variance around this rate should be extreme. These expectations, based upon the neutral theory *(58)*, are quite useful for testing predictions from the CGH *(59)*. However, such tests require population-based samples of multiple genomes per species and, unfortunately, most existing species-based genome samples are chosen to represent the diversity of clinical isolates of human pathogens and thus will often underestimate standing levels of genome diversity. The appropriate data are in the pipeline and should soon be available to permit the sort of population genomics required to address this complex and fascinating matter.

Although we are on the verge of obtaining the type and amount of genotypic data required to examine bacterial species definitions, it is important to note that there is little, if any, substantive data to support the conclusion that bacterial species do not exist. Hence, the real argument remaining is not do they exist, but rather how can they exist in the face of potentially high levels of HGT. Our job is to develop an understanding of bacterial evolution rich enough to explain this apparent paradox. The CGH provides a set of testable hypotheses from which to launch this exploration.

## 5. Conclusions

The community of bacteriologists has failed to establish a uniformly accepted definition of bacterial species. In part, this is due to the extraordinary levels of bacterial diversity and its complexity in terms of culturability, levels of observed HGT, importance in human health, and a variety of other scientific and social factors. However, we are poised to witness a synthesis of the general acknowledgment that bacteria are found in clusters of complex phenotypes (often designated as species) and the underlying genetic basis for these clusters. The Core Genome Hypothesis has, so far, provided the most compelling explanation for the apparent paradox observed for bacteria, in which the observation of stable phenotypic clusters apparently contradicts the existence of dynamic, fluid genomes. The CGH recognizes a species core genome, responsible for maintaining a species identity, and an auxiliary genome, responsible for gene transfer and rapid adaptation of strains to an ever-changing environment. The CGH argues that, even in the presence of substantial genomic fluidity, bacterial species can be rationally identified and named.

## Acknowledgments

The authors acknowledge the valuable input provided by Carla Goldstone, Chris Vriezen, and Emma White. This article was written with support from NIH R01 GM068657-01A2 and R01 A1064588-01A2 grants to MAR.

### References

1. Fox, G. E., Stackebrandt, E., Hespell, R. B., Gibson, J., Maniloff, J., Dyer, T. A., Wolfe, R. S., Balch, W. E., Tanner, R. S., Magrum, L. J., Zablen, L. B., Blakemore, R., Gupta, R., Bonen, L., Lewis, B. J., Stahl, D. A., Luehrsen, K. R., Chen, K. N., Woese, C. R. (1980) The phylogeny of prokaryotes. *Science* **209**, 457–63.
2. Olsen, G. J., Woese, C. J. (1993) Ribosomal RNA: a key to phylogeny. *FASEB J* 7, 113–23.
3. Pace, N. R., Stahl, D. A., Lane, D. J., Olsen, G. J. (1985) Analyzing natural microbial populations by rRNA sequences. *ASM News* **51**, 4–12.
4. Woese, C. R. (1987) Bacterial evolution. *Microbiological Reviews* **51**, 221–71.
5. Woese, C., Kandler, O., Wheelis, M. (1990) Towards a natural system of organisms – proposal for the domains Archaea, Bacteria, and Eucarya. *Proc Natl Acad Sci U S A* **87**, 4576–9.
6. Ludwig, W., Neumaier, J., Klugbauer, N., Brockmann, E., Roller, C., Jilg, S., Reetz, K., Schachtner, I., Ludvigsen, A., Bachleitner, M., Fischer, U., Schleifer, K., H. (1993) Phylogenetic relationships of Bacteria based on comparative sequence analysis of elongation factor Tu and ATP-synthase beta-subunit genes. *Antonie Van Leeuwenhoek* **64**, 285–305.
7. Brown, J., Douady, C., Italia, M., Marshall, W., Stanhope, M. (2001) Universal trees based on large combined protein sequence data sets. *Nat Genet* **28**, 281–5.
8. Kidwell, M. (1993) Lateral transfer in natural populations of eukaryotes. *Ann Rev Genetics* **27**, 235–56.
9. Nelson, K., Selander, R. K. (1994) Intergeneric transfer and recombination of the 6-phosphogluconate dehydrogenase gene (gnd) in enteric bacteria. *Proc Natl Acad Sci U S A* **91**, 10227–31.
10. Brown, J. R., Doolittle, W. F. (1997) Archaea and the prokaryote-to-eukaryote transition. *Microbiol Mol Biol Rev* **61**, 456–502.
11. Nesbo, C. L., L'Haridon, S., Stetter, K. O., Doolittle, W. F. (2001) Phylogenetic analyses of two "Archaeal" genes in thermotoga maritima reveal multiple transfers between Archaea and Bacteria. *Mol Biol Evol* **18**, 362–75.
12. Doolittle, W. F. (1999) Lateral genomics. *Trends Cell Biol* **9**, M5–8.
13. Doolittle, W. F., Papke, R. T. (2006) Genomics and the bacterial species problem. *Genome Biology* 7, 116.
14. Cohan, F. (2002) What are bacterial species? *Annu Rev Microbiol* **56**, 457–87.
15. Sapp, J. (2005) Microbial phylogeny and evolution: Concepts and controversies. *The Bacteriums Place in Nature*. Oxford University Press, New York.
16. Rossello-Mora, R., Amann, R. (2001) The species concept for prokaryotes. *FEMS Microbiol Rev* **25**, 39–67.
17. Shute, L.A., Gutteridge, C.S., Norris, J.R., Berkeley, R.C. (1984) Curie-point pyrolysis mass spectrometry applied to characterization and identification of selected *Bacillus* species. *J Gen Microbiol* **130**, 343–55.
18. Sneath, P., Stevens, M. (1985) A numerical taxonomic study of *Actinobacillus*, *Pasteurella*, and *Yersinia*. *J Gen Microbiol* **131**, 2711–38.

19. Barrett, S., Sneath, P. (1994) A numerical phenotypic taxonomic study of the genus *Neisseria*. *Microbiology* **140**, 2867–91.
20. Mauchline, W., Keevil, C. (1991) Development of the BIOLOG substrate utilization system for identification of *Legionella* spp. *Appl Environ Microbiol* **57**, 3345–9.
21. Kirschner, C., Maquelin, K., Pina, P., Thi, N. N., Choo-Smith, L., Sockalingum, G., Sandt, C., Ami, D., Orsini, F., Doglia, S., Allouch, P., Mainfait, M., Puppels, G., Naumann, D. (2001) Classification and identification of enterococci: A comparative phenotypic, genotypic, and vibrational spectroscopic study. *J Clin Microbiol* **39**, 1763–70.
22. Cohan, F. (2002) Sexual isolation and speciation in bacteria. *Genetica* **116**, 359–70.
23. Godoy, A. P., Ribeiro, M. L., Benvengo, Y. H., Vitiello, L., Miranda M. de C., Mendonca, S., Pedrazzoli, J., Jr (2003) Analysis of antimicrobial susceptibility and virulence factors in *Helicobacter pylori* clinical isolates. *BMC Gastroenterol* **3**, 20–26.
24. Hanage, W. P., Fraser, C., Spratt, B. G. (2005) Fuzzy species among recombinogenic bacteria. *BMC Biol* **3**, 6.
25. Priest, F., Barker, M., Baillie, L., Holmes, E., Maiden, M. (2004) Population structure and evolution of the *Bacillus cereus* group. *J Bacteriol* **186**, 7959–70.
26. Thompson, J., Pacocha, S., Pharino, C., Klepac-Ceraj, V., Hunt, D., Benoit, J., Sarma-Rupavtarm, R., Distel, D., Polz, M. (2005) Genotypic diversity within a natural coastal bacterioplankton population. *Science* **307**, 1311–3.
27. Baldwin, A., Mahenthiralingam, E., Thickett, K., Honeybourne, D., Maiden, M., Govan, J., Speert, D., Lipuma, J., Vandamme, P., Dowson, C. (2005) Multilocus sequence typing scheme that provides both species and strain differentiation for the *Burkholderia cepacia* complex. *J Clin Microbiol* **43**, 4665–73.
28. Hanage, W., Kaijalainen, T., Syrjanen, R., Auranen, K., Leinonen, M., Makela, P., Spratt, B. (2005) Invasiveness of serotypes and clones of *Streptococcus pneumoniae* among children in Finland. *Infect Immun* **73**, 431–5.
29. Stackebrandt, E., Frederiksen. W., Garrity G. M., Grimont P. A., Kämpfer P., Maiden M. C., Nesme X., Rosselló-Mora R., Swings J., Trüper H.G., Vauterin L., Ward A. C., Whitman W. B. (2002) Report of the ad hoc committee for the re-evaluation of the species definition in bacteriology. *Int J Syst Evol Microbiol* **52**, 1043–7.
30. Lindh, J., Terenius, O., Faye, I. (2005) 16S rRNA gene-based identification of midgut bacteria from field-caught *Anopheles gambiae sensu lato* and *A. funestus* mosquitoes reveals new species related to known insect symbionts. *Appl Environ Microbiol* **71**, 7217–23.
31. Drancourt, M., Berger, P., Raoult, D. (2004) Systematic 16S rRNA gene sequencing of atypical clinical isolates identified 27 new bacterial species associated with humans. *J Clin Microbiol* **42**, 2197–202.
32. Clarridge, J. R. (2004) Impact of 16S rRNA gene sequence analysis for identification of bacteria on clinical microbiology and infectious diseases. *Clin Microbiol Rev* **17**, 840–62.
33. Clayton, R., Sutton, G., Hinkle, P. J., Bult, C., Fields, C. (1995) Intraspecific variation in small-subunit rRNA sequences in GenBank: why single sequences may not adequately represent prokaryotic taxa: *Int J Syst Bacteriol* **45**, 595–9.
34. Wertz, J. E., Valletta-Goldstone, C. M., Gordon, D. M., Riley, M. A. (2003) A molecular phylogeny of enteric bacteria and implications for a bacterial species concept. *J Evol Biol* **16**, 1236–48.
35. Gordon, D. M., Fitzgibbon, F. (1999) The distribution of enteric bacteria from Australian mammals: Host and geographical effects. *Microbiology* **145**, 2663–71.
36. Holt, J. G. (1994) *Bergey's Manual of Determinative Bacteriology*, Williams and Wilkins, Baltimore, MD.
37. Maiden, M. (1998) Horizontal genetic exchange, evolution and spread of antibiotic resistance in bacteria. *Clin Infect Dis* **27**, S12–20.
38. Feil, E. J., Cooper. J. E, Grundmann, H., Robinson, D. A., Enright, M. C., Berendt, T., Peacock, S. J., Smith, J. M., Murphy, M., Spratt, B. G., Moore, C. E., Day, N. P. (2003) How clonal is *Staphylococcus aureus*? *J Bacteriol* **185**, 3307–16.
39. Whitaker, R., Grogan, D., Taylor, J. (2005) Recombination shapes the natural population structure of the hyperthermophilic archaeon *Sulfolobus islandicus*. *Mol Biol Evol* **22**, 2354–61.
40. Woodward, M., Sojka, M., Sprigings, K., Humphrey, T. (2000) The role of SEF14 and SEF17 fimbriae in the adherence of *Salmonella enterica* serotype Enteritidis to inanimate surfaces. *J Med Microbiol* **49**, 481–7.
41. Glasner, J., Perna, N. (2004) Comparative genomics of *E. coli*. *Microbiol Today* **31**.
42. Mau, B., Glasner, J., Darling, A., Perna, N. (2006) Genome-wide detection and analysis of homologous recombination among sequenced strains of *Escherichia coli*. *Genome Biology* 7, R44.
43. Milkman, R., Raleigh, E., **McKane**a, **M., Cryderman**c, **D.**, Bilodeau, P., Mcweeny, K.

(1999) Molecular evolution of the *Escherichia coli* chromosome. V. Recombination patterns among strains of diverse origin. *Genetics* **153**, 539–54.
44. Milkman, R., Jaeger, E., Mcbride, R. (2003) Molecular evolution of the *Escherichia coli* chromosome. VI. Two regions of high effective recombination. *Genetics* **163**, 475–83.
45. Edwards, S. V., Fertil, B., Giron, A., Deschavanne, P. J. (2002) A genomic schism in birds revealed by phylogenetic analysis of DNA strings. *Systematic Biology* **51**, 599–613.
46. Juhas, M., Power, P. M., Harding, R. M., Ferguson, D. J., Dimopoulou, I. D., Elamin, A. R., Mohd-Zain, Z., Hood, D. W., Adegbola, R., Erwin, A., Smith, A., Munson, R. S., Jr, Harrison, A., Mansfield, L., Bentley, S., Crook, D. W. (2007) Sequence and functional analyses of *Haemophilus* spp. genomic islands. *Genome Biol* **8**, R237.
47. Coleman, M., Sullivan, M., Martiny, A., Steglich, C., Barry, K., Delong, E., Chisholm, S. (2006) Genomic islands and the ecology and evolution of *Prochlorococcus*. *Science* **311**, 1768–9.
48. Waterfield, N., Daborn, P., Dowling, A., Yang, G., Hares, M., Ffrench-Constant, R. (2003) The insecticidal toxin makes caterpillars floppy 2 (Mcf2) shows similarity to HrmA, an avirulence protein from a plant pathogen. *FEMS Microbiol Lett* **229**, 265–70.
49. Lan, R., Reeves, P. R. (1996) Gene transfer is a major factor in bacterial evolution. *Mol Biol Evol* **13**, 47–55.
50. Feil, E. (2004) Small change: Keeping pace with microevolution. *Nat Rev Microbiol* **2**, 483–95.
51. Dobrindt, U., Reidl, J. (2000) Pathogenicity islands and phage conversion: Evolutionary aspects of bacterial pathogenesis. *Int J Med Microbiol* **290**, 519–27.
52. Karlin, S. (2001) Detecting anomalous gene clusters and pathogenicity islands in diverse bacterial genomes. *Trends Microbiol* **9**, 335–43.
53. White, P. A., Mciver, C. J., Rawlinson, W. D. (2001) Integrons and gene cassettes in the Enterobacteriaceae. *Antimicrob Agents Chemother* **45**, 2658–61.
54. Godoy, A., Ribeiro, M., Benvengo, Y., Vitiello, L., Miranda, C. M., Mendonca, S., Pedrazzoli, J. J. (2003) Analysis of antimicrobial susceptibility and virulence factors in *Helicobacter pylori* clinical isolates. *BMC Gastroenterol* **3**, 20.
55. Medini, D., Donati, C., Tettelin, H., Masignani, V., Rappuoli, R. (2005) The microbial pan-genome. *Curr Opin Genet Dev* **15**, 589–94.
56. Tettelin, H., Masignani, V., Ciesiewicz, M., Donati, C., Medini, D., Ward, N., Angiuli, S., Crabtree, J., Jones, A., Durkin, A. (2005) Genome analysis of multiple pathogenic isolates of *Streptoccocus agalactiae*: implications for the microbial "pan-genome". *Proc Natl Acad Sci USA* **102**, 13950–5.
57. Mayr, E. (1942) *Systematics and the Origin of Species*, Columbia University Press, New York.
58. Kimura, M. (1968) Genetic variability maintained in a finite population due to mutation production of neutral and nearly neutral isoalleles. *Genetic Res Camb* **11**, 247–69.
59. Fay, J., Wyckoff, G., Wu, C. (2002) Testing the neutral theory of molecular evolution with genomic data from *Drosophila*. *Nature* **415**, 1024–6.

# Chapter 22

# A Critique of Prokaryotic Species Concepts

## R. Thane Papke

. . .genera and species. . .depend on such collections of ideas as men have made, and not on the real nature of things. . .

John Locke *(1)*

## Abstract

Developments in DNA sequencing and population genetics analyses have revealed unanticipated complexity in prokaryotic variation. The observation that genetic traits are horizontally inherited at unexpected rates within and between closely related asexually reproducing lineages impacts our comprehension of prokaryotic evolution and ecology. As a result, the concepts that point to species as being discrete clusters or monophyletic lineages are at odds with most of the data, suggesting that taxon circumscription can only proceed by informed compromise, pragmatism, and subjectivity.

**Key words:** Prokaryotic species, species concepts, *Halorubrum*, haloarchaea.

## 1. Introduction

Bring up the word "species" at any gathering of professional microbiologists and a lengthy and vigorous debate will ensue. Central to the debate are two interwoven issues: First, does the current species definition(s), which among other criteria holds that different species have less than 70% DNA–DNA reassociation values, meet the challenges of modern microbiology, since it is criticized for being arbitrary, too broad, cumbersome, and indifferent to ecology and evolution *(2–4)*; and second, do evolutionary processes lead to discrete groups of individuals all having a common ancestor, such that microbial taxonomists can non-arbitrarily divide organisms into species *(4–7)*? The crux of

Maria B. Gogarten et al. (eds.), *Horizontal Gene Transfer: Genomes in Flux, vol. 532*

DOI 10.1007/978-1-60327-853-9_22 Springerprotocols.com

this manuscript will focus on the latter and argue that though diversity demonstrates central tendencies, evolutionary patterns of phenotype and genotype are often complex such that species circumscription cannot necessarily rely on discreteness of data, but requires consensus or informed subjective opinion.

Of course, species debates are not unique to microbiologists. Taxonomists who specialize in botany, zoology, or paleontology use different concepts for applying species monikers to groups, and each discipline could utilize any one of many available concepts *(8, 9)*. Probably most familiar to microbiologists is Ernst Mayr's Biological Species Concept (BSC), which can be formulated as follows: individuals belong to a species if they can produce viable offspring. The evolutionary basis behind the BSC is that in randomly mating eukaryotic populations, gene flow within populations acts as a cohesive force that prevents genetic divergence amongst its members. The slowing, biasing, or preventing of gene flow via any multitude of mechanisms promotes genetic divergence between subgroups. Eventually, subpopulations will lose the capability to exchange genes, but the historical point in time at which gene flow ceases is both theoretical and contentious, and may or may not result in species designation by human observers. Though the BSC is very successful, it fails when gene exchange occurs frequently across diverged lineages *(8)*. For instance, plants are well known for their capacity to hybridize between accepted species, which exposes the limits to the universal applicability of the BSC. Interestingly, and perhaps ironically, hybridization in plants does not just confound species circumscription, it is also considered to be a mechanism for speciation *(10)*.

Observing viable offspring is not required to know if sexual populations are mating. Mating (or its absence) can be detected by examining the distribution of alleles in a population. For instance, the random association of alleles at different loci can only occur if individuals in a population are mating indiscriminately. When the association of alleles in a population is random, they are said to be in linkage equilibrium. If, on the other hand, alleles are not randomly associated, but are structured into subpopulations or different species, then gene flow barriers are likely caused by biases in mating patterns and the population is said to be in linkage disequilibrium. However, linkage disequilibrium can be observed within randomly mating populations. In the case of prokaryotic populations, there are many possibilities: natural selection may give a temporary adaptive advantage to a specific clone, which then rises in frequency; a recent bottleneck may purge genetic diversity; there may be undetected genetic substructure such as the inclusion of geographically isolated populations into a single analysis; or sampling biases may preferentially recover specific genotypes (e.g., those associated with disease), which are actually part of a larger, more diverse "carrier" population. Therefore, interpretations of linkage disequilibrium as an indication of no or little recombination must be validated and confirmed. Moreover, important knowledge regarding the evolution of a species is gained when linkage disequilibrium is discovered in a recombining population. Linkage equilibrium, in contrast, has only one interpretation.

It is reasonable to assume that prokaryotes should form discrete clusters. They evolve by many of the same principles discovered in plants and animals. Random heritable genetic modifications create variation and natural selection shapes it. Divergence is promoted when barriers to homogenizing forces are erected and extinction can occur at any point. So, why is the term species so contentious among microbiologists? The answer might lie in the details of the mechanisms that produce variation and replication.

Prokaryotes do not reproduce in the same way as sexual eukaryotes; they replicate clonally, primarily by binary fission. Furthermore, prokaryotes do not experience homologous recombination (HR) in the same manner as eukaryotes: it is mediated by parasexual mechanisms (transformation, transduction, and conjugation) that shuttle DNA between cells in a non-reciprocal manner, and it is not required for reproduction. Recombination is typically classified as either homologous or illegitimate: the former usually results in a replacement of an existing locus with a different allele (and is the primary focus of this chapter) and the latter results in the addition of genes not previously found in a specific genome. However, it is often difficult to determine which process generated a particular observed genotype. Here, the term recombination will be used exclusively to refer to the homologous replacement of alleles.

Although prokaryotes do recombine their genes, the rate is lower than the recombination rate observed in sexually reproducing species, which undergo multiple recombinations per reproduction event. Theoretically, if sufficient genetic material is available, or if time between duplications is generously long, recombination in prokaryotes could occur multiple times between reproduction events, but because recombination in prokaryotes is not linked to reproduction, it is unlikely that any recombination will occur between reproduction events. Population genetics experiments using *Escherichia coli (11)* and *Neisseria meningitides (12)* bolstered a clonal paradigm for prokaryotes by demonstrating that recombination, as detected by multilocus enzyme electrophoresis, was too infrequent to randomize the association of alleles (i.e., break patterns of linkage disequilibrium). This observation suggested that recombination was too weak to act as a force for the cohesion of species. In the absence of significant recombination, prokaryotic diversity was proposed to be limited by a process called periodic selection *(13)*, a process by which variation in asexual populations is periodically purged when more fit mutants arise and out-compete conspecific or presumably at least same niche relatives *(14)*.

## 2. Prokaryotic Species Debate

The debate in the prokaryotic species literature focuses on the existence and meaning of phylogenetic clusters. On one hand is the concept that periodic selection is the homogenizing force of clusters, which map one-to-one with unique ecological adaptations (i.e., clusters are niche-specific). On the other hand, frequent recombination is the cohesive force of clusters, as it is for most plants and animals. In this section, I detail these two concepts and generate predictions regarding the evolutionary outcome of genetic variability. Population genetics data (mainly from haloarchaea) are evaluated to demonstrate and highlight how prokaryotic diversity tends to defy these concepts thus making species circumscription a pragmatic endeavor.

### 2.1. Periodic Selection and the Ecotype Species Concept

Periodic selection is a cohesive force that homogenizes genetic variation within populations or species when recombination rates are negligible, that is at or below the mutation rate per locus *(15)*. This suggests an evolutionary explanation for the existence of prokaryotic clusters and imparts a foundation for identifying and naming prokaryotic species *(4, 5, 15)*, which will be referred to here as the Ecotype Species Concept (EcSC). The EcSC predicts that individuals within prokaryotic species more closely resemble each other because periodic selection events on a regular basis eliminate allelic diversity within species *(15–17)*. Escape from the homogenizing force of periodic selection occurs when individuals adapt to new resources and thus avoid direct competition with conspecific relatives that occupy the old niche. This results in ecotypes, defined as a cluster of individuals with a common evolutionary history and ecological adaptations *(5, 15)*. Though different in mode, periodic selection and the formation of ecotypic species parallel gene flow barriers in sexually reproducing species: it is a homogenizing force that must be escaped for divergence to occur.

Since the EcSC is based on well-studied evolutionary forces, the variation produced and genetic relationships formed within and between lineages are predictable and testable (**Table 22.1**). The process of periodic selection should lead to the fixation of entire genomes (not individual loci) where alleles in a species are non-randomly associated (linkage disequilibrium).

**Table 22.1**
**A comparison of species concepts**

| Character | EcSC | BSC*[1] | *Halorubrum* spp. |
|---|---|---|---|
| Clusters | +[2] | + | + |
| Congruent gene trees within clusters | + | − | − |
| Congruent gene trees between clusters | + | + | − |
| Selection fixes genomes | + | − | − |
| Selection fixes alleles | − | + | + |
| Linkage disequilibrium within clusters | + | − | − |
| Clusters are niche specific | + | − | − |
| Clusters are niche non-specific | − | + | +[3] |

[1]Since the BSC* does not assume a constant recombination rate across all taxa, outcomes will vary.
[2]( + ) and ( − ) indicate presence/absence of the character.
[3]As determined by salinity pond distribution and unpublished strain phenotype analyses.

This hitchhiking process produces genetic bottlenecks, which lead to the fixation and accumulation of slightly deleterious mutations (i.e., Muller's ratchet *(18, 19)*). Under the EcSC, species are recognized in phylogenetic trees as clusters and all genes are predicted to have identical evolutionary histories. Since the periodic selection event resets the allelic diversity to zero, all loci should accumulate approximately the same number of neutral mutations.

Evidence supporting the EcSC comes mostly from work on cyanobacteria. Separate studies on hot spring cyanobacteria (*Synechococcus* spp.) and marine *Prochlorococcus* spp. demonstrate that phylogenetic clusters are correlated with unique temperature and/or light distributions (e.g., ecotypes) *(20, 21)*. However, subsequent analyses of cyanobacterial phylogeny and distributions reveal a more complex pattern than originally observed. Thermophilic *Synechococcus* spp. clusters may not reflect single niche distributions *(17)* and geographic isolation also plays a significant role *(22)*. *Prochlorococcus* spp. low-light ecotypes are paraphyletic *(23–25)*, and markers for high- and low-light adapted *Prochlorococcus* spp. ecotypes are in equal abundance at the ocean surface at some locations *(26)*. Combined, the patterns suggest that the cyanobacterial clusters are formed from organisms with diverse adaptations and that other hypotheses beyond the EcSC are required to explain the data.

### 2.2. A Biological Species Concept for Prokaryotes?

Recombination is prevalent in prokaryotes and could be a homogenizing force that acts to prevent speciation. Early on, Ravin *(27)* noted that recombination was higher within than between *Streptococcus* sp. and *Pneumococcus* sp. "taxospecies" (Ravin's term) and concluded that the observation was the result of homology between the donor and the recipient. Since then, additional experiments with different organisms including *Bacillus* sp. *(28)* and enterobacteria *(29)* have confirmed and extended these initial observations. This "distance-decay" model indicates that the probability of two sequences being exchanged by homologous recombination decreases exponentially as differences in nucleotide identity increase *(30)*.

Using the above distance-decay model as a basis for computer algorithms, recent simulations tested the possibility of recombination as a neutral hypothesis for the cohesion of prokaryotic groups or clusters *(31–33)*. Each simulation study had its own unique attributes, but they all came to the conclusion that clusters could form under different conditions and varying rates of recombination. Interestingly, it was specifically noted that under some conditions diverging clusters condensed back into a single population, indicating that recombination prevented differentiation explicitly because it was acting as a cohesive force *(32, 33)*. Furthermore, Fraser *et al.* *(32)* demonstrated that very little recombination was required for this condensing effect to take place (e.g., 0.25–2 times the per locus mutation rate), which challenges the interpretations of Cohan who previously concluded that such low recombination/mutation ratios cannot act as a cohesive force *(5, 34)*. The importance of this neutral model of prokaryotic evolution involving recombination cannot be understated: if selection as a force for cohesion does no better than clusters formed from neutral models, then selection can be rejected as a necessary cause.

Thanks to the advent of high-volume DNA sequencing and a new population genetic analysis technique called multilocus sequence analysis (MLSA, discussed below), recombination is recognized as a pervasive process occurring frequently enough to randomize the association of alleles in lineages from a wide variety of taxa and ecological niches (e.g., *Campylobacter jejuni (35)*, *Porphyromonas gingivalis (36)*, *Bradyrhizobia* spp. *(37)*, *Microcoleus* spp. *(38)*, *Sulfolobus islandicus (39)*, and *Halorubrum* spp. *(40, 41)*). Such findings are significant when viewed in the light of the above molecular and computer modeling studies, and suggest the possibility for a prokaryotic BSC. Indeed, nearly two decades ago, based on the data from *E. coli*, Dykhuizen and Green conceived of a BSC for prokaryotes in which species are composed of individuals that exchange alleles and exhibit statistically different trees for different loci whereas different species have barriers to gene exchange and between-species analyses demonstrate statistically similar trees for each locus *(42)*. To cope with the accumulating evidence for frequent gene flow between species (and higher hierarchical taxonomic units!), a revision of the BSC was proposed by Lan and Reeves *(43)*, which will be referred to here as the BSC*. The BSC* assumes a gradient of recombination capabilities and attempts to encompass clonal species by hypothesizing two general classes of genes within a species- or pan-genome, an "auxiliary" set, which is patchily distributed amongst the members and is responsible for local adaptation, and the "core", which is common to all members and are typically essential for biosynthesis pathways and central metabolism *(43)*. The core genes are thought to evolve neutrally and be vertically inherited within lineages due to recombination rates decaying with divergence. The auxiliary genes confer ecological specialization but are non-essential (e.g., pathogenicity islands) and can come and go without affecting the core of the species character. For instance, strains belong to the species *E. coli* despite possessing an enormous amount of niche variation, because they share a core set of highly similar genes. *Shigella* spp. could also be folded into the *E. coli* moniker for the same reason. The BSC* still adheres to the same rigorous standards established by Dykhizen and Green, but they are applied solely to core genes.

### 2.3. Evolutionary Outcomes of Frequent Recombination

As the BSC* is based on evolutionary principles, it leads to the predictions regarding the nature of genetic variation within and between species (**Table 22.1**). Some evolutionary outcomes are indistinguishable between the BSC* and the EcSC, such as the observation of clusters and all loci having congruent intercluster relationships in phylogenetic trees *(42)*. However, frequent recombination will cause linkage equilibrium and prevent periodic selection from purging allelic diversity *(39)* (**Fig. 22.1** and discussion below), resulting in evolutionary outcomes that are diametrically opposed to those of the EcSC. Interestingly, empirical studies of natural populations have revealed that linkage equilibrium can be attained even when the relative rates of recombination to mutation per allele are as low as two or three *(39, 41)*. These values are in line with the rates determined by the computer studies discussed above and suggest that HR in natural populations can prevent genetic divergence.

Recombination in prokaryotes, even at low rates, offers several advantages over clonality. First, repairing slightly deleterious mutations in clonal lineages happens by back mutations, while recombination provides an alternative mechanism and replaces "faulty" with "healthy" alleles from the surrounding population. Second, if two advantageous alleles arise at different loci in

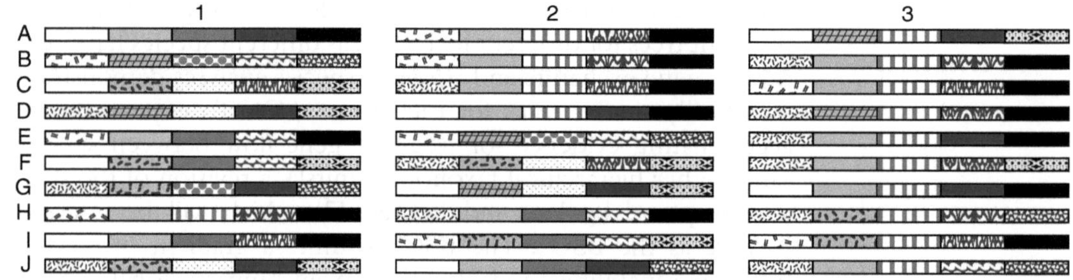


Fig. 22.1. How selection fixes a single allele in a recombining prokaryotic population over time (three panels). *Panel 1* shows a recombining population with the alleles in equilibrium. There are 10 individuals with chromosomes (**A–J**), each containing five loci (*vertical columns*). Each locus has three alleles each. The 3rd locus in chromosome H recently acquired a selectively advantageous allele (*vertical stripes*), through either mutation or recombination. In *Panel 2*, the chromosome containing the advantageous allele rises in frequency (chromosomes **A** and **B**). However, before the chromosome can reach fixation (as predicted by the EcSC), recombination randomly occurs with any other chromosome, either by any allele in the population landing in a chromosome with the *vertical stripe* allele (e.g., 2C is a descendant of 1H with a recombined allele in the 1st column/locus) or by the *vertical stripe* allele recombining into a different chromosome (e.g., 2D is a descendant of 1A). *Panel 3* shows the outcome of recombination when selection acts on a favorable allele, the *vertical stripe* allele is fixed in the population, which is still in linkage equilibrium. This scenario prevents selective sweeps and promotes gene evolution that is separate from genomes.

unrelated individuals in a clonal lineage, in the absence of recombination one will be lost due to a periodic selection event triggered by the other locus. In very large populations, advantageous mutations that are lost can be resurrected due to the sheer number of individuals and the limited number of possible mutations *(18)*. However, in a recombining lineage, both alleles can be maintained and merged into a single individual: this has recently been demonstrated in experiments with *E. coli (44)*. Thus, the rate of evolution increases substantially when two or more independently arising advantageous alleles can be fixed at any one time. Third, it generates de novo allelic diversity from the mutations that have already survived natural selection. Taken together, recombination has a huge advantage over clonality, which suggests that most prokaryotic species experience recombination to some extent.

## 3. Testing Species Concepts

### *3.1. The Haloarchaea and* Halorubrum *spp.*

Haloarchaea are the dominant organisms found in hypersaline environments *(45, 46)*. To cope with the extreme conditions of high salt concentrations, haloarchaea have many unique adaptive characteristics *(47)* including an acidic proteome, high surface-to-volume ratios and gas vesicles to migrate between oxic and anoxic environments. Perhaps, the most interesting characteristic from the perspective of suitability to test the species concepts outlined above is their well-established capabilities for frequent genetic exchange *(48–50)*. Mating haloarchaeal cells form intercellular cytosolic bridges, and during a "transiently diploid" stage *(51)* recombination occurs *(52)*. Little is known about the formation and phylogenetic breadth of intracellular cytosolic bridges, but they were also observed in *Sulfolobus* sp. *(53)*, a distantly related crenarchaeaon. Phage, which are well known for their ability to transfer genetic material, have been detected in the hypersaline environment *(54)*, isolated on several haloarchaeal genera hosts *(55)* and have recently been discovered to infect *Halorubrum* strains *(56)*.

My colleagues and I have been studying *Halorubrum* spp. as model organisms for assessing the impact of recombination in naturally occurring populations and how this trait might affect our concept of prokaryotic species *(40, 41)*. We chose to work with *Halorubrum* spp. for several reasons: (1) they were easy to cultivate from the environment, (2) they are often the dominant halophile in hypersaline communities and therefore an important taxonomic group *(57, 58)*, (3) they are not pathogens or relevant to human economics and therefore less likely to bias our interpretations (e.g., *E. coli* and *Shigella* spp. cause different diseases, are classified differently but are highly similar genetically *(59)*), (4) multiple lineages grow in the same milliliter of water and compete for the same pool of resources, (5) as extremophiles, they have a potential for geographic isolation, which might play a role in their evolution, and (6) they are Archaea, the understudied domain of life, which allows confirmation of trait universality when observed in all domains. Combined, these characteristics seemed ideal for testing species concepts.

### 3.2. EcSC vs. BSC in *Halorubrum* spp.

To assess clonality and species concepts in *Halorubrum* spp. we applied MLSA *(2, 60)* to quantify the relative contribution of recombination and mutation to the diversity observed in species. The method for binning recombination and mutation events were developed by Ed Feil and colleagues *(61–63)* and relies on sequencing multiple neutrally evolving loci (typically seven, but in our case five) from a collection of closely related strains. Neutrally evolving genes are used because selection would artificially support linkage disequilibrium despite a recombining population. The idea is to find single locus variants (SLV), which are strains that have different alleles at one locus, but otherwise share identical alleles (absolutely no nucleotide polymorphisms) at all other sequenced loci (typically six, but in our case four). If the alleles from the SLV differ by a single nucleotide polymorphism, the allele is binned into the mutation column. However, if there is more than one change, the allele is considered to have arrived in its current genetic environment via a recombination event, because it is unlikely that two mutations will occur in one locus while the others received none at all. Importantly, the method assumes a parsimonious "ancestor" determined by the sequence type (a unique allelic combination), which simultaneously has the largest number of representatives and the largest number of SLVs. This method is limited to analyzing very closely related strains, typically less than 1% or 2% divergent because of the reliance on SLV identification. After binning all of the SLVs, simple algebra is used to determine the ratio of recombination and mutation events per nucleotide or locus changed.

More recently, MLSA has been promoted as a method for species identification and circumscription *(2, 64)* and relies on phylogenetic analysis of the concatenated loci. MLSA analysis of our *Halorubrum* spp. demonstrated that strains fell into three major phylogenetic clusters (Phylogroups A, B, and C in **Fig. 22.2**). Because clustering could be explained by either one of the two homogenizing forces central to the species concepts detailed above, we examined the distribution of alleles within and between phylogroups to detect gene flow. Our analyses revealed frequent and biased patterns of recombination: alleles were randomly associated within, but structured between each of the phylogroups. Furthermore, intra-phylogroup relationships were largely unresolved. Combined, these observations largely conformed to the predictions of the BSC* rather than the EcSC (**Table 22.1**).

Rejecting the EcSC as a homogenizing force for *Halorubrum* spp. is further supported by the observation that a single allele at the bacteriorhodoprin (*bop*) locus was nearly fixed in two of the three phylogroups. The *bop* locus in Phylogroups A and B has only two polymorphic sites (**Table 22.2**), and the dominant allele was found in >85% of the strains. Other sequenced loci remained highly diverse in the populations averaging 10.5 synonymous nucleotide changes. According to neutral theory *(65)*, synonymous substitutions rates should not vastly differ across genes. Therefore, if a periodic selection event had reset the diversity to zero, all alleles would contain the same approximate number of neutral substitutions, which is not what we observed. The most reasonable explanation would be that selection drove the *bop* allele to high frequency, but recombination prevented a periodic selection event by continually spreading non-selected alleles randomly throughout the population (**Fig. 22.1**). Alternative explanations for the lack of diversity at the *bop* locus could be random chance or possibly selection in the form of codon bias. However, neither seems as satisfactory. Random chance seems unlikely given that the fixation occurred in two phylogroups (e.g., two independently arising events). On the other hand, codon bias is known to have an important selective effect on "neutral" codon positions in highly expressed genes *(66)* and may play a role here. We currently do not have any expression or codon usage data, but it is likely that the

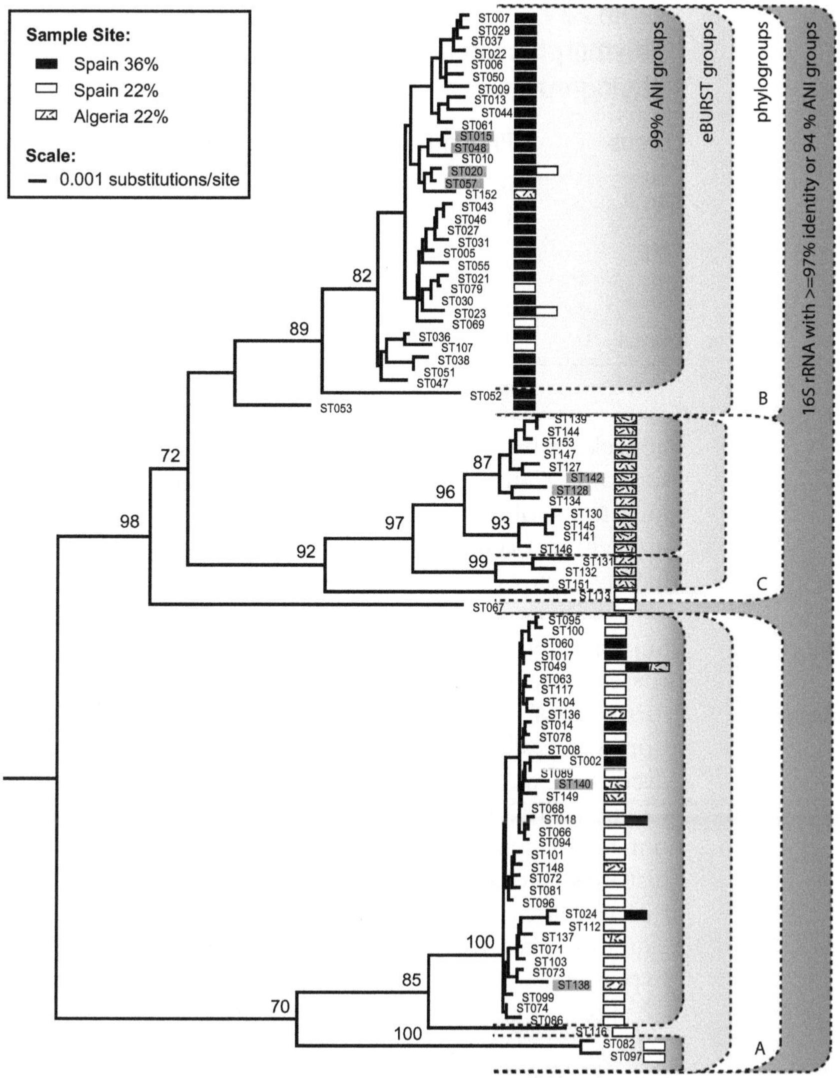


Fig. 22.2. Comparing species definitions for *Halorubrum* spp. The phylogenetic tree is modified from *(41)*. The *outer dark gray tab* reflects a clustering based on the 97% and 94% cutoffs for 16S rRNA and average nucleotide identity (ANI), respectively. These two cutoff values correlate with 70% DNA/DNA reassociation *(3, 81)*, which is the official definition used to circumscribe prokaryotic species *(82)*. Both measures consider these strains to be the same species, despite linkage equilibrium evidence that suggests at least three populations. Using the 99% ANI (*innermost tabs*), which was argued in *(83)* to reflect a more accurate representation of species, strains are divided into five groups with several strains not being assigned to clusters, indicating they belong to their own species despite having three or more alleles in common with strains in the cluster (see eBURST group tabs). The eBURST analysis (*light gray tabs*), which identifies clusters as strains that share three identical alleles with any other strain, captures three major clusters, but excludes taxa whose names are highlighted with *gray boxes*. The *gray* highlighted strains are considered "singletons" by the eBURST analysis and fall outside of "clonal complexes," despite falling well within phylogenetic clusters (for a full review of eBURST and clonal complexes, see *(62)*). Because the above objective analyses each found different members for their respective groups, an argument for the non-existence of discrete clusters can be made. Phylogroups, which are highlighted with *white tabs*, are the only subjectively and pragmatically defined clusters in the tree: careful consideration of all the data, including ecology and evolution, was used to delineate phylogroups. Sample sites for each taxon on the tree are highlighted with *black, white,* or *stippled rectangles.* No cluster containing less than four strains is composed exclusively of isolates from one site, indicating that large differences in salinity (niche) or country of origin (geography) made limited impact on the cluster circumscription.

**Table 22.2**
**Polymorphic sites per protein-coding locus within each phylogroup**

| Locus | Phylogroup A | Phylogroup B | Phylogroup C |
|---|---|---|---|
| *atpB* | 19 | 12 | 14 |
| *bop* | 2 | 2 | 20 |
| EF-2 | 3 | 8 | 13 |
| *radA* | 11 | 10 | 24 |

housekeeping genes analyzed in the study are highly expressed, suggesting that codon bias is not a factor. Furthermore, the mutations observed in *bop* are A's and T's, which would be unlikely to occur in these high G + C genes if codon biases are having their expected effect.

### 3.3. Rejecting the BSC* for Halorubrum

Superficially, each of the three *Halorubrum* spp. phylogroups discussed here seems to meet the predictions of the BSC*: they form cohesive clusters that experience high rates of recombination which allowed selection to drive a single allele (out of hundreds that were detected at all loci) to high frequency. However, Coyne and Orr *(8)* as well as Dykhuizen and Green *(42)* point out that when genes and organisms have different evolutionary histories, biological species concepts break down. Therefore, if different loci have different evolutionary histories, then the BSC* cannot explain the observed diversity and should be rejected as a hypothesis.

For prokaryotes, it has long been observed that the host cell machinery responsible for regulating recombination easily incorporates divergent homologous DNA under varying conditions *(67, 68)*, indicating that barriers to HR are extremely leaky and by extension monophyly is unlikely to be a common property of species. Phylogenetic analysis on *Halorubrum* spp. core genes revealed extensive HR between phylogroups. Approximately 8% of the protein-coding alleles demonstrated interphylogroup recombination events that were up to 20% diverged from the "typical" phylogroup allele; the 16S rRNA gene tree was completely incongruent with the concatenated tree and depending on the examined locus, Phylogroup A was sometimes most closely related to Phylogroup B, sometimes to Phylogroup C, and sometimes to the outgroup. Other lines of evidence such as splits-tree and intragenic recombination analyses also support the absence of genuine barriers to recombination among the *Halorubrum* phylogroups *(41)*, indicating that discrete clusters or biological species do not exist in the *Halorubrum* data set.

Such diversity patterns are not unique to *Halorubrum* spp.: many studies that examined closely related groups (e.g., *Neisseria* spp. *(64, 69)*, *Campylobacter* spp. *(70)*, *Prochlorococcus* spp. *(25, 71)*, enterobacteria *(59, 72)*, *Thermotoga* spp. *(73)*, and *Bradyrhizobia* spp. *(37)*) found similar types of patterns, indicating that such observations are likely the rule rather than the exception *(74)*. Several years ago, John Maynard Smith and colleagues wrote "A study of the genetic and phenotypic variation in a taxon such as *Neisseria* should be compulsory for all the philosophers who believe in the existence of natural kinds, for all cladists who believe in the universal validity of phylogenetic classification, and for all the pheneticists, whatever they believe." *(69)*. Today, it is possible to extend those comments to many more recognized groups and it is not unreasonable to conclude that there will ultimately be more "nebulous" than "discrete cluster" species.

## 4. Conclusion and Outlook

It would seem that identifying and naming species has come full circle. In the 1940s, attempts to classify prokaryotes into natural groups were renounced in favor of pragmatic approaches *(75)*. By the late 1980s, the advent of DNA sequencing and fast computers promised a new hope for classification by constructing natural rather than artificial species *(76, 77)*. Today, cheap, high-throughput DNA sequence data combined with evolution and ecology principles reveals that genetic material is readily and frequently recombined across divergent lineages and from different niches, indicating once again that a natural classification scheme for prokaryotes is in jeopardy: the data do not conform to the species models developed for plants and animals. Clusters seem apparent but, upon close examination, present many difficulties with regard to who belongs in which group (e.g., **Fig. 22.2**). The only recourse is to adopt species concepts that require pragmatic and subjective approaches which do not necessitate species to be monophyletic groups. However, when adopting such concepts, we should not conflate taxonomic species with an evolutionary group, they are not one and the same. With the benefit of hindsight it seems obvious that recombination is what confounded microbial taxonomists in the 1940s.

Microbial ecology and evolution have been transformed in the past 20–30 years by the application of molecular techniques. However, our current taxonomic procedure for species classification seems to have missed the revolution. Data from 70% DNA–DNA reassociation analyses, the gold standard of prokaryotic taxonomy, cannot be transferred from one lab to another or stored on a computer for future reference: experiments must be carried out anew every time a novel strain is cultivated. Because estimates for the number of prokaryotic species in 10 g of soil are $>10^7$, *(78)* microbiology can ill afford to maintain cumbersome taxonomic protocols that cannot be transferred or stored. MLSA, the backbone of these *Halorubrum* spp. analyses, was originally conceived as a transportable method for comparing strains of pathogenic species *(60)*, and databases were created for easy comparison of new strains against ones previously cultivated *(79)*. Applying MLSA to non-pathogenic organisms solves some of our taxonomic problems, and ideas on how to interpret and subjectively circumscribe species in the light of this technique are being developed *(2, 80)*.

To paraphrase Darwin, there is grandeur in this view of prokaryotic species. Microbiology is undergoing a metamorphosis. Insight regarding recombination and prokaryotic species is derived from thinking about diversity from a population point of view, which is similar to the way that evolutionary theory was transformed in the 20th century when natural selection was combined with genetic drift and other non-Darwinian mechanisms and ideas into a new and larger view which provided a more holistic understanding of evolution. Now, once again population-level data are challenging microbiology's ingrained ideas and a new synthesis may be on the horizon.

## Acknowledgments

I would like to thank Ford Doolittle for the rewarding discussions about species, Olga Zhaxybayeva, Peter Gogarten, Joerg Graf, and Ken Noll for critically reading this manuscript, Olga Zhaxybayeva (once again) for help with figures, and the University of Connecticut Research Foundation for financial support.

### References

1. Locke, J. (1689) *An essay concerning human understanding, p.172* David Carlisle, Boston.
2. Gevers, D., Cohan, F. M., Lawrence, J. G., Spratt, B. G., Coenye, T., Feil, E. J., Stackebrandt, E., Van De Peer, Y., Vandamme, P., Thompson, F. L., Swings, J. (2005) Opinion: Re-evaluating prokaryotic species. *Nat Rev Microbiol* **3**, 733–9.
3. Konstantinidis, K. T., Tiedje, J. M. (2005) Genomic insights that advance the species definition for prokaryotes. *Proc Natl Acad Sci U S A* **102**, 2567–72.
4. Ward, D. M., Cohan, F. M., Bhaya, D., Heidelberg, J. F., Kuhl, M., Grossman, A. (2008) Genomics, environmental genomics and the issue of microbial species. *Heredity* **100**, 207–19
5. Cohan, F. M. (2006) Towards a conceptual and operational union of bacterial systematics, ecology, and evolution. *Philos Trans R Soc Lond B Biol Sci* **361**, 1985–96.
6. Doolittle, W. F., Papke, R. T. (2006) Genomics and the bacterial species problem. *Genome Biol* 7, 116.
7. Lawrence, J. G. (2002) Gene transfer in bacteria: Speciation without species? *Theor Popul Biol* **61**, 449–60.

8. Coyne, J. A., Orr, H. A. (2004) *Speciation*, Sinauer Associates, Inc., Sunderland.
9. Wheeler, Q. D., Meier, R. (ed.) (2000) *Species concepts and phylogenetic theory: A debate.* Columbia University Press, New York.
10. Rieseberg, L. H., Willis, J. H. (2007) Plant speciation. *Science* **317**, 910–4.
11. Selander, R. K., Levin, B. R. (1980) Genetic diversity and structure in *Escherichia coli* populations. *Science* **210**, 545–7.
12. Caugant, D. A., Mocca, L. F., Frasch, C. E., Froholm, L. O., Zollinger, W. D., Selander, R. K. (1987) Genetic structure of *Neisseria meningitidis* populations in relation to serogroup, serotype, and outer membrane protein pattern. *J Bacteriol* **169**, 2781–92.
13. Levin, B. R. (1981) Periodic selection, infectious gene exchange and the genetic structure of *E. coli* populations. *Genetics* **99**, 1–23.
14. Atwood, K. C., Schneider, L. K., Ryan, F. J. (1961) Periodic selection in *Escherichia coli. Proc Natl Acad Sci U S A* **37**, 146–55.
15. Cohan, F. M. (2001) Bacterial species and speciation. *Syst Biol* **50**, 513–24.
16. Polz, M. F., Hunt, D. E., Preheim, S. P., Weinreich, D. M. (2006) Patterns and mechanisms of genetic and phenotypic differentiation in marine microbes. *Philos Trans R Soc Lond B Biol Sci* **361**, 2009–21.
17. Ward, D. M., Bateson, M. M., Ferris, M. J., Kuhl, M., Wieland, A., Koeppel, A., Cohan, F. M. (2006) Cyanobacterial ecotypes in the microbial mat community of mushroom spring (Yellowstone National Park, Wyoming) as species-like units linking microbial community composition, structure and function. *Philos Trans R Soc Lond B Biol Sci* **361**, 1997–2008.
18. Felsenstein, J. (1974) The evolutionary advantage of recombination. *Genetics* **78**, 737–56.
19. Muller, H. J. (1964) The relation of recombination to mutational advance. *Mutat Res* **1**, 2–9.
20. Moore, L. R., Rocap, G., Chisholm, S. W. (1998) Physiology and molecular phylogeny of coexisting *Prochlorococcus* ecotypes. *Nature* **393**, 464–7.
21. Ward, D. M. (1998) A natural species concept for prokaryotes. *Curr Opin Microbiol* **1**, 271–7.
22. Papke, R. T., Ramsing, N. B., Bateson, M. M., Ward, D. M. (2003) Geographical isolation in hot spring cyanobacteria. *Environ Microbiol* **5**, 650–9.
23. Rocap, G., Larimer, F. W., Lamerdin, J., Malfatti, S., Chain, P., Ahlgren, N. A., Arellano, A., Coleman, M., Hauser, L., Hess, W. R., Johnson, Z. I., Land, M., Lindell, D., Post, A. F., Regala, W., Shah, M., Shaw, S. L., Steglich, C., Sullivan, M. B., Ting, C. S., Tolonen, A., Webb, E. A., Zinser, E. R., Chisholm, S. W. (2003) Genome divergence in two *Prochlorococcus* ecotypes reflects oceanic niche differentiation. *Nature* **424**, 1042–7.
24. Rocap, G., Distel, D. L., Waterbury, J. B., Chisholm, S. W. (2002) Resolution of *Prochlorococcus* and *Synechococcus* ecotypes by using 16S-23S ribosomal DNA internal transcribed spacer sequences. *Appl Environ Microbiol* **68**, 1180–91.
25. Zhaxybayeva, O., Gogarten, J. P., Charlebois, R. L., Doolittle, W. F., Papke, R. T. (2006) Phylogenetic analyses of cyanobacterial genomes: Quantification of horizontal gene transfer events. *Genome Res* **16**, 1099–108.
26. Johnson, Z. I., Zinser, E. R., Coe, A., Mcnulty, N. P., Woodward, E. M., Chisholm, S. W. (2006) Niche partitioning among *Prochlorococcus* ecotypes along ocean-scale environmental gradients. *Science* **311**, 1737–40.
27. Ravin, A. W. (1963) Experimimental approaches to the study of bacterial phylogeny. *Am Nat* **97**, 307–18.
28. Roberts, M. S., Cohan, F. M. (1993) The effect of DNA sequence divergence on sexual isolation in *Bacillus. Genetics* **134**, 401–8.
29. Vulic, M., Dionisio, F., Taddei, F., Radman, M. (1997) Molecular keys to speciation: DNA polymorphism and the control of genetic exchange in enterobacteria. *Proc Natl Acad Sci U S A* **94**, 9763–7.
30. Lorenz, M. G., Wackernagel, W. (1994) Bacterial gene transfer by natural genetic transformation in the environment. *Microbiol Rev* **58**, 563–602.
31. Falush, D., Torpdahl, M., Didelot, X., Conrad, D. F., Wilson, D. J., Achtman, M. (2006) Mismatch induced speciation in *Salmonella*: model and data. *Philos Trans R Soc Lond B Biol Sci* **361**, 2045–53.
32. Fraser, C., Hanage, W. P., Spratt, B. G. (2007) Recombination and the nature of bacterial speciation. *Science* **315**, 476–80.
33. Hanage, W. P., Spratt, B. G., Turner, K. M., Fraser, C. (2006) Modeling bacterial speciation. *Philos Trans R Soc Lond B Biol Sci* **361**, 2039–44.
34. Cohan, F. M. (1994) The effects of rare but promiscuous genetic exchange on evolutionary divergence in prokaryotes. *Am Nat* **143**, 965–86.
35. Suerbaum, S., Lohrengel, M., Sonnevend, A., Ruberg, F., Kist, M. (2001) Allelic diversity and recombination in *Campylobacter jejuni. J*

*Bacteriol* **183**, 2553–9.
36. Frandsen, E. V., Poulsen, K., Curtis, M. A., Kilian, M. (2001) Evidence of recombination in *Porphyromonas gingivalis* and random distribution of putative virulence markers. *Infect Immun* **69**, 4479–85.
37. Vinuesa, P., Silva, C., Werner, D., Martinez-Romero, E. (2005) Population genetics and phylogenetic inference in bacterial molecular systematics: The roles of migration and recombination in *Bradyrhizobium* species cohesion and delineation. *Mol Phylogenet Evol* **34**, 29–54.
38. Lodders, N., Stackebrandt, E., Nubel, U. (2005) Frequent genetic recombination in natural populations of the marine cyanobacterium *Microcoleus chthonoplastes. Environ Microbiol* 7, 434–42.
39. Whitaker, R. J., Grogan, D. W., Taylor, J. W. (2005) Recombination shapes the natural population structure of the hyperthermophilic archaeon *Sulfolobus islandicus. Mol Biol Evol* **22**, 2354–61.
40. Papke, R. T., Koenig, J. E., Rodriguez-Valera, F., Doolittle, W. F. (2004) Frequent recombination in a saltern population of *Halorubrum. Science* **306**, 1928–9.
41. Papke, R. T., Zhaxybayeva, O., Feil, E. J., Sommerfeld, K., Muise, D., Doolittle, W. F. (2007) Searching for species in haloarchaea. *Proc Natl Acad Sci U S A* **104**, 14092–7.
42. Dykhuizen, D. E., Green, L. (1991) Recombination in *Escherichia coli* and the definition of biological species. *J Bacteriol* **173**, 7257–68.
43. Lan, R., Reeves, P. R. (2001) When does a clone deserve a name? A perspective on bacterial species based on population genetics. *Trends Microbiol* **9**, 419–24.
44. Cooper, T. F. (2007) Recombination speeds adaptation by reducing competition between beneficial mutations in populations of *Escherichia coli. PLoS Biol* **5**, e225.
45. Anton, J., Llobet-Brossa, E., Rodriguez-Valera, F., Amann, R. (1999) Fluorescence *in situ* hybridization analysis of the prokaryotic community inhabiting crystallizer ponds. *Environ Microbiol* **1**, 517–23.
46. Legault, B. A., Lopez-Lopez, A., Alba-Casado, J. C., Doolittle, W. F., Bolhuis, H., Rodriguez-Valera, F., Papke, R. T. (2006) Environmental genomics of *Haloquadratum walsbyi* in a saltern crystallizer indicates a large pool of accessory genes in an otherwise coherent species. *BMC Genomics* 7, 171.
47. Kennedy, S. P., Ng, W. V., Salzberg, S. L., Hood, L., DasSarma, S. (2001) Understanding the adaptation of *Halobacterium* species NRC-1 to its extreme environment through computational analysis of its genome sequence. *Genome Res* **11**, 1641–50.
48. Blaseio, U., Pfeifer, F. (1990) Transformation of *Halobacterium halobium*: Development of vectors and investigation of gas vesicle synthesis. *Proc Natl Acad Sci U S A* **87**, 6772–6.
49. Charlebois, R. L., Lam, W. L., Cline, S. W., Doolittle, W. F. (1987) Characterization of phv2 from *Halobacterium volcanii* and its use in demonstrating transformation of an archaebacterium. *Proc Natl Acad Sci U S A* **84**, 8530–4.
50. Zhou, M., Xiang, H., Sun, C., Tan, H. (2004) Construction of a novel shuttle vector based on an rcr-plasmid from a haloalkaliphilic archaeon and transformation into other haloarchaea. *Biotechnol Lett* **26**, 1107–13.
51. Ortenberg, R., Tchelet, R., Mevarech, M. (1999) A model for the genetic exchange system of the extremely halophilic archaeon *Haloferax volcanii*, in *Microbiology and biogeochemistry of hypersaline environments* (Oren, A., ed.) CRC Press, Boca Raton 331–8.
52. Rosenshine, I., Tchelet, R., Mevarech, M. (1989) The mechanism of DNA transfer in the mating system of an archaebacterium. *Science* **245**, 1387–9.
53. Schleper, C., Holz, I., Janekovic, D., Murphy, J., Zillig, W. (1995) A multicopy plasmid of the extremely thermophilic archaeon *Sulfolobus* effects its transfer to recipients by mating. *J Bacteriol* **177**, 4417–26.
54. Diez, B., Anton, J., Guixa-Boixereu, N., Pedros-Alio, C., Rodriguez-Valera, F. (2000) Pulsed-field gel electrophoresis analysis of virus assemblages present in a hypersaline environment. *Int Microbiol* **3**, 159–64.
55. Dyall-Smith, M., Tang, S. L., Bath, C. (2003) Haloarchaeal viruses: How diverse are they? *Res Microbiol* **154**, 309–13.
56. Sabet, S., Diallo, L., Hays, L., Dillon, J. (2007) Characterization of halophages and their hosts isolated from solar salterns in Baja Cafornia, Mexico, in: *Halophiles-2007: International Congress Exploring Life at High Salinity.* University of Essex, Colchester.
57. Walsh, D. A., Papke, R. T., Doolittle, W. F. (2005) Archaeal diversity along a soil salinity gradient prone to disturbance. *Environ Microbiol* 7, 1655–66.
58. Pasic, L., Ulrih, N. P., Crnigoj, M., Grabnar, M., Velikonja, B. H. (2007) Haloarchaeal communities in the crystallizers of two Adriatic solar salterns. *Can J Microbiol* **53**, 8–18.
59. Wirth, T., Falush, D., Lan, R., Colles, F., Mensa, P., Wieler, L. H., Karch, H., Reeves, P. R., Maiden, M. C., Ochman, H.,

Achtman, M. (2006) Sex and virulence in *Escherichia coli*: An evolutionary perspective. *Mol Microbiol* **60**, 1136–51.
60. Maiden, M. C., Bygraves, J. A., Feil, E., Morelli, G., Russell, J. E., Urwin, R., Zhang, Q., Zhou, J., Zurth, K., Caugant, D. A., Feavers, I. M., Achtman, M., Spratt, B. G. (1998) Multilocus sequence typing: A portable approach to the identification of clones within populations of pathogenic microorganisms. *Proc Natl Acad Sci U S A* **95**, 3140–5.
61. Feil, E. J., Enright, M. C., Spratt, B. G. (2000) Estimating the relative contributions of mutation and recombination to clonal diversification: A comparison between *Neisseria meningitidis* and *Streptococcus pneumoniae*. *Res Microbiol* **151**, 465–9.
62. Feil, E. J., Li, B. C., Aanensen, D. M., Hanage, W. P., Spratt, B. G. (2004) eBURST: Inferring patterns of evolutionary descent among clusters of related bacterial genotypes from multilocus sequence typing data. *J Bacteriol* **186**, 1518–30.
63. Feil, E. J., Maiden, M. C., Achtman, M., Spratt, B. G. (1999) The relative contributions of recombination and mutation to the divergence of clones of *Neisseria meningitidis*. *Mol Biol Evol* **16**, 1496–502.
64. Hanage, W. P., Fraser, C., Spratt, B. G. (2005) Fuzzy species among recombinogenic bacteria. *BMC Biol* **3**, 6.
65. Kimura, M. (1968) Genetic variability maintained in a finite population due to mutational production of neutral and nearly neutral isoalleles. *Genet Res* **11**, 247–69.
66. Sharp, P. M., Li, W. H. (1987) The codon adaptation index – a measure of directional synonymous codon usage bias, and its potential applications. *Nucleic Acids Res* **15**, 1281–95.
67. Taddei, F., Vulic, M., Radman, M., Matic, I. (1997) Genetic variability and adaptation to stress. *EXS* **83**, 271–90.
68. Matic, I., Rayssiguier, C., Radman, M. (1995) Interspecies gene exchange in bacteria: The role of SOS and mismatch repair systems in evolution of species. *Cell* **80**, 507–15.
69. Smith, J. M., Feil, E. J., Smith, N. H. (2000) Population structure and evolutionary dynamics of pathogenic bacteria. *Bioessays* **22**, 1115–22.
70. Sheppard, S. K., Mccarthy, N. D., Falush, D., Maiden, M. C. (2008) Convergence of *Campylobacter* species: Implications for bacterial evolution. *Science* **320**, 237–9.
71. Beiko, R. G., Harlow, T. J., Ragan, M. A. (2005) Highways of gene sharing in prokaryotes. *Proc Natl Acad Sci U S A* **102**, 14332–7.
72. Wertz, J. E., Goldstone, C., Gordon, D. M., Riley, M. A. (2003) A molecular phylogeny of enteric bacteria and implications for a bacterial species concept. *J Evol Biol* **16**, 1236–48.
73. Nesbϕ, C. L., Dlutek, M., Doolittle, W. F. (2006) Recombination in *Thermotoga*: Implications for species concepts and biogeography. *Genetics* **172**, 759–69.
74. O'Malley, M. A., Boucher, Y. (2005) Paradigm change in evolutionary microbiology. *Stud Hist Philos Biol Biomed Sci* **36**, 183–208.
75. Stanier, R. Y., Van Niel, C. B. (1941) The main outlines of bacterial classification. *J Bacteriol* **42**, 437–66.
76. Woese, C. R. (1987) Bacterial evolution. *Microbiol Rev* **51**, 221–71.
77. Woese, C. R., Kandler, O., Wheelis, M. L. (1990) Towards a natural system of organisms: Proposal for the domains Archaea, Bacteria, and Eucarya. *Proc Natl Acad Sci U S A* **87**, 4576–9.
78. Gans, J., Wolinsky, M., Dunbar, J. (2005) Computational improvements reveal great bacterial diversity and high metal toxicity in soil. *Science* **309**, 1387–90.
79. Chan, M. S., Maiden, M. C., Spratt, B. G. (2001) Database-driven multi locus sequence typing (MLST) of bacterial pathogens. *Bioinformatics* **17**, 1077–83.
80. Hanage, W. P., Fraser, C., Spratt, B. G. (2006) Sequences, sequence clusters and bacterial species. *Philos Trans R Soc Lond B Biol Sci* **361**, 1917–27.
81. Stackebrandt, E., Goebel, B. M. (1994) Taxonomic note: A place for DNA-DNA reassociation and 16S rRNA sequence analysis in the present species definition in bacteriology. *Int J Syst Evol Microbiol* **44**, 846–9.
82. Wayne, L. G., Brenner, D. J., Colwell, R. R., Grimont, P. A. D., Kandler, O., Krichevsky, L., Moore, L. H., Moore, W. C., Murray, R. G. E., Stackebrandt, E., Starr, M. P., Truper, H. G. (1987) Report of the ad hoc committee on reconciliation of approaches to bacterial systematics. *Int J Syst Bacteriol* **37**, 463–4.
83. Konstantinidis, K. T., Tiedje, J. M. (2005) Towards a genome-based taxonomy for prokaryotes. *J Bacteriol* **187**, 6258–64.

# Chapter 23

# What Antimicrobial Resistance Has Taught Us About Horizontal Gene Transfer

**Miriam Barlow**

## Abstract

Horizontal gene transfer (HGT) has been responsible for the dissemination of numerous antimicrobial-resistance determinants throughout diverse bacterial species. The rapid and broad dissemination of resistance determinants by HGT, and subsequent selection for resistance imposed by the use of antimicrobials, threatens to undermine the usefulness of antimicrobials. However, vigilant surveillance of the emerging antimicrobial resistance in clinical settings and subsequent studies of resistant isolates create a powerful system for studying HGT and detecting rare events. Two of the most closely monitored phenotypes are resistance to β-lactams and resistance to fluoroquinolones. Studies of resistance to these antimicrobials have revealed that (1) transformation occurs between different species of bacteria including some recipient species that were not previously known to be competent for natural transformation; (2) transduction may be playing an important role in generating novel methicillin-resistant *Staphylococcus aureus* (MRSA) strains, although the details of transferring the SCC*mec* element are not yet fully understood; (3) Resistance genes are probably moving to plasmids from chromosomes more rapidly than in the past; and (4) Resistance genes are aggregating upon plasmids. The linkage of numerous resistance genes on individual plasmids may underlie the persistence of resistance to specific antimicrobials even when use of those antimicrobials is discontinued. Further studies of HGT and methods for controlling HGT may be necessary to maintain the usefulness of antimicrobials.

**Key words:** β-lactamase, *qnr*, SCC*mec*, *parC*, penicillin-binding protein.

## 1. Introduction

The discovery and use of antimicrobials have constituted one of the most important medical advances in the history of mankind. The discovery of penicillin by Alexander Fleming in 1928 was one of the primary events that brought about the so-called antibiotic era. However, after the initial discovery of penicillin, the antibiotic

Maria B. Gogarten et al. (eds.), *Horizontal Gene Transfer: Genomes in Flux, vol. 532*

DOI 10.1007/978-1-60327-853-9_23 Springerprotocols.com

had to be purified, tested, and mass-produced. Fleming sought collaborators capable of making those necessary contributions for over 10 years before he finally garnered the interest of Ernst Chain and Howard Florey *(1, 2)*. Through the collective efforts of these scientists, sufficient penicillin became available to treat all of the allied forces suffering from wounds or infections following D-Day on June 6, 1944. It seems likely that the availability of penicillin may have been one of the deciding factors affecting the outcome of World War II *(3)*. Few would argue the importance of antimicrobials for modern society; however, just as it was difficult for Fleming to recruit interested parties to develop penicillin for clinical use, it has become increasingly difficult to find parties willing to develop new antimicrobials. While the discovery and development of new antimicrobials flourished during the 1980s and 1990s, large pharmaceutical companies are now abandoning the pursuit of antimicrobials in favor of anti-virals and drugs that treat non-infectious diseases *(4)*. Much of the reason for the lack of interest in developing novel antimicrobials is that horizontal gene exchange spreads resistance determinants so rapidly through bacterial populations. The cost of developing an antimicrobial can range from \$400 million to \$800 million over a period of >10 years, and companies find it difficult to make a profit when they develop new antimicrobials *(5, 6)* because of the rapid occurrence and spread of antimicrobial-resistance determinants *(7–9)*.

## 2. Antimicrobials Threatened by HGT

Since the discovery of penicillin, the battle between humanity and infectious diseases has been waged with the development of new antimicrobials and the emergence and horizontal spread of numerous resistance determinants. HGT has contributed to the spread of resistance to aminoglycosides *(10, 11)*, macrolides *(12–14)*, glycopeptides *(15)*, tetracyclines *(16)*, β-lactams *(17, 18)*, and fluoroquinolones *(19)*. There are already numerous review articles describing the general importance of HGT for the dissemination of those resistance determinants *(12–19)*. Therefore, this chapter is focused mainly upon cases where the study of HGT has revealed novel or previously unrealized aspects of HGT. Studies of resistance to β-lactams and fluoroquinolones have been particularly important to revealing novel aspects of HGT, so resistance to those two classes of antimicrobials is particularly emphasized.

### 2.1. β-Lactams

The β-lactams, which include penicillin, block cell wall synthesis and derive their activity from a four-membered β-lactam ring. They irreversibly bind to glycopeptide transpeptidases, also called penicillin-binding proteins (PBPs), which catalyze the formation of cross-linkages between the peptidoglycan precursors that form the cell wall of bacteria. The irreversible binding of β-lactams to PBPs destroys the functionality of PBPs. The cell walls of bacteria susceptible to β-lactams are weakened by decreased cross-linkages of the cell wall and ultimately burst due to high internal osmotic pressure. The underlying mechanism these drugs target is specific to bacteria, which means they have low toxicity to humans *(20)*. Additionally, β-lactams are effective against a broad spectrum of bacteria and have historically been used as preferred empiric treatments for numerous infectious illnesses. Transferable resistance to β-lactams is conferred either through transfer within genera with resistant PBPs that have reduced affinity for β-lactams *(21,22)* or by transfer between species and genera of β-lactamases, which hydrolyze the β-lactam ring *(18, 23, 24)*. The heavy use of β-lactams has resulted in the emergence and dissemination of numerous β-lactamases and resistant PBPs.

### 2.2. Fluoroquinolones

More recently, the fluoroquinolones have become increasingly popular antimicrobials *(25)*. The lethal effects of fluoroquinolones upon bacteria stem from their binding to DNA gyrase and topoisomerase IV, which interferes with DNA replication *(19)*. The resistance to fluoroquinolones can result from substitutions in DNA gyrase and topoisomerase IV, which reduce the susceptibility of those enzymes to inactivation by fluoroquinolones. Fluoroquinolone resistance can also result from the expression of efflux pumps, which efficiently export fluoroquinolones from the bacterium, or the expression of a *qnr* plasmidic-resistance determinant *(19, 26)*. The mechanism of *qnr* has not yet been fully elucidated; however, it is known that the Qnr protein binds to DNA gyrase and topoisomerase IV and blocks the binding of fluoroquinolones to those enzymes without completely inhibiting their activity toward DNA *(27)*. Although the fluoroquinolones have not been used as extensively or for as long as the β-lactams, resistance to them is common and increasing *(25)*.

## 3. Rapid Dissemination of Resistance by HGT

Because the vitality of bacteria critically depends upon their ability to either avoid or overcome the effects of antimicrobials, the evolutionary response time for becoming resistant to antimicrobials is extremely short. For example, within a decade, the usefulness of penicillin was seriously compromised as the horizontal spread caused a plasmidic β-lactamase to become ubiquitous among populations of *Staphylococcus aureus (8)*. The widespread occurrence of resistance to a new antimicrobial generally occurs in less than 3 years *(7–9)*, and usually includes the mobilization of a resistance determinant via horizontal transfer mechanisms. There is also some evidence (presented in a later section of this chapter), which suggests that the rate at which novel resistance determinants are mobilized in bacterial populations may be increasing.

As bacteria evolve and develop resistance to antimicrobials, clinical microbiologists closely monitor the development of resistance, and the underlying mechanisms are then determined in reference and research laboratories. The rapid response time of bacteria and the close attention paid to those responses has created an unparalleled opportunity to study bacterial evolution. Studies of the evolution of antimicrobial resistance have revealed that the accumulation of mutations in chromosomal genes can result in clinical resistance to antimicrobials, but they have also shown that the horizontal transfer of pre-existing resistance determinants is by far the most rapid and most common mechanism for the occurrence of resistance in the clinical isolates of bacteria.

It is not surprising that horizontal transfer is an effective mechanism for disseminating antimicrobial resistance among bacterial populations, because most commercially developed antimicrobials are derived from the antimicrobials naturally produced by microbes *(28–30)*. The occurrence of natural antimicrobials means that there was selective pressure for the evolution of resistance genes in the pre-antibiotic era. For every antimicrobial produced by microbes, resistance mechanisms already exist. While bacteria do not innately carry resistance determinants for all antimicrobials in their chromosomes, they are able to rapidly acquire advantageous resistance determinants via HGT from those bacteria that do carry them.

Bacteria collectively utilize all methods of HGT to acquire resistance genes, although specific species tend to specialize in the mechanism(s) they use most heavily. For example, among the *Enterobacteriaceae*, resistance genes are usually transferred on plasmids via conjugation. Among the streptococci, *Neisseriaceae*, and *Haemophilus*, resistance is disseminated via natural transformation. In *Staphylococci*, resistance genes may be disseminated both by transduction and by conjugation. The study of the acquisition of resistance determinants through each of these mechanisms has illuminated important features of each mechanism of HGT in bacteria.

## 4. Transformation

### *4.1. Transformation of Streptococci*

Transformation, or the uptake and incorporation of DNA from the environment, is an important mechanism for spreading resistance among the streptococci. *Streptococcus pneumoniae* is an important, ubiquitous pathogen. In addition to causing respiratory infections, it also causes meningitis, acute sinusitis, otitis media, septic arthritis, and other types of infections *(31)*. S. *pneumoniae* regularly acquires new DNA via natural transformation *(32)*. The competence factors underlying the natural competence of *S. pneumoniae* have been extensively studied and reviewed *(33, 34)*. The occurrence of penicillin resistance is prevalent in *S. pneumoniae (35)* and seems to be correlated with the frequency of transformation in Streptococci. It is possible that the regular occurrence of transformation within *S. pneumoniae* populations contributes to the prevalence of penicillin resistance in this pathogen.

*Streptococcus pyogenes* is also a worldwide, highly prevalent pathogen. It most frequently colonizes in the nasopharynx and causes superficial infections of the epithelium, although it can cause more serious infections as well *(36)*. While penicillin resistance is prevalent in *S. pneumoniae, S. pyogenes* is considered to be universally susceptible to penicillin despite extensive exposure to penicillin. *S. pyogenes* does not frequently undergo transformation. The infrequency of transformation in *S. pyogenes* has been named as a possible reason for its sustained susceptibility to penicillin *(37)*. However, strong selection for fluoroquinolone resistance has revealed previously unrealized aspects of transformation in *S. pyogenes.*

In recent years, fluoroquinolones have been used extensively for the treatment of infections due to streptococci. Mutations in the *gyrA* gene encoding gyrase and the *parC* subunit of topoisomerase IV result in fluoroquinolone resistance *(38)*. There are particular "hotspots" where substitutions resulting in fluoroquinolone resistance occur. They are called the "quinolone resistance determining region" (QRDR) *(39)* and are conserved across Gram-positive and Gram-negative bacteria. The occurrence of fluoroquinolone resistance in *S. pyogenes* is rare but has been documented *(38, 40, 41)*. The analysis of the resistant strains showed that point mutations in *parC* were responsible for resistance in some strains, but surprisingly HGT of the QRDR of *parC* was the best explanation for resistance in some strains *(37)*. Even more unexpectedly, the sequence of the recombined segments of DNA indicated that the uptake of DNA from a different species, *S. dysgalactiae*, was the source of the segment responsible for fluoroquinolone resistance *(37)*. Thus transformation may occur more frequently among *S. pyogenes* than previously estimated. In addition, the data also indicate that *S. pyogenes* can also uptake DNA from different species. These results demonstrate how the detection of rare events is possible when antimicrobial resistance is used as a model for studying HGT and evolution in bacteria.

### 4.2. Transformation in *Neisseria*

Another genus of bacteria that acquires new DNA via frequent horizontal gene exchange by natural transformation is *Neisseria*

*(42)*. Among the most commonly encountered and most resistant *Neisseria* species is *Neisseria gonorrhoeae*. This species is competent during all stages of growth and takes up DNA that contains a genus-specific 10 bp DNA uptake sequence (GCCGTCTGAA) *(43)*. This sequence appears about once per 1000 bp in *N. gonorrhoeoae*. Donor DNA can be released into the extracellular environment by autolysis of *Neisseria* as well as by Type IV secretion and other less characterized secretion methods *(44)*. Transformation is clearly important in *Neisseria* and may be conserved because of its role in generating surface antigen variation, which facilitates evasion of the host immune response.

Frequent transformation has effectively disseminated multiple resistance determinants among *N. gonorrhoeae*. The efficacy of fluoroquinolones *(45, 46)* and β-lactams *(47)* for treating *N. gonorrhoeae* infections has been seriously compromised by the high frequency of resistance to those drugs. β-lactam resistance in *N. gonorrhoeae* is mediated by amino acid substitutions in PBPs. In one study, a mosaic PBP-2 protein was found in β-lactam-resistant *N. gonorrhoeae* that varied from PBP-2 in susceptible bacteria by ~60 amino acid substitutions which contributed to β-lactam resistance *(48, 49)*. Mosaic PBP-2 obtained the mutations necessary for the conferral of resistance by transformation of DNA from resistant, non-pathogenic, commensal *Neisseria (48, 50)*. The species *N. flavescens* and *N. cinerea* are intrinsically more resistant to β-lactams than *N. gonorrhoeoae* and, although they are not pathogenic, they have promoted the spread and increased the difficulty of curing *N. gonorrhoeoae* infections by serving as donors of resistance.

The transformation of resistance genes in *Streptococcus* and *Neisseria* species has served as a good model for studying interspecies transformational events. It is unlikely that transformation in *S. pyogenes* would have been observable if not for the powerful selection imposed by antimicrobials upon bacteria and the close attention given to novel resistance phenotypes. It is likewise unlikely that the transfer of genes from commensal to pathogenic species of *Neisseria* would be as readily detected without the transfer of antimicrobial-resistance determinants. These examples of the transfer of resistance determinants by transformation illustrate the utility of using antimicrobial resistance as a model for studying basic microbial processes and population genetics.

## 5. Transduction

### 5.1. Transduction in *Staphylococcus Aureus*

Transduction is an important mechanism of genetic transfer among antimicrobial resistant bacteria. However, it has impacted bacteria more through the transfer of virulence factors and pathogenicity islands than the transfer of resistance genes. One of the resistant pathogens most affected by transduction is methicillin-resistant *S. aureus* (MRSA). The origin of MRSA may have resulted from HGT of the *mecA* gene, which encodes a methicillin-resistant PBP-2a protein *(51–53)* into the chromosome of *S. aureus*. The *mecA* gene is located on a mobile genetic element called the Staphylococcal Cassette Chromosome (SCC), which also contains the *ccr* genes that encode site-specific recombinases necessary for SCC mobility. The SCC*mec* cassette ranges in size from 20.9 to 66.9 kb *(54)*. The origins and transfer mechanism of SCC*mec* are not certain. The SCC has been found without the *mecA* gene in *S. hominis (55)* and *S. epidermidis (56)*, and each of these species has been suggested as possible source of the SCC that carries *mecA (55, 56)*. The *mecA* gene product shares 88% amino acid identity with the *Staphylococcus sciuri* PBP2-A *(57–59)*; however, the progenitor of *mecA* has not been determined. Additionally, the mechanism for the import of SCC*mec* into the chromosome is not clear. Among the known mechanisms of horizontal transfer, transduction has seemed the most likely for transfer of SCC*mec*. However, the segment of DNA containing this cassette is too large to fit into the heads of any of the known *S. aureus* general transducing phages. Evidence that transduction may be involved is that SCC*mec* type IV is the smallest (20.9 kb) of the known SCC*mec* elements and that it appears to be more frequently transferred in clinical *S. aureus* populations than other SCC*mec* elements *(60)*. This may indicate that it is being transduced more efficiently because of its reduced size. Currently, so little is known about the origin and transfer of SCC*mec* that it is impossible to argue strongly for transduction or any other specific transfer mechanism for this element.

There is, however, strong evidence for multiple HGT events involving SCC*mec* in *S. aureus*. For example, in one study 11 major MRSA clones were identified within five groups of related genotypes *(61)*. In this study, the authors identified multiple divergent MRSA strains with methicillin-susceptible (MSSA) strains as their nearest relatives. The relatedness of MRSA strains to multiple MSSA strains suggests multiple origins of MRSA because methicillin susceptibility is the ancestral state of *S. aureus*. It is estimated that SCC*mec* has only entered MSSA about 20 times in total *(62)*. The transfer of *SCCmec* to MSSA is extremely infrequent, but the mechanism by which SCC*mec* is transferred is continuously active. It is important that the details of this transfer are eventually understood.

One of the strongest reasons for suspecting transduction as the HGT - causing mechanism of SCC*mec* is that transduction clearly plays an important role in transmitting other genes among Staphylococci. Most strains of *S. aureus* carry between one and three prophages in their genomes *(60, 63)*. Prophages and pathogenicity islands, which are transferred by helper phages, constitute an important means by which *S. aureus* clones acquire new virulence factors. The virulence factors included on those genetic elements include staphylokinase, enterotoxins, toxic shock syndrome toxins, and the Panton-Valentine Leukocidin, which is a cytotoxin prevalent in some types of *S. aureus* infections such as abscesses and arthritis *(64)*. Bacteriophages have also been shown to transfer some resistance determinants *(65)* among *S. aureus*. Through strong selection imposed by antimicrobials, intensive monitoring of novel phenotypes by clinical microbiologists, and increased availability of genomic sequences, it is likely that what role (if any) transduction has played in generating MRSA will eventually become more clear.

## 6. Conjugation

Conjugation is the mechanism of HGT by far most responsible for broadly disseminating resistance determinants among bacterial populations. Conjugation among bacteria enables resistance genes to be transferred with increasing frequency within specific species and has resulted in many resistance determinants being disseminated broadly among distantly related species. Plasmids spread resistance determinants for virtually every type of antimicrobial that exists including but not limited to aminoglycosides, macrolides, tetracyclines, glycopeptides, flouroquinolones, and β-lactams. Some resistance determinants have been mobilized to plasmids from chromosomes, whereas others existed on plasmids for a very long time *(66)*. Among the most prevalent plasmidic resistance determinants are those that encode β-lactamases of which the TEMs, SHVs, and CTX-Ms currently cause the greatest clinical concern. The location of these resistance determinants on plasmids has perpetuated their prevalence and increased our understanding of plasmidic transmission among bacteria.

### 6.1. Increased Rate of Mobilization to Plasmids Among β-Lactamases

One of the most ominous discoveries about plasmidic resistance determinants is that the rate at which they are being mobilized from chromosomes to plasmids is increasing *(67)*. This evidence is mainly derived from Class-A β-lactamases, which are the group of β-lactamases that include the TEMs and CTX-Ms. There are numerous genes encoding Class A β-lactamases that have been mobilized from chromosomes to plasmids, and many of the mobilizations have likely occurred during the past 60 years. The evidence that suggests that the rate of mobilization is increasing comes from the *bla*$_{\text{CTX-M}}$s, which are of particular clinical interest because they are rapidly spreading through clinical populations of bacteria. Unlike the usual pattern of class A β-lactamase mobilizations in which the plasmidic alleles are all descended

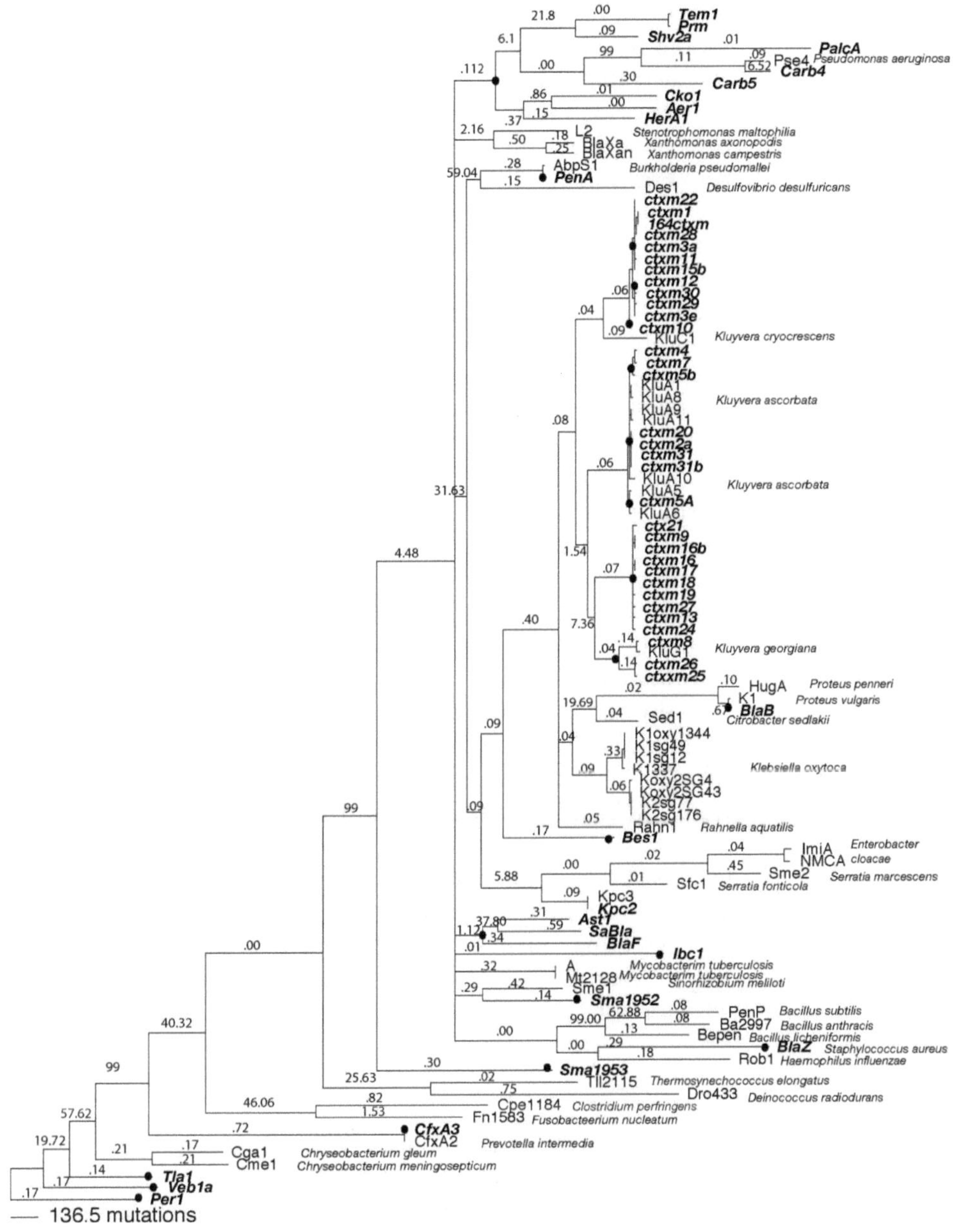


Fig. 23.1. Phylogeny of Class A β-lactamases *(67)*.This phylogeny was calculated by Bayesian inference. Number of mutations occurring along each branch are represented visually by the lengths of the branches. d$N$/d$S$ ratios for all branches except the tips are given along the lengths of the branches. *Bold* type indicates plasmidic genes. *Black dots* indicate mobilizations to plasmids. *Numbered brackets* indicate monophyletic divisions within the phylogeny.

from a single common plasmidic ancestor, evidence shows that *bla*$_{\text{CTX-M}}$s have been mobilized numerous times *(68–71)* from the chromosomes of *Kluyvera* (**Fig. 23.1** extracted from *(67)*), which is an infrequent opportunistic pathogen *(72)*. Because *Kluyvera* chromosomal genes have been found, which exactly match the sequence of plasmidic $bla_{CTX-M}$s *(68)*, many of the mobilizations have likely occurred recently. Phylogenetic analysis revealed that the $bla_{CTX-M}$s have been mobilized from the chromosomes of various *Kluyvera* species to plasmids at least eight times since they diverged from their most recent common ancestor (**Fig. 23.1**).

Among all of the class A β-lactamases, including the $bla_{CTX-M}$s, only 22 mobilizations to plasmids were found. Among the $bla_{CTX-M}$s, the ratio of mobilizations to mutations is 1 mobilization per 191 mutations. Among the remainder of the tree when the $bla_{CTX-M}$ clade is excluded from the analysis, 14 mobilizations occur with the ratio of mobilizations to mutations being 1 mobilization per 2,471 mutations. When the complete phylogeny is considered, 1 mobilization occurs per 1,870 mutations. By that comparison, the mobilization of the $bla_{CTX-M}$ genes to plasmids has occurred at an unusually high rate. This result is unlikely to be an artifact of sampling bias or clinical interest, because other class A β-lactamases have been intently studied for a longer period than the $bla_{CTX-M}$s. If any bias exists in the data, it would be the under-sampling of $bla_{CTX-M}$ mobilizations relative to other Class A β-lactamases *(67)*.

Although the reasons for the increase in the frequency of mobilization are undetermined, they may include increased frequency of plasmids among clinical and environmental populations of bacteria, and increased exposure to antimicrobials of microbes in clinical and environmental populations *(67)*. Whatever the reason, this alarming trend suggests that the time necessary for the emergence of plasmidic resistance determinants might be decreasing. Analysis of additional resistance determinants will help to establish whether this trend is generally occurring among all plasmidic resistance determinants or if it is limited to specific groups.

### 6.2. Clustering of Resistance Genes on Plasmids

Another alarming trend among bacteria has been the observation that resistance genes tend to cluster on plasmids and that acquisition of a single plasmid can therefore confer multiple drug–resistance phenotypes upon a previously susceptible strain of bacteria. As an example, a single plasmid isolated from an avian pathogenic *E. coli* isolate was described, which carried resistance determinants for aminoglycosides, sulbactam, tetracycline, quaternary ammonium compounds, silver, and tellurite *(73)*. In another example, a single clinical isolate of *Klebsiella pneumoniae* carried two plasmids that contained genes encoding the β-lactamases VIM-4, CTX-M-15, CMY-4, and TEM-1 *(74)*. These cases of several linked resistance genes are typical rather than exceptional and exemplify the enormous threat imposed by antimicrobial resistance. When bacteria acquire one resistance determinant by HGT, they often acquire several others simultaneously.

This result was explained *(75, 76)* based on a simple response of bacteria to the selective pressures imposed by antimicrobials. Bacteria will survive and reproduce in more environments if they carry numerous resistance genes. The aggregation of several resistance genes to a single plasmid will therefore be selected for in bacterial populations, because a single horizontal transfer event can confer numerous resistance capabilities upon the recipient strain. This trend has been observed numerous times and is expected to continue indefinitely.

The aggregation of resistance genes to single plasmids has likely played an important role in perpetuating resistance phenotypes even when the use of specific antimicrobials is discontinued or cycled. In numerous trials, the use of specific antimicrobials had been discontinued; however, bacteria have not responded with the expected decrease in resistance to the abandoned antimicrobial. In some instances, resistance to an antimicrobial has continued to increase even though the antimicrobial is no longer in use *(4, 77–80)*. Additionally, antimicrobial cycling studies in which the use of antimicrobials was periodically cycled in specific locations have had ambiguous results *(81, 82)*. Changes in the frequency of resistance observed during these studies are varying and usually not reproducible. It is not clear that changing the use of specific antimicrobials has any effect upon the frequency of resistance. This may result from the linkage that exists between resistance determinants which have been aggregated upon a single plasmid. Overall, the ability of one antimicrobial to select for resistance to many others may be one of the most threatening trends among bacteria because the use of all antimicrobials may have to be discontinued to observe declines in the frequencies of plasmidic resistance determinants.

## 7. Conclusions

Heavy use of antimicrobials has selected for antimicrobial resistance among bacterial populations, and horizontal transfer mechanisms have facilitated the spread of resistance determinants. Transformation is important for disseminating chromosomal resistance determinants among naturally competent bacteria. Additionally, the transfer of resistance determinants has revealed that transformation is an important mechanism of HGT in more species than previously thought. Transduction may be an important mechanism for generating new MRSA clones, although more research is needed to determine the mechanistic details of

SCC*mec* transfer. Conjugation is an important mechanism for disseminating multiple-resistance determinants simultaneously and the rate at which resistance determinants are mobilized to plasmids might be increasing. Overall, bacteria have become highly efficient at moving resistance determinants across species and throughout populations.

It is paradoxical that at a time when bacteria seem to be optimizing their capabilities to mobilize and disseminate resistance determinants, pharmaceutical companies are abandoning the development of new antimicrobials. The result of this is that the available options for treating infectious illnesses are becoming more limited. HGT underlies much of the success of bacteria and much of the discouragement of drug developers. If antimicrobials are going to remain effective treatments for infectious illnesses, it may be necessary to develop methods or drugs for inhibiting horizontal transfer of resistance genes as well as the growth of bacteria.

## References


1. Ligon, B. L. (2004) Sir Howard Walter Florey – the force behind the development of penicillin. *Semin Pediatr Infect Dis* **15**, 109–14.
2. Ligon, B. L. (2004) Penicillin: its discovery and early development. *Semin Pediatr Infect Dis* **15**, 52–7.
3. Wainwright, M. (2004) Hitler's penicillin. *Perspect Biol Med* **47**, 189–98.
4. Livermore, D. (2004) Can better prescribing turn the tide of resistance? *Nat Rev Microbiol* **2**, 73–8.
5. Shlaes, D. M. (2003) The abandonment of antibacterials: why and wherefore? *Curr Opin Pharmacol* **3**, 470–3.
6. Spellberg, B., Powers, J. H., Brass, E. P., Miller, L. G., Edwards, J. E., Jr. (2004) Trends in antimicrobial drug development: implications for the future. *Clin Infect Dis* **38**, 1279–86.
7. Clancy, J., Dib-Hajj, F., Petitpas, J. W., Yuan, W. (1997) Cloning and characterization of a novel macrolide efflux gene, mreA, from Streptococcus agalactiae. *Antimicrob Agents Chemother* **41**, 2719–23.
8. Medeiros, A. A. (1997) Evolution and dissemination of beta-lactamases accelerated by generations of beta-lactam antibiotics. *Clin Infect Dis* **24**, S19–45.
9. Maple, P., Brumfitt, W., Hamilton-Miller, J. M. (1990) A review of the antimicrobial activity of the fluoroquinolones. *J Chemother* **2**, 280–94.
10. Pantosti, A., Sanchini, A., Monaco, M. (2007) Mechanisms of antibiotic resistance in Staphylococcus aureus. *Future Microbiol* **2**, 323–34.
11. Shakil, S., Khan, R., Zarrilli, R., Khan, A. U. (2008) Aminoglycosides versus bacteria – a description of the action, resistance mechanism, and nosocomial battleground. *J Biomed Sci* **15**, 5–14.
12. Roberts, M. C. (2008) Update on macrolide-lincosamide-streptogramin, ketolide, and oxazolidinone resistance genes. *FEMS Microbiol Lett* **282**, 147–59.
13. Roberts, M. C. (2004) Resistance to macrolide, lincosamide, streptogramin, ketolide, and oxazolidinone antibiotics. *Mol Biotechnol* **28**, 47–62.
14. Roberts, M. C. (2004) Distribution of macrolide, lincosamide, streptogramin, ketolide and oxazolidinone (MLSKO) resistance genes in Gram-negative bacteria. *Curr Drug Targets Infect Disord* **4**, 207–15.
15. Willems, R. J., Bonten, M. J. (2007) Glycopeptide-resistant enterococci: deciphering virulence, resistance and epidemicity. *Curr Opin Infect Dis* **20**, 384–90.
16. Roberts, M. C. (2005) Update on acquired tetracycline resistance genes. *FEMS Microbiol Lett* **245**, 195–203.
17. Hanssen, A. M., Ericson Sollid, J. U. (2006) SCCmec in staphylococci: genes on the move. *FEMS Immunol Med Microbiol* **46**, 8–20.

18. Weldhagen, G. F. (2004) Integrons and beta-lactamases – a novel perspective on resistance. *Int J Antimicrob Agents* **23**, 556–62.
19. Robicsek, A., Jacoby, G. A., Hooper, D. C. (2006) The worldwide emergence of plasmid-mediated quinolone resistance. *Lancet Infect Dis* **6**, 629–40.
20. Livermore, D. M. (1996) Are all beta-lactams created equal? *Scand J Infect Dis Suppl* **101**, 33–43.
21. Lambert, P. A. (2005) Bacterial resistance to antibiotics: modified target sites. *Adv Drug Deliv Rev* **57**, 1471–85.
22. Hakenbeck, R., Grebe, T., Zahner, D., Stock, J. B. (1999) beta-lactam resistance in Streptococcus pneumoniae: penicillin-binding proteins and non-penicillin-binding proteins. *Mol Microbiol* **33**, 673–8.
23. Canton, R., Coque, T. M. (2006) The CTX-M beta-lactamase pandemic. *Curr Opin Microbiol* **9**, 466–75.
24. Poole, K. (2004) Resistance to beta-lactam antibiotics. *Cell Mol Life Sci* **61**, 2200–23.
25. Bhavnani, S. M., Hammel, J. P., Jones, R. N., Ambrose, P. G. (2005) Relationship between increased levofloxacin use and decreased susceptibility of Streptococcus pneumoniae in the United States. *Diagn Microbiol Infect Dis* **51**, 31–7.
26. Tran, J. H., Jacoby, G. A. (2002) Mechanism of plasmid-mediated quinolone resistance. *Proc Natl Acad Sci U S A* **99**, 5638–42.
27. Tran, J. H., Jacoby, G. A., Hooper, D. C. (2005) Interaction of the plasmid-encoded quinolone resistance protein Qnr with Escherichia coli DNA gyrase. *Antimicrob Agents Chemother* **49**, 118–25.
28. Martin, J. F., Ullan, R. V., Casqueiro, J. (2004) Novel genes involved in cephalosporin biosynthesis: the three-component isopenicillin N epimerase system. *Adv Biochem Eng Biotechnol* **88**, 91–109.
29. Martin, J. F. (2000) Molecular control of expression of penicillin biosynthesis genes in fungi: regulatory proteins interact with a bidirectional promoter region. *J Bacteriol* **182**, 2355–62.
30. Martin, J. F. (1998) New aspects of genes and enzymes for beta-lactam antibiotic biosynthesis. *Appl Microbiol Biotechnol* **50**, 1–15.
31. Lopez, R. (2006) *Pneumococcus*: the sugar-coated bacteria. *Int Microbiol* **9**, 179–90.
32. Avery, O. T., Macleod, C. M., Mccarty, M. (2000) The Classic – Studies on the chemical nature of the substance inducing transformation of pneumococcal types: induction of transformation by a desoxyribonucleic acid fraction isolated from Pneumococcus Type III (Reprinted). *Clin Orthop Relat Res Suppl* **379**, S4–8.
33. Desai, B. V., Morrison, D. A. (2006) An unstable competence-induced protein, CoiA, promotes processing of donor DNA after uptake during genetic transformation in Streptococcus pneumoniae. *J Bacteriol* **188**, 5177–86.
34. Hakenbeck, R., Balmelle, N., Weber, B., Gardes, C., Keck, W., De Saizieu, A. (2001) Mosaic genes and mosaic chromosomes: intra- and interspecies genomic variation of Streptococcus pneumoniae. *Infect Immun* **69**, 2477–86.
35. Hoffman-Roberts, H. L., Babcock, C. E., Mitropoulos, I. F. (2005) Investigational new drugs for the treatment of resistant pneumococcal infections. *Expert Opin Investig Drugs* **14**, 973–95.
36. Passali, D., Lauriello, M., Passali, G. C., Passali, F. M., Bellussi, L. (2007) Group A streptococcus and its antibiotic resistance. *Acta Otorhinolaryngol Ital* **27**, 27–32.
37. Pletz, M. W., Mcgee, L., Van Beneden, C. A., Petit, S., Bardsley, M., Barlow, M., Klugman, K. P. (2006) Fluoroquinolone resistance in invasive Streptococcus pyogenes isolates due to spontaneous mutation and horizontal gene transfer. *Antimicrob Agents Chemother* **50**, 943–8.
38. Yan, S. S., Fox, M. L., Holland, S. M., Stock, F., Gill, V. J., Fedorko, D. P. (2000) Resistance to multiple fluoroquinolones in a clinical isolate of Streptococcus pyogenes: identification of gyrA and parC and specification of point mutations associated with resistance. *Antimicrob Agents Chemother* **44**, 3196–8.
39. Piddock, L. J. (1999) Mechanisms of fluoroquinolone resistance: an update 1994–1998. *Drugs* **58**(Suppl 2), 11–8.
40. Richter, S. S., Diekema, D. J., Heilmann, K. P., Almer, L. S., Shortridge, V. D., Zeitler, R., Flamm, R. K., Doern, G. V. (2003) Fluoroquinolone resistance in Streptococcus pyogenes. *Clin Infect Dis* **36**, 380–3.
41. Reinert, R. R., Lutticken, R., Al-Lahham, A. (2004) High-level fluoroquinolone resistance in a clinical Streptoccoccus pyogenes isolate in Germany. *Clin Microbiol Infect* **10**, 659–62.
42. Koomey, M. (1998) Competence for natural transformation in Neisseria gonorrhoeae: a model system for studies of horizontal gene transfer. *APMIS Suppl* **84**, 56–61.
43. Hamilton, H. L., Dillard, J. P. (2006) Natural transformation of Neisseria gonorrhoeae: from DNA donation to homologous recombination. *Mol Microbiol* **59**, 376–85.

44. Hamilton, H. L., Schwartz, K. J., Dillard, J. P. (2001) Insertion-duplication mutagenesis of neisseria: use in characterization of DNA transfer genes in the gonococcal genetic island. *J Bacteriol* **183**, 4718–26.
45. Wang, S. A., Harvey, A. B., Conner, S. M., Zaidi, A. A., Knapp, J. S., Whittington, W. L., Del Rio, C., Judson, F. N., Holmes, K. K. (2007) Antimicrobial resistance for Neisseria gonorrhoeae in the United States, 1988 to 2003: the spread of fluoroquinolone resistance. *Ann Intern Med* **147**, 81–8.
46. Campos-Outcalt, D. (2007) Practice alert: CDC no longer recommends quinolones for treatment of gonorrhea. *J Fam Pract* **56**, 554–8.
47. Yokoi, S., Deguchi, T., Ozawa, T., Yasuda, M., Ito, S., Kubota, Y., Tamaki, M., Maeda, S. (2007) Threat to cefixime treatment for gonorrhea. *Emerg Infect Dis* **13**, 1275–7.
48. Spratt, B. G., Bowler, L. D., Zhang, Q. Y., Zhou, J., Smith, J. M. (1992) Role of interspecies transfer of chromosomal genes in the evolution of penicillin resistance in pathogenic and commensal Neisseria species. *J Mol Evol* **34**, 115–25.
49. Takahata, S., Senju, N., Osaki, Y., Yoshida, T., Ida, T. (2006) Amino acid substitutions in mosaic penicillin-binding protein 2 associated with reduced susceptibility to cefixime in clinical isolates of Neisseria gonorrhoeae. *Antimicrob Agents Chemother* **50**, 3638–45.
50. Bowler, L. D., Zhang, Q. Y., Riou, J. Y., Spratt, B. G. (1994) Interspecies recombination between the penA genes of Neisseria meningitidis and commensal Neisseria species during the emergence of penicillin resistance in N. meningitidis: natural events and laboratory simulation. *J Bacteriol* **176**, 333–7.
51. Enright, M. C. (2003) The evolution of a resistant pathogen – the case of MRSA. *Curr Opin Pharmacol* **3**, 474–9.
52. Hartman, B. J., Tomasz, A. (1984) Low-affinity penicillin-binding protein associated with beta-lactam resistance in Staphylococcus aureus. *J Bacteriol* **158**, 513–6.
53. Brown, D. F., Reynolds, P. E. (1980) Intrinsic resistance to beta-lactam antibiotics in Staphylococcus aureus. *FEBS Lett* **122**, 275–8.
54. Deurenberg, R. H., Vink, C., Kalenic, S., Friedrich, A. W., Bruggeman, C. A., Stobberingh, E. E. (2007) The molecular evolution of methicillin-resistant Staphylococcus aureus. *Clin Microbiol Infect* **13**, 222–35.
55. Katayama, Y., Takeuchi, F., Ito, T., Ma, X. X., Ui-Mizutani, Y., Kobayashi, I., Hiramatsu, K. (2003) Identification in methicillin-susceptible Staphylococcus hominis of an active primordial mobile genetic element for the staphylococcal cassette chromosome mec of methicillin-resistant Staphylococcus aureus. *J Bacteriol* **185**, 2711–22.
56. Mongkolrattanothai, K., Boyle, S., Murphy, T. V., Daum, R. S. (2004) Novel non-mecA-containing staphylococcal chromosomal cassette composite island containing pbp4 and tagF genes in a commensal staphylococcal species: a possible reservoir for antibiotic resistance islands in *Staphylococcus aureus*. *Antimicrob Agents Chemother* **48**, 1823–36.
57. Fuda, C. C., Fisher, J. F., Mobashery, S. (2005) Beta-lactam resistance in *Staphylococcus aureus*: the adaptive resistance of a plastic genome. *Cell Mol Life Sci* **62**, 2617–33.
58. Wu, S. W., De Lencastre, H., Tomasz, A. (2001) Recruitment of the mecA gene homologue of *Staphylococcus sciuri* into a resistance determinant and expression of the resistant phenotype in *Staphylococcus aureus*. *J Bacteriol* **183**, 2417–24.
59. Couto, I., Wu, S. W., Tomasz, A., De Lencastre, H. (2003) Development of methicillin resistance in clinical isolates of *Staphylococcus sciuri* by transcriptional activation of the mecA homologue native to species. *J Bacteriol* **185**, 645–53.
60. Lindsay, J. A., Holden, M. T. (2006) Understanding the rise of the superbug: investigation of the evolution and genomic variation of *Staphylococcus aureus*. *Funct Integr Genomics* **6**, 186–201.
61. Enright, M. C., Robinson, D. A., Randle, G., Feil, E. J., Grundmann, H., Spratt, B. G. (2002) The evolutionary history of methicillin-resistant Staphylococcus aureus (MRSA). *Proc Natl Acad Sci U S A* **99**, 7687–92.
62. Robinson, D. A., Enright, M. C. (2003) Evolutionary models of the emergence of methicillin-resistant Staphylococcus aureus. *Antimicrob Agents Chemother* **47**, 3926–34.
63. Iandolo, J. J., Worrell, V., Groicher, K. H., Qian, Y., Tian, R., Kenton, S., Dorman, A., Ji, H., Lin, S., Loh, P., Qi, S., Zhu, H., Roe, B. A. (2002) Comparative analysis of the genomes of the temperate bacteriophages phi 11, phi 12 and phi 13 of Staphylococcus aureus 8325. *Gene* **289**, 109–18.
64. Melles, D. C., Gorkink, R. F., Boelens, H. A., Snijders, S. V., Peeters, J. K., Moorhouse, M. J., Van Der Spek, P. J., Van Leeuwen, W. B., Simons, G., Verbrugh, H. A., Van Belkum, A. (2004) Natural population dynamics and expansion of pathogenic clones of Staphylococcus aureus. *J Clin Invest* **114**, 1732–40.
65. Schaefler, S. (1982) Bacteriophage-mediated acquisition of antibiotic resistance by

Staphylococcus aureus type 88. *Antimicrob Agents Chemother* **21**, 460–7.
66. Barlow, M., Hall, B. G. (2002) Phylogenetic analysis shows that the OXA β-lactamase genes have been on plasmids for millions of years. *J Mol Evol* **55**, 314–21.
67. Barlow, M., Reik, R. A., Jacobs, S. D., Medina, M., Meyer, M. P., Mcgowan, J. E., Jr., Tenover, F. C. (2008) High rate of mobilization for blaCTX-Ms. *Emerg Infect Dis* **14**, 423–8.
68. Rodriguez, M. M., Power, P., Radice, M., Vay, C., Famiglietti, A., Galleni, M., Ayala, J. A., Gutkind, G. (2004) Chromosome-encoded CTX-M-3 from Kluyvera ascorbata: a possible origin of plasmid-borne CTX-M-1-derived cefotaximases. *Antimicrob Agents Chemother* **48**, 4895–7.
69. Boyd, D. A., Olson, A. B., Silverman M., Mcgeer, A., Willey, B. M., Pong-Porter, V., Daneman, N., Mulvey, M. R. (2004) Identification of a progenitor of the CTX-M-9 group of extended spectrum beta-lactamases from *Kluyvera* spp. isolated in Guyana. In 44th Interscience Conference on Antimicrobial Agents and Chemotherapy; 2004 American Society for Microbiology, City.
70. Bonnet, R. (2004) Growing group of extended-spectrum beta-lactamases: the CTX-M enzymes. *Antimicrob Agents Chemother* **48**, 1–14.
71. Poirel, L., Kampfer, P., Nordmann, P. (2002) Chromosome-encoded Ambler class A beta-lactamase of Kluyvera georgiana, a probable progenitor of a subgroup of CTX-M extended-spectrum beta-lactamases. *Antimicrob Agents Chemother* **46**, 4038–40.
72. Farmer, J. J., 3rd, Fanning, G. R., Huntley-Carter, G. P., Holmes, B., Hickman, F. W., Richard, C., Brenner, D. J. (1981) Kluyvera, a new (redefined) genus in the family Enterobacteriaceae: identification of Kluyvera ascorbata sp. nov. and Kluyvera cryocrescens sp. nov. in clinical specimens. *J Clin Microbiol* **13**, 919–33.
73. Hall, R. M. (2007) Antibiotic resistance gene cluster of pAPEC-O1-R. *Antimicrob Agents Chemother* **51**, 3461–2.
74. Ktari, S., Arlet, G., Mnif, B., Gautier, V., Mahjoubi, F., Ben Jmeaa, M., Bouaziz, M., Hammami, A. (2006) Emergence of multidrug-resistant Klebsiella pneumoniae isolates producing VIM-4 metallo-beta-lactamase, CTX-M-15 extended-spectrum beta-lactamase, and CMY-4 AmpC beta-lactamase in a Tunisian university hospital. *Antimicrob Agents Chemother* **50**, 4198–201.
75. Lawrence, J. (1999) Selfish operons: the evolutionary impact of gene clustering in prokaryotes and eukaryotes. *Curr Opin Genet Dev* **9**, 642–8.
76. Lawrence, J. G. (2000) Clustering of antibiotic resistance genes: beyond the selfish operon. *ASM News* **66**, 281–6.
77. Gould, I. M. (2002) Antibiotic policies and control of resistance. *Curr Opin Infect Dis* **15**, 395–400.
78. Austin, D. J., Kristinsson, K. G., Anderson, R. M. (1999) The relationship between the volume of antimicrobial consumption in human communities and the frequency of resistance. *Proc Natl Acad Sci U S A* **96**, 1152–6.
79. Giamarellou, H., Antoniadou, A. (1997) The effect of monitoring of antibiotic use on decreasing antibiotic resistance in the hospital. *Ciba Found Symp* **207**, 76–86.
80. Monroe, S., Polk, R. (2000) Antimicrobial use and bacterial resistance. *Curr Opin Microbiol* **3**, 496–501.
81. Gerding, D. N. (2000) Antimicrobial cycling: lessons learned from the aminoglycoside experience. *Infect Control Hosp Epidemiol* **21**, S12–7.
82. Moss, W. J., Beers, M. C., Johnson, E., Nichols, D. G., Perl, T. M., Dick, J. D., Veltri, M. A., Willoughby, R. E., Jr. (2002) Pilot study of antibiotic cycling in a pediatric intensive care unit. *Crit Care Med* **30**, 1877–82.

# Chapter 24

# Potential for Horizontal Gene Transfer in Microbial Communities of the Terrestrial Subsurface


## Jonna M. Coombs



## Abstract

The deep terrestrial subsurface is a vast, largely unexplored environment that is oligotrophic, highly heterogeneous, and may contain extremes of both physical and chemical factors. In spite of harsh conditions, subsurface studies at several widely distributed geographic sites have revealed diverse communities of viable organisms, which have provided evidence of low but detectable metabolic activity. Although much of the terrestrial subsurface may be considered to be distant and isolated, the concept of horizontal gene transfer (HGT) in this environment has far-reaching implications for bioremediation efforts and groundwater quality, industrial harvesting of subsurface natural resources such as petroleum, and accurate assessment of the risks associated with DNA release and transport from genetically modified organisms. This chapter will explore what is known about some of the major mechanisms of HGT, and how the information gained from surface organisms might apply to conditions in the terrestrial subsurface. Evidence for the presence of mobile elements in subsurface bacteria and limited retrospective studies examining genetic signatures of potential past gene transfer events will be discussed.



**Key words:** Horizontal gene transfer, subsurface, conjugation, transformation, transduction.


## 1. Introduction

Much of what is currently known about HGT from an environmental standpoint involves organisms living at the Earth's surface. Exogenous plasmid isolation with *Pseudomonas* recipient cells has been used to capture conjugative plasmids from the soil and the leaf surface of plants *(1, 2)*. Reporter gene systems have been developed to monitor plasmid transfer in situ and explore plasmid host range *(3, 4)*. Predictive mathematical models have been developed to look at gene transfer on solid surfaces *(5)*. Phages,


Maria B. Gogarten et al. (eds.), *Horizontal Gene Transfer: Genomes in Flux, vol. 532*



DOI 10.1007/978-1-60327-853-9_24 Springerprotocols.com

a major reservoir of mobile DNA, have been detected in a significant fraction of the prokaryotic community and exceed prokaryotic cells in relative abundance *(6)*. Genome sequencing projects of environmental isolates have allowed direct comparisons of gene compliments from the level of strain and genus *(7, 8)* all the way up to domain *(9)*, and uncovered a multitude of genetic elements such as conjugative transposons and genomic islands which have influenced microbial evolution *(10)*.

By contrast, little is known about the transmission of genes in bacterial communities of the terrestrial and marine subsurfaces. The terrestrial subsurface has been defined as that portion of the Earth's crust located below the plant root zone (approximately 8 m) *(11)*, and the term "deep" is often used, though its exact meaning has varied from depths of 180 m and below *(12)* to below 1000 m *(13)* to 3200 m *(14)*. For this reason, it has been proposed by some researchers that the word "deep" apply specifically to isolated environments that have little to no contact with surface-associated groundwater, which do not rely on surface-derived nutrients for survival *(15)*.

The subsurface is considered a harsh environment where temperature, pressure, space, nutrient availability, water activity, and pH all have a significant impact on the survival of microorganisms *(16)*. Oxygen is available in the recharge zones of aquifers, but current evidence indicates that much of the subsurface is anaerobic, and most populations rely on fermentation, methanogenesis, or anarobic respiration using inorganic electron acceptors such as $NO_3^-$, $SO_4^{-2}$, and Fe(III) *(16)*. Physical and chemical conditions can vary widely between study sites, between sedimentary layers, and even between individual sediment aggregates at the microscale level *(17, 18)*. In addition, the subsurface can be separated into two distinct regions: (i) the unsaturated (vadose) zone, where water exists primarily as thin films adhering to soil particles and rock, and (ii) the saturated zone, which exists below the water table where rock and soil pores are filled with water rather than with gas. In both zones, microbial cells typically demonstrate starvation-survival responses such as reduced cell size, reduced culturability, and significantly impaired growth *(19, 20)*. In spite of these limitations, numerous cultivation studies have successfully obtained viable organisms from subsurface sediments even at the deepest levels, indicating that conditions are sufficient to maintain viability for at least some community members *(14, 21, 22)*. In the presence of metabolic activity, there is also the potential for HGT. This chapter will review several factors impacting HGT among bacteria in environmental settings, and describe how these factors may act as barriers or facilitators to the gene flow in subsurface microbial communities.

## 2. Factors Influencing Subsurface HGT

Genetic material subject to HGT must overcome a series of hurdles to achieve stable maintenance in a new host cell, regardless of conditions in the environment. Some of these barriers may be host related. These include host DNA restriction systems that degrade newly transferred DNA, the inability of the new host to provide proteins for the replication of extrachromosomal elements, the inability of the new host to express the acquired gene products *(23)*, deleterious effects to the host caused by the expression of newly acquired genes *(24)*, and sequence divergence between species that reduces the potential for homologous recombination with the recipient chromosome *(25)*. Other barriers may be mediated by the mobile elements themselves, such as surface exclusion mechanisms *(26)*, phage immunity *(27)*, and plasmid incompatibility *(26)*. In addition to these, physiochemical factors in the environment *(23, 28)* and cell–cell competition *(29)* may adversely affect HGT. Nonetheless, successful HGT has been documented in a number of environmental samples *(1, 2, 30)*. This section will examine the characteristics of subsurface organisms and their environment that may have implications for the acquisition of new genetic information through HGT.

### 2.1. Microbial Activity in the Subsurface

Physical and chemical factors limit the activity of microbes in the terrestrial subsurface *(16)*. Dissolved organic carbon (DOC) in pristine aquifers ranges from 0.001 to 1.1 mM with a median of 0.05 mM *(31)*. Even in the presence of organic carbon, shortages of inorganic nutrients may negatively affect microbial growth and respiration. Estimates of respiration rates in deep anaerobic aquifer sediments using geochemical modeling range from $10^{-4}$ to $10^{-6}$ mmol $CO_2$ evolved per liter of groundwater per year *(32)*, and estimates of activity in the vadose zone are even lower, at $< 10^{-10}$ mol $CO_2$ $L^{-1}$ $y^{-1}$ of porewater *(33, 34)*. Comparisons of these rates of endogenous metabolism indicate that microbial activity in the subsurface is orders of magnitude lower than in other environments such as deep ocean sediments ($10^{1}$–$10^{-1}$ mmol $CO_2$ $L^{-1}$ $y^{-1}$) *(32)*.

Microbial activity is essential to the processes mediating HGT (**Table 24.1**). Conjugation has been shown to require both ATP and proton-motive force *(35)*. The donor cell must be metabolically active in order to carry out DNA replication, pilus synthesis and retraction, and DNA translocation via type IV secretion systems *(36)*. Uptake of DNA by the recipient may not require a direct source of energy. *Agrobacterium tumefaciens* Ti plasmid has many similarities to conjugative transfer systems, and recent studies have demonstrated that the VirE protein encoded on the

**Table 24.1**
**Select ATP-utilizing components associated with three mechanisms of HGT in prokaryotes**

| HGT activity | Energy-requiring components | Example | Organism | Cell affected | Reference |
|---|---|---|---|---|---|
| **Conjugation** | | | | | |
| DNA strand separation | Helicases | TraI | *Escherichia coli* | Donor | *(128)* |
| DNA translocation | DNA-dependent ATPases | TrwB | *E. coli* | Donor | *(43)* |
| **Transduction** | | | | | |
| DNA packaging | DNA-dependent ATPases | gp16 | *Bacillus subtilis* φ29 | Donor | *(42)* |
| **Transformation** | | | | | |
| Pilus locomotion | Traffic NTPase | PilT | *Neisseria gonorrhoeae* | Recipient | *(129)* |
| DNA translocation | DNA-dependent ATPases | ComFA | *B. subtilis* | Recipient | *(45, 46)* |
| Formation of nucleoprotein complexes | ssDNA binding protein | RecN | *B. subtilis* | Recipient | *(130)* |

Ti plasmid can pull single-stranded DNA (ssDNA) into the recipient cell using the potential energy of the compacted ssDNA in the absence of an external energy source *(37)*. However, the conjugative recipient must be metabolically active in order to carry out other processes such as DNA synthesis and recombination. Transduction, a second mechanism for HGT, requires that the donor cell be active for the production and packaging of phage particles. Minimal metabolic activity is required for the infection process *(27)*, and it has been inferred from this that the new host does not need to be metabolically active due to the potential energy of the DNA packaged into the phage particle. However, recent studies indicate that the potential energy stored in the capsid DNA may not be enough to overcome the osmotic pressure of the bacterial cytoplasm *(38, 39)*, and this indicates that some additional source of energy may be necessary for phage infection. Natural transformation is a third mechanism for HGT, and the donor cell involved may be either active or inactive, as spore outgrowth *(40)* as well as cell death and lysis have been shown to release DNA into the environment. The recipient cell must maintain metabolic activity for the uptake of DNA, which requires the action of nucleases, DNA translocating motors, and proteins involved in DNA synthesis (**Table 24.1**). As will be discussed below, the pili and DNA uptake systems of transformation and conjugation are inducible, and rely on active gene expression in order to function.

Translocation of DNA across the cell membrane requires considerable energy. One of the best characterized DNA translocating motors is the FtsK-HerA-type ATPase gp16 from *Bacillus subtilis* bacteriophage φ29 *(41)*. This motor hydrolyses one ATP molecule for every two nucleotides translocated *(42)*, making viral packaging an energy-intensive process. The type IV secretion systems of conjugative plasmids such as R388 *(43)* and conjugative transposons such as Tn916 *(44)* have also been shown to have DNA translocating motors of the FtsK-HerA superfamily *(41)*, and this system is likewise present in the type IV secretion system used in natural transformation in epsilon proteobacteria. If the energy cost associated with these motors is similar to that of gp16, transport of large conjugative elements and large (>1000 kb) pieces of genomic DNA may be prohibitive in the nutrient-poor subsurface. An alternate mechanism for transformation utilizes an ATP-dependent DNA helicase of the DEAD-box family, which is found in type II secretion systems such as ComFA in *Bacillus* *(45)*. The exact mechanism and energy cost for DNA translocation for this type of helicase is unknown, although both ATP and ATP-independent proton-motive force appear to be necessary for DNA transport across the cell membrane *(46)*.

Cell growth in the subsurface has been estimated to occur at rates of up to thousands of years per division *(47)*. This has ramifications for subsurface gene transfer, as culture studies have demonstrated a dependence of HGT on active cell division. Conjugal transfer of plasmids has been shown to vary with the growth rate and phase of growth in both liquid culture *(48, 49)* and biofilms *(50)*. The highest rates of conjugal transfer occur in actively growing cultures containing exponential phase cells *(48, 50, 51)*. Likewise, the development of competence is dependent on the stage of growth and appears to be taxa-specific, as some cultures develop competence in early to mid log phase, while others become competent in stationary phase as nutrients become limiting *(26)*. The low-nutrient environments of the vadose and saturated zones might therefore favor transformation as a mechanism of gene transfer among strains residing in the subsurface. This is supported by evidence that cations which are important for the development of competence, including $Ca^{+2}$, $Sr^{+2}$, and $Mg^{+2}$ *(52)*, have been found at optimal concentrations in soil and sediment environments *(53)*.

### 2.2. Microbial Abundance and Diversity

The numbers of prokaryotes existing in the terrestrial subsurface has been estimated at $0.25–2.5 \times 10^{30}$ *(11)*. This prokaryotic abundance is greater than almost any other segment of the ecosystem, due primarily to the vastness of the terrestrial subsurface compared to other environments *(11)*. Cell numbers have been shown to decline with depth *(54, 55)*, and the numbers of prokaryotes in deeper terrestrial sediments (below 600 m) likely comprise only a small fraction ($2.6 \times 10^{28}$) of the total number of cells *(11)*. Despite this trend, even the deepest, oldest sediments that have been sampled show evidence of viable microbes *(14, 22, 55)*.

Low cell densities are a characteristic of subsurface communities. Acridine orange direct counts from subsurface materials range from a lower limit of $5.2 \times 10^3$ cells $g^{-1}$ *(22)* to an upper limit of $7 \times 10^7$ cells $g^{-1}$ *(56)*. This cell density is orders of magnitude lower than in organic-rich surface soils, where direct counts fall into the range of $3.3 \times 10^9$ cells $g^{-1}$ *(57)*. The low cell densities translate to a high degree of spatial isolation among subsurface prokaryotes, which in turn decreases the chances of HGT, especially with regard to the mechanisms that require direct cell–cell contact to occur *(24, 40)*. In addition, many of the structures involved in mediating HGT such as type IV pili and type IV secretion systems are induced by quorum sensing, either by oligopeptides *(58, 59)* or acylated homosesrine lactones *(60–62)*.

Cells can resolve the spatial isolation problem and participate in quorum sensing through the formation of cell aggregates or biofilms. These structures keep microbes in close contact with each other, and serve to trap nutrients and water and keep out toxins. In oligotrophic environments, cells tend to exist in attached states, and in the subsurface up to 90% of the cells are believed to be anchored to the solid phase *(63)*. Biofilm formation has been demonstrated on solid substrates introduced into subsurface sediments in the saturated zone *(64, 65)*, and microscopic examination of subsurface sediments has provided evidence of naturally formed microcolonies *(66)*. As biofilms have been shown to facilitate gene transfer in laboratory studies *(67)* and in the environment *(2, 50, 68)*, it is likely that bacterial microcolonies on sediment particles may act as “hotspots” for gene transfer in situ.

Currently, there is no estimation available for the number of phages that might exist in the terrestrial subsurface; however, an average of 2.6 prophages have been found in the sequenced genomes of free-living bacteria *(69)*, indicating that they are prevalent in surface environments. In the subsurface, suitable hosts may be widely distributed, and dilution of phages may dramatically reduce the probability of transduction *(27)*. It has been proposed that soil environments, which contain high diversity but also high spatial separation, and in many cases limiting nutrients and low bacterial growth rates, select for lysogenic reproduction *(70)*. This is supported by a recent study in which up to 85% of the soil microbial communities tested contained inducible phage *(71)*. There also may be a difference in viral abundance in the saturated zone vs. the vadose zone, as higher levels of phage have been found in moisture-rich environments in surface soils *(72)*. In addition, there is evidence that phages are preferentially found associated with bacterial aggregates in marine environments, where spatial distribution may interfere with transduction *(73)*. These findings suggest that perhaps the highest levels of phage in the subsurface may be detected in the saturated zone, associated with bacterial aggregates.

A variety of methods *(20, 74–78)* have been used to examine subsurface microbial communities, and all of them have detected great phenotypic and genetic diversity. The predominance of Gram-positive and Gram-negative populations appears to be correlated with the geologic composition of subsurface sediments, particularly with respect to clay content *(47, 54)*. As mentioned previously, levels of chemolithotrophic primary productivity and levels of organic carbon are low in the subsurface, and both of these factors have been found to impact the phylogenetic diversity of microbial communities *(79, 80)*, although the exact response to productivity level appears to vary for different taxa *(80)*. Another likely factor is dispersal of cells, which is limited in the subsurface. At the macroscale level, confining layers of clay and rock block or significantly reduce transport between geologic layers, and the speed of groundwater flow in the saturated zone, which ranges from m $day^{-1}$ to m $year^{-1}$ *(16)*, limits the movement of microbes. At the microscale level, diverse microhabitats develop around sediment particles, which contain different minerals. The existence of microhabitats is especially true in the vadose zone, where sporadic wetting and drying events generally limit water to thin films, which substantially reduces cell mobility and diffusion of nutrients between mineral particles *(81)*. It has been proposed for surface environments that limited diffusion of nutrients aids in the maintenance of phylogenetic diversity by reducing competition between cells *(82)*.

Community diversity affects HGT with respect to the host range of phages *(27)* and plasmids *(24)*, and the preferential uptake of signal-tagged transforming DNA *(83)*. The effect of phylogenetic diversity on the process of transformation may be especially significant. In the event of successful transfer, phylogenetic distance between donor and the host cells has a dramatic effect on the successful integration of transferred genes through recombination. The presence of homologous DNA in the chromosome results in recombination efficiencies in the range of $10^{-4}$–$10^{-5}$ transformants per gene copy *(25)*. Illegitimate recombination of acquired DNA also occurs, but at much lower efficiencies, depending on the presence ($10^{-10}$ transformants) or absence ($10^{-12}$–$10^{-13}$ transformants) of a small stretch of homologous "anchor" DNA *(84, 85)*. Even within a single species, individual strains have been shown to vary greatly in their transformation efficiencies *(86)*. Likewise, the percentage of cells in the competent state within a single population, such as might be expected within a microcolony, may vary widely due to the stage of growth or internal genetic "switches" *(87)*.

### 2.3. Stability of Extracellular DNA

The chemical heterogeneity of the subsurface environment *(16)* can influence the stability of macromolecules such as naked DNA. DNA can be stabilized and protected against the activity of DNases by $Mg^{+2}$ and $Ca^{+2}$-mediated binding to minerals *(88)*, which renders the DNA partially insoluble *(89)*. In groundwater studies, $Mg^{+2}$, $Ca^{+2}$, $Fe^{+2/+3}$, and $Mn^{+2}$ have all been shown to facilitate the adsorption of DNA to solid aquifer material *(90–92)*, providing resistance against DNase I degradation *(92)*. DNA adsorbed to soil appears to be taken up at significantly higher efficiencies in several organisms *(90, 93, 94)*, although this increase in transformation efficiency does not hold true for all microbes, particularly Gram-negatives *(90, 94)*. Phages, which are resistant to many physical and chemical factors, have also demonstrated the ability to persist for long periods of time in the environment when adsorbed onto clay minerals and other particulates *(23)*.

### 2.4. Occurrence of Mobile Genetic Elements

Mobile DNA elements are facilitators of HGT in the natural environment. The presence of these elements may act as indicators of past HGT events and provide information about the potential of bacterial strains to participate in HGT in the future. Caution must be used, however, when interpreting signatures of such elements, as the propagation and maintenance of mobile DNA may have more to do with the alteration of host fitness through gene disruption and genome re-arrangement than with any direct benefit conferred by HGT. This section will review instances in which mobilizable DNA has been detected among subsurface strains.

Plasmids have been documented in a variety of subsurface bacteria. The incidence of plasmids among cultivatable isolates from two subsurface sites varied from 1.8% of 332 bacteria *(95)* to 33% of 163 bacteria *(96)*. The two sites differed in sediment composition and therefore caution should be used in direct comparison. Plasmid incidences of 30–42% have previously been observed in bacteria from Antarctic sea ice or sediment *(97, 98)*, or 45–55% in marine bacteria existing in the neuston layer *(99)* and 24–28% from marine sediment *(100)*. Based on these studies, the occurrence of plasmids at documented subsurface sites appears to fall within the range of plasmid incidences obtained from other environments.

Subsurface strains may carry single *(101)* or multiple *(102)* plasmids, with large plasmids appearing to dominate *(96)*. The frequency of very large (>150 kb) conjugative plasmids in deep subsurface bacteria has been found to be greater than the frequency from shallower subsurface soils *(95, 96)* or from marine environments *(100, 103)*. Large subsurface plasmids frequently contain catabolic genes *(104, 105)*, which clearly enhance the metabolic flexibility of the host. Select plasmids for which sequence is available also have regions of homology to recombinases, indicating the potential for integration and excision from the host chromosome *(105)*.

While phage-like genes indicating past interactions with bacteriophage have been detected in the currently sequenced subsurface genomes (**Table 24.2**), not much is known about the prevalence of native subsurface viruses in the environment. Transport of virus particles due to groundwater flow does occur in the subsurface, as shown by studies with well-characterized bacteriophages such as MS-2 or PRD1 *(106, 107)*. In the future, metagenomic techniques such as have been employed with marine phage *(108)* may allow investigation of the phage subsurface mobilome.

Transposon prevalence has been examined using a collection of *Exiguobacterium* isolates from ancient Siberian permafrost isolated from depths of 5.5 to 43.6 m *(109)*. Transposases from nine sequence homology groups originally discovered in the genome of the subsurface permafrost isolate *Exiguobacterium* sp. 255-15 were commonly detected in other permafrost *Exiguobacterium* spp. as well as *Exiguobacterium* strains isolated from surface environments. This study demonstrates the potential for subsurface organisms to harbor transposases and their associated IS elements, although the global ubiquity of these sequences indicates that these elements may have been present prior to the burial and freezing of the *Exiguobacterium* spp. in the subsurface *(109)*, and provides no evidence of activity in permafrost.

Another putative mobile element that has been detected is a novel group I intron in the $tRNA^{Leu}$ (UAA) gene (trnL) identified in a subsurface $\gamma$-proteobacterium *(110)*. Many introns of group I are known to act as mobile elements, and transfer has previously been documented between the LSU rRNA gene of the chloroplast of *Chlamydomonas pallidostigmatica* and the LSU rRNA gene mitochondria of *Acanthamoeba castellanii (111)*. Other introns of this group such as $tRNA^{Arg}$ and $tRNA^{Ile}$ have been detected in $\alpha$- and $\beta$-proteobacteria, respectively *(112)*. The occurrence of this mobile intron in the subsurface $\gamma$-proteobacteria is the first incidence of a $tRNA^{Leu}$ (UAA) group I intron to be detected in a taxonomic group outside of the cyanobacteria. This indicates that HGT may have played a role in its distribution, although the mechanism and environment in which transfer occurred cannot be determined.

**Table 24.2**
**Completed genomes and mobile element signatures for organisms of subsurface origin**

| Organism | Site of isolation | Plasmid[1] | Phage-related genes[2] | References |
|---|---|---|---|---|
| *Dehalococcoides* sp. BAV1 | Aquifer materials | No | 11 | *(131)* |
| *Desulfovibrio desulfuricans* G20 (derived from wild-type strain G100A) | Producing oil well | pBG1 (wild-type strain only) | 29 | *(132)* |
| *Candidatus Desulforudis audaxviator MP104C* | Gold mine borehole, 2.8 km depth | No | 14 | *(133)* |
| *Exiguobacterium sibiricum* 255-15 | Russian permafrost | pEXIG01 pEXIG02 | 4 | *(109)* |
| *Geobacillus thermodenitrificans* NG80-2 | Deep subsurface oil reservoir | pLW1071 | 30 | *(101)* |
| *Geobacter uraniumreducens* Rf4 | TCE-contaminated aquifer | No | 15 | *(134)* |
| *Novosphingobium aromaticivorans* DSM 12444 | Deep subsurface, SC | pNL1 pNL2 | 42 | *(135)* |
| *Petrotoga mobilis* SJ95 | North sea oil reservoir | No | 4 | *(136)* |
| *Shewanella putrifaciens* CN32 | Subsurface rock | No | 33 | *(137)* |
| *Thermotoga petrophila* RKU-1 | Offshore oil reservoir | No | 3 | *(138)* |

[1] Based on the genome sequence information deposited in Genbank and the information available in published references.
[2] Based on the genome sequence annotation available in Genbank.

## 3. Potential for HGT in the Subsurface

### 3.1. Genome Mining

At the time of chapter submission, only 10 of the over 600 completed bacterial genome sequences originate from the organisms isolated from the subsurface (**Table 24.2**). Information from the genomes of two more subsurface organisms, *Geobacter bemidjiensis* BEM *(86)* and *Geobacter* sp. FRC-32 *(72, 87)*, are available in draft format. Although traditional measures of HGT such as G + C skew and codon bias have not currently been published for these prokaryotes, there are some hallmarks of past gene transfer/potential transfer evident in these genomes.

Four of the completed genomes contain extrachromosomal elements (**Table 24.2**). *Desulfovibrio desulfuricans* G100A, which is the wild-type strain from which strain G20 is derived, contains a cryptic non-ColE1-type 2.3 kb plasmid *(113)*. This strain has demonstrated the capacity for acting as the recipient in conjugative transfer of broad host-range IncQ plasmids *(114)*. Another subsurface organism, *Novosphingobium aromaticivorans*, contains two plasmids, the 184 kb catabolic plasmid pNL1 and a 487 kb plasmid pNL2. Plasmid pNL1 was previously sequenced *(105)*, and contains two clusters of *tnp/tra* genes for conjugative transfer. Plasmid transfer of pNL2 to a *Sphingomonas* strain has been demonstrated *(105)*. *Geobacillus thermodenitrificans* NG80-2 contains a plasmid of 58 kb. This plasmid, pLW1071, does not resemble other sequenced plasmids, including that of the related *Geobacillus kaustophilus (115)*. *G. thermodenitrificans* plasmid contains the genes necessary for replication and conjugation, with several copies of insertion sequences of the IS21 and IS982 families *(101)*. It has been proposed by Feng and coworkers that *G. thermodenitrificans*, originally native to surface soils, adapted to oil reservoir environments in part through the acquisition pLW1071, which carries genes for the degradation of hexadecane.

ORFs with a high degree of sequence homology to phage genes have been detected in all of the completed subsurface genomes (**Table 24.2**). These numbers are comparable to surface strains, although the burden of phage genes in some subsurface bacteria appears to be higher. For example, *G. thermodenitrificans* contains 30 phage-related genes compared to 10 in the genome of the related *G. kaustophilus*. *Geobacter uraniumreducens* Rf4 contains 15 phage-related genes compared to 5 in *Geobacter sulfurreducens* and 4 in *Geobacter metallireducens*. *Shewanella putrefaciens* CN-32 contains one of the highest numbers of phage-related genes out of the 13 *Shewanella* spp. currently sequenced, surpassed only by three *Shewanella* isolates from the deep sediments and waters of the Baltic Sea *(116–118)*. The prevalence of phage genes in the genomes of the subsurface strains indicates that these strains may be acquiring more temperate phages than surface organisms, they may have a higher frequency of phage-driven genetic rearrangement, or the subsurface organisms are simply slower than surface strains at purging traces of phage integration from their genomic DNA.

Of the ten subsurface genomes completed, comparative genomics to a surface microbe has been performed with one. *G. thermodenitrificans* shares 75% of its genome with its closest sequenced relative, *G. kaustophilus*, and shares an additional 11% and 3% of its genome with *Bacillus* sp. and other Firmicutes, respectively. About 2.7% of the genes in *G. thermodenitrificans* NG80-2 are only found in distantly related organisms, and indicate instances of possible HGT, including two clusters of genes involved in nitrogen utilization, which appear to have come from two different sources *(101)*. The acquisition of these genes, and of the catabolic plasmid pLW1071, has clearly increased the metabolic flexibility of this organism in environments low in nutrients.

Genomic comparison has been performed for only one of the draft subsurface sequences, *Geobacter* sp. FRC-32. Some of the same catabolic, transposase, and phage-like genes that exist on a 300-kb chromosomal island in FRC-32 were first identified in the related surface organism *Geobacter metallireducens (119)*. The genomic island is a remnant of an HGT event that may have occurred prior to the deposition of *Geobacter* FRC-32 or its ancestor in the subsurface, although given the movement of groundwater and geological weathering processes it may also be that the surface strains analyzed in this and other studies once resided in the subsurface.

### 3.2. Phylogenetic Incongruency in Select Genes from Subsurface Microorganisms

It is challenging to construct an accurate picture of HGT using retrospective approaches, reconstructing past events by looking at current sequences of DNA. There is little debate that HGT has affected the evolution of organisms currently residing in the subsurface; however, as with genomic analysis, one of the most difficult questions to answer using modern methods is whether or not HGT occurred in the subsurface.

Since groundwater quality issues are an essential component of subsurface research, much of the retrospective work that has been done with respect to subsurface HGT relates to the bioremediation of contaminated soils and aquifers. The transfer of genes pertinent to bioremediation is the subject of **Chapter 26** in this volume. One such study involves the retrospective analysis of 48 lead-resistant ($Pb^r$) subsurface isolates from an uncontaminated site along the Atlantic Costal Plain *(120)*, which detected four bootstrap-supported cases of phylogenetic incongruence among *zntA/cadA/copA* PCR amplicons. Three of these cases were supported by atypical amplicon % mol G + C content as compared to the % mol G + C content of the host genus. At least one of the potential HGT events indicated transfer of this locus among β- and γ-proteobacteria, a finding also reflected in a subsequent study using microbial genome data from surface-isolated organisms *(121)*. The finding of the same pattern in surface as well as subsurface organisms makes it impossible to pinpoint the environment in which HGT may have occurred.

Other retrospective studies of HGT have examined the transfer of genes among sulfate-reducing bacteria (SRB), common residents of subsurface environments. Phylogenetic incongruency and indel analysis of adenosine-5′-phosphosulfate reductase (*apsA*) revealed a single HGT event involving seven members of the "Syntrophobacteriaceae" *(122)*. Two of the members of this group were subsurface isolates: one from a petroleum reservoir *(123)* and the other from North Sea oil field water *(124)*. The other organisms in this group were isolated from surface methanogenic environments and man-made anaerobic bioreactors. Similar phylogenetic analysis with the *dsrAB* genes from a variety of sources *(122, 125)* detected HGT events among phylogenetic clusters of Gram-positive SRB containing both surface and subsurface strains *(122, 125, 126)*. The presence of mixed surface and subsurface isolates at potential nodes of gene transfer indicate that gene acquisition may have occurred outside the subsurface.

### 3.3. Microcosm Studies

Plasmids, particularly large plasmids that are more likely to have conjugative function, have been detected in a variety of subsurface microorganisms *(96)*. A limited number of these have demonstrated conjugative capability in a laboratory setting or in bulk soil. Often, however, the experimental setup in mating experiments and microcosms does not accurately reflect the native environment of the isolates, especially with regard to donor and recipient cell numbers and the availability of nutrients. There has been a single study involving subsurface-derived microcosms specifically designed to take place at low carbon flux under simulated subsurface groundwater leaching conditions using an *E. coli* donor as a transitory cadmium resistance ($Cd^r$) plasmid delivery strain *(127)*. The soil samples in this work originated from a contaminated aquifer, and transfer was examined with or without added Cd. Transfer of nonconjugal IncQ plasmids pMOL187 and pMOL222 and their retention in the microcosm community was demonstrated through infinite-dilution PCR (ID-PCR). However, detection of plasmid-bourne genes in microcosm eluates was low, indicating poor transfer potential of the plasmids, even in the presence of the broad host range IncPα plasmid RP4, which aids in plasmid mobilization *(127)*. This study provides evidence of plasmid persistence and transfer in subsurface-derived communities, although caution should be used in interpreting results in model ecosystems, and further work is necessary to evaluate the full potential of environmental conjugative transfer.

## 4. Conclusions/ Outlook

Several lines of evidence indicate that HGT may be possible in the terrestrial subsurface. Although the subsurface is a harsh environment, conditions such as low nutrients and ion concentrations that favor the development of competence, clay deposits supporting the persistence of bacteriophage and naked DNA, and the ability of subsurface organisms to form aggregates, which might serve as hotspots for gene transfer, indicate that HGT mechanisms may be able to operate in this remote environment. Microorganisms isolated from the subsurface contain a diverse repertoire of mobile genetic elements including plasmids, transposons, genomic islands, and phage, all of which indicate the potential to participate in HGT. Select subsurface plasmids have demonstrated the ability to be transferred via conjugation, in some cases even under simulated subsurface conditions. Nucleic acid analysis using hybridization and phylogenetic techniques have uncovered instances, which indicate that HGT has helped shape the genomes of subsurface strains. However, in all of these cases, it has been impossible to determine if these HGT events have occurred or can occur *in situ*. This poses a significant challenge to the understanding of gene flow in subsurface microbial communities. However, new methods including the fluorescence techniques developed to analyze conjugation in biofilms may be appropriate to examine HGT at the microscale level in aggregates attached to subsurface minerals. Genome amplification techniques combined with metagenomics may facilitate the analysis of the subsurface phage mobilome. Direct RNA isolation combined with quantitative PCR or microarray techniques may provide a mechanism to look at the gene expression of competence factors and components involved in conjugation in subsurface environments. Finally, the geologic history of sediments whose isolation from surface environments can be documented provides a unique opportunity to examine the evolution and speciation of microbes trapped within sedimentary layers. Further work will be necessary in order to evaluate the occurrence of HGT in the vast, heterogeneous, and isolated environment of the terrestrial subsurface, and to assess the full impact of gene transfer on subsurface microbial evolution.

## Acknowledgments

Sincere appreciation is expressed for comments and suggestions from Tamar Barkay, from the editors of this volume, and from two anonymous reviewers. This work was supported by a Faculty Development Grant from Adelphi University.

## References


1. Van Elsas, J. D., Gardener, B. B., Wolters, A. C., Smit, E. (1998). Isolation, characterization, and transfer of cryptic gene-mobilizing plasmids in the wheat rhizosphere. *Appl Environ Microbiol* **64**, 880–889.
2. Normander, B., Christensen, B. B., Molin, S., Kroer, N. (1998) Effect of bacterial distribution and activity on conjugal gene transfer on the phylloplane of the bush bean (*Phaseolus vulgaris*). *Appl Environ Microbiol* **64**, 1902–1909.
3. Mølbak, L., Licht, T. R., Kvist, T., Kroer, N., Andersen, S. R. (2003) Plasmid transfer from *Pseudomonas putida* to the indigerous bacteria on alfalfa sprouts; characterization, direct quantification, and *in situ* localization of transconjugant cells. *Appl Environ Microbiol* **69**, 5536–5542.
4. Musovic, S., Oregaard, G., Kroer, N., Sørensen, S. J. (2006) Cultivation-independent examination of horizontal transfer and host range of an IncP-1 plasmid among gram-positive and gram-negative bacteria indigenous to the barley rhisosphere. *Appl Environ Microbiol* **72**, 6687–6692.
5. Lagido, C, Wilson, I. J., Glover, L. A., Prosser, J., I. (2003) A model for bacterial conjugal gene transfer on solid surfaces. *FEMS Microbiol Ecol.* **44**, 67–78.
6. Weinbauer, M. G. (2004) Ecology of prokaryotic viruses. *FEMS Microbiol Rev* **28**, 127–181.
7. Pohlmann, A., Fricke, W. F., Reinecke, F., Kusian, B., Liesegang, H., Cramm, R., Eitinger, T., Ewering, C., Pötter, M., Schwartz, E., Strittmatter, A., Voß, I., Gottschalk, G., Steinbüchel, A., Friedrich, B., Bowien, B. (2006) Genome sequence of the bioplastic-producing "Knallgas" bacterium *Ralstonia eutropha* H16. *Nat Biotechnol* **24**, 1257–1262.
8. Makarova, K. S., Koonin, E. V. (2007) Evolutionary genomics of lactic acid bacteria. *J Bacteriol* **189**, 1199–1208.
9. Hotopp, J. C., Clark, M. E., Oliveira, D. C., Foster, J. M., Fisher, P., Torres, M. C., Giebel, J. D., Kumar, N., Ishmael, N., Wang, S., Ingram, J., Nene, R. V., Shepar, J., Tomkins, J., Richards, S., Spiro, D. J., Ghedin, E., Slatko, B. E., Tettelin, H., Werren, J. H. (2007) Widespread lateral gene transfer from intracellular bacteria to multicellular eukaryotes. *Science* **317**, 1753–1756.
10. Osborn, A. M., Böltner, D. (2002) When phage, plasmids, and transposons collide: genomic islands, and conjugative- and mobilizable-transposons as a mosaic continuum. *Plasmid* **48**, 202–212.
11. Whitman, W. B., Coleman, D. C., Wiebe, W. J. (1998) Prokaryotes: the unseen majority. *Proc Natl Acad Sci USA* **95**, 6578–6583.
12. Chandler, D. P., Brockman, F. J., Bailey, T. J., Fredrickson, J. K. (1998) Phylogenetic diversity of archaea and bacteria in a deep subsurface paleosol. *Microbial Ecol* **36**, 37–50.
13. Pedersen, K., Ekendahl, S. (1990) Distribution and activity of bacteria in deep granitic groundwaters of southeastern Sweden. *Microbial Ecol* **20**, 37–52.
14. Onstott, T. C., Moser, D. P., Pfiffner, S. M., Fredrickson, J. K, Brockman, F. J., Phelps, T. J., White, D. C., Peacock, A., Balkwill, D., Hoover, R., Krumholz, L. R., Borscik, M., Kieft, T. L., Wilson, R. (2003) Indigenous and contaminant microbes in ultradeep mines. *Environ Microbiol* **5**, 1168–1191.
15. Lovley, D. R., Chapelle, F. H. (1995) Deep subsurface microbial processes. *Rev Geophys* **33**, 365–381.
16. Fredrickson, J. K., Fletcher, M. (eds.) (2001) *Subsurface microbiology and biogeochemistry.* Wiley-Liss, Inc., Danvers, MA.
17. Colwell, F. S. (2001) Constraints on the distribution of microorganisms in subsurface environments, in *Subsurface microbiology and biogeochemistry.* (Fredrickson, J. K., Fletcher, M. eds.) Wiley-Liss, Inc., Danvers, MA, pp. 3–37.
18. Tobin, K. J., Onstott, T. C., Deflaun, M., Colwell, F., Fredrickson, J. (1999) In situ imaging of microorganisms in geologic material. *J Microbiol Meth* **37**, 201–213.
19. Amy, P. S., Durham, C., Hall, D., Haldeman, D. L. (1998) Starvation-survival of deep subsurface isolates. *Curr Microbiol* **26**, 345–352.
20. Kieft, T. L., Wilch, E., O'connor, K., Ringelberg, D. B., White, D. C. (1997) Survival and phospholipids fatty acid profiles of surface and subsurface bacteria in natural sediment microcosms. *Appl Environ Microbiol* **63**, 1531–1542.
21. Balkwill, D. L., Drake, G. R., Reeves, R. H., Fredrickson, J. K., White, D. C., Ringelberg, D. B., Chandler, D. P., Romine, M. F., Kennedy, D. W., Spadoni, C. M. (1997) Taxonomic study of aromatic-degrading bacteria from deep-terrestrial-subsurface sediments and description of *Sphingomonas aromaticivorans* sp. nov., *Sphingomonas subterranea*, sp. nov., and *Sphingomonas stygia* sp. nov. *Int J Syst Bacteriol* **47**, 191–201.

22. Zhang, G., Dong, H., Xu, Z., Zhao, D., Zhang, C. (2005) Microbial diversity in ultra-high-pressure rocks and fluids from the Chinese continental scientific drilling project in China. *Appl Environ Microbiol* **71**, 3213–3227.
23. Stotsky, G. (1989) Gene transfer among bacteria in soil, in *Gene transfer in the environment.* (Levy, S. B., Miller, R. V. eds.) McGraw-Hill Book C., New York. pp. 165–222.
24. Caumont, A. B., Jamieson, G. A., Pichuantes, S., Nguyen, A. T., Litvak, S., Dupont, C. (1996) Expression of functional HIV-1 integrase in the yeast *Saccharomyces cerevisiae* leads to the emergence of a lethal phenotype: potential use for inhibitor screening. *Curr Genetics* **29**, 503–10.
25. DeVries, J, Meier, P., Wackernagel, W. (2001) The natural transformation of the soil bacteria *Pseudomonas stutzeri* and *Acinetobacter* sp. by transgenic plant DNA strictly depends on homologous sequences in the recipient cells. *FEMS Microbiol Lett* **195**, 211–215.
26. Ippen-Ihler, K. (1989) Bacterial conjugation, in *Gene transfer in the environment.* (Levy, S. B., Miller, R. V. eds.) McGraw-Hill Book C., New York. pp. 33–72.
27. Kokjohn, T. A. (1989) Transduction: mechanism and potential for gene transfer in the environment, in *Gene transfer in the environment.* (Levy, S. B., Miller, R. V. eds.) McGraw-Hill Book C., New York. pp. 33–72.
28. Lorenz, M. G., Wackernagel, W. (1994) Bacterial gene transfer by natural genetic transformation in the environment. *Microbiol Rev* **58**, 563–602.
29. Nielsen, K. M., Townsend, J. P. (2004) Monitoring and modeling horizontal gene transfer. *Nature Biotechnol* **22**, 1110–1114.
30. Saye, D. J., Miller, R. V. (1989) The aquatic environment: consideration of horizontal gene transmission in a diversified habitat, in *Gene transfer in the environment.* (Levy, S. B., Miller, R. V. eds.) McGraw-Hill Book C., New York. pp. 33–72.
31. Leenheer, J. A., Malcolm, R. L., McKinley, P. W., Eccles, L. A. (1974) Occurrence of dissolved organic carbon in selected groundwater samples in the United States. *U S Geol Surv J Res* **2**, 361–369.
32. Chapelle, F. H., Lovley, D. R. (1990) Rates of metabolism in deep coastal plain aquifers. *Appl Environ Microbiol* **56**, 1865–1874.
33. Kieft, T. L., Phelps, J. T. (1997) Life in the slow lane: activities of microorganisms in the subsurface, in *The microbiology of the terrestrial deep subsurface.* (Amy, P. S., Haldeman, D. L. eds.) CRC Press, Boca Raton, FL. pp. 137–164.
34. Onstott, T. C., Phelps, T. J. Kieft, T. L., Colwell, F. S., Balkwill, D. L., Fredrickson, J. K., Brockman, F. J. (1998) A global perspective on the microbial abundance and activity in the deep subsurface, in *Enigmatic microorganisms and life in extreme environments: cellular origin and life in extreme habitats.* (Seckbach, J. ed.) Kluwer Publications, Norwell, MA. pp. 1–14.
35. Palmen, R., Driessen, A. J., Hellingwerf, K. J. (1994) Biological aspects of the translocation of macromolecules across bacterial membranes. *Biochim Biophys Acta* **1183**, 417–451.
36. Frost, L. S., Leplae, R., Summers, A. O., Toussaint, A. (2005) Mobile genetic elements: the agents of open source evolution. *Nature Rev. Microbiol.* **3**, 722–732.
37. Grange, W., Duckely, M., Husale, S., Jacob, S., Engel, A., Hegner, M. (2008) VirE2: a unique ssDNA-compacting molecular machine. *PLOS Biol.* **6**, e44.
38. Evilevitch, A., Lavelle, L., Knobler, C. M., Raspaud, E., Gelbart, W. M. (2003) Osmotic pressure inhibition of DNA ejection from phage. *Proc Natl Acad Sci USA* **100**, 9292–9295.
39. São-José, C., de Frutos, M, Raspaud, E., Santos, M. A., Tavares, P. (2007) Pressure built by DNA packaging inside virions: enough to drive DNA ejection in vitro, largely insufficient for delivery into the bacterial cytoplasm. *J Mol Biol* **374**, 346–355.
40. Orrego, C., Arnaud, M., Halvorson, H. O. (1978) *Bacillus subtilis* 168 genetic transformation mediated by outgrowing spores: necessity for cell contact. *J Bacteriol* **134**, 973–981.
41. Iyer, L. M., Makarova, K. S., Koonin, E. V., Aravind, L. (2004) Comparative genomics of the FtsK-HerA superfamily of pumping ATPases: implications for the origins of chromosome segregation, cell division and viral capsid packaging. *Nucleic Acids Res* **32**, 5260–5279.
42. Chemla, Y. R., Aathavan, K., Michaelis, J., Grimes, S., Jardine, P. J., Anderson, D. L., Bustamante, C. (2005) Mechanism of force generation of a viral DNA packaging motor. *Cell* **122**, 683–692.
43. Tato, I., Zunzunegui, S., de la Cruz, F., Cabezon, E. (2005) TrwB, the coupling protein involved in DNA transport during

bacterial conjugation, is a DNA-dependent ATPase. *Proc Natl Acad Sci USA* **102**, 8156–8161.
44. Flannagan, S. E., Zitzow, L. A., Su, Y. A., Clewell, D. B. (1994) Nucleotide sequence of the 18-kb conjugative transposon Tn9*16* from *Enterococcus faecalis. Plasmid* **32**, 350–354.
45. Londoño-Vallejo, J. A., Dubnau, D. (1994) Mutation of the putative nucleotide binding site of the *Bacillus subtilis* membrane protein ComFA abolishes the uptake of DNA during transformation. *J Bacteriol* **176**, 4642–4645.
46. Maier, B., Chen, I., Dubnau, D., Sheetz, M. P. (2004) DNA transport into *Bacillus subtilis* requires proton motive force to generate large molecular forces. *Nat Struct Mol Biol* **11**, 643–649.
47. Kieft, T. L., Brockman, F. J. (2001) Vadose zone microbiology, in *Subsurface microbiology and biogeochemistry.* (Fredrickson, J. K., Fletcher, M. eds.) Wiley-Liss, Inc. Danvers, MA, pp. 141–169.
48. Smets, B. F., Rittmann, B. E., Stahl, D. A. (1993) The specific growth rate of *Pseudomonas putida* PAW1 influences the conjugal transfer rate of the TOL plasmid. *Appl Environ Microbiol* **59**, 3430–3437.
49. Muela, A., Pocino, M., Arana, I., Justo, J. I., Iriberri, J., Barcina, I. (1994) Effect of growth phase and parental cell survival in river water on plasmid transfer between *Esherichia coli* strains. *Appl Environ Microbiol* **60**, 4273–4278.
50. Li, Y. H., Lau, P. C., Lee, J. H., Ellen, R. P., Cvitkovitch, D. G. (2001) Natural genetic transformation of *Streptococcus mutans* growing in biofilms. *J Bacteriol* **183**, 897–908.
51. Novotny, C.P., Lavin, K. 1971. Some effects of temperature on the growth of F pili. *J Bacteriol* **107**, 671–682.
52. Page, W. J., von Tigerstrom, M. (1979) Optimal conditions for transformation of *Azotobacter vinelandii. J Bacteriol* **139**, 1058–1061.
53. Lorenz, M. G., Reipschläger, K., Wackernagel, W. (1992) Plasmid transformation of naturally competent *Acinetobacter calcoaceticus* in non-sterile soil extract and groundwater. *Arch Microbiol* **157**, 355–360.
54. Balkwill, D. L., Murphy, E. M., Fair, D. M., Ringelberg, D. B., White, D. C. (1998) Microbial communities in high and low recharge environments: implications for microbial transport in the Vadose zone. *Microb Ecol* **35**, 156–171.
55. Kieft, T. L., Murphy, E. M., Haldeman, D. L., Amy P. S., Bjornstad, B. N., McDonald, E. V., Ringelberg D. B., White, D. C., Stair, J., Griffiths, R. P., Gsell, T. C., Holben, W. E., Boone, D. R. (1998) Microbial transport, survival, and succession in a sequence of buried sediments. *Microb Ecol* **36**, 336–348.
56. Sinclair, J. L., Ghiorse, W. C. (1989) Distribution of aerobic bacteria, protozoa, algae, and fungi in deep subsurface sediments. *Geomicrobiol J* 7, 15–32.
57. Ranjard, L., Poly, F., Combrisson, J., Richaume, A., Gourbière, F., Thioulouse, J., Nazaret, S. (2000) Heterogeneous cell density and genetic structure of bacterial pools associated with various soil microenvironments as determined by enumeration and DNA fingerprinting approach (RISA) *Microb Ecol* **39**, 263–272.
58. Ansaldi, M., Marolt, D., Stebe, T., Mandic-Mulec, I., Dubnau, D. (2002) Specific activation of the *Bacillus* quorum-sensing systems by isoprenylated pheromone variants. *Mol Microbiol* **44**, 1561–1573.
59. Solomon, J. M., Lazazzera, B. A., Grossman, A. D. (1996) Purification and characterization of an extracellular peptide factor that affects two different developmental pathways in *Bacillus subtilis. Genes Dev* **10**, 2014–24.
60. Oger, P., Farrand, S. K. (2002) Two opines control conjugal transfer of an *Agrobacterium* plasmid by regulating expression of separate copies of the quorum-sensing activator gene *traR. J Bacteriol* **184**, 1121–31.
61. McAnulla, C., Edwards, A., Sanchez-Contreras, M., Sawers, R. G., Downie, J. A. (2007) Quorum-sensing-regulated transcriptional initiation of plasmid transfer and replication genes in *Rhizobium leguminosarum* biovar *viciae. Microbiology* **153**, 2074–2082.
62. Ruhfel, R. E., Leonard, B. A. B., Dunny, G. M. (1997) Pheremone-inducible conjugation in *Enterococcus faecalis*: mating interactions mediated by chemical signals and direct contact. In *Bacteria as Multicellular Organisms* (Shapiro, J. A., Dworkin, M. eds.). Oxford University Press, Inc. New York.
63. Cozzarelli, I. M., Weiss, J. V. (2007) Biogeochemistry of aquifer systems, in *Manual of environmental microbiology*, 3rd ed. (Hurst, C. J., Crawford, R. L., Garland, J. L, Lipson, D. A., Mills, A. L., Stetzenbach, L. D. eds.) ASM Press, Washington D.C., pp. 843–859.
64. Baker, B. J., Moser, D. P., MacGregor, B. J., Fishbain, S., Wagner, M., Fry, N. K., Jackson, B., Speolstra, N., Loos, S., Takai, K., Lollar, B. S., Fredrickson, J., Balkwill, D., Onstott, T. C., Wimpee, C. F., Stahl, D. A. (2003) Related assemblages of sulphate-reducing bacteria associated with ultradeep gold mines of South Africa and deep basalt

aquifers of Washington State. *Environ Microbiol.* **5**, 267–77.

65. Boyd, E. S., Cummings, D. E., Geesey, G. G. (2007) Mineralogy influences structure and diversity of bacterial communities associated with geological substrata in a pristine aquifer. *Microb Ecol* **54**, 170–82.
66. Tobin, K. J., Onstott, T. C., DeFlaun, M. F., Colwell, F. S., Fredrickson, J. (1999) In situ imaging of microorganisms in geologic material. *J Microbiol Meth* **37**, 201–213.
67. Mølin, S., Tolker-Nielsen, T. (2003) Gene transfer occurs with enhanced efficiency in biofilms and induces enhanced stabilization of the biofilm structure. *Curr Opin Biotechnol* **14**, 255–261.
68. Licht, T. R., Christensen, B. B., Krogfelt, K. A., Mølin, S. (1999) Plasmid transfer in the animal intestine and other dynamic bacterial populations: the role of community structure and environment. *Microbiology* **145**, 2615–2622.
69. Lawrence, J. G., Hatfull, G. F., Hendrix, R. W. (2002) Imbroglios of viral taxonomy: genetic exchange and the failings of phenetic approaches. *J Bacteriol* **184**, 4891–4905.
70. Marsh, P., Wellington, E. M. H. (1994) Phage-host interactions in soil. *FEMS Microbiol Ecol* **15**, 99–107.
71. Ghosh, D., Roy, K., Williamson, KE, White, D. C., Womack, K. E., Sublette, K. L., Radosevich, M. (2008). Prevalence of lysogeny among soil bacteria and the presence of 16S rRNA and *trzN* genes in viral-community DNA. *Appl Environ Microbiol* **74**, 495–502.
72. Williamson, K. E., Radosevich, M., Wommack, K. E. (2005) Abundance and diversity of viruses in six Delaware soils. *Appl Environ Microbiol* **71**, 3119–3125.
73. Long, R., Azam, F. (2001) Microscale patchiness of bacterioplankton assemblage richness in seawater. *Aquat Microb Ecol* **26**, 101–113.
74. Kovacik, W. P. Jr., Takai, K., Mormile, M. R., McKinley, J. P., Brockman, F. J., Fredrikson, J. K., Holben, W. E. (2006) Molecular analysis of deep subsurface Cretaceous rock indicates abundant Fe(III)- and $S^0$-reducing bacteria in a sulfate-rich environment. *Environ Microbiol* **8**, 141–155.
75. Lehman, R. M., Colwell, F. S., Bala, G. A. (2001) Attached and unattached microbial communities in a simulated basalt aquifer under fracture- and porous-flow conditions. *Appl Environ Microbiol* **67**, 2799–2809.
76. Pedersen, K., Arlinger, J., Hallbeck, L., Pettersson, C. (1996) Diversity and distribution of subterranean bacteria in groundwater at Oklo in Gabon, Africa, as determined by 16S rRNA gene sequencing. *Mol Ecol* **5**, 427–436.
77. Ekendahl, S., Arlinger, J., Ståhl, F., Pedersen, K., (1994) Characterization of attached bacterial populations in deep granitic groundwater from the Stripa research mine by 16S rRNA gene sequencing and scanning electron microscopy. *Microbiology* **140**, 1575–1583.
78. DeSantis, T. Z., Brodie, E. L., Moberg, J. P., Zubieta, I. X., Piceno, Y. M., Andersen, G. L. (2007) High-density 16S rRNA microarray analysis reveals broader diversity than typical clone library when sampling the environment. *Microb Ecol* **53**, 371–383.
79. Zhou, J., Xia, B., Huang, H., Palumbo, A. V., Tiedje, J. M. (2002) Microbial diversity and heterogeneity in sandy subsurface soils. *Appl Environ Microbiol* **70**, 1723–1734.
80. Horner-Devine, M. C., Leibold, M. A., Smith, V. H., Bohannan, B. J. (2003) Bacterial diversity patterns along a gradient of primary productivity. *Ecol Lett* **6**, 613–622.
81. Stotsky, G. (1997) Soil as an environment for microbial life. In *Modern soil microbiology* (van Elsas, J. D., Trevors, J. T., Wellington, E. M. H., eds.) Marcel Dekker Inc., New York.
82. Dechesne, A., Or, D., Smets, B. F. (2008) Limited diffusive fluxes of substrate facilitate coexistence of two competing bacterial strains. *FEMS Microbiol Ecol* **64**, 1–8.
83. Danner, D.B., Deich, R. A., Sisco, K.L., Smith, H.O. (1980) An eleven-base-pair sequence determines the specificity of DNA uptake in *Haemophilus* transformation. *Gene* **11**, 311–318.
84. Hülter, N., Wackernagel, W. (2008) Double illegitimate recombination events integrate DNA segments through two different mechanisms during natural transformation of *Acinetobacter baylyi. Mol Microbiol* **67**, 984–995.
85. Meier P., Wackernagel W. (2003) Mechanisms of homology-facilitated illegitimate recombination for foreign DNA acquisition in transformable *Pseudomonas stutzeri. Mol Microbiol* **48**, 1107–1118.
86. Sikorski, J., Möhle, M., Wackernagel, W. (2002) Identification of complex composition, strong strain diversity and directional selection in local *Pseudomonas stutzeri* populations from marine sediment and soils. *Environ Microbiol* **4**, 465–476.
87. Leisner, M., Stingl, K., Rädler, J. O., Maier, B. (2007) Basal expression rate of *comK* sets a 'switching window' into the K-state of *Bacillus subtilis. Mol Microbiol* **63**, 1806–1816.

88. Greaves, M. P., Wilson, M. J. (1969) The adsorbtion of nucleic acids by montmorillonite. *Soil Bio Biochem* **2**, 257–268.
89. Joppien, G. R. (1978) Characterization of adsorbed polymers at the charged silica-aqueous electrolyte interface. *J Phys Chem* **82**, 2210–2215.
90. Chamier, B., Lorenz, M. G., Wackernagel, W. (1993) Natural transformation of *Acetobacter calcoaceticus* by plasmid DNA adsorbed on sand and groundwater aquifer material. *Appl Environ Microbiol* **59**, 1662–1667.
91. Romanowski, G., Lorenz, M. G., Wackernagel, W. (1991) Adsorption of plasmid DNA to mineral surfaces and protection against DNase I. *Appl Environ Microbiol* **57**, 1057–1061.
92. Romanowski, G., Lorenz, M. G., Wackernagel, W. (1993) Plasmid DNA in a groundwater aquifer microcosm – adsorbtion, DNAse resistance and natural genetic transformation in *Bacillus subtilis*. *Mol Ecol* **2**, 171–181.
93. Lorenz, M. G., Aardema, B. W., Wackernagel, W. (1988) Highly efficient genetic transformation of *Bacillus subtilis* attached to sand grains. *J Gen Microbiol* **134**, 107–112.
94. Lorenz, M. G., Wackernage, W. (1990) Natural genetic transformation of *Pseudomonas stutzeri* by sand-adsorbed DNA. *Arch Microbiol* **154**, 380–385.
95. Ogunseitan, O. A., Tedford, E. T., Pacia, D., Sirotkin, K. M., Sayler, G. S. (1987) Distribution of plasmids in groundwater bacteria. *J Ind Microbiol* **1**, 311–317.
96. Fredrickson, J. K., Hicks, R. J., Li, S. W., Brockman, F. J. (1988) Plasmid incidence in bacteria from deep subsurface sediments. *Appl Environ Microbiol* **54**, 2916–2923.
97. Ladd, T. I., Ventullo, R. M., Wallis, P. M., Costerton, J. W. (1982) Heterotrophic activity and biodegradation of labile and refractory compounds by groundwater and stream microbial populations. *Appl Environ Microbiol* **44**, 321–329.
98. Kobori, H., Sullivan, C. W., Shizuya, H. (1984) Bacterial plasmids in Antarctic natural microbial assemblages. *Appl Environ Microbiol* **48**, 515–518.
99. Hermansson, M., Jones, G. W., Kjelleberg, S. (1987) Frequency of antibiotic and heavy-metal resistance, pigmentation, and plasmids in bacteria of the marine air-water interface. *Appl Environ Microbiol* **53**, 2338–2342.
100. Sobecky, P. A., Mincer, T. J., Chang, M. C., Helinski, D. R. (1997) Plasmids isolated from marine sediment microbial communities contain replication and incompatibility regions unrelated to those of known plasmid groups. *Appl Environ Microbiol* **63**, 888–895.
101. Feng, L., Wang, W., Cheng, J., Ren, Y., Zhao, G., Tang, Y., Liu, X., Han, W., Peng, X., Liu, R., Wang, L. (2007) Genome and proteome of long-chain alkane degrading *Geobacillus thermodenitrificans* NG80-2 isolated from a deep-subsurface oil reservoir. *Proc Natl Acad Sci USA* **104**, 5602–5607.
102. Brockman, F. J., Denovan, B. A., Hicks, R. J., Fredrickson, J. K. (1989) Isolation and characterization of quinoline-degrading bacteria from subsurface sediments. *Appl Environ Microbiol* **55**, 1029–1032.
103. Baya, A. M., Brayton, P. R., Brown, V. L., Grimes, D. J., Russek-Cohen, E., Colwell, R. R. (1986) Coincident plasmids and antimicrobial resistance in marine bacteria isolated from polluted and unpolluted Atlantic Ocean samples. *Appl Environ Microbiol* **51**, 1285–1292.
104. Basta, T., Keck, A., Klein, J., Stolz, A. (2004) Detection and characterization of conjugative degradative plasmids in xenobiotic-degrading *Sphingomonas* strains. *J Bacteriol* **186**, 3962–3872.
105. Romine, M. F., Stillwell, L. C., Wong, K.-K., Thurston, S. J., Sisk, E. C., Sensen, C., Gaasterland, T., Fredrickson, J. K., Saffer, J. D. (1999) Complete sequence of the 184-kilobases catabolic plasmid from *Sphingomonas aromaticivorans* F199. *J Bacteriol* **181**, 1585–1602.
106. Dowd, S. E., Pillai, S. D., Wang, S., Corapcioglu, M. Y. (1998) Delineating the specific influence of virus isoelectric point and size on virus adsorption and transport through sandy soils. *Appl Environ Microbiol* **64**, 405–410.
107. Blanford, W. J., Brusseau, M. L., Jim Yeh, T. C., Gerba, C. P., Harvey, R. (2005). Influence of water chemistry and travel distance on bacteriophage PRD-1 transport in a sandy aquifer. *Water Res* **39**, 2345–2357.
108. Angly, F. E., Fels, B., Breitbart, M., Salamon, P., Edwards, R. A., Carlson, C., Chan, A. M., Haynes, M., Kelley, S., Liu, H., Mahaffy, J. M., Mueller, J. E., Nulton, J., Olson, R., Parsons, R., Rayhawk, S., Suttle, C. A., Rohwer, F. (2006) The marine viromes of four oceanic regions. *PLoS Biol* **11**, e368.
109. Vishivetskaya, T. A., Kathariou, S. (2005) Putative transposases conserved in *Exiguobacterium* isolates from ancient Siberian permafrost and from contemporary surface habitats. *Appl Environ Microbiol* **71**, 6954–6962.

110. Vepritskiy, A. A., Vitol, I. A., Nierzwicki-Bauer, S. A. (2002) Novel group I intron in the $tRNA^{Leu}$(UAA) gene of a γ-proteobacterium isolated from a deep subsurface environment. *J Bacteriol* **184**, 1481–1487.
111. Turmel, M. Mercier, J. P., Côté, V., Otis, C., Lemieux, C. (1995) Evolutionary transfer of ORF-containing group I introns between different subcellular compartments (chloroplast and mitochondrion). *Mol Biol Evol* **12**, 533–545.
112. Reinhold-Hurek, B., Shub, D. A. (1992) Self-splicing introns in tRNA genes of widely divergent bacteria. *Nature* **357**, 173–176.
113. Wall, J. D., Rapp-Giles, B. J., Rousset, M. (1993) Characterization of a small plasmid from *Desulfovibrio desulfuricans* and its use for shuttle vector construction. *J Bacteriol* **175**, 4121–4128.
114. Argyle, J. L., Rapp-Giles, B. J., Wall, J. D. (1992) Plasmid transfer by conjugation in *Desulfovibrio desulfuricans. FEMS Microbiol Lett* **73**, 255–262.
115. Takami, H., Takaki, Y., Chee, G. J., Nishi, S., Shimamura, S., Suzuki, H., Matsui, S., Uchiyama, I. (2004) Thermoadaptation trait revealed by the genome sequence of thermophillic *Geobacillus kaustophilus. Nucleic Acids Res* **32**, 6292–6303.
116. Copeland, A., Lucas, S., Lapidus, A., Barry, K., Detter, J. C., Glavina del Rio, T., Hammon, N., Israni, S., Dalin, E., Tice, H., Pitluck, S., Sims, D. R., Brettin, T., Bruce, D., Han, C., Tapia, R., Brainard, J., Schmutz, J., Larimer, F., Land, M., Hauser, L., Kyrpides, N., Mikhailova, N., Brettar, I., Klappenbach, J., Konstantinidis, K., Rodrigues, J., Tiedje, J., Richardson, P. (2007) Complete sequence of chromosome of *Shewanella baltica* OS155. Accession number NC_009052. US DOE Joint Genome Institute. Unpublished.
117. Copeland, A., Lucas, S., Lapidus, A., Barry, K., Glavina del Rio, T., Dalin, E., Tice, H., Pitluck, S., Sims, D., Brettin, T., Bruce, D., Detter, J. C., Han, C., Schmutz, J., Larimer, F., Land, M., Hauser, L., Kyrpides, N., Mikhailova, N., Brettar, I., Rodrigues, J., Konstantinidis, K., Tiedje, J., Richardson, P. (2007) Complete sequence of chromosome of *Shewanella baltica* OS185. Unpublished. Accession number NC_009665. US DOE Joint Genome Institute.
118. Copeland, A., Lucas, S., Lapidus, A., Barry, K., Detter, J. C., Glavina del Rio, T., Hammon, N., Israni, S., Dalin, E., Tice, H., Pitluck, S., Chain, P., Malfatti, S., Shin, M., Vergez, L., Schmutz, J., Larimer, F., Land, M., Hauser, L., Kyrpides, N., Lykidis, A., Tiedje, J., Richardson, P. (2007) Complete sequence of *Shewanella* sp. W3-18-1. Unpublished. Accession number NC_008750. US DOE Joint Genome Institute.
119. Butler, J. E., He, Q., Nevin, K. P., He., Z., Zhou, J., Lovley, D. R. (2007) Genomic and microarray analysis of aromatics degradation in *Geobacter metallireducens* and comparison to a *Geobacter* isolate from a contaminated field site. *BMC Genomics* **8**, 180.
120. Coombs, J. M., Barkay, T. (2004) Molecular evidence for the evolution of metal homeostasis genes by lateral gene transfer in bacteria from the deep terrestrial subsurface. *Appl Environ Microbiol* **70**, 1698–1707.
121. Coombs, J. M., Barkay, T. (2005) New findings on evolution of metal homeostasis genes: evidence from comparative genome analysis of bacteria and archaea. *Appl Environ Microbiol* **71**, 7083–7091.
122. Friedrich, M. W. (2002) Phylogenetic analysis reveals multiple lateral transfers of adenosine-5′-phosphosulfate reductase genes among sulfate-reducing microorganisms. *J Bacteriol* **184**, 278–289.
123. Rees, G. N., Grassia, G. S., Sheehy, A. J., Dwivedi, P. P., Patel, B. K. C (1995) *Desulfacinum infernum* gen. nov., sp. nov., a thermophilic sulfate-reducing bacterium from a petroleum reservoir. *Int J Syst Bacteriol* **45**, 85–89.
124. Beeder, J., Torsvik, T., Lien, T. (1995) *Thermodesulforhabdus norvegicus* gen. nov., sp. nov., a novel thermophilic suflate-reducing bacterium from oil field water. *Arch Microbiol* **164**, 331–336.
125. Klein, M., Friedrich, M., Roger, A. J., Hugenoltz, P., Fishbain, S., Abicht, H., Blackall, L. L., Stahl, D. A., Wagner, M. (2001) Multiple lateral transfers of dissimilatory sulfite reductase genes between major lineages of sulfate-reducing prokaryotes. *J Bacteriol* **183**, 6028–6035.
126. Daumas, S., Cord-Ruwisch, R., Garcia, J. L. (1988) *Desulfotomaculum geothermicum* sp. nov., a thermophilic, fatty acid-degrading, sulfate-reducing bacterium isolated with H2 from geothermal groundwater. *Antonie Van Leeuwenhoek* **54**, 165–178.
127. Smets, B. F., Morrow, J. B., Pinedo, C. A. (2003) Plasmid introduction in metal-stressed, subsurface-derived microcosms: plasmid fate and community response. *Appl Environ Microbiol* **69**, 4087–4097.
128. Dash, P. K., Traxler, B. A., Panicker, M. M., Hackney, D. D., Minkley, E. G. Jr.

(1992) Biochemical characterization of *Escherichia coli* DNA helicase I. *Mol Microbiol* **6**, 1163–1172.
129. Maier, B., Potter, L., So, M., Seifert, H. S., Sheetz, M. P. (2002) Single pilus motor forces exceed 100 pN. *Proc Natl Acad Sci USA* **99**, 16012–16017.
130. Sanchez, H., Cardenas, P. P., Yoshimura, S. H., Takeyasu, K., Alonso, J. C. (2007) Dynamic structures of *Bacillus subtilis* RecN-DNA complexes. *Nucleic Acids Res.* **36**, 110–120.
131. He, J., Ritalahti, K. M., Yang, K.-L., Koenigsberg, S. S., Löffler, F. E. (2003) Detoxification of vinyl chloride to ethane coupled to growth of an anaerobic bacterium. *Nature* **424**, 62–65.
132. Weimer, P. J., Van Kavelaar, M. J., Michel, C. B., Ng, T. K. (1988) Effect of phosphate on the corrosion of carbon steel and on the composition of corrosion products in two-stage continuous cultures of *Desulfovibrio desulfuricans. Appl Environ Microbiol* **54**, 386–396.
133. Chivian, D., Alm, E. J., Brodie, E. L., Culley, D. E., Dehal, P. S., DeSantis, T. Z., Gihring, T. M., Lapidus, A., Lin, L.-H., Lowry, S. R., Moser, D. P., Richardson, P., Southam, G., Wanger, G., Pratt, L. M., Andersen, G. L., Hazen, T. C., Brockman, F. J., Arkin, A. P., Onstott, T. C. (2008) Environmental genomics reveals a single species ecosystem deep within the Earth. Unpublished. Accession number NC_010424. US DOE Joint Genome Institute (JGI-PGF).
134. Anderson, R. T., Vrionis, H. A., Ortiz-Bernad, I., Resch, C. T., Long, P. E., Dayvault, R., Karp, K., Marutzky, S., Metzler, D. R., Peacock, A., White, D. C., Lowe, M., Lovely, D. R. (2003) Stimulating the in situ activity of *Geobacter* species to remove uranium from the groundwater of a uranium-contaminated aquifer. *Appl Environ Microbiol* **69**, 5884–5891.
135. Fredrickson, J. K., Brockman, F. J., Workman, D. J., Li, S. W., Stevens, T. O. (1991) Isolation and characterization of a subsurface bacterium capable of growth on toluene, naphthalene, and other aromatic compounds. *Appl Environ Microbiol* **57**, 796–803.
136. Lien, T., Madsen, M., Rainey, F. A., Birkeland, N.-K. (1998) *Petrotoga mobilis* sp. nov., from a North Sea oil-production well. *Int J Syst Bacteriol* **48**, 1007–1013.
137. Copeland, A., Lucas, S., Lapidus, A., Barry, K., Detter, J. C., Glavina del Rio, T., Hammon, N., Israni, S., Dalin, E., Tice, H., Pitluck, S., Chain, P., Malfatti, S., Shin, M., Vergez, L., Schmutz, J., Larimer, F., Land, M., Hauser, L., Kyrpides, N., Mikhailova, N., Romine, M. F., Fredrickson, J., Tiedje, J., Richardson, P. (2007) Complete sequence of *Shewanella putrefaciens* CN-32. Unpublished. Accession number NC_009438. US DOE Joint Genome Institute.
138. Takahata, Y., Nishijima, M., Hoaki, T., Maruyama, T. (2001) *Thermotoga petrophila* sp. nov. and *Thermotoga naphthophila* sp. nov., two hyperthermophilic bacteria from the Kubiki oil reservoir in Niigata, Japan. *Int J Syst Evol Microbiol* **51**, 1901–1909.

# Chapter 25

## Horizontal Gene Transfer and Mobile Genetic Elements in Marine Systems

**Patricia A. Sobecky and Tracy H. Hazen**

### Abstract

The pool of mobile genetic elements (MGE) in microbial communities consists of viruses, plasmids, and associated elements (insertion sequences, transposons, and integrons) that are either self-transmissible or use mobile plasmids and viruses as vehicles for their dissemination. This mobilome facilitates the horizontal transfer of genes that promote the evolution and adaptation of microbial communities. Efforts to characterize MGEs from microbial populations resident in a variety of ecological habitats have revealed a surprisingly novel and seemingly untapped biodiversity. To better understand the impact of horizontal gene transfer (HGT), as well as the agents that promote HGT in marine ecosystems and to determine whether or not environmental parameters can effect the composition and structure of the mobilome in marine microbial communities, information on the distribution, diversity, and ecological traits of the marine mobilome is presented. In this chapter we discuss recent insights gained from different methodological approaches used to characterize the biodiversity and ecology of MGE in marine environments and their contributions to HGT. In addition, we present case studies that highlight specific HGT examples in coastal, open-ocean, and deep-sea marine ecosystems.

**Key words:** Horizontal gene transfer, marine phage, virulence, plasmid, geochemical cycling, mobile genetic elements.

## 1. Introduction

Bacteria display an extraordinary phenotypic diversity with variable metabolic properties, cellular structures, and lifestyles *(1)*. It has been long assumed that the transfer of genes encoding such diverse traits acquired by transformation, conjugation,

Maria B. Gogarten et al. (eds.), *Horizontal Gene Transfer: Genomes in Flux, vol. 532*

DOI 10.1007/978-1-60327-853-9_25 Springerprotocols.com

or transduction processes were key factors in microbial community structure and function. However, the contributions of mobile genetic elements (MGEs) such as plasmids, viruses, transposons, and integron-associated gene cassettes to bacterial adaptability and diversity in aquatic and terrestrial environments are only now being fully appreciated. The significance of horizontal gene transfer (HGT) to microbial adaptation was first recognized when multiple antibiotic–resistant pathogens emerged. MGEs played a primary role in the development and dissemination of antibiotic-resistance genes and allowed bacterial populations to rapidly adapt to a strong selection pressure *(2, 3)*. The past decade of sequencing complete prokaryotic genomes now clearly shows that a large proportion of prokaryotic genes have been acquired by horizontal transfer. This evidence has been derived from sequence comparisons as well as from atypical nucleotide composition (G + C content) or patterns of codon usage. In many species, a high proportion of horizontally transferred genes can be attributed to plasmid, virus, or transposon-related sequences since remnants of MGEs are often found adjacent to these genes *(1)*. MGEs are undoubtedly an important and essential component of the marine biosphere that promote marine microbial diversification. However, due to their implication in the rapid dissemination of antibiotic-resistance and pathogenicity determinants, the vast majority of well-characterized plasmids and viruses have originated from clinical bacterial populations. Another contributing factor to the lack of information on MGEs in other ecosystems, including the marine environment, is due in part to the fact that only a minor proportion of bacteria currently are accessible to cultivation techniques *(4)*.

The oceans occupy 71% of the Earth's surface with an average depth of 4000 m and a maximum depth of approximately 11,000 m for deep-sea locales. The marine environment is the largest habitat on Earth and the oceanic water masses control global climate conditions and produce nearly half of the Earth's oxygen. Coastal near-shore estuaries and salt marshes, subject to tides and tidal flushing, are strongly influenced by terrigenous materials including freshwater runoff, sediments, organic carbon, and nutrients that affect microbial productivity. Estuaries serve as breeding grounds and habitats for a wide variety of wildlife, which are directly linked to microbial productivity and activity. In this chapter we highlight the findings and insights gained during the past decade, investigating the role of HGT and the distribution and biodiversity of MGEs present in marine microbial communities in coastal, open ocean, and deep-sea locales.

## 2. Abundance of MGEs in Marine Microbial Communities

### 2.1. Plasmids

Determinations of plasmid abundance in marine systems have most often relied on the cultivation of microbial hosts, with or without the use of selective media types. The term endogenous isolation has been coined to reflect plasmid incidence reported from a cultured microbial fraction. Numerous procedures employing a variety of cell lysis and extraction conditions have been developed for the isolation of plasmids, with many having been designed primarily for Gram-negative clinical bacteria (reviewed in *(5)*). Methods include the separation of covalently closed circular plasmid DNA from chromosomal DNA by either cesium chloride or sucrose gradient centrifugation *(6)*. However, these separation techniques are often impractical when attempting to process hundreds to thousands of marine isolates. Commercial DNA isolation kits can be used, but often yield only the best results with *Escherichia coli* strains (for which they were initially designed). Various small-scale alkaline denaturation methods *(7, 8)* and an "in-well lysis" technique for screening individual bacterial colonies developed by Eckhardt *(9)* can also facilitate processing large numbers of marine bacterial isolates. An obvious drawback to the endogenous isolation procedure is the dependency on culturing the microbial host. Such a reliance on isolation of plasmid-containing hosts undoubtedly results in a biased view of plasmid populations as the vast majority of marine bacteria are thought to be either resistant to cultivation or require intensive efforts to obtain *(10)*. Although it may be argued that a cultivation-dependent approach may not yield representative indigenous marine plasmid populations, surprising new insights into replicon diversity have resulted from the molecular characterization of endogenously isolated plasmids from coastal marine environments *(11, 12)*.

A high percentage of bacteria from marine systems (i.e., estuarine, marine water column, sediments, and salt marsh rhizosphere-associated niches) have been shown to contain plasmids with some studies detecting one or more plasmids in as much as 52% of the marine isolates examined. Although a considerable frequency of isolates carry plasmids, attempts to attribute specific traits and functions to these plasmids have proven to be more difficult. For example, Baya et al. *(13)* demonstrated that the frequency of plasmid DNA and resistance to antibiotics and toxic chemicals increased in marine bacterial isolates in closest proximity to a sewage outfall diffuser, which released pharmaceutical and industrial wastes. However, no direct correlation between plasmid presence and the observed phenotypes was obtained. Leahy et al. *(14)* were also unable to detect a correlation between hydrocarbon degradation capabilities and plasmid incidence in 242 heterotrophic sediment bacterial isolates obtained from an offshore site in the Gulf of Mexico, chronically contaminated with varying concentrations of petroleum hydrocarbons. The inability to directly correlate plasmid content and antibiotic and heavy metal resistance traits has also been reported for 30 plasmid-containing *Bacillus* isolates, representing a total of 102 plasmids, and 297 plasmid-bearing Gram-negative bacteria obtained from Canadian *(15)* and Californian *(11)* coastal marine sediments, respectively. Such findings lend support to the notion that marine plasmid populations may be encoding other beneficial and perhaps novel (phenotypic) traits.

A complementary method to endogenous isolation, i.e., exogenous isolation, has been developed to access the nonculturable microbial fraction. Two variations of the exogenous method, designated bi-parental and tri-parental isolation, have been used to obtain plasmids from microbial assemblages. Exogenous isolation eliminates the need to cultivate microbial hosts by "capturing" plasmids directly from the microbial community via mating with a selectable recipient strain (reviewed in *(16)*). In contrast to endogenous isolation, this method requires that the captured plasmid(s) either be self-transferable ($Tra^+$) or mobilizable ($Tra^-; Mob^+$), and subsequently able to replicate and express (selectable) plasmid-encoded genes in the recipient host. Conjugative plasmids conferring resistance to mercury have been exogenously isolated from microbial populations resident in marine sediments and coastal waters *(17, 18)*. For example, using *Pseudomonas putida* UWC1 as the recipient, Dahlberg et al. *(18)* were able to capture 95 marine plasmids representing 12 different structural groups from microbial communities in bulk water, biofilms, and the air–water interface. An advantage to the exogenous method is that a phenotypic trait can be readily assigned to a conjugative or mobilizable plasmid. The disadvantage is that no information is obtained regarding the identity of the marine microbial host.

### 2.2. Viruses

It is estimated that viruses outnumber bacteria and archaea by more than 15-fold, with viral densities reaching $10^6$–$10^7$ per ml of seawater in coastal environments (reviewed in *(19)*). Viral abundance generally tracks with prokaryotic abundance, decreasing further offshore and with water column depth *(20)*. Viral abundance can be determined by a number of indirect and direct methods as recently reviewed in Suttle *(21)*, and the advantages and disadvantages of these methods have been extensively reviewed *(19)*. For example, plaque assays and most probable number assays (MPN) are used when quantifying infectious viral numbers with cultured microbial hosts, while transmission electron microscopy (TEM), epifluorescence microscopy (EfM) *(22)*, and flow cytometry (FC) are used to determine total viral abundance. EfM is the most widely employed method owing to the ease of the technique relative to TEM. The nucleic acid stains, YO-PRO and SYBR Green, have greatly improved the detection of viral particles. FC is also increasing in popularity and facilitates high-throughput sampling. Marine viral diversity is determined in similar fashion to marine plasmid diversity using RFLP profiling, DNA–DNA hybridizations, and PCR amplification of viral-specific genes. As MGEs often lack common sequences that can be used to determine phylogenetic relationships such as rRNA genes, one expects that MGE diversity is likely still considerably underestimated. However, even with such limitations, patterns of viral distributions have been reported. For example, a globally distributed virus infects the picophytoplankton *Micromonas pusilla*, whereas viruses infecting *Pseudoalteromonas* spp. appear to be more geographically restricted *(23)*.

### 2.3. DNA-Based Approaches for Characterizing MGEs

The availability of sequence information for a number of well-characterized plasmid groups has greatly facilitated the development of DNA probes and PCR primers specific for conserved plasmid replication and transfer regions. For example, the broad-host-range plasmid group IncP, although detectable infrequently in culturable marine isolates *(24)*, can be amplified from marine community DNA *(11)*. Thus, when plasmid-specific primers and probes are used to screen environmental DNA samples, data on plasmid prevalence, population levels, and distribution in diverse habitats can be obtained (reviewed in *(5)*). However, as sequence information is required for primer design, our view of environmental plasmid populations is again likely to be skewed to the limited set of plasmid types for which full or partial sequences are available. Surprisingly, few studies have attempted to characterize and quantify replicon-specific sequences on temporal and spatial scales or in response to changing environmental conditions either in aquatic or in terrestrial environments. This is in contrast to the efforts characterizing the seasonal and depth distributions of aquatic viral populations (e.g., *(25)*). Techniques such as Most-Probable-Number (MPN)-PCR, quantitative PCR (QPCR), and real-time QPCR could be used for the quantification of specific marine plasmid replicon genes. In contrast to plasmid studies, the ability to physically separate viral genomes from chromosomal sequences has greatly facilitated metagenomic-based studies of viral communities. Recently, viral assemblages from four oceanic regions were sequenced *(26)*. Nearly 2 million sequences were obtained and more than 90% of the sequences had no recognizable homologs when queried against existing databases. Examples of case studies employing metagenomics approaches are presented in **Section 4**.

## 3. MGEs and Their Contribution to Host Niche Expansion

### 3.1. Diversity of Marine MGEs

During the past decade, research has revealed the unexplored diversity of MGEs in marine systems. Much of the research examining the sequence diversity of marine plasmids has been constrained by the difficulties of recovering intact plasmid DNA molecules directly from the environment. The improvement of plasmid DNA isolation kits as well as methods of isolating plasmids from chromosomal DNA using plasmid-safe DNAse *(27)* have increased the efficiency of recovering intact plasmid DNA. In addition, recent advances in genome sequencing have greatly increased the number of extrachromosomal elements characterized as a part of genome-sequencing studies. For example, the plasmids pYJ016 (48.5-kb) of *Vibrio vulnificus (28)*, pA (40.4-kb) of *Nitrosococcus oceani (29)*, and pTB1 (106.5-kb), pTB2 (69.3-kb), pTB3 (16.6-kb), and pTB4 (5.8-kb) of *Roseobacter denitrificans (30)* were all characterized as part of genome-sequencing studies. Sequence examination of completed genomes for evidence of viruses or insertion sequence (IS) elements provides information on the mobility of these MGEs as well as the diversity of associated genes. A recent study examined the diversity of MGEs among sequenced cyanobacterial and archaeal genomes and found that for some species, such as *Synechococcus* from hotsprings, IS elements are abundant, while they are absent in archaeal genomes occupying thermal vents *(31)*.

Research focused on examining the sequence diversity of widespread MGEs in the marine environment has been primarily limited to DNA *(32)* and RNA viruses *(33)*. The sequence diversity of marine plasmids has not been examined to date in a large-scale study although several ongoing studies are sequencing large numbers of plasmids from cultured hosts obtained from diverse terrestrial and aquatic environments including marine environments (E. Top, A. Summers, personal communication). The number of completed plasmid and virus sequences from marine bacteria, archaea, and eukaryotes are available in the Entrez genome of Genbank and are listed in **Table 25.1**. There are additional sequences not available in Entrez genome; however, those included in **Table 25.1** provide a cursory understanding of the relative frequency of MGEs that have been characterized from different groups. The plasmid sequences that have been characterized to date from eukaryotes range from 3.9 to 7.2 kb and are limited to several algal species including *Gracilaria* spp. *(34)* and *Porphyra* spp. *(35)*. In addition, transposable elements have been identified in the genomes of coastal and deep-sea crustaceans, with some elements exhibiting greater than 99% similarity indicating a possible role of these elements in gene

**Table 25.1**
**The number of MGE sequences characterized from marine organisms in Entrez Genome of Genbank as of February 2008**

| | Total number | Number from marine sources (% of Total) | Size range (kb) | Number from genome projects (% of marine) |
|---|---|---|---|---|
| **Plasmids** | | | | |
| Bacteria | 1,101 | 98 (9) | 1.9–821 | 55 (56) |
| Archaea | 47 | 5 (11) | 3.3–8.3 | 2 (40) |
| Eukarya | 22 | 7 (32) | 3.9–7.2 | 0 |
| **Phages*** | 471 | 42 (9) | 6.3–244.8 | 0 |
| **Viruses** | 2,498 | ND | 0.2–1,181 | ND |

*Categories as present in Entrez Genome, i.e., Phages are viruses infecting Bacteria and Archaea while Viruses are infecting Eukarya.
ND-not determined.

transfer between marine crustaceans *(36)*. Among the archaeal plasmids characterized to date are the 58.4-kb and 16.5-kb elements *of Methanocaldococcus jannaschi (37)*. The host microorganisms for plasmids and viruses that have been characterized range from *M. janaschii* of deep-sea hydrothermal vents*(37)* to *N. oceani* and *R. denitrificans (30)* in the water column *(29)*, to *Vibrio* spp. in coastal environments *(38–40)*.

Previously, methods for examining the genetic diversity of extrachromosomal elements relied on determining the structural diversity by restriction profiling or RAPD-PCR analysis *(41)*. Replicon diversity among marine plasmids has been shown to be extensive with marine plasmid incompatibility groups representing plasmid groupings unlike those previously described *(11)*. Another study examining the structural diversity of 95 marine plasmids identified 12 unique structural groups, and Southern hybridization revealed there were few similarities in gene content among the different groups *(18)*. Typically, methods such as southern hybridization have been used to examine the diversity of genes encoded by MGEs. Southern hybridization is useful for examining the overall diversity of MGEs but provides limited information in regards to the gene content. Suppression subtractive hybridization (SSH) is a method frequently used to obtain unique DNA from two DNA molecules being characterized for sequence analysis. Sequencing of entire plasmids has increased the knowledge of gene diversity and the potential contribution of cryptic plasmids to host phenotypes; however, as consistent with genome studies, many of the encoded proteins are hypothetical and often unique to the element they are encoded on. Predicted proteins encoded by MGEs are as diverse as *psbA* encoding a photosystem II protein involved in photosynthesis of *Prochlorococcus (42)* (described in more detail in **Section 4.2**) to uncharacterized proteins of the *Microscilla* plasmid pSD15 that confer the ability to degrade agar *(43)*, to a hypothetical protein on a *Micrococcus* plasmid with greatest similarity to a protein of spider silk *(44)*. The relatively unknown nature and lack of similarity of most of the proteins encoded by MGEs, which have been characterized to date, highlight the need to examine sequence diversity of more MGEs from diverse hosts and environments. Comparison of genetic diversity of MGEs from diverse hosts and niches from within the marine environment would also provide an understanding of the distribution of certain elements and their role in shaping microbial communities. For example, the viral structural protein g20 has been detected in marine systems around the world and is conserved among diverse viruses *(45)*.

Sequence analysis of extrachromosomal elements as well as entire genomes has begun to elucidate the role of certain MGEs for HGT and the contribution of their proteins to host adaptation. For example, the 62-kb multi-drug-resistance element, SXT, is a mosaic element that confers resistance to sulfamethoxazole, streptomycin, furazolidone, and trimethoprim *(46)*. Additional elements that share high identity to SXT have been identified in *V. fluvialis*, indicating the mobility of SXT from *V. cholerae* to other opportunistic *Vibrio* pathogens found in coastal environments *(47)*. The recently characterized sequence of the virus PAV1 from the deep-sea *Pyrococcus abyssi* represents the identification of a possible new family of viruses *(48)*. The 76- and 148-kb nucleotide sequences of two *Ruegeria* plasmids were shown to encode polysaccharide biosynthesis components that may allow the bacterium to adapt to changing metabolic conditions in marine sediments. *(49)*

Although research has begun to demonstrate the extent of the diversity of MGEs in marine environments and their potential for shaping the diversity of microbial populations, the majority of MGEs have not been characterized for their genetic diversity and their potential role in host adaptation. MGEs associated with *Vibrio* spp. are perhaps among the most well-characterized elements from marine bacteria. The focus of many *Vibrio* MGE studies has been on virus-like elements and the transmission of virulence-associated genes. The sequences of several small plasmids have been characterized from *V. cholerae* including pSIO1 *(50)* and pVC *(51)*, both encoding proteins with no known function. Similarly, the sequence of a 5.5-kb plasmid, pES213, isolated from *V. fischeri*, encoded hypothetical proteins and was shown to be related to similar-sized plasmids identified among other *V. fischeri* strains *(52)*. Although small extrachromosomal elements are frequently detected among marine bacteria, larger elements are often identified and have been increasingly examined for the diversity of their gene content. We recently characterized the sequences of three plasmids from *Vibrio* strains isolated from salt marsh sediment and revealed that much of the gene content of these elements is unknown *(53)*. Although many of the predicted proteins of these three plasmids had hypothetical functions, the predicted replication and partitioning proteins were similar for p09022 and p23023 and shared similarity with pKA1 of *V. cholerae*.

Marine viruses have been shown to regulate the density of the microbial population in response to changes in nutrient availability. The induction of a lysogenic *Synechococcus* cyanophage was shown to occur during nutrient limitation *(54)*, and studies have shown that phage induction is a major contributor to global geochemical cycling of nutrients in the oceans *(55)*. The following case studies are presented to highlight the remarkable contributions of HGT and MGE occurring in marine systems. In some instances, the specific MGE promoting HGT of advantageous genes is known, while in other cases it can only be inferred. Irrespective of the particular marine system examined, it is clear that HGT is a significant process that is shaping marine microbial community structure and function in unique ways that can impact microbial processes on a global scale.

### 3.2. MGEs and Pathogenicity

Sequencing of MGEs has primarily focused on characterizing the elements associated with disease-causing microorganisms; however, in recent years the sequences of a greater number of MGEs have been obtained for non-pathogenic microorganisms from marine environments. Of the MGEs associated with marine pathogens, the CTX phage of *V. cholerae* has been well characterized for its role in the transfer of cholera toxin genes *ctxAB (56)*. Also, the filamentous phage f237 *(57)* has been isolated from and associated with disease-causing strains of *V. parahaemolyticus*; however, its role in virulence of *V. parahaemolyticus* is not known. Several studies have characterized the sequences of vibriophage of diverse sizes ranging from the 8.7-kb f237 to the 244.8-kb KVP40 *(58)*. Research on the diversity of vibriophage has also involved the use of restriction analysis and hybridizations to compare vibriophage among diverse *Vibrio* hosts from coastal environments *(38, 40, 59)*. The main virulence factor of *V. parahaemolyticus* (thermostable direct hemolysins, *tdh/trh*) have been previously identified on plasmid DNA *(60)*; however, these genes are usually associated with the chromosome *(61)*. Aside from one example of a *V. parahaemolyticus* hemolysin on a plasmid, the association of host virulence factors with marine plasmids is limited to the plasmids pJM1 *(62)* and pEIB1 *(63)* of the fish pathogen *V. anguillarum*. Recently, the use of suppression subtractive hybridization (SSH) to determine unique DNA contributing to the virulence of *V. vulnificus* biotype 2 revealed three unique sequences that were associated with plasmid DNA *(64)*. The recent sequence characterization of the 68.4-kb plasmid, pR99, from a *V. vulnificus* biotype 2 strain revealed proteins encoded by this element are required for virulence in eels *(65)*.

## 4. Case Studies of HGT in Marine Niches Mediated by MGEs

### 4.1. HGT in Coastal Systems

Metagenomic-based approaches that apply large-scale sequencing efforts to assess MGE diversity, richness, and potential for the HGT of genes conferring advantageous traits to coastal marine microbial communities have revealed a remarkable wealth of novel sequences. For example, the genomic analysis of "shotgun" libraries constructed from near-shore marine viral (dsDNA) communities obtained from 200 L of surface seawater revealed a striking display of dsDNA viral diversity *(32)*. The samples were collected from two coastal sites in southern California in 2001, Scripps Pier (La Jolla, CA) and Mission Bay, CA *(32)*. The authors reported that all of the major dsDNA phage families were present in both libraries, with the marine phage Roseophage SIO1 genome being among the predominant Podoviruses detected. However, more than 65% of the sequences obtained from the libraries lacked similarity to previously reported sequences. The types of MGEs found in the two libraries did include "hits" to bacterial plasmids, insertion sequences, transposons, pathogenicity islands, and retrotransposons, which are commonly found in eukaryotic genomes. While the majority of viral-based studies have focused on dsDNA viruses, ssRNA viruses are widespread components of marine viral communities *(66)*. Recently, Culley et al. *(67)* examined the RNA virus communities from two coastal sites in British Columbia, Canada. Again, as observed for coastal dsDNA viral communities, as much as 63% of the RNA viral genomes (i.e., positive-sense ssRNA) lacked similarity to previously reported sequences. In addition, the complete genomes of ssRNA viruses recovered from the two libraries do not belong to any of the known RNA viral families *(67)*. Thus, coastal marine viral communities are themselves an untapped reservoir of diverse genetic material that carry a reservoir of MGEs that, as described in the following section, are providing a means to promote genome diversification in marine microbial communities.

Recently, a metagenomic approach targeting integron-associated gene cassette systems in coastal microbial communities was also conducted, which reinforces the notion of a pool of untapped MGE genetic diversity occurring in coastal marine habitats *(68)*. The integron/gene cassette pool was recovered from four marine sediment samples collected in Halifax, Nova Scotia, Canada, which included two sewage outfalls, a beach site, and a pristine salt marsh. The majority (~80%) of predicted proteins encoded on the sequenced cassettes appear to be novel, as no hits were obtained to the NCBI non-redundant protein database. Frequently detected cassette-encoded traits that could be identified included genes encoding multiple plasmid addiction systems, isochorismatase, acetyltransferase, and glutathione-dependent formaldehyde-activating genes *(68)*. In addition, ~23% of the gene cassettes carried putative signal peptides, presumably involved in extracellular transport processes. Thus, as concluded by the authors of this study, integrons/gene cassettes may be conferring selective advantages to their marine host(s).

A metagenomic library constructed from DNA extracted from Baltimore Harbor, MD, marine sediments was used to determine the amount of HGT in anaerobic marine microbial communities *(69)*. A total of 12 environmental fosmid clones were identified using encoded rRNA genes to infer phylogenetic incongruence of co-encoded protein coding sequences (CDSs). The authors estimated that remarkably 7–44% of the CDSs in each fosmid were acquired by HGT. For example, present on the *Flavobacteriaceae* fosmid was a large self-transferable conjugative transposon obtained from a *Bacteroides* sp. In addition, other fosmids revealed genes involved in biodegradation that were frequently detected to have undergone HGT which may be a result of the long-term impacts of pollution at this particular coastal site. In some cases, the agents of HGT were also detected, as was the case of a δ-*Proteobacterium* fosmid that encoded an integrase and transposase responsible for HGT of an acetyltransferase gene and three acyl-CoA synthase subunit genes from a β-*Proteobacterium (69)*.

Viral-like elements, denoted gene transfer agents, have recently been detected in the genome sequences of three marine isolates belonging to the *Roseobacter* clade *(70)*. Gene transfer agents (GTA), appearing exclusively in the α-proteobacteria, differ from 'true' viruses in that no negative effects are associated with their transfer. In addition, GTAs appear to function solely to move random pieces of genomic DNA between cells in a process similar to generalized transduction *(71)*. *Silicibacter pomeroyi* DSS-3 was shown experimentally to produce GTA particles, and GTA-mediated transfer was detected under controlled laboratory conditions *(70)*. Thus, HGT mediated by GTA elements could be a significant process occurring in marine ecosystems.

#### 4.2. HGT in Open Ocean Systems

Further evidence of the important contributions provided by HGT in diversifying marine microbial genomes by providing key metabolic traits is found in open ocean systems. The cyanobacteria *Prochlorococcus* and *Synechococcus* comprise the majority of the picophytoplankton fraction of open oceans and are responsible for as much as 90% of the primary productivity in oligotrophic regions of the oceans *(70, 72)*. Remarkably, it appears that marine viruses play a role in the evolution of cyanobacterial photosynthetic systems *(42, 73, 74)*. The cyanomyovirus S-PM2 infects *Synechococcus* spp. and carries *psbA* and *psbD* genes encoding for D1 and D2 proteins, part of photosystem II (PSII) *(75)*. A heterodimer of these two proteins binds pigments and cofactors for PSII photochemistry. Light-induced damage to D1 results in photo-inhibition, thereby decreasing PSII photochemistry efficiency. It has been proposed that D1 and D2 expression by S-PM2 encoded genes is a survival mechanism allowing the PSII repair cycle to continue even after host protein synthesis stops so that viral replication continues unabated *(74)*. Another gene involved in oxygenic photosynthesis, *hli*, encodes a high light-inducible protein (HLIP) that also undergoes HGT from cyanophages to their *Prochlorococcus* hosts *(42)*. During stress, HLIPs provide the photosynthetic apparatus with protection against excess excitation energy *(76)*. In addition to *hli* genes, genes encoding protein components for oxygenic photosynthetic systems have been found on the genomes of Myoviridae and Podoviridae-like marine viruses. A second case of HGT dispersal of metabolically advantageous genes in ocean systems can be seen in the horizontal transfer of genes encoding for proteorhodopsins, whose abundance correlates to their location in the photic zone of the North Pacific Subtropical Gyre *(77)*. Prokaryotic rhodopsins are retinal-binding membrane photoproteins stimulated by light. Proteorhodopsins are the energy-conserving transmembrane proton pumps that participate in light-activated proton translocation across bacterial membranes. As only a few genes encode for microbial rhodopsin-based photosystems, they are excellent candidates for dissemination by HGT among diverse marine phyla *(78)*. Surprisingly, a bacterial-like proteorhodopsin gene was discovered on an archaeal genome fragment, and heterologous gene expression in *E. coli* confirmed that the archaeal photoprotein was able to bind retinal *(77)*. These results as well as previous genome analyses from thermophilic microbial communities *(79)* and studies highlighted below are especially noteworthy as they demonstrate HGT occurring among and between the bacterial and the archaeal domains.

#### 4.3. HGT in Deep-Sea Systems

Investigations of the deep-sea and seafloor environments, cold seeps, hydrothermal vents, and surrounding seabeds, have uncovered a remarkable array of microbial life and diversity despite the fact that temperature and available energy sources represent extreme and seemingly inhospitable conditions. A comprehensive review of microbial habitats and activities in the deep-sea bed has recently been published *(80)*. The following section will highlight the examples of HGT and mobile genetic elements (MGEs) characterized from the deep-sea water column (>500 m), hydrothermal vents, and microenvironments of polychaete worms colonizing deep-sea hydrothermal chimneys.

While studies characterizing MGE distribution and diversity of plasmid and viral elements are generally lacking from deep-sea systems, Group II introns, a class of self-splicing, retrotransposable elements known to undergo HGT in soil-dwelling bacteria *(81)* have been examined. Samples of a basalt colonizing bacterial mat, and *Calyptogena magnifica* (vent clam) and *Riftia pachyptila* (vestimentiferan tubeworm) containing intracellular bacterial symbionts, were collected from a 2500-m depth hydrothermal vent community on the East Pacific Rise (EPR). Group II introns (8 total) were recovered from bacterial mat DNA only *(82)*. However, the authors were unable to identify the microbial host(s) carrying the various intron groups. In another study, the habitats of the polychaete *Alvinella pompejana* (i.e., secreted organic tubes) were collected from active black smokers at the EPR and genomic DNA was isolated for fosmid library construction *(83)*. Sequence analysis of fosmid clone Alv-FOS5 (39.8-kb), phylogenetically related to *Euryarchaeota* group II based on 16S rRNA analysis, revealed possible HGT between Bacteria and Archaea as a putative permease protein encoded on plasmids in *Nostoc* sp. was also detected on the fosmid clone *(83)*. In addition, phylogenetic incongruency and G + C content suggest that two open reading frames (ORFs) encoding putative chaperonin and sugar kinases proteins have been acquired from the *Thermococcales (83)*. Analysis of a 33.3-kb genome fragment, DeepAnt-EC39 representing Group I *Crenarchaeota* as determined by 16S rRNA analysis, recovered from Antarctic Polar Front picoplankton at 500 m depth, also indicated numerous HGT cases *(84)*. Analyses of DeepAnt-EC39 genes providing metabolic and physiological adaptations such as *menA*, encoding a menaquinone biosynthesis protein that acts as an electron carrier during anaerobic growth, indicate gene transfer between Bacteria and Archaea as the archaeal *menA* sequences were interspersed on the phylogenetic tree among bacterial sequences *(84)*.

Cultured isolates from the deep-sea also provide evidence for HGT as evidenced by phylogenetic incongruency. For example, *Halomonas variabilis*, a γ-proteobacterium, belongs to a group of microorganisms referred to as euryhaline halophiles, which are able to grow in environments with as much as 15% NaCl. Osmotic balance is controlled by ectoine, the major compatible solute vital to maintaining cell survival. Ectoine synthase, encoded by *etcC*, appears to have been acquired by HGT in *H. variabilis* strains isolated from different geographical deep-sea sites including the Mid-Atlantic Ridge, Nankai Trough, and Southern Mariana Trough *(85)*. For example, *ectC* from hydrothermal vent strains clustered with *ectC* from polar strains isolated from the Arctic Sea (3,190 m depth) and Antarctica ice-melt water. Additionally, *ectC* phylogenies were incongruent with phylogenies derived from 16S rRNA, *gyrB*, and *ectB*, suggesting HGT. *(85)*. HGT and gene fusion events have also been postulated to contribute to the survival and adaptation of a deep-sea hydrothermal vent dwelling, strict hydrogenotrophic, methanogenic archaea *(86)*. Although sulfite is known to inhibit methanogens, *Methanocaldococcus jannaschii* grows and generates methane when sulfite is the sole sulfur source *(86)*. *M. jannaschii* contains a novel coenzyme $F_{420}$-dependent sulfite reductase (Fsr), whose gene expression is induced by high levels by sulfite. The authors postulated that the unique chimeric structure of Fsr is a result of the fusion of horizontal transferred *dsrA* and *fqoF* genes, encoding a dissimilatory sulfite reductase and a $H_2F_{420}$ dehydrogenase subunit of $H_2F_{420}$–quinone oxidoreductase. HGT detected at an even greater depth, 11,000 m, has also been reported by Tamegai et al. *(87)*. *Pseudomonas* sp. strain MT-1, isolated from Mariana Trench sediments, was found to contain two genes within the *nar* gene cluster encoding for putative nitrate transporter proteins, *narK* and *narM*, which exhibited greater similarity to *narK* and *narM* genes from unrelated genera (e.g., *Halomonas, Synechococcus*) *(87)*.

## 5. Conclusions

HGT is, without question, a powerful force shaping individual marine microorganisms as well as entire marine microbial communities. By promoting the movement of genes throughout marine microbial populations, MGEs are exerting a direct effect on global ecological processes. The marine mobilome is a crucial source for rapid genome diversification, particularly when environmental stresses result in dire consequences to microbial survival and proliferation. Presently, information on the molecular functions (i.e., transfer, maintenance, host range, replication, and incompatibility) of marine plasmids is still needed to adequately assess the role of in situ plasmid-mediated HGT in marine microbial populations. Ongoing and future efforts to sequence marine plasmids and viruses and to identify and characterize marine MGE distribution and diversity will continue to reveal remarkable and unexpected aspects of gene flux in marine ecosystems.

## Acknowledgments

P. Sobecky would like to acknowledge the Office of Naval Research and the National Science Foundation (OCE-0085549) for providing prior grant support.

## References


1. Ochman, H., Lawrence, J. G., Groisman, E. A. (2000) Lateral gene transfer and the nature of bacterial innovation. *Nature* **405**, 299–304.
2. Mazel, J., Davies, J. (1999) Antibiotic resistance in microbes. *Cell Mol Life Sci* **56**, 742–54.
3. Tschape, H. (1994) The spread of plasmids as a function of bacterial adaptability. *FEMS Microbiol Ecol* **15**, 23–32.
4. Staley, J. T., Konopka, A. (1985) Measurement of *in situ* activities of nonphotosynthetic microorganisms in aquatic and terrestrial habitats. *Ann Rev Microbiol* **39**, 321–46.
5. Smalla, K., Sobecky, P. A. (2002) The prevalence and diversity of mobile genetic elements in bacterial communities of different environmental habitats: insights gained from different methodological approaches. *FEMS Microbiol Ecol* **42**, 165–75.
6. Maniatis, T., Fritsch, E. F., Sambrook, J. (1982) *Molecular cloning: a laboratory manual. Cold Spring Harbor Laboratory, Cold Spring Harbor, NY.*
7. Birnboim, H. C., Doly, J. (1979) A rapid alkaline extraction procedure for screening recombinant plasmid DNA. *Nucleic Acids* **7**, 1513–23.
8. Kieser, T. (1984) Factors affecting the isolation of ccc DNA from *Streptomyces lividans* and *Escherichia coli*. *Plasmid* **12**, 19–36.
9. Eckhardt, T. (1978) A rapid method for the identification of plasmid deoxyribonucleic acid in bacteria. *Plasmid* **1**, 584–8.
10. Giovannoni, S., Stingl, U. (2007) The importance of culturing bacterioplankton in the 'omics' age. *Nat Rev Microbiol* **5**, 820–6.
11. Sobecky, P. A., Mincer, T. J., Chang, M. C., Helinski, D. R. (1997) Plasmids isolated from marine sediment microbial communities contain replication and incompatibility regions unrelated to those of known plasmid groups. *Appl Environ Microbiol* **63**, 888–95.
12. Sobecky, P. A., Mincer, T. J., Chang, M. C., Toukdarian, A., Helinski, D. R. (1998) Isolation of broad-host-range replicons from marine sediment bacteria. *Appl Environ Microbiol* **64**, 2822–30.
13. Baya, A. M., Brayton, P. R., Brown, V. L., Grimes, D. J., Russek-Cohen, E., Colwell, R. R. (1986) Coincident plasmids and antimicrobial resistance in marine bacteria isolated from polluted and unpolluted Atlantic Ocean samples. *Appl Environ Microbiol* **51**, 1285–92.
14. Leahy, J. G., Somerville, C. C., Cunningham, K. A., Adamantiades, G. A., Byrd, J. J., Colwell, R. R. (1990) Hydrocarbon mineralization in sediments and plasmid incidence in sediment bacteria from the Campeche Bank. *Appl Environ Microbiol* **56**, 1565–70.
15. Belliveau, B. H., Starodub, M. E., Trevors, J. T. (1991) Occurrence of antibiotic and metal resistance and plasmids in *Bacillus* strains isolated from marine sediment. *Can J Microbiol* **37**, 513–20.
16. Sobecky, P. A. (2002) Approaches to investigating the ecology of plasmids in marine bacterial communities. *Plasmid* **48**, 213–21.
17. Lopez-Amoros, R., Vives-Rego, J., Garcia-Lara, J. (1997) Exogenous isolation of Hg-r plasmids from coastal Mediterranean waters and their effect on growth and survival of *Escherichia coli* in sea water. *Microbios* **92**, 3633–40.
18. Dahlberg, C., Linberg, C., Torsvik, V. L., Hermansson, M. (1997) Conjugative plasmids isolated from bacteria in marine environments show various degrees of homology to each other and are not closely related to well-characterized plasmids. *Appl Environ Microbiol* **63**, 4692–7.
19. Weinbauer, M. G. (2004) Ecology of prokaryotic viruses. *FEMS Microbiol Rev* **28**, 127–81.
20. Fuhrman, J. A. (1999) Marine viruses and their biogeochemical and ecological effects. *Nature* **399**, 541–8.
21. Suttle, C. A. (2007) Marine viruses – major players in the global ecosystem. *Nat Rev Microbiol* **5**, 801–12.
22. Wen, K., Ortmann, A. C., Suttle, C. A. (2004) Accurate estimation of viral abundance by epifluorescence microscopy. *Appl Environ Microbiol* **70**, 3862–7.
23. Wichels, A., Gerdts, G., Schutt, C. (2002) *Pseudoalteromonas* spp. phages, a significant

group of marine bacteriophages in the North Sea. *Aquatic Microb Ecol* **27**, 233–9.
24. Cook, M. A., Osborn, A. M., Bettandorff, J., Sobecky, P. A. (2001) Endogenous isolation of replicon probes for assessing plasmid ecology of marine sediment microbial communities. *Microbiology* **147**, 2089–101.
25. Cochran, P. K., Paul, J. H. (1998) Seasonal abundance of lysogenic bacteria in a subtropical estuary. *Appl Environ Microbiol* **64**, 2308–12.
26. Angly, F. E., Felts, B., Breitbart, M., Salamon, P., Edwards, R. A., Carlson, C., Chan, A. M., Haynes, M., Kelley, S., Liu, H., Mahaffy, J. M., Mueller, J. E., Nulton, J., Olson, R., Parsons, R., Rayhawk, S., Suttle, C. A., Rohwer, F. (2006) The marine viromes of four oceanic regions. *Plos Biol* **4**, 2121–31.
27. Jones, B. V., Marchesi, J. R. (2007) Transposon-aided capture (TRACA) of plasmids resident in the human gut mobile metagenome. *Nat Methods* **4**, 55–61.
28. Chen, C. Y., Wu, K. M., Chang, C., Chang, C. H., Tsai, H. C., Liao, T. L., Liu, Y. M., Chen, H. J., Shen, A. B., Li, J. C., Su, T. L., Shao, C. P., Lee, C. T., Hor, L. I., Tsai, S. F. (2003) Comparative genome analysis of *Vibrio vulnificus*, a marine pathogen. *Genome Res* **13**, 2577–87.
29. Klotz, M. G., Arp, D. J., Chain, P. S., El-Sheikh, A. F., Hauser, L. J., Hommes, N. G., Larimer, F. W., Malfatti, S. A., Norton, J. M., Poret-Peterson, A. T., Vergez, L. M., Ward, B. B. (2006) Complete genome sequence of the marine, chemolithoautotrophic, ammonia-oxidizing bacterium *Nitrosococcus oceani* ATCC 19707. *Appl Environ Microbiol* **72**, 6299–315.
30. Swingley, W. D., Sadekar, S., Mastrian, S. D., Matthies, H. J., Hao, J., Ramos, H., Acharya, C. R., Conrad, A. L., Taylor, H. L., Dejesa, L. C., Shah, M. K., O'huallachain, M. E., Lince, M. T., Blankenship, R. E., Beatty, J. T., Touchman, J. W. (2007) The complete genome sequence of *Roseobacter denitrificans* reveals a mixotrophic rather than photosynthetic metabolism. *J Bacteriol* **189**, 683–90.
31. Zhou, F., Olman, V., Xu, Y. (2008) Insertion sequences show diverse recent activities in Cyanobacteria and Archaea. *BMC Genomics* **9**, 36–47.
32. Breitbart, M., Salamon, P., Andresen, B., Mahaffy, J. M., Segall, A. M., Mead, D., Azam, F., Rohwer, F. (2002) Genomic analysis of uncultured marine viral communities. *Proc Natl Acad Sci U S A* **99**, 14250–5.
33. Culley, A. I., Lang, A. S., Suttle, C. A. (2006) Metagenomic analysis of coastal RNA virus communities. *Science* **312**, 1795–8.
34. Villemur, R. (1990) The DNA sequence and structural organization of the GC2 plasmid from the red alga *Gracilaria chilensis. Plant Mol Biol* **15**, 237–43.
35. Moon, D. A., Goff, L. J. (1997) Molecular characterization of two large DNA plasmids in the red alga *Porphyra pulchra. Curr Genetics* **32**, 132–8.
36. Casse, N., Bui, Q. T., Nicolas, V., Renault, S., Bigot, Y., Laulier, M. (2006) Species sympatry and horizontal transfer of *Mariner* transposons in marine crustacean genomes. *Mol Phylogenet Evol* **40**, 609–19.
37. Bult, C. J., White, O., lsen, G. J., Zhou, L., Fleischmann, R. D., Sutton, G. G., Blake, J. A., Fitzgerald, L. M., Clayton, R. A., Gocayne, J. D., Kerlavage, A. R., Dougherty, B. A., Tomb, J., Adams, M. D., Reich, C. I., Overbeek, R., Kirkness, E. F., Weinstock, K. G., Merrick, J. M., Glodek, A., Scott, J. L., Geoghagen, M, N. S., Weidman, J. F., Fuhrmann, J. L., Nguyen, D., Utterback, T. R., Kelley, J. M., Peterson, J. D., Sadow, P. W., Hanna, M. C., Cotton, M. D., Roberts, K. M., Hurst, M. A., Kaine, B. P., Borodovsky, M., Klenk, H., Fraser, C. M., Smith, H. O., Woese, C. R., Venter, J. C. (1996) Complete genome sequence of the methanogenic Archaeon, *Methanococcus jannaschii. Science* **273**, 1058–73.
38. Comeau, A. M., Chan, A. M., Suttle, C. A. (2006) Genetic richness of vibriophages isolated in a coastal environment. *Environ Microbiol* **8**, 1164–76.
39. Comeau, A. M., Buenaventura, E., Suttle, C. A. (2005) A persistent, productive, and seasonally dynamic vibriophage population within Pacific oysters (*Crassostrea gigas*). *Appl Environ Microbiol* **71**, 5324–31.
40. Shivu, M. M., Rajeeva, B. C., Girisha, S. K., Karunasagar, I., Krohne, G., Karunasagar, I. (2007) Molecular characterization of *Vibrio harveyi* bacteriophages isolated from aquaculture environments along the coast of India. *Environ Microbiol* **9**, 322–31.
41. Beeson, K. E., Erdner, D. L., Bagwell, C. E., Lovell, C. R., Sobecky, P. A. (2002) Differentiation of plasmids in marine diazotroph assemblages determined by randomly amplified polymorphic DNA analysis. *Microbiology* **148**, 179–89.
42. Lindell, D., Sullivan, M. B., Johnson, Z. I., Tolonen, A. C., Rohwer, F., Chisholm, S. (2004) Transfer of photosynthesis genes to and from *Prochlorococcus* viruses. *Proc NatlAcad Sci U S A* **101**, 11013–8.
43. Zhong, Z., Toukdarian, A., Helinski, D., Knauf, V., Sykes, S., Wilkinson, J. E., O'bryne, C., Shea, T., Deloughery, C., Caspi, R.

(2001) Sequence analysis of a 101-kilobase plasmid required for agar degradation by a *Microscilla* isolate. *Appl Environ Microbiol* **67**, 5771–9.
44. Zhong, Z., Caspi, R., Mincer, T., Helinski, D., Knauf, V., Boardman, K., Wilkinson, J. E., Shea, T., Deloughery, C., Toukdarian, A. (2002) A 50-kb plasmid rich in mobile gene sequences isolated from a marine micrococcus. *Plasmid* **47**, 1–9.
45. Short, C. M., Suttle, C. A. (2005) Nearly identical bacteriophage structural gene sequences are widely distributed in both marine and freshwater environments. *Appl Environ Microbiol* **71**, 480–6.
46. Waldor, M. K., Tschäpe, H., Mekalanos J. J. (1996) A new type of conjugative transposon encodes resistance to sulfamethoxazole, trimethoprim, and streptomycin in *Vibrio cholerae* O139. *J Bacteriol* **178**, 4157–65.
47. Ahmed, A. M., Shinoda, S., Shimamoto, T. (2005) A variant type of *Vibrio cholerae* SXT element in a multidrug-resistant strain of *Vibrio fluvialis*. *FEMS Microbiol Lett* **242**, 241–7.
48. Geslin, C., Gaillard, M., Flament, D., Rouault, K., Romancer, M. L., Prieur, D., Erauso, G. (2007) Analysis of the first genome of a hyperthermophilic marine virus-like particle, PAV1, isolated from *Pyrococcus abyssi*. *J Bacteriol* **189**, 4510–9.
49. Zhong, Z., Caspi, R., Helinski, D., Knauf, V., Sykes, S., O' Byrne, C., Shea, T. P., Wilkinson, J. E., De Loughery, C., Toukdarian, A. (2003) Nucleotide sequence based characterizations of two cryptic plasmids from the marine bacterium *Ruegeria* isolate PR1b. *Plasmid* **49**, 233–52.
50. Purdy, A., Rohwer, F., Edwards, R., Azam, F., Bartlett, D. H. (2005) A glimpse into the expanded genome content of *Vibrio cholerae* through identification of genes present in environmental strains. *J Bacteriol* **187**, 2992–3001.
51. Zhang, R., Wang, Y., Leung, P. C., Gu, J. D. (2007) pVC, a small cryptic plasmid from the environmental isolate of *Vibrio cholerae* MP-1. *J Microbiol* **45**, 193–8.
52. Dunn, A. K., Martin, M. O., Stabb, E. V. (2005) Characterizatioon of pES213, a small mobilizable plasmid from *Vibrio fischeri*. *Plasmid* **54**, 114–34.
53. Hazen, T. H., Wu, D., Eisen, J. A., Sobecky, P. A. (2007) Sequence characterization and comparative analysis of three plasmids isolated from environmental *Vibrio* spp. *Appl Environ Microbiol* **73**, 7703–10.
54. Mcdaniel, L., Paul, J. H. (2005) Effect of nutrient addition and environmental factors on prophage induction in natural populations of marine *Synechococcus* species. *Appl Environ Microbiol* **71**, 842–50.
55. Arrigo, K. R. (2005) Marine microorganisms and global nutrient cycles. *Nature* **437**, 349–55.
56. Waldor, M. K., Mekalanos, J. J. (1996) Lysogenic conversion by a filamentous phage encoding cholera toxin. *Science* **272**, 1910–4.
57. Nasu, H., Iida, T., Sugahara, T., Yamaichi, Y., Park, K. S., Yokoyama, K., Makino, K., Shinagawa, H., Honda, T. (2000) A filamentous phage associated with recent pandemic *Vibrio parahaemolyticus* O3: K6 strains. *J Clin Microbiol* **38**, 2156–61.
58. Miller, E. S., Heidelberg, J. F., Eisen, J. A., Nelson, W. C., Durkin, A. S., Ciecko, A., Feldblyum, T. V., White, O., Paulsen, I. T., Nierman, W. C., Lee, J., Szczypinski, B., Fraser, C. M. (2003) Complete genome sequence of the broad-host-range vibriophage KVP40: Comparative genomics of a T4-related bacteriophage. *J Bacteriol* **185**, 5220–33.
59. Comeau, A. M., Suttle, C. A. (2007) Distribution, genetic richness, and phage sensitivity of *Vibrio* spp. from coastal British Columbia. *Environ Microbiol* **9**, 1790–800.
60. Nishibuchi, M., Kaper, J. B. (1995) Thermostable direct hemolysin gene of *Vibrio parahaemolyticus*: a virulence gene acquired by a marine bacterium. *Infec Immun* **63**, 2093–9.
61. Vadivelu, J., Puthucheary, S. D., Mitin, A., Wan, C. Y., Van Melle, B., Puthucheary, J. A. (1996) Hemolysis and plasmid profiles of *Vibrio parahaemolyticus*. *Southeast Asian J Trop Med Pub Health* **27**, 126–31.
62. Di Lorenzo, M., Stork, M., Tolmasky, M. E., Actis, L. A., Farrell, D., Welch, T. J., Crosa, L. M., Wertheimer, A. M., Chen, Q., Salinas, P., Waldbeser, L., Crosa, J. H. (2003) Complete sequence of virulence plasmid pJM1 from the marine fish pathogen *Vibrio anguillarum* strain 775. *J Bacteriol* **185**, 5822–30.
63. Wu, H., Ma, Y., Zhang, Y., Zhang, H. (2004) Complete sequence of virulence plasmid pEIB1 from the marine fish pathogen *Vibrio anguillarum* strain MVM425 and location of its replication region. *J Appl Microbiol* **97**, 1021–8.
64. Lee, C., Amaro, C., Sanjuán, E., Hor, L. (2005) Identification of DNA sequences specific for *Vibrio vulnificus* Biotype 2 strains by suppression subtractive hybridization. *Appl Environ Microbiol* **71**, 5593–7.

65. Lee, C. T., Amaro, C., Wu, K. M., Valiente, E., Change, Y. F., Tsai, S. F., Chang, C. H., Hor, L. I. (2008) A common virulence plasmid in biotype 2 *Vibrio vulnificus* and its dissemination aided by a conjugal plasmid. *J Bacteriol* **190**, 1638–48.
66. Culley, A. I., Lang, A. S., Suttle, C. A. (2003) High diversity of unknown picorna-like viruses in the sea. *Nature* **424**, 1054–7.
67. Culley, A. I., Lang, A. S., Suttle, C. A. (2006) Metagenomic analysis of coastal RNA virus communities. *Science* **312**, 1795–8.
68. Koenig, J. E., Boucher, Y., Charlebois, R. L., Nesbo, C., Zhaxybayeva, O., Bapteste, E., Spencer, M., Joss, M. J., Stokes, H. W., Doolittle, W. F. (2008) Integron-associated gene cassettes in Halifax Harbour: assessment of a mobile gene pool in marine sediments. *Environ Microbiol* **doi:10:1111/j.1462-2920.20007.01524.x**.
69. Nesbo, C. L., Boucher, Y., Dlutek, M., Doolittle, W. F. (2005) Lateral gene transfer and phylogenetic assignment of environmental fosmid clones. *Environ Microbiol* 7, 2011–26.
70. Biers, E. J., Wang, K., Pennington, C., Belas, R., Chen, F., Moran, M. A. (2008) Occurrence and expression of gene transfer agent (GTA) genes in marine bacterioplankton. *Appl Environ Microbiol* **doi:10.1128/AEM.02129-07**.
71. Lang, A. S., Beatty, J. T. (2007) Importance of widespread gene transfer agent genes in alpha-proteobacteria. *Trends Microbiol* **15**, 54–62.
72. Veldhuis, M. J. W., Kraay, G. W., Vanbleijswijk, J. D., L., Baars, M. A. (1997) Seasonal and spatial variability in phytoplankton biomass, productivity, and growth in the northwestern Indian Ocean: the southwest and northeast monsoon, 1992–1993. *Deep-Sea Res Part I-Oceanographic Res Papers* **44**, 425–49.
73. Zeidner, G., Bielawski, J. P., Shmoish, M., Scanlan, D. J., Sabehi, G., Beja, O. (2005) Potential photosynthesis gene recombination between *Prochlorococcus* and *Synechococcus* via viral intermediates. *Environ Microbiol* 7, 1505–13.
74. Millard, A., Clokie, M. R. J., Shub, D. A., Mann, N. H. (2004) Genetic organization of the *psbAD* region in phages infecting marine *Synechococcus* strains. *Proc Natl Acad Sci U S A* **101**, 11007–12.
75. Mann, N. H., Cook, A., Millard, A., Bailey, S., Clokie, M. (2003) Marine ecosystems: Bacterial photosynthesis genes in a virus. *Nature* **424**, 741.
76. He, Q. F., Dolganov, N., Bjorkman, O., Grossman, A. R. (2001) The high light-inducible polypeptides in *Synechocystis* PCC6803 – Expression and function in high light. *J Biol Chem* **276**, 306–14.
77. Frigaard, N. U., Martinez, A., Mincer, T. J., Delong, E. F. (2006) Proteorhodopsin lateral gene transfer between marine planktonic Bacteria and Archaea. *Nature* **439**, 847–50.
78. Mccarren, J., Delong, E. F. (2007) Proteorhodopsin photosystem gene clusters exhibit co-evolutionary trends and shared ancestry among diverse marine microbial phyla. *Environ Microbiol* **9**, 846–58.
79. Nelson, K. E., Clayton, R. A., Gill, S. R., Gwinn, M. L., Dodson, R. J., Haft, D. H., Hickey, E. K., Peterson, L. D., Nelson, W. C., Ketchum, K. A., Mcdonald, L., Utterback, T. R., Malek, J. A., Linher, K. D., Garrett, M. M., Stewart, A. M., Cotton, M. D., Pratt, M. S., Phillips, C. A., Richardson, D., Heidelberg, J., Sutton, G. G., Fleischmann, R. D., Eisen, J. A., White, O., Salzberg, S. L., Smith, H. O., Venter, J. C., Fraser, C. M. (1999) Evidence for lateral gene transfer between Archaea and Bacteria from the genome sequence of *Thermotoga maritima*. *Nature* **399**, 323–9.
80. Jorgensen, B. B., Boetius, A. (2007) Feast and famine-microbial life in the deep-sea bed. *Nat Microbiol* **5**, 770–81.
81. Munoz, E., Villadas, P. J., Toro, N. (2001) Ectopic transposition of a group II intron in natural bacterial populations. *Mol Microbiol* **41**, 645–52.
82. Podar, M., Mullineaux, L., Huang, H. R., Perlman, P. S., Sogin, M. L. (2002) Bacterial group II introns in a deep-sea hydrothermal vent environment. *Appl Environ Microbiol* **68**, 6392–8.
83. Moussard, H., Moreira, D., Cambon-Bonavita, M. A., Lopez-Garcia, P., Jeanthon, C. (2006) Uncultured Archaea in a hydrothermal microbial assemblage: phylogenetic diversity and characterization of a genome fragment from a euryarchaeote. *FEMS Microbiol Ecol* **57**, 452–69.
84. Lopez-Garcia, P., Brochier, C., Moreira, D., Rodriguez-Valera, F. (2004) Comparative analysis of a genome fragment of an uncultivated mesopelagic crenarchaeote reveals multiple horizontal gene transfers. *Environ Microbiol* **6**, 19–34.
85. Okamoto, T., Maruyama, A., Imura, S., Takeyama, H., Naganuma, T. (2004) Comparative phylogenetic analyses of *Halomonas variabilis* and related organisms based on 16S rRNA, *gyrB*, and *ectBC* gene sequences. *Syst Appl Microbiol* **27**, 323–33.

86. Johnson, E. F., Mukhopadhyay, B. (2005) A new type of sulfite reductase, a novel coenzyme F-420-dependent enzyme, from the methanarchaeon *Methanocaldococcus jannaschii*. *J Biol Chem* **280**, 38776–86.

87. Tamegai, H., Kato, C., Horikoshi, K. (2004) Lateral gene transfer in the deep sea of Mariana trench: identification of *nar* gene cluster encoding membrane-bound nitrate reductase from *Pseudomonas* sp. strain MT-1. *DNA Sequence* **15**, 338–43.

# Chapter 26

## Horizontal Gene Transfer in Metal and Radionuclide Contaminated Soils

**Patricia A. Sobecky and Jonna M. Coombs**

### Abstract

The horizontal transfer of genes encoded on mobile genetic elements (MGEs) such as plasmids and phage and their associated hitchhiking elements (transposons, integrons, integrative and conjugative elements, and insertion sequences) rapidly accelerate genome diversification of microorganisms, thereby affecting their physiology, metabolism, pathogenicity,and ecological character. The analyses of completed prokaryotic genomes reveal that horizontal gene transfer (HGT) continues to be an important factor contributing to the innovation of microbial genomes. Indeed, microbial genomes are remarkably dynamic and a considerable amount of genetic information is inserted or deleted by HGT mechanisms. Thus, HGT and the vast pool of MGEs provide microbial communities with an unparalleled means by which to respond rapidly to changing environmental conditions and exploit new ecological niches. Metals and radionuclide contamination in soils, the subsurface, and aquifers poses a serious challenge to microbial growth and survival because these contaminants cannot be transformed or biodegraded into non-toxic forms as often occurs with organic xenobiotic contaminants. In this chapter we present cases in which HGT has been demonstrated to contribute to the dissemination of genes that provide adaptation to contaminant stress (i.e., toxic heavy metals and radionuclides). In addition, we present directions for future studies that could provide even greater insights into the contributions of HGT to adaptation for survival in mixed waste sites.

**Key words:** Horizontal gene transfer, bioremediation, efflux, radionuclide, heavy metal, subsurface.

## 1. Introduction

The treatment of hazardous mixed waste sites, especially those co-contaminated with heavy metals and radionuclides, remains one of the most costly environmental challenges currently faced by the United States and other countries. The disposal of such wastes has traditionally involved underground storage tanks, shallow

Maria B. Gogarten et al. (eds.), *Horizontal Gene Transfer: Genomes in Flux, vol. 532*

DOI 10.1007/978-1-60327-853-9_26 Springerprotocols.com

land burial pits, and trenches *(1)*. When leaks occur, these wastes come into contact with the surrounding geologic media, allowing for migration of radiological and co-occurring metal contaminants into nearby soils and groundwater. Radionuclides are unstable elements that spontaneously decay emitting radiation which can damage DNA, proteins and tissues. Metals, while necessary for metabolic processes at low concentrations, become toxic at higher concentrations and can accumulate in the food chain. The mobility and solubility of radionuclides, such as uranium (U), technetium (Tc), and metals [e.g., lead (Pb), chromium (Cr), mercury (Hg)], is dependent upon their oxidation state and chemical speciation. Microorganisms that promote the transformation and precipitation of radionuclides and metals show promise for the remediation of contaminated soils, groundwater, and waste streams *(2–5)*. Numerous studies have demonstrated the enzymatically catalyzed precipitation of insoluble phases of U *(6, 7)*, Tc *(8)*, Cr *(9, 10)*, and selenium *(11)* by microbial reduction processes. In addition to these reductive bio-precipitation processes, insoluble mineral forms of radionuclides and metals can also be formed through interactions with microbially produced sulfide *(12, 13)*, phosphate *(14–16)*, or iron through bacterial iron oxidation *(17)* in a process termed biomineralization. In contrast to microbial reductive precipitation which requires anaerobic conditions, biomineralization can occur under aerobic conditions, potentially making this process a suitable remediation strategy for radionuclides and metals in contaminated and oxygenated zones.

Metals may enter the bacterial cell cytoplasm through transport systems for essential metals *(18, 19)*. Within the cytoplasm, metals can be transformed into less toxic forms by reductases *(20–22)* or changed from organocompounds to inorganic forms through the action of lyases *(22)*. In addition, efflux systems may function to reduce harmful concentrations of metals in the cell cytoplasm. Types of efflux systems include soft-metal ($P_{IB}$-type) ATPases *(23, 24)*, cation–proton antiport *(25)*, and uniport systems powered by either electrochemical gradients or hydrolysis of ATP *(26)*. The ability of microorganisms to adapt, survive, and proliferate when exposed to a cocktail of contaminants is facilitated by effective "genome evolution" strategies. While point mutations can contribute to microbial adaptation, the horizontal dissemination of genes has proven to be more critical in promoting rapid genomic flexibility and microbial evolution *(27)*. Horizontal gene transfer (HGT) driven by mobile genetic elements (MGEs), such as plasmids *(28)*, insertion sequences *(29)*, integrons *(30)*, transposons *(31)*, and phages *(32)*, has indeed been shown to provide microbes with a wide variety of adaptive traits for microbial survival and proliferation (e.g., antibiotic and heavy metal resistance, diverse metabolic capabilities including xenobiotic degradation, and virulence factors). In this chapter we highlight cases of HGT promoting microbial adaptation to heavy metals and radionuclides. We also present other examples of selectively advantageous genes that could undergo HGT and significantly contribute to microbial survival in mixed wastes sites.

## 2. HGT in Metal-Contaminated Soils

HGT has, without a doubt, played a major role in the dissemination of metal-resistance determinants among microbial communities. Metal-resistance ($Me^R$) genes were first detected on plasmids occurring in diverse bacteria, e.g., *(33, 34)*. Subsequently, transposons have been shown to promote the acquisition of these plasmid-encoded $Me^R$ genes *(35)*. Plasmid-encoded mercury (*mer*) resistance operons, for example, are frequently associated with Class II transposons *(36,37)*, which are characterized by 36–48 bp terminal inverted repeats and which carry the genes *tnpA* and *tnpR* for transposition. These elements can often encode co-occurring antibiotic-resistance ($Ab^R$) genes *(35, 38)*, and there is increasing evidence that heavy metal contamination due to anthropogenic sources is contributing to the dissemination of $Ab^R$ genes by either co-selecting for $Ab^R$ bacteria carrying $Me^R$ genes, located on the same mobile genetic element, or by selecting for cross-resistance encoded by multidrug efflux transporters that export metals and antibiotics *(39, 40)*. The following sections will describe specific cases of HGT-mediated metal resistance in soils and the subsurface (**Sections 2.1** and **2.2**) and by the comparative sequence analysis of genes involved in arsenic resistance (**Section 2.3**).

### 2.1. HGT of mer Systems

Among the best studied metal resistance loci is the *mer* operon consisting of up to seven genes involved in the transport, catalysis, and regulation of mercury resistance ($Hg^R$) *(37)*. Exogenous plasmid isolation, developed by Hill et al. *(41)*, has been effectively used to isolate conjugative $Hg^R$ plasmids from soil bacterial populations. The method employs a genetically marked recipient strain, typically a *Pseudomonas* sp., to recover conjugative and mobilizable (i.e., not self-transferable) plasmids directly from microbial communities. While information regarding the origin of the plasmid host will not be obtained, the technique provides a means to access MGEs from the microbial community, including those bacteria that are not readily cultured. This method was subsequently used by Lilley et al. *(42)* in a pioneering study to isolate plasmids from microbial communities associated with the rhizosphere (roots) and phyllosphere (leaves) of sugar beets. Five different and novel $Hg^R$ plasmid groups were initially detected and three of these groups have been re-isolated in subsequent years. The plasmids conferred narrow-spectrum $Hg^R$ via a reductase detoxification mechanism. Exogenous isolation has also been used to examine the effect of the presence of mercury and soil types on plasmid incidence and gene-mobilizing capacity by Dronen et al. *(43)*. Of eight soils tested, only one yielded $Hg^R$ transconjugants using *P. putida* UWC1 as the recipient. However, after mercury addition to the soil, $Hg^R$ transconjugants were obtained from five of eight soils. A similar study was conducted with soil bacterial populations associated with wheat roots *(44)*. When soils were amended with mercury, novel plasmid groups conferring $Hg^R$ were recovered using *P. fluorescens* R2f, *P. putida* UWC1 and *Enterobacter cloacae* BE1 as recipients *(44)*. These studies highlight the endemic nature of MGEs conferring $Hg^R$ phenotypes among soil microbial communities.

### 2.2. HGT of $P_{IB}$-Type ATPases

There are three known mechanisms for promoting microbial heavy metal resistance or tolerance: (1) metal reduction *(6)*, (2) metal complexation *(45)*, and (3) ATP-dependent metal efflux *(46)*. The P-type ATPases are a chromosomally encoded superfamily of ion translocating proteins found in all three domains of life. P-type ATPases are essential for the transport of ions, including toxic (mono- and divalent) ions such as Cd(II), Zn(II), Cu(II), Ag(I), Cu(I), and Pb(II) from cells, acting as a "basic defense" to actively efflux heavy metal cations *(24, 47)*. The prokaryotic heavy metal translocating subgroup of P-type ATPases, the $P_{IB}$-type, detoxifies the cell cytoplasm by effluxing Cd(II), Co(II), Pb(II), Ni(II), and Zn(II) *(24, 47, 48)*. The $P_{IB}$-type ATPase genes have been found to occur on MGEs (i.e., plasmids and transposons) in both Gram-positive *(49, 50)* and Gram-negative bacteria *(51)*.

A number of studies have recently examined the sequence contexts of $P_{IB}$-type ATPase genes from bacterial isolates cultured from subsurface soils with an extensive history of radionuclide and heavy metal contamination, cultures from the deep (uncontaminated) terrestrial subsurface and sequenced genomes for evidence of HGT *(52–54)*. The bacterial isolates were obtained from contaminated soils collected from the U.S. DOE Field Research Center (ORFRC) located within the Oak Ridge National Laboratory Reservation (Oak Ridge, TN). The ORFRC subsurface represents an extreme geochemical environment that provides a number of stresses on the extant microbial community, which include low pH (e.g., $< 4$), nitrate concentrations that can exceed 100 mM, as well as co-occurring heavy metals and radionuclides (U and other actinides; http://www.esd.ornl.gov/nabirfrc) *(55)*.

Cultures from the deep, uncontaminated subsurface were obtained from the Subsurface Microbial Culture Collection (SMCC) *(56)*.

In contrast to $P_{IB}$-type ATPases genes analyzed from SMCC cultures and whole genomes, numerous $P_{IB}$-type ATPases from ORFRC isolates were determined to have unusual DNA base composition and phylogenetic incongruence *(52)*. The majority of $P_{IB}$-type ATPases (60%) obtained from ORFCR isolates exhibited some evidence of recent HGT. For example, the *zntA/cadA/pbrA*-like gene sequences amplified from two Actinobacteria isolates grouped in one of the two bifurcated nodes within the Firmicutes (**Fig. 26.1**). In addition, these isolates contained *zntA/cadA/pbrA*-like genes with a significantly lower G + C content than expected for other *Arthrobacter* spp. [35–36 mol% instead of 59–70 mol%] (**Table 26.1**). A second set of $P_{IB}$-type ATPases amplified from four Actinobacteria isolates (**Table 26.1**, **Fig. 26.1**) clustered with ZntA/ CadA/PbrA-like sequences most closely related to the second bifurcated node within the Firmicutes (**Fig. 26.1**). These four isolates also exhibited unusual G + C contents (38 mol%) (**Table 26.1**). The ATPase-related sequences amplified from three isolates, belonging to the *Firmicutes*, clustered within one distinct γ-*Proteobacteria* ZntA/CadA/PbrA-like clade (**Fig. 26.1**). All three strains contained *zntA/cadA/pbrA*-like genes with a G + C content of 58–59 mol% (**Table 26.1**). This content varied considerably from the most closely related cultured isolate, *Bacillus cereus*, which has a 32 mol% G + C. Together, such phylogenetic incongruencies and unusual G + C content provide evidence for the dissemination of horizontally acquired $P_{IB}$-type ATPase genes (**Fig. 26.1**).

### 2.3. HGT of Arsenate Reductase

Arsenic (As) occurs in four oxidation states ($As^{+5}$, $As^{+3}$, $As^{0}$, and $As^{-3}$) and can be used by diverse microorganisms for cell growth and metabolism. A global biogeochemical As cycle has recently been described *(57)* and microbial metabolic activity has been shown to play a key role in the mobilization/immobilization of arsenic compounds *(58)*. For example, As(V) can be used as an electron acceptor in anaerobic respiration by dissimilatory arsenate-respiring prokaryotes (DARPs) *(59)*. However, depending upon the geochemistry of the environment, DARP-mediated reductive mobilization of soluble As(III) to the aqueous phase can occur. DARPs occur in diverse phylogenetic lineages include archaea, Gram-positives, *alpha*-, *delta*-, and *epsilonproteobacteria (59)*. In addition to the DARP-mediated reduction of arsenic, As(V) can also be reduced enzymatically to As(III), without energy gain, as a mechanism of resistance via arsenate reductases present in many microorganisms. The arsenate

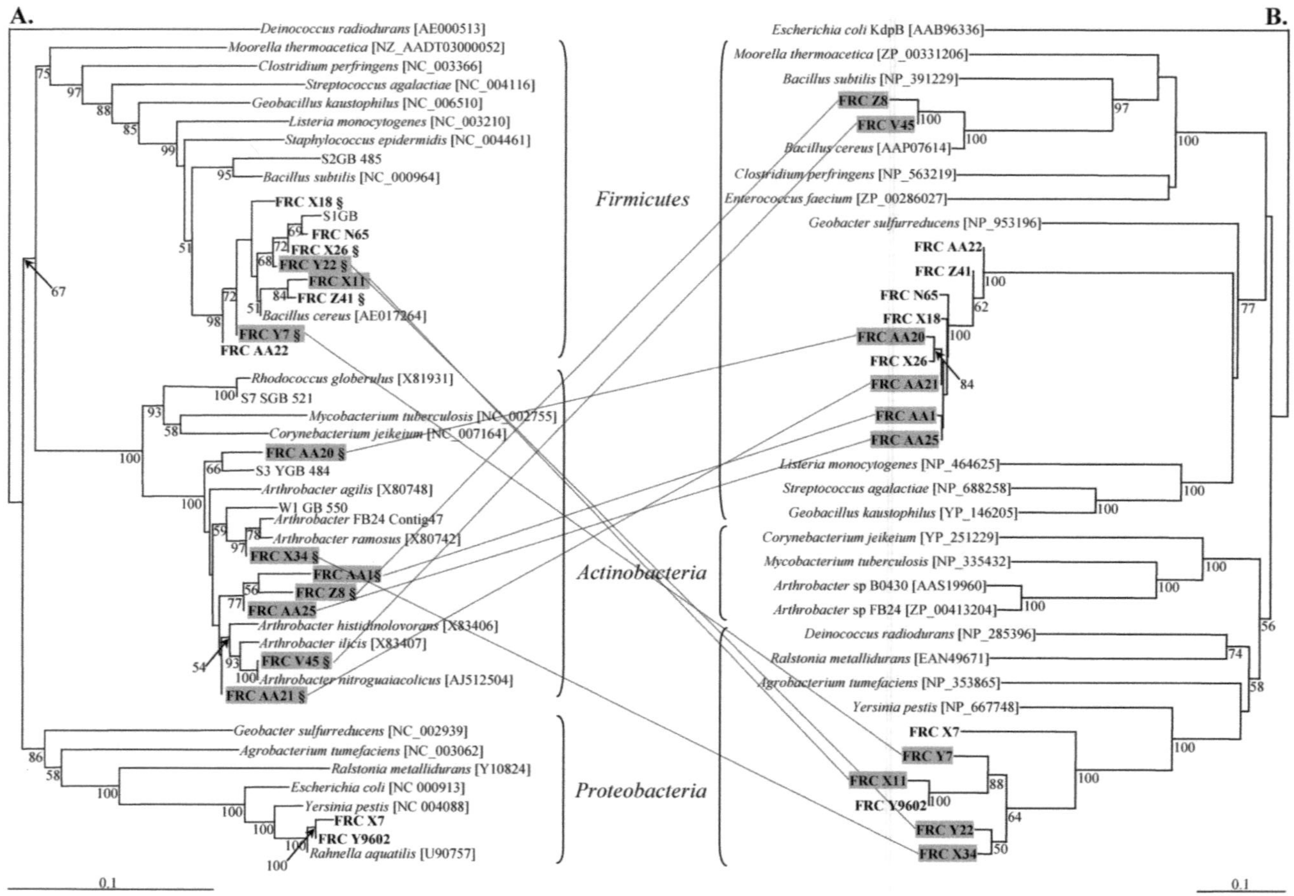


Fig. 26.1. Neighbor-joining analysis of (**A**) 16S rDNA and (**B**) *zntA/cadA/prbA*-like sequences from either subsurface FRC isolates or from completed genomes. Accession numbers are in parentheses; § denotes FRC strain containing one or more plasmids. Subsurface isolates shown in *shaded boxes* and connected by a *dotted line* are positive for horizontal acquired $P_{IB}$-type ATPases related to *zntA/cadA/prbA*-loci. Bootstrap support $> 50\%$ is shown. Scale bars for 16S rDNA and *zntA/cadA/prbA* phylogenies represent 0.1 changes per nucleotide position and 0.1 changes per amino acid position, respectively. It should be noted that *Deinococcus radiodurans* and *Geobacter sulfurreducens* indicate evidence of HGT of $P_{IB}$-type ATPases. Although previous studies have identified evidence of HGT within these species, we chose only to highlight our cultured FRC isolates that exemplified HGT through atypical base composition and phylogenetic incongruency *(54, 122, 123)*. Reprinted with permission from ASM Press.

**Table 26.1**
**Acquisition of $P_{IB}$-type ATPases genes by HGT in subsurface *Arthrobacter* and *Bacillus* isolates from mixed waste (radionuclide and heavy metal) contaminated soils**

| Genus | Strain | G + C content (%) | $P_{IB}$-type ATPase gene most closely related to |
|---|---|---|---|
| *Arthrobacter* | FRC-AA1 | 38 | *Firmicutes* |
| | FRC-AA20 | 38 | *Firmicutes* |
| | FRC-AA21 | 38 | *Firmicutes* |
| | FRC-AA25 | 38 | *Firmicutes* |
| | FRC-V45 | 36 | *Firmicutes* |
| | FRC-X34 | 59 | *γ-Proteobacteria* |
| | FRC-Z8 | 35 | *Firmicutes* |
| *Bacillus* | FRC-X11 | 59 | *γ-Proteobacteria* |
| | FRC-Y7 | 58 | *γ-Proteobacteria* |
| | FRC-Y22 | 58 | *γ-Proteobacteria* |

reductases of DARPs, however, differ from the arsenate reductases of As-resistant microorganisms *(57)*.

The *ars* operon encodes a detoxification pathway for arsenic and can be chromosomally or plasmid-encoded *(60)*. The genes for bacterial plasmid and chromosomal encoded As resistance system have been described in detail by Silver et al. *(61)*. The operon contains numerous genes including *arsC*, which encodes arsenate reductase, a soluble enzyme that reduces arsenate to arsenite *(58)*. The arsenite-specific transmembrane efflux pump, encoded by *arsB*, removes arsenite from the cell. The most frequently detected *ars* gene systems are the *arsRBC* set typically found on chromosomes *(62)* and the *arsRDABC* set encoded on plasmids such as R773 *(63)*. Phylogenetic analysis of more than 400 *arsC* gene sequences supports the role of HGT in the evolution and dissemination of arsenate reductase *(64)*. For example, evidence for HGT of *arsC* from ancestral low GC Gram-positives to Cyanobacteria as well as from Actinobacteria to *Streptococci* was detected. A recent HGT event was also proposed for the *arsC* gene encoded on plasmid pKW301 present in *Acidophilum multivorum* as the plasmid-encoded gene grouped within the Enterobacteriales *(64)*.

## 3. Potential for HGT in Radionuclide-Contaminated Soils

Radionuclides pose significant environmental contamination problems for several reasons. (i) Radionuclides are heavy metals and therefore cell exposure results in toxicity effects in addition to damage caused by radioactive decay. (ii) Radionuclides cannot be broken down or detoxified through transformation, as is possible with xenobiotics and some heavy metals, making immobilization or sequestration the most effective bioremediation method. (iii) Radionuclides often co-occur with other environmental contaminants, meaning that any surviving organisms in affected environments must be multiple-resistant. (iv) Bacteria have limited resistance mechanisms for radionuclides. Efficient DNA repair mechanisms convey resistance to the DNA damage occurring from radiation damage from radionuclides in *Deinococcus radiodurans (65)*, but these systems play no direct role in contaminant immobilization. Alternatively, dissimilatory metal-reducing bacteria (DSMR) such as *Geobacter sulfurreducens (6, 66, 67)* and sulfate-reducing bacteria such as *Desulfovibrio desulfuricans (8)* possess electron shuttle systems that function in the immobilization of radionuclides through reduction to less mobile forms [i.e., reducing highly soluble U(VI) to highly insoluble U(IV)], and are currently the model organisms for radionuclide bioremediation studies in the environment. The following sections describe the current state of knowledge with respect to the mechanisms of radionuclide and metal reduction, and provide possibilities for future studies in the field.

### *3.1. Direct Enzymatic Reduction Pathways and HGT*

Reduction of radionuclides, and of some heavy metals such as Cr(VI), is mediated in two possible ways. Indirect reduction has not yet been demonstrated in situ, but could theoretically occur when Fe(II), Mn(II), and $H_2S$ are produced by microbes during anaerobic respiration. The oxidation of these metabolic products to Fe(III), Mn(IV), and $SO_4^{-2}$ could serve to reduce materials such as U(VI) and Tc(VII). The alternative is direct enzymatic reduction. The latter process is currently not completely understood; however, periplasmic and outer membrane *c*-type cytochromes clearly play a significant role in both DSMR *(68–70)* and in sulfate-reducing bacteria such as *Desulfovibrio* spp. *(71, 72)*. Models for enzymatic reduction pathways in key organisms are starting to emerge *(73)*.

Although there is no direct evidence of HGT involving genes responsible for enzymatic reduction, work with mitochondrial-like *c*-type cytochromes (those *c*-type proteins possessing a mitochondrial-like fold) indicates that HGT of the *c*-type cytochrome does indeed occur *(74)*. One of the cyochromes involved in U(VI) reduction in *Geobacter sulfurreducens*, OmcF *(70)*, has an atypical %mol G + C content (73.3% compared to an average of 60.9% for the *G. sulfurreducens* genome) and possesses 53–59% sequence similarity to mitochondrial-like *c*-type cytochromes belonging to the $c_6$ group *(74, 75)*. The $c_6$ cytochrome group is found predominantly in cyanobacteria, where they play an important role in both photosystem I and in aerobic respiration *(76)*. An analysis of 235 bacterial genomes detected $c_6$-like cytochromes in 9 cyanobacteria, *G. sulfurreducens*, and *Nitrosomonas europaea (74)*. This patchy distribution observed with $c_6$-like cytochromes is common among other bacterial mitochondrial-like *c*-type cytochromes *(74)*, and may be an indication of HGT within the $c_6$ cytochrome group. However, caution should be used in interpreting cytochrome distribution data, as genes encoding these heme-containing proteins are prone to duplication and loss, the latter of which would also result in patchy distribution among phylogenetic groups. Future work focusing on these and other genes encoding proteins involved in radionuclide reduction, including the Ni/Fe hydrogenases of sulfate-reducing bacteria *(69, 77)* and post-translationally modified multi-heme *c*-type cytochromes (*c*-Cyt; *(73)*, should provide a more detailed picture of the potential for HGT related to radionuclide bioremediation.

## 4. HGT in Mixed Waste Sites

The term 'mixed waste' generally refers to anthropogenic contamination consisting of organic chemicals and radionuclides, although other combinations of hazardous materials are common. The mixing of more than one biohazardous compound at a waste site is significant because co-contaminants may interact with each other to facilitate or interfere with chemical transformation or transport, and variation in electron acceptors can cause the development of distinct redox zones over relatively small spatial scales *(78, 79)*. Perhaps most significant for the consideration of HGT, the presence of more than one toxic compound can alter the indigenous microbial populations *(80, 81)* and therefore can negatively influence bioremediation efforts at hazardous waste sites. Analysis of genetic elements encoding tolerance to more than one type of toxic compound may indicate the capacity of a community to adapt to mixed waste contamination in the environment and serve as a starting point for the engineering of organisms such as *Deinococcus radiodurans (82, 83)* for the bioremediation of mixed wastes. The following sections will review MGEs mediating multiple resistances and discuss one of the few retrospective studies performed with isolates from a mixed waste site.

### 4.1. Mobile Elements Encoding Genes Relevant for the Bioremediation of Mixed Wastes

Metal resistance and catabolic genes are often associated with mobile elements. This has been the subject of several excellent reviews *(35, 51, 84, 85)*. Transposons that have been sequenced from environmental samples to date appear to carry only catabolic genes (reviewed in *(86)*) or only metal-resistance genes *(35, 87–89)*. While some plasmids, such as pCAR1 and NAH7, encode only catabolic genes *(90, 91)* and others such as pMOL28 encode only metal-resistance genes *(92)*, there are a few plasmids that contain genes for both (select plasmids are presented in **Table 26.2**). Of this list, pJP4, pWW0, and pUO1 have all demonstrated conjugative ability on artificial media *(93–95)*, and mobilization of pJP4 (**Table 26.2**) has further been demonstrated in soil *(93)* and in bioreactors containing 2,4-D or a combination of 2,4-D and cadmium *(96)*. Interestingly, in a simulated mixed waste soil amended with 2,4-D and Cd, the Cd as a source of metal stress did not significantly affect transconjugate formation *(96)*. The simulated mixed-waste study mirrors the work carried out with a modified TOL plasmid, in which cells were pre-exposed to a single toxicant, and it was determined that exposure to compounds such as toluene or phenol had no statistical effect on plasmid transfer frequency *(94)*. These studies indicate that exposure to toxicants does not have a stimulatory effect on the transfer of large catabolic plasmids by conjugation.

Studies with catabolic plasmids indicate that most appear to be mosaics, containing multiple transposons or insertion sequences *(97, 98)*. This is also evident in the plasmids listed in **Table 26.2**, as all of them contain one or more IS element or full transposon. In the case of IncP-1 plasmids, many of which carry catabolic genes and/or metal-resistance genes, the site of transposon insertion appears to be important to plasmid evolution, with most integrating near the origin of replication *(99)*. Host-dependent changes in plasmid gene expression, particularly of *trb, tra, klc,* and *kle* operons have been observed a plasmid of the IncP-1beta group *(100)*. Rearrangement of catabolic plasmids following conjugation has been observed with *Sphingomonas* strains, including transfer of genes to the host chromosome and alteration of the position of gene clusters *(101, 102)*. These studies indicate that the process of gene transfer itself influences plasmid evolution, particularly with respect to adaptation to new hosts *(100)*.

### 4.2. Retrospective Analysis of HGT in Mixed Waste Sites

While retrospective studies of HGT have been carried out previously in environments contaminated with single target wastes *(103, 104)*, the only current study of HGT at a mixed waste site involves the $P_{IB}$-type ATPases described in **Section 2.2** *(52)*. An analysis of phylogenetic incongruence and %mol G + C content detected several instances of HGT among high and low

**Table 26.2**
**Examples of partially or fully sequenced plasmids encoding catabolic pathway genes and metal resistance determinants**

| Plasmid | Strain | Relevant genes | Transposons/IS elements | References |
|---|---|---|---|---|
| pJP4 | *Ralstonia eutropha* JMP134 | *tfd* I and II operons, *mer* operon | IS*1071*, IS*JP4* | *(111, 112)* |
| pRHL3 | *Rhodococcus* RH3 | *lim* genes, carveol dehydrogenase, cobalt ABC transporter | 4 genome islands, 19 transposases, intact IS element | *(113)* |
| pIJB1 | *Burkholderia cepacia* 2a | *tfd, mdc, mer* | Tn*5530* | *(114, 115)* |
| pUO1 | *Delftia acidovorans* B | *dehH1, dehH2* | Tn*Had1*, Tn*Had2*, Tn*4671*, Tn*4672* | *(95, 116, 117)* |
| pXAU1 | *Xanthobacter autotrophicus* | *dhlA*, encodes mercury resistance | ND[1] | *(118)* |
| pCAR3 | *Sphingomonas* sp. KA1 | *carR, carABC (others)* | Tn*4676* | *(90, 119)* |
| pBD2 | *Rhodococcus erythropolis* BD2 | *ipbA1A2A3A4C cadA, copA, arsABCD* | 32 ORFs similar to IS or transposase genes | *(120, 121)* |
| pWW0 | *Pseudomonas putida* | *merB, merR, xyl*, $Ni^r$ | Tn*4651*, Tn*4653* | *(98)* |

[1]ND = Not determined.

G + C Gram-positive organisms from mixed-waste bacteria, many of which carried large plasmids and were also able to withstand toxic concentrations of U(IV) at a pH of 4 *(52)*. The frequency of HGT identified by phylogenetic incongruency was higher among isolates from the contaminated site (10 instances among 28 sequenced PCR amplicons) compared to an uncontaminated site on the Atlantic Coastal Plain, Savannah River, GA, observed in a previous study (4 instances of HGT among 48 PCR amplicons) *(53)*. Although the two studies looked at different populations of microorganisms, precluding the direct comparison of results, these studies clearly indicate that dissemination of metal-resistance genes by HGT is occurring among subsurface bacteria isolated from both pristine and mixed waste sediments.

## 5. Conclusion

The occurrence of HGT at sites contaminated with organics, heavy metals, and radionuclides has a significant impact on the design and implementation of bioremediation strategies, particularly with respect to the introduction of engineered organisms, or in harnessing the indigenous microbial population for intrinsic (natural) bioremediation or biostimulation. Complete genome sequences are currently available for several organisms that are of interest with respect to the cleanup of legacy waste sites. The iron-reducing bacterium *Geobacter metallireducens* contains a 300-bp genomic island encoding genes predicted to be involved in the degradation of phenol, benzoate, and other organic compounds *(105)*. *P. putida (106)*, *Shewanella oneidensis* MR-1 *(107)*, and strain EbN1 *(108)* all contain diverse arrays of transposon related genes, inteins, and prophages. *Burkholderia xenovorans* LB400 is estimated to have acquired >20% of its genome by HGT *(109)*. In addition, the presence of integrons *(110)* and of large plasmids encoding catabolic and multiple-resistance functions (**Table 26.2**) in several strains that degrade organics, sequester metals, and precipitate radionuclides demonstrates the capacity for the mobilization of these traits in the environment. Clearly, HGT has played an important role in the past in microbes capable of carrying out bioremediation. The frequency of HGT and the prevalence of MGEs in environmental samples indicates that this mechanism of microbial evolution will continue to shape the catabolic and resistance phenotypes of soil and subsurface microorganisms in the future.

## Acknowledgments

Work in the Sobecky lab on horizontal gene transfer in the subsurface has been supported by grants from the Department of Energy.

### References

1. Department of Energy (DOE). (1995) Estimating the cold war mortgage: The 1995 baseline environmental management report (2 vol.). DOE/EM-0230.
2. Barkay, T., Schaefer, J. (2001) Metal and radionuclide bioremediation: Issues, considerations and potentials. *Curr Opin Microbiol* **4**, 318–23.
3. Stephen, J. R., Macnaughton, S. J. (1999) Developments in terrestrial bacterial remediation of metals. *Curr Opin Biotechnol* **10**, 230–3.
4. Lovley, D. R., Coates, J. D. (1997) Bioremediation of metal contamination. *Curr Opin Biotechnol* **8**, 285–9.
5. Gadd, G. M. (2000) Bioremedial potential of microbial mechanisms of metal mobilization and immobilization. *Curr Opin Biotechnol* **11**, 271–9.
6. Lovley, D. R., Phillips, E. J. P., Gorby, Y. A., Landa, E. R. (1991) Microbial reduction of uranium. *Nature* **350**, 413–6.
7. Lovley, D. R., Phillips, E. J. P. (1992) Bioremediation of uranium contamination with enzymatic uranium reduction. *Environ Sci Technol* **26**, 2228–34.
8. Lloyd, J. R., Ridley, J., Khizniak, T., Lyalikova, N. N., Macaskie, L. E. (1999) Reduction of technetium by *Desulfovibrio desulfuricans*: Biocatalyst characterization and use in a flowthrough bioreactor. *Appl Environ Microbiol* **65**, 2691–6.
9. Smith, W. L., Gadd, G. M. (2000) Reduction and precipitation of chromate by mixed

culture sulphate-reducing bacterial biofilms. *J Appl Microbiol* **88**, 983–91.
10. Turick, C. E., Graves, C., Apel, W.A. (1998) Bioremediation potential of Cr(VI)-contaminated soil using indigenous microorganisms. *Bioremediation* **2**, 1–6.
11. Tucker, M. D., Barton, L. L., Thomson, B. M. (1998) Reduction of Cr, Mo, Se and U by *Desulfovibrio desulfuricans* immobilized in polyacrylamide gels. *J Ind Microbiol Biotechnol* **20**, 13–9.
12. Lebranz, M., Druschel, G. K., Thomsen-Ebert, T., Gilbert, B., Welch, S. A., Kemner, K. M., Logan, G. A., Summons, R. E., De Stasio, G., Bond, P.L., Lai, B., Kelly, S. D., Banfield, J. F. (2000) Formation of sphalerite (ZnS) deposits in natural biofilms of sulfate-reducing bacteria. *Science* **290**, 1744–7.
13. White, C., Sharman, A. K., Gadd, G. M. (1998) An integrated microbial process for the bioremediation of soil contaminated with toxic metal. *Nat Biotechnol* **16**, 572–5.
14. Macaskie, L. E., Bonthrone, K. M., Yong, P., Goddard, D. T. (2000) Enzymically mediated bioprecipitation of uranium by a *Citrobacter* sp.: A concerted role for exocellular lipopolysaccharide and associated phosphatase in biomineral formation. *Microbiology* **146**, 1855–67.
15. Martinez, R. J., Beazley, M. J., Taillefert, M., Arakaki, A. K., Skolnick, J., Sobecky, P. A. (2007) Aerobic uranium (VI) bioprecipitation by metal-resistant bacteria isolated from radionuclide- and metal-contaminated subsurface soils. *Environ Microbiol* **9**, 3122–33.
16. Beazley, M. J., Martinez, R. J., Sobecky, P. A., Webb, S. M., Taillefert, M. (2007) Uranium biomineralization as a result of bacterial phosphatase activity: Insights from bacterial isolates from a contaminated subsurface. *Environ Sci Technol* **41**, 5701–7.
17. Banfield, J. F., Welch, S. A., Zhang, H. Z., Ebert, T. T., Penn, R. L. (2000) Aggregation-based crystal growth and microstructure development in natural iron oxyhydroxide biomineralization products. *Science* **289**, 751–4.
18. Nies, D. H., Silver, S. (1989) Metal ion uptake by plasmid-free metal-sensitive *Alcaligenes eutrophus*. *J Bacteriol* **171**, 896–900.
19. Silver, S., Keach, D. (1982) Energy-dependent arsenate efflux: The mechanisms of plasmid-mediated resistance *Proc Natl Acad Sci U S A* **79**, 6114–8.
20. Gladysheva, T. B., Oden, K. L., Rosen, B. P. (1994) Properties of the arsenate reductase of plasmid R773. *Biochemistry* **33**, 7288–93.
21. Ji, G., Silver, S. (1992) Reduction of arsenate to arsenite by the ArsC protein of the arsenic reductase operon of *Staphylococcus aureus* plasmid pI258. *Proc Natl Acad Sci U S A* **89**, 9474–8.
22. Walsh, C. T., Distefano, M. D., Moore, M., Shewchuk, L. M., Verdine, G. L. (1988) Molecular basis of bacterial resistance to organomercurial and inorganic mercury salts. *FASEB J* **2**, 124–30.
23. Rosen, B. (1999) The role of efflux in bacterial resistance to soft metals and metalloids. *In: Essays in Biochemistry, Vol 34: Metalloproteins*; S. J. Higgins and D. P. Ballou, eds., Portland Press, Ltd., London, pp 1–15.
24. Rensing, C., Ghosh, M., Rosen, B. P. (1999) Families of soft-metal-ion-transporting ATPases. *J Bacteriol* **181**, 5891–7.
25. Munkelt, D., Grass, G., Nies, D. H. (2004) The chromosomally encoded cation diffusion facilitator proteins DmeF and FieF from *Wautersia metallidurans* CH34 are transporters of broad metal specificity. *J Bacteriol* **186**, 8036–43.
26. Kuroda, M., Dey, S., Sanders, O. I., Rosen, B. P. (1997) Alternate energy coupling of ArsB, the membrane subunit of the Ars anion-translocating ATPase. *J Biol Chem* **272**, 326–31.
27. Thomas, C. M., Nielsen, K. M. (2005) Mechanisms of, and barriers to, horizontal gene transfer between bacteria. *Nat Rev Microbiol* **3**, 711–21.
28. Frost, L. S., Leplae, R., Summers, A. O., Toussaint, A. (2005) Mobile genetic elements: The agents of open source evolution. *Nat Rev Microbiol* **3**, 722–32.
29. Mahillon, J., Chandler, M. (1998) Insertion sequences. *Microbiol Mol Biol Rev* **62**, 725–74.
30. Nemergut, D. R., Martin, A. P., Schmidt, S. K. (2004) Integron diversity in heavy-metal-contaminated mine tailings and inferences about integron evolution. *Appl Environ Microbiol* **70**, 1160–8.
31. Pearson, A. J., Bruce, K. D., Osborn, A. M., Ritchie, D. A., Strike, P. (1996) Distribution of Class II transposase and resolvase genes in soil bacteria and their association with *mer* genes. *Appl Environ Microbiol* **62**, 2961–5.
32. Canchaya, C., Fournous, G., Chibani-Chennoufi, S., Dillmann, M., Brüssow, H. (2003) Phage as agents of lateral gene transfer. *Curr Opin Microbiol* **6**, 417–24.
33. Summers, A. O., Silver, S. (1972) Mercury resistance in a plasmid-bearing strain of *Escherichia-coli*. *J Bacteriol* **112**, 1228–36.
34. Silver, S., Budd, K., Leahy, K. M., Shaw, W. V., Hammond, D., Novick, R. P., Willsky, G. R., Malamy, M. H., Rosenberg, H. (1981) Inducible plasmid-determined

resistance to arsenate, arsenite, and antimony(III) in *Escherichia coli* and *Staphylococcus aureus*. *J Bacteriol* **146**, 983–96.

35. Liebert, C. A., Hall, R. M., Summers, A. O. (1999) Transposon Tn*21*, flagship of the floating genome. *Microbiol Mol Biol Rev* **63**, 507–22.
36. Grinsted, J., Delacruz, F., Schmitt, R. (1990) The Tn*21* subgroup of bacterial transposable elements *Plasmid* **24**, 163–89.
37. Barkay, T., Miller, S. M., Summers, A. O. (2003) Bacterial mercury resistance from atoms to ecosystems. *FEMS Microbiol Rev* **27**, 355–84.
38. Gilmour, M. W., Thomson, N. R., Sanders, M., Parkhill, J., Taylor, D. E. (2004) The complete nucleotide sequence of the resistance plasmid R478: Defining the backbone components of incompatibility group H conjugative plasmids through comparative genomics. *Plasmid* **52**, 182–202.
39. Baker-Austin, C., Wright, M. S., Stepanauskas, R., McArthur, J. V. (2006) Co-selection of antibiotic and metal resistance. *Trend Microbiol* **14**, 176–82.
40. Wright, M. S., Peltier, G. L., Stepanauskas, R., McArthur, J. V. (2006) Bacterial tolerances to metals and antibiotics in metal-contaminated and reference streams. *FEMS Microbiol Ecol* **58**, 293–302.
41. Hill, K. E., Weightman, A. J., Fry, J. C. (1992) Isolation and screening of plasmids from the epilithon which mobilize recombinant plasmid pd10. *Appl Environ Microbiol* **58**, 1292–300.
42. Lilley, A. K., Bailey, M. J., Day, M. J., Fry, J. C. (1996) Diversity of mercury resistance plasmids obtained by exogenous isolation from the bacteria of sugar beet in three successive years. *FEMS Microbiol Ecol* **20**, 211–27.
43. Dronen, A. K., Torsvik, V., Goksoyr, J., Top, E. M. (1998) Effect of mercury addition on plasmid incidence and gene mobilizing capacity in bulk soil. *FEMS Microbiol Ecol* **27**, 381–94.
44. Smit, E., Wolters, A., Van Elsas, J. D. (1998) Self-transmissible mercury resistance plasmids with gene-mobilizing capacity in soil bacterial populations: Influence of wheat roots and mercury addition. *Appl Environ Microbiol* **64**, 1210–19.
45. Macaskie, L. E., Empson, R. M., Cheetham, A. K., Grey, C. P., Skarnulis, A. J. (1992) Uranium bioaccumulation by a *Citrobacter* sp. as a result of enzymatically mediated growth of polycrystalline $HUO_2PO_4$. *Science* **257**, 782–84.
46. Nies, D. H. (1999) Microbial heavy-metal resistance. *Appl Microbiol Biotechnol* **51**, 730–50.
47. Nies, D. H. (2003) Efflux-mediated heavy metal resistance in prokaryotes. *FEMS Microbiol Rev* **27**, 313–39.
48. Axelsen, K. B., Palmgren, M. G. (2001) Inventory of the superfamily of P-type ion pumps in *Arabidopsis*. *Plant Physiol* **126**, 696–706.
49. Nucifora, G., Chu, L., Misra, T. K., Silver, S. (1989) Cadmium resistance from *Staphylococcus aureus* plasmid PI258 *cadA* gene results from a cadmium-efflux ATPase. *Proc Natl Acad Sci U S A* **86**, 3544–48.
50. O'Sullivan, D., Ross, R. P., Twomey, D. P., Fitzgerald, G. F., Hill, C., Coffey, A. (2001) Naturally occurring *Lactococcal* plasmid pAH90 links bacteriophage resistance and mobility functions to a food-grade selectable marker. *Appl Environ Microbiol* **67**, 929–37.
51. Mergeay, M., Monchy, S., Vallaeys, T., Auquier, V., Benotmane, A., Bertin, P., Taghavi, S., Dunn, J., Van Der Lelie, D., Wattiez, R. (2003) *Ralstonia metallidurans*, a bacterium specifically adapted to toxic metals: Towards a catalogue of metal-responsive genes. *FEMS Microbiol Rev* **27**, 385–410.
52. Martinez, R. J., Wang, Y. L., Raimondo, M. A., Coombs, J. M., Barkay, T., Sobecky, P. A. (2006) Horizontal gene transfer of $P_{IB}$-type ATPases among bacteria isolated from radionuclide- and metal-contaminated subsurface soils. *Appl Environ Microbiol* **72**, 3111–18.
53. Coombs, J. M., Barkay, T. (2004) Molecular evidence for the evolution of metal homeostasis genes by lateral gene transfer in bacteria from the deep terrestrial subsurface. *Appl Environ Microbiol* **70**, 1698–707.
54. Coombs, J. M., Barkay, T. (2005) New findings on evolution of metal homeostasis genes: Evidence from comparative genome analysis of bacteria and archaea. *Appl Environ Microbiol* **71**, 7083–91.
55. Petrie, L., North, N. N., Dollhopf, S. L., Balkwill, D. L., Kostka, J. E. (2003) Enumeration and characterization of iron(III)-reducing microbial communities from acidic subsurface sediments contaminated with uranium(VI). *Appl Environ Microbiol* **69**, 7467–79.
56. Benyehuda, G., Coombs, J., Ward, P. L., Balkwill, D., Barkay, T. (2003) Metal resistance among aerobic chemoheterotrophic bacteria from the deep terrestrial subsurface. *Can J Microbiol* **49**, 151–156.

57. Mukhopadhyay, R., Rosen, B. P., Phung, L. T., Silver, S. (2002) Microbial arsenic: From geocycles to genes and enzymes. *FEMS Microbiol Rev* **26**, 311–25.
58. Stolz, J. E., Basu, P., Santini, J. M., Oremland, R. S. (2006) Arsenic and selenium in microbial metabolism. *Ann Rev Microbiol* **60**, 107–30.
59. Oremland, R. S., Stolz, J. F. (2005) Arsenic, microbes and contaminated aquifers. *Trends Microbiol* **13**, 45–9.
60. Silver, S. (1996) Bacterial resistances to toxic metal ions – A review. *Gene* **179**, 9–19.
61. Silver, S., Phung, L. T., Rosen, B. P. (2001) Arsenic metabolism: Resistance, reduction and oxidation. *In: Environmental chemistry of arsenic*, W. T. Frankenberg Jr., ed., Marcel Dekker, Inc., New York, pp 247–72
62. Carlin, A., Shi, W. P., Dey, S., Rosen, B. P. (1995) The *ars* operon of *Escherichia coli* confers arsenical and antimonial resistance. *J Bacteriol* **177**, 981–6.
63. Chen, C. M., Misra, T. K., Silver, S., Rosen, B. P. (1986) Nucleotide-sequence of the structural genes for an anion pump – the plasmid-encoded arsenical resistance operon. *J Biol Chem* **261**, 5030–8.
64. Jackson, C. R., Dugas, S. L. (2003) Phylogenetic analysis of bacterial and archaeal arsC gene sequences suggests an ancient, common origin for arsenate reductase. *BMC Evol Biol* **3**, 18–27.
65. Zahradka, K., Slade, D., Bailone, A., Sommer, S., Averbeck, D., Petranovic, M., Lindner, A. B., Radman, M. (2006) Reassembly of shattered chromosomes in Deinococcus radiodurans. *Nature* **443**, 569–73.
66. Liu, C. X., Gorby, Y. A., Zachara, J. M., Fredrickson, J. K., Brown, C. F. (2002) Reduction kinetics of Fe(III), Co(III), U(VI) Cr(VI) and Tc(VII) in cultures of dissimilatory metal-reducing bacteria. *Biotechnol Bioeng* **80**, 637–49.
67. Lloyd, J. R., Sole, V. A., Van Praagh, C. V. G., Lovley, D. R. (2000) Direct and Fe(II)-mediated reduction of technetium by Fe(III)-reducing bacteria. *Appl Environ Microbiol* **66**, 3743–9.
68. Marshall, M. J., Beliaev, A. S., Dohnalkova, A. C., Kennedy, D. W., Shi, L., Wang, Z. M., Boyanov, M. I., Lai, B., Kemner, K. M., Mclean, J. S., Reed, S. B., Culley, D. E., Bailey, V. L., Simonson, C. J., Saffarini, D. A., Romine, M. F., Zachara, J. M., Fredrickson, J. K. (2006) c-Type cytochrome-dependent formation of U(IV) nanoparticles by *Shewanella oneidensis*. *Plos Biol* **4**, 1324–33.
69. Marshall, M. J., Plymale, A. E., Kennedy, D. W., Shi, L., Wang, Z., Reed, S. B., Dohnalkova, A. C., Simonson, C. J., Liu, C., Saffarini, D. A., Romine, M. F., Beliaev, A. S., Fredrickson, J. K. (2008) Hydrogenase- and outer membrane *c*-type cytochrome-facilitated reduction of technetium(VII) by *Shewanella oneidensis* MR-1. *Environ Microbiol* **10**, 125–136.
70. Shelobolina, E. S., Coppi, M. V., Korenevsky, A. A., Didonato, L. N., Sullivan, S. A., Konishi, H., Xu, H. F., Leang, C., Butler, J. E., Kim, B. C., Lovley, D. R. (2007) Importance of *c*-type cytochromes for U(VI) reduction by *Geobacter sulfurreducens*. *BMC Microbiol* **7**, 16.
71. Lovley, D. R., Widman, P. K., Woodward, J. C., Phillips, E. J. P. (1993) Reduction of uranium by cytochrome-c(3) of *Desulfovibrio vulgaris*. *Appl Environ Microbiol* **59**, 3572–6.
72. Payne, R. B., Casalot, L., Rivere, T., Terry, J. H., Larsen, L., Giles, B. J., Wall, J. D. (2004) Interaction between uranium and the cytochrome c(3) of *Desulfovibrio desulfuricans* strain G20. *Arch Microbiol* **181**, 398–406.
73. Shi, L., Squier, T. C., Zachara, J. M., Fredrickson, J. K. (2007) Respiration of metal (hydr)oxides by *Shewanella* and *Geobacter*: A key role for multihaem *c*-type cytochromes. *Mol Microbiol* **65**, 12–20.
74. Bertini, I., Cavallaro, G., Rosato, A. (2007) Evolution of mitochondrial-type cytochrome c domains and of the protein machinery for their assembly. *J Inorg Biochem* **101**, 1798–811.
75. Kim, B. C., Leang, C., Ding, Y. H. R., Glaven, R. H., Coppi, M. V., Lovley, D. R. (2005) OmcF, a putative *c*-type monoheme outer membrane cytochrome required for the expression of other outer membrane cytochromes in *Geobacter sulfurreducens*. *J Bacteriol* **187**, 4505–13.
76. Duran, R. V., Herv S, M., De La Rosa, M. A., Navarro, J. A. (2004) The efficient functioning of photosynthesis and respiration in *Synechocystis* sp. PCC 6803 strictly requires the presence of either cytochrome c(6) or plastocyanin. *J Biol Chem* **279**, 7229–33.
77. De Luca, G., De Philip, P., Dermoun, Z., Rousset, M., Vermeglio, A. (2001) Reduction of technetium(VII) by *Desulfovibrio fructosovorans* is mediated by the nickel-iron hydrogenase. *Appl Environ Microbiol* **67**, 4583–7.
78. Cozzarelli, I. M., Weiss, J. V. (2007) *Biochemistry of aquifer systems*. ASM Press, Washington, DC.
79. Barber, L. B., II, Thurman, E. M., Runnells, D. D. (1992) Geochemical heterogeneity in

a sand and gravel aquifer: Effect of sediment mineralogy and particle size on the sorption of chlorobenzenes. *J Contam Hydrol* **9**: 35–54.

80. Bekins, B. A., Cozzarelli, I. M., Godsy, E. M., Warren, E., Essaid, H. I., Tuccillo, M. E. (2001) Progression of natural attenuation processes at a crude oil spill site: II. Controls on spatial distribution of microbial populations. *J Contam Hydrol* **53**, 387–406.
81. Lin, Q., Zhao, H. M., Chen, Y. X. (2007) Effects of 2,4-dichlorophenol, pentachlorophenol and vegetation on microbial characteristics in a heavy metal polluted soil. *J Environ Sci Health B* **42**, 551–7.
82. Lange, C. C., Wackett, L. P., Minton, K. W., Daly, M. J. (1998) Engineering a recombinant *Deinococcus radiodurans* for organopollutant degradation in radioactive mixed waste environments. *Nat Biotechnol* **16**, 929–33.
83. Brim, H., Mcfarlan, S. C., Fredrickson, J. K., Minton, K. W., Zhai, M., Wackett, L. P., Daly, M. J. (2000) Engineering *Deinococcus radiodurans* for metal remediation in radioactive mixed waste environments. *Nat Biotechnol* **18**, 85–90.
84. Springael, D., Top, E. M. (2004) Horizontal gene transfer and microbial adaptation to xenobiotics: New types of mobile genetic elements and lessons from ecological studies. *Trends Microbiol* **12**, 53–8.
85. Nojiri, H., Shintani, M., Omori, T. (2004) Divergence of mobile genetic elements involved in the distribution of xenobiotic catabolic capacity. *Appl Microbiol Biotechnol* **64**, 154–74.
86. Wyndham, R. C., Cashore, A. E., Nakatsu, C. H., Peel, M. C. (1994) Catabolic transposons. *Biodegradation* **5**, 323–42.
87. Mindlin, S., Kholodii, G., Gorlenko, Z., Minakhina, S., Minakhin, L., Kalyaeva, E., Kopteva, A., Petrova, M., Yurieva, O., Nikiforov, V. (2001) Mercury resistance transposons of gram-negative environmental bacteria and their classification. *Res Microbiol* **152**, 811–22.
88. Bogdanova, E., Minakhin, L., Bass, I., Volodin, A., Hobman, J. L., Nikiforov, V. (2001) Class II broad-spectrum mercury resistance transposons in Gram-positive bacteria from natural environments. *Res Microbiol* **152**, 503–14.
89. Kholodii, G., Gorlenko, Z., Mindlin, S., Hobman, J., Nikiforov, V. (2002) Tn*5041*-like transposons: Molecular diversity, evolutionary relationships and distribution of distinct variants in environmental bacteria. *Microbiology* **148**, 3569–82.
90. Maeda, K., Nojiri, H., Shintani, M., Yoshida, T., Habe, H., Omori, T. (2003) Complete nucleotide sequence of carbazole/dioxin-degrading plasmid pCAR1 in *Pseudomonas resinovorans* strain CA10 indicates its mosaicity and the presence of large catabolic transposon Tn*4676*. *J Mol Biol* **326**, 21–33.
91. Sota, M., Yano, H., Ono, A., Miyazaki, R., Ishii, H., Genka, H., Top, E. M., Tsuda, M. (2006) Genomic and functional analysis of the IncP-9 naphthalene-catabolic plasmid NAH7 and its transposon Tn*4655* suggests catabolic gene spread by a tyrosine recombinase. *J Bacteriol* **188**, 4057–67.
92. Copeland, A., Lucas, S., Lapidus, A., Barry, K., Detter, J. C., Glavina Del Rio, T., Hammon, N., Israni, S., Dalin, E., Tice, H., Martinez, M., Goltsman, E., Pitluck, S., Schmutz, J., Larimer, F., Land, M., Hauser, L., Kyrpides, N., Kim, E., Mergeay, M., Bentomane, M. A., Vallaeyes, T., Michaux, A., Monchy, S., Dunn, J., Mccorkle, S., Taghavi, S., Van Der Lelie, D., Richardson, P. (2007) Complete sequence of plasmid pMOL28 of *Ralstonia metallidurans* CH34. Unpublished.
93. Neilson, J. W., Josephson, K. L., Pepper, I. L., Arnold, R. B., Digiovanni, G. D., Sinclair, N. A. (1994) Frequency of horizontal gene-transfer of a large catabolic plasmid (pJP4) in soil. *Appl Environ Microbiol* **60**, 4053–58.
94. Pinedo, C. A., Smets, B. F. (2005) Conjugal TOL transfer from *Pseudomonas putida* to *Pseudomonas aeruginosa*: Effects of restriction proficiency, toxicant exposure, cell density ratios, and conjugation detection method on observed transfer efficiencies. *Appl Environ Microbiol* **71**, 51–7.
95. Kawasaki, H., Yahara, H., Tonomura, K. (1981) Isolation and characterization of plasmid pUO1 mediating dehalogenation of haloacetate and mercury resistance in Moraxella sp. B. *Agri Biol Chem* **45**, 1477–81.
96. Newby, D. T., Gentry, T. J., Pepper, I. L. (2000) Comparison of 2,4-dichlorophenoxyacetic acid degradation and plasmid transfer in soil resulting from bioaugmentation with two different pJP4 donors. *Appl Environ Microbiol* **66**, 3399–407.
97. Schluter, A., Heuer, H., Szczepanowski, R., Forney, L. J., Thomas, C. M., Puhler, A., Top, E. M. (2003) The 64,508 bp IncP-1 beta antibiotic multiresistance plasmid pB10 isolated from a waste-water treatment plant provides evidence for recombination between members of different branches

of the IncP-1 beta group. *Microbiology* **149**, 3139–53.
98. Greated, A., Lambertsen, L., Williams, P. A., Thomas, C. M. (2002) Complete sequence of the IncP-9 TOL plasmid pWW0 from *Pseudomonas putida*. *Environ Microbiol* **4**, 856–71.
99. Sota, M., Tsuda, M., Yano, H., Suzuki, H., Forney, L. J., Top, E. M. (2007) Region-specific insertion of transposons in combination with selection for high plasmid transferability and stability accounts for the structural similarity of IncP-1 plasmids. *J Bacteriol* **189**, 3091–8.
100. Heuer, H., Fox, R. E., Top, E. M. (2007) Frequent conjugative transfer accelerates adaptation of a broad-host-range plasmid to an unfavorable *Pseudomonas* putida host. *FEM Microbiol Ecol* **59**, 738–48.
101. Basta, T., Buerger, S., Stolz, A. (2005) Structural and replicative diversity of large plasmids from Sphingomonads that degrade polycyclic aromatic compounds and xenoblotics. *Microbiology* **151**, 2025–37.
102. Basta, T., Keck, A., Klein, J., Stolz, A. (2004) Detection and characterization of conjugative degradative plasmids in xenobiotic-degrading *Sphingomonas* strains. *J Bacteriol* **186**, 3862–72.
103. Wilson, M. S., Herrick, J. B., Jeon, C. O., Hinman, D. E., Madsen, E. L. (2003) Horizontal transfer of *phnAc* dioxygenase genes within one of two phenotypically and genotypically distinctive naphthalene-degrading guilds from adjacent soil environments. *Appl Environ Microbiol* **69**, 2172–81.
104. Herrick, J. B., Stuartkeil, K. G., Ghiorse, W. C., Madsen, E. L. (1997) Natural horizontal transfer of a naphthalene dioxygenase gene between bacteria native to a coal tar-contaminated field site. *Appl Environ Microbiol* **63**, 2330–7.
105. Butler, J. E., He, Q., Nevin, K. P., He, Z. L., Zhou, J. Z., Lovley, D. R. (2007) Genomic and microarray analysis of aromatics degradation in *Geobacter metallireducens* and comparison to a *Geobacter* isolate from a contaminated field site. *BMC Genomics* **8**, 180.
106. Nelson, K. E., Weinel, C., Paulsen, I. T., Dodson, R. J., Hilbert, H., Dos Santos, V., Fouts, D. E., Gill, S. R., Pop, M., Holmes, M., Brinkac, L., Beanan, M., Deboy, R. T., Daugherty, S., Kolonay, J., Madupu, R., Nelson, W., White, O., Peterson, J., Khouri, H., Hance, I., Lee, P. C., Holtzapple, E., Scanlan, D., Tran, K., Moazzez, A., Utterback, T., Rizzo, M., Lee, K., Kosack, D., Moestl, D., Wedler, H., Lauber, J., Stjepandic, D., Hoheisel, J., Straetz, M., Heim, S., Kiewitz, C., Eisen, J., Timmis, K. N., Dusterhoft, A., Tummler, B., Fraser, C. M. (2002) Complete genome sequence and comparative analysis of the metabolically versatile *Pseudomonas putida* KT2440. *Environ Microbiol* **4**, 799–808.
107. Heidelberg, J. F., Paulsen, I. T., Nelson, K. E., Gaidos, E. J., Nelson, W. C., Read, T. D., Eisen, J. A., Seshadri, R., Ward, N., Methe, B., Clayton, R. A., Meyer, T., Tsapin, A., Scott, J., Beanan, M., Brinkac, L., Daugherty, S., Deboy, R. T., Dodson, R. J., Durkin, A. S., Haft, D. H., Kolonay, J. F., Madupu, R., Peterson, J. D., Umayam, L. A., White, O., Wolf, A. M., Vamathevan, J., Weidman, J., Impraim, M., Lee, K., Berry, K., Lee, C., Mueller, J., Khouri, H., Gill, J., Utterback, T. R., Mcdonald, L. A., Feldblyum, T. V., Smith, H. O., Venter, J. C., Nealson, K. H., Fraser, C. M. (2002) Genome sequence of the dissimilatory metal ion-reducing bacterium *Shewanella oneidensis*. *Nat Biotechnol* **20**, 1118–23.
108. Rabus, R. (2005) Functional genomics of an anaerobic aromatic-degrading denitrifying bacterium, strain EbN1. *Appl Microbiol Biotechnol* **68**, 580–7.
109. Chain, P. S. G., Denef, V. J., Konstantinidis, K. T., Vergez, L. M., Agullo, L., Reyes, V. L., Hauser, L., Cordova, M., Gomez, L., Gonzalez, M., Land, M., Lao, V., Larimer, F., Lipuma, J. J., Mahenthiralingam, E., Malfatti, S. A., Marx, C. J., Parnell, J. J., Ramette, A., Richardson, P., Seeger, M., Smith, D., Spilker, T., Sul, W. J., Tsoi, T. V., Ulrich, L. E., Zhulin, I. B., Tiedje, J. M. (2006) *Burkholderia xenovorans* LB400 harbors a multi-replicon, 9.73-Mbp genome shaped for versatility. *Proc Natl Acad Sci U S A* **103**, 15280–7.
110. Boucher, Y., Labbate, M., Koenig, J. E., Stokes, H. W. (2007) Integrons: Mobilizable platforms that promote genetic diversity in bacteria. *Trends Microbiol* **15**, 301–9.
111. Don, R. H., Pemberton, J. M. (1981) properties of 6 pesticide degradation plasmids isolated from *Alcaligenes paradoxus* and *Alcaligenes eutrophus*. *J Bacteriol* **145**, 681–6.
112. Trefault, N., De La Iglesia, R., Molina, A. M., Manzano, M., Ledger, T., Perez-Pantoja, D., Sanchez, M. A., Stuardo, M., Gonzalez, B. (2004) Genetic organization of the catabolic plasmid pJP4 from *Ralstonia eutropha* JMP134 (pJP4) reveals mechanisms of adaptation to chloroaromatic pollutants and evolution of specialized chloroaromatic degradation pathways. *Environ Microbiol* **6**, 655–68.

113. Warren, R., Hsiao, W. W. L., Kudo, H., Myhre, M., Dosanjh, M., Petrescu, A., Kobayashi, H., Shimizu, S., Miyauchi, K., Masai, E., Yang, G., Stott, J. M., Schein, J. E., Shin, H., Khattra, J., Smailus, D., Butterfield, Y. S., Siddiqui, A., Holt, R., Marra, M. A., Jones, S. J. M., Mohn, W. W., Brinkman, F. S. L., Fukuda, M., Davies, J., Eltis, L. D. (2004) Functional characterization of a catabolic plasmid from polychlorinated-biphenyl-degrading *Rhodococcus* sp strain RHA1. *J Bacteriol* **186**, 7783–95.
114. Xia, X. S., Aathithan, S., Oswiecimska, K., Smith, A. R. W., Bruce, I. J. (1998) A novel plasmid pIJB1 possessing a putative 2,4-dichlorophenoxyacetate degradative transposon Tn*5530* in *Burkholderia cepacia* strain 2a. *Plasmid* **39**, 154–9.
115. Poh, R. P. C., Smith, A. R. W., Bruce, I. J. (2002) Complete characterisation of Tn*5530* from *Burkholderia cepacia strain* 2a (pIJB1) and studies of 2,4-dichlorophenoxyacetate uptake by the organism. *Plasmid* **48**, 1–12.
116. Sota, M., Kawasaki, H., Tsuda, M. (2003) Structure of haloacetate-catabolic IncP-1 beta plasmid pUO1 and genetic mobility of its residing haloacetate-catabolic transposon. *J Bacteriol* **185**, 6741–5.
117. Sota, M., Endo, M., Nitta, K., Kawasaki, H., Tsuda, M. (2002) Characterization of a Class II defective transposon carrying two haloacetate dehalogenase genes from *Delftia acidovorans* plasmid pUO1. *Appl Environ Microbiol* **68**, 2307–15.
118. Tardif, G., Greer, C. W., Labbe, D., Lau, P. C. K. (1991) Involvement of a large plasmid in the degradation of 1,2-dichloroethane by *Xanthobacter autotrophicus. Appl Environ Microbiol* **57**, 1853–7.
119. Nojiri, H., Sekiguchi, H., Maeda, K., Urata, M., Nakai, S. I., Yoshida, T., Habe, H., Omori, T. (2001) Genetic characterization and evolutionary implications of a *car* gene cluster in the carbazole degrader *Pseudomonas* sp. strain CA10. *J Bacteriol* **183**, 3663–79.
120. Dabrock, B., Kesseler, M., Averhoff, B., Gottschalk, G. (1994) Identification and characterization of a transmissible linear plasmid from *Rhodococcus erythropolis* BD2 that encodes isopropylbenzene and trichloroethene catabolism. *Appl Environ Microbiol* **60**, 853–60.
121. Stecker, C., Johann, A., Herzberg, C., Averhoff, B., Gottschalk, G. (2003) Complete nucleotide sequence and genetic organization of the 210-kilobase linear plasmid of *Rhodococcus erythropolis* BD2. *J Bacteriol* **185**, 5269–74.
122. White, O., Eisen, J. A., Heidelberg, J. F., Hickey, E. K., Peterson, J. D., Dodson, R. J., Haft, D. H., Gwinn, M. L., Nelson, W. C., Richardson, D. L., Moffat, K. S., Qin, H. Y., Jiang, L. X., Pamphile, W., Crosby, M., Shen, M., Vamathevan, J. J., Lam, P., Mcdonald, L., Utterback, T., Zalewski, C., Makarova, K. S., Aravind, L., Daly, M. J., Minton, K. W., Fleischmann, R. D., Ketchum, K. A., Nelson, K. E., Salzberg, S., Smith, H. O., Venter, J. C., Fraser, C. M. (1999) Genome sequence of the radioresistant bacterium *Deinococcus radiodurans* R1. *Science* **286**, 1571–7.
123. Methe, B. A., Nelson, K. E., Eisen, J. A., Paulsen, I. T., Nelson, W., Heidelberg, J. F., Wu, D., Wu, M., Ward, N., Beanan, M. J., Dodson, R. J., Madupu, R., Brinkac, L. M., Daugherty, S. C., Deboy, R. T., Durkin, A. S., Gwinn, M., Kolonay, J. F., Sullivan, S. A., Haft, D. H., Selengut, J., Davidsen, T. M., Zafar, N., White, O., Tran, B., Romero, C., Forberger, H. A., Weidman, J., Khouri, H., Feldblyum, T. V., Utterback, T. R., Van Aken, S. E., Lovley, D. R., Fraser, C. M. (2003) Genome of *Geobacter sulfurreducens*: Metal reduction in subsurface environments. *Science* **302**, 1967–9.

# Chapter 27

## Horizontal Gene Transfer Between Microbial Eukaryotes


### Jan O. Andersson



### Abstract

Comparative genomics have identified two loosely defined classes of genes: widely distributed core genes that encode proteins for central functions in the cell and accessory genes that are patchily distributed across lineages and encode taxa-specific functions. Studies of microbial eukaryotes show that both categories undergo horizontal gene transfer (HGT) from prokaryotes, but also between eukaryotic organisms. Intra-domain gene transfers of most core genes seem to be relatively infrequent and therefore comparatively easy to detect using phylogenetic methods. In contrast, phylogenies of accessory genes often have complex topologies with little or no resemblance of organismal relationships typically with eukaryotes and prokaryotes intermingled, making detailed evolutionary histories difficult to interpret. Nevertheless, this suggests significant rates of gene transfer between and among the three domains of life for many of these genes, affecting a considerably diversity of eukaryotic microbes, although the current depth of taxonomic sampling usually is insufficient to pin down individual transfer events. The occurrence of intra-domain transfer among microbial eukaryotes has important implications for studies of organismal phylogeny as well as eukaryote genome evolution in general.



**Key words:** Accessory genes, core genes, gene loss, genome evolution, inter-domain HGT, intra-domain HGT, patchily distributed genes, phylogeny, protists.


## 1. Introduction

Due to the shortage of comparative genomic data covering the eukaryotic diversity, the knowledge about HGT in eukaryotic genome evolution is currently rather limited, in contrast to the situation for prokaryotes where this evolutionary mechanism have been intensively studied *(1, 2)*. However, large-scale phylogenetic analyses on genomic data (phylogenomics) have shown that several distantly related microbial eukaryotes have acquired mostly metabolic genes from prokaryotes


Maria B. Gogarten et al. (eds.), *Horizontal Gene Transfer: Genomes in Flux, vol. 532*



DOI 10.1007/978-1-60327-853-9_27 Springerprotocols.com

in their environment at an appreciable rate *(3–8)* (*see* also **Chapter 28**), suggesting an important role of gene transfer in protist genome evolution. Recent data indicate that HGT also affects animals and plants: extensive intra-domain transfer of genes between plant mitochondria has been detected *(9)*, a high-throughput genomic screening of plant-pathogenic nematodes identified a handful of HGT candidate genes *(10)*, and DNA from endosymbiotic *Wolbachia* had been transferred to their animal hosts *(11)* (*see* also **Chapter 30**). Although these findings indicate that HGT indeed occurs in genome evolution of multicellular eukaryotes, the dominance of microbial eukaryotes among the published examples of HGT affecting eukaryotes suggests that this evolutionary mechanism probably is more prominent in microbes.

In almost all large-scale studies of the role of HGT in the evolution of microbial eukaryotes, the screening procedures have been designed to detect putative transfers between prokaryotes and eukaryotes, which are easier to detect and justify than transfer between eukaryotes. Consequently, the majority of the reported gene transfers have been inter-domain events in the direction to eukaryotes *(3–8, 12)*. As more genome data are becoming available from diverse eukaryotic organisms, it is feasible to also study intra-domain gene transfer events. Indeed, indications of intra-domain HGT are frequently showing up in studies allowing for such *(8, 13)*, *see* also **Chapter 28**, indicating that microbial eukaryotes are swapping genes with each other. This chapter will focus on intra-domain HGT in non-algal microbial eukaryotes; gene transfer between eukaryotic algae, including the special case of transfer of genes from secondary or tertiary plastids to their hosts ("endosymbiotic gene transfer"), is covered in **Chapter 29**. Data suggesting that eukaryote-to-eukaryote gene transfer is common in microbial eukaryotes will be presented, and the detection methods and implications for this evolutionary process will be discussed.

## 2. Genome Data Suggest Eukaryote-to-Eukaryote HGT

Only a tiny fraction of the genes in any given genome has been found to be universally present in all studied genomes *(14, 15)*. When more restricted parts of the organismal diversity are studied, the number of conserved set of genes increases *(16, 17)*. Nevertheless, also between closely related bacteria, there is a large fraction of genes that differ between strains *(17, 18)*. Although the occurrence of such genes is much less understood for eukaryotes, many protists encode genes not universally found among eukaryotes. These patchily distributed genes are referred to as accessory

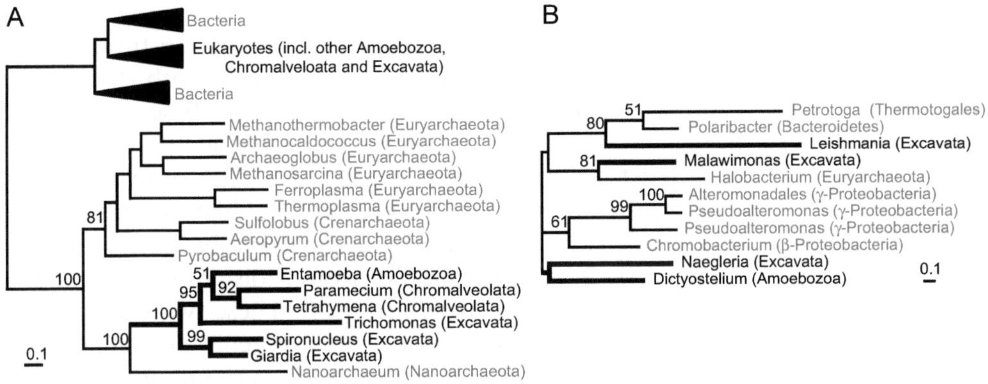


Fig. 27.1. Intra-domain HGT in core and accessory genes. (**A**) Schematic phylogeny of *alaS* coding for alanyl-tRNA synthetase, a core gene expected to be present in all eukaryotes. Most core genes rarely undergo transfer, making individual cases relatively easy to identify. Here one inter-domain transfer from archaea to excavata followed by two intra-domain transfers can be inferred, spreading the gene from archaea to excavata and then to *Entamoeba* and ciliates. Figure adapted from *(26)*. (**B**) Phylogeny of a patchily distributed gene found only in a few unrelated eukaryotes. Accessory genes appear to undergo both inter- and intra-domain HGT at comparably high frequencies. Individual transfers are typically difficult to infer because few of the relationships in the tree correspond to expected organismal phylogenies. Figure adapted from the unpublished *Dictyostelium* study described in **Section 3.3**. Eukaryotic sequences are in **boldface** with thick branches and prokaryotic sequences are in *gray font*. Numbers indicate bootstrap support values in maximum likelihood analyses. Further details on methods are found in the original publication *(26)*.

genes for the purpose of this chapter. Intriguingly, it appears that both core and accessory eukaryotic genes are subject to inter- as well as intra-domain HGT, although their patterns of transfer sometimes differ (**Fig. 27.1**).

### *2.1. Unexpected Groupings of Eukaryotes in Core Gene Phylogenies*

A general assumption in genome evolution studies of eukaryotic core genes seem to be that they rarely undergo transfer. This view appears to be an oversimplification as comparative data from diverse microbial eukaryotes are accumulating. For example, one core gene coding for elongation factor 1 alpha (EF-1α), a key protein in the translational machinery, was found to be missing in some lineages of eukaryotes. Fascinatingly, an evolutionary related, but distinct gene, coding for an EF-like protein (EFL), which likely is able to perform the same function as elongation factor 1α (EF-1α), was present in these organisms *(19)*. Follow-up studies have revealed a very complex pattern of presence of EFL in eukaryotes, almost always coupled with absence of EF-1α *(20–22)*. This punctuate distribution of EFL among eukaryotes is probably explained by a combination of multiple eukaryote-to-eukaryote gene transfer events where EF-L has replaced the canonical EF-1α in diverse eukaryotic lineages, and maintenance of both genes for relatively long evolutionary times followed by differential gene loss *(19–22)*.

A phylogenetic study of a handful of genes that aimed to determine organismal phylogenies revealed another example of intra-domain transfer of eukaryotic core genes: phylogenetic trees of genes coding for the cytoskeletal protein α-tubulin placed *Andalucia*, a jakobid-like excavate, strongly with diplomonads, parabasalids, and opisthokonts *(23)*. In trees based on the other five genes in the study *Andalucia* is found with jakobids, as expected from ultrastructural data *(24)*. The most likely explanation for these strongly incongruent results from different phylogenetic markers is that the original jakobid-like tubulin was replaced by a gene for tubulin from a distantly related eukaryote in an ancestor of *Andalucia (23)*. These two examples show that orthologous gene replacements of genes encoding proteins involved in key cellular functions, such as the translation and the cytoskeleton, have occurred between eukaryotes.

Replacements of ubiquitous genes have also occurred between prokaryotes and eukaryotes. For example, a tyrosyl-tRNA synthetase gene was transferred from Archaea to opisthokonts *(25)*, and genes coding for alanyl- and prolyl-tRNA synthetase were transferred from an archaeal lineage to an ancestor of diplomonads and parabasalids *(26)*. One of these genes, *alaS*, was likely subsequently transferred from the parabasalid lineage to ciliates and the *Entamoeba* lineage in two independent intra-domain gene transfer events (**Fig. 27.1A**). These transfers likely represent relatively ancient events, possibly between free-living ancestors of the *Entamoeba* and parabasalids that might have shared environments with ciliates. This pattern of inter-followed by intra-domain transfer appears to be rather common. Rogers and co-workers studied the evolutionary history of three eukaryotic genes coding for core carbon metabolism: transketolase (TK), glyceraldehyde-3-phosphate dehydrogenase (GAPDH), and ribulose-5-phosphate-3-epimerase (RPE) *(27)*. They found that what initially appeared to be simple inter-domain gene transfer events based on observations where single eukaryotic lineages were nested within bacterial sequences turned out to be more complicated with increased taxon sampling. In all three cases distantly related eukaryotes were found to share the same bacterial version of the gene to the exclusion of more closely related organisms, which was rationalized by transfers between eukaryotes after an initial inter-domain transfer *(27)*.

Glycolysis is a central metabolic pathway present in all extant eukaryotes, which most likely also was present in the last common eukaryotic ancestor *(28)*. This suggests that the genes coding for the enzymes necessary to perform glycolysis have been vertically inherited during evolution. However, two systematic phylogenetic analyses of glycolytic enzymes from excavate protists, the oxymonad *Monocercomonoides (29)* and *Trimastix pyriformis (30)*, identified this pathway to be an evolutionary mosaic. Although a few of the enzymes show monophyletic eukaryotes with internal relationships consistent with vertical inheritance, several enzymes have multiple origins in excavate lineages and many appear to be of bacterial origin. Again, distantly related protists are sometimes found nested together, suggesting eukaryote-to-eukaryote transfers, although in most cases the donor and recipient lineages are difficult to pinpoint due to the limited taxon sampling compared to the large number of putative transfer events *(29, 30)*. Nevertheless, these two analyses clearly suggest that both inter-domain and intra-domain gene transfer affect glycolytic genes.

### *2.2. Accessory Genes in Microbial Eukaryotes Are Affected by HGT*

Studies of closely related strains of bacteria have identified a large amount of accessory genes thought to be responsible for niche adaptation *(17, 18)*. Due to the lack of genome sequences of closely related strains, the gene content dynamics of microbial eukaryotes is much less understood. These genes are thought to code for proteins sporadically important in time and space and therefore they are frequently lost and gained over evolutionary time. If the exchange of these genes is largely uncoupled from organismal relationships, their distribution is expected to be patchy with phylogenetic relationships at odds with organismal phylogenies (**Fig. 27.1B**). Interestingly, phylogenetic studies of protist genes have often resulted in topologies showing this kind of pattern, indicating that microbial eukaryotes take part in the exchange of accessory genes. For example, studies of anaerobic or micro-aerophilic protists have shown that they have acquired genes from anaerobic prokaryotes present in their environments *(3, 5–8)*, consistent with a model where eukaryotes acquire genes from various sources in the evolutionary transition from an aerobic to an anaerobic lifestyle *(31)*.

Indications that intra-domain transfer might be important in the adaptation to anoxic environments comes from the analysis of the genes usually functioning in mitochondria, hydrogenomsomes, or mitosomes in the free-living anaerobic *Trimastix pyriformis* (excavata) *(32)*. Intriguingly, the phylogenies of both key enzymes of anaerobic metabolism, pyruvate:ferrodoxin oxidoreductase (PFO) and [FeFe] hydrogenase show *T. pyriformis* branching with the distantly related *E. histolytica*, and not the other anaerobic excavates as expected from organismal phylogenies *(33)*. This is suggestive of gene transfers of these key enzymes between the lineages leading to *E. histolytica* and *T. pyriformis (32)*.

Fungal accessory genes, for example, coding for enzymes involved in secondary metabolites, have been speculated to spread via HGT and whereby providing selective advantages to their recipients *(34, 35)*, which indeed more recently has been shown to be the case *(36–41)*. Arguably, the most striking example explains the emergence of a disease on wheat in the first half of the last century *(40)*. A highly virulent form of *Pyrenophora tritici-repentis*, which cause tan spot disease on wheat, has rapidly spread around the world after its first occurrence in the United States in 1941. This fungal species had been known to be pathogenic on wheat, but it previously caused only a mild disease. Surprisingly, a gene with 99.7% identity to the *P. tritici-repentis* gene responsible for the increased virulence was identified in the genomic sequence of *Stagonospora noderum*, another pathogen on wheat, while no virulence gene could be found in any of the more closely related species which are pathogens on other crops *(40)*. A drastic change in phenotype has also been associated with a gene transfer of the mating system of fungal genus *Stemphylium*; self-fertilization apparently has originated independently at least three times independently within the genus *(41)*. Interestingly, two of the self-fertile groups have very similar mating type loci, which indeed are monophyletic in phylogenetic analyses based on the gene sequences indicative of HGT of the mating locus within the genus *(41)*. These examples suggest that gene transfer might be relatively common between fungi *(36–41)*. Unfortunately, to my knowledge there is no systematic study to determine the impact of this evolutionary mechanism on a larger scale; a phylogenetic analysis of 4805 orthologous gene families in fungi reported that the congruence between the trees was significantly better than random, but did not try to identify individual trees incongruent with the proposed organismal phylogeny *(42)*.

Additional examples of transfer of patchily distributed genes between eukaryotic pathogens come from fungi and oomycete, which both are pathogenic for plants. Oomycetes resembles osmotrophic filamentous fungi, but belongs to a group only distantly related to fungi *(33)* (these are all eukaryotes). Thus, the similarities in lifestyle between them are likely due to convergent evolution rather than shared ancestry *(43)*. Accordingly, genes found to be shared between oomycetes and fungi to the exclusion of other eukaryotic groups are likely to represent adaptive gene exchanges of genes and may explain some of the similarities between the groups. Indeed, a dozen of shared genes have been identified and phylogenetic analyses of four of these indicated gene transfers from fungi to oomycetes *(13)*. These genes encode functions related to the utilization of rare metabolites that could be coupled to an osmotrophic lifestyle, which indirectly is coupled to the virulence of these plant pathogens. Evolutionary studies of a gene coding for cutinase represent a more direct link between HGT and pathogenicity of plants *(44)*. Cutin is the major component of the plant cuticle. Therefore, plant pathogens need to be able to degrade this insoluble lipid polyester to penetrate unwounded host tissue; functional cutinase is essential for fungal pathogenicity on plants *(45)*. This important enzyme has been identified only in oomycetes and actinobacteria in addition to fungi *(44)*. The origin of the cutinase enzyme is almost certainly more recent than the diversification of the three organismal groups where it has been detected; cutin is unique to land plants and probably evolved as an adaptation to life on land *(46)*. Indeed, phylogenetic analyses suggest that this enzyme has spread via gene transfers, suggesting that the pathogenicity of those fungi and oomycete, at least to some extent, is the result of HGT of an accessory gene *(44)*.

In the examples above, the functions of the products of the patchily distributed genes could be coupled to the lifestyles of the organisms, suggesting that gene acquisitions confer advantages in the environments of the recipient organisms. Relatively recent gene transfers between the lineages that share the genes are reasonable scenarios, although details of the transfers are not

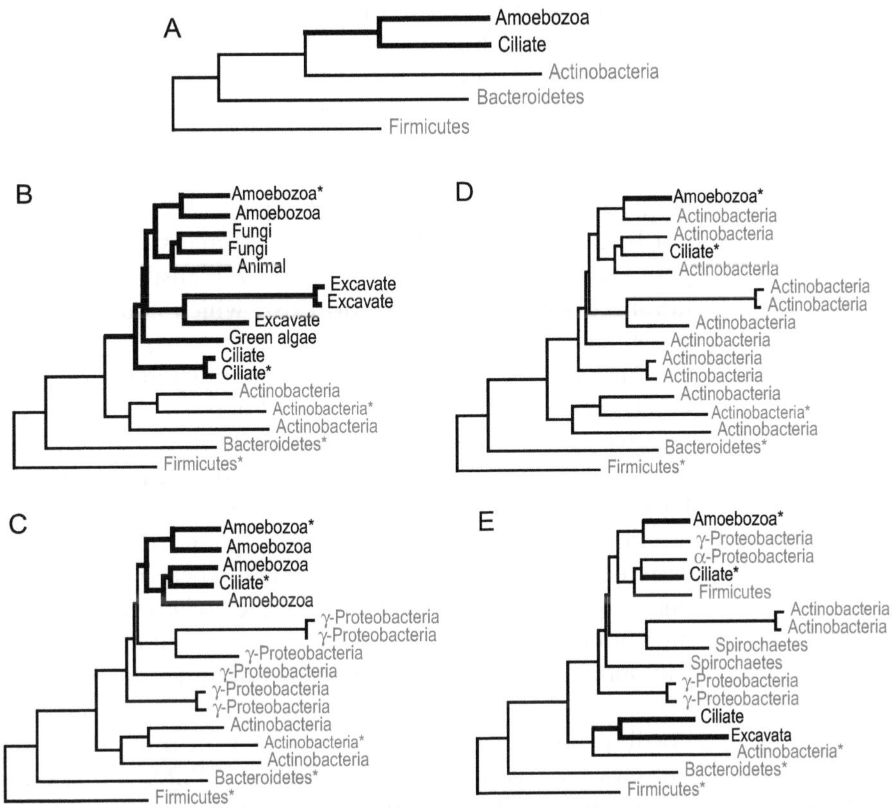


Fig. 27.2. Alternative explanations for patchily distributed genes showing indications of intra-domain transfer. (**A**) Hypothetical phylogeny of a gene found in two eukaryotic and three bacterial species in the available datasets. The grouping of the eukaryotic sequences suggests transfer between eukaryotes. Increased taxonomic sampling may clarify the evolutionary history of the gene family. Several alternative scenarios are possible: (**B**) the genes are detected in a wide diversity of eukaryotes, indicating that the pattern observed initially was due to gene losses; (**C**) one of the eukaryotic lineages are nested within the other, indicating intra-domain HGT; (**D**) both eukaryotic sequences are nested within a bacterial group, indicating two independent inter-domain HGT events: or (**E**) the complexity of the topology is increasing making individual HGT events difficult to identify even with a broader taxonomic sampling. *Asterisks* indicate sequences present in (**A**). Eukaryotic sequences are in **boldface** with thick branches and prokaryotic sequences are in *gray font*.

always straightforward to infer. Similar results are often found for genes without any well-characterized function (**Figs. 27.1B** and **27.2A**). Distantly related eukaryotes, typically from different eukaryotic supergroups, are found as sisters in trees of patchily distributed proteins putatively indicating eukaryote-to-eukaryote transfers. The distribution of the gene is usually sporadic also in prokaryotes where the individual species that do encode the gene typically are often only distantly related (**Figs. 27.1B** and **27.2A**).

## 3. Detecting HGT Between Eukaryotes Using Phylogenetic Methods

In order to identify donor and recipient lineages for gene transfer events, the organisms that received the gene should be nested within a natural group of descendants of the donor lineage in the gene phylogeny *(47)* (**Fig. 27.1A**). Furthermore, the recipients are expected to represent a monophyletic organismal group where all members encode the gene, while organisms that diverged before the transfer event lack the gene. It is very rare that these requirements are met for patchily distributed proteins (**Figs. 27.1B** and **27.2A**). Since donor and recipient lineages cannot be determined, it could be argued that such phylogenies are poor indicators of HGT; a number of plausible alternative evolutionary scenarios actually do exist, which could produce such topologies (**Fig. 27.2B–F**).

An obvious problem with using phylogenetics to test for HGT events is that all tree reconstructions methods are prone to error to some extent, and especially single-gene phylogenies are problematic due to the small number of positions available. However, the major signals in a dataset should be reliable if the protein is of a suitable length, the sequence divergences are not extreme, and the alignment is robust. There is indeed no obvious reason to believe that phylogenies of proteins with limited distributions in the tree of life are more problematic than more widespread proteins showing similar alignment properties.

### *3.1. Different Evolutionary Scenarios May Result in Intra-Domain Transfer Topologies*

Possibly, a more severe complication for the interpretation of phylogenies such as the one shown in **Fig. 27.2A** is that only a very tiny fraction of the sequence diversity in nature has yet been sampled, as emphasized by a recent metagenomic project of ocean water *(48)*. Thus, increased sampling of a protein family should possibly shed light on its evolutionary history. If improved taxon sampling results in a tree where most of the eukaryotic diversity is represented with the expected internal relationships (**Fig. 27.2B**), the sisterhood of the distantly related eukaryotes (**Fig. 27.2A**) were probably due to losses of the gene in branches leading to organisms sampled in genome projects. On the other hand, it may turn out that the addition of more data results in a phylogeny where recipients and donors of HGT events might be identified. For example, in **Fig. 27.2C** ciliates likely acquired the gene from a member of Amoebozoa, in agreement with the eukaryote-to-eukaryote transfer putatively inferred from **Fig. 27.2A**. However, interpretations of the details of transfers may change with increased taxon sampling. An apparent intra-domain gene transfer (**Fig. 27.2A**) could actually be revealed to be the result of two independent inter-domain events from closely related prokaryotic lineages, in this case both from the actinobacterial lineage (**Fig. 27.2D**). These three examples show that more sequence data indeed could improve the understanding of the detailed evolutionary events underlying the observed phylogenies.

Unfortunately, in many cases increased sampling may not be as successful; more sequences could lead to phylogenetic trees that are more complicated to interpret using the accepted organismal relationships as a guide. In **Fig. 27.2E** the original ciliate and amoebozoan sequences are nested with prokaryotic sequences from different groups; it is impossible to determine the events leading to this subtree although a direct intra-domain transfer between ancestors of the two eukaryotes (**Fig. 27.2A**) appears unlikely in the light of more data. In a different part of the tree a ciliate sequence is found as a sister to an excavate sequence, which, again, could be interpreted as a putative intra-domain transfer (**Fig. 27.2E**). Only the most similar prokaryotic sequences are showing any resemblance with organismal relationships. Although no HTG event can be pinpointed in such a tree, gene transfer has likely played a major role in the distribution of the gene family. However, much denser taxonomic sampling would most likely be needed to obtain a more detailed understanding.

### 3.2. Increased Taxon Sampling Often Results in More HGT Events

The sampling of the diversity of all domains of life is steadily improving, suggesting that these alternative scenarios could be tested. Indeed, additional taxonomic sampling may lead to conclusions involving an increased number of gene transfer events. For example, intensified studies of the relationships between EF1-α and EFL in microbial eukaryotes have led to a more complex picture *(20–22)* than in the initial report *(19)*. Similarly, phylogenetic studies of glycolytic enzymes from a broader diversity of protists show complex phylogenies suggestive of multiple gene transfer events *(29, 30)*, while an earlier study based on a narrower sampling of eukaryotes dismissed this possibility *(49)*.

Four genes originally found to have been introduced in the genomes of the unrelated micro-aerophilic intestinal parasites *E. histolytica* and *Giardia lamblia (3)* were re-analyzed, combining the usage of more advanced models of sequence evolution and a considerable broadening of the taxon sampling *(50)*. The number of putative inter-domain transfers increased for all four genes; the majority of the sampled relatives of *E. histolytica* and *G. lamblia* (*Mastigamoeba* and *Trichomonas*, respectively) showed up in distinct regions of the trees in a pattern similar to **Fig. 27.2E**. Interestingly, unrelated eukaryotic sequences often clustered together in the trees suggestive of intra-domain transfers *(50)*. However, there are also examples where increased taxon sampling and/or the application of more appropriate models in phylogenetic analyses have disproved earlier suggestions of HGT events affecting eukaryotes *(51–53)*.

### 3.3. Intra-Domain Transfers in Phylogenomic Studies

Most large-scale studies of gene transfer affecting eukaryotes have addressed inter-domain transfers *(3–8, 31)* likely because these are easier to detect than intra-domain eukaryotic transfers, especially in the absence of complete genome sequences from a diversity of eukaryotes. However, a test of the effect of increased sampling on inferences of HGT from one of these was performed by Alsmark and co-workers and is reported in **Chapter 23**. They re-analyzed 96 HGT candidate genes detected in the *E. histolytica* genome *(5)* and added recently published prokaryotic and eukaryotic sequences to the earlier datasets. Most trees still supported a HGT scenario, although five topologies contradicted the initial HGT hypothesis. On the other hand, in nine trees the *E. histolytica* sequence now clustered with one of the added eukaryotic sequences (*Trichomonas* in the majority of the cases), suggestive of intra-domain gene transfer events. Very similar results were found when multiple eukaryotes nested with prokaryotes were allowed in the screening process in a phylogenomic study of the diplomonad *Spironucleus salmonicida (8)*. Interestingly, in 20% of the cases where the diplomonad gene had been acquired via HGT from prokaryotes, subsequent gene transfers between diplomonads and in most cases members of Amoebozoa were found *(8)*. These results suggest that some gene transfers currently reported as inter-domain HGT may be revealed to represent transfer between eukaryotes in future analyses with a denser sampling of the eukaryotic diversity, if the possibility of such transfers is acknowledged.

Using a different approach to analyze the evolution of patchily distributed genes, proteins present in the cellular slime mold *Dictyostelium discoideum* and a limited number of other taxa (including at least one prokaryotic species) were identified (unpublished observation); 49 protein families were detected containing on average 42 sequences using arbitrarily chosen criteria for the maximum number of taxa allowed in each dataset. The phyletic distribution of these conflicted strongly with organismal phylogenies; 25 proteins are shared with the distantly related soil-dwelling amoeboflagellate *Naegleria* (excavata), while only two are present in the more closely related *Entamoeba*. The vast majority of the protein families indeed showed unexpected topologies in phylogenetic analyses; eukaryotes were polyphyletic in 85% of the trees with unrelated eukaryotes often grouping together (unpublished observation). The analyses of the identified proteins suggest that the majority, if not all, have been distributed via inter- and intra-domain gene transfers affecting diverse microbial eukaryotes. The observation that the proteins tend to be shared between organisms in soil suggests that the gene exchange occurred in this environment; transfer of accessory genes within specific environments has previously been observed in prokaryotes *(54)*.

## 4. Concluding Remarks

In an influential 1992 review, Russell Doolittle and colleagues stressed the importance of taking various supporting lines of evidence into consideration when evaluating putative cases of HGT, such as anomalous positions in phylogenetic analyses and the possibility of putative donor and recipients to meet in the environment *(47)*. They concluded their review (3 years before the first complete genome sequence of any cellular organism) that "horizontal gene transfer should be considered carefully and claimed with caution" *(47)*. It might seem discouraging that individual cases of HGT still often is difficult to prove or disprove despite the fact that we now have a massive amount of genomic data along with improved phylogenetic methods. In fact, many transfers of core genes are only manifested in anomalous results in phylogenetic analyses without any obvious overlaps in lifestyles of the descendants of donors and recipients (**Fig. 27.1A**). However, the availability of complete genome sequences provide us with an additional criterion that can be used to evaluate putative HGT events of accessory genes; absences of genes can be scored with confidence enabling evaluation of gene transfer scenarios against gene duplications in combination with differential losses. Furthermore, the strength of the support of HGT increases when a putative acquisition of an accessory gene is possible to couple with a modification of the lifestyle of the organism, for example, a change in virulence or metabolic properties.

The occurrence of intra- in addition to inter-domain HGT affecting protists has implications for studies of eukaryotic organismal phylogenies. The finding that also core genes, such as tubulin and EF-1α that have been extensively used as phylogenetic markers *(55)*, may have undergone eukaryote-to-eukaryote HGT *(19–23)* suggests that no gene can be assumed to be immune to HGT. Accordingly, concatenation of genomic sequences for use in phylogenetic studies should be performed with caution *(56)*.

Fortunately, there are efforts to develop methods for screening of individual genes to check for incongruities in phylogenetic signals between markers *(57, 58)*.

The presence of a gene with a foreign origin in two organismal groups to the exclusion of other putatively related groups could be due to an introduction via HGT in a common ancestor. Such rare evolutionary events have been suggested to be useful in determining organismal relationships *(59)*, *see* also **Chapter 7**. For example, acquisitions of prokaryotic genes by eukaryotes have been used in support of the existence of specific eukaryotic organismal groupings *(25, 26)*. These inferences rely on proper identification of donor and recipient lineages. A relatively high rate of intra-domain gene transfer within eukaryotes would be problematic since the sharing of a foreign gene could be due to a recent gene transfer between two groups rather than the introduction in their common ancestor.

More generally, any phylogenetic study involving microbial eukaryotes needs to take into account that genes may be transferred between lineages. For example, the presence of a gene in genomes of distantly related organisms belonging to different eukaryotic super-groups is not necessarily due to the presence of the gene in the common ancestor; HGT could distribute the gene to different parts of the tree maybe resulting in niche adaptation in the recipient. It is indeed very likely that more examples where eukaryotes have converted to similar lifestyles partly via gene acquisitions, as outlined for plant pathogens, anaerobic protists, and soil amoebae, will be revealed. This gene exchange will certainly involve both inter- and intra-domain transfers; HGT among eukaryotic-specific accessory genes is currently very poorly studied.

A rapid increase of comparative genomic data covering the eukaryotic diversity is expected in the near future, both from metagenomic and from genomic projects. New sequencing technologies should facilitate sequencing of several isolates of the same species of protists. Such approaches should provide insight into the evolutionary dynamics and plasticity of protist genomes. Combining sequence data from these different sources with knowledge of the biology of the organisms should provide us with exciting new insights into the genome dynamics and evolution of microbial eukaryotes.

## Acknowledgments

I thank two anonymous reviewers and the editors for constructive comments and suggestions. The author is supported by a grant from the Swedish Research Council (VR).

## References


1. Boucher, Y., Douady, C. J., Papke, R. T., Walsh, D. A., Boudreau, M. E. R., Nesbø, C. L., Case, R. J., Doolittle, W. F. (2003) Lateral gene transfer and the origins of prokaryotic groups. *Annu Rev Genet* **37**, 283–328.
2. Gogarten, J. P., Townsend, J. P. (2005) Horizontal gene transfer, genome innovation and evolution. *Nat Rev Microbiol* **3**, 679–87.
3. Andersson, J. O., Sjögren, Å. M., Davis, L. A. M., Embley, T. M., Roger, A. J. (2003) Phylogenetic analyses of diplomonad genes reveal frequent lateral gene transfers affecting eukaryotes. *Curr Biol* **13**, 94–104.
4. Huang, J., Mullapudi, N., Lancto, C. A., Scott, M., Abrahamsen, M. S., Kissinger, J. C. (2004) Phylogenomic evidence supports past endosymbiosis, intracellular and horizontal gene transfer in *Cryptosporidium parvum*. *Genome Biol* **5**, R88.
5. Loftus, B., Anderson, I., Davies, R., Alsmark, U. C. M., Samuelson, J., Amedeo, P., Roncaglia, P., Berriman, M., Hirt, R. P., Mann, B. J., Nozaki, T., Suh, B., Pop, M., Duchene, M., Ackers, J., Tannich, E., Leippe, M., Hofer, M., Bruchhaus, I., Willhoeft, U., Bhattacharya, A., Chillingworth, T., Churcher, C., Hance, Z., Harris, B., Harris, D., Jagels, K., Moule, S., Mungall, K., Ormond, D., Squares, R., Whitehead, S., Quail, M. A., Rabbinowitsch, E., Norbertczak, H., Price, C., Wang, Z., Guillen, N., Gilchrist, C., Stroup, S. E., Bhattacharya, S., Lohia, A., Foster, P. G., Sicheritz-Ponten, T., Weber, C., Singh, U., Mukherjee, C., El-Sayed, N. M., Petri, W. A., Jr., Clark, C. G., Embley, T. M., Barrell, B., Fraser, C. M., Hall, N. (2005) The genome of the protist parasite *Entamoeba histolytica*. *Nature* **433**, 865–8.
6. Ricard, G., Mcewan, N. R., Dutilh, B. E., Jouany, J. P., Macheboeuf, D., Mitsumori, M., Mcintosh, F. M., Michalowski, T., Nagamine, T., Nelson, N., Newbold, C. J., Nsabimana, E., Takenaka, A., Thomas, N. A., Ushida, K., Hackstein, J. H., Huynen, M. A. (2006) Horizontal gene transfer from bacteria to rumen ciliates indicates adaptation to their anaerobic carbohydrates rich environment. *BMC Genomics* **7**, 22.
7. Carlton, J. M., Hirt, R. P., Silva, J. C., Delcher, A. L., Schatz, M., Zhao, Q., Wortman, J. R., Bidwell, S. L., Alsmark, U. C., Besteiro, S., Sicheritz-Ponten, T., Noel, C. J., Dacks, J. B., Foster, P. G., Simillion, C., Van De Peer, Y., Miranda-Saavedra, D., Barton, G. J., Westrop, G. D., Muller, S., Dessi, D., Fiori, P. L., Ren, Q., Paulsen, I., Zhang, H., Bastida-Corcuera, F. D., Simoes-Barbosa, A., Brown, M. T., Hayes, R. D., Mukherjee, M., Okumura, C. Y., Schneider, R., Smith, A. J., Vanacova, S., Villalvazo, M., Haas, B. J., Pertea, M., Feldblyum, T. V., Utterback, T. R., Shu, C. L., Osoegawa, K., De Jong, P. J., Hrdy, I., Horvathova, L., Zubacova, Z., Dolezal, P., Malik, S. B., Logsdon, J. M., Jr., Henze, K., Gupta, A., Wang, C. C., Dunne, R. L., Upcroft, J. A., Upcroft, P., White, O., Salzberg, S. L., Tang, P., Chiu, C. H., Lee, Y. S., Embley, T. M., Coombs, G. H., Mottram, J. C., Tachezy, J., Fraser-Liggett, C. M., Johnson, P. J. (2007) Draft genome sequence of the sexually transmitted pathogen *Trichomonas vaginalis*. *Science* **315**, 207–12.
8. Andersson, J. O., Sjögren, Å. M., Horner, D. S., Murphy, C. A., Dyal, P. L., Svärd, S. G., Logsdon Jr, J. M., Ragan, M. A., Hirt, R. P., Roger, A. J. (2007) A genomic survey of the fish parasite *Spironucleus salmonicida* indicates genomic plasticity among diplomonads and significant lateral gene transfer in eukaryote genome evolution. *BMC Genomics* **8**, 51.
9. Richardson, A. O., Palmer, J. D. (2006) Horizontal gene transfer in plants. *J Exp Bot* **58**, 1–9.
10. Scholl, E. H., Thorne, J. L., Mccarter, J. P., Bird, D. M. (2003) Horizontally transferred genes in plant-parasitic nematodes: a high-throughput genomic approach. *Genome Biol* **4**, R39.
11. Hotopp, J. C., Clark, M. E., Oliveira, D. C., Foster, J. M., Fischer, P., Torres, M. C., Giebel, J. D., Kumar, N., Ishmael, N., Wang, S., Ingram, J., Nene, R. V., Shepard, J., Tomkins, J., Richards, S., Spiro, D. J., Ghedin, E., Slatko, B. E., Tettelin, H., Werren, J. H. (2007) Widespread lateral gene transfer from intracellular bacteria to multicellular eukaryotes. *Science* **317**, 1753–6.
12. Andersson, J. O. (2005) Lateral gene transfer in eukaryotes. *Cell Mol Life Sci* **62**, 1182–97.
13. Richards, T. A., Dacks, J. B., Jenkinson, J. M., Thornton, C. R., Talbot, N. J. (2006) Evolution of filamentous plant pathogens: gene exchange across eukaryotic kingdoms. *Curr Biol* **16**, 1857–64.
14. Ciccarelli, F. D., Doerks, T., Von Mering, C., Creevey, C. J., Snel, B., Bork, P. (2006) Toward automatic reconstruction of a highly resolved tree of life. *Science* **311**, 1283–7.
15. Charlebois, R. L., Doolittle, W. F. (2004) Computing prokaryotic gene ubiquity: rescuing the core from extinction. *Genome Res* **14**, 2469–77.

16. Lerat, E., Daubin, V., Ochman, H., Moran, N. A. (2005) Evolutionary origins of genomic repertoires in bacteria. *PLoS Biol* **3**, e130.
17. Kettler, G. C., Martiny, A. C., Huang, K., Zucker, J., Coleman, M. L., Rodrigue, S., Chen, F., Lapidus, A., Ferriera, S., Johnson, J., Steglich, C., Church, G. M., Richardson, P., Chisholm, S. W. (2007) Patterns and implications of gene gain and loss in the evolution of *Prochlorococcus*. *PLoS Genet* **3**, e231.
18. Welch, R. A., Burland, V., Plunkett, G., 3rd, Redford, P., Roesch, P., Rasko, D., Buckles, E. L., Liou, S. R., Boutin, A., Hackett, J., Stroud, D., Mayhew, G. F., Rose, D. J., Zhou, S., Schwartz, D. C., Perna, N. T., Mobley, H. L., Donnenberg, M. S., Blattner, F. R. (2002) Extensive mosaic structure revealed by the complete genome sequence of uropathogenic *Escherichia coli*. *Proc Natl Acad Sci U S A* **99**, 17020–4.
19. Keeling, P. J., Inagaki, Y. (2004) A class of eukaryotic GTPase with a punctate distribution suggesting multiple functional replacements of translation elongation factor 1α. *Proc Natl Acad Sci U S A* **101**, 15380–5.
20. Ruiz-Trillo, I., Lane, C. E., Archibald, J. M., Roger, A. J. (2006) Insights into the evolutionary origin and genome architecture of the unicellular opisthokonts *Capsaspora owczarzaki* and *Sphaeroforma arctica*. *J Eukaryot Microbiol* **53**, 379–84.
21. Noble, G. P., Rogers, M. B., Keeling, P. J. (2007) Complex distribution of EFL and EF-1alpha proteins in the green algal lineage. *BMC Evol Biol* **7**, 82.
22. Gile, G. H., Patron, N. J., Keeling, P. J. (2006) EFL GTPase in cryptomonads and the distribution of EFL and EF-1α in chromalveolates. *Protist* **157**, 435–44.
23. Simpson, A. G. B., Perley, T. A., Lara, E. (2008) Lateral transfer of the gene for a widely used marker, alpha-tubulin, indicated by a multi-protein study of the phylogenetic position of *Andalucia* (Excavata). *Mol Phylogenet Evol* in press.
24. Lara, E., Chatzinotas, A., Simpson, A. G. (2006) Andalucia (n. gen.) – the deepest branch within jakobids (Jakobida; Excavata), based on morphological and molecular study of a new flagellate from soil. *J Eukaryot Microbiol* **53**, 112–20.
25. Huang, J., Xu, Y., Gogarten, J. P. (2005) The presence of a haloarchaeal type tyrosyl-tRNA synthetase marks the opisthokonts as monophyletic. *Mol Biol Evol* **22**, 2142–6.
26. Andersson, J. O., Sarchfield, S. W., Roger, A. J. (2005) Gene transfers from Nanoarchaeota to an ancestor of diplomonads and parabasalids. *Mol Biol Evol* **22**, 85–90.
27. Rogers, M. B., Watkins, R. F., Harper, J. T., Durnford, D. G., Gray, M. W., Keeling, P. J. (2007) A complex and punctate distribution of three eukaryotic genes derived by lateral gene transfer. *BMC Evol Biol* **7**, 89.
28. Fothergill-Gilmore, L. A., Michels, P. A. (1993) Evolution of glycolysis. *Prog Biophys Mol Biol* **59**, 105–235.
29. Liapounova, N. A., Hampl, V., Gordon, P. M., Sensen, C. W., Gedamu, L., Dacks, J. B. (2006) Reconstructing the mosaic glycolytic pathway of the anaerobic eukaryote *Monocercomonoides*. *Eukaryot Cell* **5**, 2138–46.
30. Stechmann, A., Baumgartner, M., Silberman, J. D., Roger, A. J. (2006) The glycolytic pathway of *Trimastix pyriformis* is an evolutionary mosaic. *BMC Evol Biol* **6**, 101.
31. Andersson, J. O. (2006) Genome evolution of anaerobic protists: metabolic adaptation via gene acquisition, in *Genomics and Evolution of Microbial Eukaryotes* (Katz, L. A., Bhattacharya D., ed.) Oxford University Press, Oxford 109–22.
32. Hampl, V., Silberman, J. D., Stechmann, A., Diaz-Trivino, S., Johnson, P. J., Roger, A. J. (2008) Genetic evidence for a mitochondriate ancestry in the 'amitochondriate' flagellate *Trimastix pyriformis*. *PLoS ONE* **3**, e1383.
33. Adl, S. M., Simpson, A. G., Farmer, M. A., Andersen, R. A., Anderson, O. R., Barta, J. R., Bowser, S. S., Brugerolle, G., Fensome, R. A., Fredericq, S., James, T. Y., Karpov, S., Kugrens, P., Krug, J., Lane, C. E., Lewis, L. A., Lodge, J., Lynn, D. H., Mann, D. G., Mccourt, R. M., Mendoza, L., Moestrup, O., Mozley-Standridge, S. E., Nerad, T. A., Shearer, C. A., Smirnov, A. V., Spiegel, F. W., Taylor, M. F. (2005) The new higher level classification of eukaryotes with emphasis on the taxonomy of protists. *J Eukaryot Microbiol* **52**, 399–451.
34. Rosewich, U. L., Kistler, H. C. (2000) Role of horizontal gene transfer in the evolution of fungi. *Annu Rev Phytopathol* **38**, 325–63.
35. Walton, J. D. (2000) Horizontal gene transfer and the evolution of secondary metabolite gene clusters in fungi: an hypothesis. *Fungal Genet Biol* **30**, 167–71.
36. Patron, N. J., Waller, R. F., Cozijnsen, A. J., Straney, D. C., Gardiner, D. M., Nierman, W. C., Howlett, B. J. (2007) Origin and distribution of epipolythiodioxopiperazine (ETP) gene clusters in filamentous ascomycetes. *BMC Evol Biol* **7**, 174.
37. Khaldi, N., Collemare, J., Lebrun, M. H., Wolfe, K. H. (2008) Evidence for horizontal

transfer of a secondary metabolite gene cluster between fungi. *Genome Biol* 9, R18.

38. Temporini, E. D., Vanetten, H. D. (2004) An analysis of the phylogenetic distribution of the pea pathogenicity genes of *Nectria haematococca* MPVI supports the hypothesis of their origin by horizontal transfer and uncovers a potentially new pathogen of garden pea: *Neocosmospora boniensis. Curr Genet* **46**, 29–36.
39. Slot, J. C., Hibbett, D. S. (2007) Horizontal transfer of a nitrate assimilation gene cluster and ecological transitions in fungi: a phylogenetic study. *PLoS ONE* **2**, e1097.
40. Friesen, T. L., Stukenbrock, E. H., Liu, Z., Meinhardt, S., Ling, H., Faris, J. D., Rasmussen, J. B., Solomon, P. S., Mcdonald, B. A., Oliver, R. P. (2006) Emergence of a new disease as a result of interspecific virulence gene transfer. *Nat Genet* **38**, 953–6.
41. Inderbitzin, P., Harkness, J., Turgeon, B. G., Berbee, M. L. (2005) Lateral transfer of mating system in *Stemphylium. Proc Natl Acad Sci U S A* **102**, 11390–5.
42. Fitzpatrick, D. A., Logue, M. E., Stajich, J. E., Butler, G. (2006) A fungal phylogeny based on 42 complete genomes derived from supertree and combined gene analysis. *BMC Evol Biol* **6**, 99.
43. Latijnhouwers, M., De Wit, P. J., Govers, F. (2003) Oomycetes and fungi: similar weaponry to attack plants. *Trends Microbiol* **11**, 462–9.
44. Belbahri, L., Calmin, G., Mauch, F., Andersson, J. O. (2008) Evolution of the cutinase gene family: evidence for lateral gene transfer of a candidate *Phytophthora* virulence factor. *Gene* **408**, 1–8.
45. Kolattukudy, P. E., Rogers, L. M., Li, D., Hwang, C. S., Flaishman, M. A. (1995) Surface signaling in pathogenesis. *Proc Natl Acad Sci U S A* **92**, 4080–7.
46. Kenrick, P., Crane, P. R. (1997) The origin and early evolution of plants on land. *Nature* **389**, 33–9.
47. Smith, M. W., Feng, D. F., Doolittle, R. F. (1992) Evolution by acquisition: the case for horizontal gene transfers. *Trends Biochem Sci* **17**, 489–93.
48. Yooseph, S., Sutton, G., Rusch, D. B., Halpern, A. L., Williamson, S. J., Remington, K., Eisen, J. A., Heidelberg, K. B., Manning, G., Li, W., Jaroszewski, L., Cieplak, P., Miller, C. S., Li, H., Mashiyama, S. T., Joachimiak, M. P., Van Belle, C., Chandonia, J. M., Soergel, D. A., Zhai, Y., Natarajan, K., Lee, S., Raphael, B. J., Bafna, V., Friedman, R., Brenner, S. E., Godzik, A., Eisenberg, D., Dixon, J. E., Taylor, S. S., Strausberg, R. L., Frazier, M., Venter, J. C. (2007) The Sorcerer II global ocean sampling expedition: expanding the universe of protein families. *PLoS Biol* **5**, e16.
49. Canbäck, B., Andersson, S. G. E., Kurland, C. G. (2002) The global phylogeny of glycolytic enzymes. *Proc Natl Acad Sci U S A* **99**, 6097–102.
50. Andersson, J. O., Hirt, R. P., Foster, P. G., Roger, A. J. (2006) Evolution of four gene families with patchy phylogenetic distribution: influx of genes into protist genomes. *BMC Evol Biol* **6**, 27.
51. Richards, T. A., Hirt, R. P., Williams, B. A., Embley, T. M. (2003) Horizontal gene transfer and the evolution of parasitic protozoa. *Protist* **154**, 17–32.
52. Dyall, S. D., Yan, W., Delgadillo-Correa, M. G., Lunceford, A., Loo, J. A., Clarke, C. F., Johnson, P. J. (2004) Non-mitochondrial complex I proteins in a hydrogenosomal oxidoreductase complex. *Nature* **431**, 1103–7.
53. Hrdy, I., Hirt, R. P., Dolezal, P., Bardonova, L., Foster, P. G., Tachezy, J., Embley, T. M. (2004) *Trichomonas* hydrogenosomes contain the NADH dehydrogenase module of mitochondrial complex I. *Nature* **432**, 618–22.
54. Beiko, R. G., Harlow, T. J., Ragan, M. A. (2005) Highways of gene sharing in prokaryotes. *Proc Natl Acad Sci U S A* **102**, 14332–7.
55. Baldauf, S. L., Roger, A. J., Wenk-Siefert, I., Doolittle, W. F. (2000) A kingdom-level phylogeny of eukaryotes based on combined protein data. *Science* **290**, 972–7.
56. Delsuc, F., Brinkmann, H., Philippe, H. (2005) Phylogenomics and the reconstruction of the tree of life. *Nat Rev Genet* **6**, 361–75.
57. Leigh, J. W., Susko, E., Baumgartner, M., Roger, A. J. (2008) Testing congruence in phylogenomic analysis. *Syst Biol* **57**, 104–15.
58. Planet, P. J. (2006) Tree disagreement: measuring and testing incongruence in phylogenies. *J Biomed Inform* **39**, 86–102.
59. Huang, J., Gogarten, J. P. (2006) Ancient horizontal gene transfer can benefit phylogenetic reconstruction. *Trends Genet* **22**, 361–6.

# Chapter 28

## Horizontal Gene Transfer in Eukaryotic Parasites: A Case Study of *Entamoeba histolytica* and *Trichomonas vaginalis*

**U. Cecilia Alsmark, Thomas Sicheritz-Ponten, Peter G. Foster, Robert P. Hirt, and T. Martin Embley**

### Abstract

Over the past few years it has become apparent that horizontal gene transfer (HGT) has played an important role in the evolution of pathogenic prokaryotes. What is less clear is the exact role that HGT has played in shaping the metabolism of eukaryotic organisms. The main problems are the reliable inference of HGT on a genomic scale as well as the functional assignment of genes in these poorly studied organisms. We have screened the completed genomes of the protists *Entamoeba histolytica* and *Trichomonas vaginalis* for cases of HGT from prokaryotes. Using a fast primary screen followed by a conservative phylogenetic approach, we found 68 and 153 recent cases of HGT in the respective organisms. The majority of transferred genes that fall into functional categories code for enzymes involved in metabolism. We found a broad range of prokaryotic lineages represented among the donors, but organisms that share similar environmental niches with *E. histolytica* and *T. vaginalis*, such as the gut and the vaginal mucosa, dominate.

**Key words:** *Entamoeba histolytica, Trichomonas vaginalis*, genome-wide analysis, phylogeny, metabolic genes, donor lineages, sampling.

## 1. Introduction

HGT plays a significant role in prokaryotic genome evolution, contributing up to ~20% of the content of a given genome *(1)*. HGT thus provides an efficient means of gaining new phenotypes, such as resistance to antibiotics and new physiological and metabolic capabilities, permitting or facilitating adaptation to new ecological niches *(2–4)*. More recently, data from microbial eukaryotes suggest that HGT also plays a role in

Maria B. Gogarten et al. (eds.), *Horizontal Gene Transfer: Genomes in Flux, vol. 532*

DOI 10.1007/978-1-60327-853-9_28 Springerprotocols.com

eukaryotic genome evolution, particularly among protists that can feed on bacteria *(1, 5–10)*. The sequencing of the genomes of two anaerobic parasites of humans, *Entamoeba histolytica (10)* and *Trichomonas vaginalis (11)*, has recently been completed. The availability of these data provides an excellent opportunity to investigate the role of HGT in the evolution of parasitic protists and in shaping eukaryotic genomes more generally. *Trichomonas vaginalis*, an anaerobic, flagellated parabasalid, is the most common, non-viral, sexually transmitted parasite of humans in the developed and developing world, mainly affecting women by colonising the urethra and the vaginal mucosa *(12)*. *Entamoeba histolytica* is a microaerophilic phagotrophic parasite living in the human gut *(13)*, an environment that is rich in microorganisms and where HGT is thought to be common between bacteria *(14)*. These two human parasites belong to distinct phylogenetic groups and provide ideal model systems for investigating prokaryote-to-eukaryote HGT.

## 2. Methods

### 2.1. Primary Screen

We set up a screening pipeline to detect HGT from a prokaryote donor to the ancestors of *T. vaginalis* and *E. histolytica*, respectively. For all analyses we used a previously published suite of python scripts and modules called PyPhy *(15)*, which we modified and adapted for large eukaryotic genome projects (SpyPhy, http://www.cbs.dtu.dk/staff/thomas/pyphy/spyphy.html). As many genes appear to be present in multiple copies, we first carried out a cluster analysis to identify putative paralogues using Blastclust (length coverage threshold = 0.9, score average threshold = 75, and this length coverage threshold must be met for both neighbours) from the BLAST package *(16)*. The initial proteomes were thus reduced to a number of clusters or singletons at the level of $\geq$ 75% identity over $\geq$ 90% of the gene length for cluster members. A single randomly chosen representative of each cluster was used as a seed for BLASTP searches against a database obtained by merging TrEMBL and Swissprot. Sequences above an empirically determined given cutoff value (40% similarity over a minimum of 70% of the protein length), and all cluster members were aligned using ClustalW *(17)*. The resulting alignments were automatically processed (allowed gap positions = half, minimum length of a block = 2, maximum number of contiguous non-conserved positions = 20) using GBLOCKS *(18)* to remove sequence positions where the inference of positional homology was ambiguous. Bootstrap (100 replicates) consensus p-distance trees were made using a PAUP pipeline from edited alignments of all proteins, for which there were sufficient homologues ( $\geq 3$) in SwissProt and TrEMBL to make trees. The trees were analysed to identify those trees where the nearest neighbour to the input gene was a prokaryote. As an additional primary screen for putative HGT, we identified all proteins for which a prokaryote was the top BLAST hit. These initial screens identified about 1000 candidate HGT for each organism. After inspection of alignments, BLAST outputs, tree support values and sequence identities, cases of potential HGT were retained for more detailed phylogenetic analyses. During the manual inspection we discarded short versions of genes already represented, genes that show similarity to a prokaryote orthologue due to a biased amino-acid composition (for example, LRR-repeats) and genes where only a short domain shows similarity to a prokaryote gene.

### 2.2. Secondary Screen

Each candidate for HGT was analysed by MrBayes *(19)* using the WAG matrix, a gamma correction for site rate variation and a proportion (pinvar) of invariable sites. The analyses were run for 600,000 generations and sampled every 100 generations, with the first 2000 samples discarded as a conservative burn-in. A consensus tree was made from the remaining samples. Because posterior probabilities, the support values used by Bayesian analysis to indicate confidence in groups, have been criticised *(20)*, we also use bootstrapping to provide an additional indication of support for relationships. Each data set was bootstrapped (100 replicates) and used to make distance matrices under the same evolutionary model as in the Bayesian analysis, using custom (P4) software *(21)*. Trees were made from the distance matrices using FastME *(22)* and a bootstrap consensus tree made using P4.

### 2.3. Evaluation of Trees

Our two-step screen is aimed at detecting recent HGT from a prokaryote donor to an ancestor of *Entamoeba* or *Trichomonas*. By recent HGT, we include cases where the query genes from *T. vaginalis* or *E. histolytica* were clustered inside well-supported prokaryotic groups and/or were separated from any other eukaryotes in the Bayesian tree, by at least two well-supported (posterior probabilities $\geq 0.95$, bootstrap $\geq 70$) nodes. In cases when tree topologies were more weakly supported but still suggested a possible HGT, we scrutinised bootstrap partition tables for partitions where the query sequence clustered with another eukaryote. If no such partitions were found we considered that gene also to be a putative HGT. We also considered a case to be a putative HGT if no other eukaryote contained the gene in question as determined by BLAST and HSSP scores falling below our thresholds (see below). We focused on recent HGT because they represent the most robust, least controversial, and easily detected examples of HGT. Our screen was not designed to detect more ancient gene transfers such as those from the mitochondrial endosymbiont *(23)* or those occurring at the base of major eukaryotic clades (*see* **Chapter 7**).

## 3. Results

### 3.1. Prevalence of HGT

For *E. histolytica* a total of 5740 trees were made and 548 of these were selected from the primary screen as representing potential HGT. These included all of the genes that had previously been published as HGT for *Entamoeba (8)*. The 548 candidate genes were then processed through the secondary screen to make better trees using a Bayesian approach. From the 548 trees HGT was inferred in 96 cases using the stringent criteria we applied. The remaining 452 trees are not discussed further here, apart from commenting that their topologies rarely reflected the relationships among taxa which are depicted in the universal rRNA tree of life.

In *Trichomonas vaginalis* we found 153 genes where prokaryote-to-eukaryote HGT, from diverse prokaryotes, is supported either by phylogenetic trees or by the presence of a gene in *Trichomonas* and prokaryotes, but in no other eukaryote sampled. We identified 76 genes where the most straightforward interpretation of trees indicated that HGT from a prokaryote to an ancestor of *T. vaginalis* had occurred. We also identified 77 genes where *Trichomonas* and diverse prokaryotes possess a particular gene, but we were unable to find a convincing homologue among other eukaryotes. To infer putative homology, we used a structure-based criterion called the HSSP score *(24, 25)*. The HSSP approach was designed to identify protein homologues in the so-called twilight zone of 20–35% amino acid identity *(24)*. The HSSP curve plots the number of aligned residues and percent residue identity to identify a length dependent threshold above which homology is likely, based upon comparisons of known homologous protein structures (i.e. true positives). In our analysis we identified eukaryotic and prokaryotic proteins as putative homologues when they showed HSSP scores with a distance of $\geq 10$ from this threshold value. At this distance from threshold, the frequency of false positives is effectively zero *(24)*.

### 3.2. Where Do the Genes Come from?

Based upon our trees certain prokaryote lineages are favoured as potential donors for genes now residing on the *Trichomonas vaginalis* and *Entamoeba histolytica* genomes. Thus, in *Trichomonas*, of 42 unrooted trees where we could identify a nearest neighbour to the *Trichomonas* gene, 15 of these trees placed the *Trichomonas* gene next to a member of the Bacteroidetes phylum

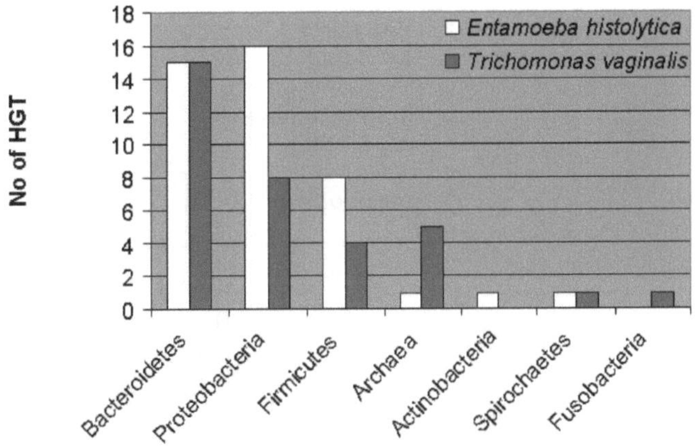


Fig. 28.1. **Taxonomic distribution of putative donors of genes to *Entamoeba histolytica* and *Trichomonas vaginalis*.** Among the 76 trees strongly supporting an HGT event from a prokaryotic donor, a total of 42 trees recovered the *Trichomonas vaginalis* sequence(s) as the nearest *neighbour* to a single species or a coherent prokaryotic taxon with $\geq 70\%$ bootstrap support and $\geq 0.95$ posterior probability. For *Entamoeba histolytica*, 34 of 68 HGT gave a tree with a nearest neighbour by the same criteria. In both species, the majority of such nearest neighbours comprised one or more members of the Bacteroidetes. Assuming a root outside of the specific *Trichomonas* or *Entamoeba*-prokaryote(s) split, these trees support the notion that *Trichomonas* or *Entamoeba* received the gene from an ancestor shared with that prokaryote lineage. The trees that did not resolve the position of *Trichomonas vaginalis / Entamoeba histolytica* sequences among prokaryotic homologues are not represented in the chart.

(**Fig. 28.1**). A similar bias towards the Bacteroidetes as potential donors was observed (**Fig. 28.1**) for candidate HGT for *Entamoeba histolytica (10)*. In 15 well-resolved trees, *Entamoeba* was recovered next to a member of the Bacteroidetes/Chlorobi group. Bacteroidetes are abundant anaerobic members of the human intestinal flora *(14)* where *Entamoeba* typically resides, and most trichomonads are also associated with the digestive tract and adjacent mucosa *(12)*. For example, Bacteria from the digestive tract are commonly present in the urogenital *(26)* tract where *Trichomonas vaginalis* typically resides. Members of the Bacteroidetes/Chlorobi and *Fusobacterium* (a pathogen colonising the oral cavity) groups are all obligate anaerobes. The observed bias for donor lineages is thus consistent with the idea that anaerobic prokaryotic and eukaryotic cohabitants of the human mucosa are sharing genes *(27)*. **Figure 28.2** shows an intriguing example where *Trichomonas vaginalis* clusters with members of the Bacteroidetes/Chlorobi with maximum support values, and *Entamoeba* clusters with *Fusobacterium*.

### 3.3. What Kinds of Genes Are Being Transferred?

Most of the transferred genes that can be assigned to a functional category are enzymes involved in metabolism. Furthermore, the genes transferred are diverse and affect many different metabolic pathways; HGT has thus profoundly influenced the

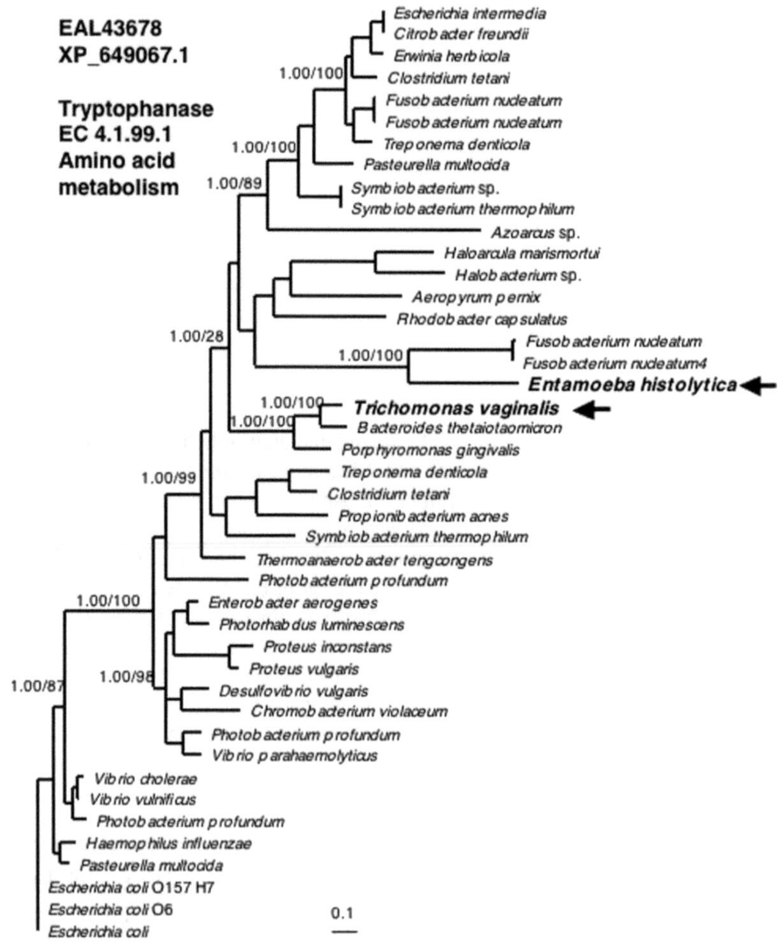


Fig. 28.2. **Example of a phylogenetic tree supporting two strong inferences of HGT.** Consensus MrBayes tree with support values corresponding to the posterior probabilities of the Bayesian analysis followed by the corresponding bootstrap support value of the equivalent maximum likelihood distances analysis (see methods). The tree suggests that the *Entamoeba* tryptophanase was acquired by HGT from a relative of the anaerobic bacterial genus *Fusobacterium*. By contrast, the *Trichomonas vaginalis* gene appears to have a separate origin by HGT from a relative of the anaerobic *Bacteroides* group. The *scale bar* represents 10% of inferred sequence divergence. Both the GenBank and RefSeq accession numbers are given for the *Entamoeba* entry. The EC number is also shown.

evolution of the *Trichomonas* and *Entamoeba* genomes and their metabalomes. Mapping the HGT onto a schematic of *Entamoeba* metabolism *(10)* indicates that HGT has affected some important pathways including iron-sulfur cluster biosynthesis, amino acid metabolism, and nucleotide metabolism (**Fig. 28.3**). In *Trichomonas*, metabolic pathways affected include salvage pathways, amino acid metabolism, synthesis of lipophosphoglycan, and many more (**Fig. 28.4**). For *E. histolytica*, 45% of the inferred transferred genes are hypothetical or unclassified proteins and in

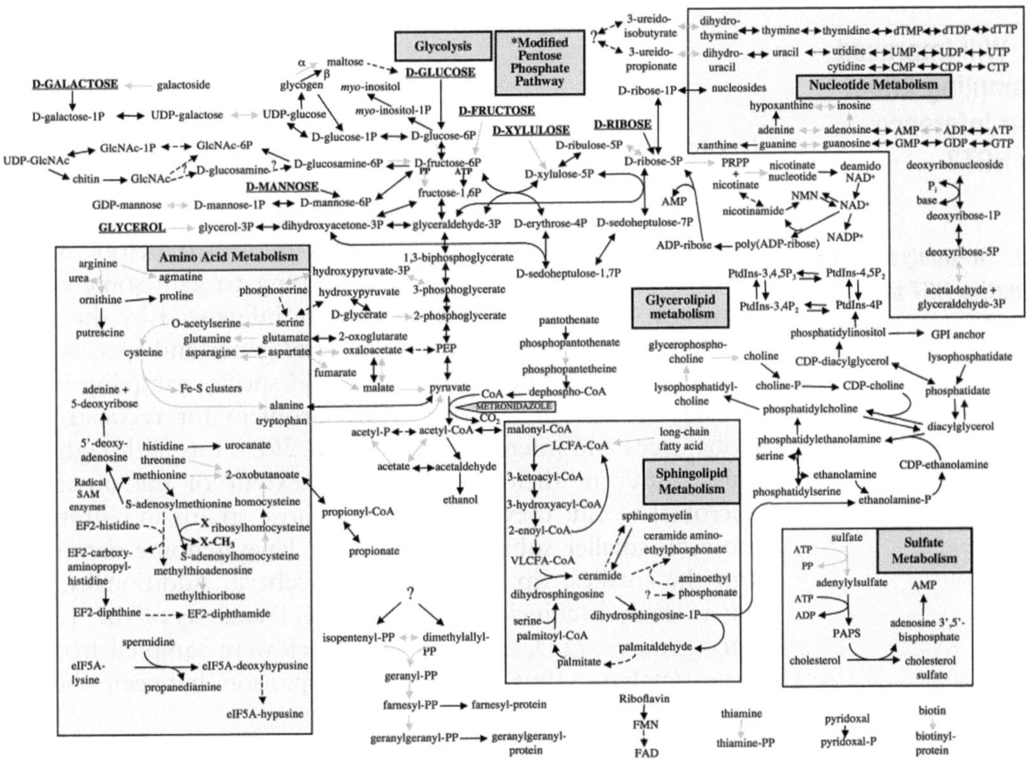


Fig. 28.3. **Predicted metabolic pathways of *E. histolytica* based on the analysis of its genome showing inferred HGT** *(9)*. Glycolysis and fermentation are the major energy generation pathways. *Bold grey arrows* represent enzymes encoded by genes that are among the 96 candidates for HGT into the *E. histolytica* genome. *Broken arrows* indicate enzymes for which no gene could be identified from the genome data, although the activity is thought to be present. The *framed arrow points* to the target of Metronidazole, the major drug for treatment of amoebic liver abscess. Abbreviations: PEP, phosphoenolpyruvate; GlcNAc, $N$-acetylglucosamine; LCFA, long chain fatty acid; VLCFA, very long chain fatty acid; PRPP, phosphoribosyl pyrophosphate; GPI, glycosylphosphatidylinositol; PAPS, phosphoadenosine phosphosulfate.

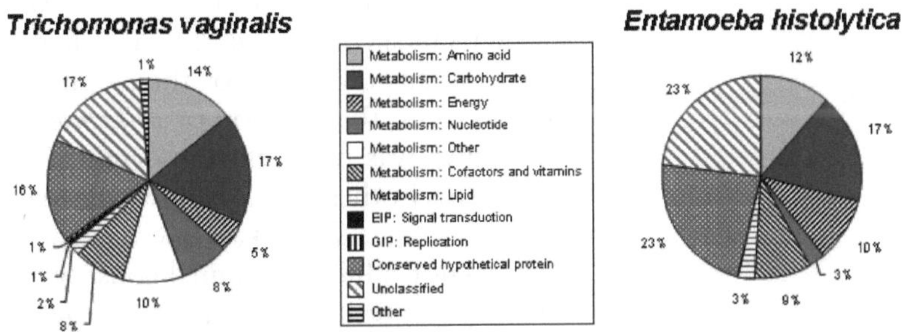


Fig. 28.4. **Functional categories among 153 *Trichomonas vaginalis* and 68 *Entamoeba histolytica* candidate HGT**. Distribution of functional annotation from the KEGG database among candidate HGT. The % values were rounded up.

*T. vaginalis* the corresponding value is 33%. These values may simply reflect the observation that for *Entamoeba histolytica* and *Trichomonas vaginalis* around 30% of the proteins predicted from the genome sequence are also hypothetical or unclassified.

## 4. Will Improved Sampling Affect Our Inference of HGT?

### *4.1. Re-analysis of 96 Putative HGT to Entamoeba histolytica*

In order to shed some light on how taxon sampling might affect the inference of HGT, we reanalysed the 96 putative HGTs we identified in the *Entamoeba* genome paper *(10)*, adding data from eukaryotic and prokaryotic genomes published before August 2005 *(27)*. In doing so, we were hoping to gain some insight into how our previous inferences were influenced by the sparse sampling of eukaryotic and prokaryotic genes and species available at the time. Such sparse gene and species sampling is, and is likely to remain, a very serious problem for reconstructing global trees and inferring HGT *(8, 28, 29)*. Thus, although ecologists differ in their claims for the extent of the unsampled microbial world, they all agree that those strains in culture and the even smaller subset for which we have genome data represent the smallest tip of a very large iceberg. Additionally, since this re-analysis coincided with our HGT analysis of the *T. vaginalis* genome *(11)*, taxa and sequences were sampled from the same database, thus facilitating comparison between the two genomes.

A total of 41 HGTs remain as strongly supported as before based upon the original criteria. For the remaining 56 tree topologies, support for recent HGT into the *Entamoeba* lineage is not as strong as before. For 27 of these 56 trees, where previously there were two strongly supported nodes separating *Entamoeba* from other eukaryotes, the hypothesis of HGT is now supported by only one well-supported node. However, close scrutiny of the bootstrap partition tables for these trees revealed that, as before, there are no trees in which *Entamoeba* is found together with another eukaryote. Thus, HGT still remains the best hypothesis to explain 68 (70%) of the original 96 topologies. In a further 14 trees, the position of *Entamoeba* among prokaryotes and eukaryotes was not well supported. The taxonomic sampling of eukaryotes in these trees is now very patchy and the trees do not resemble consensus eukaryotic relationships. Thus, although the trees do not fulfil our original conservative criteria for HGT, they do not provide strong support for the alternative hypothesis that the *Entamoeba* genes were vertically inherited from a common ancestor shared with other eukaryotes.

### *4.2. Eukaryote-to-Eukaryote Transfers?*

In nine of the re-sampled trees, *Entamoeba* either clustered with a single newly published eukaryotic sequence, or we could not reject such a relationship among mainly prokaryotic sequences. Six of these nine trees recovered *Entamoeba* and *Trichomonas* together, and two trees grouped *Entamoeba* with the diatom *Thalassiosira* (for example, *see* **Fig. 28.5**). Such trees are not easy to explain by simple vertical inheritance for the species concerned

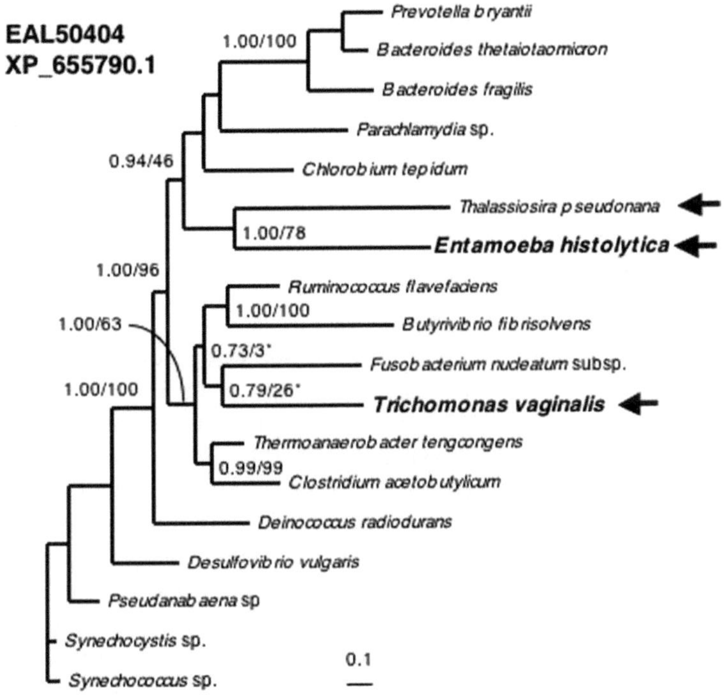


Fig. 28.5. **Some trees show unusual relationships for eukaryotes that are not easy to explain within the framework of the current consensus for eukaryotic phylogeny.** *Entamoeba* and the diatom *Thalassiosira* are closely related in the tree for glutamine synthase (EC 6.3.1.2), but no external data suggest that these two eukaryotes are closely related. One possible explanation is eukaryote-to-eukaryote HGT. *Trichomonas vaginalis* also contains a homologue of glutamine synthase, but in this case it clusters weakly with *Fusobacterium*. The *scale bar* represents 10% of inferred sequence divergence. Both the GenBank and RefSeq accession numbers are given for the *Entamoeba* entry.

within the framework of the current consensus for eukaryotic relationships *(30–32)*. Similar unusual topologies have been previously reported for other eukaryotes *(5)* (compare **Chapter 27**). The explanations advanced to explain the absence of the gene in other eukaryotes, include massive gene loss from eukaryotic lineages, or HGT between the eukaryotes concerned. *Entamoeba* can phagocytose both eukaryotes and prokaryotes, and it has been suggested that HGT between eukaryotes, after one of them had acquired the gene from a prokaryote, could explain such peculiar tree topologies and sparse eukaryotic sampling *(5)*. The fact that six among nine entries recover a relationship between *Entamoeba* and *Trichomonas*, the other mucosal pathogen of humans discussed here, is consistent with this idea. Recent large-scale analyses already support the hypothesis that prokaryotes from the same environment may share a set of niche-specific genes *(33, 34)*.

For five trees, the gene now appears to be present in several eukaryotes from different taxonomic groups and the trees cannot exclude a common origin for all eukaryotic sequences. Thus, for about 5% of the original trees, the simplest hypothesis to explain the observed pattern for the *Entamoeba* gene is no longer HGT, but vertical inheritance from a common ancestor shared with other eukaryotes.

## 5. Conclusions

The results of our study demonstrate that HGT has played a significant role in the evolution of the *Trichomonas* and *Entamoeba* genomes. The majority of functionally categorised HGTs are enzymes involved in metabolism, thus affecting various metabolic pathways. This is consistent with previously published continual transfer hypotheses, such as the complexity hypothesis *(35)*, which posit that HGT of genes involved in processing a single substrate are more likely to be transferred than those genes encoding proteins that interact with many other cellular components, such as proteins found in the ribosome. Furthermore, phylogenetic analyses imply that a broad range of donors have contributed the acquired genes with a bias towards prokaryotes that share the same ecological niche as *Entamoeba* and *Trichomonas.* Given that our screen was designed to detect only "recent" HGTs, the cases we record may be only the tip of a much larger historical iceberg *(1)* of transfers. The results are thus consistent with ideas that prokaryote to eukaryote transfers have occurred continually throughout eukaryotic history *(6)*.

## Acknowledgments

This work was supported by a Wellcome grant awarded to TME, RPH and UCMA. We would like to thank Dr Simon Harris for advice on HSSP analysis.

### References

1. Doolittle, W. F., Boucher, Y., Nesbo, C. L., Douady, C. J., Andersson, J. O., Roger, A. J. (2003) How big is the iceberg of which organellar genes in nuclear genomes are but the tip? *Philos Trans R Soc Lond B Biol Sci* **358**, 39–57.
2. Koonin, E. V., Makarova, K. S., Aravind, L. (2001) Horizontal gene transfer in prokaryotes: quantification and classification. *Annu Rev Microbiol* **55**, 709–42.
3. Lawrence, J. G. (2005) Horizontal and vertical gene transfer: the life history of pathogens. *Contrib Microbiol* **12**, 255–71.
4. Ochman, H., Lawrence, J. G., Groisman, E. A. (2000) Lateral gene transfer and the nature of bacterial innovation. *Nature* **405**, 299–304.
5. Andersson, J. O. (2005) Lateral gene transfer in eukaryotes. *Cell Mol Life Sci* **62**, 1182–1197.

6. Doolittle, W. F. (1998) You are what you eat: a gene transfer ratchet could account for bacterial genes in eukaryotic nuclear genomes. *Trend Genet* **14**, 307–11.
7. Lawrence, J. G. (2005) Common themes in the genome strategies of pathogens. *Curr Opin Genet Dev* **6**, 584–8.
8. Richards, T. A., Hirt, R. P., Williams, B. A., Embley, T. M. (2003) Horizontal gene transfer and the evolution of parasitic protozoa. *Protist* **154**, 17–32.
9. Andersson, J. O., Sjögren, A. M., Horner, D. S., Murphy, C. A., Dyal, P. L., Svärd, S. G., Logsdon, J. M. Jr, Ragan, M. A., Hirt, R. P., Roger, A. J. (2007) A genomic survey of the fish parasite *Spironucleus salmonicida* indicates genomic plasticity among diplomonads and significant lateral gene transfer in eukaryote genome evolution. *BMC Genomics.* **14**, 8–51.
10. Loftus, B., Anderson, I., Davies, R., Alsmark, U. C., Samuelson, J., Amedeo, P., Roncaglia, P., Berriman, M., Hirt, R. P., Mann, B. J., Nozaki, T., Suh, B., Pop, M., Duchene, M., Ackers, J., Tannich, E., Leippe, M., Hofer, M., Bruchhaus, I., Willhoeft, U., Bhattacharya, A., Chillingworth, T., Churcher, C., Hance, Z., Harris, B., Harris, D., Jagels, K., Moule, S., Mungall, K., Ormond, D., Squares, R., Whitehead, S., Quail, M. A., Rabbinowitsch, E., Norbertczak, H., Price, C., Wang, Z., Guillen, N., Gilchrist, C., Stroup, S. E., Bhattacharya, S., Lohia, A., Foster, P. G., Sicheritz-Ponten, T., Weber, C., Singh, U., Mukherjee, C., El-Sayed, N. M., Petri, W. A., Jr., Clark, C. G., Embley, T. M., Barrell, B., Fraser, C. M., Hall, N. (2005) The genome of the protist parasite *Entamoeba histolytica*. *Nature* **433**, 865–8.
11. Carlton, J. M., Hirt, R. P, Silva, J. C., Delcher, A. L., Schatz, M., Zhao, Q., Wortman, J. R., Bidwell, S. L., Alsmark, U. C., Besteiro, S., Sicheritz-Ponten, T., Noel, C. J., Dacks, J. B., Foster, P. G., Simillion, C., Van de Peer, Y., Miranda-Saavedr, D., Barton, G. J., Westrop, G. D., Müller, S., Dessi, D., Fiori, P. L., Ren, Q., Paulsen, I., Zhang, H., Bastida-Corcuera, F. D., Simoes-Barbos, A., Brown, M. T., Hayes, R. D., Mukherjee, M., Okumura, C. Y., Schneider, R., Smith, A. J., Vanacova, S., Villalvazo, M., Haas, B. J., Pertea, M., Feldblyum, T. V., Utterback, T. R., Shu, C. L., Osoegawa, K., de Jong, P. J., Hrdy, I., Horvathova, L., Zubacova, Z., Dolezal, P., Malik, S. B., Logsdon, J. M. Jr., Henze, K., Gupta, A., Wang, C. C., Dunne, R. L., Upcroft, J. A., Upcroft, P., White, O., Salzberg, S. L., Tang, P., Chiu, C. H., Lee, Y. S., Embley, T. M., Coombs, G. H., Mottram, J. C., Tachezy, J., Fraser-Liggett, C. M., Johnson, P. J. (2007) Draft genome sequence of the sexually transmitted pathogen *Trichomonas vaginalis. Science.* **315**, 207–12.
12. Nanda, N., Michel, R. G., Kurdgelashvili, G., Wendel, K. A. (2006) Trichomoniasis and its treatment. *Expert Rev Anti Infect Ther* **1**, 125–35.
13. Petri, W. A. Jr., (2000) Protozoan parasites that infect the gastrointestinal tract. *Curr Opin Gastroenterol.* **1**, 18–23.
14. Shoemaker, N. B., Vlakamis, H., Slyers, A. A. (2001) Evidence for extensive resistance gene transfer among *Bacteroides* spp. and among *Bacteroides* and other genera in the human colon. *Appl Environ Microbiol* **67**, 561–8.
15. Sicheritz-Ponten, T., Andersson, S. G. (2001) A phylogenomic approach to microbial evolution. *Nucleic Acids Res* **29**, 545–52
16. Altschul, S. F., Madden, T. L., Schaffer, A. A., Zhang, J., Zhang, Z., Miller, W., Lipman, D. J. (1997) Gapped BLAST and PSI-BLAST: a new generation of protein database search programs. *Nucleic Acids Res* **25**, 3389–402.
17. Thompson, J. D., Higgins, D. G., Gibson T. J. (1994). Clustal W: improving the sensitivity of progressive multiple sequence alignment through sequence weighting, position-specific gap penalties and weight matrix choice. *Nucleic Acids Res* **22**, 4673–80.
18. Castresana, J. (2000) Selection of conserved blocks from multiple alignments for their use in phylogenetic analysis. *Mol Biol Evol* **17**, 540–52
19. Huelsenbeck, J. P., Ronquist, F. (2001) MRBAYES: Bayesian inference of phylogenetic trees. *Bioinformatics* **17**, 754–55.
20. Cummings, M. P., Handley, S. A., Myers, D. S., Reed, D. L., Rokas, A., Winka, K. (2003) Comparing bootstrap and posterior probability values in the four-taxon case. *Syst Biol* **52**, 477–87.
21. Foster, P. G. P4. http://www.bmnh.org/~pf/p4.html
22. Desper, R., Gascuel, O. (2004) Theoretical foundation of the balanced minimum evolution method of phylogenetic inference and its relationship to weighted least-squares tree fitting. *Mol Biol Evol* **21**, 587–98.
23. Timmis, J. N., Ayeliffe, M. A., Huang, C. Y., Martin,W. (2004) Endosymbiotic gene transfers: organelle genomes forge eukaryotic chromosomes. *Nat Rev Genet* **5**, 123–35.
24. Rost, B., (1999) Twilight zone of protein sequence alignments. *Protein Eng* **12**, 85–94.
25. Nair, R., Rost, B. (2002) Sequence conserved for subcellular localization. *Protein Sci* **11**, 2836–47.
26. Reid, G., Bocking, A. (2003) The potential for probiotics to prevent bacterial vagi-

nosis and preterm labor. *Am J Obstet Gynecol.* **189**, 1202–8.
27. Clark, C. G., Alsmark, U. C. M., Hofer, M., Saito-Nakano,Y., Ali, V., Marion, S., Weber, C., Mukherjee, C., Bruchhaus, I., Tannich, E., Leippe, M., Sicheritz-Ponten, T., Foster, P. G. Samuelson, J., Noël, C. J., Hirt, R. P., Embley, T. M., Gilchrist, C. A., Mann, B. J., Singh, U., Ackers, J. P., Bhattacharya, S., Bhattacharya, A., Lohia, A.,Guillén, N., Duchêne, M., Nozaki, T., Hall, N. (2007) Structure and Content of the *Entamoeba histolytica* Genome. *Adv Parasitol,* **65**, 51–190.
28. Andersson, J. O., Doolittle, W. F., Nesbo, C. L. (2001) Genomics. Are there bugs in our genome? *Science* **292**, 1848–50.
29. Salzberg, S. L., White, O., Peterson, J., Eisen, J. A. (2001) Microbial genes in the human genome: lateral transfer or gene loss? *Science* **292**, 1903–6.
30. Baldauf, S. L. (2003) The deep roots of eukaryotes. *Science* **300**, 1703–6.
31. Embley, T. M., Martin, W. (2006) Eukaryotic evolution, changes and challenges. *Nature.* **440**, 623–30.
32. Adl, S. M., Simpson, A. G., Farmer, M. A., Andersen, R. A., Anderson, O. R., Barta, J. R., Bowser, S. S., Brugerolle, G., Fensome, R. A., Fredericq, S., James, T. Y., Karpov, S., Kugrens, P., Krug, J., Lane, C. E., Lewis, L. A., Lodge, J., Lynn, D. H., Mann, D. G., McCourt, R. M., Mendoza, L., Moestrup, O., Mozley-Standridge, S. E., Nerad, T. A., Shearer, C. A., Smirnov, A. V., Spiegel, F. W., Taylor, M. F. (2005) The new higher level classification of eukaryotes with emphasis on the taxonomy of protists. *J Eukaryot Microbiol.* **52**, 399–451.
33. Beiko, R. G., Harlow, T. J., Ragan, M. A. (2005) Highways of gene sharing in prokaryotes. *Proc Natl Acad Sci U S A* **102**, 14332–7.
34. Mira, A., Pushker, R., Legault, B. A., Moreira, D., Rodriguez-Valera, F. (2004) Evolutionary relationships of *Fusobacterium nucleatum* based on phylogenetic analysis and comparative genomics. *BMC Evol Biol* **4**, 50.
35. Jain, R., Rivera, M. C., Lake, J. A. (1999) Horizontal gene transfer among genomes: the complexity hypothesis. *Proc Natl Acad Sci U S A* **96**, 3801–6.

# Chapter 29

# Role of Horizontal Gene Transfer in the Evolution of Photosynthetic Eukaryotes and Their Plastids

Patrick J. Keeling

## Abstract

Plastids are the organelles derived from a cyanobacterium through endosymbiosis. Unlike mitochondria, plastids are not found in all eukaryotes, but their evolution has an added layer of complexity since plastids have moved between eukaryotic lineages by secondary and tertiary endosymbiotic events. This complex history, together with the genetic integration between plastids and their host, has led to many opportunities for gene flow between phylogenetically distinct lineages. Some intracellular transfers do not lead to a protein functioning in a new environment, but many others do and the protein makeup of many plastids appears to have been influenced by exogenous sources as well. Here, different evolutionary sources and cellular destinations of gene flow that has affected the plastid lineage are reviewed. Most horizontal gene transfer (HGT) affecting the modern plastid has taken place via the host nucleus, in the form of genes for plastid-targeted proteins. The impact of this varies greatly from lineage to lineage, but in some cases such transfers can be as high as one fifth of analyzed genes. More rarely, genes have also been transferred to the plastid genome itself, and plastid genes have also been transferred to other non-plant, non-algal lineages. Overall, the proteome of many plastids has emerged as a mosaic of proteins from many sources, some from within the same cell (e.g., cytosolic genes or genes left over from the replacement of an earlier plastid), some from the plastid of other algal lineages, and some from completely unrelated sources.

**Key words:** Endosymbiotic gene transfer, endosymbiotic gene replacement, plastid-targeting, endosymbiosis.

## 1. Introduction

Mitochondria and plastids arose by the endosymbiotic uptake and retention of an alpha-proteobacterium and a cyanobacterium, respectively *(1)*. These endosymbionts were reduced in complexity and substantially integrated with their host, primarily though

Maria B. Gogarten et al. (eds.), *Horizontal Gene Transfer: Genomes in Flux, vol. 532*

DOI 10.1007/978-1-60327-853-9_29 Springerprotocols.com

the transfer of genes to the host genome and the targeting of proteins back to the endosymbiont. The processes of reduction and integration were seemingly fairly similar since the resulting two organelles now share a number of characteristics in common *(2)*, but the subsequent evolution of the two differed in several interesting ways. Mitochondria arose through a single endosymbiosis and the organelle or some derivative of it has been identified in virtually all eukaryotes (the few cases where it is not yet known for certain are most likely due to lack of evidence rather than absence), so it is now thought to have originated in the common ancestor of all known extant eukaryotes (**Fig. 29.1**), *(3)*. The evolution of plastids differed in two interesting ways. Like mitochondria, plastids also arose through a single endosymbiosis; there was some debate about the common origin of canonical plastids, and there are other cyanobacterial endosymbionts with very narrow taxonomic distribution, which may be sufficiently integrated to consider them plastids *(4–8)*, but for the purposes of this discussion these exceptions will not be considered further. However, unlike mitochondria, this endosymbiosis took place well after the origin and diversification of extant eukaryotes so that the lineage containing the endosymbiont consists only of three groups of algae, glaucophytes, red algae, green algae, and land plants *(9)*. Moreover, after the establishment of the plastid in these lineages, the organelle spread to other eukaryotes by further rounds of endosymbiosis. In "secondary" endosymbiosis, either a red or green alga was itself taken up and integrated into a new host and once again converted into an organelle (**Fig. 29.1**) *(10)*. Secondary endosymbiosis is known to have involved both green and red algae, so it must have taken place at least twice. Two lineages contain green secondary plastids, euglenids, and chlorarachniophytes, and both plastid and nuclear gene phylogenies support the conclusion that they acquired their plastids independently *(11)*. Red algal secondary plastids are known from cryptomonads, haptophytes, heterokonts, dinoflagellates, and apicomplexa. Despite this wider variety of groups and much debate (e.g. *(12–15)*), the current consensus is that these plastids originated from a single endosymbiosis event in the ancestor of these lineages, the so-called chromalveolates *(16)*. In addition to secondary endosymbiosis, additional events of endosymbiosis have taken place in dinoflagellate algae (**Fig. 29.1**). Here, the ancestral secondary red algal plastid has been lost or reduced several times and in a few lineages a new plastid has been acquired *(10, 17)*. When the new plastid is acquired from a secondary alga, they are referred to as tertiary plastids (these include plastids derived from cryptomonads, haptophytes and heterokonts); and when they are derived from a primary alga, they are called serial secondary plastids (this includes a single case where a green algal plastid has been taken up *(18)*). Of course, plastid gain is a long-term process, so

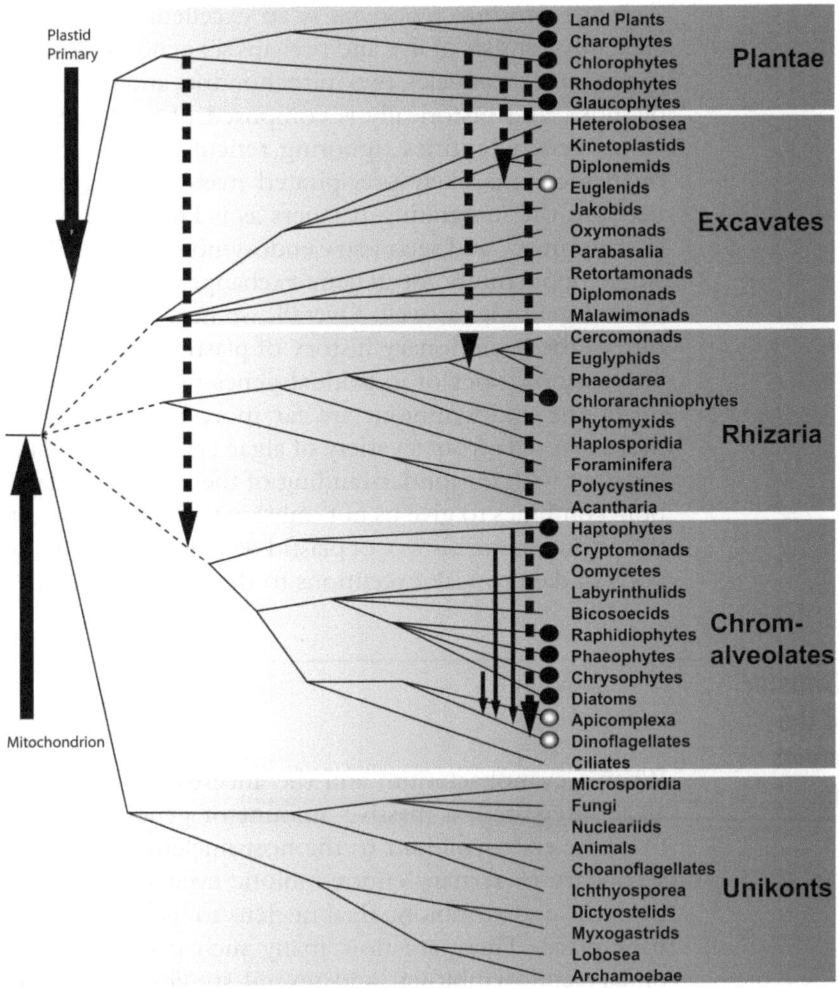


Fig. 29.1. Hypothetical tree of eukaryotes showing relationship of major lineages, including those discussed in text, and the many endosymbiotic events that explain modern plastid diversity. Many eukaryotic lineages (mostly protistan) are clustered into five major lineages, or supergroups, which are *shaded* and *named* to the *right*. Unknown relationships are indicated by polytomes, and supergroups of still contentious monophyly are indicated by the *dashed lines* at their base. Primary endosymbiosis (origin of mitochondria and primary plastids) is indicated by a *thick, solid arrow*. Secondary endosymbioses involving red or green algae are indicated by *slightly thinner, dashed arrows*. Tertiary endosymbioses in dinoflagellates are indicated by *thin, solid arrows*. Figure redrawn from *(64)*.

there are many intermediates in the spectrum between symbiont and organelle. There are also many lineages for which there is no evidence for a plastid, but which phylogenetically might be predicted to have once contained one. These may have lost the organelle or only lost photosynthesis, resulting in a cryptic plastid that has not yet been detected; in most cases the distinction cannot be made.

The evolution of photosynthetic lineages may therefore involve layers upon layers of endosymbiosis, making them genetically complex cells. The diatom plastid-containing dinoflagellate

*Kryptoperidinium foliaceum* is an excellent example *(19–21)*: it currently consists of five and perhaps six genome-containing compartments (two nuclei, two mitochondria, and one or perhaps two plastids), and historically is composed of no less than ten individual genetic entities (ignoring reticulations in the tree). Each of these events likely precipitated massive movements of DNA between the integrating partners as is known to have happened in the primary and secondary endosymbioses, and likely provided ample opportunity for genetic exchanges among the other component genomes as well. Overall, we now have a fairly clear picture of the evolutionary history of plastids *(10, 17)*, but the evolutionary histories of individual genes related to either the plastid or the endosymbiont are far more uncertain. Nevertheless, molecular data from a variety of algae are now abundant and this, together with our understanding of the history of plastid acquisition, combines to give us not only a set of specific expectations for the phylogenetic history of plastid derived genes but also a decent body of data to seek exceptions to those expectations.

## 2. Gene Transfer Related to the Establishment of Plastids

When a cyanobacterium and the ancestor of plants and primary algae integrated, a massive amount of gene transfer took place from the endosymbiont to the host nucleus *(1)*. Further, during secondary or tertiary endosymbiotic events, large-scale transfers from the endosymbiont algal nucleus to its new host nucleus also took place. There are now many such genes known from secondary endosymbioses, and several studies of how the proteins are targeted back to the organelle *(22–27)*. In the case of tertiary plastids there is comparatively little data, and the situation is more complex because all known tertiary plastids are in dinoflagellates, and therefore both the host and the endosymbiont had a plastid or plastid-containing ancestry (*see* **Section 3.2**. for more discussion on this). While these transfers are well characterized and certainly are important, they need to be distinguished from HGT in general because the circumstances surrounding these transfers are very different, as are their implications. Here, we will use the terms Endosymbiotic Gene Transfer (EGT) and Endosymbiotic Gene Replacement (EGR) to distinguish these special subsets of transfers, as described below.

### *2.1. Endosymbiotic Gene Transfer (EGT): Transfer of Genes Whose Products Are Targeted Back to the Organelle*

The plastid genome encodes only a small fraction of the genes needed for plastid function *(2)*. Most plastid proteins are encoded by host nuclear genes, and most of these were transferred from the cyanobacterium to the host nucleus in the ancestor of primary plastids (**Fig. 29.2-A**). The protein products of these genes are translated in the host cytoplasm and post-translationally targeted back to the organelle using a specific import pathway (**Fig. 29.2-1**), *(28)*. In secondary plastids, these genes were transferred yet again, in this case from the nucleus of the primary alga to the nucleus of the secondary one, and in tertiary plastids the same set of events occurred once more. Regardless of the number of times these genes are transferred, however, the environment in which their protein product functions remains unchanged, so while these transfers might affect the evolution of the gene (e.g. by moving to a compartment with different mutation rates), they do not so much generate novel functions or combinations of functions, but rather are a new way to organize information. They should therefore be distinguished from HGTs that yield novel combinations of proteins. These Endosymbiotic Gene Transfers (EGT) are undisputedly an important process, but will not be discussed further here.

### *2.2. Endosymbiotic Gene Replacement (EGR): Transfer of Genes Whose Products Are Not Targeted Back to the Organelle*

In addition to the transfer of hundreds of genes for proteins destined to be targeted back to the plastid, it is now hypothesized that many other cyanobacterial genes made their way into the host genome whose products do not function in the plastid today (**Fig. 29.2-2**) *(29)*. This process, referred to as Endosymbiotic Gene Replacement (EGR), in one way could be seen as a subclass of HGT because these genes have the potential to introduce new functions and combinations of functions. However, in another way they are like EGT because they come en masse from a single source, as opposed to a slower trickle of new genes from a wide variety of sources. In some cases transferred genes can be subjected to both EGT and EGR, since differentially expressed proteins can function in either compartment (aminoacyl-tRNA synthetases provide a great example as an entire class of proteins where dual targeting is common: *(30)*). However, the full potential impact of EGR had not been obvious until recent analyses of plants concluded that the contribution of the cyanobacterium to the plant lineage was substantial *(31)*. Several examples of EGR between two eukaryotes also seem to have occurred in the secondary endosymbiosis *(32)*. Interestingly, however, a large-scale analysis of another primary plastid-containing lineage, the glaucophytes, did not reach the same conclusion *(33)*. Here, many genes of cyanobacterial origin were found, but the protein products of most were concluded to be plastid-targeted. Whether the original estimates of EGR were too high, or alternatively why two systems seem to differ when they are products of the same endosymbiosis remains to be seen, and a detailed analysis of many of the cyanobacterial genes thought to encode cytosolic proteins also remains to be performed.

### 2.3. The Contribution of Chlamydial Genes to the Origin of Plastids

While the plastid may be derived from a cyanobacterium, recent large-scale analyses of plastid-targeted proteins have suggested that there is also an unexpected contribution from Chlamydiales. Chlamydiales are a group of pathogenic bacteria that have no known relationship to photosynthesis or plastids. Nevertheless, an unusual number of plastid-targeted proteins have been shown to share a closer relationship to chamydial homologs than they do to cyanobacterial homologs *(34–36)*. This was first described in plants, and was originally interpreted as either HGT or as a possible link between the plant host lineage and Chlamydiales *(34)*.

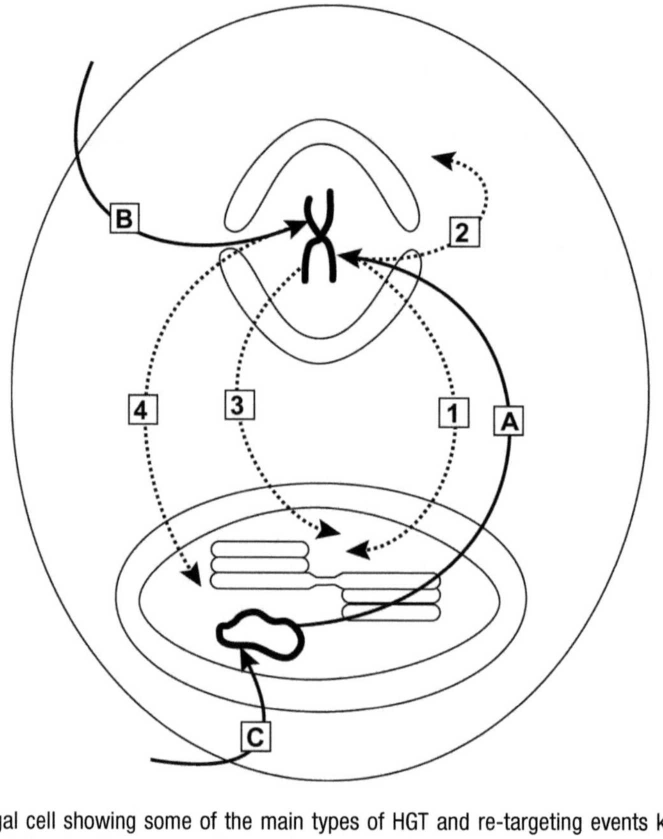


Fig. 29.2. Schematic of a primary algal cell showing some of the main types of HGT and re-targeting events known to have affected the protein makeup of the plastid and host. Gene transfers are indicated by *solid lines* labeled with letters, whereas protein-targeting is indicated by *dashed lines* labeled with numbers. (**A**) is plastid to nucleus EGT, which is widespread and common. Many protein products of these genes are targeted back to the plastid (**1**), and so the environment of the protein is unchanged. Others (**2**) may take up a function in the cytoplasm, and there is some indication that this might be common in some lineages as well. In contrast to this, the protein products of a few ancestrally nuclear genes were found to be targeted to the plastid (**3**), suggesting this process can work in both directions. (**B**) is gene transfer from an exogenous source to the nucleus. This has been well documented for numerous proteins in several algal lineages. The majority of such transfers are expected to result in cytosolic proteins (not shown), but many of the cases that are best described are proteins that are subsequently targeted to the plastid (**4**). Lastly, (**C**) is gene transfer from an exogenous source directly to the plastid genome. Such events have been documented, but are relatively rare. Not shown are the mitochondria, which could participate in the same kinds of events, or the added complexity of secondary or tertiary endosymbiosis, where genes can move from one eukaryote or its plastid to the other.

However, some of the same proteins are also found in a range of algae, and in at least some cases the genes encode plastid-targeted proteins. This was suggested to reflect a close relationship between Chlamydiales and cyanobacteria *(37)* rather than HGT;, however, this cannot explain many of the genes, where plant and algal homologs are robustly related to Chlamydiales to the exclusion of cyanobacteria *(38, 39)*. Most recently, these genes have been suggested to be relicts of a chlamydial presence in the host lineage that undertook the original primary endosymbiosis. It has been suggested that the presence of a chlamydial parasite in this host might have actually facilitated the transition of the cyanobacterium from a symbiont to an organelle, perhaps by providing essential transporters that were unlikely to have been present in a free-living cyanobacterium *(35, 39)*.

## 3. HGT of Genes for Plastid-Targeted Proteins

The establishment of the plastid-targeting system and the apparent ease of acquiring targeting peptides (inferred from the facts that transit peptide sequences are not highly conserved, and more importantly that they have been acquired in their thousands, repeatedly by several lineages) meant that HGT to either the plastid genome itself or to the nuclear genome could affect the plastid proteome. Here some cases will be reviewed where HGT to the nucleus or plastid, or EGT between nuclei have impacted the plastid proteome.

### *3.1. HGT of Plastid Targeted Proteins in Several Algal Systems*

Plastid-targeted proteins make an attractive class of proteins for detecting HGT in algae because they are relatively well sampled, and there is also a clear phylogenetic expectation that they are related to cyanobacteria-derived homologs present in green or red algal nuclear genomes. In contrast, host nuclear genes from many algal lineages are poorly sampled across eukaryotic diversity, and the closest relative of the host is often either not well known or even more poorly sampled *(10, 17)*. It is probably for these reasons, and not some intrinsic tendency of genes for plastid-targeted proteins to be transferred, that many of the first cases of HGT in algae came from genes for plastid-targeted proteins (**Fig. 29.2B**). Two groups where this is especially well examined are the chlorarachniophytes and the dinoflagellates. Chlorarachniophytes have a green algal plastid, and an analysis of expressed sequence tags (ESTs) from the model species *Bigelowiella natans* revealed that almost 20% of the genes for recognized plastid-targeted proteins were derived by HGT *(22)*. The majority of these were derived from other algae, several from red algae or lineages with red algal plastids, but others were derived from streptophyte green algae (some might be derived from chlorophyte green algae too, but these cannot be detected since the *B. natans* plastid is itself derived from a chlorophyte). A small fraction of these genes (glyceraldehyde-3-phosphate dehydrogenase and ribulose-5-phosphate-3-epimerase) were not related to the plastid lineage at all, but rather to other bacterial lineages. Interestingly, a closely related, plastid-targeted ribulose-5-phosphate-3-epimerase has recently been found in distantly related algae with red secondary plastids *(40)*, indicating a more complex history of this transfer. The chlorarachniophytes are myxotrophic, and so it was suggested they acquired genes from their food, and the tendency for the proteins to originate from algae was most likely due to the fact that many plastid proteins only exist in phototrophs *(22)*. Similar surveys in dinoflagellates also revealed slightly fewer cases of HGT, again mostly genes derived from other algae *(23, 24, 32, 41)*. The dinoflagellate plastid is derived from a red alga, so in this case the genes most clearly originating from HGT came from green algae, with a few genes coming from other sources.

Rather than analyzing all the plastid-targeted proteins from a random sample (like ESTs), other analyses have focused on the evolutionary history of all proteins in a certain pathway. For example, the analysis of the heme biosynthetic pathway in red algae and red algal–derived plastids shows some genes have the expected phylogenetic origin, whereas others seem to be derived from the host lineages, others still are proteobacterial and inferred to be derived from the mitochondrion *(42)*. Indeed, in apicomplexan parasites this process seems to have gone one step further, as the reactions themselves are distributed among various cellular compartments *(43)*. Similarly, the shikimate pathway for aromatic amino acid biosynthesis is composed of plastid-targeted proteins that are derived not only from cyanobacteria, but also at least two other bacterial lineages, so that the majority of the pathway appears to be proteins derived from HGT *(44)*.

Many other individual cases of plastid-targeted proteins being involved in HGT have been described, but only a few will be reviewed here. In the dinoflagellates one interesting example involves a gene fusion where the context of the source gene can be inferred. Here, the dinoflagellates and their sister lineages *Oxyrrhis marina* acquired an AroB gene that is closely related to an AroB paralog found only in a subgroup of cyanobacteria and not normally in plastids *(45)*. Immediately downstream of this gene in the cyanobacterial genomes is an *O*-methyl transferase of unknown function, and in dinoflagellates this gene has been fused to the 3′ end of AroB. The *O. marina* fusion protein is cytosolic, but in several other dinoflagellates the protein encodes a plastid-targeting leader. Interestingly, the AroB and *O*-methyl transferase moieties have been split up again in one species, and here the

*O*-methyl transferase has acquired its own plastid-targeting leader, suggesting it acquired some function in the plastid while attached to AroB.

Another interesting example is GAPDH, where the plastid-targeted protein in most dinoflagellates is only distantly related to that of cyanobacteria and other plastids, and instead appears to be derived from a duplication of the cytosolic GAPDH *(46, 47)*. In the dinoflagellate genera *Pyrocystis* and *Akashiwo*, however, a second, apparently plastid-targeted GAPDH has been found. This protein is of the cyanobacterial type, but they are specifically related to plastid-targeted homologues from euglenid algae, which have green algal plastids as opposed to the red algal plastids of dinoflagellates *(48, 49)*. This is even more intriguing since another GAPDH has also been found in another subgroup of dinoflagellates that is related to a different class of euglenid GAPDH again, suggesting a complex history of transfers between these two distantly related algal groups *(48)*.

An apparently more ancient transfer has been described for the plastid-targeted fructose-bisphosphate aldolase of chromalveolates, perhaps originally incorporated into the cell as a cytosolic enzyme, and becoming plastid-targeted after a gene duplication event *(50)*. This case is noteworthy because, along with a similar event involving GAPDH *(47, 51)*, it lends support to the single origin of the plastid in these lineages.

Overall, the impact of HGT on plastid proteins in algae is significant in some groups, but uneven. On one hand, some genes appear to be more prone to transfer than others, for example, GAPDH, where plastid-targeted proteins have arisen by HGT in at least some dinoflagellates and the chlorarachniophytes *(10, 48, 49)*, or glucose-6-phosphate isomerase *(52)* where a variety of transfer events appear possible. One the other hand, many of the cases described here affect the same or overlapping groups of algae (in particular chlorarachniophytes and dinoflagellates), whereas other groups show little or no evidence of HGT affecting plastid proteins. The same set of proteins analyzed in the chlorarachniophyte *B. natans* was also analyzed in the green alga *Chlamydomonas reinhardtii* and, in contrast to *B. natans*, none showed any evidence for transfer in *C. reinhardtii (10)*. In addition, most of these cases are apparently relatively recent and only affect one subgroup or perhaps a few species of the algae in question (one exception being plastid-targeted fructose-bisphosphate-aldolase).

### 3.2. Plastid-Plastid EGT in Algae with Redundant Plastids

As described in **Section 2**, the origin of tertiary plastids involves the movement of a great many genes from the nucleus of a secondary alga to the nucleus of its new host. This process is not unlike the origin of secondary plastids, but one very major difference is that all known tertiary plastids are found in dinoflagellates, which means both the host and the incoming secondary alga have (or had) a plastid of distinct evolutionary origin. In most cases no data for plastid-targeted proteins are available from these complex systems, but EST surveys have been carried out in two related genera with haptophyte-derived plastids, *Karlodinium micrum* and *Karenia brevis (25, 53)*. Since these organisms only contain the haptophyte-derived plastid, the null expectation would be that all or nearly all their plastid-targeted proteins should also be derived from the haptophyte, but their presence in the dinoflagellate nucleus raises the intriguing possibility that the original dinoflagellate plastid may have contributed proteins too. Indeed, the *K. micrum* survey revealed that about one-third of all phylogenetically analyzable plastid-targeted proteins were leftovers from the dinoflagellate plastid, and were re-targeted to the new haptophyte plastid, suggesting the proteome of this organelle is substantially chimeric *(25)*. Analysis of *K. brevis* ESTs also revealed this, and showed that proteins from other sources were also present, as with other dinoflagellates *(53)*. Interestingly, the targeting peptides in these organisms are different from those of either dinoflagellates or haptophytes *(25)*, and it is possible that this played a role in the mixing of these proteomes. As the targeting system shifted emphasis from the characteristics common to the ancestral haptophyte plastid, a dinoflagellate-derived protein might have been as likely to co-evolve with the import system as were the haptophyte-derived proteins. It is therefore plausible that being well adapted to the environment (which would presumably favor haptophyte proteins) was not the only or even the main determinant of which protein was retained in the new plastid *(25)*. Whatever the case may be, these are special cases of EGR where new combinations of plastid proteins were the outcome.

### 3.3. HGT of Plastid Proteins to Prokaryotes

While the subject of this review is the impact of HGT on plastids and their hosts, it is worthwhile to discuss whether plastids have also been the donor in HGT events. Of course, many of the events described in the preceding sections involve plastid-derived genes moving between plastid lineages (e.g., plastid-targeted proteins moving between algal groups), and plastid-derived proteins taking up a function in the cytosol would have the same overall effect as an HGT event, but what about plastid genes moving to lineages and new environments? Direct transfer of plastid DNA between plants has been documented, but it has not been shown that the genes moved to the plastid genome of the recipient, or that they are expressed *(54)*. It has been demonstrated that plastid DNA could be incorporated into the genome of the naturally competent soil bacterium *Acinetobacter* under experimental conditions *(55)*, but whether such transfers have occurred in nature is uncertain. The presence of *Chlamydia*-like genes in plants and algae, which led to the idea that a *Chlamydia*-like symbiont participated in the origin of plastids (**Section 2.3**) could also, in theory, be due to the transfer of plastid-derived genes to *Chlamydia*. This has been suggested in the case of *fabI* *(38)*, but in no case is the relationship of these genes to cyanobacteria so clear as to distinguish this possibility from a *Chlamydia*-to-plastid transfer *(39)*. One case involving an apparently functional gene transfer that is interesting for several reasons is the transfer of plastid-targeted FBA from red algae to the cyanobacteria *Prochlorococcus* and *Synechococcus* *(56)*. This is partly interesting because the recipient of the plastid gene is a cyanobacterium, which is ironic, but also because the gene has inserted beside the ancestral cyanobacterial FBA in the genome and its distribution within the genera where it is found is highly irregular, altogether suggesting a very complex history. This case is further complicated by the fact that the plastid gene in question is not actually ancestrally cyanobacterial either. Early in the evolution of the primary plastid lineage the host, cytosolic FBA duplicated and replaced the plastid-targeted analog (they are different and non-homologous classes of FBA) in the common ancestor of red and green algae *(57)*. Accordingly, the gene transferred to the cyanobacterium was not originally for a plastid-targeted protein.

## 4. HGT to Plastid Genomes

### 4.1. Direct Transfer of Genes Between Plastid Genomes

So far only nucleus-encoded genes, presumed to be cyanobacterial and presumed to be plastid targeted, have been discussed in any detail, but the plastid also has its own genome. However, HGT involving the plastid genome directly appears to be comparatively rare (**Fig. 29.2C**), in contrast to what has been found in plant mitochondria *(58)*. Many plastid genomes apparently encode unique proteins with no detectable similarity to any other protein, but the majority of these are likely not derived from HGT. One case that does appear to be transferred involves a mobile element found in the plastid-encoded *psbA* of *Euglena myxocylindracea* and a psychrophilic *Chlamydomonas* *(59, 60)*. In both species, phylogenetic analysis of a mobile group II intron showed it to be comparatively distant to group II introns in other plastid genomes, and instead closely related to group II introns from cyanobacteria *(59)*. The discovery of this intron in the primary green plastid of a *Chlamydomonas* and the secondary green plastid of a *Euglena* may be taken to suggest its origin in plastids is ancient, but the intron is absent in close relatives of both, and indeed in many other plastid lineages, and homing introns should be relatively likely to move. Two other cases of transfer into the plastid genome stand out as well, *rpl36* and *dnaX* *(61, 62)*. There are two distinct and distantly related paralogs of *rpl36* in bacteria, and plastid genomes typically contain only one of these, embedded within a conserved operon of ribosomal proteins. In the cryptomonads and haptophytes, however, the same operon contains an *rpl36* at the same position, but unusually it is a member of the other paralogous family *(61)*. This is significant for a number of reasons, first it is interesting that the incoming paralogue was inserted at the same position as the ancestral *rpl36*, and second it is a strong evidence for a relationship between the plastids of haptophytes and cryptomonads, which has since been supported by nuclear gene phylogenies *(13, 14)*. DnaX is a component of DNA polymerase responsible for repair. Neither DnaX nor any other DNA polymerase subunit has been found in a plastid genome previously, with the exception of several closely related species of the cryptomonad *Rhodomonas (62)*. Significantly, other cryptomonad plastid genomes lack the enzyme and it is not phylogenetically related to cyanobacteria, indicating it originated recently by HGT from some unidentified bacterial lineage.

## 5. Conclusion/ Outlook

The plastid proteome is much more complex than simply a residual collection of cyanobacterial proteins, and its relationship to the host is also more complicated than originally envisioned. The close connection between the two lineages has provided them with ample opportunity for proteins to move into a new environment, and this has happened many times. Similarly, genes for plastid-targeted proteins have moved between algal lineages many times, some genes showing a propensity to do this. The plastid genome seems resistant to HGT, so most of the changes to its proteome have occurred via the host nucleus, but even plastid genomes have been invaded by new genes on a few occasions. Overall, the plastid proteome is doubtlessly a variable mosaic of proteins from several sources, although the degree to which this is the case is still unclear. Genome-wide surveys in recent years have radically changed our view of the plastid proteome, so perhaps the biggest change we can expect in the years to come will emerge from analyzing plastid proteomics with an eye for invasion of foreign genes *(63)*. Such data will reveal which proteins are actually located in the organelle, something we infer with varying degrees of confidence now but do not really know in most cases. Without a more reliable catalog of plastid proteins, it is difficult to conclude what actually was the contribution of the host to the plastid proteome and to be certain of the plastid location of non-cyanobacterial genes derived from HGT.

## Acknowledgments

Work in the Keeling lab on HGT is supported by a grant from the Natural Sciences and Engineering Research Council of Canada and a grant to the Centre for Microbial Diversity and Evolution from the Tula Foundation.

### References

1. Gray, M. W., Burger, G., Lang, B. F. (1999) Mitochondrial evolution. *Science* **283**, 1476–81.
2. Gray, M. W. (1999) Evolution of organellar genomes. *Curr Opin Genet Dev* **9**, 678–87.
3. Williams, B. A. P., Keeling, P. J. (2003) Cryptic organelles in parasitic protists and fungi. *Adv Parasitol* **54**, 9–67.
4. Nowack, E. C., Melkonian, M., Glockner, G. (2008) Chromatophore genome sequence of *Paulinella* sheds light on acquisition of photosynthesis by eukaryotes. *Curr Biol* **18**, 410–8.
5. Keeling, P. J., Archibald, J. M. (2008) Organelle evolution: What's in a name? *Curr Biol* **18**, R345–7.
6. Palmer, J. D. (2003) The symbiotic birth and spread of plastids: How many times and whodunit? *J Phycol* **39**, 1–9.
7. Larkum, A. W., Lockhart, P. J., Howe, C. J. (2007) Shopping for plastids. *Trends Plant Sci* **12**, 189–95.
8. Keeling, P. J. (2004) The diversity and evolutionary history of plastids and their hosts. *Am J Bot* **91**, 1481–93.
9. Rodriguez-Ezpeleta, N., Brinkmann, H., Burey, S. C., Roure, B., Burger, G., Loffelhardt, W., Bohnert, H. J., Philippe, H., Lang, B. F. (2005) Monophyly of primary photosynthetic eukaryotes: Green plants, red algae, and glaucophytes. *Curr Biol* **15**, 1325–30.
10. Archibald, J. M., Keeling, P. J. (2002) Recycled plastids: A green movement in eukaryotic evolution. *Trends Genet* **18**, 577–84.
11. Rogers, M. B., Gilson, P. R., Su, V., Mcfadden, G. I., Keeling, P. J. (2007) The complete chloroplast genome of the chlorarachniophyte *Bigelowiella natans*: Evidence for independent origins of chlorarachniophyte and euglenid secondary endosymbionts. *Mol Biol Evol* **24**, 54–62.
12. Burki, F., Shalchian-Tabrizi, K., Minge, M., Skjaeveland, A., Nikolaev, S. I., Jakobsen, K. S., Pawlowski, J. (2007) Phylogenomics reshuffles the eukaryotic supergroups. *PLoS ONE* **2**, e790.
13. Hackett, J. D., Yoon, H. S., Li, S., Reyes-Prieto, A., Rummele, S. E., Bhattacharya, D. (2007) Phylogenomic analysis supports the monophyly of cryptophytes and haptophytes and the association of rhizaria with chromalveolates. *Mol Biol Evol* **24**, 1702–13.
14. Patron, N. J., Inagaki, Y., Keeling, P. J. (2007) Multiple gene phylogenies support the monophyly of cryptomonad and haptophyte host lineages. *Curr Biol* **17**, 887–91.
15. Leigh, J. W., Susko, E., Baumgartner, M., Roger, A. J. (2008) Testing congruence in phylogenomic analysis. *Syst Biol* **57**, 104–15.
16. Cavalier-Smith, T. (2004) Chromalveolate diversity and cell megaevolution: Interplay of membranes, genomes and cytoskeleton, in*Organelles, genomes and eukaryotic evolution* (Hirt, R. P., Horner D., ed.) Taylor and Francis, London 71–103.
17. Delwiche, C. F. (1999) Tracing the thread of plastid diversity through the tapestry of life. *Am Nat* **154, Supplement**, S164–77.
18. Watanabe, M. M., Suda, S., Inouye, I., Sawaguchi, I., Chihara, M. (1990) *Lepidodinium viride* gen. et sp. nov. (Gymnodiniales, Dinophyta), a green dinoflagellate with a chlorophyll *a*- and *b*-containing endosymbiont. *J Phycol* **26**, 741–51.
19. Chesnick, J. M., Hooistra, W. H., Wellbrock, U., Medlin, L. K. (1997) Ribosomal RNA analysis indicates a benthic pennate diatom ancestry for the endosymbionts of the dinoflagellates *Peridinium foliaceum* and *Peridinium balticum* (Pyrrhophyta). *J Eukaryot Microbiol* **44**, 314–20.
20. Inagaki, Y., Dacks, J. B., Doolittle, W. F., Watanabe, K. I., Ohama, T. (2000) Evolutionary relationship between dinoflagellates bearing obligate diatom endosymbionts: Insight into tertiary endosymbiosis. *Int J Syst Evol Microbiol* **50 Pt 6**, 2075–81.
21. Imanian, B., Keeling, P. J. (2007) The dinoflagellates *Durinskia baltica* and *Kryptoperidinium foliaceum* retain functionally overlapping mitochondria from two evolu-

tionarily distinct lineages. *BMC Evol Biol* **7**, 172.
22. Archibald, J. M., Rogers, M. B., Toop, M., Ishida, K., Keeling, P. J. (2003) Lateral gene transfer and the evolution of plastid-targeted proteins in the secondary plastid-containing alga *Bigelowiella natans*. *Proc Natl Acad Sci USA* **100**, 7678–83.
23. Bachvaroff, T. R., Concepcion, G. T., Rogers, C. R., Herman, E. M., Delwiche, C. F. (2004) Dinoflagellate expressed sequence tags data indicate massive transfer of chloroplast genes to the nuclear genome. *Protist* **155**, 65–78.
24. Hackett, J. D., Yoon, H. S., Soares, M. B., Bonaldo, M. F., Casavant, T. L., Scheetz, T. E., Nosenko, T., Bhattacharya, D. (2004) Migration of the plastid genome to the nucleus in a peridinin dinoflagellate. *Curr Biol* **14**, 213–8.
25. Patron, N. J., Waller, R. F., Keeling, P. J. (2006) A tertiary plastid uses genes from two endosymbionts. *J Mol Biol* **357**, 1373–82.
26. Patron, N. J., Waller, R. F., Archibald, J. M., Keeling, P. J. (2005) Complex protein targeting to dinoflagellate plastids. *J Mol Biol* **348**, 1015–24.
27. Ralph, S. A., Van Dooren, G. G., Waller, R. F., Crawford, M. J., Fraunholz, M. J., Foth, B. J., Tonkin, C. J., Roos, D. S., Mcfadden, G. I. (2004) Tropical infectious diseases: Metabolic maps and functions of the *Plasmodium falciparum* apicoplast. *Nat Rev Microbiol* **2**, 203–16.
28. Mcfadden, G. I. (1999) Plastids and protein targeting. *J Eukaryot Microbiol* **46**, 339–46.
29. Martin, W., Stoebe, B., Goremykin, V., Hansmann, S., Hasegawa, M., Kowallik, K. V. (1998) Gene transfer to the nucleus and the evolution of chloroplasts. *Nature* **393**, 162–5.
30. Duchene, A. M., Giritch, A., Hoffmann, B., Cognat, V., Lancelin, D., Peeters, N. M., Zaepfel, M., Marechal-Drouard, L., Small, I. D. (2005) Dual targeting is the rule for organellar aminoacyl-trna synthetases in *Arabidopsis thaliana*. *Proc Natl Acad Sci USA* **102**, 16484–9.
31. Martin, W., Rujan, T., Richly, E., Hansen, A., Cornelsen, S., Lins, T., Leister, D., Stoebe, B., Hasegawa, M., Penny, D. (2002) Evolutionary analysis of *Arabidopsis*, cyanobacterial, and chloroplast genomes reveals plastid phylogeny and thousands of cyanobacterial genes in the nucleus. *Proc Natl Acad Sci USA* **99**, 12246–51.
32. Li, S., Nosenko, T., Hackett, J. D., Bhattacharya, D. (2006) Phylogenomic analysis identifies red algal genes of endosymbiotic origin in the chromalveolates. *Mol Biol Evol* **23**, 663–74.
33. Reyes-Prieto, A., Hackett, J. D., Soares, M. B., Bonaldo, M. F., Bhattacharya, D. (2006) Cyanobacterial contribution to algal nuclear genomes is primarily limited to plastid functions. *Curr Biol* **16**, 2320–5.
34. Stephens, R. S., Kalman, S., Lammel, C., Fan, J., Marathe, R., Aravind, L., Mitchell, W., Olinger, L., Tatusov, R. L., Zhao, Q., Koonin, E. V., Davis, R. W. (1998) Genome sequence of an obligate intracellular pathogen of humans: *Chlamydia trachomatis*. *Science* **282**, 754–9.
35. Tyra, H. M., Linka, M., Weber, A. P., Bhattacharya, D. (2007) Host origin of plastid solute transporters in the first photosynthetic eukaryotes. *Genome Biol* **8**, R212.
36. Royo, J., Gimez, E., Hueros, G. (2000) CMP-KDO synthetase: A plant gene borrowed from gram-negative eubacteria. *Trends Genet* **16**, 432–3.
37. Brinkman, F. S., Blanchard, J. L., Cherkasov, A., Av-Gay, Y., Brunham, R. C., Fernandez, R. C., Finlay, B. B., Otto, S. P., Ouellette, B. F., Keeling, P. J., Rose, A. M., Hancock, R. E., Jones, S. J., Greberg, H. (2002) Evidence that plant-like genes in *Chlamydia* species reflect an ancestral relationship between Chlamydiaceae, cyanobacteria, and the chloroplast. *Genome Res* **12**, 1159–67.
38. Ryall, K., Harper, J. T., Keeling, P. J. (2003) Plastid-derived type II fatty acid biosynthetic enzymes in chromists. *Gene* **313**, 139–48.
39. Huang, J., Gogarten, J. P. (2007) Did an ancient chlamydial endosymbiosis facilitate the establishment of primary plastids? *Genome Biol* **8**, R99.
40. Rogers, M. B., Watkins, R. F., Harper, J. T., Durnford, D. G., Gray, M. W., Keeling, P. J. (2007) A complex and punctate distribution of three eukaryotic genes derived by lateral gene transfer. *BMC Evol Biol* **7**, 89.
41. Waller, R. F., Patron, N. J., Keeling, P. J. (2006) Phylogenetic history of plastid-targeted proteins in the peridinin-containing dinoflagellate *Heterocapsa triquetra*. *Int J Syst Evol Microbiol* **56**, 1439–47.
42. Obornik, M., Green, B. R. (2005) Mosaic origin of the heme biosynthesis pathway in photosynthetic eukaryotes. *Mol Biol Evol* **22**, 2343–53.
43. Sato, S., Clough, B., Coates, L., Wilson, R. J. (2004) Enzymes for heme biosynthesis are found in both the mitochondrion and plastid of the malaria parasite *Plasmodium falciparum*. *Protist* **155**, 117–25.
44. Richards, T. A., Dacks, J. B., Campbell, S. A., Blanchard, J. L., Foster, P. G., Mcleod,

R., Roberts, C. W. (2006) Evolutionary origins of the eukaryotic shikimate pathway: Gene fusions, horizontal gene transfer, and endosymbiotic replacements. *Eukaryot Cell* **5**, 1517–31.

45. Waller, R. F., Slamovits, C. H., Keeling, P. J. (2006) Lateral gene transfer of a multigene region from cyanobacteria to dinoflagellates resulting in a novel plastid-targeted fusion protein. *Mol Biol Evol* **23**, 1437–43.
46. Fagan, T., Woodland Hastings, J., Morse, D. (1998) The phylogeny of glyceraldehyde-3-phosphate dehydrogenase indicates lateral gene transfer from cryptomonads to dinoflagellates. *J Mol Evol* **47**, 633–9.
47. Fast, N. M., Kissinger, J. C., Roos, D. S., Keeling, P. J. (2001) Nuclear-encoded, plastid-targeted genes suggest a single common origin for apicomplexan and dinoflagellate plastids. *Mol Biol Evol* **18**, 418–26.
48. Takishita, K., Ishida, K., Maruyama, T. (2003) An enigmatic GAPDH gene in the symbiotic dinoflagellate genus *Symbiodinium* and its related species (the order Suessiales): Possible lateral gene transfer between two eukaryotic algae, dinoflagellate and euglenophyte. *Protist* **154**, 443–54.
49. Fagan, T. M., Hastings, J. W. (2002) Phylogenetic analysis indicates multiple origins of chloroplast glyceraldehyde-3-phosphate dehydrogenase genes in dinoflagellates. *Mol Biol Evol* **19**, 1203–7.
50. Patron, N. J., Rogers, M. B., Keeling, P. J. (2004) Gene replacement of fructose-1,6-bisphosphate aldolase (FBA) supports a single photosynthetic ancestor of chromalveolates. *Eukaryot Cell* **3**, 1169–75.
51. Harper, J. T., Keeling, P. J. (2003) Nucleus-encoded, plastid-targeted glyceraldehyde-3-phosphate dehydrogenase (GAPDH) indicates a single origin for chromalveolate plastids. *Mol Biol Evol* **20**, 1730–5.
52. Grauvogel, C., Brinkmann, H., Petersen, J. (2007) Evolution of the glucose-6-phosphate isomerase: The plasticity of primary metabolism in photosynthetic eukaryotes. *Mol Biol Evol* **24**, 1611–21.
53. Nosenko, T., Bhattacharya, D. (2007) Horizontal gene transfer in chromalveolates. *BMC Evol Biol* **7**, 173.
54. Park, J. M., Manen, J. F., Schneeweiss, G. M. (2007) Horizontal gene transfer of a plastid gene in the non-photosynthetic flowering plants *Orobanche* and *Phelipanche* (Orobanchaceae). *Mol Phylogenet Evol* **43**, 974–85.
55. De Vries, J., Herzfeld, T., Wackernagel, W. (2004) Transfer of plastid DNA from tobacco to the soil bacterium *Acinetobacter* sp. by natural transformation. *Mol Microbiol* **53**, 323–34.
56. Rogers, M. B., Patron, N. J., Keeling, P. J. (2007) Horizontal transfer of a eukaryotic plastid-targeted protein gene to cyanobacteria. *BMC Biol* **5**, 26.
57. Gross, W., Lenze, D., Nowitzki, U., Weiske, J., Schnarrenberger, C. (1999) Characterization, cloning, and evolutionary history of the chloroplast and cytosolic class I aldolases of the red alga *Galdieria sulphuraria*. *Gene* **230**, 7–14.
58. Bergthorsson, U., Adams, K. L., Thomason, B., Palmer, J. D. (2003) Widespread horizontal transfer of mitochondrial genes in flowering plants. *Nature* **424**, 197–201.
59. Sheveleva, E. V., Hallick, R. B. (2004) Recent horizontal intron transfer to a chloroplast genome. *Nucleic Acids Res* **32**, 803–10.
60. Odom, O. W., Shenkenberg, D. L., Garcia, J. A., Herrin, D. L. (2004) A horizontally acquired group II intron in the chloroplast psbA gene of a psychrophilic *Chlamydomonas*: In vitro self-splicing and genetic evidence for maturase activity. *RNA* **10**, 1097–107.
61. Rice, D. W., Palmer, J. D. (2006) An exceptional horizontal gene transfer in plastids: Gene replacement by a distant bacterial paralog and evidence that haptophyte and cryptophyte plastids are sisters. *BMC Biol* **4**, 31.
62. Khan, H., Parks, N., Kozera, C., Curtis, B. A., Parsons, B. J., Bowman, S., Archibald, J. M. (2007) Plastid genome sequence of the cryptophyte alga *Rhodomonas salina* CCMP1319: Lateral transfer of putative DNA replication machinery and a test of chromist plastid phylogeny. *Mol Biol Evol* **24**, 1832–42.
63. Van Wijk, K. J. (2004) Plastid proteomics. *Plant Physiol Biochem* **42**, 963–77.
64. Keeling, P. J., Palmer, J. D. (2008) Horizontal gene transfer in eukaryotic evolution. *Nat Rev Genet* **9**, 605–18.

# Chapter 30

## Role of Horizontal Gene Transfer in the Evolution of Plant Parasitism Among Nematodes

**Makedonka Mitreva, Geert Smant, and Johannes Helder**

### Abstract

Horizontal gene transfer (HGT) implies the non-sexual exchange of genetic material between species – in some cases even across kingdoms. Although common among Bacteria and Archaea, HGTs from pro- to eukaryotes and between eukaryotes were thought to be extremely rare. Recent studies on intracellular bacteria and their hosts seriously question this view. Recipient organisms could benefit from HGT as new gene packages could allow them to broaden or change their diet, colonize new habitats, or survive conditions that previously would have been lethal.

About a decade ago, plant parasitic nematodes were shown to produce and secrete cellulases. Prior to this, animals were thought to fully depend on microbial symbionts for the breakdown of plant cell walls. This discovery prompted Keen and Roberts *(1)* to hypothesize that the ability of nematodes to parasitize plants was acquired by HGT from soil bacteria to (ancestral) bacterivorous nematodes. Since the identification of the first nematode cellulases, many more plant cell wall–degrading enzymes (CWDE) have been identified in a range of plant parasitic nematode species.

Here we discuss a number of criteria that can be used to underpin an HGT claim. HGT requires close physical contact between donor and recipient, and this could be achieved in, for example, a symbiont–host, or a trophic relationship. The former type of relationship was indeed shown to potentially result in the transfer of genetic material (e.g., *Brugia malayi* and *Wolbachia*). However, currently known endosymbionts of nematodes may not be the source of CWDEs. Remarkably, all cellulases discovered so far within the order Tylenchida belong to a single glycoside hydrolase family (GHF5). A range of soil bacteria harbours GHF5 cellulases, but of course nothing can be said about the gene content of soil bacteria at the time HGT took place (if at all). We suggest that characterisation of cellulases (and other CWDEs) and their genomic organisation in more basal (facultative) plant parasitic Tylenchida is needed to find out if CWDEs were indeed acquired via HGT from bacteria. A more complete picture about the evolution of CWDEs among plant parasitic Tylenchida will require a detailed characterisation of two – so far – fully unexplored basal suborders, Tylenchina and Criconematina. Finally, we performed a computational high-throughput identification of potential HGT candidates (including ones unrelated to CWDEs) in plant parasitic nematodes using a genomics approach.

**Key words:** Nematodes, plant parasitic nematodes, endosymbiont, horizontal gene transfer, lateral gene transfer, cell wall–degrading enzymes, endo-glucanase, cellulase, bioinformatics, genomics.

Maria B. Gogarten et al. (eds.), *Horizontal Gene Transfer: Genomes in Flux, vol. 532*

DOI 10.1007/978-1-60327-853-9_30 Springerprotocols.com

## 1. Introduction

Nematodes are probably the most numerous animals on Earth. They are estimated to include more than one million species *(2)*, and are present worldwide in marine, freshwater and terrestrial habitats. The most well-known representative is *Caenorhabditis elegans*; the first animal of which the genome was completely sequenced *(3)*. *C. elegans* is a bacterivorous soil-inhabiting nematode, and other members of the speciose phylum Nematoda show a wide range of trophic ecologies. Apart from bacterial feeders, fungivorous nematodes, plant parasites, predators, omnivores, and parasites of invertebrates and vertebrate animals, including humans, are represented. Among plant parasitic nematodes, cyst (*Globodera*, **Fig. 30.1**, and *Heterodera* spp.) and root knot (*Meloidogyne* spp.) nematodes are most notorious for causing major damage to crops such as soybean, potato, and sugar beet, resulting in about 80 billion dollars in damage annually *(4)*. Human parasitic nematodes have a major, long-term impact (directly and indirectly) on human health and cause substantial suffering, particularly in children. The four most prevalent soil-transmitted species (*Ascaris lumbricoides, Trichuris trichura* and the hookworms, *Necator americanus* and *Ancylostoma ceylanicum*) alone infect nearly 3 billion people *(5)*. Furthermore, there are species that in combination with their endosymbiont

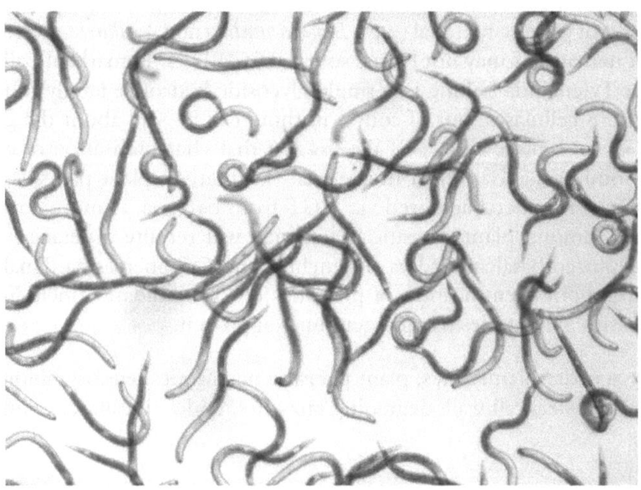

Fig. 30.1. Infective juveniles of the potato cyst nematodes (*Globodera rostochiensis*) each measuring about 0.5 cm in length. A hallmark of plant-parasitic nematodes is the black protrusible needle-like structure, named the oral stylet, at the head end of the nematode.

*Wolbachia* are causal agents of river blindness (*Onchocerca volvulus*), or elephantiasis (*Brugia malayi, Wucheria bancrofti*).

It has been hypothesized that the ability of nematodes to parasitize plants was acquired by HGT from soil bacteria to bacterivorous nematodes *(1)*. Nematodes as a group are suitable to study this hypothesis because within the phylum Nematoda plant parasitism arose independently multiple times *(6, 7)*, and because this feeding type is associated with a number of known, relatively well-characterized genes. In this chapter we concentrate on the origin of cellulases, enzymes that are used by plant parasitic nematodes to depolymerise plant (or oomycete) cell walls.

## 2. Plant Parasitism Arose Independently Multiple Times Within the Phylum Nematoda

The ribosomal DNA cistron and – more specifically an approximately 1,700 bp part thereof coding for the small subunit (SSU) of the ribosomal RNA – was shown to be very useful for the deduction of phylogenetic relationships within the phylum Nematoda. On the basis of 53 small subunit (SSU) ribosomal DNA sequences, Blaxter et al. *(6)* subdivided the phylum Nematoda into five major clades. Feeding types such as plant and animal parasitism were suggested to have arisen multiple times. More recently, Holterman et al. *(7)* analyzed 339 nearly full-length SSU rDNA sequences from the phylum Nematoda, and these data suggested a subdivision of the Nematoda into 12 clades, with the most basal clade dominated by representatives of the orders Enoplida and Triplonchida. Most likely, plant parasitism arose independently within the (sub) orders Triplonchida (Clade 1), Dorylaimida (Clade 2), and Tylenchina (Clade 12). This conclusion is supported by differences in stylet structure. – Stylets are protrusible hollow needles used by the nematode to puncture plant cell walls (**Fig. 30.1**). Plant parasitic members of the Triplonchida – members of the family Trichodoridae – are equipped with an onchiostyle, whereas the plant parasitic Dorylaimida and Tylenchina use an odonto (often referred to as "spear") or a stomatostyle (often referred to as "stylet"), respectively. The analysis of a relatively variable part of the large subunit (LSU) rDNA sequence suggests that there could be three independent plant parasite lineages within the order Dorylaimida: the Longidoridae family, and two smaller ones consisting of representatives of the genera *Pungentus* and *Longidorella (8)*. Hence, if we want to study the mechanisms underlying the evolution of plant parasitism, nematodes are interesting as the same phenomenon occurred perhaps five or even more times.

## 3. Detecting Horizontal Gene Transfer in Eukaryotes

Vertical gene transfer describes the transmission of traits (via genetic material) from parents to their offspring. Horizontal (or lateral) gene transfer (HGT), on the other hand, is a process of nonsexual exchange of genes between genomes of different species, and in some cases even across kingdom boundaries. HGT is a well-recognized phenomenon in the evolution of Bacteria and Archaea *(9–11)*. However, for animals, the role of HGT during their evolution is largely unknown, except in the context of mobile genetic elements and ancient acquisition of organelle genes *(12)*. There are no strict objective rules about how to test whether a given gene was acquired from another non-related organism via HGT. Nevertheless, a combination of methods can be proposed that together may – to a large extent – lift limitations of the individual approaches. Andersson *(13)* proposed a combination of phylogenetic methods (**Section 3.1**), the analysis of the distribution pattern (**Section 3.2**), and habitat overlap between inferred donor and recipient (**Section 3.3**). Furthermore, the evolutionary significance of the identified HGT candidate requires demonstration of the longevity and integration into the host biology *(14)*.

### 3.1. Phylogenetic Methods

Similarity searches based on the primary sequence level (e.g. BLAST, FASTA) may reveal that a given gene shows highest similarity to genes from phylogenetically non-related organisms. Further phylogenetic analysis of significantly matching sequences may then expose an unexpected disparity only between certain clades in the gene tree and the corresponding consensus organismal tree. Such results could be a first indication for horizontal gene transfer. However, similarity searches alone are rarely accepted as a basis for a HGT claim as there are a number of potential pitfalls, including:

- The current phylogeny is incorrect; the "non-related organism" is in fact much more related than suggested.
- "Best sequence similarity match" – even among dissimilar genes or predicted proteins, there will by definition be a set that is labeled as being most similar. This phenomenon may be especially prominent within classes of proteins that have overall high levels of conservation due to functional constraints.
- Skewed or insufficient sampling ("tip of the iceberg effect"). Mainly due to the publication of a very substantial number of bacterial genomes, there is a reasonable chance that a given unusual eukaryotic gene appears most closely related to prokaryotic genes because matching homologues from

related organisms are not yet in the database. For a long time the nematode GHF5 cellulases as a monophyletic group stood out as nearest neighbours of bacterial homologues. But with the availability of more closely related cellulases from other Metazoa, this biased view had to be revised (**Section 4.2.2**).

The outcome of the similarity search may in fact reflect imbalances in the original database (e.g. *(15)*). This can be further illustrated by a simple key word search of the NCBI's Genome database (November, 2007); a search using the key word "Bacteria [ORGN]" yielded 2,094 bacterial genomes and using "Eukaryota [ORGN] NOT mitochondrion" yielded 956 eukaryotic (non-mitochondrion, of which 447 are Metazoa) genomes. Another example is the humane genome that initially was reported to contain more than 100 genes that were suggested to be of bacterial origin *(16)*. Soon after, most cases were refuted after sampling more related sequences in a wider range of taxonomic groups *(17)*.

### *3.2. Analysis of the Distribution Pattern*

The absence of a putatively "foreign" gene in closely related organisms could be an indication of HGT. Absence of such a gene can most convincingly be shown if whole genome sequences are available. However, this argument is used infrequently because (currently) only in a very few cases are closely related whole genome sequences available. For example, although there are nearly 30 nematode genome projects completed, planned or in progress (*see* **Note 1**), the results of only nine of these are currently available; and all are non-plant parasitic species. The majority of the plant parasitic nematode sequence data are Expressed Sequence Tags (ESTs) (144,954 ESTs from 15 species, *see* **Note 2**), therefore the public databases are even limited in representation and coverage of plant parasitic nematode transcriptomes.

### *3.3. Habitat Overlap Between Donor and Recipient*

In the most optimal case – i.e. the data set genuinely reflects the diversity of a given gene (all major taxonomic groups are adequately sampled) – a similarity search could reveal the identity of a (fairly) close living representative of the donor. HGT presupposes physical contact and a potential case of HGT would be strengthened if: (i) it is plausible that the putative donor and recipient lived in the same habitat; (ii) a trophic relationship exists between the donor and the recipient, e.g. the putative donor could have served as food for the putative recipient *(18)*; or, (iii) a symbiotic relationship exists between the putative donor and its recipient (e.g. the α-proteobacterium *Wolbachia pipientis*, a maternally inherited endosymbiont widespread among arthropods and presumably nematodes).

## 4. Detecting Horizontal Gene Transfer in Parasitic Nematodes

### 4.1. Symbionts and Symbiont-Dependent HGT in Parasitic Nematodes

Recently, a number of genomes from animal parasitic nematodes known to be infected by *Wolbachia* were screened for the presence of bacterial fragments *(19)*. In the case of the filarial nematode *Brugia malayi, Wolbachia* is needed for the completion of its life cycle, and screening of the *B. malayi* genome revealed 249 contigs with *Wolbachia* sequences. Sequence comparison with functional *Wolbachia* genes suggested that these inserts were non-functional. This case is not unique among nematodes; screening of the dog heartworm (*Dirofilaria immitis*) genome revealed another *Wolbachia* insertion. These findings, together with various similar examples from insects, prompted the authors to suggest that HGT from bacteria to animals is not necessarily as rare as previously supposed. Their research included a single plant pest, the aphid *Acyrthosiphon pisum*. This biting-sucking insects feeds on the phloem of a range of monocotyledons, and was infected with *Wolbachia*, but no *Wolbachia* traces were detected in its genome.

The frequency by which *Wolbachia* infections are found among animal parasitic nematodes made us wonder whether such infections could also occur among plant parasitic nematodes. *Wolbachia* belongs to the Anaplasmataceae in the Rickettsiales, an order of small proteobacteria. Most described members of the Rickettsiales are known as endosymbionts. In 1973, Shepherd et al. *(20)* described the presence of bacterium-like organisms living intracellularly in the cyst nematode species *Globodera rostochiensis* and *Heterodera goettingiana*. These bacteria-like organisms were particularly abundant in the reproductive system, and resembled – according to the authors – a symbiont found in the leafhopper *Helochara communis (21)*. More recently, the ultrastructure of these putative endosymbionts was suggested to relate to another bacterial endosymbiont, namely *Cardinium (22)*. Just like *Wolbachia, Cardinium* (phylum Bacteriodetes) is associated with alternations in the mode of reproduction such as parthenogenesis or feminization. However, in the case of the obligatory amphimictic cyst nematode species mentioned above, no alternation in their mode of reproduction was observed. For one particular nematode species, the soybean cyst nematode *Heterodera glycines*, the endosymbiont was identified as *Paenicardinium endonii* (93% 16S rDNA sequence identity to *Cardinium hertigii*; *(23)*).

The association between plant parasitic nematodes and bacterial endosymbionts is not restricted to the order Tylenchida, as they have also been found in *Xiphinema americanum* (order Dorylaimida) (e.g. *(24)*). Contrary to *H. glycines, X. americanum* reproduces by parthenogenesis, and males are extremely rare. In this case parthenogenesis is associated with the presence of a bacterial endosymbiont from the phylum Verrucomicrobia, namely *Xiphinematobacter (25)*.

Hence, just like animal parasitic nematodes, plant parasites are occasionally associated with endosymbionts. Only two bacterial species have been identified so far, *P. endonii* (phylum Bacteroidetes) in *H. glycines*, which apparently has no impact on the mode of reproduction, and the completely unrelated endosymbiont in the plant parasitic *X. americanum* (members of the genus *Xiphinematobacter*, phylum Bacteroidetes) that is associated with parthenogenesis. Within the order Tylenchida, frequent shifts from amphimixis to parthenogenesis are observed (e.g. in the genera *Meloidogyne* and *Pratylenchus*), and it remains to be established whether these shifts are associated with the presence of endosymbionts.

Detailed screening of the *Heterodera glycines* genome, and genome of the *Xiphinema americanum*, could reveal inserts from *Paenicardinium endonii* and *Xiphinematobacter*, respectively. In the case of cyst nematodes, the presence of a bacterial endosymbiont seems to have no effect at all on pathogenicity, host preferences or reproduction *(20)*, whereas *Xiphinematobacter* was suggested to induce thelytokous parthenogenesis in *Xiphinema (25)*. Hence, we have no indications for a role of these bacteria as putative bacterial donor(s) from which plant parasitic nematodes could have acquired the genes encoding for plant cell wall–degrading enzymes.

### 4.2. Symbiont-Unrelated HGT in Nematodes

#### 4.2.1. Plant Parasitic Nematodes Produce Cell Wall–Degrading Proteins Themselves

Until about a decade ago, mechanical puncturing of the cell wall by the stylet or spear was thought to be the mechanism underlying plant (root) penetration. In 1998, Smant and co-workers *(26)* discovered that the potato and soybean cyst nematodes (*G. rostochiensis* and *H. glycines* (order Tylenchida)) produce and secrete $\beta$-1,4-endoglucanase (cellulases). This finding constituted the starting point of a series of papers reporting a range of plant cell wall–degrading enzymes (CWDE) from plant parasitic nematodes, including pectate lyases *(27, 28)*, exo-polygalacturonase *(29)*, endo-$\beta$-1,3-glucanase *(30)*, endo-xylanases *(31)*, expansins *(32)* and cellulose-binding proteins *(33, 34)*. Nematodes are devoid of plant cell wall–like structures, and hence it seems safe to state that plant cell wall penetration by parasitic nematodes is the result of mechanical weakening and local depolymerization.

The finding of these CWDEs came as a surprise: without exception these cell wall–degrading proteins had never been detected before in animals. Hundreds of CWDE-encoding genes have been cloned from a wide taxonomic range of plants, fungi and bacteria. As compared to CWDE-encoding genes from plants and fungi, the nematode genes show remarkably high similarities to their prokaryotic equivalents. No CWDE-encoding genes have been reported in the free-living nematodes *C. elegans* and *C. briggsae* (completely sequenced in 1998 *(3)* and 2003 *(35)*, respectively). Phylogenetic analyses suggest that plant parasitic nematodes arose from fungivorous ancestors, which had evolved from bacterivores. In 1998, Keen and Roberts *(1)* wrote: "The bet is that the nematodes borrowed the cellulase genes from microorganisms at some point in their evolution. It is quite possible that microbes in the digestive system of non-parasitic nematodes may have been the source of cellulase and other genes that converted the nematodes to a parasitic habit". This hypothesis is exciting, and at the same time hard to prove. Although it is realized that a simple BLAST output with nematode sequences followed by a reasonable number of prokaryotic equivalents is only suggestive for HGT (e.g. *(36)*), it can be a practical starting point (**Section 4.3**).

Cellulases are a good object to focus on when investigating HGT in phytopathogens, because they are widely produced by pathogens that attack plants (including bacteria and fungi, *(37)*) but not found in pathogens that attack animals. Therefore, in the following section we will discuss the origin of cellulases in plant parasitic nematodes and how much they satisfy the requirements/criteria used to detect and test possible cases of HGT in the evolution of plant parasitism within the phylum Nematoda.

#### *4.2.2. On the Origin of Cellulases in Plant Parasitic Nematodes*

Among the cell wall–degrading enzymes produced by nematodes (**Section 4.2.1**), the cellulases (EC 3.2.1.4) are by far best documented. Cellulases are produced by many plant pathogenic and plant-feeding organisms *(38)* and are a rather diverse group of enzymes as illustrated by the fact that cellulases can be found in 14 different glycoside hydrolase families (GHF 5, 6, 7, 8, 9, 10, 12, 26, 44, 45, 48, 51, 61, 74; http://afmb.cnrs-mrs.fr/CAZY/). The cellulases that have been identified and characterized from plant parasitic Tylenchida (the order including most of the major plant sedentary endoparasitic nematodes such as *Meloidogyne*, *Heterodera* and *Globodera* and migratory parasitic species *Pratylenchus penetrans*) are from glycoside hydrolase family 5 (GHF5). Since the discovery of the first GHF5 nematode cellulase in 1998 *(33)*, there has been a strong overrepresentation of bacterial cellulases in the GHF5 family. While cellulases from other eukaryotes were classified into other GH families and analysis indicate that some derive from an ancient ancestor *(39, 40)* the nematode cellulases consistently showed significant similarity (BLAST E-values $< 10^{-10}$) to homologues from a range of bacteria, including plant-pathogens. Similarity searches and subsequent phylogenetic analysis resulted in seemingly credible trees in which nematode cellulases clustered as nearest neighbours of certain groups of bacterial cellulases. Hence, it was hypothesized that GHF5 genes have been acquired via HGT from bacteria. The only other eukaryotic cellulases next to the nematode genes in GHF5 originated from different yeast species. However, ten years ago genomics was in its infancy and, except for a few model organisms, systematic searches into the transcriptomes of many organisms have just recently begun. Thus, there was reasonable doubt as to whether GHF5 is a class specific mostly to bacterial and nematode enzymes. The number of accessions in GHF5 from other organisms has grown steadily ever since, giving new insight in the evolution of cellulases in this family (**Fig. 30.2**). At present enzymes significantly matching nematode GHF5 cellulases from plants, fungi, beetles, dinoflagellates, protist symbionts of termites, and diverse classes of bacteria are in the CAZY database. The inclusion of a richer diversity of organisms in the most up-to-date phylogenies of GHF5 shows that the nematode cellulases no longer constitute a monophyletic group together with bacterial cellulases, suggesting that the HGT hypothesis

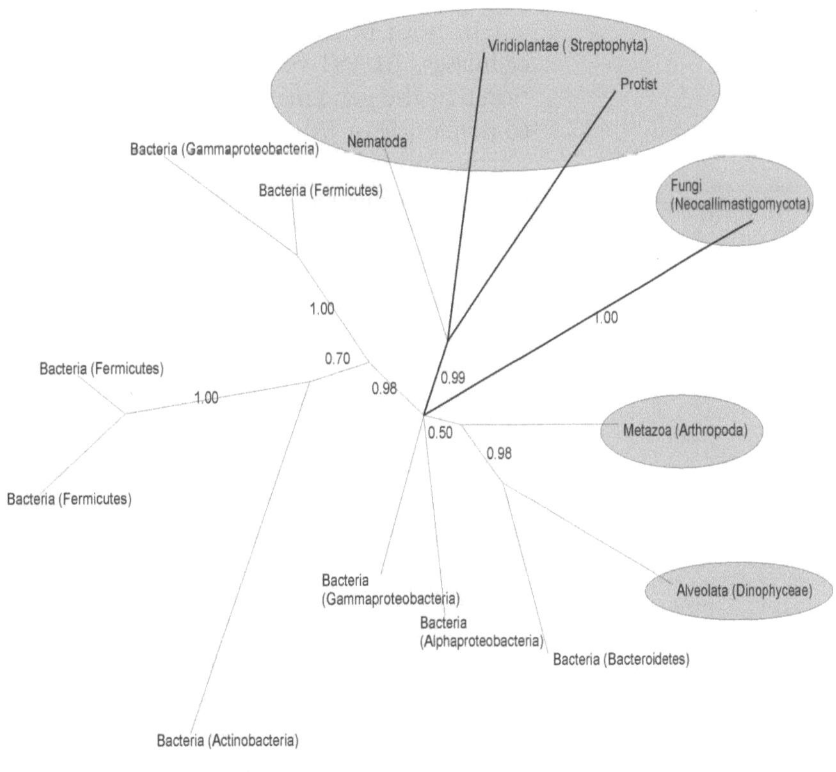


Fig. 30.2. Major groups of GHF5 cellulases with significant similarity to nematode cellulases. The phylogeny is a summarizing tree in which each group is monophyletic and assembled from taxa out the 135 homologous protein sequences in GHF5. The encircled taxa are cellulases from eukaryotes, and the values in the tree represent posterior probabilities of the basal nodes as calculated in a Bayesian analysis. Note that the nematode cellulases as a group are not positioned within a single monophyletic cluster together with bacterial cellulases.

may not hold true much longer. The work on the GHF5 cellulases demonstrates the inherent risk of misjudgements of phylogenies based on skewed datasets due to poor sampling ("tip of the ice berg" in **Section 3.1**). Future research will likely discover more GHF5 genes from yet unexplored organisms and the phylogeny of GHF5 cellulases may undergo more significant changes accordingly.

Contrary, the pine wood nematode *Bursaphelenchus xylophilus* (Clade10, order Aphelenchida; (7)), a pathogenic species that is unique in its ability to feed on live trees and fungi, harbours GHF45 cellulases *(41)*. GHF45 cellulases have been found in fungi, bacteria, protists, and a very small number of animals *(42)* (see http://afmb.cnrs-mrs.fr/CAZY/). The ancestor of *B. xylophilus* is likely a non-pathogenic *Bursaphelenchus* (which are solely fungal feeders), and the pathogenic *B. xylophilus* GHF45 phylogenetically is more closely related to the fungal cellulases than to ones found in insects. Hence, it was hypothesized that *B. xylophilus* cellulases are of fungal origin, and acquired by HGT *(41, 43)*.

In both of the above-mentioned cases, GHF5 and GHF45 cellulases, BLAST-based similarity searches were the first indication that the gene may have been acquired by HGT from bacteria to nematode or from fungi to nematode (**Sections 3.1** and **4.3**). Furthermore, using similarity searches it has been suggested that GHF45 cellulases are absent from root knot and cyst nematodes, and similarly GHF5 cellulases have not been detected in *B. xylophilus* (this observation was also supported by the inability to amplify GHF5 cellulases using degenerate primers). While the GHF5 cellulases have been detected in five root knot and four cyst nematode species, the absence of sequence data from the other two completely unexplored suborders of the order Tylenchida and from other *Bursaphelenchus* species (*see* **Note 3**) precludes using the distribution pattern to strengthen the HGT claim (**Section 3.2**). Furthermore, because full genomes are not available, we are not able to make definitive statements about the presence or absence of a particular gene, but the observations to date strongly suggest that in the case of Tylenchida and Aphelenchida the different overlapping habitat and trophic ecology between the donor and the recipient contributed to independent, parallel evolutionary trajectories, i.e. acquiring different cellulases to perform a common function, which is the ability to digest cellulose and parasitize plants (**Section 3.3**). Hence, independent lineages of plant parasitic nematodes investigated so far harbour cellulases from different GH families. Representatives of the individual lineages only contain cellulases that belong to a single, apparently lineage-typical GH family. Characterization of the cellulase content of basal members of these lineages will tell us whether acquisition via HGT is a likely scenario.

***4.3. Computational High-Throughput Identification of HGT Candidates in Plant Parasitic Nematodes***

To date most of the potential HGT acquired genes in plant parasitic nematodes are the result of the characterization of individual genes. Rarely, a specific search has been performed to identify HGT candidates *(44, 45)*. Additional lines of evidence are needed to (dis)confirm a HGT case, such as endogenous localization of the transcript, experimental demonstration of *trans*-spliced leader on the mRNA, or presence of introns in the gene, just to name a few. However, it is also widely accepted that a primary sequence similarity search can be a practical starting point to identify HGT candidates. To demonstrate the latter we have used 60,643 translated EST contig consensus sequences (www.nematode.net; *(46)*), representing 45,583 genes (EST clusters; *(47)*) from 13 plant parasitic nematode species as queries to make the homology assignments to 2 phylogenetically distinct databases (*see* **Note 4**): (i). Non-redundant GenBank (built November 2007) with all plant parasitic nematode and bacterial sequences removed (E) (**Table 30.3**), and (ii) all bacterial protein sequences in GenBank (B) (built November 2007). Of the 45,583 plant parasitic nematode sequences, 47% shared homology to known proteins. Of the 4,545 that had hits to the database containing both eukaryotic and bacterial sequences (E&B), 384 EST clusters had a higher percent identity to the bacterial compared to the eukaryotic sequence. Furthermore, 151 clusters matched the bacteria-only database (using the specified cut-off, *see* **Note 4**). We identified the previously postulated HGT genes such as different CWDEs, the nodL and other hits to genes of rhizobial origin (validating the method). While more sophisticated algorithms to detect HGT have started to become available *(48)*, our approach, in addition to rediscovering previously identified genes, also identifies many new candidates for HGT (*see* **Note 5**). However, we have to note that based on the limitations described in **Sections 3.1** and **4.2.2** gene products from this category are likely to shift toward and expand the E&B category as additional complete genomes become available.

The bacterial sequences matched by the 151 nematode ESTs clusters originated from 110 unique bacterial species. Most of the bacterial species that appeared as potential donors of the HGT candidates were members of the phylum Proteobacteria, followed by Actinobacteridae and low G + C Firmicutes. Similarly, in most of the reported cases (**Section 4.2**), individual enzyme families are represented in numerous bacterial species from different divisions and these enzymes show various degrees of sequence similarity with each other. For example, cellulases from GHF5 can be found in *Erwinia chrysanthemi* and *E. carotovora* (both are plant pathogenic Proteobacteria), *Bacillus licheniformis* (Firmicutes, a saprophytic soil bacterium) and *Streptomyces* (Actinobacteria).

Therefore, genes encoding cellulases were indeed donated to bacterivorous or fungivorous nematodes (**Section 4.2.2**) by soil-borne prokaryotes, and there is probably no necessity for taxonomic relatedness among the donors.

## 5. Notes

1. **Table 30.1** lists nematode taxa for which a genome project is completed, in progress or planned (at the time this book was written).
2. **Table 30.2** gives the number of ESTs from plant parasitic nematode species registered in the dbEST division of GenBank at the time this book was written.
3. The order Tylenchida comprises four suborders: Tylenchina, Hexatylina, Criconematina and Hoplolaimina. Except for the order Hexatylina that includes mainly insect parasites, these suborders embrace predominantly facultative and obligate plant parasitic nematodes. We hypothesize that most members of the Tylenchina, Criconematina and Hoplolaimina harbour genes encoding cell wall–degrading enzymes. The genera *Meloidogyne, Heterodera, Globodera* and *Pratylenchus* all belong to the Hoplolaimina, the most distal suborder *(7)*. The two remaining suborders, Tylenchina and Criconematina, have so far not been sampled. A closer investigation of members of these suborders should reveal whether these plant parasites harbour GHF5 cellulases, or cellulases from other GH families. If GHF5 cellulases are present among, for example, facultative plant parasitic representatives of the Tylenchina, these sequences could be compared with other cellulases from non-nematodes. Hence, the phylogenetic analysis of a few distal members of the Hoplolaimina and the careful exon/intron analysis cannot give rise to conclusions about HGT at the level of the order Tylenchida as done by Ledger et al. *(49)*. At most this analysis gives a first glimpse of GHF5 evolution among members of the suborder Hoplolaimina. For a robust picture, more (basal) members of the Hoplolaimina should be included.
4. **Table 30.3** lists BLAST matches of plant parasitic EST clusters per database. WU-BLAST sequence comparison *(50, 51)* was performed. Homologies were reported for matches with at least 50 bits and 40% identity. Internal PERL-based scripts were used to summarize the data on cluster (gene) level and to do taxonomy parsing using the NCBI's taxonomy structure.
5. Of the 151 genes with bacterial matches only, 87 got assigned InterPro id *(52)*. Focusing on the Gene Ontology

**Table 30.1**
**Nematode taxa for which a genome project is completed, in progress or planned**

| Genome coverage | Species | Clade[a] | Clade[b] | Trophic ecology[c] | Status[d] |
|---|---|---|---|---|---|
| Full genome | | | | | |
| | *Caenorhabditis elegans* | V | 9 | B | C |
| Draft genome (6–8x coverage) | | | | | |
| | *Ancylostoma caninum* | V | 9 | V | P |
| | *Ascaris lumbricoides* | III | 8 | V | P |
| | *Brugia malayi* | III | 8 | H | C |
| | *Caenorhabditis branneri* | V | 9 | B | P |
| | *Caenorhabditis briggsae* | V | 9 | B | C |
| | *Caenorhabditis japonica* | V | 9 | B | C |
| | *Caenorhabditis remanei* | V | 9 | B | C |
| | *Cooperia oncophora* | V | 9 | V | P |
| | *Dictyocaulus viviparus* | V | 9 | V | P |
| | *Haemonchus contortus* | V | 9 | V | P |
| | *Heterorhabditis bacteriophora* | V | 9 | I | P |
| | *Meloidogyne hapla* | IV | 12 | P | P |
| | *Meloidogyne incognita* | IV | 12 | P | P |
| | *Heterodera glycines* | IV | 12 | P | C |
| | *Globodera pallida* | IV | 12 | P | P |
| | *Necator americanus* | V | 9 | H | P |
| | *Nematodirus battus* | V | 9 | V | P |
| | *Nippostrongylus brasiliensis* | V | 9 | V | P |
| | *Oesophagostomum dentatum* | V | 9 | V | P |
| | *Onchocerca volvulus* | III | 8 | H | P |
| | *Ostertagia ostertagii* | V | 9 | V | P |
| | *Pristionchus pacificus* | V | 9 | A-O-P | C |
| | *Strongyloides ratti* | Iva | 10 | V | P |
| | *Teladorsagia circumcincta* | V | 9 | V | P |

(continued)

**Table 30.1 (continued)**

| Genome coverage | Species | Clade[a] | Clade[b] | Trophic ecology[c] | Status[d] |
|---|---|---|---|---|---|
| | *Trichinella spiralis* | I | 2 | V | C |
| | *Trichostrongylus colubriformis* | V | 9 | V | P |
| | *Trichuris trichura* | I | 2 | H | P |
| Low genome coverage ( < 2x) | | | | | |
| | *Ancylostoma caninum* | V | 9 | V | C |
| | *Ascaris suum* | III | 8 | V | C |
| | *Pristionchus entomophagous* | V | 9 | A-O-P | C |
| | *Pristionchus naupasi* | V | 9 | A-O-P | C |

[a] Nematode phylogeny, Blaxter et al. *(6)*.
[b] Nematode phylogeny Holterman et al. *(7)*.
[c] Food source: B, Bacteriovore; A-O-P, Algivore-Omnivore-Predator, I, Insect-associated bacteriovore; H, human parasite, V, vertebrate parasite, P, Plant parasite.
[d] Status of genome sequencing project: C, completed, P, in progress or planned.

*(53)* organizing principle molecular function hydrolase activity, hydrolyzing *O*-glycosyl compounds (GO:0004553), we were able to rediscover characterized cellulases and identify new cellulases. While all endo-glucanases clustered together in one protein family (using Tribe-MC *(54)* which relies on the Markov cluster algorithm MCL that simulates flow in a protein similarity graph and assigns complete protein sequences into families based on density and strength between them), there was a *Meloidogyne hapla* putative endo-glucanase (MH10227.cl) that clustered separately from all the other glucanases. This sequence had a best hit to an endo-1,4 beta-glucanase from *Xylella fastidiosa*; a xylem-dwelling, insect-transmitted, gamma-proteobacterium that causes diseases in many plants *(55)*. Obviously, because full genomes are not available, we are not able to make definitive statements about the interspecific presence or absence of a particular gene (**Section 3.2**). Alternatively, the partial sequence may preclude detection of a better homology to other, more similar genes. Another example was the *X. index* EST cluster (XI03644.cl), which encodes a putative cellulase homologous to GHF12 cellulases (based on hits to IPR013319 and IPR002594). This suggests that cellulases present in plant parasitic Dorylaimids belong to a glycoside hydrolase family different from GHF5 (distal Tylenchida) or GHF45

**Table 30.2**
**Number of ESTs from plant parasitic nematode species registered in the dbESTs division of GenBank (November 2007)**

| Order[a] | Nematode species | ESTs | Feeding habit[b] |
|---|---|---|---|
| Tylenchina (Clade 12) | | | |
| | *Globodera pallida* | 4378 | Sen |
| | *Globodera rostochiensis* | 11851 | Sen |
| | *Heterodera schachtii* | 2818 | Sen |
| | *Heterodera glycines* | 24444 | Sen |
| | *Meloidogyne arenaria* | 5018 | Sen |
| | *Meloidogyne chitwoodi* | 12218 | Sen |
| | *Meloidogyne hapla* | 24452 | Sen |
| | *Meloidogyne incognita* | 20334 | Sen |
| | *Meloidogyne javanica* | 7587 | Sen |
| | *Meloidogyne paranaensis* | 3710 | Sen |
| | *Ditylenchus africanus*[c] | 4847 | Men |
| | *Radopholus similis* | 6206 | Men |
| | *Pratylenchus penetrans* | 1928 | Men |
| | *Pratylenchus vulnus* | 5812 | Men |
| Dorylaimida (Clade 2) | | | |
| | *Xiphinema index* | 9351 | Mec |
| Triplonchida (Clade 1) | | | |
| | None | – | – |

[a]Nematode phylogeny according to Holterman et al. *(8)*.
[b]Definitions: SEn: *Sedentary endoparasites*: Enter the root system and feed from highly modified cells. Lose the ability to move and require an active feeding site. MEn: *Migratory endoparasites*: Enter the root system and feed from cells as they migrate. MEc: *Migratory ectoparasites*: Remain outside the root and feed on epidermal cells or on cells deeper in the root. Retain the ability to move to new feeding sites.
[c] Peanut pod nematode – facultative plant parasite – can also be reared on fungal cultures such as *Aspergillus* sp., *Botrytis* sp., etc.

(Aphelenchida) (**Section 4.2.2**). While this gene has been previously reported *(43)*, more firm conclusions require a full-length sequence, gene structure analysis and biochemical characterisation of the identified *X. index* transcript.

**Table 30.3**
**BLAST matches of plant parasitic EST clusters per database**

| Species | ESTs | Contigs | Clusters | Similarity match | | |
|---|---|---|---|---|---|---|
| | | | | E[a] | E & B | B[b] |
| *Globodera pallida* | 4378 | 2919 | 2227 | 767 | 238 | 6 |
| *Globodera rostochiensis* | 11851 | 2496 | 2348 | 991 | 267 | 2 |
| *Heterodera glycines* | 24442 | 12121 | 8547 | 3014 | 861 | 42 |
| *Heterodera schachtii* | 2818 | 1466 | 1441 | 560 | 172 | 7 |
| *Meloidogyne arenaria* | 5018 | 3272 | 3163 | 1146 | 310 | 9 |
| *Meloidogyne chitwoodi* | 12218 | 5824 | 3555 | 1284 | 300 | 11 |
| *Meloidogyne hapla* | 24452 | 11072 | 6624 | 2278 | 568 | 20 |
| *Meloidogyne incognita* | 20334 | 9019 | 5559 | 1971 | 522 | 18 |
| *Meloidogyne javanica* | 7587 | 4030 | 3881 | 1573 | 353 | 14 |
| *Meloidogyne paranaensis* | 3710 | 1914 | 1873 | 745 | 210 | 2 |
| *Pratylenchus penetrans* | 1928 | 482 | 415 | 172 | 64 | 0 |
| *Radopholus similis* | 1154 | 713 | 707 | 244 | 82 | 1 |
| *Xiphinema index* | 9351 | 5315 | 5243 | 1936 | 598 | 19 |
| Total | 129241 | 60643 | 45583 | 16681 | 4545 | 151 |

[a] E, eukaryotic database is the GenBank nr database with plant parasitic nematode and bacterial species removed (includes Archaea).
[b] B, all bacterial sequences.

## Acknowledgments

GS is supported by the Vernieuwingsimpuls grant from the Dutch Science Foundation (NWO) and the European Commission FP6 project BIOEXPLOIT. MM is in part supported by the National Institute of Health-NIAID grant AI46593.

### References

1. Keen, N. T., Roberts, P. A. (1998) Plant parasitic nematodes: Digesting a page from the microbe book. *Proc Natl Acad Sci U S A* **95**, 4789–90.
2. Lambshead, P. J., Brown, C., Ferrero, T., Hawkins, L., Smith, C., Mitchell, N. (2003) Biodiversity of nematode assemblages from the region of the clarion-clipperton fracture zone, an area of commercial mining interest. *BMC Ecology* **3**, 1.
3. The *C. elegans* Sequencing Consortium (1998) Genome sequence of the nematode *C. elegans*: A platform for investigating biology. *Science* **282**, 2012–8.
4. Agrios, G. N. (2005) *Plant pathology*, Academic Press, New York.
5. WHO (2005) Deworming for health and development. Report of the third global meeting of the partners for parasite control. World Health Organization, Geneva.

6. Blaxter, M. L., De Ley, P., Garey, J. R., Liu, L. X., Scheldeman, P., Vierstraete, A., Vanfleteren, J. R., Mackey, L. Y., Dorris, M., Frisse, L. M., Vida, J. T., Thomas, W. K. (1998) A molecular evolutionary framework for the phylum Nematoda. *Nature* **392**, 71–5.
7. Holterman, M., Van Der Wurff, A., Van Den Elsen, S., Van Megen, H., Bongers, T., Holovachov, O., Bakker, J., Helder, J. (2006) Phylum-wide analysis of SSU rDNA reveals deep phylogenetic relationships among nematodes and accelerated evolution toward crown clades. *Mol Biol Evol* **23**, 1792–800.
8. Holterman, M., Rybarczyk, K., Van Den Elsen, S., Van Megen, H., Mooyman, P., Santiago, R. P., Bongers, T. O. M., Bakker, J., Helder, J. (2008) A ribosomal DNA-based framework for the detection and quantification of stress-sensitive nematode families in terrestrial habitats. *Mol Ecol Res* **8**, 23–34.
9. Smith, M. W., Feng, D. F., Doolittle, R. F. (1992) Evolution by acquisition: The case for horizontal gene transfers. *Trends Biochem Sci* **17**, 489–93.
10. Koonin, E. V., Makarova, K. S., Aravind, L. (2001) Horizontal gene transfer in prokaryotes: Quantification and classification. *Annu Rev Microbiol* **55**, 709–42.
11. Lawrence, J. G., Hendrickson, H. (2003) Lateral gene transfer: When will adolescence end? *Mol Microbiol* **50**, 739–49.
12. Brown, J. R. (2003) Ancient horizontal gene transfer. *Nat Rev Genet* **4**, 121–32.
13. Andersson, J. O. (2005) Lateral gene transfer in eukaryotes. *Cell Mol Life Sci* **62**, 1182–97.
14. Blaxter, M. (2007) Symbiont genes in host genomes: Fragments with a future? *Cell Host Microbe* **2**, 211–3.
15. Koski, L. B., Golding, G. B. (2001) The closest BLAST hit is often not the nearest neighbor. *J Mol Evol* **52**, 540–2.
16. International Human Genome Sequencing Consortium (2001) Initial sequencing and analysis of the human genome. *Nature* **409**, 860–921.
17. Stanhope, M. J., Lupas, A., Italia, M. J., Koretke, K. K., Volker, C., Brown, J. R. (2001) Phylogenetic analyses do not support horizontal gene transfers from bacteria to vertebrates. *Nature* **411**, 940–4.
18. Ford Doolittle, W. (1998) You are what you eat: A gene transfer ratchet could account for bacterial genes in eukaryotic nuclear genomes. *Trends Genet* **14**, 307–11.
19. Hotopp, J. C. D., Clark, M. E., Oliveira, D. C. S. G., Foster, J. M., Fischer, P., Torres, M. C. M., Giebel, J. D., Kumar, N., Ishmael, N., Wang, S., Ingram, J., Nene, R. V., Shepard, J., Tomkins, J., Richards, S., Spiro, D. J., Ghedin, E., Slatko, B. E., Tettelin, H., Werren, J. H. (2007) Widespread lateral gene transfer from intracellular bacteria to multicellular eukaryotes. *Science* **317**, 1753–6.
20. Shepherd, A. M., Clark, S. A., Kempton, A. (1973) Spermatogenesis and sperm ultrastructure in some cyst nematodes, *Heterodera* spp.. *Nemaiologica* **19**, 551–60.
21. Chang, H. P., Musgrave, A. J. (1972) Multiple symbiosis in a leafhopper, *Helochara communis* Fitch (cicadellidae: Homoptera): Envelopes, nucleoids and inclusions of the symbiotes. *J Cell Sci* **11**, 275–93.
22. Zchori-Fein, E., Perlman, S. J. (2004) Distribution of the bacterial symbiont *Cardinium* in arthropods. *Mol Ecol* **13**, 2009–16.
23. Noel, G. R., Atibalentja, N. (2006) '*Candidatus* Paenicardinium endonii', an endosymbiont of the plant-parasitic nematode *Heterodera glycines* (Nemata: Tylenchida), affiliated to the phylum Bacteroidetes. *Int J Syst Evol Microbiol* **56**, 1697–702.
24. Adams, R. E., Eichenmuller, J. J. (1963) A bacterial infection of *Xiphenema americanum. Phytopathology* **53**, 745.
25. Vandekerckhove, T. T. M., Willems, A., Gillis, M., Coomans, A. (2000) Occurrence of novel verrucomicrobial species, endosymbiotic and associated with parthenogenesis in *Xiphinema americanum*-group species (Nematoda, Longidoridae). *Int J Syst Evol Microbiol* **50**, 2197–205.
26. Smant, G., Stokkermans, J. P., Yan, Y., De Boer, J. M., Baum, T. J., Wang, X., Hussey, R. S., Gommers, F. J., Henrissat, B., Davis, E. L., Helder, J., Schots, A., Bakker, J. (1998) Endogenous cellulases in animals: Isolation of beta-1, 4-endoglucanase genes from two species of plant-parasitic cyst nematodes. *Proc Natl Acad Sci U S A* **95**, 4906–11.
27. Popeijus, H., Overmars, H. A., Jones, J. T., Blok, V. C., Goverse, A., Helder, J., Schots, A., Bakker, J., Smant, G. (2000) Degradation of plant cell walls by a nematode. *Nature* **406**, 36–7.
28. Kudla, U., Milac, A.-L., Qin, L., Overmars, H., Roze, E., Holterman, M., Petrescu, A.-J., Goverse, A., Bakker, J., Helder, J., Smant, G. (2007) Structural and functional characterization of a novel, host penetration-related pectate lyase from the potato cyst nematode *Globodera rostochiensis. Mol Plant Pathol* **8**, 293–305.
29. Jaubert, S., Laffaire, J. B., Abad, P., Rosso, M. N. (2002) A polygalacturonase of animal origin isolated from the root-knot nematode *Meloidogyne incognita. FEBS Lett* **522**, 109–12.

30. Kikuchi, T., Shibuya, H., Jones, J. T. (2005) Molecular and biochemical characterization of an endo-beta-1,3-glucanase from the pinewood nematode *Bursaphelenchus xylophilus* acquired by horizontal gene transfer from bacteria. *Biochem J* **389**, 117–25.
31. Mitreva-Dautova, M., Roze, E., Overmars, H., De Graaff, L., Schots, A., Helder, J., Goverse, A., Bakker, J., Smant, G. (2006) A symbiont-independent endo-1,4-beta-xylanase from the plant-parasitic nematode *Meloidogyne incognita. Mol Plant Microbe Interact* **19**, 521–9.
32. Qin, L., Kudla, U., Roze, E. H. A., Goverse, A., Popeijus, H., Nieuwland, J., Overmars, H., Jones, J. T., Schots, A., Smant, G., Bakker, J., Helder, J. (2004) Plant degradation: A nematode expansin acting on plants. *Nature* **427**, 30.
33. Ding, X., Shields, J., Allen, R., Hussey, R. S. (1998) A secretory cellulose-binding protein cDNA cloned from the root-knot nematode (*Meloidogyne incognita*). *Mol Plant Microbe Interact* **11**, 952–9.
34. Gao, B., Allen, R., Davis, E. L., Baum, T. J., Hussey, R. S. (2004) Molecular characterisation and developmental expression of a cellulose-binding protein gene in the soybean cyst nematode *Heterodera glycines. Int J Parasitol* **34**, 1377–83.
35. Stein, L. D., Bao, Z., Blasiar, D., Blumenthal, T., Brent, M. R., Chen, N., Chinwalla, A., Clarke, L., Clee, C., Coghlan, A., et al. (2003) The genome sequence of *Caenorhabditis briggsae*: A platform for comparative genomics. *PLoS Biol* **1**, E45.
36. Andersson, J. O. (2005) Lateral gene transfer in eukaryotes. *Cell Mol Life Sci* **62**, 1182–97.
37. Walton, J. D. (1994) Deconstructing the cell wall. *Plant Physiol* **104**, 1113–8.
38. Lynd, L. R., Weimer, P. J., Van Zyl, W. H., Pretorius, I. S. (2002) Microbial cellulose utilization: Fundamentals and biotechnology. *Microbiol Mol Biol Rev* **66**, 506–77.
39. Davison, A., Blaxter, M. (2005) Ancient origin of glycosyl hydrolase family 9 cellulase genes. *Mol Biol Evol* **22**, 1273–84.
40. Lo, N., Watanabe, H., Sugimura, M. (2003) Evidence for the presence of a cellulase gene in the last common ancestor of bilaterian animals. *Proc Biol Sci* **270 Suppl 1**, S69–72.
41. Kikuchi, T., Jones, J. T., Aikawa, T., Kosaka, H., Ogura, N. (2004) A family of glycosyl hydrolase family 45 cellulases from the pine wood nematode *Bursaphelenchus xylophilus. FEBS Lett* **572**, 201–5.
42. Henrissat, B., Bairoch, A. (1993) New families in the classification of glycosyl hydrolases based on amino acid sequence similarities. *Biochem J* **293**, 781–8.
43. Jones, J. T., Furlanetto, C., Kikuchi, T. (2003) Horizontal gene transfer from bacteria and fungi as a driving force in the evolution of plant parasitism in nematodes. *Nematology* 7, 641–6.
44. Mccarter, J., Dautova Mitreva, M., Martin, J., Dante, M., Wylie, T., Rao, U., Pape, D., Bowers, Y., Theising, B., Murphy, C. V., Kloek, A. P., Chiapelli, B. J., Clifton, S. W., Bird, M. D., Waterston, R. (2003) Analysis and functional classification of transcripts from the nematode *Meloidogyne incognita. Genome Biol* **4**, R26: 1–19.
45. Scholl, E. H., Thorne, J. L., Mccarter, J. P., Bird, D. M. (2003) Horizontally transferred genes in plant-parasitic nematodes: A high-throughput genomic approach. *Genome Biol* **4**, R39.
46. Wylie, T., Martin, J., Dante, M., Mitreva, M., Clifton, S. W., Chinwalla, A., Waterston, R. H., Wilson, R. K., Mccarter, J. P. (2004) Nematode.Net: A tool for navigating sequences from parasitic and free-living nematodes. *Nucleic Acids Res* **32**, D423–6.
47. Mitreva, M., Mccarter, J. P., Martin, J., Dante, M., Wylie, T., Chiapelli, B., Pape, D., Clifton, S. W., Nutman, T. B., Waterston, R. H. (2004) Comparative genomics of gene expression in the parasitic and free-living nematodes *Strongyloides stercoralis* and *Caenorhabditis elegans. Genome Res* **14**, 209–20.
48. Li, Z., Wang, L., Zhong, Y. (2005) Detecting horizontal gene transfer with t-rex and rhom programs. *Brief Bioinform* **6**, 394–401.
49. Ledger, T. N., Jaubert, S., Bosselut, N., Abad, P., Rosso, M.-N. (2006) Characterization of a new beta-1,4-endoglucanase gene from the root-knot nematode *Meloidogyne incognita* and evolutionary scheme for phytonematode family 5 glycosyl hydrolases. *Gene* **382**, 121–8.
50. Altschul, S. F., Gish, W., Miller, W., Myers, E. W., Lipman, D. J. (1990) Basic local alignment search tool. *J Mol Biol* **215**, 403–10.
51. Gish, W. (1996–2002) http://blast.wustl.edu.
52. Zdobnov, E. M., Apweiler, R. (2001) Interproscan – an integration platform for the signature-recognition methods in interpro. *Bioinformatics* **17**, 847–8.
53. The Gene Ontology Consortium (2000) Gene ontology: Tool for the unification of biology. *Nat Genet* **25**, 25–9.
54. Enright, A. J., Van Dongen, S., Ouzounis, C. A. (2002) An efficient algorithm for large-scale detection of protein families. *Nucl Acids Res* **30**, 1575–84.

55. Van Sluys, M. A., De Oliveira, M. C., Monteiro-Vitorello, C. B., Miyaki, C. Y., Furlan, L. R., Camargo, L. E. A., Da Silva, A. C. R., Moon, D. H., Takita, M. A., Lemos, E. G. M., Machado, M. A., Ferro, M. I. T., Da Silva, F. R., Goldman, M. H. S., Goldman, G. H., Lemos, M. V. F., El-Dorry, H., Tsai, S. M., Carrer, H., Carraro, D. M., De Oliveira, R. C., Nunes, L. R., Siqueira, W. J., Coutinho, L. L., Kimura, E. T., Ferro, E. S., Harakava, R., Kuramae, E. E., Marino, C. L., Giglioti, E., Abreu, I. L., Alves, L. M. C., Do Amaral, A. M., Baia, G. S., Blanco, S. R., Brito, M. S., Cannavan, F. S., Celestino, A. V., Da Cunha, A. F., Fenille, R. C., Ferro, J. A., Formighieri, E. F., Kishi, L. T., Leoni, S. G., Oliveira, A. R., Rosa, V. E., Jr., Sassaki, F. T., Sena, J. A. D., De Souza, A. A., Truffi, D., Tsukumo, F., Yanai, G. M., Zaros, L. G., Civerolo, E. L., Simpson, A. J. G., Almeida, N. F., Jr., Setubal, J. C., Kitajima, J. P. (2003) Comparative analyses of the complete genome sequences of Pierce's disease and citrus variegated chlorosis strains of *Xylella fastidiosa*. *J Bacteriol* **185**, 1018–26.

# Index

## B



## C

## D

## F



## G

## J



## K



## L



## M

## Q

## R



## S

## W



## X



## Y



## Z

MIX
Papier aus verantwortungsvollen Quellen
Paper from responsible sources
FSC® C105338

If you have any concerns about our products,
you can contact us on
**ProductSafety@springernature.com**

In case Publisher is established outside the EU,
the EU authorized representative is:
**Springer Nature Customer Service Center GmbH**
**Europaplatz 3, 69115 Heidelberg, Germany**

Printed by Libri Plureos GmbH
in Hamburg, Germany

MIX
Papier aus verantwortungsvollen Quellen
Paper from responsible sources
FSC® C105338

If you have any concerns about our products,
you can contact us on
**ProductSafety@springernature.com**

In case Publisher is established outside the EU,
the EU authorized representative is:
**Springer Nature Customer Service Center GmbH**
**Europaplatz 3, 69115 Heidelberg, Germany**

Printed by Libri Plureos GmbH
in Hamburg, Germany